FIREFIGHTER'S HANDBOOK

BASIC ESSENTIALS OF FIREFIGHTING

Delmar is proud to donate a portion of the proceeds from this book to NFAAA.

FIREFIGHTER'S HANDBOOK

BASIC ESSENTIALS OF FIREFIGHTING

DELMAR
THOMSON LEARNING

Australia Canada Mexico Singapore Spain United Kingdom United States

Firefighter's Handbook: Basic Essentials of Firefighting
Thomson Delmar Learning

Vice President, Technology and Trades SBU:
Alar Elken

Editorial Director:
Sandy Clark

Acquisitions Editor:
Alison S. Weintraub

Developmental Editor:
Jennifer A. Thompson

Marketing Director:
Cyndi Eichelman

Channel Manager:
Bill Lawrensen

Marketing Coordinator:
Mark Pierro

Production Director:
Mary Ellen Black

Production Editor:
Barbara L. Diaz

Art & Design Specialist:
Rachel Baker

Technology Project Manager:
Kevin Smith

Editorial Assistants:
Jennifer Luck
Stacey Wiktorek

COPYRIGHT 2004 by Thomson Delmar Learning. Thomson and Delmar Learning are trademarks used herein under license.

Printed in the United States of America
1 2 3 4 5 XX 07 06 05 04

For more information contact Delmar Learning
Executive Woods
5 Maxwell Drive, PO Box 8007,
Clifton Park, NY 12065-8007
Or find us on the World Wide Web at
www.delmarlearning.com

ALL RIGHTS RESERVED. Portions of this work © 2000. No part of this work covered by the copyright hereon may be reproduced in any form or by any means—graphic, electronic, or mechanical, including photocopying, recording, taping, Web distribution, or information storage and retrieval systems—without the written permission of the publisher.

For permission to use material from the text or product, contact us by
Tel.   (800) 730-2214
Fax   (800) 730-2215
www.thomsonrights.com

Library of Congress Cataloging-in-Publication Data:

Firefighter's handbook : Basic essentials of firefighting.
   p. cm.
   Includes bibliographical references and index.
   ISBN 1-4018-3582-1
   1. Fire extinction—Handbooks, manuals, etc.
I. Delmar Publishers. II. Title.
TH9151.F49 2005
628.9′25—dc22
                              2003066267

ISBN: 1-4018-3582-1

NOTICE TO THE READER

Publisher does not warrant or guarantee any of the products described herein or perform any independent analysis in connection with any of the product information contained herein. Publisher does not assume, and expressly disclaims, any obligation to obtain and include information other than that provided to it by the manufacturer.

The reader is expressly warned to consider and adopt all safety precautions that might be indicated by the activities herein and to avoid all potential hazards. By following the instructions contained herein, the reader willingly assumes all risks in connection with such instructions.

The publisher makes no representation or warranties of any kind, including but not limited to, the warranties of fitness for particular purpose or merchantability, nor are any such representations implied with respect to the material set forth herein, and the publisher takes no responsibility with respect to such material. The publisher shall not be liable for any special, consequential, or exemplary damages resulting, in whole or part, from the readers' use of, or reliance upon, this material.

CONTENTS

Dedication xiv
Foreword xv
About Our Authors xvi
Preface xx
Acknowledgments xxviii
NFPA 1001 Correlation Guide xxxi
Job Performance Requirement Correlation Guide xxxiii

CHAPTER 1 OVERVIEW OF THE HISTORY, TRADITION, AND DEVELOPMENT OF THE AMERICAN FIRE SERVICE

Objectives 4
The Mission of the Fire Service 4
Roots in the Past 4
   Ancient Beliefs 5
   Recorded History 6
   Early History and Symbols of the Fire Service 7
   The Middle Ages 7
   Early American History 8
   The Civil War 13
   The Industrial Revolution 14
   The Beginning of the Twentieth Century 15
Technology, Transition, and Times of Change 16
The Effects of World War II 17
Modernization of the Fire Service 17
The Fire Service of Today 18
Lessons Learned 19
Key Terms 19
Review Questions 20
Additional Resources 20

CHAPTER 2 FIRE DEPARTMENT ORGANIZATION, COMMAND, AND CONTROL

Objectives 23
Introduction 23
Fire Department Organization 23
   The Business of Fire Protection 24
   Mission Statement 24
   Organizational Structure 24
The Firefighter 26
The Company 28
   The Engine Company 29
   The Truck Company 29
   The Rescue Company 29
   Specialty/Combination Units 31
   Emergency Medical Services 32
   The Chief Officers 32
Additional Fire Department Functions 32
   Fire Prevention and Life Safety 32
   Training 33
   Emergency Medical Services 34
   Apparatus Maintenance and Purchasing 34
   Special Operations 34
Regulations, Policies, Bylaws, and Procedures 34
   Regulations 34
   Policies 34
   Bylaws 35
   Procedures 35
Allied Agencies and Organizations 35
Incident Management 36
   Command and Control 36
Incident Management System (IMS) 37
   Five Major Functions of an Incident Management System 39
Lessons Learned 42
Key Terms 42
Review Questions 44
Additional Resources 44

CHAPTER 3 COMMUNICATIONS AND ALARMS

Objectives 47
Introduction 47
Communications Personnel 48
The Communications Facility 49
Computers in the Fire Service 51
Receiving Reports of Emergencies 51
   Methods of Receiving Reports of Emergencies 53
Emergency Services Deployment 58
Traffic Control Systems 62
Radio Systems and Procedures 63
Arrival Reports 67
Mobile Support Vehicles 68
Records 68
Lessons Learned 70
Key Terms 71
Review Questions 72
Endnote 72
Additional Resources 73

CHAPTER 4
FIRE BEHAVIOR

Objectives 76
Introduction 76
Fire Triangle, Tetrahedron, and Pyramid 77
Measurements 77
Chemistry and Physics of Fire 78
Sources of Heat 82
   Chemical 83
   Mechanical 83
   Electrical 83
   Nuclear 85
Combustion 85
Oxygen and Its Effect on Combustion 86
Vapor Pressure and Vapor Density 86
   BLEVE 88
Boiling Point 89
Flammable and Explosive Limits 90
The Burning Process—Characteristics
  of Fire Behavior 91
   Ignition Stage 91
   Growth Stage 91
   Fully Developed Stage 92
   Decay Stage 92
Modes of Heat Transfer 94
   Conduction 94
   Convection 95
   Radiation 96
Thermal Conductivity of Materials 97
Physical State of Fuels and Effect on Combustion . . 98
   Solid 98
   Liquid 98
   Gas 98
Theory of Fire Extinguishment 99
Unique Fire Events 99
   Thermal Layering 99
   Rollover and Flashover 99
   Backdraft 100
Classes of Fire 100
Lessons Learned 103
Key Terms 103
Review Questions 105
Endnotes 105
Additonal Resources 105

CHAPTER 5
FIREFIGHTER SAFETY

Objectives 108
Introduction 108
Safety Issues 108
   Firefighter Injury and Death Causes 108
   Safety Standards and Regulations 109
   Accident Prevention 110
The Safety Triad 111
   Procedures 111
   Equipment 112
   Personnel 114
Firefighter Safety Responsibilities 117
   The Department 117
   The Team 118
   The Individual Firefighter 119
Lessons Learned 120
Key Terms 121
Review Questions 121
Endnote 122
Additional Resources 122

CHAPTER 6
PERSONAL PROTECTIVE CLOTHING AND ENSEMBLES

Objectives 125
Introduction 125
Personal Protective Equipment Factors 126
   Standards and Regulations 126
Types of Personal Protective Equipment 127
   Ensembles 127
   Miscellaneous PPE Components 134
Care and Maintenance of Personal
  Protective Equipment 136
Personal Protective Equipment
  Effectiveness: "Street Smarts" 137
   Good PPE Habits and Attitude 137
   Streetsmart Suggestions 138
Lessons Learned 139
Key Terms 139
Review Questions 139
Additional Resources 140

CHAPTER 7
SELF-CONTAINED BREATHING APPARATUS

Objectives 143
Introduction 143
Conditions Requiring Respiratory Protection . . . 145
   Oxygen-Deficient Environments 145
   Elevated Temperatures 145
   Smoke 145
Effects of Toxic Gases and Toxic
  Environments 146
   Carbon Monoxide 146

Legal Requirements for Self-Contained
   Breathing Apparatus Use................. 148
   Title 29 Code of Federal Regulations,
      Section 1910.134........................ 148
   NFPA 1500: Standard on Fire Department Occupational
      Safety and Health Program................. 148
Limitations of Self-Contained
   Breathing Apparatus................... 148
   SCBA Design and Size..................... 150
   Limitations of the SCBA User............... 150
Types of Self-Contained Breathing
   Apparatus............................. 151
   Open-Circuit Self-Contained Breathing Apparatus....... 152
   Closed-Circuit Self-Contained Breathing Apparatus..... 158
   Open-Circuit Supplied Air Respirators................. 158
Donning and Doffing Self-Contained
   Breathing Apparatus................... 158
   General Considerations................... 159
   Storage Case............................ 159
   Seat-Mounted Apparatus.................. 162
   Compartment or Side-Mounted Apparatus........... 164
   Donning the SCBA Face Piece................ 164
   Removing/Doffing the SCBA Unit............. 166
Self-Contained Breathing Apparatus
   Operation and Emergency Procedures...... 168
   Safe Use of SCBA........................ 168
   Operating in a Hostile Environment............ 170
   Restricted Openings...................... 170
   Emergency Procedures.................... 171
Inspection and Maintenance of Self-Contained
   Breathing Apparatus................... 172
   Daily Maintenance....................... 172
   Monthly Maintenance.................... 172
   Annual and Biannual Maintenance.............. 172
   Changing SCBA Cylinders.................. 172
   Servicing SCBA Cylinders.................. 176
Lessons Learned.......................... 180
Key Terms................................ 180
Review Questions......................... 181
Endnote.................................. 182
Additional Resources...................... 182

CHAPTER 8
PORTABLE FIRE EXTINGUISHERS

Objectives............................... 185
Introduction............................. 185
Fire Classification and Risk................. 185
   Class A................................ 185
   Class B................................ 185
   Class C................................ 186
   Class D................................ 186
   Class K................................ 186

Types of Fire Extinguishers................. 187
   Types of Extinguishing Agents.............. 187
   Kinds of Extinguishers.................... 188
Rating Systems for Portable Extinguishers..... 193
   Class A................................ 193
   Class B................................ 194
   Class C................................ 194
Limitations of Portable Extinguishers........ 194
Portable Extinguisher Operation............. 195
Care and Maintenance of Portable
   Extinguishers......................... 197
Inspection Requirements................... 197
Lessons Learned.......................... 198
Key Terms................................ 198
Review Questions......................... 199
Endnotes................................. 199
Additional Resources...................... 199

CHAPTER 9
WATER SUPPLY

Objectives............................... 203
Introduction............................. 203
Sources of Water Supply................... 203
   Groundwater........................... 203
   Surface Water.......................... 204
   Mobile Water Supply Apparatus............ 204
   Tanks, Ponds, and Cisterns................ 205
Water Distribution Systems................. 205
Fire Hydrants............................ 206
   Wet Barrel............................. 206
   Dry Barrel............................. 207
   Dry Hydrant........................... 208
   Specialty Hydrants...................... 209
Valves Associated with Water
   Distribution Systems.................. 209
Rural Water Supply....................... 210
   Portable Water Tanks.................... 210
   Tender Operation....................... 211
Pressure Associated with Water
   Distribution Systems.................. 212
Testing Operability and Flow of Hydrants..... 213
Determining Static, Residual, and Flow Pressures.. 214
Obstructions and Damage to Fire
   Hydrants and Mains.................... 215
Lessons Learned.......................... 216
Key Terms................................ 217
Review Questions......................... 218
Endnotes................................. 218
Additional Resources...................... 218

CHAPTER 10
FIRE HOSE AND APPLIANCES

Objectives	221
Introduction	221
Construction of Fire Hose	221
Care and Maintenance of Fire Hose	223
Types of Hose Coupling	225
Care and Maintenance of Couplings	226
Hose Tools and Appliances	227
Coupling and Uncoupling Hose	228
Hose Rolls	230
Straight/Storage	230
Single Donut	231
Twin or Double Donut	231
Hose Carries	234
Drain and Carry	234
Shoulder Loop Carry	234
Single-Section Street Drag	236
Hose Loads	236
Accordion Load	237
Flat Load	237
Horseshoe Load	239
Finish Loads and Preconnected Loads	243
Flat Load, Minuteman Load, and Triple-Layer Load	245
Stored Hose Loads/Packs	248
Wildland Firefighting Hose Loads	248
Advancing Hoselines— Charged/Uncharged	251
Into Structures	253
Up and Down Stairs	254
Using a Standpipe System	255
Working Hose Off Ladders	257
Establishing a Water Supply Connection	261
From Hydrants	262
From Static Water Supplies	264
Extending Hoselines	265
Replacing Sections of Burst Hose	268
Hose Lay Procedures	269
Forward Lay	270
Reverse Lay	270
Split Lay	271
Deploying Master Stream Devices	271
Service Testing of Fire Hose	274
Lessons Learned	276
Key Terms	276
Review Questions	278
Endnotes	278
Additional Resources	278

CHAPTER 11
NOZZLES, FIRE STREAMS, AND FOAM

Objectives	281
Introduction	281
Definition of Fire Stream	281
Nozzles	281
Solid Tip or Stream	282
Fog	283
Straight Stream	285
Special Purpose	286
Playpipes and Shutoffs	286
Nozzle Operations	287
Operating Hoselines	288
Small-Diameter Handlines	288
Medium-Diameter Handlines	289
Master Stream Devices	289
Stream Application, Hydraulics, and Adverse Conditions	289
Direct, Indirect, and Combination Attack	289
Basic Hydraulics, Friction Loss, and Pressure Losses in Hoselines	290
Adverse Conditions That Affect Fire Streams	295
Types of Foam and Foam Systems	295
Foam Characteristics	295
Classification of Fuels	296
Class A	296
Class B	297
Application of Foam	297
Fog Nozzle versus Foam Nozzles	300
Lessons Learned	304
Key Terms	304
Review Questions	306
Endnotes	306
Additional Resources	306

CHAPTER 12
PROTECTIVE SYSTEMS

Objectives	309
Introduction	309
Detection Systems	309
People or Manual Systems	309
Heat Detectors	310
Smoke Detectors	311
Gas Detectors	313
Flame Detectors	313
Sprinkler Systems	314
Sprinklers and Life Safety	314

Sprinkler Head Design and Operation 315
Types of Sprinkler Systems 318
    Wet Pipe Systems. 318
    Dry Pipe Systems. 319
    Deluge Systems 320
    Preaction Systems 321
    Residential Systems 321
Sprinkler System Connections and Piping 321
Control Devices for Sprinkler Systems. 324
Returning Sprinkler Systems to Service 325
Standpipe Classifications 329
Standpipe System Connections and Piping. . . . 330
Alarms for Standpipes and Sprinklers 331
Other Protective Systems 331
    Local Application and Hood Systems. 332
    Total Flooding Systems 332
Fire Department Operations
    with Protective Systems 333
    Standpipe Operations. 334
    Sprinkler System Operations 335
    Detector Activation Operations 335
    Operations for Other Protective Systems 336
Lessons Learned 336
Key Terms 337
Review Questions 338
Endnotes 338
Additional Resources 338

CHAPTER 13
BUILDING CONSTRUCTION

Objectives 341
Introduction 341
Building Construction Terms and Mechanics 342
    Types of Loads. 342
    Imposition of Loads 344
    Forces 345
Structural Elements. 346
    Beams 346
    Columns 346
    Walls 347
    Connections 347
Fire Effects on Common Building
    Construction Materials. 347
    Wood 348
    Steel. 349
    Concrete 349
    Masonry. 350
    Composites 350
Types of Building Construction 352
    Type I: Fire-Resistive. 353
    Type II: Noncombustible 354
    Type III: Ordinary 354

    Type IV: Heavy Timber 355
    Type V: Wood Frame 356
    Other Construction Types. 357
    Relationship of Construction Type to Occupancy Use 359
Collapse Hazards at
    Structure Fires 360
    Trusses. 361
    Void Spaces 362
    Roof Structures 362
    Stairs 363
    Parapet Walls 363
    Collapse Signs 364
    Buildings under Construction. 364
    Preparing for Collapse 364
Lessons Learned 366
Key Terms 366
Review Questions 368
Additional Resources 368

CHAPTER 14
LADDERS

Objectives 371
Introduction 371
Ladder Terminology 371
    Parts of a Ladder 371
Ladder Companies 373
Types of Truck-Mounted Ladders 373
    Aerial Ladder. 373
    Tower Ladder. 375
    Articulating Boom Ladder 376
Types of Ground or Portable Ladders 376
    Straight Ladder 376
    Extension Ladder 377
    Roof or Hook Ladder 378
    Folding Ladder 379
    A-Frame Combination Ladder 379
    Pompier Ladder 381
Use and Care of Portable or Ground
    Ladders 381
Maintenance, Cleaning, and Inspection. 382
    Cleaning Ladders 383
Ladder Safety 383
Ladder Uses 383
    Access 384
    Rescue 384
    Stability 384
    Ventilation 384
    Bridging 384
    Elevated Streams 384
    Elevated Work Position 384
Ladder Selection. 384
    Butt Section 385
    Fly Section. 385

Special Uses 387
    Removal of Numerous Victims 387
    Chute with a Tarp 388
    Over a Fence 388
    Elevated Hose Streams 388
    Portable Pool 388
    Barrier 389
    Support 389
    Hoist Point 389
    Ventilation Fan Supports 390
Safety 390
    Overhead Obstructions 390
    Climbing Path 391
    Ground Considerations 391
    Ladder Load 391
    Working Off a Ladder 392
Miscellaneous Ladder Information 393
    Ladder Storage 393
    Apparatus Ladder Storage 393
    Ladder Apparatus Parking 393
    Ladder Painting 394
    Certification and Testing Procedures 395
Ladder Skills 395
    Carrying Ladders 395
Raising Skills 399
    Raising Ladders 399
    Rung and Beam Raises 400
    Leg Lock 408
    Carrying Tools 408
    Mounting and Dismounting 410
    Engaging the Hook on a Hook Ladder 412
    Roof Ladder Deployment 412
    Hoisting Ladders by Rope 412
Lessons Learned 415
Key Terms 415
Review Questions 416
Additional Resources 416

CHAPTER 15
ROPES AND KNOTS

Objectives 419
Introduction 419
Rope Materials and Their Characteristics 419
    Natural Materials 419
    Synthetic Materials 420
Construction Methods and
    Their Characteristics 422
    Laid (Twisted) 422
    Braided 422
    Braid-on-Braid 423
    Kernmantle 423
Primary Uses 423
    Utility 423
    Firefighting and Rescue Uses 424

Fire Service Knots 424
    Nomenclature of Rope and Knots 425
    Knots 426
Inspection 439
Maintenance 441
    Cleaning 441
    Storage 444
Rigging for Hoisting 449
    Specific Tools and Equipment 449
    Securing a Rope between Two Objects 452
Lessons Learned 457
Key Terms 457
Review Questions 458
Additional Resources 458

CHAPTER 16
RESCUE PROCEDURES

Objectives 461
Introduction 461
Hazards Associated with
    Rescue Operations 461
Search of Burning Structures 461
    Primary Search 465
    Secondary Search 465
Victim Removal, Drags, and Carries 467
    Carries 468
    Drags 469
    Backboard, Stretcher, and Litter Uses 478
Extrication from Motor Vehicles 482
    Tools and Equipment 482
    Scene Assessment (Size-Up) 486
    Establishment of Work Areas 486
    Vehicle Stabilization 487
    Patient Access 488
    Disentanglement 489
    Patient Removal 489
    Scene Stabilization 490
Specialized Rescue Situations and Tools 490
    Vertical Rescue 491
    Water Rescue 492
    Ice Rescue 494
    Structural Collapse Rescue 494
    Trench and Below-Grade Rescue 497
    Confined Space Rescue 498
    Rescue from Electrical Situations 498
    Industrial Entrapment Rescue 499
    Elevator and Escalator Rescue 500
    Farm Equipment Rescue 505
Lessons Learned 505
Key Terms 505
Review Questions 506
Additional Resources 507

CHAPTER 17
FORCIBLE ENTRY

Objectives 510
Introduction 510
   Knowledge. 510
   Skill 511
   Experience 511
Forcible Entry Tools 511
   Striking Tools 512
   Prying and Spreading Tools 513
   Cutting Tools 515
   Pulling Tools 518
   Special Tools 518
Safety with Forcible Entry Tools 519
   Rotary and Chain Saws 520
   Carrying Tools 520
   Hand Tools 521
Maintenance of Forcible Entry Tools 521
Construction and Forcible Entry 522
   Door Construction 522
   Types of Doors. 522
   Locks. 526
   Additional Security Devices. 530
Methods of Forcible Entry 530
   Conventional 531
   Through-the-Lock Forcible Entry. 536
   Operating Lock Mechanisms 538
   Lock Variations 539
Windows. 540
   Forcible Entry of Windows 540
   Glazing 542
   Types of Windows 543
Breaching Walls and Floors. 544
   Techniques for Breaching Walls 544
   Techniques for Breaching Floors 545
Tool Assignments 546
Lessons Learned 546
Key Terms. 546
Review Questions 547

CHAPTER 18
VENTILATION

Objectives 550
Introduction 550
Principles, Advantages, and Effects
   of Ventilation. 550
Heat, Smoke, and Toxic Gases 551
Considerations for Proper Ventilation 551
Fire and Its By-Products 554
Flashover 556
Backdraft (Smoke Explosion). 556
Rollover 560
What Needs to Be Vented? 560
   Voids and Compartments 561
   Cocklofts 561
   Horizontal and Vertical Voids. 562
Air Movement. 562
Types of Ventilation. 563
   Natural. 563
   Mechanical 563
Mechanics of Ventilation. 567
   Vertical Ventilation. 567
   Horizontal Ventilation 567
Ventilation Techniques 568
   Break Glass 568
   Open Doors 569
   Effects of Glass Panes 569
   Rope and a Tool 570
   Hook or Pike Pole 570
   Iron or Halligan 570
   Ax 571
   Portable Ladder 571
   Negative Pressure Ventilation. 572
   Positive Pressure Ventilation 573
Roof Ventilation 575
   Expandable Cut 575
   Center Rafter Cut (Louver) 577
   Triangular Cut 579
   Trench Cut or Strip Cut 579
   Inspection Cut 581
   Smoke Indicator Hole 581
Safety Considerations. 581
   Will Ventilation Permit the Fire to Extend? 581
   Will the Escape Route Be Cut Off? 581
   Will Ventilation Endanger Others? 581
   Work in Teams. 582
   Proper Supervision. 582
Obstacles to Ventilation 583
   Access 583
   Security Devices 583
   Height 583
   Poor Planning 584
   Personnel Assignment 584
   Unfamiliar Building Layout 584
   Ventilation Timing 584
   Cut a Roof—Open a Roof 585
Factors Affecting Ventilation. 585
   Partial Openings. 585
   Partially Broken Windows 585
   Screens 586
   Roof Material. 586
   Dropped or Hanging Ceilings. 587
   Building Size. 587
   Weather 588
   Opening Windows 588

Lessons Learned 590
Key Terms 590
Review Questions 590
Additional Resources 590

CHAPTER 19
FIRE SUPPRESSION

Objectives 593
Introduction 593
Elements of Fire Control 593
   Structural Fire Components and
      Considerations 593
   Ground Cover Fire Components and Considerations 595
   Vehicular Fire Components and
      Considerations 599
   Flammable Liquids Fire Components and Considerations . 601
   Flammable Gas Fire Components and Considerations 602
   Process of Fire Extinguishment 603
   Proper Stream Selection 604
Tactical Considerations 606
   Residential Occupancies 611
   Business and Mercantile Occupancies 615
   Multistory Occupancies 616
   Below-Ground Structures or Basements 618
   Structures Equipped with Sprinklers
      or Standpipes 619
   Exposure Fires 620
   Nonstructural Fires 620
Lessons Learned 628
Key Terms 628
Review Questions 629
Additional Resources 630

CHAPTER 20 SALVAGE, OVERHAUL, AND FIRE CAUSE DETERMINATION

Objectives 633
Introduction 633
Salvage Tools and Equipment 633
   Salvage Covers 634
   Floor Runner 634
   Water Vacuum 635
   Miscellaneous Salvage Tools 636
Maintenance of Tools and Equipment
   Used in Salvage 636
   Salvage Cover Folds and Rolls 637
Salvage Operations 639
   Safety Considerations 639
   Stopping Water Flowing from
      Sprinkler Heads 642
   Methods of Protecting Material Goods 642
   Arranging of Furnishings and Salvage Cover Deployment . 642
   Water Removal 644
Salvage Operations in Sprinklered Buildings .. 647
   Post Indicator Valve and Outside Screw
      and Yoke Valve 647
   Sprinkler Stops 648
   Salvage Operations Lessons Learned 648
Overhaul Tools and Equipment 648
   Common Tools 649
   Carry-All 649
Overhaul Operations 649
   Overhauling Roofs 650
   Electronic Heat Sensors 650
   Revisits of the Involved Structure 650
   Debris Removal 651
   Overhaul Operations Lessons Learned 651
Fire Cause Determination Concerns 652
   Preservation of Evidence 653
   Basics of Point of Origin Determination .. 653
   Fire Cause Determination Lessons Learned . 654
Securing the Building 654
Lessons Learned 655
Key Terms 655
Review Questions 656
Additional Resources 656

CHAPTER 21 PREVENTION, PUBLIC EDUCATION, AND PRE-INCIDENT PLANNING

Objectives 659
Introduction 659
Administration of the Fire Prevention
   Division 659
Fire Company Inspection Program 660
   Equipment 660
   Preparation for Inspections 661
   Conducting the Inspection 661
   Typical Violations 662
   Concluding the Inspection 674
   Reinspections 677
Home Inspections 677
Fire and Life Safety Education 679
   Fire and Life Safety Program Presentations .. 680
   Forms of Fire and Life Safety Programs ... 681
Pre-Incident Management Process 684
   Deciding to Preplan 685
   Site Visit 685
   Diagrams 685
   Seek Input from Others 685
   The Finished Document 687
Lessons Learned 687
Key Terms 687
Review Questions 688
Additional Resources 688

CHAPTER 22
EMERGENCY MEDICAL SERVICES

Objectives 691
Introduction 691
Roles and Responsibilities of an
   Emergency Care Provider 691
     Key Responsibilities 692
     Legal Considerations for Emergency Care Providers 694
     Interacting with Emergency Medical
       Services Personnel 694
Safety Considerations 695
     Analyzing the Safety of the Emergency Scene 695
     Firefighter Physical and Mental Health 696
     Infection Control 696
Assessing a Patient 700
     Performing an Initial Assessment 700
     Vital Signs and the Focused History
       and Physical Exam 704
     Patient Findings 707
Cardiopulmonary Resuscitation/AED 707
Bleeding Control and Shock Management 709
     Internal and External Bleeding 709
     Caring for Patients with Internal Bleeding 710
     Caring for Patients with External Bleeding 710
     Types of Wounds Requiring First Aid 712
     What Is Shock? (Hypoperfusion) 713
     Recognizing the Signs and Symptoms of Shock
       (Hypoperfusion) 713
     Caring for Patients in Shock 713
Emergency Care for Common Emergencies 714
     Trouble Breathing 714
     Chest Pain 714
     Medical Illnesses 714
     Allergic Reactions 714
     Thermal Burns 715
     Chemical Burns 716
     Poisoning 717
     Fractures and Sprains 718
Lessons Learned 718
Key Terms 719
Review Questions 720
Additional Resources 720

CHAPTER 23
FIREFIGHTER SURVIVAL

Objectives 723
Introduction 723
Incident Readiness 723
     Personal Protective Equipment 723

     Personal Accountability 725
     Fitness for Duty 726
Safety at Incidents 728
     Team Continuity 728
     Orders/Communication 729
     Risk/Benefit 730
     Personal Size-Up 730
     Rehabilitation 730
     Rapid Intervention Teams 732
Firefighter Emergencies 733
     Rapid Escape 733
     Rapid Escape Steps 734
     Lost, Trapped, and Injured Firefighters 735
     Post-Incident Survival 736
Lessons Learned 738
Key Terms 739
Review Questions 739
Endnotes 739
Additional Resources 740

CHAPTER 24
TERRORISM AWARENESS

Objectives 743
Introduction 743
Types of Terrorism 747
Potential Targets 748
Indicators of Terrorism 750
HAZMAT Crimes 752
     Clandestine Labs 752
Incident Actions 756
General Groupings of Warfare Agents 757
     Nerve Agents 757
     Incendiary Agents 758
     Blister (Vesicants) 759
     Blood and Choking Agents 759
     Irritants (Riot Control) 759
     Biological Agents and Toxins 760
     Radioactive Agents 761
     Other Terrorism Agents 761
Detection of Terrorism Agents 762
Federal Assistance 762
Lessons Learned 764
Key Terms 764
Review Questions 765
Endnote 765
Additional Resources 765

Glossary **767**
Acronyms **783**
Other Fire Science Titles from Delmar **784**
Index **789**

DEDICATION

Dedicated to the courageous firefighters and emergency responders who have given of themselves the greatest sacrifice, their lives. On September 11, 2001, the fire and emergency service community changed forever, and as we continue on we are left with the scar of this day and the tears of many loved ones left behind. We share in the heartache of the loss of every single firefighter and emergency responder on that day and others. Let their lives shine on in the dedication and bravery of those left to respond when the tones drop, the bells ring, and the sirens blare.

This text is also dedicated to the driving force behind the continuation of firefighter heritage, the sharing of wisdom and experience, and the art of discovery and learning—trainers and educators. Every single classroom session, practical scenario, and review session directly affects the quality of response the fire service provides. Never underestimate the power of positive change the training and education community holds.

In honor and support of all Fire Service Educators, we are privileged to announce that Delmar Learning, a Thomson company, will donate a portion of the proceeds to the National Fire Academy Alumni Association (NFAAA) for every copy of *The Firefighter's Handbook* we sell. The NFAAA was selected as the sole recipient of this contribution because of the similarities of our missions and our belief that NFAAA makes a positive difference in the education, safety, and welfare of firefighters.

And to every firefighter who has touched the life of someone in need and made a positive difference—you are truly the epitome of human compassion and selflessness. Don't ever stop caring.

FOREWORD

So much has changed in our country and our world since the first edition of *The Firefighter's Handbook* was published back in 2000. Of course the tragic events of September 11, 2001, have caused many changes in the lives of America's firefighters and how we do business. Sure, we used to talk about terrorism—but not with the urgency and realism that permeates today's discussions. The study of weapons of mass destruction (WMD) was an emerging specialty and today is mandatory reading for every firefighter. The changes are not limited to terrorism and WMD but have spilled over into rapid intervention team tactics, firefighter survival, and the value and hazards of live fire training.

However, some things have not changed. The value and importance of preparedness and training for every possible scenario has not changed. The principle of treating and serving the people who rely on us with respect and dignity has not changed. The idea that firefighters are held to a higher standard of service and duty has not changed. Even in the face of widespread changes in the world and in firefighting, there are some key principles that stay the same. These principles make firefighting a proud and honorable tradition and tie firefighters to the communities they serve.

This second edition of *The Firefighter's Handbook* is a reflection of the new world that today's firefighters live in. We have updated and revised many chapters and sections relating to the new threat of terrorism and the part that the fire service plays in handling this heightened threat. Other areas of study have also been improved and expanded such as self-contained breathing apparatus, firefighter safety, rescue procedures, and firefighter survival. All of the areas that we all deal with on a much more frequent basis than terrorism have received the same attention and have been noticeably improved. All of this has been done to keep you, the first line of defense, as technically prepared and tactically ready as possible. It has become apparent that every firefighter in every city and town in America must be able to respond quickly and effectively to any and every emergency that arises there. Every effort has been made to create a new and updated handbook that can be used as a reference for chief and company officers, training manual for company officers and firefighters, and textbook for candidates preparing to enter the most difficult and rewarding profession, firefighting.

John Salka, Jr.
Battalion Chief
New York City Fire Department

ABOUT OUR AUTHORS

The expertise, dedication, and passion of our contributing authors have created a text that determines a standard of excellence in the education of our nation's firefighters.

To continue in this standard of excellence for the second edition of *The Firefighter's Handbook: Essentials of Firefighting and Emergency Response,* our authors have dedicated their time to ensure that the book remains current to the 2002 edition of *NPFA Standard 1001,* as well as the changing landscape of the fire service world. Thanks to our revising authors Andrea Walter, David Dodson, Dennis Childress, Chris Hawley, and Marty Rutledge, as well as to the outstanding authors who provided the foundation of this textbook: Ron Coleman, Thomas J. Wutz, Willis T. Carter, Frank J. Miale, T. R. (Ric) Koonce III, Robert F. Hancock, Robert Morris, Geoff Miller, and Donald C. Tully.

Andrea A. Walter

Author of Chapter 22, Emergency Medical Services. Revising author of 1, 2, 3, and 4.

Andrea A. Walter is a firefighter with the Washington Metropolitan Airports Authority and a member and former officer of the Sterling Volunteer Rescue Squad. Walter has been active in the fire and emergency services community for many years, serving as the Manager of the Commission on Fire Accreditation International for the International Association of Fire Chiefs and assisting in a variety of projects with the National Volunteer Fire Council, Women in the Fire Service, and the United States Fire Administration. She has over fifteen years of experience in the fire and emergency services. In addition to being an author for this text, Walter also took on the expanded role of serving as the project's Content Editor. She is also an author and the Content Editor for Delmar Learning's *First Responder Handbook: Fire Service Edition* and the *Law Enforcement Edition.*

Ronny J. Coleman

Author of Chapter 1, Overview of the History, Tradition, and Development of the American Fire Service.

Chief Coleman is a nationally and internationally recognized member of the fire service who formerly served as the Chief Deputy Director, Department of Forestry and Fire Protection, and as California State Fire Marshal. He has served in the fire service for thirty-eight years. Previously he was Fire Chief for the Cities of Fullerton and San Clemente, California, and was the Operations Chief for the Costa Mesa Fire Department. Chief Coleman possesses a Master of Arts Degree in Vocational Education from Cal State Long Beach, a Bachelor of Science Degree in Political Science from Cal State Fullerton, and an Associate of Arts Degree in Fire Science from Rancho Santiago College. He has served in many elected positions in professional organizations, including President, International Association of Fire Chiefs; Vice President, International Committee for Prevention and Control of Fire; and President, California League of Cities, Fire Chiefs Department. He is the author of *Going For Gold*, Delmar Thomson Learning.

Thomas J. Wutz

Author of Chapter 2, Fire Department Organization, Command, and Control; Chapter 7, Self-Contained Breathing Apparatus.

Chief Wutz has been involved in the fire service for more than thirty years in both volunteer and military fire departments. He is currently Chief of Fire Services, New York State Office of Fire Prevention and Control. In this position, his duties include supervision and management of the state's outreach training program, delivered by 230 instructors assigned to fifty-seven counties. In addition, he is responsible for curriculum development and implementation of new training programs, New York State's Wireless 9-1-1 program, and state fire mobilization and mutual plan and response. On completion of a twenty-eight-year career, he recently retired as Fire Chief of the 109th Airlift Wing, New York Air National Guard. Chief Wutz is also a member of the faculty in the Fire Science Program at Schenectady County Community College, Schenectady, New York, and a firefighter with the Midway Fire Department, Town of Colonie, New York.

Willis T. Carter

Author of Chapter 3, Communications and Alarms.

Chief Willis Carter has been a member of the fire service for over thirty years. He began his career in 1972 as a firefighter with the Shreveport Fire Department, and for the past twenty-five years has served as the Chief of Communications for the department. Carter is responsible for the management and operations of the Fire Communications Center, which serves as the Primary Public Safety Answering Point (PSAP) for the Caddo Parish, Louisiana, 9-1-1 system.

In addition to his work in the fire service, he is active at the national level. He has served as past president of the Louisiana Chapter of the Association of Public Safety Communications Officials (APCO) and currently serves as Executive Council representative for the state of Louisiana. His other work with APCO includes serving as Chairperson for the Membership Task Force, and as a member of APCO Project 37 Team (Telecommunicator

Certification Program). He also serves as a member of APCO Bulletin Editorial Advisory Board. In addition to his work with APCO, he is a member of the International Fire Chief's Association and the National Emergency Number Association. He is also an Assessment Team Leader for the Commission on Law Enforcement Accreditation (CALEA). Carter led the effort by the Shreveport Fire Department Communications Center to become the first Public Safety Communications Center in the state of Louisiana to achieve accredited status through CALEA.

Frank J. Miale

Author of Chapter 4, Fire Behavior; Chapter 14, Ladders; and Chapter 18, Ventilation.

Miale, a Battalion Chief with over thirty years in the FDNY, recently retired. A twenty-five-year active member in his local Volunteer Lake Carmel Fire Department, he maintains a busy role as treasurer and training instructor. A former high school teacher, he holds two Bachelor of Science degrees with several concentrations in Education, Biology, and Fire Administration. During his career in the FDNY, he taught at the NYC fire academy, participated in the introduction of a communication system using apparatus-mounted computers, and headed a special Emergency Command Unit while an active line officer. Formerly the Training Officer for the 27th Battalion in the FDNY, he taught many ladder company and ventilation courses throughout the country. His career was spent primarily in busy ladder companies in Brooklyn, Harlem, and the South Bronx sections of New York City prior to promotion to Chief Officer. He is the recipient of nine awards for courage and valor, including two department medals from the FDNY, and has been published many times in *WNYF, Fire Command,* and *Fire Service Today.*

David W. Dodson

Author of Chapter 5, Firefighter Safety; Chapter 6, Personal Protective Clothing and Ensembles; and Chapter 23, Firefighter Survival. Revising author of Chapters 5, 6, 9, 10, 11, 12, 13, 16, and 23.

Dodson is a twenty-four-year fire service veteran. He started his fire service career with the U.S. Air Force. He served at Elmendorf AFB in Alaska and spent two years teaching at the USAF Fire School. After the USAF, Dodson spent almost seven years as a Fire Officer and Training/Safety Officer for the Parker Fire District in Parker, Colorado. He became the first Career Training Officer for Loveland Fire and Rescue in Colorado and rose through the ranks, including time as a HAZMAT Technician, Duty Safety Officer, and Emergency Manager for the city. He accepted a Shift Battalion Chief position for the Eagle River Fire District in Colorado before starting his current company, *Response Solutions,* which is dedicated to teaching firefighter safety and practical incident handling. Chief Dodson has served on numerous national boards including the NFPA Firefighter Occupational Safety Technician Committee and the International Society of Fire Service Instructors (ISFSI). He also served as president of the Fire Department Safety Officers' Association. In 1997, Dodson was awarded the ISFSI "George D. Post Fire Instructor of the Year." He is also the author of *Fire Department Incident Safety Officer,* published by Delmar Learning, a Thomson Company.

T. R. (Ric) Koonce, III

Author of Chapter 8, Portable Fire Extinguishers; Chapter 9, Water Supply; Chapter 10, Fire Hose and Appliances; Chapter 11, Nozzles, Fire Streams, and Foam; and Chapter 12, Protective Systems.

Koonce is an Assistant Professor and Program Head of Fire Science Technology at J. Sargeant Reynolds Community College in Richmond, Virginia. He is a retired Battalion Chief with the Prince George's County (Maryland) Fire Department and has over thirty years of fire service experience. He is an adjunct instructor for the Virginia Department of Fire Programs. He holds two associate degrees, a Bachelor of Science degree in Fire Service Management from University College of the University of Maryland, and a Certificate of Public Management from Virginia Commonwealth University.

Robert F. Hancock
Author of Chapter 15, Ropes and Knots; and Chapter 16, Rescue Procedures.

Hancock is Assistant Chief/Administration with Hillsborough County Fire Rescue in Tampa, Florida, a department that services an area of 931 square miles and a population of over 600,000 with 615 career personnel and 205 volunteers and a $42.7 million budget. He was hired in November 1974 as a firefighter and was promoted through the ranks to his present position in October 1993. He was awarded an Associate of Science Degree in Fire Science, with honors, from Hillsborough Community College. He graduated from the Executive Fire Officers Program at the National Fire Academy and has been certified as an instructor with the State of Florida since 1983. Hancock is chairman of the Florida Fire Chiefs' Disaster Response Communications Sub-Committee, charged with identifying short- and long-term solutions to the disaster response communication issue statewide. He is Rescue Series Editor, a contributing author for Delmar Thomson Learning, and a member of Florida EDACS PS User's Group, serving as President for 1999.

Robert Morris
Author of Chapter 17, Forcible Entry.

Morris is a thirty-year veteran of the New York City Fire Department and has been assigned to some of the busiest fire companies in New York City, including Ladder Company 42, Engine 60 in the Bronx, and Rescue Company 3 in Manhattan. After serving in the Bronx and Harlem, he served as Company Commander of Ladder Company 28. Captain Morris is currently Company Commander of Rescue Company 1 in Manhattan. He is an Instructor with the Connecticut State Fire Academy, the New York City Fire Department Institute, a national lecturer, and a Contributing Editor for *Firehouse Magazine*. Captain Morris is the recipient of seventeen meritorious awards, including three department medals.

Dennis Childress
Author of Chapter 19, Fire Suppression.

Childress is with the Orange County Fire Authority in Southern California and has been in the fire service for just over thirty-five years. He is a Certified Chief Officer with the state of California, and he holds an Associate of Arts degree in Fire Science and a Bachelor of Science degree in Fire Protection Administration. He holds a seat on the Board of Directors for the Southern California Fire Training Officers Association, chairs the California State Firefighters Association Health and Safety Committee, and sits on the Statewide Training and Education Advisory Committee for the State Fire Marshal's office. He is a principal member of the NFPA 1500 Fire Service Occupational Safety and Health Committee and the NFPA 1561 Standard for Emergency Services Incident Management System Committee. He has authored a number of articles in fire service publications over the years, and he has also been an instructor in Fire Command and Management in the California State Fire Training System for over twenty years as well as an instructor for the National Fire Academy. He is the original author of the *Workbook to Accompany Firefighter's Handbook: Essentials of Firefighting and Emergency Response.*

Geoff Miller
Author of Chapter 20, Salvage, Overhaul, and Fire Cause Determination.

Miller is a twenty-nine year veteran of the fire service and is currently a line Battalion Chief with the Sacramento Metropolitan Fire District in California. Previous assignments have included four years as the district's Training Officer, ten years as a line Captain, and two years as an Inspector. He has been involved in several California Fire Fighter I and II curriculum development workshops as well as participating on the rewrite of Fire Command 1A and 1B. He was also on the IFSTA Material Review Committee for Fire Department Company Officer, third edition. He is happily married, has two daughters, and lives in El Dorado Hills, California.

Donald C. Tully

Author of Chapter 21, Prevention, Public Education, and Pre-Incident Planning.

Tully is a member of the Orange County, California, Fire Authority. With thirty years in the fire service, he has also been a Division Chief/Fire Marshal in Buena Park and Westminster, California, for ten years, and a Fire Technology Instructor at Santa Ana College, California. He is Past President of the Orange County Fire Prevention Officers' Association and was a member of IFSTA's Fire Investigation Committee. He also served as a member of NFPA Committees 1221 (CAD Dispatch and Public Communications) and 72 (Fire Alarms), and as a member of the California State Fire Marshal Committees on Fire Sprinklers and Residential Care Facilities (ad hoc committees). He is a California State Certified Chief Officer, Fire Officer, Fire Investigator, Fire Prevention Officer, and Fire Service Instructor and Technical Rescue Specialist.

Chris Hawley

Author and revising author of Chapter 24, Terrorism Awareness.

Hawley is a retired Fire Specialist with the Baltimore County Fire Department. Prior to this post, he was assigned as the Special Operations Coordinator and was responsible for the coordination of the Hazardous Materials Response Team and the Advanced Technical Rescue Team along with the team leaders. He has served on development teams for local, state, and federal projects, including the National Fire Academy, and provides HAZMAT and terrorism response training nationwide. Hawley is also the author of four Delmar Learning textbooks, including *Hazardous Materials Response and Operations, Hazardous Materials Incidents, Haz-Mat Air Monitoring and Detection Devices,* and *Fire Department Response to Sick Buildings.* He is currently the owner of FBN Training, which provides a wide variety of emergency response training, including hazardous materials, confined space, technical rescue, and emergency medical services, as well as consulting services to emergency services and private industry.

Marty L. Rutledge

Revising author of Chapters 7, 8, 14, 15, 20, and 21.

Rutledge is a Firefighter/Engineer, ARFF Specialist, and EMS Program Manager for Loveland Fire and Rescue in Loveland, Colorado. He is a member of the Fire Certification and Advisory Board to the Colorado Division of Fire Safety, as well as serving as the State First Responder program coordinator. He is also a member of the Colorado State Fire Fighter's Association and has over thirteen years of fire and emergency services experience in both volunteer and career ranks. Rutledge has authored and served as technical expert for a supplementary firefighter training package for Delmar Learning's *The Firefighter's Handbook* and as co-author for Delmar Learning's *First Responder Handbook: Fire Service Edition* and *First Responder Handbook: Law Enforcement Edition.*

PREFACE

Welcome new recruits, volunteers, and experienced firefighters to *The Firefighter's Handbook: Basic Essentials of Firefighting.*

The Firefighter's Handbook is a comprehensive guide to the basic principles and fundamental concepts involved in firefighting, emergency medical services, and terrorism response as defined by the 2002 edition of the *National Fire Protection Association 1001 Standard for Firefighter Professional Qualifications.* The text can be used by both new and experienced firefighters to study the basic skills required to perform a wide variety of firefighting and emergency service activities.

As a firefighter you will make a difference in the lives of many. Use your knowledge, practice your skills, and above all—be safe.

Development of This Text

The Firefighter's Handbook was created to fill a void in the firefighting and emergency service education system. Through the dedication of our authors, content and technical reviewers, as well as our Fire Advisory Board members, this new edition of *The Firefighter's Handbook* continues to remain up to date with the changing landscape of the fire service world. This text is designed to reflect revisions to the requirements set forth in the 2002 edition of *NFPA Standard 1001* and presents the information in a realistic and challenging way. The content is written in a clear and concise manner, and step-by-step photo sequences illustrate the need-to-know Job Performance Requirements that are so critical to hands-on training. A special emphasis on safety and the development of critical thinking skills through featured text boxes ensures that both aspiring and experienced firefighters have the knowledge they need to effectively respond to fires and other emergencies.

New for the release of the 2002 edition of the *NFPA Standard 1001 The Firefighter's Handbook: Basic Essentials of Firefighting* retains all the features of the original text, but excludes the coverage of hazardous materials. This is an excellent choice for fire departments, academies, and schools in which hazardous materials are taught in a separate course with separate learning materials.

The *Basic Essentials of Firefighting* textbook meets the requirements of the *NFPA Standard 1001* when taught in conjunction with a hazardous materials course that meets the requirements of *NFPA Standard 472,* Awareness and Operations levels.

Organization of This Text

The Firefighter's Handbook: Basic Essentials of Firefighting consists of twenty-four chapters, including coverage of terrorism. All the essential information—from the history of the fire service to the governing laws and regulations, from the use of apparatus and equipment to the practice of procedures, from understanding fire behavior and building construction to effective planning and prevention measures—is covered in this text. The chapters are set up to deliver a straightforward, systematic approach to training, and each includes an outline, objectives, introduction, lessons learned, key terms, review questions, and a list of additional resources. Also included at the front of the book is an NFPA 1001 Correlation Guide and Job Performance Requirement Guide that correlates the requirements outlined in the Standard to the content of *The Firefighter's Handbook* by chapter and page references. These resources can be used as a quick reference and study guide.

The Firefighter's Handbook: Basic Essentials of Firefighting contains many new updates and additional information to address the needs of the fire service today:

- *Safety:* Safety—when responding to incidents in apparatus, and while performing scene assessment on vehicle accidents—is thoroughly covered. New sections on unique fire events and "reading smoke" encourage firefighters to apply an understanding of basic fire behavior when responding to structural fires. In addition, expanded content on the two in/two out rule educates firefighters on how to rescue their own in emergencies.

- *Current Technology:* This text offers information on the latest technology, including information on up-to-date communication systems and a new section on thermal imaging cameras.

- *Building Construction:* Chapter 13 was completely revised to address new building structures and additional considerations in structural collapse. Also included is a special section dedicated to the expert in the field, Francis L. Brannigan.

- *Ladders:* Chapter 14 was thoroughly revised to reflect a variety of procedures utilizing ladders in rescue situations.

- *Terrorism:* In light of the events of September 11, 2001, and the worldwide terrorist attacks that followed, Chapter 24 has been thoroughly revised to reflect the latest threats to our nation.

- *New Photos:* The inclusion of new photos brings the text up to date with the latest in apparatus, tools, equipment, and procedures in the fire service.

- *New Feature Text Boxes:* Additional text boxes and featured articles provide helpful tips, advice on safety, and important information for firefighters.

FEATURES

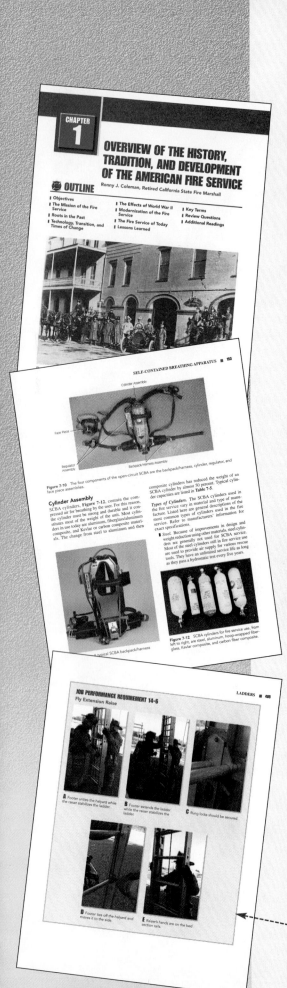

The Firefighter's Handbook contains a number of features that set it apart from other basic fire service texts. What is unique about this text is that it offers a realistic approach to the world of firefighting and emergency response. It is comprehensive in coverage, including all the need-to-know information, but presents the content in a clear and concise manner, so it is easy to read, follow, and understand. What is essential is not only acquiring the knowledge, but more importantly, putting it into practice. *The Firefighter's Handbook* emphasizes this fact through the use of step-by-step photo sequences and text descriptions to illustrate critical Job Performance Requirements. Firefighters throughout the nation also continue to contribute their stories to the book, bringing insight, advice, and experience to the text.

The Firefighter's Handbook continues to recognize that today's firefighters must respond to more than just fire emergencies. Firefighters are expected to effectively respond to medical emergencies as well as terrorism incidents. In response, the textbook includes a chapter on emergency medical services, as well as a chapter discussing terrorist activities. Tips for safety that are vital to survival and critical to the success of the operation on the scene of an incident are integrated throughout the chapters.

In addition, *The Firefighter's Handbook* provides instructors with handy reference tools and supplemental materials that alleviate those heavy preparation commitments. A new emphasis on the modularized lesson plans in the Instructor's Curriculum Kit and Instructor's Curriculum CD-ROM for departments provides instructors with flexible lecture outlines that will meet individual department or academy schedules.

How to Use This Text

The following suggests how you can best utilize the features of this text to gain competence and confidence in learning firefighting essentials.

NFPA Standard 1001/472 Correlation Guide

This grid provides a correlation between *The Firefighter's Handbook* and the requirements for Firefighter I and II, as stipulated by the *NFPA Standard for Firefighter Professional Qualifications,* 2002 Edition, Chapters 5 and 6. These sections from the Standard are correlated to the textbook chapters and are referenced by page numbers.

Job Performance Requirement Correlation Guide

This grid provides an outline of the Job Performance Requirements stipulated by the *NFPA Standard for Firefighter Professional Qualifications,* 2002 Edition, Chapters 5 and 6, and the corresponding step-by-step photo sequences illustrated in *The Firefighter's Handbook.* Additional supplemental skills are also included to encourage practice of essential skills for on-the-job training. A quick reference for studying and reviewing important hands-on skills, this grid correlates the JPR to the textbook chapters and references the page numbers.

Job Performance Requirements

Step-by-step photo sequences illustrating important procedures are integrated throughout the chapters. These are intended to be used as a guide in mastering the job performance requirement skills and to serve as a review guide reference.

Street Stories

Each chapter opens with a personal experience written by noted contributors from across the nation. These personal accounts draw you into the minds of those who wrote them and help to reinforce to you why the subsequent chapter can make all the difference in the world.

Streetsmart Tips

As is true of any profession, sometimes experience can be the best teacher. These tips are power-packed with a wide variety of hints and strategies that will help you to become a streetsmart firefighter.

Firefighter Facts

These boxes offer a detailed snapshot of facts based on firefighting history, experience, and recorded data to provide thought-provoking information.

Notes

The Note feature highlights and outlines important points for you to learn and understand. Based on key concepts, this "must know" material is an excellent study resource.

Safeties and Cautions

As a firefighting professional, you will face situations in which you will need to react immediately in order to ensure your safety and the safety of those with you. These tips emphasize when and how to react safely.

Lessons Learned

The Lessons Learned feature summarizes the main points presented in the chapter and is ideal for review purposes.

Review Questions

At the end of every chapter, the review questions assist you with the learning process and help you develop the necessary critical thinking skills.

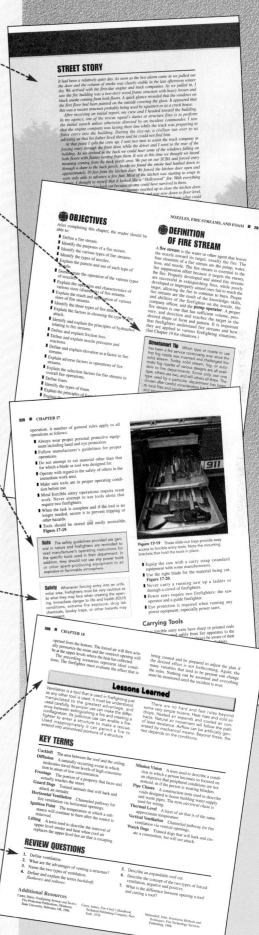

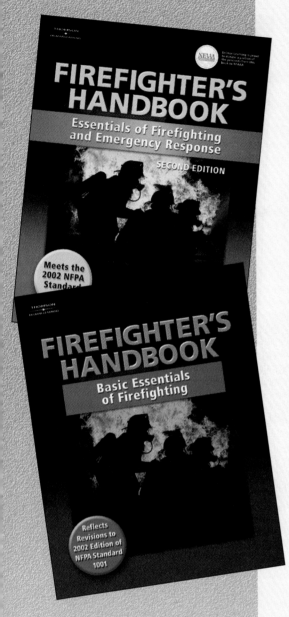

Supplemental Curriculum Package

The Firefighter's Handbook

This text was created not only as a stand-alone manual for firefighters, but as a special package of materials for the full instructional experience. The supplement package provides a variety of tools for students and instructors to enhance the learning experience.

Instructor's Curriculum Kit

The *Instructor's Curriculum Kit* is designed to allow instructors to run programs according to the standards set by the authority having jurisdiction where the course is conducted. It contains the information necessary to conduct Firefighter I, Firefighter II, hazardous materials awareness, and hazardous materials operations courses. It is divided into sections to facilitate its use for training:

Administration: Provides the instructor with an overview of the various courses, student and instructor materials, and practical advice on how to set up courses and run skill sessions.

Modularized Lesson Plans: Are ideal for instructors, whether they are teaching at fire departments, academies, or longer-format courses. Each section presents learning objectives, recommended time allotment, equipment and reading assignments for each lesson and outlines key concepts presented in each chapter of the text with coordinating PowerPoint slides, textbook readings, and Job Performance Requirement and Supplement Skill sheets.

Equipment Checklist: Offers a quick guide for ensuring the necessary equipment is available for hands-on training.

Job Performance Requirement and Supplemental Skills Sheets: Outline important skills that each candidate must master to meet requirements for certification and provide the instructor with a handy checklist. Also includes a *Job Performance Requirement Correlation Guide,* which cross-references the Standard with the *Job Performance Requirements* outlined in *The Firefighter's Handbook.* These guides can be used for quick reference when reviewing important skills and for studying for the Firefighter I and II certification exam.

Progress Log Sheets: Provide a system to track the progress of individual candidates as they complete the required skills.

Quick Reference Guides: Contain valuable information for instructors. Included are three grids: *NFPA Standard 1001/472 Correlation Guide* used to cross-reference *The Firefighter's Handbook* with standards 1001/472, *New Edition Correlation Guide* used to cross-reference the revisions between the first and second editions of *The Firefighter's Handbook,* and a *Comparison Guide* that correlates *The Firefighter's Handbook, Second Edition,* to the *IFSTA/Essentials, Fourth Edition,* text.

Answers to Review Questions: Included for each chapter in the text.

Additional Resources: Offer supplemental resources for important information on various topics presented in the textbook.

Also includes: Instructor's Curriculum CD-ROM

Order #: 1-4018-3576-7

Instructor's Curriculum CD-ROM

Available in the *Instructor's Curriculum Kit* and as a separate item, the *Instructor's Curriculum CD-ROM* ensures a complete, electronic teaching solution for *The Firefighter's Handbook: Essentials of Firefighting and Emergency Response, Second Edition*. Designed as an integrated package, it includes the following:

- *PowerPoint Presentations* outline key concepts from each chapter, and contain graphics and photos from the text, as well as video clips, to bring the content to life.
- A *Testbank* containing hundreds of questions helps instructors prepare candidates to take the written portion of the certification exam for Firefighter I and II.
- A searchable *Image Library* containing hundreds of graphics and photos from the text offers an additional resource for instructors to enhance their own classroom presentations or to modify the PowerPoint provided on the CD-ROM.
- *Complete Curriculum* available in Word to allow instructors to add their own notes or revise to meet the requirements of the Authority Having Jurisdiction.

Order #: 1-4018-7175-5

Student Workbook

This is helpful in the classroom setting as a guide for study and a tool for assessing progress. The workbook consists of questions in multiple formats, including new and revised questions to support the second edition of *The Firefighter's Handbook*.

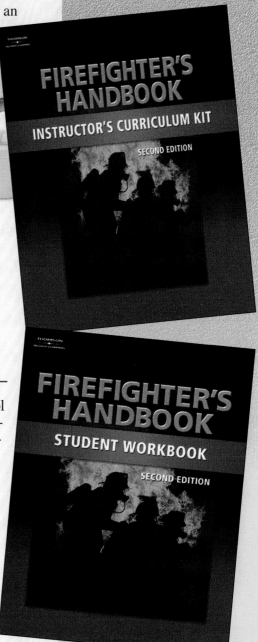

ACKNOWLEDGMENTS

The Firefighter's Handbook: Basic Essentials of Firefighting, remains true to the tradition of Delmar to remain dedicated to the individuals we serve—among them, both aspiring and experienced firefighters. However, we would not be able to accomplish this without the contributions of many professionals whose passion, commitment, and hard work have helped shape a book of which we all can be proud.

Delmar and the contributing authors would like to extend a special thanks to those who served on the Editorial Review Board, as well as those who are members of our Fire Advisory Board. Their expertise and objectivity has provided invaluable insight:

Mike Arnhart, Deputy Chief, High Ridge Fire District, High Ridge, Missouri

Francis Brannigan, (Building Construction chapter) SFPE & Fellow, Port Republic, Maryland

Kevin Barkley, Fire/HazMat Coordinator, Saratoga County Emergency Services, Galway, New York

Robert Bettenhausen, Chief Fire Marshal, Village of Tinley Park, Tinley Park, Illinois

George Braun, Lieutenant, Gainesville Fire Rescue, Gainesville, Florida

Tony Calorel, Senior Instructor, Burlington County Emergency Services Training Center, West Hampton, New Jersey

Steve Chickerotis, Chief of Training, Chicago Fire Department, Chicago, Illinois

Michael J. Connors, Assistant Fire Chief, Naperville Fire Department, Naperville, Illinois

Lee Cooper, Fire Service Specialist, Wisconsin Indianhead Technical College, New Richmond, Wisconsin, and President, Fire Instructors Association of Minnesota, Bloomington, Minnesota

Richard W. Davis, Lieutenant, Loveland Fire and Rescue, Loveland, Colorado

Peter Evers, Captain, Auburn City Fire Department, Auburn, California

Tom Feierabend, Director of Fire Technology, Mount San Antonio College, Walnut, California

Herald Good, Associate Instructor, Virginia Department of Fire Programs, Newport News, Virginia

Craig Hanna, Training Chief, Sioux Falls Fire Rescue, Sioux Falls, South Dakota

Attila Hertelendy, Instructor, University of Nevada Fire Science Academy, Carlin, Nevada

Al Ionnone, Director of Fire Technology, American River College, Sacramento, California

Kim Johnson, MBA, Fitness Trainer, 24-Hour Fitness

John Kingyens, Training Safety Officer, Sarnia Fire Rescue, Canada

Kent D. Neiswender, Supervisor, Office of Training and Certification, New Jersey Division of Fire Safety, Trenton, New Jersey

Bob Leigh, Battalion Chief, Aurora Fire Department, Aurora, Colorado

Ron Marley, Fire Technology Director, Shasta College, Redding, California

Pat McAulliffe, Director of Fire Science, Collin Community College Fire Academy, McKinny, Texas

Peter McBride, Safety Battalion Chief, Ottawa Fire, Canada

David P. Pritchett, Director Georgia Fire Academy, Forsyth, Georgia, and Training Captain, City of Jackson, Jackson, Georgia

Taylor Robertson, District/Training Chief, City of Eugene Fire and EMS Department, Eugene, Oregon

Chris Reynolds, Battalion Chief, Hillsborough County Fire Rescue, Tampa, Florida

John J. Salka, Jr., Battalion Chief, New York City Fire Department

Bob Sanborn, Captain, Bowling Green Fire Department, Bowling Green, Kentucky

Randy Scheerer, Battalion Chief, Newport Beach Fire Department, Newport Beach, California

R. Peter Sells, Chief Training Officer, Toronto Fire Services, Canada

Billy Shelton, Curriculum and ARFF Manager, Virginia Department of Fire Programs, Richmond, Virginia

William Shouldis, Deputy Chief, Philadelphia Fire Department, Philadelphia, Pennsylvania

Fred C. Windisch, Fire Chief, Ponderosa VFD, Houston, Texas, and Chairman Volunteer Chief Officers Section, International Association of Fire Chiefs

We would also like to recognize those individuals who contributed content:

Mike West (Thermal Imaging Article/Photos), Lieutenant, South Metro Fire and Rescue, Instructor, SAFE-IR, Colorado

Tom Wutz (Original Author/Building Construction Chapter), Deputy Chief of Fire Services, New York State Office of Fire Prevention and Control

The majority of photographs shown in this book are the result of numerous days of photo shoots at various locations. A special acknowledgment is owed to four very patient firefighters who facilitated numerous days of shooting. Our appreciation is extended to Kevin P. Terry, Assistant Chief, Fuller Road Fire Department, Patrolman, Town of Colonie Police Department, and New York Regional Team-1 (NYRRT-1) Logistics Liaison; Steven M. Leonardo, Past Chief, Shaker Road, Loudonville Fire Department, Patrolman, Town of Colonie Police Department, and New York Regional Team-1 (NYRRT-1) Rescue Specialist; Warren E. Carr, Jr., Past Chief, S.W. Pitts Hose Company, Inc., Latham, New York, and New York Regional Response Team-1 (NYRRT-1) Team Leader; and Mike Kelleher, Troy Fire Department, Troy, New York, and New York Regional Response Team-1 (NYRRT-1) Team Manager. Your knowledge and willingness were invaluable.

Credit is also owed to the many departments, models, and photographers who shared their time, expertise, equipment, and photographs with us: Fuller Road Fire Department, Albany, New York; S.W. Pitts Hose Company, Inc., Latham, New York; Troy Fire Department, Troy, New York; Albany Fire Department, Albany, New York; Metropolitan Washington Airports Authority, Washington, DC; Loveland Fire and Rescue, Loveland, Colorado; Poudre Fire Authority, Fort Collins, Colorado; South Metro Fire and Rescue, Greenwood Village, Colorado; Hillsborough County Fire Rescue, Tampa, Florida; Baltimore County Fire Department, Baltimore, Maryland; Fairfax County Department of Fire and Rescue, Fairfax, Virginia; Sterling Park Rescue Squad and Sterling Volunteer Fire Department, Sterling, Virginia; Ashburn Volunteer Fire and Rescue Department, Ashburn, Virginia, and Loudoun County Department of Fire Rescue Services, Leesburg, Virginia. Our photographers included Michael Dzaman, Dzaman Photography, Latham, New York; Rick Fulford, Cockeysville, Maryland; Rick Michalo, Brentt Sporn, California Fire Photos, Anaheim, California; and Captain Pete Evers, City of Auburn Fire Department, California.

We also would like to recognize those who participated in our focus groups, reviewed various material, or were just there to answer questions: Randy Napoli, Chief, Bureau of Fire Standards and Training, Division of State Fire Marshal, Ocala, Florida; Dave Edmunds, Sarasota County Technical Institute, Sarasota, Florida; Mike Brackin, State Fire Academy, Jackson, Mississippi; David Pritchett, Georgia Fire Academy, Forsyth, Georgia; David Herndon, Georgia Fire Academy, Forsyth, Georgia; Claude Shew, North Carolina Fire and Rescue Commission, North Carolina; David Fultz, Louisiana State Fire Academy, Baton Rouge, Louisiana; Larry McCall, Florida State Fire Academy, Ocala, Florida; Timothy Dunkle, Pennsylvania State Fire Academy, Pennsylvania; Gregory Kirt, Michigan Firefighters Training Council, Lansing, Michigan; Ron Coleman, Elk Grove, California; Michael Richwine, Department of Forestry and Fire Protection, Ione, California; Michael Ridley, Elk Grove Community Services District Fire Department, Elk Grove, California; Mark Lewandowski, Connecticut Fire Academy, Windsor

Locks, Connecticut; John Pangborn, Jersey City Fire Department, Jersey City, New Jersey; Doug Hall, Red Rocks Community College, Red Rocks, Colorado; Michael Forgy, Fairfax Volunteer Fire Department, Fairfax City, Fairfax, Virginia.

We also give our sincere appreciation to those who shared with us their stories that you see at the beginning of each chapter: Andrea Walter, Lead Author; Mike Smith, District of Columbia Fire and EMS Department, Washington, DC; Mike Kelleher, Troy Fire Department, Troy, New York; Fred Windisch, Ponderosa Volunteer Fire Department, Houston, Texas; Randy Sheerer, Newport Beach Fire Department, Newport Beach, California; Gordon Sachs, Fairfield Fire and EMS, Fairfield, Virginia; Mike Gala, FDNY, New York, New York; Chief Bernard Lach (ret.) Torrington Fire Department, Torrington, Connecticut; Michael Arnhart, High Ridge Fire Protection District, High Ridge, Missouri; Battalion Chief Billy Goldfelder, Loveland Fire Department, Loveland, Ohio; Peter F. Kertzie, Buffalo Fire Department, Buffalo, New York; Paul LePore, Long Beach Fire Department, Long Beach, California; Lieutenant Michelle Steele, Miami-Dade Fire Rescue, Miami, Florida; Richard Arwood (ret.), Memphis Fire Department, Memphis, Tennessee; Battalion Chief Frank Montagna, FDNY, New York, New York; William Shouldis, Philadelphia Fire Department, Philadelphia, Pennsylvania; James P. Smith, Philadelphia Fire Department, Philadelphia, Pennsylvania; Michael Ramsey, Oran Fire Protection District, Oran, Iowa; James Angle, Palm Harbor Fire Department, Palm Harbor, Florida; Mary K. Marchone, Montgomery County Fire and Rescue Services; John J. Salka, FDNY, New York, New York; Rob Schnepp, Alameda County Fire Department, San Leandro, California; Mike Callan, Callan & Company, Ltd., Middlefield, Connecticut; Jan R. Dunbar, Sacramento Fire Department, Sacramento, California; Lieutenant Julius Stanley, Chicago Fire Department, Chicago, Illinois; Frank Docimo, Turn of River Fire Department, Turn of River, Connecticut; Joseph DeFrancesco, Madison County Fire and Rescue, Madison, New York; David Mitchell, Fayettville Fire Department, Fayettville, Arkansas; Greg Noll, Hildebrand & Noll Associates, Lancaster, Pennsylvania; Greg Socks, Washington County Special Operations, Hagerstown, Maryland; Tom Creamer, Worcester Fire Department, Worcester, Massachusetts; and Rick Townsend, Sierra Vista Fire Department, Sierra Vista, Arizona.

And to those we rarely take the time to recognize because this is their job, a special thanks. The Delmar Learning team developed, produced, and marketed *The Firefighter's Handbook* setting an example for not only getting the job done, but having the creativity and fortitude to go above and beyond. Our appreciation to Alison Weintraub, Mary Ellen Black, Jennifer Thompson, Rachel Baker, Toni Hansen, Barbara Diaz, Jennifer Luck, Erin Coffin, Mark Pierro, Fair Huntoon, Cindy Eichelman, Sandy Clark, and Alar Elken.

How to Contact Us

At Delmar Learning, listening to what our customers have to say is the heart of our business. If you have any comments or feedback on *The Firefighter's Handbook,* you can e-mail us at firescience@delmar.com or fax us at 518-881-1262, Attention: Fire Rescue Editorial.

For additional information on other titles that may be of interest to you, or to request a catalog, see our listing on pages 959–963, or visit www.firescience.com.

NFPA 1001 CORRELATION GUIDE

Entrance Requirements

NFPA 1001 Section	NFPA 1001 Description	*Firefighter's Handbook* Chapter Reference	*Firefighter's Handbook* Page Reference
4.3	Emergency Medical Care	22	696, 707, 709

Firefighter I

NFPA 1001 Section	NFPA 1001 Description	*Firefighter's Handbook* Chapter Reference	*Firefighter's Handbook* Page Reference
5.1	General	Not Applicable	
5.1.1	General	24–30	See hazmat grid.
5.1.1.1	General	2, 5, 15	23, 26, 24, 34, 35, 109, 419, 423, 424, 439, 449
5.1.1.2	General Skill Requirements	2, 6, 13, 15, 20	137, 449, 432, 428, 433, 427, 430, 427
5.2	Fire Department Communications	3	51, 58, 63
5.2.1	Fire Department Communications	3	51, 63
5.2.2	Fire Department Communications	3	51
5.2.3	Fire Department Communications	3	63
5.3	Fireground Operations	4, 7-20, 23	74, 141, 721
5.3.1	Fireground Operations	7	158, 168, 145, 150, 172, 170
5.3.2	Fireground Operations	5, 23	120, 723
5.3.3	Fireground Operations	16, 19	624, 486
5.3.4	Fireground Operations	13, 17, 18	339, 522, 511, 530
5.3.5	Fireground Operations	7, 16, 19, 23	725, 733, 168, 591, 459
5.3.6	Fireground Operations	14	369
5.3.7	Fireground Operations	11, 19	279, 622
5.3.8	Fireground Operations	10, 11, 19	219, 279, 603, 627
5.3.9	Fireground Operations	7, 14, 16, 17	170, 383, 459, 511
5.3.10	Fireground Operations	4, 10, 11, 19	74, 219, 279, 591
5.3.11	Fireground Operations	4, 14, 18, 19	79, 369, 548, 591
5.3.12	Fireground Operations	4, 13, 14, 15, 18	79, 339, 369, 449, 591
5.3.13	Fireground Operations	11, 20	279, 631
5.3.14	Fireground Operations	12, 20	307, 631
5.3.15	Fireground Operations	9, 10	201, 219
5.3.16	Fireground Operations	4, 8	100, 183
5.3.17	Fireground Operations	Instructor's Guide	Instructor's Guide

NFPA 1001 Section	NFPA 1001 Description	Firefighter's Handbook Chapter Reference	Firefighter's Handbook Page Reference
5.3.18	Fireground Operations	19	591
5.3.19	Fireground Operations	19	595
5.4	Rescue Operations	Not Applicable in FF I	—
5.5	Prevention, Preparedness, Maintenance	21	657
5.5.1	Prevention, Preparedness, Maintenance	21	657
5.5.2	Prevention, Preparedness, Maintenance	21	657
5.5.3	Prevention, Preparedness, Maintenance	7, 14, 15, 17, 20	172, 382, 439, 441, 511, 636
5.5.4	Prevention, Preparedness, Maintenance	10	223, 226

Firefighter II

NFPA 1001 Section	NFPA 1001 Description	Firefighter's Handbook Chapter Reference	Firefighter's Handbook Page Reference
6.1	General	Not Applicable	—
6.1.1	General	Not Applicable	—
6.1.1.1	General Knowledge Requirements	2, 5	37, 106
6.1.1.2	General Skill Requirements	2	37
6.2	Fire Department Communications	3	45
6.2.1	Fire Department Communications	3	68
6.2.2	Fire Department Communications	3	45
6.3	Fireground Operations	4, 9-11, 13, 14, 16-21	74, 201, 339, 369, 459
6.3.1	Fireground Operations	11, 19	279, 625
6.3.2	Fireground Operations	10, 11, 13, 16, 17, 18, 19	219, 279, 339, 459, 508, 548, 591
6.3.3	Fireground Operations	4, 19	86, 88, 625
6.3.4	Fireground Operations	20	652
6.4	Rescue Operations	16	459
6.4.1	Rescue Operations	16	482
6.4.2	Rescue Operations	16	490
6.5	Prevention, Preparedness, Maintenance	9, 10, 12, 17, 21	201, 274, 307, 511, 657
6.5.1	Prevention, Preparedness, Maintenance	9, 12, 21	201, 307, 657
6.5.2	Prevention, Preparedness, Maintenance	17	511
6.5.3	Prevention, Preparedness, Maintenance	10	274
6.5.4	Prevention, Preparedness, Maintenance	9	213

JOB PERFORMANCE REQUIREMENT CORRELATION GUIDE

Designation	No.	Description	NFPA 1001 References	Firefighter's Chap. Ref.	Handbook Page Ref.
Supplemental Skill	6-1	Don Protective Clothing	5.1.1.2, 5.3, 6.3, 6.4	6	N/A
JPR	7-1	Donning Self-Contained Breathing Apparatus, Over the Head Method	5.3.1, 5.3, 6.3	7	161
JPR	7-2	Donning Self-Contained Breathing Apparatus, Coat Method	5.3.1, 5.3, 6.3	7	163
JPR	7-3	Donning Self-Contained Breathing Apparatus, Seat Method	5.3.1, 5.3, 6.3	7	165
JPR	7-4	Donning Self-Contained Breathing Apparatus Face Piece	5.3.1, 5.3, 6.3	7	167
JPR	7-5	Self-Contained Breathing Apparatus, Daily Inspection	5.3.1, 5.3, 6.3	7	173
JPR	7-6	SCBA Cylinder Replacement Procedure	5.3.1, 5.3, 6.3	7	175
JPR	7-7	SCBA Cylinder Replacement Procedure, Firefighter Wearing SCBA	5.3.1, 5.3, 6.3	7	176
JPR	7-8	Servicing an SCBA Cylinder Using a Cascade System	5.3.1, 5.3, 6.3	7	178
JPR	7-9	Servicing an SCBA Cylinder Using a Compressor System	5.3.1, 5.3, 6.3	7	179
JPR	8-1	Operation of Portable Fire Extinguisher	5.3.16	8	196
JPR	10-1	Coupling Hose—One-Person Foot-Tilt Method	5.3, 6.3	10	229
JPR	10-2	Coupling Hose—One-Person Over-the-Hip Method	5.3, 6.3	10	229
JPR	10-3	Coupling Hose—Two-Person Coupling Method	5.3, 6.3	10	230
JPR	10-4	Uncoupling Hose with Spanners	5.3, 6.3	10	230
JPR	10-5	Uncoupling Hose—One-Person Knee Press	5.3, 6.3	10	231
JPR	10-6	Storage Hose Roll	5.5.4	10	232
JPR	10-7	Single-Donut Hose Roll (Option 1)	5.5.4	10	232
JPR	10-8	Single-Donut Hose Roll (Option 2)	5.5.4	10	233

Designation	No.	Description	NFPA 1001 References	Firefighter's Handbook Chap. Ref.	Firefighter's Handbook Page Ref.
JPR	10-9	Twin-Donut Hose Roll	5.5.4	10	233
JPR	10-10	Hose Drain and Carry	5.5.4	10	234
JPR	10-11	Hose Carry—Shoulder Loop	5.5.4	10	235
JPR	10-12	Hose Drag	5.5.4	10	236
JPR	10-13	Accordion Hose Load	5.5.4	10	238
JPR	10-14	Advancing an Accordion Load	5.3, 6.3	10	239
JPR	10-15	Demonstrate a Flat Hose Load	5.5.4	10	240
JPR	10-16	Advancing a Flat Hose Load from a Supply Bed	5.3, 6.3	10	241
JPR	10-17	Demonstrate a Horseshoe Hose Load	5.5.4	10	242
JPR	10-18	Advancing a Horseshoe Hose Load	5.3, 6.3	10	243
JPR	10-19	Demonstrate a Reverse Horseshoe Load for an Attack Line	5.5.4	10	244
JPR	10-20	Advancing a Flat Load from a Preconnected Bed	5.3, 6.3	10	246
JPR	10-21	Loading a Minuteman or Slot Load	5.5.4	10	247
JPR	10-22	Advancing the Minuteman Load	5.3, 6.3	10	248
JPR	10-23	Perform the Triple-Layer Load	5.5.4	10	249
JPR	10-24	Advancing the Triple-Layer Load	5.3, 6.3	10	250
JPR	10-25	Modified Gasner Bar Pack	5.3.19, 5.5.4	10	252
JPR	10-26	Advancing a Charged Hoseline Up a Stairwell	5.3.10	10	255
JPR	10-27	Advancing an Uncharged Hoseline Over a Ladder	5.3.10	10	259
JPR	10-28	Advancing an Uncharged Hoseline Over a Ladder at Entry Point of Building	5.3.10	10	260
JPR	10-29	Soft Sleeve Hydrant Connection	5.3.10	10	264
JPR	10-30	Hard Sleeve Hydrant Connection	5.3.15	10	265
JPR	10-31	Assemble and Connect Equipment for Drafting	5.3.15	10	266
JPR	10-32	Extending a Hoseline with a Break-Apart Nozzle	5.3.10, 5.3, 6.3	10	267

Designation	No.	Description	NFPA 1001 References	Firefighter's Handbook Chap. Ref.	Firefighter's Handbook Page Ref.
JPR	10-33	Extending a Hoseline Using a Hose Clamp	5.3.10, 5.3, 6.3	10	268
JPR	10-34	Wildland Hose Advancing and Extension	5.3.19	10	269
JPR	12-1	Using "Stops" to Stem the Flow of Water from a Sprinkler Head	5.3.14	12	327–328
JPR	14-1	Ladder Suitcase Carry	5.3.6, 5.3.12, 5.3.13	14	396
JPR	14-2	Shoulder Carry	5.3.6, 5.3.12, 5.3.13	14	397
JPR	14-3	Flat Ladder Carry	5.3.6, 5.3.12, 5.3.13	14	398
JPR	14-4	Two-Person Rung Raise	5.3.6, 5.3.12, 5.3.13	14	401
JPR	14-5	Two-Person Beam Raise	5.3.6, 5.3.12, 5.3.13	14	403–404
JPR	14-6	Fly Extension Raise	5.3.6, 5.3.12, 5.3.13	14	405
JPR	14-7	Lowering a Ladder into a Building	5.3.6, 5.3.12, 5.3.13	14	406
JPR	14-8	One-Person Ladder Raise	5.3.6, 5.3.12, 5.3.13	14	407
JPR	14-9	Use Ladder Leg Lock	5.3.6, 5.3.12, 5.3.13	14	408
JPR	14-10	Carry Tools Up and Down Ladder	5.3.6, 5.3.12, 5.3.13	14	409
JPR	14-11	Securing/Heeling a Ladder	5.3.6, 5.3.12, 5.3.13	14	411
JPR	14-12	Engaging the Hooks on a Roof Ladder	5.3.6, 5.3.12, 5.3.13	14	413
JPR	14-13	Hoisting Ladders	5.1.1.1	14	414
JPR	15-1	Half Hitch Around an Object	5.1.1.2	15	427
JPR	15-2	Tie Overhand Safety	5.1.1.2	15	427
JPR	15-3	Tie a Clove Hitch in the Open	5.1.1.2	15	428
JPR	15-4	Tie a Clove Hitch Around an Object	5.1.1.2	15	429
JPR	15-5	Tie a Becket Bend Knot	5.1.1.2	15	430
JPR	15-6	Tie a Double Becket Bend	5.1.1.2	15	431
JPR	15-7	Tie a Bowline Knot	5.1.1.2	15	432
JPR	15-8	Tie a Figure Eight Knot	5.1.1.2	15	434
JPR	15-9	Tie a Follow-Through Figure Eight	5.1.1.2	15	435
JPR	15-10	Tie a Figure Eight Knot on a Bight	5.1.1.2	15	436
JPR	15-11	Tie a Rescue Knot	5.1.1.2	15	437–438
JPR	15-12	Tie a Water Knot	5.1.1.2	15	439
JPR	15-13	Coiling a Rope	5.5.3	15	446
JPR	15-14	Rope Storage Bag	5.5.3	15	448

Designation	No.	Description	NFPA 1001 References	Firefighter's Handbook Chap. Ref.	Firefighter's Handbook Page Ref.
JPR	15-15	Hoist an Ax	5.3.12	15	450
JPR	15-16	Pike Pole Hoist	5.3.12	15	451
JPR	15-17	Hoist a Charged Hoseline	5.3.12	15	452–453
JPR	15-18	Hoist an Uncharged Hoseline	5.3.12	15	453–454
JPR	15-19	Hoisting Small Equipment	5.3.12	15	454
JPR	15-20	Hoisting a Ladder	5.3.12	15	455
JPR	15-21	Rope Between Two Objects	5.1.1.2	15	456–457
JPR	16-1	Firefighter's Carry	5.3.9	16	468
JPR	16-2	Extremity Carry	5.3.9	16	470
JPR	16-3	Seat Carry	5.3.9	16	471
JPR	16-4	Blanket Drag	5.3.9	16	472
JPR	16-5	Clothing Drag	5.3.9	16	473
JPR	16-6	Webbing Sling Drag	5.3.9	16	474
JPR	16-7	Sit and Drag Method	5.3.9	16	475
JPR	16-8	Firefighter's Drag	5.3.9	16	476
JPR	16-9	Rescue of a Firefighter Wearing SCBA	5.3.9	16	477
JPR	16-10	Placing a Patient on a Blackboard	5.3.9	16	478–479
JPR	16-11	Placing a Patient on an Ambulance Stretcher	5.3.9	16	481
JPR	17-1	Conventional Door Opening Away from Team	5.3.4, 6.3.2	17	534
JPR	17-2	Conventional Door Opening Toward Team	5.3.4, 6.3.2	17	535–536
JPR	17-3	"Through the Lock" Wrenching Lock	5.3.4, 6.3.2	17	537
JPR	17-4	"Through the Lock" K Tool	5.3.4, 6.3.2	17	538
JPR	18-1	Horizontal Ventilation from Above	5.3.11	18	570
JPR	20-1	Salvage Cover Roll	5.3.14	20	638
JPR	20-2	Preparing a Folded Salvage Cover for a One-Firefighter Spread	5.3.14	20	640
JPR	20-3	Preparing a Folded Salvage Cover for a Two-Firefighter Spread	5.3.14	20	641
JPR	20-4	Salvage Cover Shoulder Toss	5.3.14	20	644
JPR	20-5	Salvage Cover Balloon Toss	5.3.14	20	645
JPR	22-1	Removing Gloves	4.3	22	699

CHAPTER 1

OVERVIEW OF THE HISTORY, TRADITION, AND DEVELOPMENT OF THE AMERICAN FIRE SERVICE

Ronny J. Coleman, Retired California State Fire Marshall

 OUTLINE

- Objectives
- The Mission of the Fire Service
- Roots in the Past
- Technology, Transition, and Times of Change
- The Effects of World War II
- Modernization of the Fire Service
- The Fire Service of Today
- Lessons Learned
- Key Terms
- Review Questions
- Additional Resources

Photo courtesy of Marysville Fire Department (California)

STREET STORY

When I began my work in the fire and emergency services, I was confused by the many traditions, symbols, and practices that have come straight out of history. In these high-tech times, when industry technology seems to change almost weekly, it seemed odd to find these traditions still alive and well in the fire service.

As I learned more about the history of the fire service, I realized how valuable these pieces of history are to what we do every day. It is from this history that the fire service derives its pride, honor, integrity, and courage. The symbolism ties our modern-day practices to those of the early firefighters and emergency responders. It provides firefighters and emergency responders with a unique sense of belonging, perspective, unity, and promise for the future.

The history of firefighting and emergency response also gives modern-day firefighters the safety and protection that we need to survive and thrive in the industry. This history, including knowledge of the incidents in which firefighters have fallen in the line of duty, provides us with the information and expertise we need to ensure that firefighters can perform their duties safely with safer personal protective equipment, technological advances, and improved operational practices.

The fire service has changed dramatically over the centuries, but some key elements still remain strong. Would Ben Franklin ever have expected that fire departments in America would be responding to terrorism and hazardous materials incidents and training firefighters on how to protect themselves from infectious diseases? No, of course not. But he would have expected that whatever firefighters do, it would be the best that they could.

The types of emergency calls may have changed with society, but our dedication to protecting our communities and our brother and sister firefighters has not. I hope it never does.

The history of the fire service helps modern firefighters learn from the successes and failures of the past and unites us in a common tradition of serving those in need.

—Street Story by Andrea A. Walter, Lead Author, *Firefighter's Handbook*

OBJECTIVES

After completing this chapter, the reader should be able to:

- Describe the role of the firefighter in the fire service.
- Define the importance of the mission of the fire service and the purpose of a mission statement.
- Identify the major events that have altered the history of the fire service.

THE MISSION OF THE FIRE SERVICE

In general, the tasks of firefighters are the same the world over: to save lives and property from fire. However, not all fire departments approach that goal in the same manner. Many variables are considered by the agencies that provide fire protection. Some agencies only protect airports, whereas others do structural firefighting and provide emergency medical services (EMS). Fire departments that provide EMS may only perform basic life support (BLS), whereas others may provide advanced life support (ALS).

The term *mission,* used in the context of modern fire department management, is usually synonymous with an agency's purpose for existence, or even its legal authority to act in a certain manner. A **mission statement** is a written declaration by a fire agency describing the things that it intends to do to protect its citizenry or customers. When we look at the variety of situations under which the fire service functions, we begin to understand why one mission statement will not cover all agencies. The size and organizational structure of fire departments range from small volunteer companies that protect only a few hundred souls to metropolitan departments that protect millions.

Many fire agencies have a written mission statement. Firefighters should understand their departments' mission statements and how they contribute to their departments' successes. Mission statements vary from very general to very specific. Once written and posted, it is the responsibility of every individual in that department to attempt to achieve its goals. Each and every firefighter, from the beginning of civilization to the modern fire service, has contributed something to the accomplishment of an agency's mission. That is one of the legacies of the fire service.

ROOTS IN THE PAST

Somewhere, lost in the ancient past, is the name of the very first firefighter. That person was probably thrust into the job of fighting fire out of sheer desperation, to protect himself, his family, or his possessions. It was a courageous act then, and it has remained a courageous act as the fire problem has passed from those obscure ancient events to the streets of small towns and villages, suburban cities, metropolitan areas, airports, harbors, marinas, and wildlands, all of which are protected by contemporary firefighters.

> **Note** What is important to recognize is that the task of being a firefighter is an old, yet constantly evolving occupation.

The fire service contains a great deal of tradition. These traditions are often in conflict with the constant changes required in the fire service to keep up with the task of combating fire in a modern society. Being a firefighter today is much different from the past, but practically everything firefighters and fire service organizations do to save lives and protect property is rooted in events from the past. That is why it is important to look at the legacy and heritage of the fire service for the fundamental reasons firefighters do many of the things they do.

> **Firefighter Fact** The fire service of today is a direct result of an evolution in the methodology, technology, and responsibility of a service that has been vital to communities since the beginning of civilization.

Present-day firefighters need to recognize that what they do today is going to be the basis for the processes future firefighters will use. Understanding the history of the fire service is a lot like climbing a ladder, as shown in **Figure 1-1**. The ladder has to be well grounded in order to be stable, and there must be strong beams to support the ladder. There are rungs to stand on as the ladder is climbed. The two beams of the fire service's history are courage and commitment. The bottom rungs represent the experience and knowledge gained from past years. The rung we are standing on is the present. The ones that are beyond

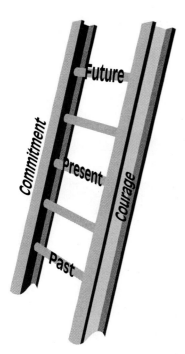

Figure 1-1 Understanding the history of the fire service is like climbing a ladder.

our reach are the future. The ladder is being raised to achieve the goal of saving lives and protecting property. We anticipate that we will always be climbing that ladder in an unending quest to achieve that goal.

The history, traditions, and development of the fire service is a pageant that covers a great deal of time. Significant events have shaped the development of fire service technology and methods. Individual people have actually shaped fire protection philosophy through their actions. Fire service history is a colorful and exciting cavalcade of personalities, tools, equipment, and apparatus that have been used to combat both the ordinary fire and the **conflagration**. The legacy and heritage of the fire service is rich with experience.

Someone once made a joke that the fire service is "two hundred years of tradition, unhampered by progress." Nothing could be more inaccurate. For the last 350 years, the fire service in this country has advanced on all fronts. What has not changed a great deal is the nature of individuals who have chosen to follow a fire and emergency service career, either as a paid, career firefighter or as a volunteer. Granted, in the last few decades there have been more changes in the firefighters' working environment than at any other time in the history of firefighting. But a brief review of history reveals that the process of change has been there for a long time. We just have to search for the stepping stones and recognize that they are incremental steps of change.

> **Note** The fire service has evolved into what it is today because it is a service for people. What communities need and want from their fire service determine what firefighters should or could be providing them. That is what makes the role of the firefighter so exciting and fulfilling. It is a job that not everyone can do. It is not a job that remains static over time. It is a constantly challenging and stimulating environment where the past collides with the future in the actions of those who serve today.

Although many fire agencies have their own mission statement, that does not mean they are all different in all ways. Some factors are common to all agencies. For example, fire agencies all have an organizational structure, an inventory of facilities, apparatus, equipment, and methods for responding to emergencies. They almost all conduct a variety of programs to achieve their goal of protecting life and property. These programs, in general, fall into the categories of fire and emergency operations, fire prevention, arson investigation, training, emergency medical services, communications, and maintenance to name a few. These elements of the fire service are all founded on events and decisions from the past.

Ancient Beliefs

The invention of fire was a turning point in the human race's quest for safety and security. In actuality, humans did not invent fire at all. Fire is a natural phenomenon that results from a number of natural occurrences like lightning or volcanic activity. Even spontaneous combustion requires no human help. What the human race did was to recognize the value of fire for heating their food and warming their homes or living area, and then eventually using it to forge metals and create engines of commerce and weapons of war. Although the human race did not invent fire, the greatest initial challenge was to learn to manage fire and to prevent it from destroying its user.

Records from ancient history clearly indicate that fire was considered a force of nature that was to be feared. The Greeks believed that fire was given to common humans as a gift by one of their gods, Prometheus. A number of religions regarded fire and flame as a deity. The ancient myth of the Phoenix bird was based on the consequences of fire. The Phoenix, as shown in **Figure 1-2**, reportedly lived for 1,000 years. When it reached the end of its life cycle, the bird reportedly built a nest of wooden sticks and

Figure 1-2 The Phoenix rising from the ashes.

set fire to it. When the ashes cooled, a small worm emerged from the ruins of the fire and turned, once again, into a beautiful, peacock-like bird that lived again for another 1,000 years. This legend was based on the idea that fire was a destroyer of monumental power, but also a giver of life. The concept of a resurrection from the flames created the idea that out of death and destruction comes hope.

As human philosophy evolved, the human race continued to experiment and improve on the use of fire as a tool instead of worshipping it as a sign from the gods. Archaeologists have proven not only that ancient cultures used fire, but also that fire destroyed their homes, consumed their property, and in general did a great deal of harm to early villages and towns.

Recorded History

The Romans worshipped a goddess, Vesta, who was the protector of the hearth fire. In the cities of ancient Rome, Vesta was the only goddess with no statue. Instead, a flame was kept burning on her altar constantly. The Romans worshipped fire and respected its impact, for they suffered a lot of fires in their city. Hero of Alexandria, a Roman leader, was credited with creating the first fire pump, which was based on a syringe-like design that squirted water when the plunger was pushed, **Figure 1-3**.

The first record of a truly organized fire department began with actions taken by the Roman Empire to protect their capital. In 22 B.C. a group of individuals, called magistrates, was responsible for maintaining a watch over slaves during the night. Naturally, they were in a good position to report the outbreak of fire. Wealthy Roman citizens expanded that group to become what was called the "familia publica," which were organized along military lines, but they were not very efficient. After a major fire occurred in Rome in A.D. 6 a permanent fire brigade, called the "Cohortes Vigilum," or Vig-

Figure 1-3 An early European hand-operated pump based on the ancient Roman design.

iles, was established. They were housed in large barracks and toured the streets at night looking for fires.

The Emperor Augustus created seven cohorts, which were seven military units, each under the command of a tribune, and the groups were under a prefect, or an officer of equestrian rank. The prefect was an important person in the Roman hierarchy. Each unit initially had 500 men. Later this was changed to a group of 1,000 men. The Vigiles were further broken down into a variety of specialists. There were men who worked the pumps and men with grappling hooks who tore off roofs and tore down walls to get to the fires. There were also individuals who directed the fire stream provided by the pump-like device. In spite of the presence of this brigade, Rome burned many more times during the next several centuries. Many Roman leaders attempted to use fire regulations to control the buildings that were breeding these destructive fires. Despite these losses nothing slowed down humans' use of fire. Its use was expanded in areas of commerce and industry, such as for the mining and smelting of ore.

With the collapse of the Roman Empire, the Vigiles passed out of use. Europe was not to see any dedicated fire forces for another 1,000 years. This did not mean that no one was fighting fires, but it did mean that there were no organized or collective efforts. The legacy left by the Romans was the idea that firefighting required organizational effort, that there were specialized tasks to be performed, and that a command and control structure, similar to the military, was a good model.

Early History and Symbols of the Fire Service

The next organized groups of individuals devoted to the saving of lives and protection of property arose during the Crusades. Bands of knights going to the Holy Land to search for the Holy Grail were organized into "orders." One such order was entitled the Order of Saint John of Hospitaliers. This band of knights dedicated themselves to the treatment of serious wounds caused on the battleground.

During this same time period there was an order called the "Knights of Malta." These brave knights did not go into combat to kill, but to save. They developed a reputation for saving lives by serving as stretcher-bearers for the victims of battles. After a battle was over they carried the victims back to the crude battlefield hospitals, often using shields as stretchers. In the age of the Crusades, one of the only ways to tell friend from foe was by a symbol on the uniform and another on the shield carried by soldiers. These symbols were used to denote that a person had the training and the responsibility to act on behalf of the order they represented. The symbol for the Order of Saint John was a red cross. The symbol for the Knights of Malta was the Maltese cross. Both emblems became symbolic for saving lives and property. Today, these symbols still exist in the fire and emergency services, **Figure 1-4**.

The Middle Ages

In the Middle Ages, fire was a serious threat to every city. History books are full of stories of the devastation of major European cities in the years between the end of the Roman Empire and the beginning of the Middle Ages. During the next ten centuries, technological advances were created all around the globe, but they were not widely available to European cultures. For example, in China, fire pumps were created that were based on the syringe-like concept mentioned earlier. The Chinese also created fire brigades to protect their largest cities. Because of limited contact between Europe and the Orient, these advances were not copied.

The English, from about A.D. 1100 to 1600, devised a whole set of regulations about fire. For example, in 1189, a law required that homeowners had to have ladders prepared at their homes to help their neighbors. Other laws required buckets and barrels of water to be kept handy. **Arson** remained a serious crime during this period, even though the punishment for arson was to be burned alive.

Fires continued to devastate, even after the creation of the concept of fire prevention. London burned in 1666. Originally labeled the Pudding Lane fire for the street on which it began, it has since been called the Great Fire of London. Unfortunately, even with the proliferation of business and industry in London, the concept of fire protection had been allowed to lag.

The community leaders who were creating the commerce, business, and industrial growth were also placing a new emphasis on the need to eliminate conflagration. The next 200 years would see the creation of almost all of the basic institutions that provide fire protection in the modern age. These components would consist of the organization of fire departments, the creation of a fire insurance industry, and the rise of technology to both prevent and combat fires.

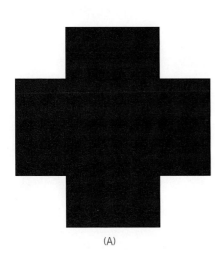
(A)

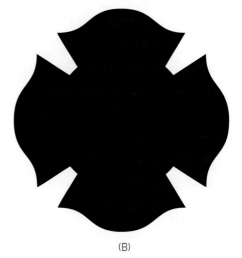
(B)

Figure 1-4 Symbols taken from history: (A) red cross and (B) Maltese cross.

Early American History

Most of the settlers of this country's colonies were of English, Irish, German, French, or other European background. Therefore, the first steps to providing fire protection were based on the practices from these same countries. Among the very first laws put on the books in the New World were ones dealing with fire prevention. Among those regulations were prohibitions on the construction of wooden chimneys and limitations on the candle-making process. Most of the early settlements were constructed of wooden buildings and were extremely vulnerable to fire spread. Therefore, it became important for the colonists to prevent fires from happening rather than try to combat them.

History books contain many stories of the early settlements in New England that burned to the ground, leaving their inhabitants exposed to the elements. Although fire protection in these early years consisted mostly of active fire prevention, it was not always effective. Fire suppression was essentially done by bucket brigades, in which buckets of water were passed down a line of people from the source of the water to the fire.

The King of England authorized the firm of Ryley and Mabb to create a new idea called "insurance" in 1637. This unique idea was far from successful at the time. Little or no attention was paid to the idea until the Great Fire in 1666. Nonetheless, the insurance companies went into business and marked the occupancies they protected with signs on sheets of metal telling the firefighters which company held the insurance policy on the building. These were called **firemarks**, as shown in **Figure 1-5**, and were used in the colonies by the early settlers.

The development of fire suppression capability became a pressing need. As early as 1647 the city of Boston sent an order to London to purchase a fire pump. Unfortunately, the pump did not arrive in time for the fire of 1653 that almost destroyed the town. That fire was fought by the bucket brigades. During the colonization period, most cities realized that they were entirely on their own against fire, and they began to prepare for it by developing some specific strategies to cope in their individual communities. Fire protection was initially viewed as a local problem. Because of the close relationship with Europe, the exchange of ideas was soon flowing. Boston, in the early 1700s, designated specific people to serve as **fire wardens**. These individuals were not unlike the Romans who walked the streets at night looking for fire. They were given big wooden rattles, as shown in **Figure 1-6**, and told to make as much noise as they could if they found a fire. They were soon labeled the "Rattlewatch." By 1718, these concepts had jelled into the idea of creating **fire societies**. Fire societies consisted of groups of people who voluntarily banded together to deal with a community's fire problems. It was to

Figure 1-5 Firemarks were used to show which insurance company protected a home or building. *(Courtesy CIGNA Museum and Art Collection)*

Figure 1-6 Examples of rattles used by the Rattlewatch. *(Courtesy CIGNA Museum and Art Collection)*

be the beginning of one of the American fire service's most colorful ages, the era of the volunteer firefighter. Many of these volunteer firefighters were also destined to play an increasingly important role in history as founding fathers of the American political system.

Firefighter Fact Benjamin Franklin, then a businessman from Philadelphia, began his influence over the early American fire service by being a cofounder of the Union Fire Company in 1736. Supposedly, he had witnessed the Boston Towne House fire, one of the largest conflagrations in the 1700s. He modeled the Union Fire Company after Boston's Fire Society. Franklin, known for his adventures in politics, science, publishing, and other fields, is considered by many to be America's first fire chief, **Figure 1-7**.

As the need to develop a fire service grew in this country, Franklin was joined by many other key historical figures such as George Washington, Thomas Jefferson, John Hancock, and Samuel Adams. These individuals were involved in the creation of fire companies or the acquisition of fire equipment from Europe.

The colonies in New England suffered terrible losses during the next few decades. A review of history books reveals that almost every major city was struck by serious fires, many reaching conflagration size. Boston, New York, and Philadelphia each experienced severe losses of life and property several times.

Around 1752, Benjamin Franklin, once again following European practice, started up his own insurance company, the Philadelphia Contributionship, in concert with other Philadelphians. One of the ideas he borrowed from Europe was the concept of placing firemarks on the outside of the structure to let fire suppression crews know that the contents were insured. In England, individual engine companies

Figure 1-7 Benjamin Franklin, commonly referred to as "America's first fire chief." *(Courtesy CIGNA Museum and Art Collection)*

were hired to protect the insured homes and buildings of certain insurance companies. The fire marks were important indicators because they allowed the fire companies to know whether or not they should extinguish the fire in a home or not! These metal signs were made from lead castings and had the company's logo on the surface. They were attached to the building that had the policy. In Philadelphia the Contributionship's directors saw the value of fire marks and continued the practice even though volunteer fire companies extinguished fires regardless of insurance. The fire mark of The Philadelphia Contributionship, four clasped hands on a wood shield, is known as the Hand in Hand, **Figure 1-8**. Interestingly, most of the early insurance companies are still active, but many of their fire marks were removed from houses during the Revolutionary War in order to make bullets out of the lead.

Fire was used as a weapon of war in the colonies during the two wars fought over independence. Major fires occurred in most of the cities when they were occupied by British military forces, but because many volunteers had joined the rebel military forces themselves, there were few firefighters to cope with local fires. When the war ended, almost all of the fire equipment that existed prior to the war had been either destroyed or badly damaged. Early America, which needed to build a federal government, also had to spend some time in rebuilding the fire service. New companies were formed, volunteers were recruited, and the remains of the cities destroyed by fire were replaced by new homes, new businesses, and an increasing population.

> **Note** With the close of the war for independence, the American fire service entered a new age. With the end of one century, fire protection had become a public need. With the beginning of the new century, fire was becoming an important economic consideration. Fire protection went from being the concern of only the larger cities to being a concern for smaller communities as well. By the beginning of the 1800s, there were very few settlements that did not have a volunteer fire company, and most even had some form of pumping apparatus powered by volunteers.

The American fire service began to evolve away from European practices, customs, and styles. The idea of a distinctive American fire service began to emerge. Changes began to occur, but not without traces of the legacies of the past. For example, many of the volunteer fire companies were organized around the ethnicity of the community they served. The Hibernia Fire Company consisted mostly of those of Irish descent. The hat for the Hibernia Fire Company is shown in **Figure 1-9**. There were also African-American fire companies, including one from 1815, which had to quickly disband. Some required a person to pay extensive dues and were therefore only open to the wealthy. There were volunteer fire companies for just about every ethnic, social, and political orientation in society.

The volunteer fire companies were often made up of soldiers who had fought in the war. When the smoke of battle faded, many young men converted their sense of community service from combating Redcoats to combating fire. Among the traits that attracted the volunteer were the

Figure 1-8 Fire mark for the Philadelphia Contributionship for the Insurance of Houses from Loss by Fire, organized 1752. *(Courtesy of The Philadelphia Contributionship for the Insurance of Houses from Loss by Fire)*

Figure 1-9 Firehat illustrating the Hibernia Fire mark. *(Courtesy of The Philadelphia Contributionship for the Insurance of Houses from Loss by Fire)*

excitement of firefighting and the camaraderie of the organization.

From the beginning of the 1800s to about 1850 many advances were made in fire protection, but almost all of the emphasis was on fire suppression. What was most significant in this era was the development of larger, more powerful hand pumpers, as shown in **Figure 1-10**, and the development of the American fire helmet. Although the hand pumper owed its origins to England's early models, American versions were more elaborate and, for lack of a better term, more flamboyant. Volunteer fire companies vied with each other to see who could design and build the most powerful and yet beautiful machine.

The American fire helmet was developed by a leather hat maker named Andrew Gratacap. The original hats were shaped like top hats and had the name of the fire company painted on them. Later, the idea of a short front brim and an extended back brim was incorporated into the design to assist in water runoff. Firefighters also copied the large front piece off the Hessian soldier military helmets to create the distinguished American fire helmet profile. Europeans tended to use military-style helmets, so the American fire helmet began to be a symbol of the departure from the past, **Figure 1-11**.

The time period from 1800 to 1850 was one of great expansion and development in American society. The thirteen tiny colonies that created the United States were soon dwarfed by the states that

Figure 1-11 An early American fire helmet. *(Courtesy CIGNA Museum and Art Collection)*

were carved out of the rest of the country. City after city was established. The country grew quickly as western expansion resulted in the entire continent being settled in about 100 years. Cities like Chicago and Detroit became major trading centers. The growth of the fire service during this period was more along the lines of expansion of numbers of volunteer personnel and new fire departments and less along the lines of technological advance. Volunteer fire companies of the era became heavily involved in local politics, and individual fire companies became known as social clubs for younger, often aggressive and boisterous men.

This led to a time in the fire service that has been called the "Rowdies and Rum" era. During this era,

Figure 1-10 This early hand pumper, called the "Red Rover," was used by the Howard Engine Company in New York. *(Courtesy CIGNA Museum and Art Collection)*

manually operated hand pumpers required a tremendous number of people to keep a fire stream in operation for any lengthy period of time. Some larger pumpers required the service of more than 100 individuals for the period it took to control a large building fire. Ladders at this time were limited in height to about four stories and were hauled to the scene of fires by a team of men pulling on a long tongue of rope.

Firefighter Fact With an increase in the sociopolitical nature of volunteer fire companies, and in some situations, outright physical competition between fire companies, there was often more violence between fire companies than confrontation with the fire. Incidents often occurred that involved exchanges of blows between firefighters over who had the right to be on a specific fire. There were even cases of volunteer companies lying in wait to ambush another company on the way to a fire.

On the other hand, this era was also responsible for the creation of the image of the firefighter as being an important part of society. Currier and Ives produced a series of lithographs during this time that are still popular today. This series, which focused on firefighters and firefighting activities, included scenes at the firehouse, scenes of responding fire apparatus, and large fire scenes. These pictures also chronicled the changes that occurred during that period, showing both hand-drawn and horse-drawn firefighting equipment. One of the prints that demonstrates an image that resulted in a symbol was a print showing a **foreman**, which was a leader of a fire company, holding a **speaking trumpet** aloft, **Figure 1-12**. The foreman's trumpet was originally used to allow the crew boss to shout orders over the noise of the firefighting activity. The trumpet became the icon of authority within the fire service.

The mid-1800s was a period of both social and economic change. Invention after invention was brought forth to both simplify and complicate life. The railroad was created, printing presses became widely available, petroleum was discovered, and America's fire problems continued to become more complicated. Not all fire companies evolved in parallel to the need.

Arson was a common crime in this era. It was even called the "working man's vengeance" because arson was used as a retribution against unfair employers. Authorities searched for a measure that would cut down on the crime. In 1856,

Figure 1-12 A reproduction of a Currier and Ives lithograph entitled "Rushing to the Conflict," published in 1858. *(Courtesy CIGNA Museum and Art Collection)*

Philadelphia Mayor Richard Vaux, a former foreman of a fire company, established a special arson investigation unit. Although rewards for arsonists had been offered as early as 1723 in Boston, specially trained individuals who could track down arsonists did not come into being until it became clear that arson was a crime against society instead of against an individual.

One event that significantly affected the fire service was a riot by firefighters at a fire in Cincinnati, Ohio, in the 1850s. The city leaders were so distressed at the behavior of their volunteers that they instituted the first full-time, paid firefighting force in 1853. To cope with the need to provide the necessary energy to pump the water previously pumped manually by the volunteers, they provided a new technology—the steam-powered fire engine. The volunteers were outraged.

Nonetheless, during the next decade, city after city began to realize that their fire protection needs now required a fire department with paid firefighters. The conflict over creating a new way of fighting fire was a double-edged sword. On the one hand, moving to full-time paid firefighters on fire apparatus removed a lot of the political turmoil from the ranks, but caused a great deal of conflict for local officials. Adopting steamer technology was a boost to the development of fire streams to match the growing size of buildings. This technology provided the power to reach higher and larger buildings, but required a lot more commitment of community resources, especially in the way of funding.

> **Firefighter Fact** It should be noted that this transition to paid forces was certainly not universal. In fact, many communities chose to retain their volunteer fire forces. Many volunteer fire departments began to embrace the technology being used by the paid fire service agencies. Today, the volunteer fire force still constitutes more than half of the fire service in the United States.

The Civil War

The Civil War started at a crucial time in the history of the American fire service. While the fire service had a background in paramilitary organizational structures, the Civil War played a key part in reestablishing the significance of command and control. It also created the environment for the establishment of many of the organizational practices for a paramilitary organization.

Earlier in this chapter it was noted that the Revolutionary soldier returned to become a volunteer; this was not true of the Civil War. The fire service had provided a huge reserve of candidates to become soldiers in the Civil War, including entire units, for example, the 11th New York Fire Zouaves, the 73rd New York Infantry, and the 72nd Philadelphia Fire Zouaves. A monument dedicated to the 73rd New York Infantry, shown in **Figure 1-13**, stands on the battlefield in Gettysburg, Pennsylvania, the site of a crucial Civil War battle.

For the most part, the Civil War did not provide any specific advances in firefighting technology itself. Because thousands of volunteer firefighters went off to that war, on both sides, and were either killed or so severely injured that they could no longer serve their communities, when the war was over most of the country's experienced firefighters were gone.

Shortly after the Civil War was over, hundreds of veterans returned to communities looking for jobs and some opportunities for community service. Those cities that were converting their volunteer forces over to paid personnel saw the military veterans as excellent choices. They had demonstrated their courage in combat, were familiar with the technology of the emerging industrial age, and were well disciplined.

Figure 1-13 Located at the battlefield in Gettysburg, Pennsylvania, this monument is dedicated to the 73rd New York Infantry. *(Photo by Michael O. Forgy)*

Another factor that played into this time period was that a new skill was needed in using steam fire apparatus. Most of the steamers were so heavy they could not be pulled by firefighters alone. Horses had been introduced to provide the muscle to move the steamers. Many soldiers, experienced in use of horses for pulling artillery batteries and operating logistical support wagon trains, went right into service driving teams of horses to haul steamers.

One of the most recognizable symbols of the fire service today was introduced at that time: the firehouse dog—specifically the Dalmatian, as shown in **Figure 1-14**. Although firefighters had had mascots in firehouses for years, the Dalmatian was not there as a pet. The breed was originally developed as a companion for horses, to calm them and chase away any small animals in their path. Dalmatians trotting alongside a team of horses, was already a common sight in other commercial applications. It is very unlikely that the firefighters who brought these dogs into the stations would have predicted that fifty years later, the horses would be gone and the dogs would still be there!

The Civil War veteran brought many new ideas into the firehouse. Among these were the concept

Figure 1-14 The Dalmatian has become a symbol of the fire service in America.

of military rank structure, command and control similar to infantry tactics of the war, and even the coloration and design of firefighters' uniforms. Prior to the war, uniforms looked more like costumes, with color and flair being the lead criteria. Red, green, and yellow were commonly used in volunteer uniforms. After the war, uniforms tended to be more military-like, almost exclusively blue in color. The use of gold for top-level insignias and silver for lower level ranks came from the military system. The speaking trumpet went out of use as a tool and became a symbol of authority on collar brass. Prior to the war a person in charge of a group of firefighters was commonly called a foreman. After the war, terms such as lieutenant, captain, sergeant, and battalion were borrowed from the military and became common in the fire service.

The end of the Civil War also brought a new wave of professionalism and organizational structure to the fire service. In a few short years, the fire chiefs, who were often called fire engineers, had banned together to create the National Association of Fire Engineers (which was to become the International Association of Fire Chiefs) in 1873. A host of state and local fire chiefs' organizations were started at about the same time.

None of the advances in either methodology or technology would prevent the continuous exposure to large losses from fires. Among the most famous fires of this era were the Great Chicago Fire and the Peshtigo (Wisconsin) forest fire (1871). Both fires were major catastrophes and neither the presence of courageous firefighters nor powerful steamers prevented the loss of lives and property. Fire was proving to be a greater adversary than anticipated.

The Industrial Revolution

The 1870s through the turn of the century has often been called the "Industrial Revolution." America moved from a farming society of a large number of small towns and villages into a world leader in industry and commerce. Part of that revolution was the requirement that advances were needed to deal with fire.

> **Firefighter Fact** One of the largest industries that took hold in the United States was the textile industry. Cotton from the South was sent to the North to be made into cloth to sell to the rest of the world. In New England, these textile factories were burning to the ground with regularity when a young man named Henry Parmalee invented a device called an **automatic sprinkler system**. Sprinkler systems were installed in factory after factory and the losses began to decrease.

A new organization was formed consisting of individuals that were involved in designing and installing automatic sprinklers. They called themselves the National Fire Protection Association (NFPA). They took the task of forming this new concept into a "standard," which was given the number 1. It was the first, but certainly not the last, national fire code of the United States.

As the technology of industry increased, the technology of firefighting tended to increase too. Almost all of the basic research in **fire engineering** today was started during this period. The first **fire alarm** systems were created in the 1870s. Much of the basic information on **fire hydraulics** was done during this time. The first **aerial apparatus**, a spring-loaded device, was designed by San Francisco firefighter Daniel Hayes. The fire service of this era was quick to adopt the new ideas coming from the inventors and innovators of the industrial world. The steam-powered fire pump was one of the best examples, as shown in **Figure 1-15**. Steam power was not invented for the fire service. It was adapted for use in the fire service when it was needed. Its primary purpose was to be an inexpensive source of power to operate machinery. Steam engines ran trains and boats and operated machinery in factories. The trend that was created during the industrial age was to link the technological advances of the firefighter with the advances of society in general. At the outset of this phenomenon, the time frame for adoption was just slightly behind the creation of the concept in general society. It has never been that closely linked since.

Figure 1-15 An example of an early steam-powered fire engine. *(Courtesy CIGNA Museum and Art Collection)*

The Beginning of the Twentieth Century

The nation's fire problem did not go away as a result of technological innovations. To the contrary, it got more complicated. Career firefighters recognized many of these changes in working conditions. They banned together to form a labor union, the International Association of Fire Fighters (IAFF), near the turn of the century.

Although the capacity to fight fires was increased by technology, the very equipment, materials, and processes used to modernize the country created new fire problems. Fire continued to devastate cities and towns. In the late 1890s and into the turn of the century, fire took a terrible toll. There were so many large fires that the insurance industry almost went bankrupt. As a result they formed a group called the National Board of Fire Underwriters (NBFU) and started to evaluate the level of fire defense in different cities to see which ones were ready to deal with conflagration and which were not. This concept would have a lasting effect on the fire service.

As a result, fire departments and insurance organizations began to place more emphasis on the subject of fire prevention. Fire and building codes began to place emphasis on such things as fire-resistive construction techniques and fire control processes. A series of large loss of life fires during this era brought the subject of fire prevention into the forefront of fire science. Many fires, such as the Iroquois Theater, the Triangle Shirtwaist, and the *General Slocum* ferry fires, took the headlines away from courageous firefighting efforts and placed an emphasis on eliminating the reason for the fires occurring in the first place. This has sometimes been called the "theory of catastrophic reform."

> **Note** Almost all of the advances in fire prevention that have created the fire and building codes of today were created during this time frame. Simple concepts such as adequate exits, sprinkler protection for high-risk occupancies, exit requirements for assembly occupancies, and limitations on processes that used flammable or hazardous materials were all created in a short time frame by the leadership of the fire service of the time.

During the 1920s, Fire Chief Ralph Scott of the Los Angeles City Fire Department expressed a concern about the quality and quantity of fire service training. It had been almost forty years since the concept of having a training system for firefighters had been created and he suggested that it needed an overhaul. Chief Scott, who was the president of the International Association of Fire Chiefs, created an Education Committee and made himself the chair. While participating in a regional conference on trade and education he had learned of federal legislation that funded vocational training. Scott requested and received money to write a job analysis for the occupation of a firefighter.

Most of the metropolitan fire departments had training programs in place at the time. What was lacking, however, was an organized, systematic listing of tasks, knowledge, and skills based on current practices and needs in the fire service. The document that was produced by this project was called Firefighting Bulletin Number 155, series 44, Federal Board of Vocational Education, 1931. This was the landmark document on which almost all subsequent fire service training and education has been based. The report was validated by representatives from fire agencies all over the United States. It created the minimum standards for the training of entry-level fire personnel and created a body of knowledge that resulted in the industrial education system recognizing that firefighting was a separate and distinct occupation. Since then, other studies have been conducted and vast improvements have been made in the standards, but they all started with this bulletin.

TECHNOLOGY, TRANSITION, AND TIMES OF CHANGE

What was really remarkable about the first quarter of the twentieth century was the speed with which change impacted the fire service. In fact, the speed of change increased rapidly. The result was that during this era there were often many different levels of technology in the fire service at the same time. For example, the steamer era did not end with the creation of the internal combustion engine. Steamers were in service on a parallel track for many years after gasoline engines became available. Steamers had to be hauled with horses, and when the internal combustion engine first came along it was not used to pump water, but rather to replace the horses. This time period saw fire stations that had both horse-drawn and motorized fire apparatus in the stations next to one another.

Prior to World War I, the internal combustion engine was a novelty. After the war it became the power source of choice. Initially horses were replaced on steamers, not aerial apparatus. Aerials were still cranked up by hand or required spring power to elevate. One of the more unique motorized pieces of equipment was the early adaptation of soda-acid extinguisher capacity to small vehicles. They were often sent ahead to try to perform an initial attack on small fires.

A whole generation of fire apparatus arose that consisted of hybrids of steam power and internal combustion. During that era there were hose wagons to haul hose, and they worked very well with horses. Once the internal combustion engine was adapted to not only haul the apparatus, but to operate the pump too, hose wagons were then incorporated into engine companies. Once the horsepower of combustion engines improved, water tanks were added to the chassis. Prior to this time, several pieces of equipment had to be assembled to fight a fire.

The internal combustion engine made it possible to combine three fire apparatus into just one. The same vehicle could carry water, have a pump, and also carry the amount of hose and equipment required to apply the water. Basically this is the design of even the most modern of pumpers, the **triple combination engine company**, as shown in **Figure 1-16**.

Figure 1-16 Triple combination engine companies can carry water and hoses and other equipment as well as pump the water. *(Owned and photographed by William Killen)*

As tools were added to the apparatus, they became larger and more complicated. Aerial ladders were converted from horse drawn with the advent of hydraulic systems. What is interesting to note about this transition is that the fire service improved quickly on its ability to get to the fire, but did not make many advances in its capacity to fight the fire once on scene. The control panel of a steamer in the 1880s describes the pump as having a capacity of 1,250 gallons per minute (gpm) at 150 pounds per square inch (psi). That is about the average capacity for fire engines today.

A review of the apparatus designs from the beginning of the twentieth century to current times demonstrates that the fire service tends to adopt any technological improvement as soon as it has proven its dependability. Fads and trends are not as important to the firefighter as reliability and predictability. New ideas must go through a test of both time and experience to demonstrate their survivability. As a result, even with hundreds of millions of dollars being devoted to designing and building fire apparatus, the contemporary fire service is usually on the trailing edge of technology instead of the leading edge.

THE EFFECTS OF WORLD WAR II

World War II also had its effects on the American fire service. Once again, large numbers of firefighters were called to serve their country. Another impact was not as obvious, but was far reaching. The war accelerated the need to deal with fire. Fire was still a weapon of war, and the United States devoted a fairly large amount of resources to prepare for it. In Europe, the fire service had already suffered terribly in the war, on both sides of the conflict. In the United States, several military projects resulted in findings that would influence the postwar fire service. One was the research that resulted in the development of the indirect attack method. This work, performed by Fire Chief Lloyd Layman, was distributed to the fire service after the war in the form of two books, *Firefighting Tactics and Strategy* and *Attacking and Extinguishing Interior Fires*. Research into fighting flammable liquid fires resulted in improved foams for use by the fire service. Personal protective clothing used for firefighting was also improved. Fire nozzle technology was improved drastically with the development of advanced versions of fog nozzles.

No records were kept of the number of firefighters that lost their lives in World War II, but it can be anticipated that the fire service suffered a loss of both experience and knowledge as a result of the conflict. Firefighters who did return home after World War II to reenter the department and the veterans that returned home trying to get jobs in the fire service both had a frame of reference that would affect the fire service for generations. This group saw the fire service as being even more paramilitary than ever in the past.

Innovation after innovation was created from the war effort. The military buildup also resulted in an increased need for domestic fire service agencies to protect new factories and increased residential developments. Many innovations for the fire service emerged: the availability of radios for fire engines to improve communications and of diesel engines to improve road performance. These innovations and the growth of the fire service placed new demands on the fire service.

From the 1950s to the end of the twentieth century, the fire service saw more changes than in the previous 200 years. Most of the hardware and basic components of the service have evolved fairly slowly. The process of accelerating change has more to do with the actual duties of the firefighter and fire agency staffing. For example, the role of EMS has grown rapidly in the fire service; many departments are much more involved in **hazardous materials** response, **Figure 1-17**, than ever before; the role of **search and rescue** has grown in view of major earthquakes and structural collapses; and fire service agencies are becoming more and more involved as first responders in **terrorism** incidents. More importantly, the fire service has also evolved into a profession that embraces gender and ethnic diversity. Members of the fire service of today tend to think of themselves as being the "modern fire service"—but then so has every other generation.

MODERNIZATION OF THE FIRE SERVICE

As used in this text, the term *modernization* is a process, not a place in time. The process of change has been constant in the fire service. What has not changed has been the basic mission of firefighters. When it comes to describing modernization in this text, we are talking only about the present. What is being done today is necessary to keep the fire service modern. This gives rise to two concepts that all firefighters should be aware of as they enter the service: information half-life and technological obsolescence. Information half-life is how long it takes for 50 percent of the information a firefighter has to become obsolete, inaccurate, or ineffective. Information half-life is not the same for all generations of firefighters.

Figure 1-17 The fire service has expanded into many areas, including hazardous materials response.

A member of the fire service in Ben Franklin's day would have used the same tools that his grandfather had used. Moreover, his grandchildren would not have used much more effective equipment. The information half-life in the 1700s for the fire service was about 100 years. The information half-life in the last 100 years has decreased almost every generation. While there is no scientific study to determine what the half-life is today, it is a fact that the body of knowledge regarding fire protection is changing rapidly. It can be anticipated that fire science information will change a great deal more in the future than it has in past generations.

The second phenomenon is called *technological obsolescence*. This term means that any given technology will only be useful for a certain period of time before it is replaced by another. Most technology goes through an initial stage of being controversial. Then it becomes superior to the one that it replaces. After it serves a period of time as "state of the art," it too has the possibility of becoming less than adequate. Eventually it can become old-fashioned and is replaced. Steamers were once superior to hand pumpers, internal engines replaced horses, diesel replaced gasoline, and so on. All obsolete technology becomes a source for museums and collections of memorabilia. All status quo technology is just one invention away from obsolescence.

The lesson to be learned here is that the fire service needs to be alert to any indications that its current technology is not doing the job and be prepared to adopt any improvements. It is equally important that firefighters do everything they can to be competent and effective in the knowledge and skills needed to perform the job of firefighter safely.

THE FIRE SERVICE OF TODAY

Although fire protection remains essentially a function of local government, the second half of the twentieth century saw efforts to create a national focus on the fire problem in the United States and bring together training and information sources at state and national levels to improve the business of fire protection. In the 1970s the federal government recognized that the country's fire problem was changing and commissioned a panel of exports to study it. This group, called the National Commission on Fire Protection and Control, published a document called *America Burning*. As a result of that document a lot of positive change occurred, including the creation of the United States Fire Administration and the development of the National Fire Academy.

In the twenty-first century, the fire service continues to grow and develop. There are an estimated 30,020 fire departments in the United States with an estimated 1,078,300 firefighters. The actual numbers vary because new communities are created that require fire protection, and sometimes several fire organizations band together as a result of mergers or consolidations. Many fire departments are still totally volunteer, while others are growing larger due to population increases and increased density in some urban areas. In some areas, fire and rescue services may be provided by private, for-profit companies. Fire departments are increasingly complex in this day and age, with a diversification in the services they provide. In addition, fire departments are also more gender and ethnically diverse, offering new insights into the way they provide services to their communities.

According to the United States Fire Administration, a fire department in this country responds to a fire approximately every eighteen seconds. Even with the incredible advancements in fire service tactics, training, and technology, the United States still has the fourth highest fire death rate of all industrialized countries, with someone dying in a fire an average of every two hours and someone injured by fire approximately every twenty-three minutes. Fire kills more Americans than all natural disasters combined. America is still burning, and the fire service continues to race to meet these ever-evolving challenges.

The late twentieth century and the early twenty-first century also laid a new obstacle before the fire service, the threat of terrorism in this country. The attacks on the World Trade Center and the Pentagon on September 11, 2001, shed new light on the fire service by demonstrating that firefighters are not immune from the horrible acts of terrorists. Although this event will remain fresh in every American's mind, it is very possible that the events of that day will shape the fire service in new and different ways well into the future.

The fire service of today has a very diverse, complicated system of delivery, as a result of the contributions of all past generations. The basic mission of saving lives and protecting property has not changed. Only the manner and means to cope with the problem have changed. Nonetheless, the fire service of today still contains a great deal of the past in its inventory of tools, equipment, and methods. If we go back to the beginning of this chapter and revisit the term mission, it may be possible to reconcile the past with the present. The fire service is today what it has become from past experiences.

Lessons Learned

What does the future hold for firefighters as they begin their journeys? In one sense a firefighter's career is like climbing that ladder we described earlier in the chapter. There are a lot more rungs to be climbed as firefighters pursue their careers. And, while we stand on only one rung at a time, we must be able to place our hands on those rungs that are higher.

No one should attempt to predict the future without giving due credit to the past. That is what this chapter has been all about. The future will contain a lot of challenges and opportunities. The future will contain some changes that will be difficult to deal with, just as they were for previous generations. The future will contain a requirement that firefighters continue to learn and develop skills that did not even exist the day they entered the fire service, not unlike those from the past.

> **Ethics** What is predictable about the future is that firefighters will likely be called on to deal with whatever dilemma or disaster strikes the community they serve. They must be prepared.

The motto of the Roman fire brigade was "Semper Vigilans" or "Always Vigilant." It was good enough for then. It is good enough for today.

KEY TERMS

Aerial Apparatus   Fire apparatus using mounted ladders and other devices for reaching areas beyond the length of ground ladders.

Arson   A malicious fire or fires set intentionally by humans for vengeance or profit.

Automatic Sprinkler System   A system of devices that will activate when exposed to fire, connected to a piping system that will supply water to control the fire. Typically, an automatic sprinkler system is also supported by firefighters when they arrive on the scene.

Conflagration   A large and destructive fire.

Fire Alarm   Notification to the fire department that a fire or other related emergency is in progress, which results in a response.

Fire Engineering   The study of fire, fire behavior, fire extinguishment, and suppression.

Fire Hydraulics   The principles associated with the storage and transfer of water in firefighting activities.

Fire Societies   Groups of people who voluntarily banded together to deal with a community's fire problems.

Fire Wardens   Designated community individuals who walked the streets at night looking for fire and carrying large wooden rattles with which to signify a found fire.

Firemark   Signs on sheets of metal telling firefighters which company held the insurance policy on a home or building.

Foreman Individual designated as the leader of an early fire company; a predecessor to the modern title of fire chief.

Hazardous Materials Chemicals that are flammable, explosive, or otherwise capable of causing death or destruction when improperly handled or released.

Mission Statement A written declaration by a fire agency describing the things that it intends to do to protect its citizenry or customers.

Search and Rescue Attempts by fire and emergency service personnel to coordinate and implement a search for a missing person and then effect a rescue.

Speaking Trumpet Trumpet used by a foreman or crew boss to shout orders above the noise of firefighting activities.

Terrorism Acts of violence that are arbitrarily committed against lives or property and intended to create fear and anxiety.

Triple Combination Engine Company Fire apparatus that can carry water, pump water, and carry hose and equipment.

REVIEW QUESTIONS

1. What is a mission statement? Why do fire and emergency services departments have mission statements? What is the mission statement for your department?
2. What is a firemark? What was the purpose of the firemark in early American history?
3. What was the purpose of the "Rattlewatch"?
4. What contributions did Benjamin Franklin make to the American fire service? What other American historical figures played a role in the development of the American fire service?
5. List some symbols present in today's fire service and describe their historical significance. For example, why is the Dalmatian considered a "firehouse dog"?
6. What do you think the future of the American fire service will look like in 5 years? In 10 years? In 100 years?
7. Discuss what effects you think terrorism has had on the fire service.

Additional Resources

Bureau of Alcohol, Tobacco, and Firearms, Arson and Explosives National Repository http://www.atf.treas.gov/aexis2/index.htm

Goudsblom, Johan, *Fire and Civilization.* Penguin Press, New York, 1992.

Great Fires of America. Country Beautiful Corporation, Waukesha, WI, 1973.

Gurka, Andrew G., *Hot Stuff Firefighting Collectibles.* L-W Book Sales, Gas City, IN, 1994.

National Center for Injury Prevention and Control http://www.cdc.gov/ncipc

National Institute for Occupational Safety and Health, Fire Fighter Fatality Investigation and Prevention Program http://www.cdc.gov/niosh/firehome.html

National Fire Information Council http://www.nifc.org

National Interagency Fire Center http://www.nifc.gov

Smith, Dennis, *History of Firefighting in America.* Dial Press, New York, 1978.

United States Department of Transportation, Office of Hazardous Materials Safety http://hazmat.dot.gov

United States Fire Administration http://www.usfa.fema.gov

CHAPTER 2

FIRE DEPARTMENT ORGANIZATION, COMMAND, AND CONTROL

Thomas J. Wutz, Midway Fire Department and New York State Office of Fire Prevention and Control

OUTLINE

- Objectives
- Introduction
- Fire Department Organization
- The Firefighter
- The Company
- Additional Fire Department Functions
- Regulations, Policies, Bylaws, and Procedures
- Allied Agencies and Organizations
- Incident Management
- Incident Management System (IMS)
- Lessons Learned
- Key Terms
- Review Questions
- Suggested Readings
- Additional Resources

STREET STORY

Winters in Washington, D.C., can be fickle. One year the temperature could be 70 degrees during January, or it could easily be 17 degrees with three feet of snow on the ground. On January 13, 1982, it was the latter. When people woke up that morning to blowing snow and sub-teen temperatures, they may have expected rough weather but little did they realize that by afternoon rush hour the city would experience a major subway crash in the tunnel with seven fatalities and that Air Florida flight 90 would collide, with seventy-nine souls on board, into the 14th Street Bridge that links Washington with Virginia across the Potomac River.

Concepts such as command and control and tools such as the Incident Management System (IMS) were not readily employed in 1982. The incident was under the command of a senior fire department officer. The outcome would depend heavily on the abilities of that officer because he was responsible for all aspects of command. I was a young officer in command of a ladder company that day. Before the day was over I would have the fortune of responding to both the subway and the airplane crashes. My clearest recollection of the incidents that day is that there was a great amount of confusion. Many times orders were issued only to be countermanded or withdrawn. A jurisdiction dispute existed between many of the responding entities. The number of casualties was compounded by the fact that many of the dead and injured had to be removed from the wreckage in the subway and at the 14th Street Bridge. The aircraft itself went into the water with no flames but it took most of the victims with it. The weather was the final trump card played against us that day, making traveling and responding extremely difficult. This was a very taxing and difficult operational day for anyone to handle and should not reflect on the commanders that fateful day, but recently a panel of chief officers who had experience in the implementation of IMS and the concepts of command and control revisited that day in the form of an exercise. No facts were given except for the pertinent data of resources available, time of day, weather, and the scene as reported by the first arriving units. Needless to say, the outcome from a command point of view was much more successful this second time. The IMS can be considered one of the high points in tool development for the fire service of the twentieth century.

The fire service continues to develop tools, and concepts continue to evolve as the need dictates. Some of these tools, such as the IMS, will be used well into the next millennium; that is, if we embrace and use them wisely. The leaders of the fire service must ensure that those who follow us maintain an open mind and continue to educate themselves in order to push the concept of command and control to levels where the safety of our people will be ensured and our operations will be successful in protecting civilians and property.

—Street Story by Mike Smith, Battalion Fire Chief, District of Columbia Fire and EMS Department

OBJECTIVES

After completing this chapter, the reader should be able to:

- Describe a typical fire department organization and mission statement.
- Define the functions of a firefighter and list the common tasks a firefighter must be able to perform.
- Explain what a standard operating procedure is and list five general areas covered by SOPs.
- List five rules and regulations of an organization and describe how they apply to the firefighter.
- List and define the five major components of an incident management system.
- Describe duties and responsibilities in assuming and transferring command within the incident management system.
- List five allied agencies that assist with fire department operations and describe their functions.

INTRODUCTION

The American fire service has a history and tradition that is more than 250 years old, dating back to Ben Franklin who organized the Union Fire Company in 1736. That basic organizational structure is still in use today. Depending on the size and needs of a community, a fire department will consist of a number of companies in one or a number of stations throughout a community. The number of stations depends on the geography of the response area, travel time, and distance.

Companies are usually divided into functions such as engine, ladder, or truck companies and specialized units such as rescue or hazardous materials companies. In addition, many fire departments provide either basic or advanced life support emergency medical services. The organization is designed to establish a clear division of work assignments:

- To assign responsibility for a specific response area
- To assign responsibility for a specific activity
- To eliminate duplication of work and confusion
- To establish an adequate level of equipment and personnel response to control the emergency scene or incident.

This chapter describes the many different roles a firefighter may have in the organization of a fire department. The firefighter is the basic unit of the organization, and a well-trained firefighter is essential for the department to accomplish its mission.

FIRE DEPARTMENT ORGANIZATION

Fire departments, as shown in **Figure 2-1**, are like any other organization in that they have a reason for existing and a structure for operations. The mission

Figure 2-1 A typical fire department with apparatus and facilities. It is necessary to look into the fire department's organization to understand how it functions.

statement should communicate the reason for being, and the organizational structure defines the chain of command and authority.

The Business of Fire Protection

Providing fire protection and life safety services is a business. Like businesses of all sizes and types, the fire service strives to provide the highest quality services as effectively and efficiently as possible while working to meet customer or community expectations. The fire service also works to meet these goals while protecting the safety of firefighters, emergency responders, and the public. Also similar to businesses, the fire and emergency services have adopted a quality mind-set and a focus on customer service in the provision of their emergency and nonemergency services.

Mission Statement

Each fire department should have a mission statement. This will provide the members of the organization with a clear and defined purpose of the type and level of service the department will provide. Also, the mission statement must be specific so the public—the people depending on the entire department for help—know what to expect from the organization and understand what duties firefighters can and cannot perform.

> **Streetsmart Tip** The fire department must communicate its mission statement to the public. A few ways this can be accomplished are during fire prevention inspections and fire safety information presentations to community groups.

The mission statement provides a clear and concise statement of what the organization intends to do or accomplish. The statement should contain the type of services or product to be delivered and who will receive this service or product.

Sample Mission Statement 1
The Midway Fire Department is organized to deliver fire prevention, fire suppression (extinguishment), and rescue services to the citizens of its protection area. This will include response to conduct vehicle extrication, hazardous materials mitigation, and basic life support emergency medical services.

The sample mission statement describes what the Midway Fire Department does. The mission statement does not limit what an organization will do; it provides the focus for the service it will provide to the people in its community. This is especially true for an emergency response organization when a person calls for help.

If the Midway Fire Department decides to expand its services to include additional activities, its mission statement would read as follows:

Sample Mission Statement 2
The Midway Fire Department is organized to deliver fire prevention, life safety, fire suppression (extinguishment), and rescue services to the citizens of its protection area. This will include response to conduct vehicle extrication, hazardous materials mitigation, confined space rescue, advanced life support emergency medical services, disaster response, and fire life safety code enforcement.

Again the services provided by this organization are very specific, and the firefighters and public know what to expect from the fire department. The second mission statement shows that the Midway Fire Department has decided to provide advanced life support, confined space rescue, disaster response, and code enforcement services.

Organizational Structure

To accomplish its tasks, a fire department must have some type of organizational structure. This structure may be as complex as that shown in **Figure 2-2** or as simple as that of **Figure 2-3**. Both figures show a structure for internal organization and how functions and responsibilities are delegated in the organization. Neither of these shows the receiver of the fire department's services, the citizens they protect, or how the fire department is connected to the community. Another way of defining the fire department organization could look like **Figure 2-4**. This structure shows circles of interdependence the organization has with its community. The fire department depends on the community and governing body for support. In turn, the community depends on the fire department for services and the governing body depends on the department to provide those services. The firefighter is the key position in the organization and is essential for the department to fulfill its mission.

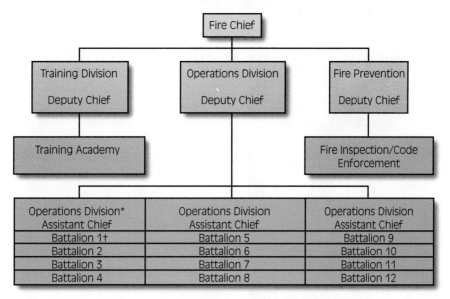

*Each division has one heavy rescue company.
† Battalion consists of four engine companies and one ladder company.

Figure 2-2 The organizational structure for a medium to large fire department shows the division of work assignments and chain of command. Most large departments will have a structure like this with separate divisions for suppression, training, and fire prevention.

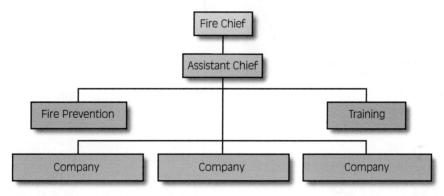

Figure 2-3 The organizational structure for a small fire department will have a more direct chain of command and structure. In this size of organization, one individual may be responsible for a number of functions.

Figure 2-4 Different from most organizational charts, this organizational structure shows the interdependence of the community, fire department, governing body, and the firefighters.

Figure 2-5 The firefighter is the individual who makes a department operate.

THE FIREFIGHTER

A firefighter is an individual trained to perform the function of fire prevention and suppression. Firefighters must possess both the knowledge and skills to be able to perform safely and effectively at an emergency incident, **Figure 2-5.**

Depending on the mission of the fire department or organization, there may be other areas that firefighters must be knowledgeable about in addition to firefighting. A firefighter may be required to be an **emergency medical technician (EMT)** or **paramedic (EMT-P)**, a **hazardous materials technician**, a **rescue specialist**, **Figure 2-6,** a fire investigator, or a fire prevention officer. Many other positions are available in the firefighting community, and as firefighters begin their training they can start establishing personal and career goals. The only limitation to advancement and personal growth is a failure to take advantage of training and learning opportunities.

There are many organizations in the fire and emergency service community that assist in the determination of the necessary qualifications firefighters need to perform their jobs. The **National Fire Protection Association (NFPA)**, through its professional qualification committee, has established a number of training standards for various department positions. These standards address firefighters, fire officers, hazardous materials responders, rescue specialists, and driver/operators to name a few. In addition, many state and local agencies have established standards and qualification requirements for firefighting and emergency response occupations and volunteer positions. All firefighters must be familiar with the standards in their jurisdictions and determine what types of training and certification are necessary to maintain a position as a firefighter.

> **Firefighter Fact** Two national organizations provide certification of state and local firefighter training programs in accordance with NFPA standards. These are the National Board on Fire Service Professional Qualifications and the International Fire Service Accreditation Congress (IFSAC).

Through a consensus development process, the NFPA provides the *Standard for Fire Fighter Professional Qualifications,* also known as **NFPA 1001**, **Figure 2-7.** This standard outlines the minimum knowledge, skills, and abilities firefighters must have at two different levels, Firefighter I and Firefighter II. In addition to knowledge and skills gained through training and experience, firefighters must also be in good physical condition to perform the demanding and strenuous tasks involved in firefighting and emergency response.

> **Firefighter Fact** According to the United States Fire Administration, 102 firefighters died in the line of duty in 2002. Of these line-of-duty deaths, 31 (nearly one-third of the total) were from heart attacks. Regular exercise, physical conditioning, and a healthy lifestyle can help save firefighters' lives and reduce the number of line-of-duty deaths from heart attacks.

Listed here are the typical requirements an individual trained in structural fire suppression would be expected to meet:

- Know the department's organizational structure and operating procedures.
- Perform all duties safely.
- Know the department's response area or district, including streets and hazards.

FIRE DEPARTMENT ORGANIZATION, COMMAND, AND CONTROL ■ 27

(A)

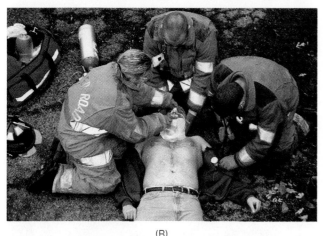

(B)

(C)

Figure 2-6 Some positions available to firefighters are (A) rescue specialist, (B) paramedic, and (C) hazardous materials technician.

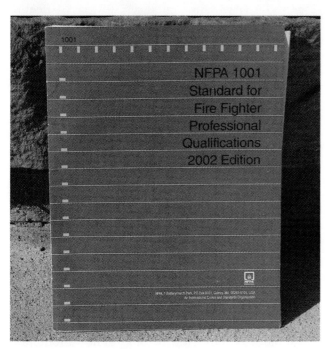

Figure 2-7 National Fire Protection Association 1001, *Standard for Fire Fighter Professional Qualifications*, provides the basis for firefighter training.

- Maintain firefighting equipment, especially personal protective equipment.
- Respond to alarms of fire as a member of a trained unit/company.
- Use self-contained breathing apparatus and personal protective equipment.
- Rescue people endangered by the fire.
- Use fire department tools to conduct forcible entry, ventilation, and fire extinguishment at structure, vehicle, and ground cover fires.
- Conduct overhaul operations at a fire scene.
- Conduct fire prevention inspections.
- Present fire safety information to the community.
- Provide basic life support activities such as CPR or control of bleeding.

Depending on the organization's mission statement and responsibilities, the firefighter may also be required to perform additional duties or functions. A sampling of these and the corresponding training standard are shown in **Table 2-1** and **Figure 2-8**.

Additional Duty Training Standards

POSITION DESCRIPTION	TRAINING STANDARD
Fire apparatus driver/operator	NFPA 1002
Airport firefighter	NFPA 1003
Public safety communications telecommunicator	NFPA 1061
Paramedic/emergency medical technician	State or local standards

TABLE 2-1

THE COMPANY

Historically the basic unit of a fire department is the **company**, a team of firefighters with apparatus assigned to perform a specific function in a designated response area. A company is typically designated as an engine, ladder, truck, squad, or rescue depending on the type of apparatus, equipment, and firefighters on board. The company officer, often a captain, is responsible for all activities of the company. The company officer may have lieutenants or other officers assigned to assist with company tasks and objectives.

Streetsmart Tip Depending on history, traditions, and/or geography, the titles used to designate a position in the fire department may vary from those presented in this text. Not all organizations have the position of lieutenant and some use a military designation such as sergeant or private. Firefighters should be familiar with the rank structure and titles used in their own organization.

(A)

(B)

(C)

Figure 2-8 Firefighters may be required to perform additional duties as (A) dispatcher, (B) airport firefighter, or (C) apparatus driver/operator.

Company officers are supervisory-level positions and are responsible for both firefighters and administrative duties, **Figure 2-9**. Requirements for the position of company officer are addressed in a separate training standard, NFPA 1021, *Standard for Fire Officer Professional Qualifications*.

Just as firefighters and company officers have standards that help in determining the knowledge, skills, and abilities to complete their jobs, so does the apparatus they ride to emergency incidents. NFPA 1901, *Standard for Automotive Fire Apparatus*, outlines minimum equipment, specifications,

FIRE DEPARTMENT ORGANIZATION, COMMAND, AND CONTROL

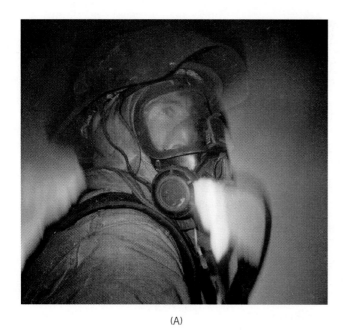
(A)

(B)

Figure 2-9 Company officers perform both (A) firefighting and (B) administrative duties.

and capabilities for various types of firefighting apparatus. **Table 2-2** outlines some of the basic requirements for common types of fire apparatus.

The Engine Company

The **engine company**, as shown in **Figure 2-10,** is organized to provide firefighters who deliver water at the fire scene, deploy hoselines, and attack and extinguish fires in vehicles and structures.

An engine company often uses an apparatus referred to as an engine or pumper. NFPA 1901 states that a pumper should have a permanently mounted fire pump with a capacity of at least 750 gallons per minute (gpm). NFPA 1901 also states that a pumper should carry no less than 300 gallons of water. Other specific hose and ladder requirements are listed in **Table 2-2.**

In some organizations, the engine company may also perform rescue or squad functions, as detailed later in this section. This increases the amount of equipment necessary on the apparatus and may result in a different company designation, such as a rescue engine.

The Truck Company

The primary functions of the **truck company**, as shown in **Figure 2-11** (also known as a ladder company), are to carry firefighters for forcible entry, search and rescue, ventilation, provision of ladders, securing of utilities, and overhaul functions at a fire scene.

NFPA 1901 states that the apparatus used by truck companies, an aerial fire apparatus, should have a minimum of 115 feet of ground ladders, including at least one attic ladder, two straight ladders with folding roof hooks, and two extension ladders. Aerial fire apparatus may or may not be equipped with a fire pump. Aerial fire apparatus are also required to carry an extensive list of hand and power tools to accomplish the many tasks that are associated with truck company operations.

In addition to traditional truck company operations, in some jurisdictions, truck companies may also perform the functions of a rescue company, as detailed in the next subsection. Regardless of apparatus, all fire departments must be able to perform the functions associated with the truck company.

The Rescue Company

The **rescue company** (also known as a squad or squad company) is shown in **Figure 2-12.** It is organized to provide specially trained firefighters and specialized rescue equipment at an incident scene. The apparatus designated as a rescue company carries tools to conduct forcible entry and search and rescue at fire operations. Most rescue companies also carry specialized tools to conduct vehicle extrications, confined space rescue, rope

NFPA 1901: Standard for Automotive Fire Apparatus

OVERVIEW OF BASIC APPARATUS REQUIREMENTS

PUMPER

Fire pump with minimum rated capacity of 750 gpm.

Water tank with a minimum of 300 gallons.

Ground ladders, to include at least one attic ladder, one extension ladder, and one straight ladder with roof hooks.

A minimum of 15 feet of soft suction hose or 20 feet of hard suction hose.

A minimum of 800 feet of 2 1/2-inch or larger diameter fire hose.

A minimum of 400 feet of 1 1/2-inch, 1 3/4-inch, or 2-inch fire hose.

AERIAL FIRE APPARATUS

A minimum of 115 feet of ground ladders, to include at least one attic ladder, two straight ladders with folding roof hooks, and two extension ladders.

Apparatus may or may not have a fire pump and water tank.

MOBILE WATER SUPPLY APPARATUS

A minimum of 1,000-gallon water tank.

Apparatus may or may not have a fire pump. If the apparatus does have a pump, there are minimum requirements for the fire hose that must be carried.

QUINT

A permanent fire pump with a minimum rated capacity of 1,000 gpm.

Water tank with a minimum of 300 gallons.

A minimum of 85 feet of ground ladders, to include at least one attic ladder, one extension ladder, and one straight ladder with roof hooks.

A minimum of 15 feet of soft suction hose or 20 feet of hard suction hose.

A minimum of 800 feet of 2 1/2-inch or larger diameter fire hose.

A minimum of 400 feet of 1 1/2-inch, 1 3/4-inch, or 2-inch fire hose.

TABLE 2-2

Figure 2-10 The engine company provides the firefighters with the tools, equipment, and hose necessary for fire suppression operations.

FIRE DEPARTMENT ORGANIZATION, COMMAND, AND CONTROL 31

Figure 2-11 The ladder or truck company is designed to carry firefighters and for forcible entry, search and rescue of building occupants, ventilation, ladder provision, and overhaul functions at a fire scene.

Figure 2-12 The rescue company is designed to carry specialized equipment for rescue operations and may provide support equipment such as light towers.

rescue, and other forms of technical rescue operations. These specialized functions depend on local requirements and the mission of the organization.

Specialty/Combination Units

Because the fire service has a great amount of diversity in the types of services provided to communities, and an equally great diversity in the way these services are delivered, there are a wide variety of unique fire and emergency service apparatus that are used in departments worldwide.

Combination units are typically a blend of two major company functions into one piece of apparatus. For example, many departments use a **quint** to deliver fire suppression services. A quint is a combination of an engine and ladder company and can perform either function as necessary on a fire scene. According to NFPA 1901, a quint must have a permanently mounted fire pump and water tank, an area to store hose, an aerial device (ladder or platform) with a permanently mounted waterway, and at least 85 feet of ground ladders.

Another type of specialty unit, a mobile water supply apparatus, is used in areas where a water supply is not present, such as hydrants or natural water sources. Mobile water supply apparatus, also known as tankers or tenders, carry a minimum of 1,000 gallons of water per NFPA 1901, and they may or may not have a fire pump and fire hose.

Specialty units provide unique types of services at emergency scenes. Examples of specialty units include wildland fire or brush fire response units, **aircraft rescue and firefighting (ARFF)** apparatus, light and air units, hazardous materials units, mass casualty response units, and water rescue and firefighting apparatus. These units are specially equipped to handle unique emergency situations, and the personnel who operate on these units are specially trained.

Emergency Medical Services

Many fire departments provide either basic life support (BLS) or advanced life support (ALS) **emergency medical services**. This may be an additional duty assigned to an existing engine, truck, or rescue company or a specific unit designated as a BLS/ALS response unit. Also, fire departments may operate ambulances to provide transport services.

The Chief Officers

The chief of the department, as shown in **Figure 2-13**, is the person ultimately responsible for the operations and administration of the fire department. To carry out these responsibilities, the chief may have a number of deputy, division, assistant, or battalion chiefs.

The rank structure and position designation depend on the size, needs, and in some cases the history of an individual fire department. Not all fire departments will have all of the positions listed next because some designations serve the same function:

- *Chief of department:* Responsible for all department functions.
- *Deputy chief:* A staff position designated by the chief of department for a specific function, such as personnel, training, or administration. Deputies usually have the authority to act in the absence of the chief.
- *Assistant chief:* Similar to a deputy chief; the assistant may be a staff position responsible for a functional area. In some departments, an assistant chief may be a higher rank than a deputy chief.
- *Division chief:* Usually an operational position responsible for a large geographic area or a number of battalions.
- *Battalion/district chief:* An operational position responsible for a specific number of companies.

The number of officers in any department depends on the size of the organization and the necessity to maintain unity of command, span of control, and division of task responsibilities. Chief officers are management-level positions and requirements for these positions are addressed in a separate training standard, NFPA 1021, *Standard for Fire Officer Professional Qualifications*.

ADDITIONAL FIRE DEPARTMENT FUNCTIONS

Because of the complex issues and tasks facing a fire department, many additional functions are assigned to its operations. Some of these, such as training or fire prevention, are necessary for day-to-day operations. Additional sections may be established to manage complex issues such as hazardous materials, urban/technical search and rescue, water rescue, or the delivery of emergency medical services.

Fire Prevention and Life Safety

One of the prime missions of all fire departments is the prevention of fires. Preventing fire not only reduces the risk to a community both in lives and property but also reduces the exposure of firefighters to a very dangerous and hazardous duty. The organization of a fire prevention office is usually divided into two functions: code enforcement/inspection services and fire/life safety education. A chief-level officer usually heads the fire prevention office.

Fire department personnel, as shown in **Figure 2-14**, should conduct fire prevention inspections. This provides the organization with increased knowledge of the hazards present in a community and provides the opportunity to develop prefire plans.

Public fire/life safety education activities are usually assigned to the fire prevention office and include fire prevention education and fire survival programs. In addition to code enforcement, fire prevention programs are designed to prevent the start of

Figure 2-13 Chief officers are responsible for many functions, including command of emergency operations.

FIRE DEPARTMENT ORGANIZATION, COMMAND, AND CONTROL 33

Figure 2-14 The most important function of the fire prevention staff is to conduct inspections to prevent fires or ensure early control by fire suppression systems.

a fire. If an area were to experience a high number of fires starting because of careless disposal of smoking materials, an education program on proper disposal would be delivered to the public. Fire survival programs educate the public on what to do after a fire has started. An example of these activities includes education on installation and maintenance of smoke detectors or **exit drills in the home (EDITH)**.

Staff assigned to fire prevention/fire inspection duties should meet the requirements as addressed in NFPA 1031, *Standard for Professional Qualifications for Fire Inspector and Plan Examiner.*

Training

> **Street Smart Tip** Training must be a continuing function in all fire departments.

Training begins with basic firefighter or probationary training, as shown in **Figure 2-15**, and continues with proficiency training as new tools, equipment, or techniques become available. Depending on the size, organizational structure, and other responsibilities, a chief-level officer usually heads the training division or group. Regardless of size, all fire departments must have a designated training officer or division.

The training division administers and documents all training activities for the department and will usually have a number of instructors assigned to present specific subjects. This division may also be responsible for conducting mandatory employee

Figure 2-15 Training must be a continuing function in all fire departments regardless of size or area served.

safety training required by the Occupational Safety and Health Administration (OSHA). This program requires firefighters to be trained to perform assigned tasks in a safe manner. Some of the required topics for firefighting are respiratory protection, response safety, and workplace safety.

Training officers and instructors should meet the requirements for these positions as addressed in NFPA 1041, *Standard for Fire Service Instructor Professional Qualifications*.

Emergency Medical Services

As mentioned previously in this chapter, and depending on the size of the organization or its jurisdiction, the EMS function may be a separate division within the fire department with a chief-level officer responsible for its activities.

Apparatus Maintenance and Purchasing

Large departments may have a fire apparatus maintenance or repair shop that is usually responsible for all vehicle repairs, maintenance, and purchasing. Depending on local policy this organization may be headed by a fire department officer or a nonuniform staff member knowledgeable in the areas of vehicles, heavy equipment, and fire apparatus.

Special Operations

Depending on the size of its community or potential hazards present, a fire department may establish a special operations section, **Figure 2-16**, which will deliver or support services such as hazardous materials mitigation, high-rise operations, air operations, confined space rescue, and swift water or ice rescue. A chief-level officer usually heads these specialized functions.

Individuals assigned to these areas must have specific knowledge and skills for their specialty and this usually requires many additional hours of training.

REGULATIONS, POLICIES, BYLAWS, AND PROCEDURES

All organizations must have regulations, policies, bylaws, and procedures to operate in an effective manner. This is especially true in a fire department to ensure an adequate and effective emergency response. These regulations are implemented and used to establish the daily and emergency operations of the organization.

Figure 2-16 Trench rescue is one of many specialized operations requiring additional equipment and training.

Regulations

Regulations are the rules that determine how an organization operates and are usually established by top-level management or a governing body. Regulations address topics such as time and attendance, uniform, and use of department equipment and vehicles. These are usually very specific in nature and may carry some type of disciplinary action for violation.

In addition, state or federal government agencies, such as OSHA, may establish regulations governing fire department activities. These usually deal with such issues as health and safety, antidiscrimination laws, and work hours.

> **Note** Regulations established by the governing body of a fire department are internal in nature and design, whereas regulations established by state or federal agencies have the force of law, with some penalty or fine charged for violations.

Policies

Policies are formal statements or directives established by fire department managers to provide guidance for decision making. Policies will usually be general in nature and provide a framework for administering the day-to-day department activities. Topics addressed by policies include types of response levels, staffing of companies, and the release of information at an incident.

Bylaws

Volunteer fire departments provide a large amount of fire protection, both in the United States and Canada. These departments may be organized as previously described in this chapter, as independent corporations, or as some combination of both types of organizational structure. A fire corporation will usually be organized as a not-for-profit organization with a governing body of a president and vice president, or a board of directors. The board of directors or membership will establish bylaws in place of regulations or policies on how the department business structure is organized. Generally, fire and emergency operations are not covered as part of the bylaws.

Procedures

Standard operating procedures (SOPs) provide specific information and instructions on how a task or assignment is to be accomplished.

SOPs are established so that all members of a department will perform the same function with the same level of uniformity. Also, these procedures are generally tactical in nature because, in most instances, they address emergency operations. In addition, all SOPs must be distributed to all department members. Some topics addressed by SOPs include cleaning and use of protective equipment, apparatus response policies, use of specific tactics or tools, and how to establish a water supply at a fire scene.

A note on terminology: This text uses the term *standard operating procedure* to describe documentation outlining actions or procedures to be followed. Many departments use the term *standard operating guideline (SOG)* to describe this documentation. SOGs are often thought of as providing guidance to company officers and firefighters when operating on emergency scenes as opposed to outlining strict parameters for actions. Each department is different in its use of SOP and SOG terminology and even different in its interpretation of the flexibility of the documents. It is up to the individual firefighter to understand what terms are used by his or her department.

There are a variety of ways in which SOPs are developed and documented. In general, the following should be true:

- SOPs address the who, what, when, where, and how of a topic.
- Firefighter safety is the first consideration in all procedures.
- SOPs are brief, clear, and concise.
- Lengthy SOPs are broken down into smaller sections.

Sample Standard Operating Procedure

Subject: Emergency Vehicle Response   Section: 303
Date: 01/01/2001
Purpose: To provide requirements for emergency response of Delmar Fire Department apparatus and vehicles.
Scope: This procedure applies to all drivers/operators of Delmar Fire Department vehicles.

1. Apparatus will not exceed the posted speed limit at any time.
2. Seat belts must be used by all personnel riding on apparatus.
3. No riding is allowed on the rear step area nor standing in the crew compartment while the vehicle is in motion.
4. Apparatus must come to a complete stop before attempting to proceed in any of the following conditions:
   - Intersection against a red traffic light
   - All railroad grade crossings
   - Blind intersections
   - Stop signs
   - Any condition when the operator cannot account for all traffic lanes
5. A spotter shall be located at the rear of the apparatus at any time the vehicle moves in reverse.
6. Operators are responsible for any traffic citation issued to them.

Figure 2-17 A sample standard operating procedure.

- SOPs cover dynamic topics and must be reviewed often, at least once every three years.
- A method for changing SOPs exists if they should become obsolete, inaccurate, or ineffective.

Figure 2-17 shows a sample of an SOP that might be found in a typical fire service organization.

ALLIED AGENCIES AND ORGANIZATIONS

During the course of any operation, the fire department and firefighter interact with many different people and organizations. Depending on the local jurisdiction, a few of these agencies could be police/law enforcement, public works or highway department, utility companies, environmental conservation/protection agencies, and private business. Depending on the nature of the incident, these organizations may be assisting with the operation or they may be receiving fire department services.

- *Police:* Provide assistance with traffic management and scene security at all incidents. May be the assisting or lead agency in fire investigations, bomb threats, or terrorism incidents.
- *Public works:* Provide assistance with traffic management and maintain public water system for fire protection. May maintain fire station facilities and provide assistance in securing materials (earth, sand) for the control of hazardous material incidents.
- *Utility companies:* Secure building and street utilities during or after an incident.
- *Environmental protection:* Depending on the incident they may be a lead or assisting agency at a hazardous materials spill.
- *Private business:* May provide supplies for emergency operations or may be the receiver of fire department services.

INCIDENT MANAGEMENT

Fire departments respond to millions of emergency incidents every year. These incidents are the operations for which a firefighter trains. Some incidents require a single company; some require multiple companies. Extremely large incidents such as wildfires or natural disasters may require mutual aid assistance from outside the authority having jurisdiction and involve fire agencies on a national basis.

> **Note** Mutual aid or assistance agreements are prearranged written agreements of the type and amount of assistance one jurisdiction will provide to another in the event of a large-scale fire or disaster. Mutual aid plans may include the fire department resources from a town, county, or entire state. The key to understanding mutual aid is that it is a reciprocal agreement. Many fire departments use automatic mutual aid to provide for immediate dispatch to an incident. This may be to supplement staffing or equipment required for high hazard facilities such as a hospital.

Command and Control

To manage incidents, firefighters must understand the concept of command and control and how it is applied at the emergency incident. The basic tactical unit at any incident is the company or single resource. As the size or complexity of the incident grows, additional companies and resources respond. To maintain command and control, the command officer must have the ability to manage effectively the increasing number of companies operating at an incident. This is accomplished by using **unity of command** and span of control.

Unity of command means having one designated leader or officer in charge of an operation, company, or single resource. Every operating resource has a single designated supervisor. On the initial response this could be a company officer or a firefighter. All units responding to the incident report to command and receive instructions on operations. As the incident becomes larger, command may transfer to a high ranking officer. This is accomplished by briefing the new command on the incident, actions taken, and units operating.

Span of control refers to the number of resources (people, companies, etc.) that any one person supervises. For emergency operations the span of control is usually three to seven, with the optimum number being five. This allows for effective management and safety of all operations.

Assuming Command

The first unit arriving at an incident will usually assume command of the incident and give a status report. This is accomplished with a radio announcement, for example: "Engine 438 arriving, establishing Maple Avenue Command. We have a two-story wood frame residential structure with smoke showing from the second floor. Two occupants are reported in the structure." This advises all other responding units of a fire condition in an occupied residential building. Generally, SOPs will dictate the tasks assigned to other responding units.

> **Streetsmart Tip** Local and regional policies may dictate the format of the radio status report, and firefighters should be familiar with this format and structure. In addition, the locality or region may have special terminology for radio status reports, such as location designations. In some areas, the address side of a structure may be considered side 1, whereas in other areas this may be considered side A. Firefighters should be familiar with these specific requirements and local terminology.

Transfer of Command

Transfer of command is the process of briefing an individual of equal or higher experience or authority so that person can assume command of an incident. As noted earlier, the first arriving unit will usually assume command of an incident and as the incident becomes more complex, the command structure will expand and the incident commander (IC) will usually change. The information necessary for a complete transfer of command is known as a status or situation status report and must include the following information:

- What has happened or the type of incident
- Incident action plan (strategy and tactics)
- Current units/resources operating, status, and location
- Civilians injured or trapped
- Firefighters injured or trapped
- Actions that have been accomplished
- Ability to control the incident with available resources

Transfer of command should occur during a face-to-face meeting, but under extreme conditions transfer may be accomplished by radio or telephone. Transfer can only be accomplished when the person assuming command is at the incident scene. Also, the individual assuming command must acknowledge receipt of the information, confirm its accuracy, and announce the transfer of command to avoid any confusion with the previous IC. The more complex an incident is, the more complex this briefing will be, possibly involving additional staff or technical experts.

INCIDENT MANAGEMENT SYSTEM (IMS)

The **incident management system** is exactly what the name implies: a systematic approach for the command, control, and management of an emergency incident. It provides a command structure and designated responsibilities for the functions that must be addressed to stabilize any incident, from a room and contents fire to a large-scale natural disaster.

The concept of incident command, as shown in **Figure 2-18**, has evolved over the years and is based on the FireScope Project from California. As a result of numerous and large wildland fires, the California Department of Forestry developed the system to provide a unified system for use by all agencies responding to an incident. As it has evolved, the application of the IMS has taken on many names such as Fire Ground Command, Incident Command System, National Fire Academy (U.S. Fire Administration), or the Incident Command System, National Wildfire Coordinating Group.

Regardless of the name of the system used in a particular jurisdiction, to function properly, it must contain the following components:

- **Common terminology**: The designation of a term that is the same throughout an IMS.

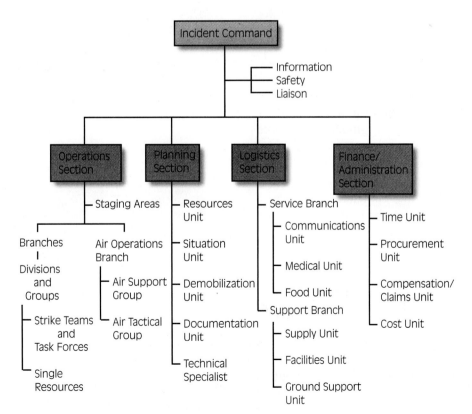

Figure 2-18 This chart of a typical incident management system shows the modular organization necessary to manage an incident.

- **Modular organization**: The ability to start small and expand if an incident becomes more complex.
- **Integrated communications**: The ability of all units or agencies to communicate at an incident. This may be as simple as assigning a tactical radio channel to an incident or developing a communications plan for use by many agencies at a large-scale incident.
- **Consolidated incident action plans**: The strategic goals to eliminate the hazard or control the incident. All companies or agencies working an incident must have the same objective.
- **Span of control**: The ability of one individual to supervise a number of other people or units. At an emergency incident the range for safe supervision is usually three to seven people or units. The ideal number is five.
- **Designated incident facilities**: These may be as simple as a command post established at a chief's vehicle **Figure 2-19A**, or complex enough to include a staging area, rehabilitation area for firefighters, feeding facilities, and office space for other agencies. A specially designed mobile command post is shown in **Figure 2-19B**. In some instances it is possible to designate these facilities prior to an incident.
- **Resource management**: Common designators are used for all resources assigned to an incident. Resources are assigned and managed in one of the following ways:
  - *Single resource:* Personnel or vehicle and required equipment.
  - *Task force:* A task force is any combination of single resources within the span of control guidelines, assembled for a common task or assignment. A structural assignment task force may consist of three engines, two trucks, and one rescue company.
  - *Strike team:* A combination of a specific number of units of the same kind and type. A water supply strike team would be five Type 1 engine companies.

Note Two key concepts that firefighters and officers must remember for all incidents are that the incident commander maintains responsibility for safe control of an incident and that the IC may delegate authority for a function or task but never responsibility. Also, any function or task that the IC does not delegate remains with the IC to accomplish.

(A)

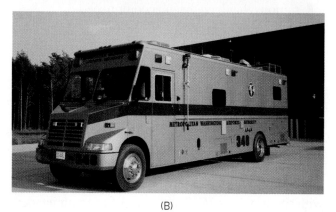

(B)

Figure 2-19 Depending on an incident's complexity, a command post may be as simple as (A) the command officer's vehicle or (B) a specially designed mobile command post.

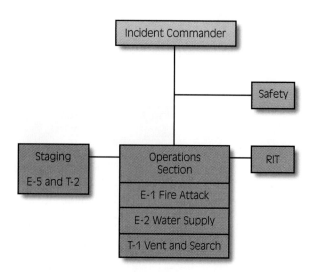

Figure 2-20 Command organization for a small structure fire incident.

As shown in **Figure 2-20**, the command structure for a small structure fire (such as a single-family residential fire) involves an IC who has designated a safety officer and retained responsibility for operations. The IC is always responsible for all functions and unassigned tasks. The IC, as operations chief, has established a staging area and designated a fire attack group, a water supply group, a ventilation/search group, and a **rapid intervention crew (RIC)**. These firefighter rescue teams may also be known as a **firefighter assist and search team (FAST)** or a **rapid intervention team (RIT)**.

Five Major Functions of an Incident Management System

An incident command or management system has five major functional areas, four of which may or may not be established depending on incident complexity. All incidents will always have an incident commander, who may establish the operations, planning, logistics, and finance/administration sections. When these positions are established, an individual is designated as the chief and reports directly to the IC. In addition, the IC may establish three command staff positions—safety officer, liaison officer, and public information officer—to assist with the incident management. These positions are not considered under the span of control guidelines. During incidents involving long **operational periods** or other agencies these functions should be established. The managerial level of the organization structure used should follow a format similar to that shown in **Figure 2-21**.

Command

The incident commander is responsible for developing the **strategic goals** for control of an incident. This may be a chief-level officer, captain, or firefighter, depending on the staffing of the fire department that responds and who arrives first. The IC has the authority to request and assign resources to an incident, the power to establish functional areas to control the incident, and the responsibility for the safety of all responders. The IC manages the incident, not the **tactics**.

Operations

The operations section chief is responsible for implementing the tactical assignments to meet the strategic goals established by the IC. The operations chief reports directly to the IC. Depending on the size and complexity of the incident, the operations section may be a single unit or a number of **branches** or **divisions**.

Staging

Staging is part of the operations section under the direct control of the operations chief with the assistance of the staging area manager. All apparatus and personnel assigned to staging are committed to the incident and must be available for deployment within three minutes.

Planning

The planning section chief is responsible for the development of the incident action plan. This plan is based on evaluation of the information and particulars of an incident. The planning section is also responsible for tracking the status of all resources used at an incident.

The planning section provides the IC with situation status report updates. These reports detail what

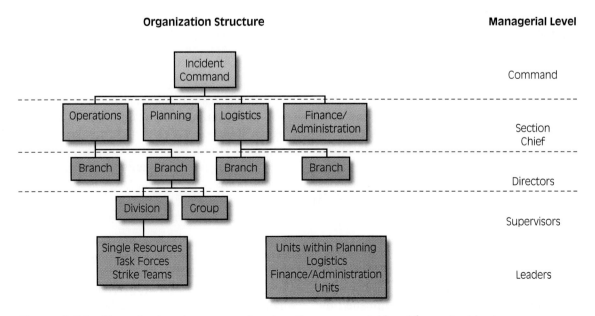

Figure 2-21 Organizational structure showing the managerial level for an incident.

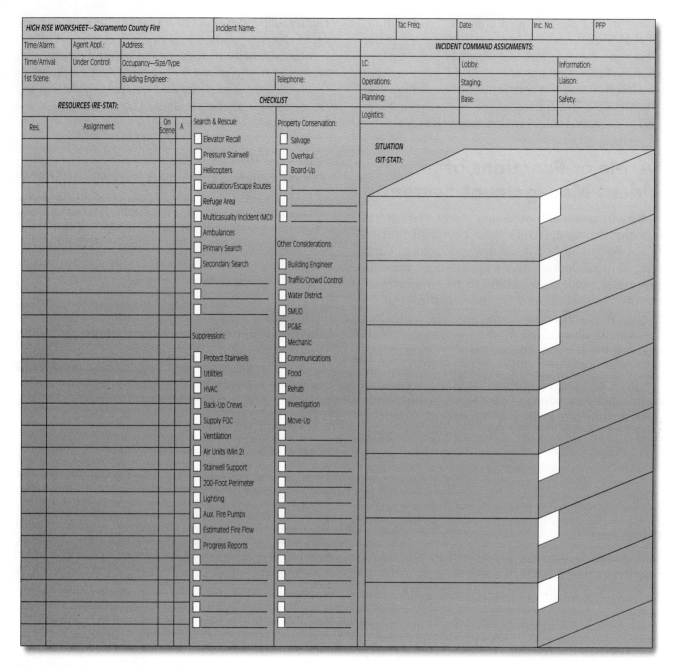

Figure 2-22 Tactical worksheets provide the incident commander with a guide for managing an incident.

has happened at the incident, injuries, how the incident is being controlled, resources in use, and suggested actions for stabilization of the incident. For large-scale incidents or incidents of long duration, the planning section will develop an incident action plan for approval and use by the IC. These plans are usually for operational periods of eight to twelve hours in length.

Logistics

The logistics section chief is responsible for securing the facilities, services, equipment, and materials for an incident. Usually this is accomplished by using a support and service section. The support section is responsible for medical support for incident responders, communication, and food services. The services section is responsible for supplies (food, medical, and specialized extinguishing agent), facilities, and the resources to deliver these items.

Finance/Administration

The finance/administration section chief is responsible for documenting cost of materials and personnel for the incident. This is an important function, especially when the potential for reimbursement exists from areas such as state or federal disaster declarations and insurance companies. Generally this func-

FIRE DEPARTMENT ORGANIZATION, COMMAND, AND CONTROL

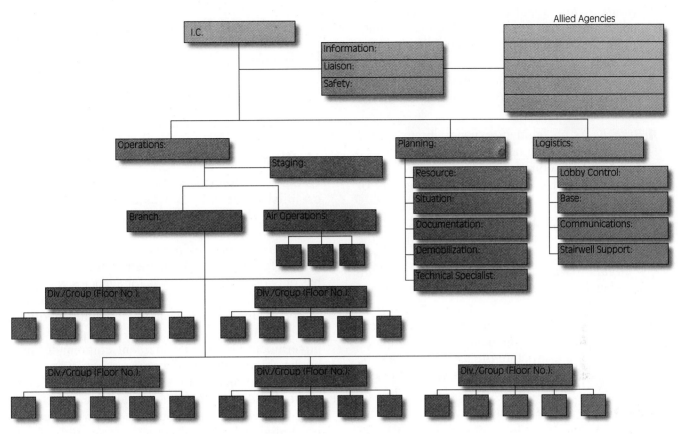

Figure 2-22 Continued

tion will be established for large-scale or long-duration incidents such as natural disasters.

Command Staff Positions

An incident commander has the ability to create staff positions to assist with an incident. These individuals report directly to the IC.

Safety Officer

The safety officer is responsible for the safety of all responders at an incident. The incident safety officer reports directly to the IC and has the authority to stop any activities that pose an immediate danger to incident responders. All other safety issues are channeled through the IC. For large-scale incidents the safety officer will develop an incident safety plan to be incorporated as part of the incident action plan.

Liaison Officer

The liaison officer is responsible for communication and contact between other agencies that respond. These may include environmental protection, police, highway or transportation, and state and federal government agencies. This position is especially important for long-duration, hazardous materials, or mass casualty incidents.

Streetsmart Tip The PIO position is very important for incidents involving evacuation of people or an unusual event. The news media is a valuable resource for conducting evacuation operations. Another benefit from this is the development of a relationship with the media for publicizing other events such as fire safety education or recruitment campaigns.

Firefighter Fact To implement an incident management system many fire departments use tactical worksheets like that shown in **Figure 2-22**.

Public Information Officer

The public information officer (PIO) is responsible for providing factual and accurate information concerning an incident to the news media. Only one information officer is appointed for each incident.

Incident Command Designations

- *Division:* The division is responsible for all operations within an assigned geographic area. At structural incidents a division may be designated for each floor or level of a building.

- *Group:* A functional designation to conduct a specific task such as search and rescue or ventilation. A group can operate in coordination with any division or sector.
- *Section:* The organizational level with responsibility for a major functional area of the incident.
- *Branch:* As an incident expands, branches are established to maintain span of control over a number of divisions, sectors, or groups. A branch must have at least two divisions or groups.
- *Type:* A resource that has specific capabilities or equipment. As an example, a Type 1 engine company has a minimum 1,000-gpm pump, 400-gallon water tank, 1,200 feet of 2½-inch hose, and a 500-gpm master stream appliance with four firefighters.
- *Task force:* Any combination of single resources assembled for an assignment. As an example, a task force for a structure fire assignment may be three engines, two ladder trucks, and one rescue company.
- *Strike teams:* Designation for a set number of resources of the same type and kind. A water supply strike team would be five Type 1 engine companies.
- *Crew:* A specific number of firefighters with an assigned task, but usually without apparatus. A ventilation crew may consist of six firefighters with tools but not assigned to a truck company.

> **Streetsmart Tip** Incident command terms and designations may vary by jurisdiction. For example, "sector" and "division" are interchangeable terms used for tactical assignments. Firefighters must know the terms used in their organization!

Unified Command

One of the benefits of any IMS is the ability to establish what is known as a **unified command** structure. The unified command structure is used to manage an incident involving multiple response agencies or when multiple jurisdictions have responsibility for control of an incident. Unified command has only one IC, but it allows for agencies with jurisdiction to be part of the command structure or team.

For example, a natural disaster may involve a large number of agencies from multiple levels of government. The initial incident commander may be the chief of the fire department; however, due to the complex nature of the incident, law enforcement, building code, and social service officials may be needed for legal reasons or for their expertise. Individuals from these agencies may be designated deputy ICs or may assume command as incident problems and priorities involve their specific discipline. Regardless of how complex or large an incident is, the incident will have one and only one IC. The use of unified command structure requires knowledge and training before the incident!

Lessons Learned

In the past few years, five notable incidents have resulted in the line-of-duty death of sixteen firefighters. The investigation report from each incident cited the lack of an organized and effective fire department command structure as one of the factors resulting in these deaths. To survive on the fire scene, firefighters must know the roles and responsibilities of the personnel, how their fire department command structure works, and be part of that command structure.

KEY TERMS

Aircraft Rescue and Firefighting (ARFF) Of or pertaining to firefighting operations involving fixed or rotary wing aircraft.

Branch The command designation established to maintain span of control over a number of divisions, sectors, or groups.

Common Terminology The designation of a term that is the same throughout an IMS.

Company A team of firefighters with apparatus assigned to perform a specific function in a designated response area.

Consolidated Incident Action Plan The strategic goals to eliminate the hazard or control the incident.

Division Command designation responsible for operations within an assigned geographic area.

Emergency Medical Services The delivery of prehospital medical treatment.

Emergency Medical Technician (EMT) An individual trained and certified to provide basic life support emergency medical care.

Engine Company The unit designation of a group of firefighters assigned to a piece of apparatus designed to deliver water to the fire scene.

Exit Drills in the Home (EDITH) A fire survival program to encourage people to practice fire drills from their home or residence.

Firefighter Assist and Search Team (FAST) A company designated to search for and rescue trapped or lost firefighters. May also be called a rapid intervention team (RIT).

Hazardous Materials Technician An individual trained to meet the requirements of CFR OSHA 1910.120, *Technician Level for Hazardous Materials Response.*

Incident Management System (IMS) An organized, systematic method for the command, control, and management of an emergency incident.

Integrated Communications The ability of all units or agencies to communicate at an incident.

Modular Organization The ability to start small and expand if an incident becomes more complex.

Mutual Aid or Assistance Agreements Prearranged written agreements of the type and amount of assistance one jurisdiction will provide to another in the event of a large-scale fire or disaster. The key to understanding mutual aid is that it is a reciprocal agreement.

National Fire Protection Association (NFPA) A not-for-profit membership organization that uses a consensus process to develop model fire prevention codes and firefighting training standards.

NFPA 1001 *Standard for Fire Fighter Professional Qualifications,* a national consensus training standard establishing the job performance requirements of tasks to be performed by firefighters.

Operational Period The time frames for operations at an incident. At large-scale or complex incidents these will usually be eight- to twelve-hour time frames.

Paramedic (EMT-P) An individual trained and certified to provide advanced life support emergency medical care, including drug therapy.

Quint A combination fire service apparatus with components of both an engine company and a truck company.

Rapid Intervention Crew (RIC) See *Rapid Intervention Team.*

Rapid Intervention Team (RIT) A company designated to search for and rescue trapped or lost firefighters. Depending on location, may also be called a FAST.

Rescue Company The unit designation of a group of firefighters assigned to perform specialized rescue work and/or tactics and functions such as forcible entry, search and rescue, ventilation, and so on.

Rescue Specialist A firefighter with specialized training and experience in areas such as high angle rope rescue, confined space, trench, or structural collapse rescue.

Span of Control The ability of one individual to supervise a number of other people or units. The normal range is three to seven units or individuals, with the ideal being five.

Staging Part of the operations section where apparatus and personnel assigned to the incident are available for deployment within three minutes.

Standard Operating Procedure (SOP) Specific information and instruction on how a task or assignment is to be accomplished.

Strategic Goals The overall plan developed and used to control an incident.

Tactics The specific operations performed to satisfy the strategic goals for an incident.

Truck Company The unit designation of a group of firefighters assigned to perform tactics and functions such as forcible entry, search and rescue, ventilation, and so on.

Unified Command The structure used to manage an incident involving multiple response agencies or when multiple jurisdictions have responsibility for control of an incident.

Unity of Command One designated leader or officer to command an incident.

REVIEW QUESTIONS

1. Write a mission statement for a fire department including the level of services provided and to whom the services will be delivered.
2. Draw an organizational diagram for a fire department including the chain of command and who is responsible for training, fire prevention, apparatus maintenance, and fire scene operations.
3. List five functions or tasks performed by firefighters in a fire department.
4. Explain the reason for standard operating procedures.
5. List five areas addressed by standard operating procedures.
6. Explain where to find the rules and regulations for a particular firefighting organization and identify who is responsible for developing and enforcing them.
7. List five rules and regulations of a firefighting organization and describe how they apply to the firefighter.
8. Research a jurisdiction's mutual aid agreement and list the apparatus, specialized equipment, and personnel available for response.
9. List and define the five major components of an incident management system.
10. Describe duties and responsibilities of assuming and transferring command for the IMS used by a particular organization.
11. List and describe the functions of five allied agencies that assist a fire department.

Additional Resources

Diamantes, David, *Fire Prevention: Inspection & Code Enforcement*. Delmar Learning, a part of the Thomson Corporation, Clifton Park, NY, 1998.

Incident Command System National Fire Academy (available through state training agencies).

Incident Command System National Training Curriculum I-100 and I-200 courses (available through state training agencies).

International Fire Service Accreditation Congress, http://www.ifsac.org

Klinoff, Robert, *Introduction to Fire Protection*. Delmar Learning, a part of the Thomson Corporation, Clifton Park, NY, 1997.

National Board on Fire Service Professional Qualifications, http://www.npqs.win.net

Smoke, Clinton H., *Company Officer*. Delmar Learning, a part of the Thomson Corporation, Clifton Park, NY, 1999.

United States Fire Administration, http://www.usfa.fema.gov

CHAPTER 3

COMMUNICATIONS AND ALARMS

Willis Carter, Chief, Shreveport Fire Department

OUTLINE

- Objectives
- Introduction
- Communications Personnel
- The Communications Facility
- Computers in the Fire Service
- Receiving Reports of Emergencies
- Emergency Services Deployment
- Traffic Control Systems
- Radio Systems and Procedures
- Arrival Reports
- Mobile Support Vehicles
- Records
- Lessons Learned
- Key Terms
- Review Questions
- Endnote
- Additional Resources

Photo courtesy of Scot Smith, Smith Photographic, Shreveport, Louisiana

STREET STORY

Several years ago I had the unique opportunity to spend time working in the 9-1-1 communications center. It was a very eye-opening experience for me since I had always been a firefighter in the field. Spending time in the center allowed me to realize the role that communications personnel play in the overall fire and emergency services system.

As a firefighter, I would get frustrated when calling dispatch on the phone for alarm numbers and times and the dispatcher would say, "We are busy—can you call back?" I would think to myself, "How long can it take to give me times and a number? I have to get these reports done, and the dispatchers are holding me up." When responding to emergency calls, if the dispatch information was not completely accurate, the field personnel would often blame the dispatchers and communications personnel. We often did not consider that maybe the caller did not provide accurate information.

After spending time in the communications center, I realized a new appreciation for what communications personnel do on a daily basis. While it may be frustrating to be told to call back for information, I realized that dispatchers were dealing with units calling in from twenty-four other stations. While they were dealing with these phone calls, they also had to answer and dispatch 9-1-1 emergency calls, answer inquiries from the media calling about incidents, and of course talk to people calling 9-1-1 for nonemergency information. I also came to understand that the communications personnel often have difficulty in obtaining accurate information from callers. Many times this is because the callers are upset, frantic, or just do not have good information to pass on to the communications center. The information given to the responding units is only as good as the information available to the dispatchers.

My experience in the communications center taught me a great deal about the role of the communications personnel. They are truly the critical link between the public and the fire rescue system. They are a dedicated group who try to do the best job possible, even under difficult circumstances and situations. They take the 9-1-1 calls and dispatch the units and then try to anticipate the needs of the units throughout the incident.

Do not make the same mistake that I did. Visit your communications center and experience what they do for yourself. And when they ask you to call back because they are busy, just do it with a smile.

—Street Story by James Angle, Chief, Palm Harbor Fire Department, Palm Harbor, Florida

OBJECTIVES

After completing this chapter, the reader should be able to:

- Demonstrate the proper method of answering a nonemergency administrative call.
- Demonstrate the proper method of answering an emergency call and effectively obtaining full and complete information, and promptly relaying that information to the communications center.
- Demonstrate the proper method of operating a mobile radio.
- Demonstrate the proper method of operating a portable radio.
- Complete a basic incident report.

INTRODUCTION

"Nine-one-one. What is your emergency?"

"Oh, please send help! There has been a terrible accident!"

"What is the location of the accident?"

"It's at the intersection of Main and...."

Terrifying emotionally wrenching scenarios such as this unfold thousands of times each day in **emergency communications centers** all across America. To get the information emergency responders need to do their jobs effectively, **telecommunicators** must be prepared to communicate effectively with citizens and relay accurate information to first responders.

In this chapter we discuss many aspects of fire service communications. It is safe to say that throughout history, effective communications have been essential to all successful endeavors. Public safety communications is certainly no exception to this. What is **communications**? Although communications takes on many forms, conveyance of information through the medium of speech is considered one of the most universal functions. It is estimated that interactions involving speech account for 74 percent of our communications time.[1] Yet effective communication is still a challenge for most of us in today's busy and complex workplace.

Note From the standpoint of fire service emergency communications center operations, the communications process must include four basic elements: information from the caller must be *received, understood, recorded* accurately, and *communicated* to emergency responders.

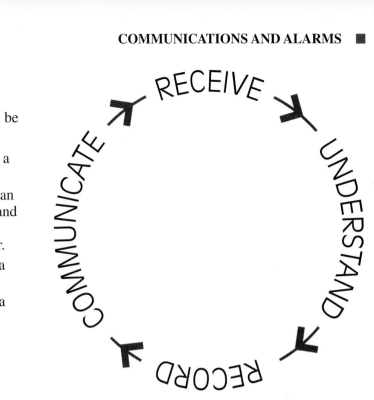

Figure 3-1 The communications process must be complete and clearly understood in order to be effective.

The communications process is illustrated in **Figure 3-1**. Reports from citizens can usually be categorized as true emergencies, perceived emergencies, nonurgent reports, or requests for information.

Despite our knowledge of the importance of complete communications, in many cases complete communication does not occur between individuals. Ironically, it has not been until recently that many fire service leaders have come to realize the importance of good, reliable communications and the impact that successful communication can have on the delivery of the services provided by their departments. As a result of a greater awareness of the need to maintain effective communications, many fire service leaders have started taking proactive measures to ensure that the quality of the communications process is maintained. These measures include:

1. Teaching communications skills to employees.
2. Upgrading communications systems.
3. Incorporating modern technology into the daily operational routines of their departments.

While it takes many special skills and various personnel to provide total service, the communications center is the heart of the operations. As a result, fire service leaders are putting more emphasis on having qualified staff managing and operating emergency dispatch systems. Due to this increased awareness, greater emphasis is being placed on the selection and training of the telecommunicators who staff communications centers in larger departments, and also for those smaller volunteer departments who rely on volunteers to perform these duties.

> **Firefighter Fact** A communications center may also be referred to in some instances as a PSAP, public safety answering point.

COMMUNICATIONS PERSONNEL

Although the names may have changed and the roles and responsibilities expanded, the basic role of the telecommunicator has remained remarkably the same. The primary role of the telecommunicator is to receive emergency requests from citizens, evaluate the need for a response, and ultimately sound the alarm that starts first responders on their way to the scene of an emergency.

Receiving calls is only the first challenge facing telecommunicators. This is the beginning of what can be a very complex process of gathering information and deploying emergency apparatus, personnel, and equipment. A myriad of work goes on behind the scenes of communications centers. Telecommunicators provide a variety of ancillary services to support field operations. In some cases a telecommunicator serves as the public relations officer for the department. After all, a telecommunicator is the first person to speak to the caller in need of emergency services. A telecommunicator is often the first person contacted by the news media when updates are required. Some departments issue departmental news and updates from their communications center, whereas others require that this type of information be distributed by the chief or the person acting as the public information officer. In either case, the news media can be the department's best friend or worst enemy; they should be handled with care.

In some instances local protocols may dictate that after initial contact by the citizen, the telecommunicator will maintain contact with the caller, enabling the telecommunicator to provide valuable **prearrival instructions**. As a result of the expanding role of the telecommunicator, fire service leaders are reconsidering the complexity of emerging technology and the importance of the *individuals* hired to operate these systems.

> **Firefighter Fact** A telecommunicator's job has been accurately defined by some as hours of boredom punctuated by moments of terror.

The NFPA 1061 standard, *Standard for Professional Qualifications for Public Safety Telecommunicator,* clearly outlines behavioral characteristics or traits that a person should possess to be a viable candidate for a position in the emergency communications center. Among these traits are the ability to perform multiple tasks, the ability to make decisions based on common sense and standard values, the ability to maintain composure in high stress conditions, and the ability to remember details and recall information easily. Candidates must also be able to exercise voice control, including the ability to maintain balanced tone, modulation, volume, and inflection while communicating. Only a tried, trusted, and complete training program will ensure that telecommunicators are armed with the knowledge and skills necessary to do the job.

A quality training program must be followed by a comprehensive work performance evaluation program to ensure fire service leaders that their communications centers are staffed by well-trained personnel who are able to apply the training they receive and develop the skills necessary to perform the required tasks. One organization that has been a leader in the field of communications training for many years is the **Association of Public Safety Communications Officials–International, Inc. (APCO)**. APCO has established minimum training standards that are being adopted by many states throughout the country. Initial certification, through a tested program coupled with a strong continuing education program, is paramount to the efficient and effective operations of a modern communications center.

An adequate staffing level at communications centers is the next most significant concern of fire service leaders. Most communications center managers will readily attest that there are never enough personnel when a **mass casualty** or **multiple-alarm incident** occurs. Communications managers rely on historical data to produce staffing models that closely resemble actual staffing requirements. By using some widely accepted formulas, managers are able to closely replicate staffing needs.

COMMUNICATIONS AND ALARMS ■ 49

Note The NFPA states that "Ninety-five percent of alarms shall be answered within 30 seconds, and in no case shall the initial operator's response to an alarm exceed 60 seconds." It additionally recommends that "the dispatch of the appropriate fire services shall be made within 60 seconds after the completed receipt of an emergency alarm."

Firefighter Fact The first national emergency telephone number, 999, was introduced in Great Britain in 1930.

To meet this requirement, some departments with limited resources and low call volumes need only ensure that at least one person is on duty at all times to answer emergency phones. Regardless of size, fire departments must provide well-trained personnel to serve in the telecommunicator role because those individuals who answer calls from citizens have a direct impact on the overall response time of the agency.

THE COMMUNICATIONS FACILITY

The importance of staffing levels and having properly trained personnel in a communications center in order to process emergency calls within the time prescribed by the NFPA has been discussed. Now we turn our attention to the places where these telecommunicators work. If the facility is intended to house emergency communications operations, there are several important considerations.

Communications centers have many different configurations. Some are served by **9-1-1** systems, and some rely on a regular seven-digit telephone number to receive calls. Many serve only a single agency, whereas others have joint authority over police, fire, emergency medical services, and even utilities. Many communications centers are nothing more than a small alcove in a larger administrative office, with only one person responsible for communications duties for a single agency such as police. That same person might also book prisoners, answer phones, serve walk-up customers, and perform other combinations of job functions as assigned. Another example is a small volunteer fire department that has only one person on duty to answer calls and dispatch personnel. This is common in jurisdictions with low call volume activity. As shown in **Figure 3-2**, other facilities can be larger and constructed expressly for the purpose of serving as an emergency communications center. These are generally found in larger metropolitan areas with high volumes of emergency responses.

Regardless of the size or type of facility, all serve the one common goal of receiving and disseminating information that can be both emergency and nonemergency in nature.

The NFPA also has standards (NFPA 1221) for location and construction of emergency communi-

Figure 3-2 Exterior view of a 9-1-1 emergency communications center that houses fire/EMS, police, and sheriff communications operations.
(Courtesy of Caddo Parish 9-1-1)

cations centers. Modern emergency communications facilities are typically constructed in areas where there is little risk of damage either by natural or man-made hazards. Communications centers should not be constructed in low-lying areas prone to flooding or along coastlines prone to hurricanes or high winds. In like terms, most agree that they should be located in areas of limited traffic and limited exposure to man-made hazards such as terrorism, arson, and so on. The NFPA suggests that communications centers that are built closer than 150 feet to existing structures should receive special attention regarding protection from collapse. Therefore, parks and other open areas are well-suited locations for these types of facilities.

Special attention should also be given to protecting the facility from vandalism and the effects of civil disturbance. Only authorized personnel should be permitted access to the communications center. Communications centers should have a limited number of outside windows, and all outside entrances should be monitored by security systems. The facility should also be monitored by a fire alarm system that complies with **NFPA 72,** National Fire Alarm Code.

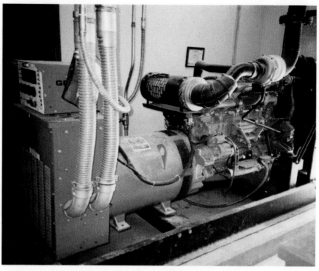

Figure 3-3 The emergency communications center is protected from power outages by use of backup power generators. Generators of this type can provide continuous power to operate the communications center during periods of power outages. *(Photo courtesy of Caddo Parish 9-1-1)*

> **Caution** Every communications facility should be supported by a backup location in the event that the primary facility encounters problems that render it inoperable and result in evacuation.

Emergency communications centers should be equipped with backup power supplies that also meet the requirements set forth by the NFPA. An example of a backup power generator is shown in **Figure 3-3**. In some cases, backup power generators must be able to maintain the operations of a large communications center for prolonged periods of time. Some emergency generators are fueled by diesel fuel, which is stored on site as shown in **Figure 3-4**. It is also important to note that liquid petroleum or natural gas may fuel some power-generating equipment. Most backup power supplies are designed with electrical load distribution, which provides separation of power between emergency and nonemergency circuits. In this type of arrangement, where multiple generators are in use, they are typically installed with a switching device that automatically transfers the *emergency* loading to the first generator to achieve operating speed.

Uninterruptible power supplies (UPS) are also used in some of the more modern communications centers. These units use battery power to provide uninterrupted power to the critical communications and computer equipment used in the communications center. Properly designed and regularly tested power backup systems that employ automatic switching devices that do not interrupt the opera-

Figure 3-4 A 500-gallon supply of fuel is maintained to ensure that uninterrupted power can be provided for a minimum of twenty-four hours, and for longer periods if necessary. Liquefied petroleum gas and natural gas (not shown) may also be used to power backup generators. *(Photo courtesy of Caddo Parish 9-1-1)*

COMMUNICATIONS AND ALARMS

Figure 3-5 UPS systems such as this help maintain uninterrupted service should a power outage occur.

Figure 3-6 Computer-aided dispatch workstations such as this one are found in many communications centers that have high activity levels.

tions are extremely important, especially when modern computerized systems are in use. A large UPS system is shown in **Figure 3-5**.

COMPUTERS IN THE FIRE SERVICE

The proliferation of new and more technically advanced computer hardware, along with the development of highly specialized computer software designed especially for use by public safety agencies, has made it possible for many departments to incorporate the use of computers into their communications systems. Many fire departments have found that the use of **computer-aided dispatch (CAD)** systems, **Figure 3-6**, has made it easier to handle increased call volumes.

The computer can keep track of the location of active incidents and which units have been assigned to respond and therefore help telecommunicators to better manage their resources. At the same time, computers can create and store valuable records on each incident and other departmental activities. Computers have greatly improved the way in which statistical analyses of departmental activity are performed. Additionally, computers can provide remote locations with access to a variety of information that is stored in a main **database** that can be accessed when needed. This information may include maps, hazardous materials information, prefire plans, policies and procedures, mutual aid instructions, or any other information that may be useful to firefighters. This type of information is often stored on **mobile data computers** in all types of fire apparatus and **command vehicles**. Computers also allow fire personnel to access off-site databases that may be beneficial either for training or incident mitigation.

RECEIVING REPORTS OF EMERGENCIES

The call-taking process consists of receiving a report, interviewing, and referral or dispatch composition. **Figure 3-7** provides a visual representation of the various stages of call processing. Whether a highly trained telecommunicator or a firefighter is performing this duty, speed is very important during the interview portion of the call-taking process. Telecommunicators must be able to prioritize incoming calls in order to ensure that the most important call gets the fastest attention. Incoming telephone calls should be answered in the following priority: (1) 9-1-1 and other emergency lines, (2) direct lines, and (3) business or administrative lines. Telecommunicators should answer incoming lines promptly and determine if the caller has an emergency. If not, and another emergency line is ringing, the telecommunicator should put the caller on hold and answer the other incoming call. However, speed is of little consequence if accurate and complete information is not obtained from the caller and relayed to emergency response personnel.

When receiving reports of emergencies by telephone, telecommunicators should always speak clearly and slowly with good volume. If questions to the caller have to be repeated, valuable time is

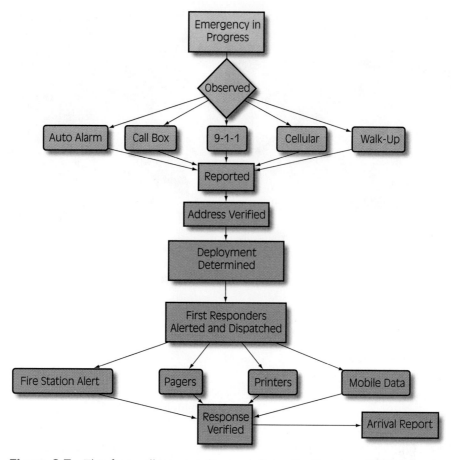

Figure 3-7 This figure illustrates the work flow of call processing by a public safety telecommunicator.

wasted. When speaking with a citizen who is reporting an emergency, the telecommunicator's voice should project authority and knowledge. The telecommunicator should use plain, everyday language at all times when speaking to the public and maintain a polite and friendly tone. It is never appropriate to argue with a caller. The citizen is the "customer."

Time is of the essence when answering and processing requests for service. The 9-1-1 and other emergency lines should always be the top priority. When answering incoming lines, the telecommunicator should always identify the department and ask, "Do you have an emergency?" or "What is your emergency?" It is critically important for the telecommunicator to control the conversation in order to obtain the necessary information in the least amount of time. In many cases the caller knows what has happened but finds it difficult to relay key elements of the situation to emergency service providers. The key information must be extracted from the caller, even in cases where a caller is hysterical or in severe emotional distress. A calm and authoritative tone of voice is helpful. In most cases, asking short, specific questions (name, address, telephone number, location) is the most effective tactic.

In the case of nonemergency calls, the telecommunicator should make every effort to accommodate the requests of the callers, which includes transferring them to another agency that may be better suited to provide assistance. It is important from a public relations standpoint to always offer to transfer the caller if the telephone system is capable of this function. Prior to transferring the call, the telecommunicator should provide the number to the caller in case the transfer process is interrupted or fails.

> **Streetsmart Tip** A professional telecommunicator who speaks with confidence and authority will be able in most cases to gain the trust of the caller, who will then be more apt to follow the instructions issued by the call taker.

From time to time a caller may hang up, be disconnected, or simply drop the telephone before providing all of the needed information. This could be the result of a medical problem, an equipment failure, or simply the fact the caller realized he or she dialed a wrong number.

> **Caution** Telecommunicators should diligently follow up on all incomplete calls.

How this is done varies greatly from one department to another depending on the type of telephone system available, but the basic fundamentals remain the same.

When processing an emergency call, telecommunicators should always attempt to obtain the following minimum information:

- Location of the emergency
- Nature of emergency
- Callback number
- Caller's location and situation

Without a doubt the most important piece of information is the *location* of the incident. *What* is happening is of little importance if the call taker doesn't know *where* to send help. Once the caller has provided the location, the telecommunicator should attempt to secure any additional information that the caller can provide with respect to landmarks such as crossstreets or easily recognizable buildings, if this can be done safely.

Life safety is of primary importance. Therefore, the call taker should always try to determine if the caller is in danger. If so, the call taker should provide prearrival instructions in accordance with local protocols.

> **Caution** Getting and verifying callback numbers of the caller reporting the emergency is extremely important in the event it is necessary to recontact the caller in order to obtain additional information concerning the incident.

This information can be relayed to field units via radio as it is received and updated. Callers should be asked to provide all the information they can about an incident as long as they can do so *safely*.

In addition to a callback number, the caller's proximity to the location of the incident should be noted. This information is sometimes beneficial when attempting to locate incidents by following the directions of the caller. It is helpful if the first telecommunicator and the first responders are able to visualize the location of the caller in proximity to the locations of the incident. In most dispatch systems, once sufficient address and incident type information is verified, deployment of emergency apparatus and personnel can be initiated. As a rule, the average citizen will only find a need to report an emergency once in a lifetime.

Therefore, in emergency situations citizens may be very excited and unable to formulate and convey clear and concise information. A telecommunicator providing clear, calm instructions can reduce the normal panic associated with this activity and thereby effect faster and more accurate responses.

> **Streetsmart Tip** The caller is the telecommunicator's set of eyes and ears at the emergency scene.

The information callers provide can be extremely valuable but only if the call taker is able to ask the right questions that will generate meaningful responses. After deployment of the appropriate responders, communications center personnel should retrieve any pre-incident plans, SOPs, hazardous conditions, or any other information that may be available on site at the communications center and relay all pertinent information to the emergency responders.

In the case of emergency medical calls, much more information may be requested from the caller in accordance with **emergency medical dispatch** protocols. Emergency medical dispatch systems are designed to meet the needs of specific jurisdictions by providing the telecommunicators with a set of standard questions and predetermined actions used to evaluate the situation. A response to an emergency medical incident must be based on a telecommunicator's ability to recognize and react to the "symptoms" being displayed by the victim, rather than an attempt on the part of the telecommunicator to make a "diagnosis" of the injury or illness.

Methods of Receiving Reports of Emergencies

Reports of emergencies can be received in a variety of ways. We briefly discuss some of the most common means in this section:

- Conventional telephones
- Wireless or cellular telephones
- Emergency call boxes
- Automatic alarms

- TDD equipment for the hearing impaired
- Still alarms or walk-ups

Receiving Reports by Telephone

Throughout the country, conventional residential and business telephones are the most commonly used method of reporting emergencies. However, the use of cellular telephones is becoming much more popular as they become available to larger cross sections of citizens, as discussed in the next subsection.

According to the **National Emergency Number Association** (NENA), nearly 93 percent of the population of the United States is covered by some type of 9-1-1, and 95 percent of 9-1-1 coverage is enhanced 9-1-1. Approximately 96 percent of the geographic United States is covered by some type of 9-1-1.

> **Caution** To reduce confusion, 9-1-1 should always be pronounced "nine-one-one," not "nine-eleven."

Incidents have been reported in which a panicked citizen has cited problems locating the "eleven" on the telephone touch pad. Regardless of whether **basic 9-1-1** or **enhanced 9-1-1** is used, these systems provide access to emergency services via the use of a simple, easy-to-remember three-digit number.

In addition to voice communication, enhanced 9-1-1 service provides emergency communications centers with the telephone number and address from which the call is originating. This decreases the time necessary to determine the caller's location when callers are either unwilling or unable to provide this information. This feature eliminates the need to trace or research telephone company records to associate an address with a telephone number. Again, features like this provide valuable assistance quickly and effectively. They are extremely important to emergency call takers, especially in cases where the caller is very excited, the caller is not familiar with an area, or for other reasons the caller is unable to communicate clearly the location of an emergency.

> **Caution** When considering the value of 9-1-1, also be reminded that the caller may not be calling from the address of the emergency, so address information must be verified.

Both basic and enhanced 9-1-1 service are available through conventional business and residential telephone service. In addition to conventional or wired telephones, wireless communications devices are growing more popular as a means of communicating emergencies.

Receiving Reports via Cellular Telephones

Wireless systems, such as cellular telephones, are proving very beneficial to the process of reporting emergencies. They are widely available to a large cross section of citizens and are very mobile. Any 9-1-1 calls initiated by a citizen using a cellular telephone are routed to a predetermined answering point for processing. In some areas the state highway patrol or state police is designated to receive all 9-1-1 cellular calls for a specific area or jurisdiction. In other areas of the country, 9-1-1 calls that originate from cellular telephones are routed to a communications center that may serve either a single agency or multiple agencies and even multiple jurisdictions.

Although the use of cellular telephones is proving very beneficial, it also has some negative points. The use of cellular telephones has caused significant increases in communications center call volumes. For example, on today's busy highways, one accident usually generates multiple reports by motorists using cellular telephones to access 9-1-1. Cellular callers are also more likely not to know their exact location, because they may be traveling along busy highways in an unfamiliar area. In these types of situations, landmarks and direction of travel are very important to the telecommunicator. In the case of small departments, the increase in the number of calls generated by cellular phones can have a budgetary impact. Also, in some cases, the telecommunicator will be criticized unfairly for not getting "accurate" information when in fact some of the things discussed earlier in this chapter are the cause for inaccurate information. Government-mandated upgrades in technology require cellular manufacturers to provide a means by which the location of wireless callers can be determined and provided to the emergency call taker. This greatly reduces the risk of location errors during the processing of emergency reports received via cellular phones.

New technologies created for automobiles provide citizens with specialized communication systems built into vehicles that will contact emergency services with the press of a button or when certain safety systems, such as air bags, are activated within the vehicle. This new wireless emergency reporting technology does not provide much information about the nature of the emergency; however, using satellite technology it can provide the communications center with an exact position of the vehicle reporting the emergency. As this system is used and tested, its popularity in new automobiles may grow in the future.

Receiving Reports via Municipal Fire Alarm Systems

Municipal fire alarm systems are those systems that allow a coded or voice message to be generated from an alarm box typically located in highly visible, easily accessible areas that are open to the general public. Systems of this type came into use during the late 1800s and many are still in use today with few upgrades and modifications.

> **Firefighter Fact** **Emergency call boxes** were first installed in the United States in 1852 and are still used in many parts of the country as a means of reporting emergencies.

According to the Boston Fire Department, the first emergency call box was placed in operation in Boston, Massachusetts, on April 28, 1852. This system is still in operation today and has seen only one upgrade since its installation. While Boston finds it beneficial to use a system of this type, some cities have discontinued them largely due to high rates of false alarms. As seen in **Figure 3-8**, call boxes can be operated via "hardwired" systems or they can be of the wireless solar-powered variety that can be installed in remote locations not serviced by electrical power or conventional telephone lines. An example of this type of technology is shown in **Figure 3-9**.

Some call boxes simply transmit a preset identification number to the communications center without providing a means for the reporting party to communicate verbally with the communications center. Others are equipped with signal switches that allow the caller to select the type of emergency being reported by pressing the appropriately labeled button, **Figure 3-10**.

Figure 3-8 "Fire alarm" boxes of this type came into being during the late 1800s and some are still in use today. *(Photo courtesy of Shreveport Fire Department)*

Figure 3-9 More modern technology is being used as solar-powered cellular call boxes go up in areas with limited access to power and communications networks. This type of call box supports voice communications with emergency communications centers. *(Photo courtesy of Caddo Parish 9-1-1)*

Figure 3-10 Some call boxes are equipped with signal switches that allow the caller to select the type of emergency being reported.

Receiving Reports via Automatic Alarm Systems

There are two types of public alarm systems as defined by the NFPA. A **Type A reporting system, Figure 3-11,** is defined by NFPA as "a system in which an alarm from a fire alarm box is received and is retransmitted to fire stations either manually or automatically." A **Type B reporting system, Figure 3-12,** is defined by NFPA as "a system in which an alarm from a fire alarm box is automatically transmitted to fire stations and, if used, to outside alerting devices." Properly designed and installed automatic fire alarm systems can be the key element to any building's overall safety. Automatic alarm monitoring systems are typically comprised of five common types:

1. *Local protective signaling system:* An alarm system operating in the protected premises.
2. *Auxiliary protective signaling system:* An alarm system that utilizes a municipal fire alarm box to transmit a fire alarm from a protected property to a fire communications center.
3. *Remote station protective signaling system:* An alarm system that connects a protected premise over leased telephone lines to a remote station such as a fire communications center.
4. *Central station protective signaling system:* An alarm system that connects a protected premise to a privately owned monitoring site that monitors the lines constantly for any indication of fire or other trouble signals.
5. *Proprietary protective signaling system:* An alarm system that protects contiguous or non-contiguous properties with common ownership from a location on the protected property.

Alarm systems typically consist of a system of sensors designed to detect smoke, heat, or a combination of both and also manual stations that can be activated by occupants in the event a fire is detected.

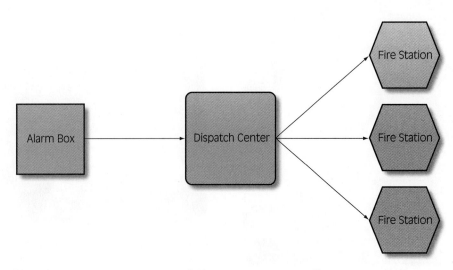

Figure 3-11 Type A municipal alarm systems typically transmit an alarm from a call box to a communications center where the alarm is retransmitted to emergency responders.

COMMUNICATIONS AND ALARMS 57

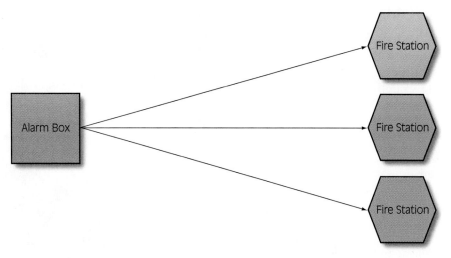

Figure 3-12 Type B municipal alarm systems typically transmit alarms directly from a call box to first responders.

Detecting devices protect specific and limited areas. Initiating the signal for any fire alarm system is accomplished by many means: manual pull boxes, heat detectors, and smoke detectors. **Figure 3-13** shows a manual pull box, and **Figure 3-14** shows both a heat and a smoke detector. Fire alarms are covered in greater detail later in Chapter 12.

Figure 3-13 Manual pull box stations are used to activate local alarms.

Figure 3-14 Heat detectors are designed to activate fire alarms and automatic extinguishing systems when ambient room temperatures reach certain levels. Smoke detectors monitor air quality and can also activate automatic alarm systems. This photo shows both in a typical installation.

Receiving Reports via TDD

Telecommunications devices for the deaf (**TDDs**) are becoming much more widely used by individuals with hearing disabilities. The **Americans with Disabilities Act** (ADA) entitles citizens to equal service from public agencies.

> **Firefighter Fact** ADA regulation 35.162 states: "The legislative history of Title II specifically reflects congressional intent that public entities must ensure that telephone emergency services, including 9-1-1 services, be accessible to persons with impaired hearing and speech through telecommunication technology."

Therefore, communications centers are required to remain ready to receive calls via specialized equipment designed to allow citizens to communicate with telecommunicators through the use of a keyboard using text messages instead of voice communications, **Figure 3-15**.

Figure 3-15 Devices such as this assist public safety agencies in communicating with citizens who have hearing impairments.

Receiving Reports via Still Alarm or Walk-Ups

From time to time citizens may report an emergency directly to the personnel at the fire station. Receiving complete and accurate information is just as important in these cases as those that are received at the communications center by telecommunicators. The section titled Receiving Reports of Emergencies (earlier in this chapter) provides information that can be applied to deal with a citizen who is reporting an emergency in person. **Figure 3-16** shows a firefighter relaying information to the communications center that was received from a citizen who actually stopped at the fire station and reported an emergency. Although the protocols of different departments may vary with respect to how these types of reports are handled, it is important to always obtain as much information as possible from the person reporting the emergency. Once this is done, and following departmental protocols, the person taking the report should immediately activate the emergency notification system and relay all of the information to the communications center. How specific emergency notification systems are activated is covered in departmental protocols. The same is true in the case of notifying the communications center that a report of an emergency has been received. Some departments may use **ringdown circuits**, **base radio**, or **mobile radio** to communicate with their communications center. It

Figure 3-16 A firefighter relays information from the fire station to the communications center via direct telephone circuit.

is very important for firefighters always to notify the communications center of location and assignment when reports are received directly from citizens. Telecommunicators can then ensure that any necessary support and assistance are provided.

EMERGENCY SERVICES DEPLOYMENT

Once an emergency is recognized and subsequently reported or relayed to the communications center, the next step is to determine what action must be taken. A variety of methods is used to accomplish this throughout the country; however, some elements of the process are essential regardless of jurisdiction or geographic location. Identification of the situation, address verification, and unit selection must occur in order for telecommunicators to deploy the appropriate types and numbers of emergency responders.

As stated earlier, the most important information to obtain from the caller is the address. However, to deploy the most effective emergency response, the telecommunicator must also know the nature of the emergency. Emergency response organizations typically identify the most common types of situations and preassign a standard response to each such situation. An example would be a fire department that predetermines the routine response to all single-family dwelling fires will be three engines, one ladder, and one district chief. Based on this plan, the telecommunicators have baseline criteria from which to develop a **deployment plan**. In the modern fire service, deployment plans are based not only on apparatus types, but also take into consideration what equipment is carried on the apparatus, the number of personnel, and their skill levels.

The process of deploying apparatus and personnel varies greatly from one department to another. For instance, in smaller communities with a low volume of emergency response activity, a manual **run card system** may be sufficient to manage the emergency response deployment process. Manual run card systems, similar to the one shown in **Figure 3-17**, are typically comprised of a card file containing street and location information relating to a jurisdiction and predetermined unit assignments for each location.

As discussed earlier, jurisdictions with high volumes of activity may require automated systems to assist in the deployment process. CAD systems are widely used today and provide a very sophisticated method of assessing resources and making recommendations for deployment of equipment and personnel.

Regardless of which method of deployment is used, whether it is accomplished manually or through the use of a CAD system, a predetermined deployment of apparatus must exist. An example of a deployment table is shown in **Figure 3-18**.

The basic elements of the deployment process remain the same in either the manual or automated systems: Verify the location and nature of the emergency and determine what resources are available.

> **Firefighter Fact** Enhanced 9-1-1 systems provide number identification and location information even in those cases where computer-aided dispatch is not being used.

STREET NAME: Delmar Ave.   BOX NUMBER: 1128
BLOCK RANGE FROM: 1000 TO: 1500

BLOCK	ENGINES	TRUCKS	RESCUE	D/CHIEF	A/CHIEF	MEDIC	OTHER
1000	1, 5, 4, 7, 10	1, 7, 10	1, 9	1, 2, 3	1	1, 4, 5	
1100	1, 5, 4, 7, 10	1, 7, 10	1, 9	1, 2, 3	1	1, 4, 5	
1200	1, 5, 4, 7, 10	1, 7, 10	1, 9	1, 2, 3	1	1, 4, 5	
1300	1, 4, 5, 7, 10	1, 7, 10	1, 9	1, 2, 3	1	1, 4, 5	
1400	1, 5, 4, 7, 10	1, 7, 10	1, 9	2, 1, 3	1	1, 5, 4	
1500	1, 5, 4, 7, 10	1, 7, 10	1, 9	2, 1, 3	1	1, 5, 4	

STREET NAME: Delmar Ave   BLOCK RANGE FROM: 1000 TO: 1500

Figure 3-17 Run cards show the response assignment of a variety of apparatus to specific street addresses.

TYPE	DISP. CODE	DESCRIPTION	RESPONSE
Fire	01	Single Company Response	1E
	11	Structure Fire/Alarm (Residential)	2E-1D-1R-2T
	21	Structure Fire/Alarm (Commercial)	3E-1D-1R-1T
	51	Hospitals	4E-2D-2R-2T
EMS	08	Medical (Noncritical)	1E-(COLD)
	18	MVA	1M-1E

Legend: E — Engine or Pumper
D — Chief Officer
R — Rescue Unit
T — Truck or Aerial
M — Medic Unit
(Cold) — Nonemergency

Figure 3-18 Deployment tables such as the example shown are used to identify the appropriate apparatus response for the type of dispatch.

Once telecommunicators receive a report of an emergency situation, verify the location, and determine the appropriate deployment scheme, the next step is to notify the emergency responders. As is the case with deployment plans, this process also varies greatly from agency to agency. Volunteer departments may rely on personal pagers or use an automatic telephone system that "rings" multiple telephones on a common circuit and, in some cases, a system of sirens to alert them of an emergency. **Figure 3-19** shows a volunteer wearing a personal pager.

A variety of **home alerting devices** are also available. These are used by some agencies that operate either an all-volunteer or combination volunteer and paid department. A common example of this type of alerting device is shown in **Figure 3-20**.

In departments where fire stations are staffed twenty-four hours a day, some type of **fire station alerting system** is usually employed. Fire station alerting can be accomplished in a variety of ways, but should always comply with NFPA standards. In some systems a voice message is transmitted from the communications center to a fire station via the use of a vocal alarm system. These systems typically operate either via some type of control unit connected to leased telephone circuits or a radio transmitter. **Figure 3-21** shows a receiver that would be located at the fire station. The telecommunicator, using the control device, which may be similar to the **encoder** shown in **Figure 3-22**, decides the appropriate fire stations to notify and activates the system. Normally some type of distinctive tone is transmitted via a public address system within the fire station to alert personnel of an incoming message.

This alert tone is followed by voice instructions over the PA from the telecommunicator. Some fire station alerting systems perform additional functions such as turning on selected lights in the fire station, opening apparatus bays, turning off appliances, and controlling traffic signals. Some systems are capable of a *zoned* alert that can notify only specified areas of a station such as those stations that house both fire and EMS units and personnel.

Departments that utilize CAD systems can enhance this method by installing "tear and run" printers in each fire station. These printers provide the responders with a hardcopy printout showing details of the incident and location. A printer of this type is shown in **Figure 3-23**.

Some departments also employ mobile data terminals and mobile data computers in their dispatch and deployment process. These units allow dispatch information to be transmitted directly to the apparatus on a display screen or, in some cases, directly to mobile printers. Modern communications equipment such as this can provide two-way confidential information flows between the communications center and emergency responders. **Figure 3-24** shows a **mobile**

COMMUNICATIONS AND ALARMS 61

Figure 3-19 Some departments use personal pagers to alert personnel of the need to respond to emergencies.

Figure 3-21 This is yet another device that is used to alert emergency responders. This unit is equipped with relays that activate station lights and open station doors in preparation for a response. *(Photo courtesy of Shreveport Fire Department)*

Figure 3-20 This device is typically found in fire stations and also in some private homes and is used to call out emergency responders.

Figure 3-22 Encoders such as this are used to control many paging systems.

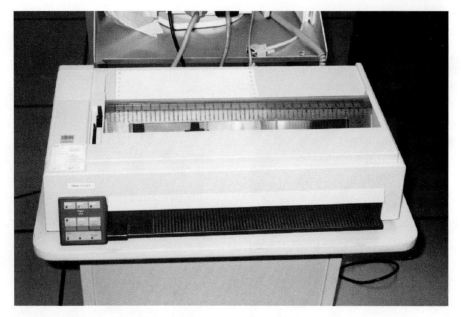

Figure 3-23 Printers such as these are used to relay incident information to first responders at the time of dispatch. They are typically referred to as "tear and run" printers. *(Photo courtesy of Caddo Parish 9-1-1)*

Figure 3-24 Mobile data terminals such as this one are used by some departments and provide information to emergency responders while en route to incidents. Unit status can also be managed with systems of this type. *(Photo courtesy of Shreveport Fire Department)*

data terminal widely used in the late 1980s and mid- to late 1990s, and **Figure 3-25** shows a more modern mobile data computer, which is replacing the older model mobile data terminals in some departments.

Regardless of the type of fire station alerting system in use, it is critically important that accurate information pertaining to an emergency situation be transmitted from the communications center to first responders in a clear, concise, and expedient manner.

Again, speed and accuracy are of the utmost importance.

Streetsmart Tip Remember that the response times of first responders are directly impacted by the amount of time required for telecommunicators to receive information, verify location, determine deployment, compose a dispatch, and transmit this information to first responders.

TRAFFIC CONTROL SYSTEMS

Caution Thousands of collisions resulting in tragic accidents involving emergency response vehicles are evidence of the dangerous nature of emergency responses.

COMMUNICATIONS AND ALARMS 63

Figure 3-25 Mobile data computers such as this one are replacing the less powerful mobile data terminal as shown in Figure 3-24. Units such as this are capable of storing information on board the apparatus for easy retrieval. *(Photo courtesy of Shreveport Fire Department)*

Careful planning and use of the technology described in this section can create a safer emergency response path without undue or prolonged disruption of normal traffic flows. However, there is no better way for emergency responders to reduce traffic accidents than by exercising prudent judgment and applying safe driving practices at all times.

To speed the response of emergency responders as well as reduce stress and increase the safety of both emergency responders and the general public, some jurisdictions utilize various types of emergency preemption systems to control traffic signals and provide a safe transition to a priority right-of-way for emergency vehicles. Systems of this type are designed to recognize an emergency response vehicle and actually allow it to change the traffic control signals on its route to allow clear passage for the emergency responders. A variety of these systems are in operation today and each uses slightly different technology. Some operate using radio-frequency while others communicate with remote signal detection devices through the use of microwave or laser technology mounted on the vehicle.

RADIO SYSTEMS AND PROCEDURES

Once apparatus and personnel are deployed to emergency situations, the function of fire communications personnel then becomes that of providing support for the field units deployed. The primary link between the communications center and field units is the radio system. The use of radios in the fire service serves to carry both verbal and digital messages. Radio systems have various components; however, every system must have at minimum a base station and antenna capable of transmitting at a power or signal strength necessary to provide coverage to all parts of a jurisdiction.

The radio-frequencies that have commonly been used by the fire services are VHF low band frequencies, 33 to 46 MHz; the VHF high band frequencies 150 to 174 MHz; and the UHF frequencies, 450 to 460 MHz. These frequency ranges have provided reliable fire service communications for many years. However, as the result of growth, the fire service has experienced severe difficulties when attempting to add frequencies to their radio systems. The **Federal Communications Commission (FCC)** closely monitors frequency allocations, and in some cases additional frequencies in the ranges mentioned are simply not available. As the result of a need for additional frequency spectrum, other frequencies have been approved for use by the fire service. The 800-MHz frequency range is being used successfully by some departments for voice communications. Voice communications via the 800-MHz frequency range are limited by the FCC to systems using trunking technology. This type of radio system is discussed next.

One common radio system used by fire departments is a simplex system that uses only one frequency to transmit outgoing messages and to receive incoming messages, **Figure 3-26**. The advantage is simplistic design, resulting in decreased system cost. The primary disadvantage of systems of this type is the limited range and interference between multiple

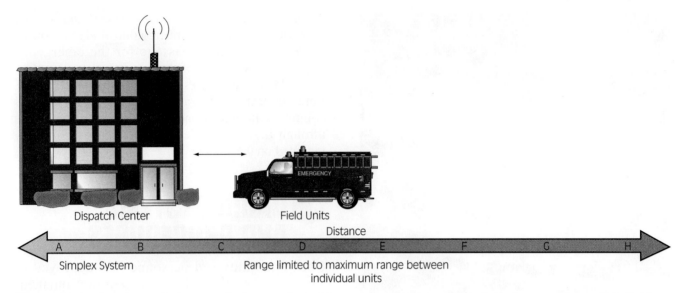

Figure 3-26 Simplex radio system designs such as this one are reliable and relatively inexpensive to install. However, they are limited with respect to range of operations.

units in the same system attempting to access the base station simultaneously.

A more complex system, shown in **Figure 3-27**, is a duplex system that uses two frequencies per channel, transmitting outgoing messages on one and receiving incoming messages on the other. This system uses base repeater stations whereby a fixed control station transmits an outgoing message that is received at a repeater site and retransmitted to mobile and portable units in the field. The benefits of systems of this type are more range and the elimination of the self-interference found in simplex systems. The disadvantages are more complex system design, the need for multiple frequencies, system cost, and the ongoing maintenance costs associated with the system.

Multisite trunking systems as depicted in **Figure 3-28** use computer processors that make the most efficient use of radio spectrum. Multiple transmitters operating on different channels are controlled by microprocessors that sense available channels and reallocate their use as needed. In duplex systems of this type, user transmissions are transmitted on one frequency, received by the field unit on another, and retransmitted back to the communications center on the other. The user does not notice the fact that the system is "changing" frequencies with each transmission. However, this more efficient allocation of radio-frequency resources allows the use of fewer frequencies by individual agencies. Several agencies can operate

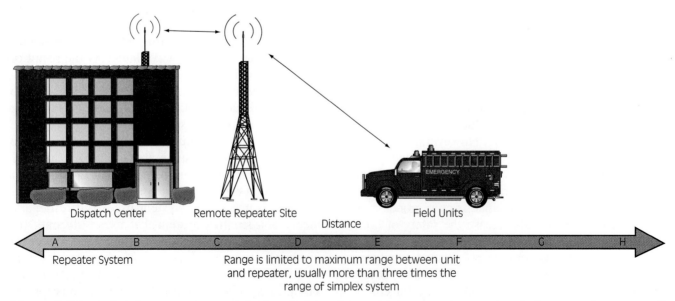

Figure 3-27 A slightly more advanced duplex design using multiple transmitters extends the operating range of the simplex system shown in Figure 3-26.

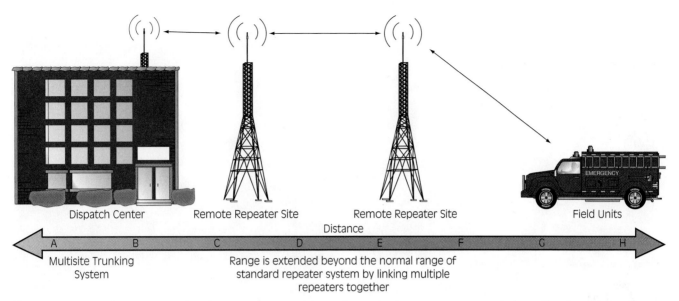

Figure 3-28 Multisite trunked radio systems provide perhaps the best coverage and also offer direct benefits associated with the most efficient use of radio resources.

simultaneously on the same trunked radio system and not interfere with each other. The benefits of systems of this type are expanded range, more efficient frequency use, and the ability of multiple agencies to operate on one system. Disadvantages are more complex system design and overall higher system cost.

Streetsmart Tip The proper operation of the radio system, regardless of type, is of primary importance if information to and from the scene of the emergency is to be relayed in a timely and accurate manner.

Proper radio discipline is very important during active incidents.

Caution On-scene personnel must listen before they talk. Routine or nonemergency traffic should not interfere with emergency communications. Transmissions should occur when airwaves are clear.

When using any nontrunked two-way radio, it is important to depress and hold the "push to talk" button at least two seconds before talking to avoid **clipping** the first part of the message. The same is true at the end of the message. The user should always pause briefly before releasing the "push to talk" button to avoid clipping the end of a message. Before keying any radio, users should know what they are going to say. The golden rule is to be brief but be concise. Firefighters should avoid touching any radio antenna during transmission to avoid burns that can result from radio-frequency energy.

Firefighter Fact Some radios are equipped with a time-out feature that forces brevity by interrupting transmissions that are longer than the preset duration allowed. This feature also prevents long-term open microphones from accidental keying.

The user should never attempt to operate a radio while eating or chewing anything and should never use slang or jargon when transmitting a message over the radio. The message must be clearly understood. During large-scale incidents, radio economy is important. When using mobile radios, the user should speak clearly (not shout) across the microphone, as shown in **Figure 3-29**, as opposed to speaking directly into the microphone as pictured in **Figure 3-30.**

When using portable radio equipment, the unit should be held perpendicular to the ground with the antenna facing skyward to allow for better radio wave distribution. The user should speak across the microphone as shown in **Figure 3-31**. The use of a portable radio that is located in a "radio pocket" or belt clip impairs the performance of the unit. **Figure 3-32** shows the improper position of the portable radio when transmitting. In transmitting on both mobile and portable radios holding the microphone one to two inches from the mouth and at a 45° angle is important for a clear transmission.

Figure 3-29 This figure shows the proper use of a mobile radio microphone.

Figure 3-31 In this figure, the user has positioned the portable radio properly and is speaking across the microphone.

Figure 3-30 Improper use of a mobile microphone. The radio microphone should not be held directly in front of the operator's mouth when transmitting.

Figure 3-32 This figure shows the improper positioning of a portable radio.

Microphones should never be left on the seat or in other locations where they might accidentally be keyed up. Microphones and portable radios should be placed in their appropriate storage locations or in a protected location. Local agency protocols and policy will dictate how radios are to be operated in a particular jurisdiction. However, the following is a simple illustration of the "call up" method that is used by some agencies:

Telecommunicator: "Engine One, this is dispatch."
Engine One: "Dispatch, this is Engine One."
Telecommunicator: "Engine One, respond to 1234 Main Street for a trash fire."
Engine One: "Dispatch, Engine One is responding for a trash fire at 1234 Main Street."
Engine One: "Dispatch, Engine One is on the scene and establishing command. We have a small trash fire in the rear of the building, no exposures, no assistance needed."
Telecommunicator: "Engine One, dispatch copies."

Ten codes are a set of radio signals preceded by the number "10" that make up a predetermined message. Considerable debate exists as to whether radio codes or clear speech should be used, but in most cases that decision is made based on the needs of each specific agency. Use of radio signal codes provides a somewhat more confidential and cryptic means of communicating. However, such codes must be learned and remembered and are not usually standardized from department to department, making communications during multijurisdictional responses somewhat problematic.

Clear speech, on the other hand, is exactly what the name implies. Clear speech is used to convey information and issue instructions. The use of clear speech eliminates much of the confusion associated with the use of radio codes and, although somewhat lengthier, in most cases is easily understood by all. However, even in the "clear speech" environment, the same phrase may carry a different meaning from agency to agency. For example, "in-service" for some agencies means that a crew is *going to work at an incident*. On the other hand, in some agencies "in-service" means that the crew is *available* for an assignment. Thus, it is important not to assume anything during radio communications.

Another important function associated with the radio systems is the issuance of emergency evacuation signals to on-scene personnel who may be subject to imminent danger. Some departments use the radio system to transmit electronic *tones* that are intended to attract the attention of firefighters and alert them of the need to evacuate to a safe area.

This works relatively well in most cases. However, a typical emergency scene is very noisy and not everyone on the scene has a portable radio. Another system used by some departments employs the use of apparatus air horns. Three bursts on the air horn indicate the need to evacuate. Regardless of the type of system used, it is critically important that firefighters learn it and be familiar with how it is used. Firefighter safety is paramount in all operations.

The use of two-way radios has grown greatly over the years. Routine administrative traffic, which is necessary in the daily operations of a department, should never be allowed to interfere with emergency operations. In departments that have access to multiple radio channels, emergency operations should be assigned to a separate channel dedicated for use on that scene only. This greatly improves the ability for incident commanders to communicate with on-scene personnel and also minimizes the threat of interference from some other source.

Radios and radio systems are evolving as the advances in technology grow. Newer radio systems can identify radios assigned to a particular apparatus to help in tracking transmissions. As a result of the technology, some radios incorporate an emergency feature that alerts the emergency communications center when a crew is in trouble or encountering an emergency. Firefighters should spend time familiarizing themselves with the radio system in use in their department and the many features associated with the equipment and the system.

> **Note** In summary, firefighters must consider several important factors when transmitting messages across mobile or portable radios. First, the information provided to an emergency communications center or other emergency service units must be accurate, clear, and complete. The radio transmissions must also be within the time parameters established by the department or local jurisdiction if such policies and procedures exist. Firefighters should consult their department or jurisdiction's policies on time parameters for radio communications, specific terminology, and proper designations for units operating on a radio system.

ARRIVAL REPORTS

Until field units arrive on the scene, for all practical purposes, the communications officer is the incident commander. The first unit arriving on the scene should establish command and give a brief arrival report.

Note The initial incident commander remains in command of the incident until command is either assumed by a higher ranking officer or transferred to another officer.

This arrival report contains pertinent, but brief, information about the on-scene conditions. The arrival report should be given using clear, precise, language. The report should contain, at minimum, the following information: (1) the correct address, (2) a situation evaluation, (3) where the emergency is located in the building, (4) some information about the building as well as its potential occupants, (5) a request for any other agency support such as law enforcement, (6) the location of the on-scene command post, and (7) the identity of the incident commander. Arrival reports may also contain a very brief action plan for the incident.

In the case of most routine emergency medical incidents, this much detail is not necessary. However, in the event of a major accident or other incidents involving multiple patients the same type of detailed information should be provided along with any additional information such as a quick assessment of the number of patients and general types of injuries involved.

Status reports are very important to the overall success of any major scene operations. Most fire command officers agree that the first status report from the field should be made no longer than ten minutes into the incident, and every ten to fifteen minutes thereafter until the situation is brought under control. Communications centers may implement SOPs that call for "time marking" incidents at important thresholds for improved documentation and reporting purposes. Some CAD mobile data computer systems perform this function automatically.

MOBILE SUPPORT VEHICLES

Major events involving fire and EMS sometimes call for the use of **mobile support vehicles** (MSVs) or mobile communications units. These vehicles greatly enhance the overall effectiveness of the communications system in use at the scene of major incidents. Coordination of the communications process is absolutely necessary in order to manage large-scale operations effectively. MSVs provide an on-scene command post from which operations can be directed. The need to deploy vehicles of this type is usually determined by the size of the incident and the projected duration of activities. MSVs are normally highly specialized vehicles designed exclusively for use as on-scene command posts. MSV size depends greatly on the jurisdiction that it serves. The vehicle should have radios that allow communications on all of the frequencies that may be in use by the command agency and any additional jurisdictions that may be called in for mutual aid. Telephone service to vehicles such as these can be provided by the local telephone company via temporary connections or through the use of cellular technology. The MSV should be equipped to operate on both battery power as well as standard 120-V current. **Figure 3-33** shows an example of an MSV that is the result of modifications to a public transit bus. Conversions such as these, although often time consuming, are for the most part a much less expensive way to incorporate the use of this type of vehicle into the operations of a department. Custom-built units from the factory are also available, as shown in **Figure 3-34.** This particular unit was custom built to serve both fire and police operations through a cooperative agreement between the services.

Figure 3-33 The vehicle pictured here is the result of a conversion project performed on a city transit system bus. The resulting mobile support vehicle cost much less than a custom-built factory unit.

RECORDS

Complete and accurate communications center records should be maintained on all responses. It is considered routine practice in most communications centers to record all emergency telephone and radio traffic either in some type of manual log book or on magnetic or digital recording devices for future reference. Examples of both a tape logging device and a digital recording device are shown in **Figure 3-35** and **Figure 3-36**, respectively.

COMMUNICATIONS AND ALARMS 69

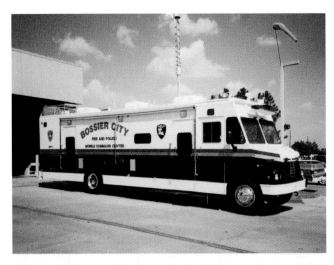

Figure 3-34 Unlike the vehicle shown in Figure 3-33, the vehicle shown here is a custom-built mobile support vehicle. *(Photo courtesy of Bossier City Fire Department)*

Figure 3-36 Digital logging recorders such as this one are taking the place of magnetic tape loggers in some modern communications centers.

Figure 3-35 Magnetic tape logging recorders are used by some communications centers to record telephone and radio activity. *(Photo courtesy of Shreveport Fire Department)*

In most states fire reports are considered public record and as such are available to newspaper reporters, insurance adjusters, lawyers, and others. Records of this type speak for the department. A well-written report will be very valuable when years later a firefighter is called to testify at a deposition regarding legal action resulting from an incident.

The following is the minimum information that should be recorded and maintained as the legal record of an incident. This information can be used to complete a more detailed National Fire Incident Report or other appropriate reports approved for use by a particular state or jurisdiction.

- *Time call received:* The time the telecommunicator received the call from the caller.
- *Units dispatched:* The unit numbers of the units dispatched to the incident.
- *Dispatch times:* The time of the initial dispatch and the times of any additional units subsequently dispatched.

Inc. Number: 9807670				Under Control Time: 10:20		
Address: 1234 Delmar Ave						
Location: City Hall				Dispatch Code:		
Complaint: Fire Alarm (21)						
Reported By: Joe Smith				Call Back Number: 555-5555		
UNIT	I/ROUTE	ARRIVE	TO HOSP	AT HOSP	COMP	I/SERVE
E01	1005	1007			1025	—
E04	1005	1008			1020	—
E05	1005	1006			1020	—
D01	1005	1015			1020	—
T01	1005	1007			1020	—
R01	1005	1007			1020	—

Figure 3-37 A manual incident card that is used to maintain a record of event history.

- *Arrival times:* The arrival times of all units at the scene and the unit initiating the incident management system.
- *Command post information:* Commander, location, and so on.
- *Requests:* For example, for additional fire department units or for other agencies or services.
- *All clear time.*
- *Under control time.*
- *Back in service times for all units.*

Figure 3-37 shows an example of a "manual" dispatch card that is used to record incident information. This type of logging information is typically found in smaller departments and also as backup systems for departments that rely on CAD and other automatic systems.

Lessons Learned

Whether a telecommunicator in a large emergency communications center or a firefighter answering an emergency call at a fire station, the manner in which calls are answered and information processed has a direct impact on citizens' impressions of the department. In the case of emergency incidents, information that is accurately collected and rapidly transmitted to first responders is paramount to the successful resolution of the incident. The ability to answer incoming calls quickly, gain control of the call, and, in some instances, calm the caller is a very important aspect of this process. Remember that the telecommunicator is the first person "on the scene." Utilize the caller's eyes and ears. A telecommunicator's knowledge and skills, combined with the ability to make wise use of all available resources, will play a vital role in the successful outcome of an emergency incident.

KEY TERMS

9-1-1 Emergency telephone number that provides access to the public safety services in the community, region, and, ultimately, nation.

Americans with Disabilities Act Public law that bars discrimination on the basis of disability in state and local services. Enacted in 1990.

Association of Public Safety Communications Officials–Int., Inc. (APCO) International not-for-profit organization dedicated to the advancement of public safety communications. Membership is made up of public safety professionals from around the world.

Base Radio Radio station that contains all of the antennas, receivers, and transmitters necessary to transmit and receive messages.

Basic 9-1-1 Telephone system that automatically connects a person dialing the digits "9-1-1" to a predetermined answering point through normal telephone service facilities. Number and location information is not normally provided in *basic* systems.

Clipping Term associated with the use of two-way radios that is used to describe instances when either the first part of a message or the last part of a message is cut off as the result of either speaking before pressing the transmit key or releasing the transmit key prior to the end of a transmission.

Command Vehicle Typically used by operations chief officers in the fire service.

Communications Sending, giving, or exchanging of information.

Computer-Aided Dispatch Computer-based automated system that assists the telecommunicator in assessing dispatch information and recommends responses.

Database Organized collection of similar facts.

Deployment Plan Predetermined response plan of apparatus and personnel for specific types of incidents and specific locations.

Emergency Call Box System of telephones connected by private line telephone, radio-frequency, or cellular technology usually located in remote areas and used to report emergency situations.

Emergency Communications Center Facility either wholly or partially dedicated to being able to receive emergency and, in some instances, nonemergency reports from citizens. Centers such as these are sometimes referred to as fire alarm, headquarters, dispatch, or a public safety answering point (PSAP).

Emergency Medical Dispatch System designed for use by telecommunicators to assist them in evaluating patient symptoms using predetermined criteria and responses.

Encoder Device that converts an "entered" code into paging codes, which in turn activate a variety of paging devices.

Enhanced 9-1-1 Similar in nature to basic 9-1-1 but with the capability to provide the caller's telephone number and address.

Federal Communications Commission Government agency charged with administering the provisions of the Communications Act of 1934 and the revised Telecommunications Act of 1996 and is responsible for nonfederal radio-frequency users.

Fire Station Alerting System System used to transmit emergency response information to fire station personnel via voice and/or digital transmissions.

Home Alerting Devices Emergency alerting devices primarily used by volunteer department personnel to receive reports of emergency incidents.

Mass Casualty EMS incidents that involve more than five patients.

Mobile Data Computer Communications device that, unlike the mobile data terminal, does have information processing capabilities.

Mobile Data Terminal Communications device that in most cases has no information processing capabilities.

Mobile Radio Complete receiver/transmitter unit that is designed for use in a vehicle.

Mobile Support Vehicle Vehicle designed exclusively for use as an on-scene communication center and command post.

Multiple-Alarm Incident Involves the response of additional personnel.

National Emergency Number Association Not-for-profit organization founded in 1982 and made up of more than 6,000 members. The association fosters technical advancement, availability, and implementation of a universal emergency telephone number system.

NFPA 72 National Fire Alarm Code.

Prearrival Instructions Self-help instructions intended to enhance the overall safety of the citizen until first responders arrive on the scene.

Ringdown Circuits Telephone connection between two points. Going "off-hook" on one end of the circuit causes the telephone on the other end of the circuit to "ring" without having to dial a number.

Run Card System System of cards or other form of documentation that provides specific information on what apparatus and personnel respond to specific areas of a jurisdiction.

TDD Device that allows citizens to communicate with the telecommunicator through the use of a keyboard over telephone circuits instead of voice communications.

Telecommunicator Individual whose primary responsibility is to receive emergency requests from citizens, evaluate the need for a response, and ultimately sound the alarm that sends first responders to the scene of an emergency.

Type A Reporting System System in which an alarm from a fire alarm box is received and retransmitted to fire stations either manually or automatically.

Type B Reporting System System in which an alarm from a fire alarm box is automatically transmitted to fire stations and, if used, to outside alerting devices.

REVIEW QUESTIONS

1. Throughout history, what element has been essential to all of successful endeavors?
2. List four basic elements of communications.
3. What is the primary role of the telecommunicator in relation to fire and EMS apparatus?
4. What is the most important part of call processing for a telecommunicator or firefighter?
5. List the order in which ringing lines should receive priority.
6. When speaking with citizens who are reporting emergencies, what should the call taker's voice project?
7. When processing a nonemergency call, and an emergency line rings, what steps should the telecommunicator take?
8. Telecommunicators or others who answer emergency telephone lines must extract key information from individuals who are, in some instances, hysterical and in severe emotional distress. How do telecommunicators extract information in such a case?
9. What is the most important information to learn from a caller reporting an emergency?
10. What other information is important in the event it becomes necessary to recontact the caller?
11. What are emergency medical dispatch protocols designed to accomplish?
12. Name the five most common technologies used to report emergencies.
13. What is the primary benefit of basic or enhanced 9-1-1 service?
14. What sector of our society would have the most need for TDD equipment?
15. What is a preassigned standard response to a specific situation called?
16. When operating radio equipment, emergency personnel should always do what before talking?
17. How should a user speak when using a mobile radio?
18. True or false: Routine administrative radio traffic always gets priority attention.
19. In most states, fire reports are considered _____ .
20. True or false: The actions of the person answering an emergency telephone call are vital to the successful outcome of the emergency incident.

Endnote

1. *NFPA 1061*. National Fire Protection Association, Quincy, MA.

Additional Resources

Basic Telecommunicator Course. Association of Public Safety Communications Officials, South Daytona, FL. http://www.apcointl.org

Bukowski, Richard W., and Robert J. O'Laughlin, *Fire Alarm Signaling Systems.* National Fire Protection Association, Quincy, MA, and Society of Fire Protection Engineers, Boston, MA, 1997.

Bunker, Merton W., Jr., *National Fire Alarm Code Handbook, 1996 Edition.* National Fire Protection Association, Quincy, MA.

Fire Department Communications Manual, A Basic Guide to System Concepts and Equipment. U.S. Fire Administration, Federal Emergency Management Agency, Washington, DC, 1995.

National Emergency Number Association http://www.nena9-1-1.org

Nathonal Academies of Emergency Dispatch http://www.emergencydispatch.org

NFPA 1061. National Fire Protection Association, Quincy, MA.

NFPA 1221. National Fire Protection Association, Quincy, MA.

Pivetta, Sue, *9-1-1 Emergency Communications Manual,* 3rd ed. Kendall/Hunt Publishing Company, Dubuque, IA, 1997.

Project 33, Telecommunicator Training Standards. Association of Public Safety Communications Officials, South Daytona, FL, August 1996.

CHAPTER 4

FIRE BEHAVIOR

Frank J. Miale, Retired Battalion Chief, New York City Fire Department and Lake Carmel Volunteer Fire Department

OUTLINE

- Objectives
- Introduction
- Fire Triangle, Tetrahedron, and Pyramid
- Measurements
- Chemistry and Physics of Fire
- Sources of Heat
- Combustion
- Oxygen and Its Effect on Combustion
- Vapor Pressure and Vapor Density
- Boiling Point
- Flammable and Explosive Limits
- The Burning Process—Characteristics of Fire Behavior
- Modes of Heat Transfer
- Thermal Conductivity of Materials
- Physical State of Fuels and Effect on Combustion
- Theory of Fire Extinguishment
- Unique Fire Events
- Classes of Fire
- Lessons Learned
- Key Terms
- Review Questions
- Endnotes
- Additional Resources

STREET STORY

It had been a relatively quiet day. As soon as the box alarm came in we pulled out the door and the column of smoke was clearly visible in the late afternoon winter sky. We arrived with the first-due engine and truck companies. As we pulled in, I saw the fire building was a two-story wood frame structure with heavy brown and black smoke coming from both floors. A quick glance revealed that the windows on the first floor had been painted on the outside covering the glass. It appeared that this was a vacant structure probably being used by squatters or as a crack house.

After receiving an initial report, my crew and I headed toward the building. In my agency, one of the rescue squad's duties at structure fires is to perform the initial search unless otherwise directed by an incident commander. I saw that the engine company was laying their line while the truck was preparing to force entry into the building. During the size-up, a civilian ran over to us advising us that his father lived there and he could not find him.

At that point I split the crew up. I sent two men to assist the truck company in forcing entry through the front door, while the driver and I went to the rear of the building. As we arrived in the back we could hear some of the windows falling on both floors with flames venting from them. It was at this time we thought we heard moaning coming from the back porch area. We put on our SCBA and forced entry through a door to the back porch. Inside we found the smoke had banked down to approximately 10 feet from the kitchen door. We forced the kitchen door open and were only able to advance a few feet. Most of the kitchen was starting to erupt in flames. I thought to myself that it looked like a "Hollywood" fire. With everything burning, we started to back out because no one could have survived in there.

As we backed out of the kitchen my partner reached up to close the kitchen door. As he was doing this, the smoke became jet black and was now down to floor level. The temperature became unbelievably hot. I thought to myself "I hope what could happen doesn't," but it did. The porch flashed over. It was incredible. It was like having a giant orange flashbulb go off in your face. The entire porch was in flames. As we turned and headed back toward the door, we kept the wall to the left because, on the way in, the wall had been on our right. We came to the back wall and turned right to where the door should have been. What we didn't realize was that the inside panel of the door was covered by plywood with no doorknobs, and it was secured with latches on the inside because the resident was concerned about break-ins. When we could not find the door, I actually became afraid that we would not get out alive. My partner made a Mayday call giving our location while I searched for the door. The pain from the heat was like being stung by a thousand bees. At one point I just wanted to stand up and try to run where I thought the door was but I told myself "Stand up and you've really had it. Stay calm, stay together, and they'll get us out." I thought of my son who was only six weeks old at the time, my partner's kids, and our wives, and I said to myself, "I'm not dying in this place."

We stayed together in the area by the rear wall, when suddenly I felt someone grab me by the air-pack harness and throw me out the door. As I landed on the ground I turned and saw my partner come flying out the door. Thankfully, one of the captain/paramedics had seen us enter the building and heard our Mayday call.

Because we were wearing our full protective ensemble—boots, bunker pants, hoods, coats, gloves, helmets, and SCBA—we were fortunate enough to have sustained only minor burns. Our training also saved us by prompting us to create a search pattern that led us back to the door, make a Mayday call, and stay together.

—Street Story by Mike Kelleher, Captain, Troy Fire Department, Troy, New York

OBJECTIVES

After completing this chapter, the reader should be able to:

- Describe the chemistry and physics of fire.
- Identify the sources of heat.
- Describe the characteristics of fire.
- Describe the effect of oxygen on fire.
- Define combustion.
- Describe vapor pressure and vapor density.
- Describe the meaning of flammable and explosive limits.
- Describe the three types of heat transfer.
- Describe the significance of the thermal conductivity of materials.
- Describe fuel types and their effect on combustion.
- Describe the basis for the theory of fire extinguishment.
- Identify the classes of fire and methods of extinguishment.
- Explain thermal layering, flashover, and backdraft.

INTRODUCTION

Humans have been familiar with fire all of their lives. From their earliest memories, people have been warned against its dangers, entertained by its explosions and sound, comforted by its warmth, frightened by its power, mystified by its characteristics, and assisted by its light. In the history of mankind, fire has played a major role as a tool in the development of society. Sometimes an ally, sometimes an enemy, much has been learned about it, especially in the last thirty years, **Figure 4-1**.

So what is **fire**? Burning, also called **combustion**, is a simple chemical reaction. It is described as "a rapid, persistent chemical change that releases heat and light and is accompanied by flame, especially the exothermic oxidation of a combustible substance."[1] It is also referred to as a process of "rapid oxidation with the development of heat and light,"[2] and as "a reaction that is a continuous combination of a fuel (reducing agent) with certain elements, prominent among which is oxygen in either a free or combined form (oxidizing agent)."[3]

As more research into the mystery of combustion is accomplished, more is understood.

> **Firefighter Fact** Once thought to be simply a gift from the gods, combustion is now understood to be a complex chemical reaction.

When fighting a foe, one of the best weapons one can have is knowledge and understanding of the enemy. Military training and warfare are predicated on such information. Entire units in a military operation are devoted to the mission of uncovering, analyzing, and developing intelligence to explain the needs, capabilities, and probable actions of the enemy. The information learned dictates the battle plan. Firefighters need to know

Figure 4-1 Since its discovery, fire has been considered both an ally and an enemy.

about the behavior of fire as one of the elements in understanding the enemy. This chapter examines and discusses the process of burning. Once we know what causes fire to begin, grow, and spread, the means employed to extinguish it will become more understandable.

FIRE TRIANGLE, TETRAHEDRON, AND PYRAMID

The combustion process was once depicted as a triangle with three sides. Each side represented an essential ingredient for fire. Heat, fuel, and oxygen were thought to be the essential elements. As the scientific study of fire progressed, it became evident that a fourth ingredient was necessary. That fourth element was the actual chemical reaction that permitted flame propagation. A new four-sided figure was created to represent the essential ingredients for fire, the **fire tetrahedron**. The fire tetrahedron, **Figure 4-2**, is a pyramid shape describing the heat, fuel, oxygen, and chemical reaction necessary for combustion.

The four elements of the fire tetrahedron must be present in order to support the combustion process. Therefore, removing one or more of the elements in the fire tetrahedron will result in an end to the combustion process. Basic firefighting strategies are based on this principle of removing an element of the fire tetrahedron to halt the combustion process and put out the fire. An example of this is a simple fire in a pan on the stove. Placing a lid on the pan removes the oxygen source, thereby stopping the combustion process and putting out the fire.

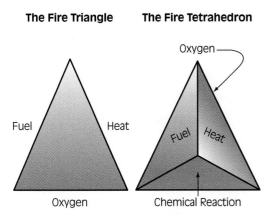

Figure 4-2 The old and new ways of visualizing the combustion process, the fire triangle and the fire tetrahedron.

MEASUREMENTS

In the field of firefighting, knowing units of measurements is essential. For example, firefighters need to know that a fire stream putting out 175 gallons per minute of water is releasing 175 times the amount of water that a stream putting out 1 gallon per minute is releasing. Furthermore, it is essential for a firefighter to know that a fire that is generating many thousands of **British thermal units (Btus)** will need a fire flow output of hundreds of gallons per minute to extinguish the fire. Measurement is an important part of firefighting.

Some types of measurement are part of our everyday language such as minutes, gallons, liters, watts, feet, centimeters, pounds per square inch, horsepower, degrees, and many others, **Table 4-1** and **Figure 4-3**.

Firefighters use terms for heat, electricity, volume, length, energy output, concentrations, and weight, an example of which is shown in **Figure 4-4**. To a firefighter, it is important to know the forms of measurement that describe these elements and understand the limits of each as it relates to safety of the firefighter and the team. For example, a 1,000-gallon pump on an engine is capable of putting out about 1,000 gallons per minute. If a particular hoseline requires a water flow of 250 gallons per minute to be

Measurements	
ENGLISH SYSTEM	**METRIC SYSTEM**
Length	
Inches	Millimeters
Feet	Centimeters
Yards	Meters
Miles	Kilometers
Volume	
Ounces	Milliliters
Pints	Liters
Quarts	
Gallons	
Weight	
Ounces	Grams
Pounds	Kilograms
Tons	

TABLE 4-1

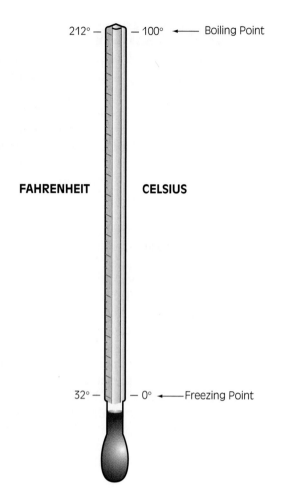

Figure 4-3 Several types of scales are used to measure heat. The two most common are the Fahrenheit and Celsius scales.

Note Different types of measurements are used in this text. The English measurement system and the metric system **(Table 4-1)** can be converted from one to the other. The key to metric measurement is in the terms that precede the measurement. *Milli-*, *centi-*, and *kilo-* are terms that mean thousandth, hundredth, and thousand, respectively. A meter is a measurement of length. (For example, a millimeter is one thousandth of a meter, a centimeter is one hundredth of a meter, and a kilometer is 1,000 meters.) The English version is a little more complicated, and memorization of the terms and multiples is required. (For example, there are 12 inches in a foot and 3 feet in a yard.)

effective, then the 1,000-gallon pumper could feed water to four of these hoses. A fifth hoseline of the same size would be beyond the capability of this pumper. A separate water supply source would be required to deliver the proper streams.

CHEMISTRY AND PHYSICS OF FIRE

The universe is made of a substance that is referred to as **matter**. Matter is defined as "something that occupies space and can be perceived by one or more senses." Some forms of matter are so small that they cannot actually be touched by a human hand, felt by the human sense of touch, or even seen with the human eye. But matter is the basis for the existence of the universe.

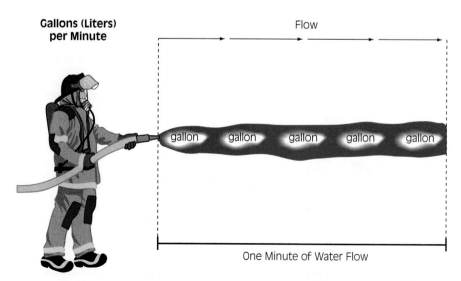

Figure 4-4 A way to measure water flow is by gallons (liters) per minute.

As matter interacts, substances are formed, changed, and destroyed. Matter doesn't disappear, it merely changes form. For example, a piece of iron that is left outside in the weather for years will eventually deteriorate into a form of dust and then disappear. Although it may seem as if the substance that makes up the iron has been destroyed, it has actually just changed form.

A house can be disassembled piece by piece and, at some point, simply become a pile of wood, shingles, nails, masonry, glass, roofing, metal, wire, pipes, and ducts. These components make up the house, and each component can be broken down further.

The wood contained in the house is made up of a material called *cellulose*. Because the wood was once a living organism, it is made up of millions of cells. The cells, in turn, are made up of compounds. A **compound** is composed of **molecules**, which are in turn composed of atoms of two or more elements in chemical combination, **Figure 4-5**. These compounds are **organic**, which refers to substances that are or were once living organisms. Organic substances generally contain the elements of carbon, hydrogen, and oxygen in their makeup. Examples of organic substances are hair, gasoline, wood, and plastics. It should be understood that something can be organic although it has never been alive itself, but is made up of chemicals that were once alive. For example, a plastic substance, although never actually alive, is made from oil more commonly referred to as **hydrocarbons**. The oil is left over from the decay, stabilization, and collection of millions of years of prehistoric life-forms that pooled into oil fields and has been tapped and drawn from wells. Once alive, the material is now made into nonliving materials by natural chemical actions.

Organic compounds are made up of many different forms of substances. These substances are various forms of chemicals that exist in the universe. Through combinations of these chemicals, substances are formed. Chemicals are made up of combinations of molecules. Molecules are joined and separated by bonding actions that use a form of electricity as energy.

Molecules are made up of atoms joined in various combinations. Atoms are bound together in much the same way that molecules join. In fact, it is the molecules with atoms that have loose ends that combine with other molecules with atoms with loose ends when conditions are right. The combination and separation of these molecules and atoms provide the basis for what causes oxidation or combustion.

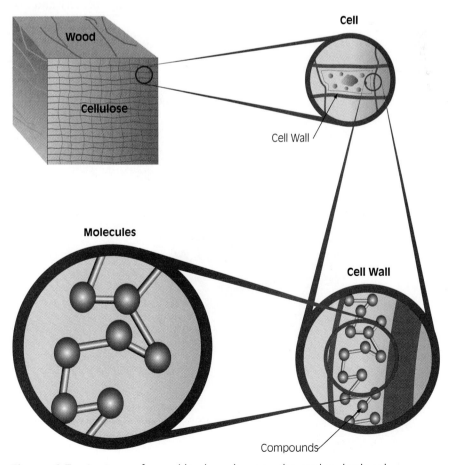

Figure 4-5 A piece of wood broken down to the molecular level.

Atoms are made up of still smaller particles called protons, neutrons, and electrons. The nucleus, or core, of the atom is made up of the protons and neutrons. The electrons encircle or orbit the nucleus, creating an electron cloud around the protons and neutrons, **Figure 4-6**.

> **Note** The numbers of protons and neutrons and the electrons circling them in various configurations are the determining factors in what properties that element will exhibit.

Atoms lacking electrons will tend to be quick to link up and form molecules. Atoms that are stable will tend to be those that have a satisfactory balance between the electron/proton ratio.

The term *organic* was described as a substance that is or was once living. In the universe, there are also substances that were never alive and the term used to describe that type of substance is **inorganic**. For example, iron, sulfur, granite, quartz, and silicon are all examples of inorganic substances. They are generally termed *minerals* and for the most part do not contribute to the combustion process.

The differences between organic and inorganic substances become important when discussing the chemistry of fire. As a rule of thumb, only organic materials will burn. There are some exceptions, which are discussed in a later section.

The term **bond** describes the "atomic glue" that holds molecules together. It is in this bond that fire (or combustion) has its origins. When molecules and atoms are joined, a certain amount of heat is absorbed into the bond as part of the mechanism that keeps the elements together. This is a process known as an **endothermic reaction**, **Figure 4-7**. This means that heat is absorbed when the bond is created. When a bond is broken, an **exothermic reaction** occurs, **Figure 4-8**. This type of reaction causes the release of heat. The energy released comes in the form of heat and light. If the release is rapid enough to sustain a continuous reaction, we see it as fire and feel the release of the heat. When this occurs, combustion is taking place. Fighting a fire is actually a process of breaking up a chain reaction. A look at the chemical reaction of combustion will help lead to a better understanding of what is achieved when fires are fought and extinguished.

A hydrocarbon is a compound that is made up of at least two elements, hydrogen and carbon. Chemical compounds can often be deciphered by their names. Looking closely at the hydrocarbon itself, one can see the two elements described in the name: *hydro* (hydrogen) and *carbon* (carbon). The ability

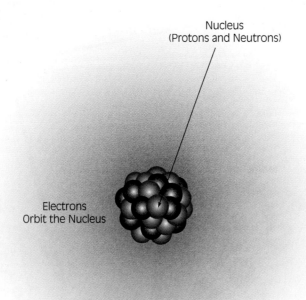

Figure 4-6 An atom is made up of protons, neutrons, and electrons.

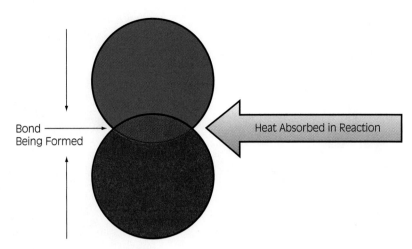

Figure 4-7 An endothermic reaction.

to recognize a chemical compound can tip off a firefighter as to what extinguishment agent to employ.

Organic compounds all contain some form of hydrocarbons. The number of carbon atoms in combination with the number of hydrogen atoms will determine the properties of the substance and, more importantly, how it will react under varying conditions.

To understand the process of combustion, a basic knowledge of chemistry is helpful. Imagine a jigsaw puzzle that has common joints fashioned in such a way that only certain pieces can be joined with other pieces. In **Figure 4-9**, the hydrocarbon elements are joined to form the compound methane. In its bonds between the atoms, the substance methane holds forces that, when split apart, will emit heat and light.

Most chemicals that are stable will maintain their form unless something presents itself to change it. An **oxidizer** acts as a catalyst in the breakdown of otherwise stable molecules. An oxidizer possesses a chemical property that can pull apart a molecule and break apart the bond that previously existed. The emission of light and, more importantly, heat then causes the chemical to break apart other compounds, letting loose more light and heat, which causes other bonds to break apart, and so on. If the process is able to continue a self-sustaining chemical reaction, combustion results. One example of an oxidizer is oxygen. The unique characteristic of these elements is that they all have the same chemical properties in attraction and combination with other chemicals. It is this property that makes oxidizers react in a similar fashion.

> **Streetsmart Tip** The key to the chemicals that promote and support combustion is found in the suffix of the word (**Table 4-2**). It is important to watch for chemicals that end in *-ines*, *-ates*, and *-ides*. In chemistry, these endings identify oxidizers. Firefighters must be aware of these chemicals and the potential effect on combustion when discovered at a fire scene or during an inspection.

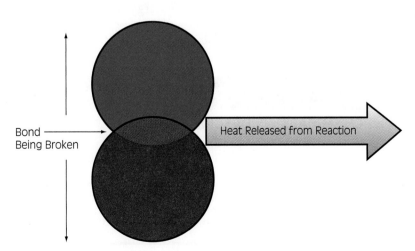

Figure 4-8 An exothermic reaction.

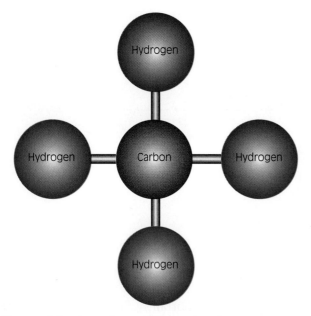

Figure 4-9 A methane (CH_4) molecule.

Figure 4-10 Rusting metal is a common example of the oxidation process.

If the reaction is slow, a gradual deterioration of the material will occur over time almost without notice. An example of an oxidizing process is iron that is undergoing a rusting process, **Figure 4-10**. The oxygen of the atmosphere combines with the properties of the iron and gradually pulls out the bonds that are keeping the iron atoms together. Although not visible, heat and light are emitted during this process. If sensitive monitoring instruments were affixed to the piece of iron, it would be possible to measure the heat and light.

The oxidation process is much more evident during combustion. The presence of oxygen in combination with heat from a previous bond break causes a rapid chain reaction to take place. Because of the speed at which the reaction takes place, the light is visible and the heat emitted can be perceived by the senses. This is the process of combustion.

> **Firefighter Fact** In oxidizing reactions, such as combustion, electrons are transferred from one atom to another. In these reactions, there are reducing agents and oxidizing agents. Reducing agents lose electrons in the chemical reaction. In the fire tetrahedron, fuel is considered the reducing agent in the chemical reaction of fire.

SOURCES OF HEAT

Heat is an energy form that, in essence, powers the universe. It is heat that forges atomic-level synthesis of compounds, and it is heat that is released when those bonds are separated. Heat, because it is an energy force, can be neither created nor destroyed. It is merely the recognizable physical manifestation of energy as it changes from one form to another.

Chemicals and Their Names

SUBSTANCE	CHEMICALS	DESCRIPTION	SUFFIX
Chlor*ine*	Cl	Chlorine by itself (Cl)	-ine
Sodium Chlor*ide*	Na Cl	Chlorine with sodium (Na)	-ide
Sodium Chlor*ite*	Na Cl O_2	Chlorine with sodium and two oxygens (O)	-ite
Sodium Chlor*ate*	Na Cl O_3	Chlorine with sodium and three oxygens	-ate

TABLE 4-2

FIRE BEHAVIOR

The phenomenon of heat comes from four basic sources, all of which are important to understand because they can all be initiators of fire. Although some sources of heat are very common and are experienced every day, other sources are not well understood. The four sources of heat are chemical, electrical, mechanical, and nuclear.

Chemical

The most common of the four sources of heat that firefighters deal with on a regular basis is the chemical reaction that releases heat as a by-product. Anything that burns does so through a chemical reaction in which heat is released as the bonds of the molecules break down. Anything from organic cellulose to petrochemicals will burn through a process called **pyrolysis**, which is decomposition or transformation of a compound caused by heat. The decomposition is the breaking down of the compound.

> **Caution** The transformation of a compound is the reconstruction of the molecules to form another compound. It is important to understand that in the process of transformation, chemicals can be created that are extremely hazardous to health. It is also essential to possess a respect for the invisible by-products of burning. In addition to heat and light, products of combustion include water vapor, carbon particles, carbon monoxide, sulfur dioxide, and hydrogen cyanide. These products of combustion can cause damage to and diseases of the lung tissue from the heat, toxic chemicals, and particulate matter. The use of protective breathing apparatus in fires is required because of this danger, **Figure 4-11**.

Figure 4-11 The breakdown or transformation of substances can release hazardous chemicals. This is one reason why firefighters wear self-contained breathing apparatus.

becomes important when inspections for fire hazards are taking place. When extinguishing fire, ensuring that the source of the heat has been stopped becomes significant. Although a fire can be extinguished even if the heat source is still pumping in heat, the likelihood of keeping the fire extinguished is reduced.

Electrical

As a source of heat, electricity is probably the most recognized. Electrical power is used to heat homes and cook food. Nowhere is electricity as a heat

Mechanical

Another form of heat source is mechanical. Friction causes heat that can reach levels hot enough to ignite surrounding combustible material. The buildup of heat from friction is often the cause of fire in machinery. Because heat from friction can be produced whenever any rubbing or compression occurs, **Figure 4-12**, industrial processes that employ this type of action have means by which to carry the heat away from the source. These heat transfer mediums can be simply water or a chemical solution designed in various forms to be a coolant.

The breakdown of this cooling mechanism will cause the two materials rubbing against each other to heat up to a point where the surrounding materials can ignite. Understanding how this can result in a fire

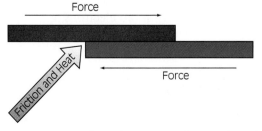

Figure 4-12 Heat from friction can be produced whenever any rubbing or compression occurs.

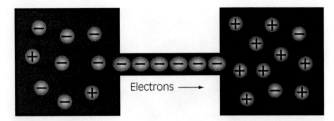

Figure 4-13 Electricity is simply a flow of electrons from a place where there are many to a place where they are lacking.

source more evident than in the home where space heaters with glowing orange-red metallic strips radiate heat into a room.

An electrical current can generate heat. Electricity is simply a flow of electrons from a place where there are electrons to a place where they are lacking, **Figure 4-13**. The place where electrons are in abundance is referred to as being *negatively charged*. Conversely, the place where electrons are lacking is referred to as being *positively charged*. It is the nature of electricity to attempt to equalize charges by sending electrons from where they are in abundance to where they are lacking. This is most commonly done by way of an **electrical conductor**. An electrical conductor permits the flow of electrons from one place to the other. When this electrical flow occurs, one electron does not exactly jump from one atom to the other. What occurs is more like a relay of electrons where one jumps onto an atom, which releases another to move to the next atom. When electrons are traveling along a conductor, the atoms that make up that conductor trade off their electrons from one atom to the next. During this activity, billions of electrons are jumping and, inevitably, collisions occur. When the collisions send the electrons astray, they collide with molecules and cause them to break apart. With the breakdown of the molecule, the energy that held the molecule together is broken and the bond is released as heat. Depending on the density of the conductor and the insulation surrounding it, the heat generated differs greatly. If a substance heats up easily, it is generally not a good conductor because the greater the amount of heat, the more inefficient is the action of the electrons milling about, causing much confusion at the atomic level. An efficient conductor will have less heat because more electrons are busy being transferred and there are fewer collisions.

> **Firefighter Fact**   A general rule of thumb is that the heavier the material, the more efficient it is as a conductor. For example, metal is a good conductor, but cotton is not.

It is not simply the weight of the substance that determines this "rule of thumb" of conduction capability; it is also determined by **density**. Usually, the more dense a substance is, the better its conductivity. Conduction is discussed in greater detail later in this chapter.

Electrical energy is a heat source. As firefighters, it is important to recognize forms of electrical energy. Obviously, lightning, arcing, and wiring outlets in the home are sources but also included in this classification are static electricity and induction heating. Static electricity occurs when dissimilar materials are rubbed, scraped, or suddenly joined or separated. This action creates heat in the form of mechanical heat, but it also creates a different potential in electrical charges due to the tearing apart of the surface at the molecular level. Electrons from one substance are taken by another and when enough of them collect, they will attempt to equalize by jumping the gap in the form of a static charge.

> **Firefighter Fact**   Static electricity is a small but powerful electrical charge that can emit temperatures of more than 2,000°F. Although this heat level is high as far as a temperature reading goes, it dissipates very quickly and offers no particular hazard to ordinary combustibles. If in the presence of a flammable gas, however, the small charge can ignite the gas with explosive results.

With one molecule breaking apart from the heat, more heat is released and that causes other molecules to break apart in like fashion. The resulting chain reaction will continue until all of the fuel is spent. If there is enough sustained heat, ordinary surrounding combustibles will have been exposed to the heat long enough to break down and generate their own heat for combustion.

The other type of induction heat can be found in a microwave oven. In this case, waves of alternating electrical energy subject the food to exposure. The current waves, called *microwaves,* alternate direction at high speeds resulting in molecular bombardment of the substances' internal electrons attempting to join the flow. The resultant collisions of the electrons release heat energy as the molecular bonds break apart and heat the surrounding material.

> **Streetsmart Tip**   As firefighters, knowing the process of combustion will provide a direction to the solution. A fire in a microwave oven requires the removal of outside power before any further internal action can be taken.

FIRE BEHAVIOR ■ 85

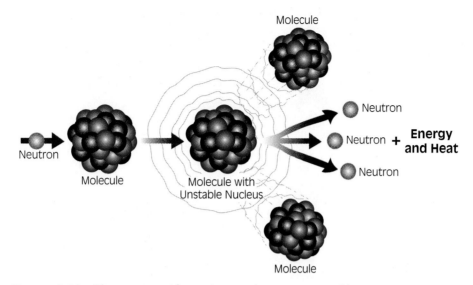

Figure 4-14 The process of creating nuclear energy and heat.

Nuclear

The last heat source is nuclear, **Figure 4-14**. Used in the atomic bomb, nuclear power has been hotly debated for years as to whether it is a safe source of energy. Essentially, all forms of atomic energy that we use are first converted to heat. The means by which nuclear energy generates heat is essentially the same as for the other forms already described. Radioactive materials are very unstable and are constantly breaking down as they seek to form a stable molecular composition. In that process, atomic particles are randomly flying in all directions. When encased in a heavy shield called a *core* made of very dense materials such as lead, the particles remain confined within the radioactive material. When activity is desired, the radioactive material, usually in the form of rods, is pulled out of the protective surrounding core, and a controlled amount of energy is permitted to be transferred into the surrounding water. This heat release into the water then turns to steam. The steam then turns turbines that generate electricity. As the need for steam is increased or reduced, the rods are extracted from or reinserted into the protective core. When fully inserted, the nuclear bombardment is confined.

As firefighters, little can be done to safely fight a fire that has been caused by a nuclear heat source. In the nuclear accident that occurred at Chernobyl (Ukraine) in 1986, a meltdown of the core occurred as a result of the water level being too low to accept the heat from the rods. The heat then melted the core and exposed the rods. Once exposed, the rods have no controlling mechanism and runaway reaction results, appropriately called a *meltdown*.

If a nuclear heat source causes a fire, there is little to do but protect uninvolved areas, evacuate, and let more skilled technicians handle the emergency. On-site personnel are trained to act in such an emergency. The real danger inherent in a nuclear fuel fire results from exposure to radiation. Without proper protection, a firefighter can sustain a serious long-term illness or even a fatal injury from radiation exposure.

In discussing the source of heat energy, one common thread winds through the subject. In all cases, the heat source is generally the initiator of the fire that ignites surrounding combustibles, be they solid, liquid, or gas. Once removed as a heat source, the extinguishment of the fire is simply a matter of removal of the oxygen or fuel or a breakup of the flame's chain reaction.

COMBUSTION

Combustion is often confused with the term *fire*. It is not important to make the distinction a critical element at this level of learning, but it is important to know that there is a difference on a chemical level. Fire is a chemical reaction that is a self-sustaining process that emits light and heat as a by-product of that reaction.

In combustion, the released heat energy is reinvested in the process, causing the continued reaction to occur repeatedly. Unchecked, with involvement of greater access to fuel, oxygen, and heat, the growth will accelerate. The products of combustion, **Figure 4-15**, can be hazardous and deadly to firefighters.

86 ■ CHAPTER 4

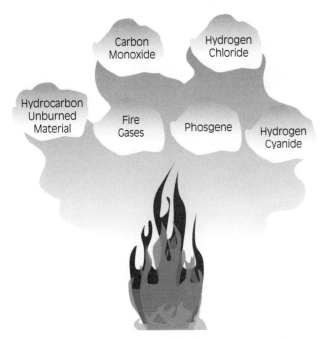

Figure 4-15 The products of combustion can be deadly.

Figure 4-16 The speed of combustion is affected by the amount of oxygen present.

OXYGEN AND ITS EFFECT ON COMBUSTION

Oxygen is an important element in the sicence of fire. It acts as a catalyst to the combustion process. The chemical reaction that occurs during combustion, called oxidation, is the process of oxygen bonding to other elements. The amount of oxygen present affects the process of oxidation, thus affecting the process of combustion.

Firefighter Fact The concentration of oxygen in the atmosphere is important to the combustion process. Oxygen occurs normally in the Earth's atmosphere in a concentration of approximately 21 percent. Combustion can occur in oxygen concentrations of 14 percent or greater. In concentrations less than 14 percent, there is not enough oxygen present to support combustion.

The amount of oxygen present can also affect the rate at which oxidation occurs. It can be very slow, as is the case with rusting, or it can occur very quickly, as is the case in combustion. With a diminished amount of oxygen, the combustion process is slowed. With an abundance of oxygen, the chemical reaction is accelerated, **Figure 4-16**. In addition, the presence of oxygen can affect a material's combustibility. Higher concentrations of oxygen can cause some materials to ignite spontaneously or permit materials to burn that would not burn under otherwise normal oxygen concentrations.

Streetsmart Tip Oil in the presence of high levels of oxygen will ignite spontaneously. For this reason, any pipe, gauge, or fitting that carries oxygen posts the warning to "use no oil." Nomex, a material that is used as a fire protection component in fire protection equipment, will ignite and burn in high levels of oxygen.

VAPOR PRESSURE AND VAPOR DENSITY

Pressure is defined as the application of continuous force by one body on another body that it is touching. In terms of liquid or gas, it is the amount of force applied to the surrounding container or in the open atmosphere at a given altitude. *Vapor pressure* is the measurable amount of pressure being exerted by a liquid substance as it converts to a gas and exerts pressure against a confined container, **Figure 4-17**.

A liquid is a collection of molecules that occupy the same space in a fluid state. In normal molecular activity, these molecules are constantly circulating and colliding with either one another or the sides of the vessel in which they are contained. However, when the molecules circulate beyond the upper boundary of the sur-

FIRE BEHAVIOR

Figure 4-17 Vapor pressure. Container on left has less vapor pressure than container on right.

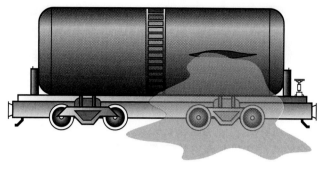

Vapor Density Greater than 1

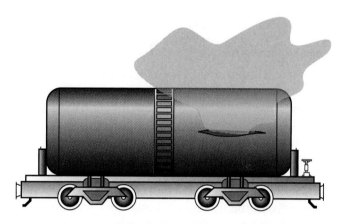

Vapor Density Less than 1

Figure 4-18 Vapor density. Vapors lighter than air rise and those heavier than air sink.

face, they escape into the space above the liquid. Some return to the liquid in their random paths. The molecules that escape and do not return tend to fill the surrounding area with the molecules of that liquid in a gaseous state. This is called **diffusion**.

If the weight of the gas is lighter than the surrounding atmosphere, the molecules will continue to escape and return with more molecules rising and escaping than returning. If this were to continue, eventually total **evaporation** of the liquid would occur. If the gas were heavier than air, the molecules would remain in the area between the upper surface of the liquid and the top of the open vessel. At the point where there are as many molecules being liberated from the liquid as being reabsorbed, it is referred to as being in a state of **equilibrium**. Of course, if the gas were lighter than air, this state would never be achieved because the molecules would be constantly escaping to higher levels. The rate of molecular escape or evaporation is highly dependent on many factors, most notably temperature and pressure. The liberation of molecules can be curtailed if the pressure of the surrounding atmosphere is higher than normal because it might exceed the liquid's ability to diffuse. Or, if the temperature is lower and the molecular activity is reduced, a reduction in the number of molecules being liberated from the liquid could result.

Air pressure changes at different altitudes and must be considered when examining this area of study. At sea level, air pressure is 14.7 psi (pounds per square inch). If a vapor pressure exceeds 14.7 psi, it will escape if it is at sea level.

Vapor density describes the weight of a gas as compared to normal air and is identified as a number. A normal concentration of air at sea level (14.7 psi) is designated as "1" when measured at 0°C (32°F). Gases that weigh less than the same volume of air will be lighter and tend to rise and their vapor density is described in terms of a number less than 1. If the gas is heavier than air, the number will be greater than 1, **Figure 4-18**. The importance of this to the firefighter is knowing that a particular gas that is heavier than air (a vapor density of more than 1) will tend to collect in low areas such as depressions, basements, cellars, sewers, ravines, or the like, **Table 4-3**. Normal air pressure can act as a cap to an open vessel. This is where the terms *vapor pressure* and *vapor densities* are often confused.

Firefighter Fact Vapor density can be a valuable number to a firefighter, **Table 4-3**. The vapor density of a substance can display whether it is heavier or lighter than air. Propane, with a vapor density of 1.6, will tend to collect in low areas, depressions, or cellars. Methane, on the other hand, with a vapor density of 0.6, is lighter than air and will rise and tend to mix and dilute with atmospheric air.

Vapor pressure is the force exerted on the sides of a closed container. Vapor density is a function of the weight of the gas. Although different, they are similar

Vapor Density of Normal Atmospheric Air at Sea Level Is Designated as 1

	VAPOR DENSITY	EFFECT
Propane	1.6	Will sink and collect at low points
Methane	0.6	Will rise

TABLE 4-3

in some respects. The pressure that exists between the top of the liquid and the sides of the container vessel occurs as a result of the liquid's conversion to a gas and is the vapor pressure. Conventional gauges can measure this pressure, and determinations can be made from those data if the vapor pressure poses a threat. In a sense, a container that is open at the top is capped at the top by the weight of air acting as a lid.

If the liquid vaporizes and creates its respective vapor pressure, which turns out to be a pressure that is less than the ambient air (14.7 psi), the weight of the air will keep the substance "capped" in the open vessel. In this case, the vapor density of the substance would be less than air and the vapor pressure not enough to overcome the weight of the air holding it in the open vessel.

Conversely, if the vapor pressure is greater than the ambient air, the vapor will escape and not be measurable at all because no pressure is being created. The gas escapes as it evaporates through the "cap" of the weight of the air. In this case the vapor pressure is greater than the 14.7 psi of air. However, higher altitudes will exhibit different air pressures, so although numbers will be different, the principles will remain the same. Therefore, at higher altitudes, vapor escape will occur at lower vapor pressure levels. Evaporation of a liquid will occur more quickly and diffusion will be accelerated.

Note Density—the weight of a material—is dependent on the number of molecules and atoms that occupy a given volume. The more molecules crushed into a given volume, the denser it is. As an added factor, the heavier each individual molecule is, the heavier the substance will be. So a heavily loaded volume with a bunch of heavy molecules will result in a weighty material. The more molecules there are, the easier it is for them to collide with one another and pass on the heat energy. The more dense the material, the greater its ability to conduct heat.

BLEVE

A **boiling liquid expanding vapor explosion (BLEVE)** occurs when the vessel holding liquid ruptures as a result of pressure being exerted on its sides when the liquid it holds boils and the resulting pressure exceeds the container's ability to hold it, **Figure 4-19**. This usually occurs when heat is

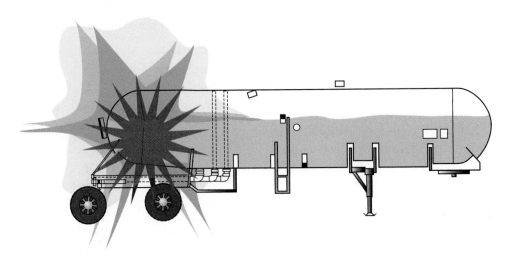

Figure 4-19 A BLEVE.

applied to the closed container, causing the liquid inside to boil. The vapor pressure in the closed container is increased until it reaches a point where the vessel can no longer withstand the pressure exerted against it. When the vessel fails, the ensuing release of vapor and liquid can be very violent, causing the explosive release of tremendous forces.

> **Firefighter Fact** The most common example of a BLEVE is popcorn. The liquid inside the hard kernel shell heats, boils, and exerts pressure against its container until the shell fails. The result is a cooked kernel that has escaped its container while the pressure inside has escaped and equalized.

Even if the liquid in a vessel is not flammable, the container rupture could still be violent, resulting in a force that can send container fragments great distances accompanied by a shock wave. If the liquid is flammable, the container failure will exhibit a fireball that adds in a fire extension element to the problem. If the liquid is a hazardous material, still another set of variables must be addressed.

BOILING POINT

All materials obey the laws of nature and exist in one of three forms or states: solid, liquid, or gaseous. Pressure and temperature will affect the state of matter, **Figure 4-20**. As a point of reference, the words *normal* and *ambient* are often used to describe the conditions under which a substance exists. To standardize the understanding of scientific description, 70°F was established as the reference point for temperature and 14.7 psi for pressure. This set of parameters is termed *normal*. When a substance being described is referred to as being in its normal state, it is understood that it is at that temperature and pressure.

If the pressure remains the same and the temperature changes, the form of the substance can change. Similarly, a change in pressure while the temperature remains the same can cause a change in the state of the substance. Gases compressed to liquids under great pressure generate heat, and liquids that boil absorb heat. When a material absorbs heat, it usually does so at its boiling point.

The state of conversion from a liquid to a gas under normal atmospheric conditions is called *evaporation*. Evaporation is simply the state where a liquid's molecules, which are constantly sending off molecules into the surrounding area and accepting them as they randomly return, wind up with a greater loss rate than recovery rate. Eventually, the entire supply of the liquid's molecules will escape into the surrounding atmosphere. This phenomenon occurs as long as the surrounding atmospheric pressure is less than that of the gas above the liquid. If the liquid is heated, the molecules move about more quickly, crash into the sides of the container vessel, and increase pressure on the vessel. When the molecules collide into the sides, they rebound into the liquid, eventually escaping over the surface of the liquid. The increased activity of the molecules over the surface increases the pressure of that vapor. When the pressure exerted overcomes the surrounding atmospheric pressure, then the boiling point of that liquid has been reached. Water is the liquid state and steam is the gas state, **Figure 4-21**.

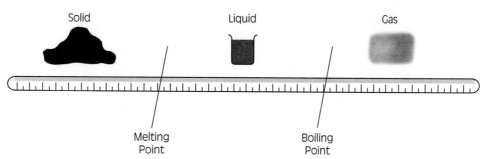

Figure 4-20 States of matter, in many cases, are temperature dependent.

Figure 4-21 Water turns to steam at the boiling point.

FLAMMABLE AND EXPLOSIVE LIMITS

A combustible material must meet certain requirements before it can oxidize. Depending on the molecular makeup of the compound, the presence of oxygen is required at different concentrations. This relative concentration is described in terms of percentages.

If there are 100 gallons of a liquid made up of 50 gallons of water, 35 gallons of detergent, and 15 gallons of a degreasing solution, we could say that the percentages of each substance in the total mixture are represented as 50 percent water, 35 percent detergent, and 15 percent degreaser. The number 100 was used for simplification. Whatever the total volume happens to be, the percentage of the components of that mixture can be described as a percentage of the total. A percentage is determined by dividing the amount of the substance to be measured by the total amount in the mixture, **Figure 4-22**.

The same is true for gases. Because combustion can only take place when a substance is a gas, the levels of the gases are described in terms of percentages. Through chemical analysis and scientific testing, it has been determined that gases can only ignite when certain concentrations of that substance are present in air. If enough combustible gas is not present, it is said that the mixture is too lean to burn. If there is too much gas, it is said to be too rich to burn. When a concentration of a gas falls into the range where it can ignite, it is said to be within its **flammable** or **explosive limits**.

Flammable limits can change depending on the temperature. The limits can contract or expand depending on the surrounding conditions. Comparing the flammable or explosive limits of one substance with another will assist the firefighter in understanding the relative volatility of that substance. Referring to **Table 4-4**, compare the flammable limits of gasoline, natural gas, and carbon monoxide. For example,

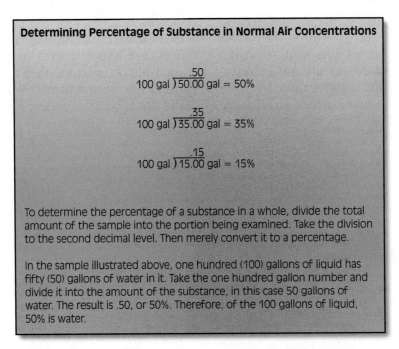

Figure 4-22 Determining the percentage of a substance in normal air concentrations.

Flammable Limits of Some Materials

	LOWER	UPPER
Acetone	2.6	12.8
Butane	1.9	8.5
Kerosene	0.7	5
Natural gas	6.5	17
Gasoline (92 octane)	1.5	7.6
Carbon monoxide	12.4	74

TABLE 4-4

a concentration of carbon monoxide at 37 percent is very significant if the reading was taken in a 50,000 square foot undivided building space. On the other hand, a reading of 37 percent in a 1 cubic foot container is much less dangerous and might only emit a slight "pop" if ignited, whereas in the 50,000 square foot building space, an explosion could lift the roof off its supports and blow out concrete walls.

> **Firefighter Fact** Flammable limits of the materials in **Table 4-4** demonstrate how different the flammable limits of different substances can be. Especially notable are the upper and lower limits of carbon monoxide, a substance found at every fire. It will ignite at a much greater range than natural gas.

Specialized instruments have been developed to display the measurements of flammable and explosive limits. Through the use of these instruments, firefighters can better evaluate their surroundings, have a better understanding of the enemy, and be better prepared to engage in firefighting activities safely.

THE BURNING PROCESS— CHARACTERISTICS OF FIRE BEHAVIOR

The process of burning occurs in clearly defined stages. By recognizing the different stages, a firefighter can better understand the process of burning and fight the fire at different levels and with different tactics and tools. As is true in any combat situation, knowledge of the enemy's needs and practices leads to the use of different tactics and practices to attack that enemy successfully.

> **Note** Burning materials follow a specific sequence of events from start to finish. Interrupt any of the sequential steps and fire can be minimized or extinguished entirely. Older texts refer to these stages as the incipient stage, the smoldering stage, and the free-burning stage. Today's firefighting science has redefined the stages of fire burn. They are the ignition, growth, fully developed, and decay stages, **Figure 4-23**.

Ignition Stage

When a substance begins to heat up, it liberates gases that can burn. The preliminary heating usually comes from an outside source such as a match or spark from a fire already burning. Heating can also occur through conduction, convection, or radiation from another fire or heat source. At some point, the amount of heat being created exceeds the amount of heat that is dispersed and the combustion feeds itself the necessary heat to sustain a burn.

When the necessary ingredients of a self-sustaining chemical reaction are present, ignition occurs. **Ignition** is that point where the need for outside heat application ceases and the ability for the material to sustain combustion comes from the heat generation of the material itself.

The ignition stage of the fire is, very simply, the point at which the four elements in the fire tetrahedron come together, materials reach their ignition temperatures, and a fire is started. At the ignition stage, the fire is typically very small and limited in area.

Growth Stage

From the point of ignition, fire begins to grow. Starting out as a spark or a small flame, other combustibles heat up, liberate flammable gases, and ignite, spreading the chain reaction to other flammables and resulting in an increase in size. (The growth stage was formerly considered the incipient and smoldering stages.)

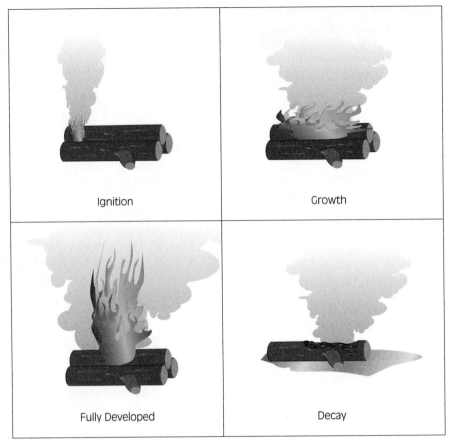

Figure 4-23 Four stages of fire.

The speed of the growth and ultimately the size of the fire are dependent on several factors:

- *Oxygen supply:* The amount of oxygen will have a direct effect on the speed of growth and the size of the fire. Any limitation of the oxygen supply will curtail the growth and can even result in extinguishment.
- *Fuel:* The size of the fire will naturally depend on the amount of fuel available to burn.
- *Container size:* In a structure, the container would be the surrounding walls and obstructions. A large container would permit dissipation of heat and slow the growth of the fire. In an open and unconfined area it even serves to inhibit fire ignition. With a container, heat can radiate back down and back into a space, further heating up uninvolved fuel sources.
- *Insulation:* Heat that is radiated back into unburned areas will accelerate growth. If the container wall and ceiling are insulated, the amount of heat kept trapped in the container and not permitted to escape serves to further "reinvest" the heat produced back into the fire formula.

Fully Developed Stage

This stage is recognized as the point at which all contents within the perimeter of the fire's boundaries are burning. In a structure this would mean the entire contents of a room. In an outside fire, it would mean all combustible material within the fire's furthest reach. The fully developed stage is regulated by one of two features. In a structure, the speed and extent of a fully developed stage is controlled or regulated by the amount of air that can be introduced or supplied to the fire area, making it an air-dependent fire. In an outside fire, because the amount of air is unlimited, the amount of fuel will dictate the size of the fire, making it a fuel-dependent fire. The fully developed phase is where a phenomenon called flashover can occur, **Figure 4-24.** Flashover is detailed later in this chapter.

Decay Stage

When the point at which all fuel has been consumed is reached, the fire will begin to diminish in size. Ultimately, the fire will extinguish itself when the fuel or oxygen supply is exhausted. Obviously,

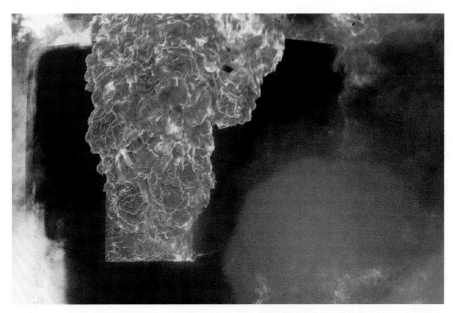

Figure 4-24 A flashover occurs suddenly when all or most of the combustible materials reach their ignition temperature nearly simultaneously.

this can take a considerable period of time. The bright array of flames will diminish, becoming a series of separate flame fronts, **Figure 4-25**. Then the flames will disappear until only glowing embers are visible. Eventually, those too will disappear. What is witnessed in a fireplace from ignition to final extinguishment is the same series of events that occurs on a much larger scale in a structure fire or an open wildfire. As the decay stage progresses, even a fire that was air controlled will become a fuel-dependent fire as the amount of fire decreases. At some point the amount of fire will be small enough to be fully supplied by the air supply. At that point the fire's future will depend on the amount of fuel left.

These factors will dictate the tactics that will be employed when fighting a fire. In some cases, such as a fire in the hold of a ship, cutting off the oxygen is usually the tactic of choice. If a ship's watertight spaces can be sealed, the supply of oxygen can be cut off. Ultimately, this will cause the fire to self-extinguish. In fighting a wildfire, use of a fire break to cut off access to additional fuel might be the choice of attack.

Figure 4-25 The decay stage.

Note The type of fire, the characteristics of the burn, and the manner in which fuel is being supplied are the determining factors in deciding the best attack.

MODES OF HEAT TRANSFER

Heat is a by-product of combustion that is of significant importance to the firefighter. It is heat that causes fire to sustain its combustion and, more importantly, to extend. When heat given off as a product of combustion is exposed to an unheated substance, certain changes occur that can make the new substance a contributing factor in extending a fire. Therefore, knowing how heat is transmitted from one place to another is the first step in knowing how to control the extension of fire, one of the first steps in extinguishing it.

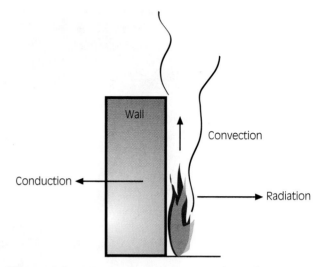

Figure 4-26 Examples of heat transfer in fire.

Caution Heat, as a by-product of the combustion process, can be dangerous to firefighters, causing dehydration and heat exhaustion as well as burns. Each of these conditions can be mild to severe and ultimately life threatening. The amount of energy released as heat over a certain period of time is called a *heat release rate*. Heat release rate is influenced by the quantity and type of burning materials. The greater the heat release rate for a particular fire, the greater the danger to firefighters for heat-related injuries.

The three modes by which heat transfers its energy from one substance to another are through conduction, convection, and radiation, **Figure 4-26**. The type of substance being heated, the distance covered by the material being heated, and the ability of the substance to retain the heat will be factors in the spread of fire.

Conduction

When a hot object transfers its heat, conduction has taken place. The transfer could be to another object or to another portion of the same object. Combustion occurs on the molecular level. When an object heats up, the atoms become agitated and begin to collide with one another. A chain reaction of molecules and atoms, like a wave of energy, occurs and causes the agitated molecules to pass the heat energy to areas of nonheat. As the heat increases, so do the waves of energy passing to the cooler areas, **Figure 4-27**. As more agitation occurs, the substance heats up and unless the heat energy can be dissipated in a **heat sink**, the overall and internal temperature of the substance increases until it reaches its boiling point if it is a liquid, or its ignition temperature if it is a solid.

When examining conduction as a heat transfer vehicle, it is important to understand that heat is conducted through materials at different rates. The rate of heat transfer will depend on several factors, the most significant being density. If a given volume of any substance is weighed, we can determine its relative density.

The more dense a material, the better a conductor it will be. Because density is a function of weight, the heavy substances are generally better conductors. Metals, being among the most dense in the universe, are therefore generally better conductors of heat, and the heavier the metal, the better the ability to transfer heat. Concrete, a dense material made up of rock, sand, and cement, will also conduct heat through it with enough transfer ability to ignite combustibles. Essentially, the ability of a material to transfer heat is dependent on that material's ability to keep the heat energy accumulating faster than it can be dissipated.

If the denser materials are better conductors, then conversely, the less dense materials are better insulators. Lighter materials whose ignition temperature is very high become the best insulators and, therefore, the poorest conductors. Mineral wool, essentially a form of rock spun into a web-like material with a lot of air space between fibers, is one of the best insulators. Because heat energy uses the collision chain

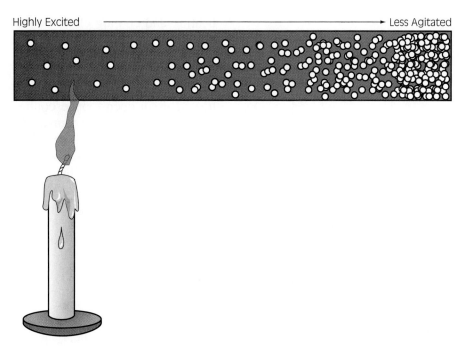

Figure 4-27 Conduction. A metal bar is heated at one end, causing the molecular activity that transfers the heat.

reaction to pass heat along, it stands to reason that energy will dissipate in the absence of collisions. There is no conduction whatsoever in a vacuum because there are no molecules or atoms to contribute to the collision process. The fewer such molecules, the poorer a conductor the material will be. Therefore, besides the noncooperation of the material in the mineral wool, the air spaces between the strands of mineral wool serve to limit conduction further, making it an excellent insulator and a poor conductor.

The relative ability of a substance to be a good or poor conductor is a function of time and application of heat. If the application of heat to a substance over a given period of time exceeds its ability to dissipate or shed that heat, the substance will begin to heat up. If it is a poor conductor because of its light density, it will be able to shed the heat without having its internal temperature rise. If it is a good conductor, it will absorb the heat and transfer that heat to distant locations internally. A piece of heated steel will take a long time to cool because the heat energy must come to the surface and dissipate in the less dense air. Mineral wool, conversely, will cool almost as quickly as the heat source is removed because of its excellent ability to dissipate heat, owing to its light density.

Convection

Air that is hotter than its surroundings rises. Air that is cooler than its surroundings sinks. Air is made up of many molecules floating about freely. Even so, it still has weight. Some molecules are made up of the same element. For example, oxygen in its natural state will combine with another oxygen atom to form a stable oxygen molecule. In a given volume, air at a given temperature will have the same density. When heated, as in conduction theory, the molecules become agitated and begin to collide with one another. In the process, the molecules are demanding more space to accommodate the vibrations and they push into one another as they seek that space, **Figure 4-28.** When that happens, the density of a given volume is reduced and it weighs less. Because it weighs less, it rises until it reaches equilibrium—the level at which the weight is the same as the surrounding atmosphere. On the way up, it mixes with cooler air around the perimeter of the column. Once there, the agitation reaches its maximum. As the molecules calm down, they start to return to their original density and become less buoyant. As the buoyancy decreases, that particular volume of air drops. If it is heated again by the rising heat energy, its agitation increases again. If it falls outside the **thermal plume**, it drops down to the bottom of the plume.

In theory, an unobstructed plume that is unaffected by any other influences such as horizontal air current would behave in the mushroom fashion. In a structure, the flow and rate of convection would be accelerated when the heated air is confined and must seek a "cooler" space after meeting a vertical obstruction such as a ceiling. The "mushroom"

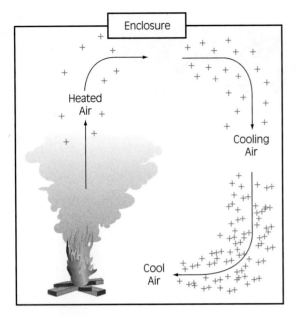

Figure 4-28 Convection. As air is heated, its molecules become excited and seek more space.

heating and cooling effect contributes to horizontal spread of fire at upper levels, **Figure 4-29**.

Convection currents created by heat that occur in the air also occur in liquids in the same fashion. The process of evaporation is accelerated in a liquid heated in an open-topped container. The molecules that are heated tend to bounce off the sides and reintroduce the energy back into the liquid. The reintroduced energy then causes more molecules to collide until the energy forces them over the top of the container where they escape. This is also the same phenomenon that causes pressure in a closed container to rise. In the case of the closed container, there is no escape and the reintroduced energy is kept in the contained vessel.

Radiation

The last form of heat transfer occurs by radiation. As discussed earlier, heat energy can be transmitted directly when molecules collide with one another and cause the waves of heat energy to travel. It stands to reason that in the absence of molecules, such as would be the case in a vacuum, heat energy would not be able to travel. However, this is not the

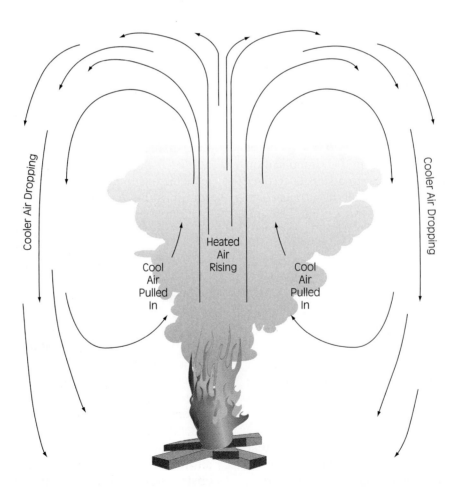

Figure 4-29 Fire plume.

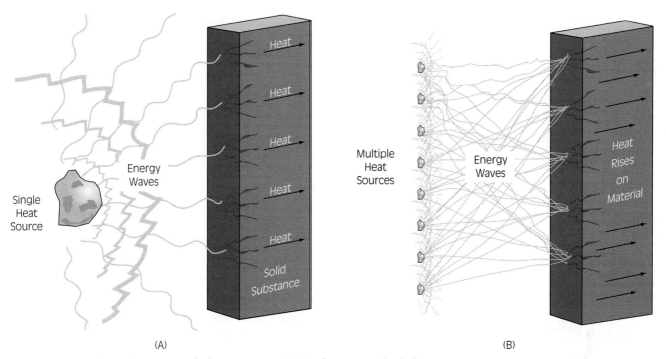

Figure 4-30 (A) Radiation, single heat source. (B) Radiation, multiple heat sources.

case. When combustion occurs, light is produced. Light travels by way of lightwaves. These lightwaves range from ultraviolet to infrared. Contained in the lightwaves are electromagnetic forces that can actually travel across vacuums and deposit themselves on remote objects. An example of this is the sun and its ability to heat the Earth.

Fire produces the same infrared lightwaves and, with enough concentration, can permit fire to jump from the source to a distant object, heat it up, and, if intense enough, cause it to ignite, **Figure 4-30**. Several factors need to be in place before this can occur: the source must be strong enough to sustain the bombardment of the lightwaves; the recipient must be able to absorb the heat energy; and the object must be able to build up the heat without it dissipating away through its own heat sink.

Radiation is a major contributor to flashover, in which heat buildup at upper levels in a compartment radiates heat down into the room. Eventually the objects in the room at lower levels approach and reach their respective ignition temperatures through this mode.

Firefighters must cope with this phenomenon because fire can extend through radiation as much as through conduction or convection. Water is used to absorb the heat and carry it away so it cannot serve to heat surrounding materials. Water absorbs the accumulating heat from the exposed objects, thereby depriving them from reaching ignition temperature. By controlling this element of the fire tetrahedron, the fire is prevented from extending. To a limited extent, water can also be employed to combat conduction extension. In that case, the water carries away the heat, keeping the unburned or unheated portion of the material from reaching the ignition temperature or permitting it to transfer the heat to another substance.

THERMAL CONDUCTIVITY OF MATERIALS

All matter will conduct heat. The ability of a material to conduct thermal energy depends on its density. Because heat transmission is actually a transfer of agitation from one molecule to another, for heat to be conveyed from one place to another, molecules must be present. The less dense a material, the more difficult it is for heat to be transferred through it. In the absence of all molecules—a vacuum—there is no transmission of heat. The more dense a material is, the greater the thermal conductivity potential. If a substance has a great deal of open space between its molecules, the molecules can become agitated without transferring that agitation to other molecules. The same holds true for material in a larger scope beyond the molecular level. Materials such as cellulose or mineral wool, which have voids in their makeup, contain a great deal of space, and these spaces serve to insulate the heat from being transferred from the heated side of the material to the unheated side.

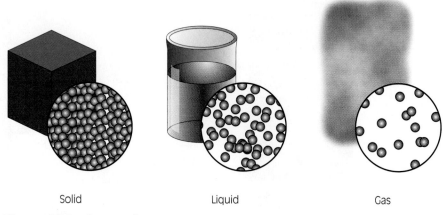

Figure 4-31 States of matter.

PHYSICAL STATE OF FUELS AND EFFECT ON COMBUSTION

Matter can be found in one of three states: solid, liquid, or gas. To take part in the combustion process, most fuels must be in their gaseous state. Although this is not the case in every single situation, in most cases, a molecular free-floating state must exist before oxidation reactions can take place. In many cases, wood for example, the solid matter is turned to gas and it combusts almost simultaneously. Knowledge of a material in its physical state and what state it is required to be in before combustion takes place is of paramount interest to the firefighter. The physical state can affect combustion, **Figure 4-31**.

Solid

The molecules in a solid material are packed closely together. That criterion gives the material its density and, to a great extent, its physical state. The molecules are bound together by bonds that maintain the form and shape. When heat is applied, the molecules become agitated and begin to collide with one another. In some cases, this causes the molecules to break apart and free up the components of the material. Among some of the molecules being broken apart are those that will readily combine with oxygen and become oxidized. The others are left to free float for the time being. As the bonds with oxygen are formed, a heat-producing exothermic reaction occurs, and the released heat causes the free-floating molecules to develop a greater affinity for the oxygen. These newly formed by-products of oxidation then release even more heat, and the chain reaction called combustion is under way.

Heat that is produced causes additional molecules to break away from the solid physical state, become a gas, and combust. The additional heat causes more reactions to occur, and a free-burning self-sustained chain reaction results. In the case of the solid material, the applied heat is absorbed into the mass and dissipates to some degree. When the amount of heat applied exceeds the ability of the mass to dissipate that heat, chemical breakdowns occur and the self-sustaining burning reaction results and continues until it either is interrupted or runs out of fuel.

Liquid

In a burning liquid, the ability to burn is dependent on the substance's ability to place its molecules into suspension. A liquid cannot burn unless it is in suspension, also referred to as **atomization**. That liquid will then become engaged in the self-sustaining combustion process if it has the affinity for oxygen and will permit oxidation. Not all liquids possess that property. Some will boil, evaporate, and never engage in combustion. Water is a prime example.

Similar to a solid, a liquid will act as a heat sink and dissipate the introduced heat into the cooler areas of the liquid. When the entire pool of liquid is heated or the ability to dissipate is overcome by the application of heat, a rise in temperature of the liquid will occur, resulting in boiling or **flash point** ignition. In most cases, there is a physical state transformation from solid to liquid to gas before combustion can take place.

Gas

A gas apart from the other two states is essentially primed for combustion. Its physical property is in a ready-made state that will permit a chemical reaction.

Awareness of these properties and their place in the combustion process is another weapon in the arsenal for fighting fire. Knowing that keeping a liquid cool enough to prevent **vaporization** or that by excluding oxygen ignition can be minimized or even prevented becomes a valuable piece of information when dealing with an incident such as a tank truck fire or a liquid fuel spill.

> **Note** Remember that as a substance changes from one state to another, heat is either given off (exothermic reaction) or absorbed (endothermic reaction).

We see heat absorbed when water turns to steam, and a carbon dioxide extinguisher will actually create snow and collect frost on the cylinder wall when the liquid is released as a gas. Essentially, when pressure is applied heat is released. If enough pressure is applied to a gas, it turns to a liquid (e.g., carbon dioxide gas to liquid carbon dioxide). Liquid CO_2 is used in carbon dioxide extinguishers. When additional pressure is applied, the liquid turns to a solid with the release of additional heat (liquid carbon dioxide to dry ice). Conversely, when the solid becomes a liquid, it absorbs great quantities of heat (ice to water) and again, when the liquid becomes a gas (water to steam).

The heat-absorbing capabilities of the water-steam conversion are employed when extinguishing fire in ordinary combustibles. Water robs the fire of its heat and reduces the heat level so that self-sustaining combustion is no longer possible.

THEORY OF FIRE EXTINGUISHMENT

The way to stop a fire is to remove one of its essential ingredients. Heat, fuel, oxygen, and a continuing chemical reaction are the four elements in the fire tetrahedron. Remove one of them and the fire will collapse.

UNIQUE FIRE EVENTS

There are several unique events that can occur in a fire within a compartment or structure that firefighters must be familiar with, including thermal layering, rollover, flashover, and backdraft.

Thermal Layering

Hot air and gases from fire rise and will continue to rise until they reach equilibrium, or balance, with the surrounding atmosphere. **Thermal layering** occurs when the gases produced by fire stratify into layers based on their temperatures. The hottest gases rise to the highest point in the compartment or room, and the gases with the lowest temperatures stay closer to the bottom of the container, or nearer to the floor.

When confined to a structure, the hottest of the gases and air will accumulate near the ceiling of the compartment or room, and as the amount of by-products from fire increases, the gases, heated air, and smoke will **bank down** until they can find an escape route.

Thermal layering is an important concept for firefighters because it affects how to enter and function in a room or area that is on fire or where the fire has just been extinguished. Firefighters must stay low to the floor in attacking structural fires because the hottest gases rise and stay at the highest layers in the compartment.

Rollover and Flashover

As the contents of a closed compartment or room burn, the superheated gases stratify and thermal layering occurs. These superheated gases can ignite in the uppermost thermal layers and cause the fire to travel across the top of the compartment. This is known as **rollover**, or flameover. This situation can be dangerous for firefighters because it can occur quickly and cause the fire to travel over the top of the firefighting team and sometimes get behind the team, affecting the firefighters' egress point from the compartment. Also, a rollover can cause the fire to spread to unaffected areas of the compartment, expanding the amount of area involved.

As a fire continues to burn in a compartment, more and more heated gases are generated, and the hottest layer of gases becomes larger, expanding down to the container. These superheated gases raise the temperature of all the unburned contents of the compartment. When the gases and fire have brought the entire contents of a container to their ignition temperature, a flashover can occur. In a **flashover**, the entire contents of the compartment ignite almost simultaneously, generating intense heat and flames. This rapid change in the compartment occurs in seconds and is a very dangerous and often deadly occurrence for firefighters. The tactics and skills covered in this text will assist firefighters in strategies to avoid flashover situations through the extinguishment of fire and ventilation of the compartment.

Backdraft

The fire tetrahedron shows that there are four elements that must be present to support combustion: fuel, oxygen, heat, and a continuing chemical reaction. A fire burning in a closed container, such as a room in a structure, can use all the available oxygen in the compartment, slowing the combustion process and the fire. The introduction of a source of oxygen to the compartment, such as opening a door or window to a room, will bring together the four elements of the fire tetrahedron again, causing the room to violently burst into flames. A **backdraft** is a dangerous situation. Firefighters should look for the signs of a potential backdraft situation when they are on the scene of fires in closed areas or structures. Signs of backdraft are covered in Chapter 18 of this text.

CLASSES OF FIRE

To better assist both the firefighter and the nonfirefighter alike, fires have been classified into different types based on the substance burning. This is primarily done so that the correct extinguishing agent can be applied to the fire by the firefighter.

- *Class A:* Class A type fires are made up of ordinary combustibles such as cellulose, rubber, or plastic. Combustibles such as paper, wood, cloth, rubber, and other organic solids including petrochemical solids (plastics) make up this class.
- *Class B:* Class B type fires are fueled by liquids, gases, or grease-type fuels. Oil, gasoline, alcohol, and other liquids are the more common types found in this class of fuel.
- *Class C:* Class C type fires are basically fueled by electricity. In this case, the electricity is actually the heat source that propagates the fire and often communicates to other fuels of the Class A or B type to sustain the burning process. If no other fuels are involved, merely shutting off the flow of electricity is enough to extinguish the fire.
- *Class D:* Class D, a less common fire type, is fueled by metals. A particular class of heavy metals, which can be identified on the periodic table of the elements and found mostly in the alkali metal group, will burn. Most common metals in the group are magnesium, titanium, zirconium, sodium, and potassium.

> **Firefighter Fact** In 1998 a new classification of fire, Class K, was created to cover fires in combustible cooking fuels such as vegetable or animal oils and fats. Class K fires are similar to Class B fires, although the oils and fats have some special characteristics. Class K fires and extinguishing agents will be covered in detail in Chapter 8.

Because each type of fuel has different burning characteristics, the method of extinguishing them differs, **Figure 4-32**.

The Class A fire, made up of what are considered normal combustibles, is extinguished by cooling the fire. The application of water cools the fire by absorbing heat as water is converted to steam. When enough of the heat is removed, the temperature of the fire is lowered below the ignition temperature of the substance and extinguishes the fire.

In a Class B fire, the fuel is a liquid, grease, or gas. As the temperature of ignition is approached, the liquid fuel vaporizes into gas and ignites. The gaseous materials are already in that state. The application of a smothering agent is used to prevent oxygen from getting to the fuel and propagating the chain reaction of fire by removing the oxygen element of the fire tetrahedron. In this case, the fire collapses due to a lack of oxygen.

A Class C fire, fueled by electricity, is overcome by the removal of the flow of electricity—the source and sustainer of heat. In this case the removal of the fuel, electricity, is the action taken to break down the fire tetrahedron and put the fire out. Keep in mind that in most cases an electrical fire is only the initiator. The fire then communicates to surrounding combustibles and can become a combination Class A or B in addition to Class C fire. Once the electrical power flow is removed, it becomes a straight Class A or B fire and is extinguished accordingly. In other cases, removal of the electrical current will extinguish the fire without any further effort. So here removal of what essentially is the fuel, one element of the tetrahedron, serves as the mechanism by which Class C fires are extinguished.

The Class D heavy metal fire is a chemical reaction fire. Almost all metals will burn if conditions are correct. Although some merely oxidize very slowly as in metal rusting, others will burn with great violence producing high heat and brilliant light. Among the combustible metals of significance are thorium, titanium, plutonium, hafnium, lithium, magnesium, zirconium, zinc, uranium, sodium, potassium, and calcium.

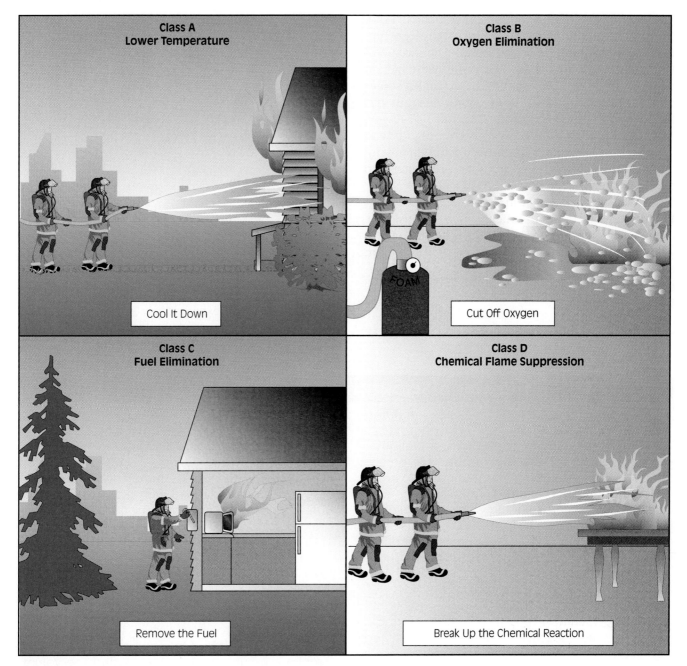

Figure 4-32 Four methods of extinguishing fires that closely follow the classification assigned to the material involved.

The combustible heavy metals differ somewhat in their reactions under fire. In some cases, the mere presence of water will cause a violent reaction, releasing heat and brilliant light. In other cases, the mere presence of air will cause the reaction. Some metals will burn so hot that the water molecule will actually be broken down into its component hydrogen and oxygen atoms. Then the hydrogen burns away in the presence of oxygen creating a brilliant and hot fire. Each metal's characteristics must be evaluated. Fortunately, these metals are not found in great abundance in normal occupancies. They are usually found in industrial processes and their presence should be known in advance to responding firefighters.

Other metals are also subject to burning but usually only when the material is in fine shaving form. For example, steel will not ignite easily, but a piece of steel wool will ignite rapidly when flame is applied.

While the classes of fires are essentially predicated on the fuel type, the classification follows the same track as the extinguishment method. And, as with most things in the field of firefighting, nothing is absolute. For example, a heavy metal fire can be extinguished without the use of chain reaction–breaking chemicals. For example, sometimes a huge volume of water can extinguish a magnesium fire, whereas in small applications, water will accelerate the chain reaction. Study and exposure will acquaint the novice firefighter with the basics. Experience will turn that firefighter into a knowledgeable veteran.

Classed primarily as a means by which to identify the type of extinguishment process to use, the extinguishment agents are labeled by several codes. The classification symbols are identified by letter, shape, and color and are attached to fire extinguishers for better recognition, identification, and utilization, **Figure 4-33**.

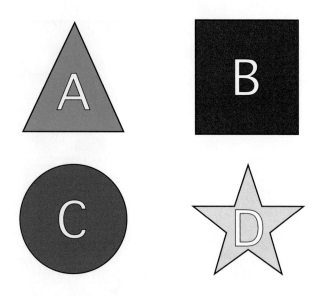

Figure 4-33 Fire extinguisher classification symbols may be displayed by shape, color, and/or letter.

READING SMOKE AT STRUCTURAL FIRES
Contributed by David Dodson

When responding to structural fires, a firefighter can apply an understanding of basic fire behavior by "reading smoke." Smoke issuing from multiple openings of a structure is often the only clue as to what a fire is doing within the building. Reading smoke can help firefighters discover clues about the location of the fire within a building as well as the severity of the fire and the potential for a hostile fire event such as flashover or smoke explosion/backdraft.

Typically, firefighters view smoke as "light" or "heavy." While this is fine for a rapid radio report, it is not descriptive enough when trying to understand what is actually happening with a fire. Smoke from a structural fire has four attributes that must be analyzed. These are *volume*, *velocity*, *density*, and *color*. All four of these factors of smoke are comparative—each opening that issues smoke is compared to the others—thus painting a story about the fire behavior and the location of fire within the building.

Volume

Smoke volume is an indicator of the amount of fuels that are "off-gassing" within a given space. In itself, smoke volume tells very little about the fire. A hot, clean-burning fire will emit very little visible smoke—yet a hot, fast-moving fire in a poorly ventilated building will show a tremendous volume of smoke. Dampened material will burn slowly and emit lots of smoke (typically a lighter color). Low-mass fuels and plastics emit large volumes of smoke with little heat. A large volume of smoke may indicate that the container holding the fire (room, area, or other portion of a structure) is full and needs ventilation.

Velocity

The "speed" that smoke leaves a building is referred to as velocity. In actuality, smoke velocity is an indicator of pressure that has built up within the building. Only two things can create smoke pressure: heat and restricting the volume of smoke within a container (a room or building). Velocity caused by heat will typically rise and slow gradually after it leaves the building. Velocity caused by restricted volume will immediately slow and balance with the outside atmosphere. If the velocity of the smoke leaving an opening is agitated or turbulent (some call this boiling or rolling smoke), a flashover is likely to occur. This flow is caused by the rapid expansion of the smoke due to heat. In other words, the box cannot hold the rapid expansion and therefore a turbulent flow of smoke is created. In these cases, the structure must be ventilated and cooled. Persons in these conditions have little chance of surviving due to smoke toxicology and thermal exposure.

Density

Incomplete burning causes smoke density. Some refer to this as smoke thickness. Smoke density is indicative of the amount of fuel that is laden within the smoke (particulates, gases, and aerosols). The greater the smoke density, the more likely a hostile fire event, such as flashover or rapid fire spread, can occur. In essence, the thicker the smoke, the more spectacular the flashover or fire spread. Thick, black smoke within a compartment reduces the chance of life sustainability due to the toxicology of the smoke. A few breaths of thick, black smoke will render a victim unconscious and cause death within minutes.

Color

For single-fuel fires, smoke color may indicate the type of material burning. In typical residential and commercial fires, it is rare that a single fuel source is emitting smoke. Smoke color can, however, tell the firefighter what stage of burning is taking place or where the fire is within a building. Virtually all solid materials will emit a white "smoke" when first heated. This white smoke is actually moisture. As a material dries out and breaks down, the color of the smoke will change. Natural materials will change to tan or brown, while plastics and painted surfaces will turn to gray. Gray is a result of moisture and hydrocarbons (black) mixing. All materials will eventually off-gas a black smoke—flame contact will cause materials to give off black smoke right away. As smoke leaves a fuel that is ignited, it heats up other materials and the moisture from those objects can cause black smoke to turn gray. As smoke travels, carbon content from the smoke will deposit along surfaces and objects. This also lightens the smoke color.

By combining these smoke attributes, some basic observations about the fire can be made before firefighters even enter a structure. Smoke velocity and color differences from opening to opening help firefighters find the location of the fire. Faster/darker smoke is closer to the fire seat, whereas slower/lighter smoke is farther away. If smoke from multiple openings is a constant color and velocity, firefighters should start thinking that the fire is deep-seated within the building. In these cases, the smoke has traveled some distance or has been pressure-forced through close doors or seams (walls/concealed spaces) prior to leaving the building. "Black fire" is a good phrase to describe smoke that is high volume, turbulent velocity, ultradense, and black. Black fire is a sure sign of impending autoignition and flashover.

Other factors influence smoke and may cause a false read. Wind, thermal balance, fire streams, ventilation openings, and sprinkler systems change the appearance of smoke. These four simple factors—volume, velocity, density, and color—can help firefighters refine their ability to read smoke and ultimately help protect their own safety and predict what fires will do next within a building.

Lessons Learned

An understanding of fire behavior is a fundamental requirement for anyone who has aspirations in the field of firefighting. Like the chemist who must know and understand measurements, or the tennis player who must know the rules of the game and employ them in a strategy, so must the firefighter know and understand what to expect fire will do based on information obtained on arrival at a fire scene. Because so many factors will be unknown in the initial stages of a fire, such as structural components, contents in the structure, effect of wind or weather, and other elements, it is incumbent on the firefighter to be armed with as much knowledge as possible about fire behavior so trends of fire spread in a given situation can provide the basis for strategy and tactics to be employed while more information is obtained.

KEY TERMS

Atomization   The separation of atoms and molecules into an unconnected state where they are in suspension rather than in liquid form.

Backdraft   A sudden, violent reignition of the contents of a closed container fire that has consumed the oxygen within the space when a new source of oxygen is introduced.

Bank Down   A condition in which the heat, smoke, and fire gases have reached the uppermost level in a compartment and, instead of continuing up, begin to push down from the ceiling toward the floor.

Boiling Liquid Expanding Vapor Explosion (BLEVE)   Describes the rupture of a container when a confined liquid boils and creates a vapor pressure that exceeds the container's ability to hold it.

Bond   A substance or an agent that causes two or more objects or parts to bind.

British Thermal Unit (Btu)   A measurement of heat that describes the amount of heat required to raise 1 pound of water 1°F.

Combustion   The chemical action in which heat and light are produced and the heat is used to maintain the chemical chain reaction to continue the process.

Compound   A combination of substances joined in a chemical bond that exists in a proportional amount and cannot be separated without chemical interaction.

Density   The mass per unit volume of a substance under specified conditions of pressure and temperature.

Diffusion   A process in which liquids or gases mix with one another by natural processes stemming from molecular movement.

Electrical Conductor   Any material that will permit electricity to flow through it.

Endothermic Reaction   A chemical reaction that absorbs heat.

Equilibrium   When referring to gas or liquids, a state where a balance has occurred in mixture or weight.

Evaporation   A process in which the molecules of a liquid are liberated into the atmosphere at a rate greater than the rate at which the molecules return to the liquid. Ultimately the liquid becomes fully airborne in a gaseous state.

Exothermic Reaction   A chemical reaction that liberates heat.

Explosive Limits   A concentration of a gas or liquid that is not too rich or too lean to ignite with force.

Fire   Also called combustion, it is the chemical action in which heat and light are produced and the heat is used to maintain the chemical chain reaction to continue the process.

Fire Tetrahedron   Four-sided pyramid-like figure describing the heat, fuel, oxygen, and chemical reaction necessary for combustion.

Flammable Limits   The concentration level of a substance at which it will burn.

Flash Point   The temperature at which a liquid will liberate a flammable gas.

Flashover   A sudden event that occurs when all the contents of a container reach their ignition temperature nearly simultaneously.

Heat Sink   The term used to denote a place where heat is drained away from a source.

Hydrocarbon   Any of numerous organic compounds, such as benzene and methane, that contain only carbon and hydrogen.

Ignition   The point at which the need for outside heat application ceases and a material sustains combustion based on its own generation of heat.

Inorganic   A substance that is not of any living organism.

Matter   Something that occupies space and can be perceived by one or more senses; a physical body, a physical substance, or the universe as a whole. Something that has mass and exists as a solid, liquid, or gas.

Molecule   The smallest particle into which an element or a compound can be divided without changing its chemical and physical properties; a group of like or different atoms held together by chemical forces.

Organic   A substance derived from living organisms.

Oxidizer   A catalyst in the breakdown of molecules.

Pyrolysis   Decomposition or transformation of a compound caused by heat.

Rollover   A phenomenon where the burning of superheated gases from fire extends into the top areas of the compartment in the upper thermal layers.

Thermal Layering   The stratification of gases produced by fire into layers based on their temperature.

Thermal Plume   A column of heat rising from a heat source. A fully formed plume will resemble a mushroom as the upper level of the heat plume cools, stratifies, and begins to drop outside the rising column.

Vaporization   The process in which liquids are converted to a gas or vapor.

REVIEW QUESTIONS

1. Why is knowing the sources of heat important to the firefighter?
2. What are the four modes of heat sources and why are they important?
3. Does an increasing span for a flammable limit denote a more hazardous material?
4. What are the three modes of heat transfer and why are they important?
5. What are the three states in which fuel may be found?
6. What is the basis for classification of fire into classes?
7. What is the significance of thermal balance and imbalance in a fire?

Endnotes

1. The American Heritage® Dictionary (CD-ROM), based on *The American Heritage® Dictionary of the English Language,* Third Edition © 1992 by Houghton Mifflin Company. Licensed from and portions © 1994 INSO Corporation.
2. *Fire Protection Handbook,* 18th ed., pp. 4–9. National Fire Protection Association, Quincy, MA, 1997.
3. *Fire Protection Handbook,* 18th ed., pp. 2–21. National Fire Protection Association, Quincy, MA, 1997.

Additional Resources

Bettelheim, Frederick, William H. Brown, and Jerry March, *Introduction to General, Organic, and Biochemistry,* 7th ed., Brooks/Cole, a part of the Thomson Corporation, Belmont, CA, 2004.

Bevelacqua, Armando S., *Hazardous Materials Chemistry,* Delmar Learning, a part of the Thomson Corporation, Clifton Park, NY, 2001.

Casey, James F., *The Fire Chief's Handbook,* Reuben H. Donnelley Corp., New York, 1978.

Cracolice, Mark S. and Edward I. Peters, *Introductory Chemistry: An Active Learning Approach,* 2nd ed., Brooks/Cole, a part of the Thomson Corporation, Belmont, CA, 2004.

Fire Protection Handbook, 18th ed., National Fire Protection Association, Quincy, MA, 1997.

Joesten, Melvin, D., James L. Wood, and Mary E. Castellion, *World of Chemistry: Essentials,* 3rd ed. Brooks/Cole, a part of the Thomson Corporation, Belmont, CA, 2004.

Klinoff, Robert, *Introduction to Fire Protection,* Delmar Learning, a part of the Thomson Corporation, Clifton Park, NY, 1997.

Lowe, Joseph D., *Wildland Firefighting Practices,* Delmar Learning, a part of the Thomson Corporation, Clifton Park, NY, 2001.

Masterton, William L. and Cecil N. Hurley, *Chemistry: Principles and Reactions,* 5th ed., Brooks/Cole, a part of the Thomson Corporation, Belmont, CA, 2004.

McMurry, John E., *Organic Chemistry,* 6th ed., Brooks/Cole, a part of the Thomson Corporation, Belmont, CA, 2004.

Quintiere, James G., *Principles of Fire Behavior,* Delmar Learning, a part of the Thomson Corporation, Clifton Park, NY, 1998.

Skoog, Douglas A., Donald M. West, James F. Holler, and Stanley R. Crouch, *Fundamentals of Analytical Chemistry,* 8th ed., Brooks/Cole, a part of the Thomson Corporation, Belmont, CA, 2004.

Whitten, Kenneth W., Raymond E. Davis, Larry M. Peck, and George Stanley, *General Chemistry,* 7th ed., Brooks/Cole, a part of the Thomson Corporation, Belmont, CA, 2004.

CHAPTER 5

FIREFIGHTER SAFETY

David Dodson, Lead Instructor, Response Solutions, Colorado

OUTLINE

- Objectives
- Introduction
- Safety Issues
- The Safety Triad
- Firefighter Safety Responsibilities
- Lessons Learned
- Key Terms
- Review Questions
- Endnote
- Additional Resources

STREET STORY

I was at work one day when the phone rang. The caller stated that my son Scott, who happens to be both a career and volunteer firefighter, was injured at a fire on his career job. It is the phone call a parent always dreads. I immediately had a rush of adrenaline, an increase in respiration, and became flushed. I asked, "How bad?" The caller informed me that luckily the injuries appeared to be minor.

Of course I made a phone call or two to verify the information and to explain the situation to my wife. I was informed that Scott had some burns on his shoulders and hand, where there would be some scarring, and that he would miss a shift or two, but he would be O.K. "O.K.?" I guess that term takes on a new definition to a father who is 45 miles away and worried!

I left work early to coincide with Scott's arrival at home. He had large bandages around his shoulders and his hand, and he was beginning to feel some pain. We talked about the fire, and he told me it was a typical garden apartment-type fire and that his company had been assigned on automatic mutual aid. He and his lieutenant were in search mode because residents reported an eleven-year-old was trapped. Their backup line was right behind them, but it was not quite ready and they were; seconds became hours!

During their search, crawling on hands and knees to stay below the heat and smoke as they had learned, Scott felt something very hot on the back of his right hand, a sharp pain. He turned to his lieutenant and both of them shouted at the same time, "Let's get out of here now. It is too hot!" They exited quickly.

They were "greeted" at the door, and their coats were smoking and discolored around the shoulders. The firefighter at the door began patting the smoking gear, stating it was "on fire." The air barrier was now gone, and both the lieutenant and Scott felt the heat, big time! They both ended up with second-degree burns on both shoulders, and the hand burn resulted from the inside Nylon label on the gloves melting, even though they were compliant gloves!

The good news is that the properly designed and worn personal protective clothing prevented serious injuries, maybe even third-degree, debilitating burns. The coats, pants, and gloves could be replaced, but the scar tissue is still there. Scott received a very nice leather jacket from the PPE outer shell material manufacturer, and the damaged PPE pictures can be seen at major fire service conferences across the country.

As Scott's parents, it was a difficult day for both of us, and emotions were running high. I think sometimes we forget how our choice to be a member of the fire service can have such an impact on our family members. His mother was very concerned about the inevitable scarring and strongly suggested that he really needed to quit doing that! Scott looked her in the eye and with deep conviction in his voice stated, "Mom, it's my job." Kind of made me shiver. But I know one thing. Scott and his lieutenant learned a lot that day—about safety, about PPE, and that on the fireground you never know what might come next. As he shares what he learned with other firefighters, foremost on his mind is what he might teach to his son, Cody, the next generation of Windisch firefighters.

—Street Story by F. C. (Fred) Windisch, Fire Chief, Ponderosa VFD, Houston, Texas

OBJECTIVES

After completing this chapter, the reader should be able to:

- Define risk management.
- List the leading causes of death and injury in the fire service.
- List the NFPA standards that affect and pertain to firefighter occupational safety.
- List the five components that make up the accident chain.
- List the three key components of the safety triad.
- Discuss the difference between formal and informal procedures.
- Name the three factors that influence the equipment portion of the safety triad.
- Name the three factors that influence the personnel portion of the safety triad.
- Name the three partners that work together to achieve firefighter safety.

INTRODUCTION

Simply put, the firefighting profession is one of significant **risk**—that is, one filled with the potential to be seriously injured or killed. Typically, this potential for death has created the aura that firefighters are hero types. Unfortunately, the "hero badge" has led to unnecessary injuries and deaths. Today's firefighter understands that certain risks have no tangible benefit. In simple terms, this understanding is called **risk management**. Firefighters practice risk management when they ask questions of risk/benefit. Why risk a team of firefighters for a building that is basically lost? Why cross over a collapse-zone barrier tape just to get a better angle for a fire stream?

To better understand risk management and to improve firefighter safety, the firefighter needs to look at the common causes of injuries and deaths associated with firefighting. Additionally, an understanding of the forces that combine to make firefighting safer is imperative. With this knowledge, the firefighter can fill the "hero" role with intelligence and wisdom rather than recklessness.

SAFETY ISSUES

One way to keep firefighter injury and death events to a minimum is to understand what events and circumstances typically lead to injury.

> **Firefighter Fact** Firefighter injury and death information is invaluable and has, in many notable cases, prevented additional injuries. For example, the Hackensack (NJ) Fire Department lost five firefighters after a roof collapsed on interior firefighters battling an attic fire in an automobile service shop. This shop had a bowstring-trussed roof. Following the incident and subsequent investigations, recommendations were made by various fire service organizations to help prevent a similar incident.
>
> One Oregon chief publicly thanked the fire service organizations for their recommendations—he experienced a similar situation and withdrew his firefighters ten minutes before the roof collapsed.

A study of injury causes has inspired fire and safety professionals to create standards and regulations to help prevent injuries and deaths. These standards and regulations directly affect some of the training and tactics the fire service employs today. The firefighter who understands simple accident prevention steps is actually helping the fire service address safety issues.

Firefighter Injury and Death Causes

A study of firefighter injury and death statistics reveals that approximately one-half of all duty deaths and injuries occur at the incident scene. The other half is split between training, response to/returning from an incident, and "other" duties. This helps explain *where* firefighters get hurt and killed. The next question is what *caused* the injuries and deaths. **Figure 5-1** shows the leading causes of death, and **Figure 5-2** shows the leading causes of injuries.

Firefighter death statistics also track the nature of the injury that caused the death. These statistics have shown that heart attacks (as a result of stress) are the leading type of death-producing injury. Internal trauma, crushing injuries, and asphyxiation follow heart attacks. Data suggest that firefighter duty deaths are not noticeably decreasing—even with advances in equipment, training, and uniform procedures over the past decade. According to the U.S. Fire Administration, firefighter fatalities as a result of fire-related causes (burns, asphyxiation, and structural collapse) have actually increased.[1] This trend suggests that even more emphasis needs to be placed on firefighter safety during actual fires.

FIREFIGHTER SAFETY

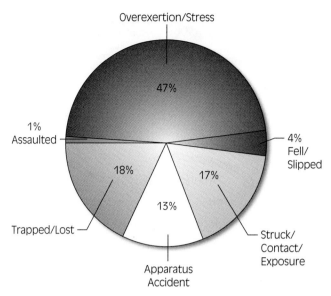

Figure 5-1 Firefighter deaths by immediate cause, 1990–2000. *(Source: U.S. Fire Administration, April 2002, Publication FA-220)*

> **Note** Unlike firefighter fatalities, firefighter injuries have not noticeably decreased—meaning, the fire service needs to increase accident prevention and risk management practices.

Safety Standards and Regulations

In 1970, the William Stieger Act was passed by congress and signed into law by President Nixon. This act created the **Occupational Safety and Health Administration (OSHA)**, which is part of the Department of Labor. OSHA is responsible for the enforcement of

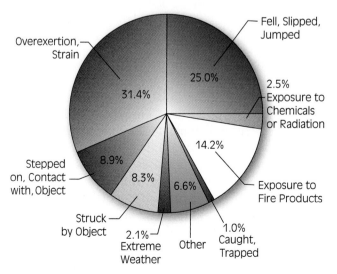

Figure 5-2 Firefighter injuries by cause—2000 sample. *(Source: NFPA, NFPA Journal, Nov/Dec 2001)*

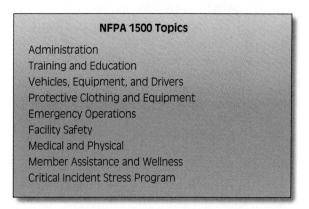

Figure 5-3 NFPA 1500, *Standard on Fire Department Occupational Safety and Health Program*, covers a multitude of topics.

safety-related regulations in the workplace. These regulations are part of the **Code of Federal Regulations (CFR)**. OSHA CFRs outline specific processes and procedures to be followed to help reduce the chance of injury. Interesting enough, most fire departments and public agencies were exempted from complying with OSHA CFRs when they were first developed. This created a double standard for occupational safety. In the 1980s, states began to address this issue by creating state OSHA plans that were compulsory for public agencies. For the first time, fire departments were obligated to follow OSHA CFRs adopted by an individual state. Unfortunately, the spontaneous nature of firefighting did not fit well into defined OSHA procedures and processes. Further, the firefighter injury and death trends of the 1970s and 1980s suggested that firefighting was the *most dangerous profession* in America. With this in mind, fire service representatives came together and helped write a comprehensive occupational health and safety standard for the fire service.

NFPA 1500, *Standard on Fire Department Occupational Safety and Health Program,* outlines consensus requirements and procedures for the safety of fire service personnel—much like OSHA CFRs do for industry. **Figure 5-3** shows the topical areas addressed by NFPA 1500. This standard was written to help fire departments address a whole host of safety issues.

> **Note** Currently, OSHA looks at NFPA standards as a guideline to address issues not directly covered by CFRs.

This OSHA/NFPA alliance is furthering the importance and accountability placed on firefighter safety. A classic example of this alliance is the requirement

NFPA Safety-Related Standards		
1500	F.D.	Occupational Safety and Health Program
1521	F.D.	Safety Officer
1561		Emergency Services Incident Management System
1581	F.D.	Infection Control Program
1582		Comprehensive Occupational Medical Program for F.D.
1583		Health Related Fitness Programs for Firefighters

Figure 5-4 The NFPA 1500 series specifically addresses firefighter safety and wellness issues.

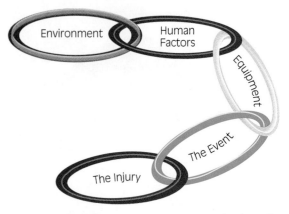

Figure 5-5 Accident prevention is simply the efforts expended to break the accident chain.

and clarification given to safe initial fire attack operations. The two-in/two-out rule is a concept underscored in OSHA 1910.134 and NFPA 1500. In essence, the rule states that before firefighters engage in interior fire attack operations, they must have two equipped and ready firefighters who form a team to make that attack. Additionally, two equipped and ready firefighters must be outside to monitor the progress of the interior crew and serve as an initial rapid intervention team should the interior firefighters experience an emergency. It is also important to note the outside team will likely need to perform some support activities (incident command or pump operations). The persons performing these tasks must be immediately available to rescue the interior team. These two are not considered a dedicated rapid intervention team—as more resources arrive, a dedicated rapid intervention team is required. This concept is covered more specifically in Chapters 19 and 23.

NFPA has also created other standards that address safety. It can be argued that all NFPA standards address safety, but specifically, NFPA has created a 1500 series that focuses on safety issues, **Figure 5-4**. The **National Institute of Occupational Safety and Health (NIOSH)** has also entered the firefighter safety equation. NIOSH does not set regulations or standards—they write recommendations based on the investigation of firefighter fatalities. In the 1990s, President Clinton signed legislation that mandated the investigation of incident-related firefighter fatalities. NIOSH is tasked with preparing recommendations so that other fire departments can take preventive steps to reduce the chance of an injury or fatality.

Accident Prevention

An **accident** can be defined as the result of a series of events and conditions that lead to an unsafe situation resulting in injury and/or property damage. This series of events and conditions is called the **accident chain**, **Figure 5-5**. The accident chain is made up of five components:

1. *The environment:* Included here are physical surroundings such as weather, surface conditions, access, lighting, and physical barriers.
2. *Human factors:* This includes human and social behaviors, training (or lack of training), fatigue, fitness, and attitudes.
3. *Equipment:* Included here are apparatus, personal protective equipment (PPE), maintenance and serviceability, proper application, and equipment limitations.
4. *The event:* The event is the intersection of the foregoing components. Something or someone had to bring those components together in such a way to create the unsafe situation.
5. *The injury:* This last part of the chain deals with the actual injury (or property damage) associated with the accident. A "near miss" or "close call" is an accident without injury or physical damage.

The key to accident prevention is awareness of these five components and the ability to "break the chain" prior to the last link. Any action designed to break the accident chain is known as an **intervention**. Intervention is typically a *reactive* action. Any strategy designed to reduce the potential of creating an accident chain is known as **mitigation**. Mitigation is typically a *proactive* action. Knowing the difference between mitigation and intervention is important because these terms are being used more and more frequently in all aspects of the fire service, **Figure 5-6**.

Most often, firefighters and fire officers are instrumental in preventing accidents through intervention.

FIREFIGHTER SAFETY 111

Intervention (Reactive)	vs.	Mitigation (Proactive)
• Incident Safety Officers • Collapse Zones • Rehab		• Protective Equipment • Training • Physical Fitness

Figure 5-6 Accident prevention relies on *proactive* and *reactive* efforts.

> **Safety** A firefighter's awareness of surroundings, proficiency in tasks, appropriate use of equipment, and, most importantly, positive attitude about safety will ultimately prevent injuries.

THE SAFETY TRIAD

Most fire service operational environments are made up of three key components: procedures, equipment, and personnel. To mitigate injuries, each of these components needs to be addressed. This mitigation effort is called the safety triad, **Figure 5-7**.

Procedures

As stated, the three basic components or elements of the safety triad come together to form an operational environment like the one found at working structure fires or multicar extrications. If the operation is to be "safe," each component needs to be addressed.

Procedures or processes are the structures from which all activity at an incident begins. The first-arriving engine at a fire alarm activation will most likely follow a set of procedures or start a series of processes to investigate the alarm. Procedures can be classified as *formal* or *informal*. Formal procedures are those that are in writing as standard operating procedures (SOPs). In some departments, formal procedures are derived from standard evolutions or lesson plans. These evolutions and lessons can be drilled periodically, on a rotating basis, to ensure that a crew's response to a given situation is appropriate. Many of these procedures and evolutions have been written down to help ensure consistency and, therefore, a greater margin of safety.

Informal procedures are those processes and operations that are obviously part of the routine of a given department, but may not be written. Informal procedures are passed on through new member training as well as day-to-day operations. One example of an informal procedure is the practice of placing a full-

(A)

(B)

(C)

Figure 5-7 The safety triad includes (A) procedures, (B) equipment *(Courtesy of Richard W. Davis)*, and (C) personnel.

face hood across bunker boots so that the firefighter must don the hood before putting on the boots. Another example is the company officer who talks through assignments at the start of each shift or when new members are assigned to the crew.

1. 1.1 Incident Command System
2. Emergency Ground Operations
   2.1 Rapid action group
   2.2 Gas/odor investigation
   2.3 Auto alarms
   2.4 Train fires
   2.5 Vehicle fire
   2.6 Fires at postal facilities
   2.7 Emergent driving procedure
   2.8 Kaneb pipeline response
   2.9 Volunteer and fire apparatus placement for motor vehicle accidents (MVAs)
   2.10 Operations involving Thompson Valley Ambulance
   2.11 Minimal staffing for interior firefighting
   2.12 Foreground formation and activation of companies
   2.13 Standard fire attack procedures/dwelling fires
3. Alarm Levels/Dispatching
   3.1 City alarm level assignments
   3.2 Rural alarm level assignments
   3.3 Fire resource officer
   3.4 Fire alarm panel operation and response policy
   3.5 Mutual/automatic aid agreement
   3.6 Staffing considerations during adverse weather conditions
   3.7 Cancellation procedures for emergency medical service (EMS) and MVA incidents
4. Hazardous Materials
   4.1 Hazardous materials operations
5. Emergency Medical Services
   5.1 Duties for non-EMS-certified personnel
6. Aircraft Rescue and Firefighting (ARFF)
   6.1 ARFF standby policy
7. Technical Rescue and Special Operations
   7.1 Vehicle extrication
   7.2 Rope
   7.3 Trench
   7.4 Collapse rescue
   7.5 Confined space
   7.6 Farm equipment and industrial rescue
   7.7 Loveland dive rescue standard operating procedures
   7.8 Use of Civil Air Patrol

Figure 5-8 Sample SOP index.

> **Safety** Both formal and informal procedures can be responsible for the overall safety of a department.

> **Safety** Making equipment safe is addressed in three ways: equipment selection, equipment inspection and maintenance, and equipment application.

Typically, SOPs are arranged in a topical format, **Figure 5-8**. Individual SOPs are formatted for easy reading and reference, **Figure 5-9**.

Equipment

In the past few years, the fire service has seen the release of vast amounts of new equipment designed uniquely for improved "safety." Equipment helps makes an operation more safe but is arguably the least important factor in the safety triad of procedures, equipment, and personnel.

Each of these is important as it relates to safety.

Equipment Selection

Most critical equipment used in firefighting (PPE, ladders, hose, etc.) is designed and built to meet NFPA standards. NFPA standards are written with safety paramount to the topic. The fire department that purchases equipment that meets NFPA standards is buying equipment that has a certain safety element built into it. **Figure 5-10** lists common equipment that is designed to improve safety.

> **Purpose:**
> To establish policy and direction to all department members regarding minimal staffing and resource allocation for safe and aggressive interior structural firefighting.
>
> **Responsibility:**
> It is the responsibility of all officers and firefighters engaged in firefighting operations to adhere to this policy. The Incident Commander is accountable for procedure included within this policy.
>
> **Procedure:**
> 1. This policy is applicable to situations where the Incident Commander (IC) has made a tactical decision to initiate an **offensive fire attack**, by firefighters, inside the structure. Additionally, tactical firefighting assignments that expose firefighters to an atmosphere that is immediately dangerous to life and health (IDLH) dictate the application of this policy.[1]
>
> 2. Prior to initiating interior fire attack or exposure of firefighters to an IDLH atmosphere, a **minimum of four (4) firefighters shall assemble on scene**.[2] These four members shall utilize a "two-in, two-out" concept.
>
> 3. The **"two-in"** firefighters that enter the IDLH atmosphere shall remain as partners in close proximity to each other, generally fulfilling the operational role as the **FIRE ATTACK GROUP**. As a minimum, the **"two-in"** firefighters entering the IDLH atmosphere shall have full PPE, with SCBA and PASS devices engaged, and have between them a two-way portable radio, forcible entry tool, and flashlight or lantern.
>
> [1] An **IDLH atmosphere** can be defined as an atmosphere that would cause immediate health risks to a person who did not have **Personal Protective Equipment (PPE)** and/or **Self-Contained Breathing Apparatus (SCBA)**. This includes smoke, fire gases, oxygen deficient atmospheres, or hazardous materials environments. For Loveland Fire and Rescue application, an IDLH atmosphere can be further defined as an environment that is **suspected** to be IDLH, has been **confirmed** to be IDLH, or **may rapidly become** IDLH. The use of full protective equipment including an activated SCBA and an armed PASS device is mandatory for anyone working in or near an IDLH atmosphere.
>
> [2] The firefighters must be SCBA qualified and capable of operating inside fire buildings without immediate supervision.

Figure 5-9 Sample SOP format.

Firefighter Safety Equipment

PPE
Accountability Tags
Nomex/PBI Clothing
EMS Gloves/Masks
SCBA
Goggles

Apparatus
Enclosed Cabs
Headsets
Automatic Vehicle Locator
Reflective Trim
Swing-Down Tool Trays

Tools
Multi-Gas Detectors
Two-Way Radios
Infrared Cameras
Disposable EMS Adjuncts
Rehab Kits

Figure 5-10 Fire departments spend thousands of dollars on equipment designed to enhance firefighter safety.

Equipment Inspection and Maintenance

For equipment to be safe, it must be inspected and maintained. Firefighters spend many hours each week ensuring that their equipment is clean, properly functional, and ready for rough service during critical situations, **Figure 5-11**.

Because many different firefighters may use and maintain a given piece of equipment, complete documentation of repairs and maintenance is essential. Further, a complete set of guidelines is often developed or adopted for essential equipment. These guidelines include:

- Selection
- Use

Figure 5-11 Ensuring that equipment is clean, functional, and well maintained is essential to firefighter safety.

- Cleaning and decontamination
- Storage
- Inspection
- Repairs
- Criteria for retirement

Equipment Application

Choosing the right tool for a given job is paramount for safety. It is imperative to use equipment in the manner for which it was designed. Using an ax as a prying tool can cause the handle to shatter. At times, the ingenuity and inventiveness of firefighters facing an unusual situation places equipment in a position for which it may not be intended. Extra care and insight are required here.

> **Safety** Firefighters must always prepare for the worst—including equipment failure.

Personnel

Human factors are often cited as the cause of injuries and deaths. To mitigate these injuries and deaths, the safety triad must address personnel issues. A firefighter's training, fitness/health, and attitude all factor into the safety equation.

Training

> **Note** Without a doubt, the single most important step that can be taken to reduce firefighter injuries is that of regular training. It is widely accepted that once a person enters the fire service, the training *never* ends.

Unfortunately, many firefighters become comfortable in their knowledge or position and let the basics slip away unknown. Each and every skill, knowledge item, or behavior that is learned in basic academy should be drilled, tested, or confirmed every year. In a perfect example, the New York City Fire Department (FDNY) launched a huge "back-to-the basics" program after a series of firefighter duty deaths within a year. The FDNY line leadership felt that many of the deaths could have been avoided had firefighters and fire officers practiced the "basics." Many departments utilize company drills on a twice-monthly basis in order to ensure that the "basics" are practiced, **Figure 5-12**.

Because regular training is the single most important step in firefighter safety, the firefighter must strive to retain the information and skills that are presented in training sessions. Here are some strategies for firefighters to help retain essential skills and information and thereby improve safety:

- *Always take notes and keep handouts.* Establish a library and/or filing system for the volumes of notes and information collected during fire service tenure. These notes will provide a handy reference for future recall and assist the firefighter in imprinting information.
- *Envision the application of all training.* After a training session, firefighters should visualize themselves using the skill or information on an actual emergency or scenario. This can be achieved alone or as part of a group. Firefighters should ask themselves what building, situation, or series of events will place them in a position to use the given training.
- *Acknowledge that skills will diminish over time if they are not used.* An old fire service adage says firefighters need to spend 95 percent of their training time practicing for what they do

FIREFIGHTER SAFETY

Figure 5-12 Basic skills need to be practiced on a regular basis.

Essential Training Subjects for Increased Incident Safety	
Subject	**Degree of Understanding**
• Personal Protective Equipment	Mastery
• Accountability Systems	Mastery
• Company Formation and Team Continuity	Mastery
• Fire Behavior and Phenomena	Proficient
• Incident Management Systems	Proficient
• Apparatus Driving	Proficient Under Stress
• Fitness and Rehabilitation	Practitioner

Figure 5-13 Firefighter safety is directly related to the training of essential topics.

5 percent of the time. Arguments can be made on which training subjects or behaviors are most important to safe operations; however, a common list can be created based on firefighter injury and death statistics. **Figure 5-13** lists training subjects that directly affect incident safety—if these subjects are practiced and appropriately applied, incident operations will become safer. An expectation of the depth of understanding and the methods needed to achieve this are also suggested.

Firefighter Fitness and Health

The safety and well-being of any firefighter increases with the health of the individual firefighter. Much has been written on the benefits of healthy firefighters—most of which centers on *physical* health.

> **Caution** "Stress" continues to lead the list of causes of firefighter duty deaths and is a significant contributor in injuries.

To handle the inherent stress of firefighting, each firefighter's body must be accustomed to and capable of handling stress. Additionally, firefighters need to protect themselves from, and prevent the spread of, communicable diseases and infections. Keys to improving physical health, and therefore firefighter safety, include these:

- *Annual health screening for all firefighters and line officers.* These physicals should also include vaccination and immunization offerings, blood screening, and stress testing.
- ***Work hardening*** *and mandatory ongoing fitness programs.* A personal devotion to physical fitness is essential to firefighter safety. Fitness includes improving flexibility, strength, and cardiovascular endurance. Work hardening is the effort and physical training put forth to better perform physical tasks without overstressing or injuring an individual. The firefighter who addresses these fitness areas will breeze through any "fit-for-duty" agility tests that a fire department may require as a condition of employment or association. Improper lifting techniques and slip and fall accidents are two of the most common causes of injury on duty. *Work hardening helps prevent these injuries through strength, balance, and coordination improvement,* **Figure 5-14.**
- *Firefighter fueling (nutrition) education.* Nutrition is important not only to lifestyle changes (like losing excessive fat) but also to incident readiness.

> **Caution** Diet fads and meal-replacement aids may help an individual lose weight but often interfere with maintaining energy during working incidents.

Figure 5-14 Work hardening must become a habit—leading to increased safety and professional tenure.

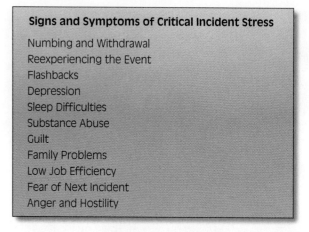

Figure 5-15 Critical incident stress is inevitable and unpreventable. CISM can be addressed through peer support and formal debriefings.

This is especially true with the sporadic nature of the firefighter's energy demands. Chapter 23, Firefighter Survival, more clearly defines appropriate fueling of the firefighter.

Attention to *physical health* is indeed important, but *mental health* is also important to firefighter safety. Keys to addressing mental health include the following:

- *Training and understanding of critical incident stress management (CISM)*. Knowing the signs and symptoms of critical incident stress is important, **Figure 5-15**. Most departments have developed a process to have a **critical incident stress debriefing (CISD)** team activated or available for unusual events or at the request of one or more responders.
- *Making an employee assistance program (EAP) available*. An EAP offers individuals a confidential approach to dealing with situations and problems that can affect job performance. Drug and alcohol dependence, depression, worker relationships, and job stress are a few of the issues that can be addressed through an EAP.

Attitude

Of all the "people" factors affecting safety, attitude is the hardest to address. Many factors affect the attitude of a given individual, not to mention the fact that attitudes are dynamic. Here are some factors that affect safety attitudes:

- *The fire department's safety "culture."* The culture of an organization is reflected in the ideas, skills, and customs that are passed through generations. If safety has not been given a high priority by administrators and members, a poor safety attitude will be reflected in daily activities. The inverse is also true.
- *The fire department's history.* A department that has experienced a firefighter duty-related death or serious injury is more likely than not to have increased safety awareness. Likewise, some departments have taken from others' experiences—this is known as having a **vicarious experience**. Realizing that a potential for injury or death exists is an important first step in developing a proactive safety attitude.
- *The example set by others.* Firefighters and line officers display their safety attitudes in what they do rather than what they say. These examples will shape the attitudes of others, **Figure 5-16**.

Ethics It is best to shape a positive safety attitude early in a career rather than later. Likewise, the exciting period just following a promotion is ideal for setting in good habits.

FIREFIGHTER SAFETY   117

Figure 5-16 Developing a positive safety attitude and practicing safe habits will demonstrate safe examples to others.

The following are steps a firefighter can take to create a positive safety attitude:

- *Practice good habits.* Each and every incident response and training activity results in opportunities to practice and reinforce good habits. PPE use, readiness, team formation, and assignment follow-through are all items that lead to a more positive safety attitude. These habits will serve the firefighter well when the incident becomes more complex or dangerous.
- *Learn from others.* Each month, trade magazines and investigative reports outline circumstances that have led to a firefighter injury or death. These serve as a constant reminder that an injury can happen. Firefighters can learn these lessons vicariously. Significant fatality and injury reports can be found at the NIOSH Web site, http://www.cdc.gov/niosh.
- *Be vigilant.* Inactivity, complacency, and overconfidence have been cited as primary factors in many firefighter fatalities. Often, merely visualizing an injury will help firefighters regain focus and obtain vigilance in checking their safety attitude.

FIREFIGHTER SAFETY RESPONSIBILITIES

Firefighter safety is dependent on the efforts of everyone. Specifically, responsibility for firefighter safety rests in one of three areas: the department itself (administration), the working team, and the individual firefighter, **Figure 5-17**. Knowing these responsibilities—especially where individuals fit—will help achieve firefighter safety. Firefighter safety is achieved if, and only if, all "partners" hold up their end of the safety partnership.

The Department

The responsibility for firefighter safety ultimately rests with the department's leadership. This is an incredible weight to carry. Fire chiefs must impose rules, regulations, and expectations in order to address the issue of firefighter safety. Additionally, a department's leadership must obtain budget appropriations to purchase equipment and apparatus that will ultimately help make firefighting safer. Often, a fire chief will appoint a health and safety officer to specifically address firefighter safety issues. In a nutshell, NFPA 1500 outlines what firefighter safety components are needed as part of an occupational safety and health program.

To hold up the department's end of the safety partnership, the fire chief, health and safety officer, or other administrative officer should create a health and safety committee, develop standard procedures and guidelines, and implement a risk management plan. Additionally, the department should research and purchase safety equipment, as well as develop and deliver hazard awareness training. The following paragraphs address each of these department responsibility areas.

Creation of a Health and Safety Committee

This group is usually made up of personnel from all levels. The focus here is to identify and create solutions for safety issues through a committee process.

Development of SOPs

Defining a proper and expected level of procedure or behavior is an important step in addressing operations that can cause injury. It is common to have safety SOPs that address these areas:

- PPE
- Firefighter injuries
- Training safely

Figure 5-17 Firefighter safety is dependent on all partners holding up their responsibilities: (A) administration, (B) teams, and (C) individual firefighters.

- Zoning
- Evacuation
- Emergent driving
- Slip and fall hazards
- Self-contained breathing apparatus (SCBA) use and care
- Accountability systems
- Incident command
- Riding apparatus
- Lifting heavy objects
- Power tool use
- Infection control
- Hazard reporting.

Implementation of a Risk Management Plan

NFPA 1500 requires that the department develop, implement, and train all firefighters on a risk management plan. This plan is developed based on local needs and includes expectations from the community. Typically, though, the risk management plan acknowledges that certain risks are inherent in firefighting. These risks can be categorized, however. An example of this risk/benefit philosophy follows:

- Significant risk to the life of a firefighter shall be limited to those situations where the firefighter can potentially save endangered lives.
- Situations endangering valued property shall cause firefighters to take a calculated and weighted risk.
- Where no life or valued property can be saved, no risk shall be taken by firefighters.

Research and Purchase of Appropriate Apparatus and Equipment

Firefighting requires rugged, specially designed equipment that ensures a certain level of reliability and safety for the user. NFPA standards outline some of the requirements that make apparatus and equipment safer. This equipment is often very expensive and requires a significant effort to obtain funds.

Development and Delivery of Hazard Awareness Training

Virtually all training is designed to help firefighters operate in a safe manner. The training effort must also inform firefighters of the hazards they may face on any particular type of incident.

The Team

The department as a whole cannot be effective in its firefighter safety effort without the support of a team approach to handling incidents. Therefore, the team that handles an incident must hold up its part of the safety partnership. This team obviously includes the individual firefighter. To ensure safety, the team should follow these procedures:

1. Utilize an **incident management system (IMS)**. The IMS should detail lines of authority and communication, be reasonable in its span of control, and include an action plan. Company officers, sector officers, or group supervisors need to communicate progress and status. Orders need to be clear and accomplished within an acceptable risk environment.

2. Work together and remain intact. The "buddy" concept is imperative for firefighter safety.

> **Caution** The separation of members within a team is a contributing factor to firefighter fatalities.

3. Look after each other. *Team members who continuously watch each other will find and address conditions that will lead to injury.* Aggressive rehabilitation will provide hydration, rest, and nourishment—thereby increasing the time the team can perform safely. At the team level, issues of PPE and SCBA use or misuse should be addressed. Likewise, the team must watch each other for signs of fatigue, overaggressiveness that is dangerous, and freelancing.

The Individual Firefighter

The individual firefighter holds the final key to making the safety partnership work. The following paragraphs outline *how* a firefighter can contribute to individual safety and the safety of the team.

Be Ready

> **Streetsmart Tip** A firefighter's readiness includes not only wearing PPE and SCBA as appropriate, but also being ready mentally and physically.

Reporting to duty or an incident with a physical limitation will increase everyone's danger. Firefighters working while sick or under the influence of certain over-the-counter and prescribed drugs can experience less than clear thinking as well as injury. Likewise, mental barriers and "baggage" will likely impair a firefighter's ability to think and act.

Understand and Act within the Chain of Command

For an incident management system to work, each individual must fill a role and not operate outside of it. Each person is responsible to someone else. Crossing these lines of responsibility causes confusion—a deadly proposal. In addition to understanding the chain of command, firefighters must clearly understand their orders (tasks). It is good practice to repeat the orders that have been given and to ask for clarification if those orders are not clear.

Perform as Trained

Incident task needs may place a firefighter in a position to perform a skill or task that the individual may or may not have been trained for. This should not be allowed to happen.

> **Safety** If a firefighter has not been trained for a task that is required, the firefighter should tell the direct on-scene supervisor.

The incident scene is rarely the place for on-the-job training of critical skills. Firefighters should practice individual skills—knot tying, ladder raising, and tool use can deteriorate over time. Additionally, the department has outlined SOPs to be followed—firefighters should do so.

Do Not Freelance

Fire and rescue work is a team effort. A leader guides a team and the leader must follow the incident commander's action plan. Working alone or outside the action plan endangers individuals and the team, **Figure 5-18**.

Figure 5-18 Freelancing endangers individuals and the team. *Never work alone!*

Use an Incident Engagement Checklist

At the individual level, firefighter safety can be achieved if firefighters practice a standard approach to incidents—that is, doing things in a standard way all the time. Using an engagement checklist, **Figure 5-19,** helps achieve a standard approach to incidents.

> **Safety** Riding the fire apparatus is one of the most common dangers the firefighter experiences. Routinely, firefighters are injured or killed while riding to and from incidents. This may seem incongruent to a profession dedicated to preserving life. A firefighter can minimize the chance of injury by following a few simple steps. The first is to always wear a seat belt! While most departments have purchased apparatus with fully enclosed cabs, some firefighters still "jump on" the tailboard when the fire apparatus is shuttling people during training drills and hose-laying activities. The firefighter must resist the temptation to ride the tailboard; this is considered a major safety violation for most fire departments. Finally, if a firefighter rides in fire apparatus that is not equipped with a fully enclosed crew compartment (open jump seat), it is important to use eye and ear protection as well as any installed safety bar or restraint device. In all cases, it is common practice for the firefighter to remain seated and belted until the officer or driver gives instructions to dismount.

> **Incident Engagement Checklist**
> - Don personal protective equipment appropriate for the response.
> - Mount apparatus using handholds.
> - Buckle seat belt, don headset—report you are ready.
> - Listen to radio reports and listen for details.
> - Mentally run through tasks that might be required for the type of incident.
> - Don SCBA or collect tools, gloves, etc.., if appropriate and if this can be done with seat belt on.
> - Upon arrival, listen for orders.
> - Unbuckle only when orders have been given.
> - Prior to leaving the apparatus, look out the window and make sure the path is clear. Watch for traffic when dismounting.
> - Close all doors and compartments after retrieving tools and/or equipment.
> - Scan the environment for overhead wires, trip/fall hazards, and unusual circumstances.
> - Know who you are reporting to—make sure your accountability tag/passport is processed.
> - Proceed with orders.
> - Make sure you can be seen and heard by other firefighters at all times during an incident.

Figure 5-19 Firefighters should perform a mental "incident engagement checklist" for every response.

Lessons Learned

The issue of firefighter safety is dependent on many factors. The fact that roughly half of injuries and deaths occur at incident scenes demonstrates that firefighting is a dangerous profession. The majority of these deaths and injuries are a result of stress. To help prevent injuries and death, the fire service is required to follow federal regulations called OSHA CFRs. In addition, NFPA standards, like NFPA 1500, have been written and adopted to further the firefighter safety effort.

Accident prevention is actually the process of mitigating hazards or intervening with the accident chain. Within any operational environment, the safety triad is used to address issues. To make sure the safety triad is effective, a partnership between the department administration, the working teams, and the individual firefighter is formed. Each of these partnerships carries equal weight in creating a safe atmosphere. It is then the individual's challenge to develop safe habits and be ever mindful of the events and conditions that cause injury.

KEY TERMS

Accident The result of a series of events and conditions that lead to an unsafe situation resulting in injury and/or property damage.

Accident Chain A series of events and conditions that can lead to or have led to an accident. These events and conditions are typically classified into five areas: environment, human factors, equipment, events, and injury.

Code of Federal Regulations (CFR) The documents that include federally promulgated regulations for all federal agencies.

Critical Incident Stress Debriefing (CISD) A formal gathering of incident responders to help defuse and address stress from a given incident.

Critical Incident Stress Management (CISM) A process for managing the short- and long-term effects of critical incident stress reactions.

Employee Assistance Program (EAP) A defined program that offers professional mental health and other health services to employees.

Incident Management System (IMS) An expandable management system used to deal with a myriad of incidents. Helps achieve the highest level of accountability and effectiveness for incident handling. Limits span of control and provides a framework to help make tasks manageable.

Intervention The act of intervening; to come between as an influencing force. Typically a reactive action.

Mitigation Actions taken to eliminate a hazard or make a hazard less severe or less likely to cause harm. Typically a proactive action.

National Institute for Occupational Safety and Health (NIOSH) A federal institute tasked with investigating firefighter fatalities and making recommendations to prevent reoccurrence.

Occupational Safety and Health Administration (OSHA) The federal agency, under the Department of Labor, that is responsible for employee occupational safety.

Risk The chance of injury, damage, or loss; hazard.

Risk Management The process of minimizing the chance, degree, or probability of damage, loss, or injury.

Vicarious Experience A shared experience by imagined participation in another's experience.

Work Hardening A phrase given to the effort and physical training designed to prepare an individual to better perform the physical tasks that are expected of the individual. Work hardening is key in preventing injuries resulting from typical firefighting tasks.

REVIEW QUESTIONS

1. Define risk management and the concept of risk/benefit.
2. List the leading causes of death and injury in the fire service.
3. List the NFPA standards that affect and pertain to firefighter occupational safety.
4. List the five components that make up the accident chain and discuss the difference between mitigation and intervention.
5. List and briefly describe the three key components of the safety triad.
6. Explain the difference between formal and informal procedures.
7. Name the three factors that influence the equipment portion of the safety triad.
8. Name the three factors that influence the personnel portion of the safety triad.
9. Name the three partners that work together to achieve firefighter safety and give examples of how the individual firefighter can contribute to firefighter safety.

Endnote

1. U.S. Fire Administration, Firefighter Fatality Retrospective Study, FA-220, USFA, Washington, DC, April 2002.

Additional Resources

Angle, James S., *Safety in the Emergency Services*. Delmar Learning, a part of the Thomson Coproration, Clifton Park, New York, 1999.

Dodson, David W., *Fire Department Incident Safety Officer*. Delmar Learning, a part of the Thomson Coproration, Clifton Park, New York, 1999.

National Institute for Occupational Safety and Health: http://www.cdc.gov/niosh– to obtain firefighter fatality reports and recommendations.

NFPA 1500: Standard on Fire Department Occupational Safety and Health Program, National Fire Protection Association, Quincy, MA, 2002.

Risk Management Practices for the Fire Service, FA-166. United States Fire Administration, Emmitsburg, MD, 1996.

CHAPTER 6

PERSONAL PROTECTIVE CLOTHING AND ENSEMBLES

David Dodson, Lead Instructor, Response Solutions, Colorado

 OUTLINE

- Objectives
- Introduction
- Personal Protective Equipment Factors
- Types of Personal Protective Equipment
- Care and Maintenance of Personal Protective Equipment
- Personal Protective Equipment Effectiveness: "Street Smarts"
- Lessons Learned
- Key Terms
- Review Questions
- Additional Resources

STREET STORY

As the department's training officer, I served as the safety officer on all major incidents. One afternoon while working in my office at our suppression headquarters station, I heard Engine 4 dispatched to a vegetation fire. Engine 4 arrived on scene to report approximately one acre involved, with the fire rapidly spreading. They requested two additional engines. The paramedics I was working with decided to drive down the street to observe, as we were on the opposite side of the involved area. I decided to ride with them. I did not plan on this being a major incident and did not take any protective clothing with me.

As we approached the area across from the fire, we noticed that the fire had spread across the ravine and was now threatening homes and utility services. An engine had laid lines, and one firefighter was attempting to put a 2½-inch line into service to protect the exposures. The incident commander was calling for two strike teams. The paramedics were given an assignment to help extend lines to the right flank. I jumped in to assist with the exposure line. My protective equipment consisted of slacks, a dress shirt, a tie, and shoes. Within minutes we were exposed to heat and smoke thick enough to cause difficulty breathing. Through the smoke an engineer brought me his turnout coat and helmet. He later gave me his turnout boots and stood at the pump panel in stockinged feet.

Not only was I completely ineffective as a safety officer, I was a safety hazard myself! By not having my protective gear, I created a safety hazard for myself and took away safety gear from another firefighter. The few minutes it would have taken for me to grab my gear would have been more than worthwhile. Our protective clothing is as essential to us as a scalpel is to a surgeon. The lesson I learned was never to respond to an incident without my safety gear and never to enter the hazard environment without it. Fortunately, no one was injured and the fire was extinguished with no loss of structures.

—Street Story by Randy Scheerer, Battalion Chief, Newport Beach Fire Department, Newport Beach, California

PERSONAL PROTECTIVE CLOTHING AND ENSEMBLES

OBJECTIVES

After completing this chapter, the reader should be able to:

- Describe the role of personal protective equipment for firefighters.
- Define the relationship between PPE and national standards and regulations.
- List the components and unique elements of structural, proximity, and wildland PPE ensembles.
- Describe a serviceability inspection of structural PPE.
- Describe the conditions and damage that render structural PPE unserviceable.
- Given a structural PPE ensemble, appropriately don the ensemble within one minute.
- Demonstrate a team check following PPE donning.

INTRODUCTION

Firefighters and emergency medical providers respond to incidents that are often immediately dangerous to life and health. The vernacular for this is IDLH, **Figure 6-1**. Often, the difference between injury and safety is determined by the responder's personal protective equipment (PPE). It is important to note, however, that PPE provides a minimum level of protection and should be considered the *last resort* of protection for firefighters and emergency responders operating at an incident. Proper fire streams, **zoning**, and sound tactics and procedures should provide a greater measure of safety for teams, especially in IDLH atmospheres. Simply stated, PPE is the *first* thing a firefighter puts on to deal with an incident and the *last* thing taken off when the incident is over.

Personal protective equipment for the firefighter can take on many forms. Clothing, helmets, hoods, gloves, harnesses, **personal alert safety systems (PASS)**, and self-contained breathing apparatus (SCBA) are just some of examples of the PPE used by firefighters. Further, defined collections of PPE make up ensembles that are designed for specific firefighter hazards. These ensembles include structural, proximity, and wildland PPE. Other ensembles may include technical rescue, ice rescue, and swift-water rescue gear. Each of these ensembles is discussed later in this chapter. Ensembles such as EMS infection control and hazardous materials PPE are used by many fire departments. It is important to note that each ensemble or piece of protective equipment is designed to meet a minimum level of safety and each has specific limitations that govern performance of the equipment.

> **Caution** Failure to operate within a PPE component's designed limitations can lead to an injury, illness, or perhaps death of the user.

Pushing the limitations of a PPE component is not the only way PPE-related injuries occur. Injuries and illnesses have been suffered by the firefighter who fails to properly don and secure PPE—usually because the wearer was trying to create a "macho" image or skipped complete donning in the haste to perform a task.

This chapter starts with a discussion of the factors that influence PPE design and use, including national standards and regulations. Next, the various

Figure 6-1 A hostile fire within a structure creates an IDLH environment. Personal protective equipment can help the firefighter work in an IDLH environment.

types of firefighter PPE are outlined followed by care and maintenance of PPE. This chapter then discusses how firefighters can effectively use PPE and keep their first and last defense intact. SCBA—although an important component of PPE—is discussed in Chapter 7. Other PPE items such as those for emergency medical operations and hazardous materials operations are discussed in later chapters.

PERSONAL PROTECTIVE EQUIPMENT FACTORS

Firefighter personal protective equipment has evolved significantly during the past two decades. The reasons for these recent improvements are varied. A study of the history of injuries, evolution of materials, and sound risk management practices have led to vast improvements in protective equipment. Modern PPE has been developed as a result of the direct efforts of labor groups (most notably, the International Association of Firefighters, IAFF), other fire service membership associations, equipment manufacturers, and government entities such as NASA. These efforts usually come together as a result of a consensus process that establishes minimum standards. The National Fire Protection Association (NFPA) provides the forum for this consensus building. Although NFPA standards certainly influence PPE design and use, the federal government has also become involved in the PPE equation through the development of regulations and guidelines.

Standards and Regulations

The NFPA has developed numerous standards for firefighter protective equipment and ensembles, **Figure 6-2**. Typically, these standards cover component parts, manufacturers' quality assurance and labeling requirements, design requirements, user information, performance requirements, and test methods. The specific NFPA standards listed in **Figure 6-2** primarily address the design, performance, and manufacturing of PPE. Additionally, NFPA has developed an important PPE "use" standard. NFPA 1500, *Standard on Fire Department Occupational Safety and Health Program,* dedicates a whole chapter to the use, care, and maintenance of many forms of protective equipment.

All equipment worn by a firefighter should meet current applicable standards. Firefighters

NFPA Standards That Address PPE and Ensembles	
1500	Fire Department Occupational Safety and Health Program
1971	Protective Ensemble for Structural Firefighting
1975	Station/Work Uniforms for Firefighters
1976	Protective Clothing for Proximity Firefighting
1977	Protective Clothing and Equipment for Wildland Firefighting
1981	Open-Circuit Self-Contained Breathing Apparatus for the Fire Service
1982	Personal Alert Safety Systems
1983	Life Safety Rope and System Components
1991	Vapor-Protective Hazardous Ensembles/Materials Emergencies
1999	Protective Clothing for Medical Emergency Operations

Figure 6-2 NFPA develops consensus standards that address PPE.

should check PPE components for a conspicuous and permanently attached label that verifies that a component meets an applicable NFPA standard, **Figure 6-3**.

The federal government is also involved in PPE use. Of primary importance is the involvement of the Occupational Safety and Health Administration (OSHA). OSHA is responsible for the development and enforcement of regulations—namely, the Code of Federal Regulations (CFRs)—that govern safe work practices. At the firefighter level, this involvement essentially means that failure to use appropriate PPE in certain applications (e.g., IDLH atmospheres) means that a federal regulation has been violated, leading to a potentially expensive fine and disciplinary actions against the firefighter and the department. Other government and ancillary agencies that are involved in protective equipment issues include the Environmental Protection Agency, Centers for Disease Control and Prevention, American National Standards Institute, American Society for Testing and Materials, and the National Institute for Occupational Safety and Health.

PERSONAL PROTECTIVE CLOTHING AND ENSEMBLES

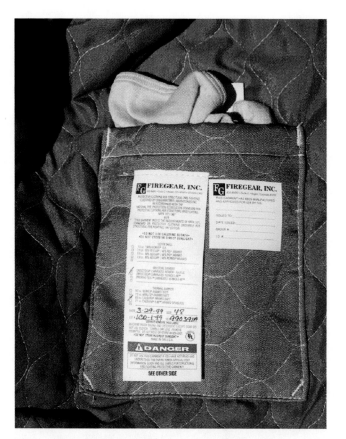

Figure 6-3 NFPA-compliant PPE components should have a permanently affixed label.

TYPES OF PERSONAL PROTECTIVE EQUIPMENT

Ensembles

Within the fire service, many personal protective ensembles have been developed. The NFPA has developed minimum ensemble standards for structural firefighting PPE, proximity PPE (commonly used for aircraft rescue and firefighting), and wildland PPE. NFPA also addresses specialized ensembles that exist for technical rescue such as Urban Search and Rescue (USAR). The important point to underscore is the word *ensemble*. Frequently, firefighters short-circuit the concept of an "ensemble" in the use of their PPE. An action as simple as forgetting to secure the top fastener of a protective coat violates the concept of an ensemble and can lead to an injury.

> **Safety** To be effective, all components of a PPE ensemble must be utilized as recommended by the manufacturer. Failure to complete an ensemble can cause an injury.

Structural Ensemble

The protective ensemble for structural firefighting includes all of the components listed and shown in **Figure 6-4**. For the sake of clarification, NFPA defines structural firefighting as "the activities of rescue, fire suppression, and property conservation in buildings, enclosed structures, aircraft interiors, vehicles, vessels, or like properties that are involved in a fire or emergency situation."

Structural PPE is commonly referred to as **bunkers**. The term *bunkers* was originally associated with short boots and protective pants that

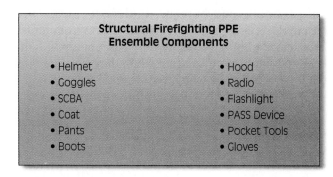

Structural Firefighting PPE Ensemble Components

- Helmet
- Goggles
- SCBA
- Coat
- Pants
- Boots
- Hood
- Radio
- Flashlight
- PASS Device
- Pocket Tools
- Gloves

Figure 6-4 A full structural firefighting ensemble includes more than the NFPA minimum required components.

duty firefighters would only wear at night—donned in the bunk room of a fire station. During the day, these same firefighters wore long coats and three-quarter pull-up boots—a practice no longer allowed by NFPA 1500 due to the inadequate protection offered to the wearer. (Many serious burns have been experienced by firefighters using this antiquated type of clothing.) Today, many terms are used to refer to the structural firefighting ensemble. These terms include the previously mentioned *bunkers* as well as *turnouts* and *structural gear.* Some highlights of the structural ensemble are discussed next.

Coats and Trousers. The heart of structural PPE is the coat and pant combination. Both components rely on a layered protection system that includes a fire-resistive outer shell, vapor barrier, and thermal barrier.

> **Safety** All three layers of the pant and trouser assembly must be intact—failure to wear the entire layer system may cause injury or death during fire suppression operations.

The three layers help the coat and pant meet thermal protective criteria—insulation that minimizes the chance that the wearer will be burned. This criterion is called **thermal protective performance (TPP).** The minimal TPP for structural coats and pants is 35. In simple terms, this means the wearer has 35 seconds of protection before a second-degree burn is likely to be sustained when exposed to flames. Dynamic influences such as dirty gear, moisture (including perspiration), and fabric compression (from an SCBA) can reduce the TPP of the clothing.

The coat and trouser combination has reflective trim to increase the visibility of the wearer to others, **Figure 6-5.** Flaps, enclosures, wristlets, and fastening devices are all designed to seal the ensemble and to provide a protective interface with gloves, hoods, boots, and helmets. Failure to "seal" the ensemble invites injury in hazardous environments. Trousers should have heavy-duty suspenders to help keep the pants from sagging when they become wet with perspiration or water.

Shoulder padding, knee reinforcements, and various types of pockets and accessory attachments can also be found on coats and trousers. Some firefighters have had webbing sewn into their structural trousers to create a "sit harness" for certain rescue situations. This customizing should only be performed by a certified manufacturer that understands and complies with NFPA standards for both struc-

Figure 6-5 Reflective trim increases firefighter visibility to others during low-light activities.

tural PPE and life safety system components (NFPA 1983, *Standard on Fire Service Life Safety Rope and System Components*).

Helmets. The classic firefighter's helmet was designed to help shed water and to prevent hot embers from falling down on the firefighter's neck, back, and ears. This classic design remains in most styles of helmets. Newer helmet safety features exceed those of yesteryear in many ways and include impact resistance; thermal insulation; earflaps for a layered interface with hoods, coats, and SCBA face pieces; chin straps; and clear or tinted face shields and/or eye protection accessories. The chin strap is an important part of the helmet—it must be utilized to take full advantage of the helmet's safety feature. Failure to utilize the chin strap is an invitation to injury or loss of the helmet during critical operations.

Gloves. Hand protection is essential in the structural ensemble. Gloves meeting NFPA standards for structural firefighting must provide thermal protection as well as protection from cuts, punctures, and scrapes. Unfortunately, dexterity is almost always reduced when wearing structural gloves. This is a common complaint of firefighters.

> **Streetsmart Tip** Two things can improve dexterity with gloves: good fit and practice. It is important to note that dexterity when wearing gloves increases with practice—thanks to muscle memory.

The simple act of practicing hands-on tool use with gloves on can increase muscle memory and reduce the frustration of lost dexterity.

Footwear. The firefighter's choices for approved and effective footwear are growing. Now firefighters can choose from traditional rubber-like boots or leather-type boots. Each has advantages and disadvantages, **Figure 6-6**. Both must meet NFPA standards for foot protection.

Protective Hoods. An important interface that creates an encapsulating link to the firefighter's helmet, coat, and SCBA face piece is the protective hood. Hoods are made of fire-resistive, form-fitting cloth that protects the face, ears, hair, and neck in areas not covered by the helmet, earflaps, and coat collar, **Figure 6-7**. It is important to note that structural protective hoods have a TPP less than that of a structural coat (20 for a hood versus 35 for a coat). Additionally, hoods are designed to be worn *over* the straps of an SCBA face piece. This practice helps protect the SCBA face piece straps and helps keep the hood from binding.

Miscellaneous Structural PPE. While not necessarily covered by the NFPA standards, many seasoned firefighters feel that certain accessories are as much a part of the full PPE ensemble as the coat, pants, and helmet. These items include primary eye protection (goggles or safety glasses), hearing protection, PASS devices, and pocket tools, **Figure 6-8**.

Proximity Ensemble

The proximity firefighting PPE ensemble is most often associated with aircraft rescue and firefighting (ARFF). Proximity gear utilizes an aluminized coating to help reflect radiant heat, **Figure 6-9**.

Common Rubber Boot
- Easy to Don
- Excellent Water Repellency
- Easy Decontamination
- Inexpensive
- Sloppy Fit

Leather Pull-Up Boot
- Lightweight
- Durable
- Comfortable
- Minimal Ankle Support

Leather Lace-Up Boot
- Tight Fit
- Ankle Support
- Durable
- Expensive

Figure 6-6 Firefighters have choices for structural firefighting footwear—each with its own advantages and disadvantages.

Figure 6-7 A protective hood provides an interface layer that links the coat collar, helmet flaps, and opening for an SCBA mask.

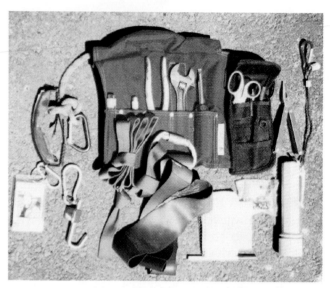

Figure 6-8 Personal protective equipment can also include many tools that firefighters pocket.

Although similar in many ways to structural PPE, the proximity gear must meet more stringent heat reflection and wearer insulation standards.

> **Note** It is important to note that proximity PPE is *not* designed for fire entry (to be totally enveloped by fire).

The aluminized fabric allows the wearer to get closer to fires that emit extreme radiant heat such as that found with petroleum-based fires. Bulk fuel facilities, airports, and certain chemical plants are complexes where the assigned fire attack teams may be required to wear proximity ensembles.

In addition to the aluminized fabric, proximity PPE features full face shields that are coated with an anodized gold material to help create a "mirrored" reflective surface. Without this special coating, the wearer could receive radiant facial burns, and the face shield could quite possibly melt.

Wildland Ensemble

Wildland firefighting conditions are unique in that firefighting operations are often outdoors, often require prolonged physical effort, and are typically accomplished when ambient temperatures are high. Fighting wildland fires with a structural ensemble can invite strained necks, heat stroke, and sprained ankles. Wildland PPE (called **brush gear** by many) addresses the specific needs of the wildland firefighter: It is lightweight and provides breathability, firm ankle support, and hot ember protection, **Figure 6-10**.

The wildland PPE ensemble is designed to be worn over undergarments. These undergarments (long-sleeve T-shirt, pants, and socks) should be 100 percent cotton or of a fire-resistive material, **Figure 6-11**.

> **Caution** Synthetic material should never be worn under wildland PPE—most synthetic materials can melt and cause severe burns to the wearer.

Other highlights of the wildland PPE ensemble are discussed next.

Lightweight Jacket/Shirt and Trousers. These garments are typically made of a fire-resistive material or a treated cotton. Wool has also been used for wildland PPE garments. Once again, these protective elements need to be worn over undergarments in order to increase thermal insulation.

PERSONAL PROTECTIVE CLOTHING AND ENSEMBLES 131

(A)

(B)

Figure 6-9 Proximity firefighting ensembles can utilize (A) a special helmet or (B) a full hood. Either can interface with an SCBA. Note the gold-anodized visors.

Figure 6-10 Wildland PPE is lightweight, but still provides protection from hot, flying brands.

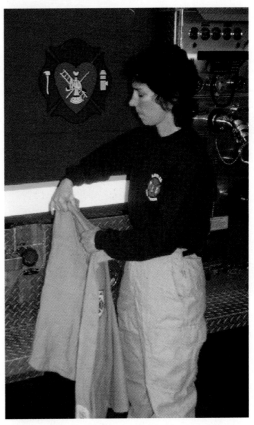

Figure 6-11 Wildland PPE includes cotton or fire-resistive undergarments. Long-sleeve T-shirts are a *must!*

Footwear. Lace-up leather boots that rise well above the ankle (8 to 10 inches) help protect the wearer from cuts, snakebites, and burns. Additionally, a good fitting, tightly laced boot can help prevent ankle sprains and reduce foot fatigue.

Fire Shelter. A **fire shelter** is another unique component in the wildland PPE ensemble, **Figure 6-12**. The fire shelter must be carried in a case that protects the aluminized fabric from being crushed, yet still allows the shelter to be deployed quickly. A fire shelter is a last-resort protective device for firefighters caught or trapped in an environment where a firestorm or blow-up is imminent.

Web Gear. Although not listed in the NFPA standard for wildland PPE, **web gear** is essential for the wildland firefighter. Web gear consists of a belt (with or without shoulder support straps) that can carry a fire shelter, water bottles, flares (fusees), a radio, and other assorted gear to help the wildland firefighter, **Figure 6-13**. Often, a web gear setup includes a detachable day sack that can carry meals and maps, or even overnight sleeping gear and personal items.

Ice Rescue Ensemble

In areas where frozen lakes and recreation ponds exist, the fire department may offer an ice rescue service. The ice rescue ensemble includes a buoyant, insulated suit that protects the wearer from submersion and the freezing cold water should the rescuer break through the ice. The suit has a form-fitting face seal and hood to minimize the amount of water that can enter the suit. Typically one-piece, the suit has sealed gloves and boots built in. To complete the ensemble, the suit should be worn with a simple chest harness and lightweight helmet with face screen (similar to a football face mask), **Figure 6-14A**.

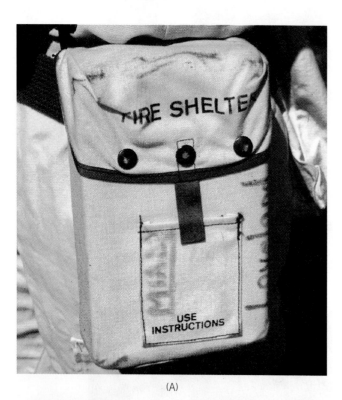

(A)

(B)

Figure 6-12 (A) A wildland fire shelter is an essential PPE component. (B) Training for rapid deployment can help save a firefighter's life.

Figure 6-13 Wildland web gear is designed to keep personal items, water, flares, and the fire shelter within easy reach.

Figure 6-14 Specialized ensembles include (A) the technical rescue ensemble, (B) the ice rescue ensemble, and (C) the swift-water ensemble. *(Courtesy of Ron Marley)*

Technical Rescue Ensemble

Rescue operations such as confined space, collapse, rope, and trench do not necessarily require full structural PPE. In these cases, a lightweight yet durable ensemble is often used to protect rescuers. Typically, this ensemble consists of a durable pant and overshirt (or coverall) coupled with lace-up leather boots, tight-fitting durable gloves, lightweight helmet and eye protection, and a harness, **Figure 6-14B**.

Swift-Water Ensemble

The use of structural PPE near swift water can introduce *more* danger to the firefighter if swept into the stream. While specific intense training can teach firefighters how to "float" in structural PPE, it is not a preferred choice for swift-water environments. The swift-water ensemble includes a typical work uniform overlaid with a personal flotation device (PFD), harness with throw-rope bag, lightweight helmet with face cage, and no-slip gloves **Figure 6-14C**.

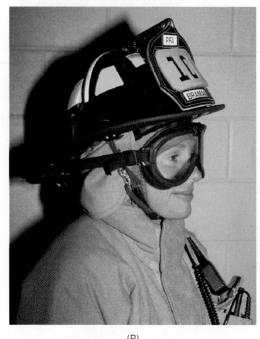

(A)   (B)

Figure 6-15 Different types of eye protection are available for firefighters. (A) Wearing safety glasses under a face shield improves eye protection. (B) A face shield alone may not protect the eyes.

Miscellaneous PPE Components

Firefighters use many different types of PPE components to increase their personal safety in various hazardous environments. The following is a list of some of these items.

Eye Protection

Firefighters use many different forms of eye protection. Goggles, safety glasses, and wrap-around shields are examples. An SCBA face shield can provide primary eye protection although a structural helmet face shield may not. The helmet face shield—because of its wide facial opening—may not prevent windblown particulate or liquid splashes from getting into the eyes, **Figure 6-15**.

Hearing Protection

Many typical rescue and firefighting operations can expose firefighters to noise levels above 100 dB, which can cause hearing damage (including permanent damage) after a short-duration exposure. For this reason, firefighters should always have rapid access to hearing protection. Hearing protection takes on many forms—foam plugs, rigid earmuffs, and headsets are common examples. Many fire apparatus are equipped with a technology that combines hearing protection/intercom/radio microphone into a single headset. These combination systems make communication within the cab of a responding apparatus efficient and effective, **Figure 6-16**.

Figure 6-16 Combination headsets protect the wearer's hearing and also increase communication effectiveness between crew members and other responders.

PERSONAL PROTECTIVE CLOTHING AND ENSEMBLES

Personal Alert Safety System (PASS)

A PASS device is a small, motion-sensitive unit that is typically battery powered and includes a loud audible warning signal. Some devices include a small flashing beacon or strobe. The PASS device, when worn on a firefighter, senses the firefighter's motion. As long as the device senses motion, the alarm will not sound. Inactivity for thirty seconds causes the device to send a "chirp" or other reminder to the wearer. If the wearer fails to move, the device will go into alarm mode and emit a loud noise to signal or warn other firefighters that the wearer may be in trouble. Most new PASS devices are an integral part of the SCBA unit, although some fire departments issue PASS devices to individual firefighters, **Figure 6-17**. The biggest problem with PASS devices results when wearers simply forget to turn their units on—this simple mental lapse has contributed to numerous firefighter fatalities. The National Fire Protection

(A)

(B)

(C)

Figure 6-17 PASS devices can help save a firefighter's life—but they *must* be activated. (A) A manual PASS device *must* be armed by the wearer. (B and C) An integrated PASS device on an SCBA, which is activated when the wearer opens the SCBA bottle.

Association requires that PASS devices integrated with SCBA activate automatically when the SCBA air supply is turned on.

> **Safety** Arming a manual PASS device can save a firefighter's life!

Work Uniform

Firefighters assigned to emergency response duty are often required to wear a uniform—a uniform that meets NFPA standards. NFPA 1975, *Standard on Station/Work Uniforms for Fire Fighters,* outlines manufacturers' guidelines and requirements for work uniforms. The underlying message is that a work uniform can add a protective measure to firefighters engaged in support activities or station duties. A work uniform that meets NFPA standards is *not* designed to protect the wearer from IDLH atmospheres, but can add another layer of reasonable protection under wildland, proximity, and structural ensembles.

CARE AND MAINTENANCE OF PERSONAL PROTECTIVE EQUIPMENT

The key to maintaining personal protective equipment in a high state of readiness is simple: Follow the specific instructions given by the manufacturer. NFPA requires manufacturers to clearly label care instructions for cleaning each piece of equipment. In addition, manufacturers should provide the user with specific instructions and information that address the following:

- Safety considerations
- Limitations of use
- Marking recommendations and restrictions
- Warranty information
- Sizing/adjustment procedures
- Recommended storage practices
- Inspection frequency and details
- Donning and doffing procedures
- Interface issues

Specific to cleaning and maintenance, the manufacturer will provide cleaning instructions and precautions, including a warning that the user should not wear equipment that is not thoroughly clean and dry.

> **Caution** Wearing clothing that is not dry can lead to reduced thermal resistance and to burns.

Equipment exposed to biological and chemical contaminants should be decontaminated in accordance with manufacturers' instructions, **Figure 6-18**. NFPA 1581, *Standard on Fire Department Infection Control Program,* requires that clothing be cleaned every six months as a minimum. Obviously, PPE that is dirty or contaminated should be cleaned immediately. Fire departments should have specific cleaning equipment or a cleaning service that can clean equipment in accordance with manufacturers' guidelines. Washing structural or wildland gear along with linens or other household items should be forbidden because cross-contamination can result.

Finally, personal protective equipment must be routinely inspected and, when appropriate, retired and disposed of as suggested by the manufacturer. This last maintenance check is best accomplished through a team process that involves the individual firefighter, his or her officer, and the organization. This partnership process creates checks and balances that help ensure that PPE is serviceable for the user.

Figure 6-18 Structural and wildland clothing should be washed in accordance with manufacturers' recommendations. Dedicated washers that are used for PPE *only* can help lower cross-contamination of other washables.

PERSONAL PROTECTIVE EQUIPMENT EFFECTIVENESS: "STREET SMARTS"

To maximize the effectiveness of all PPE, firefighters must develop "automatic behaviors." Simple steps can help achieve these behaviors. This final—and perhaps most important—section of the chapter contains suggestions to help firefighters develop good habits that will reduce their chance of injury.

Good PPE Habits and Attitude

Fire departments spend literally thousands of dollars equipping firefighters with PPE for the hazards they may face. Unfortunately, many firefighters are still injured—all too often because the individual firefighter has failed to properly utilize the PPE. This unfortunate result can be eliminated simply by good PPE habits and a positive attitude toward safety. The proper PPE attitude starts with a simple concept:

> **Safety** Firefighters should don all PPE necessary for the potential worst case scenario.

Granted, this approach may lead to "overdressing" for an incident. In these cases, the firefighters' company officer, incident commander, or incident safety officer may allow firefighters to "dress-down." The inverse, however, is unacceptable. Firefighters who take a "wait and see" attitude to decide what level of PPE they are going to need have set themselves up to shortcut their PPE ensemble if, on arrival at the scene, the situation requires immediate lifesaving actions. It is better to be prepared first—then act.

Another key to proper and effective use of PPE is the development of good habits that include fast, proper, and complete donning of the appropriate PPE ensemble, **Figure 6-19**. Unfortunately, many firefighters begin shortcutting their use and completeness of PPE donning as days, weeks, months, and years of firefighting experience are accrued. This is one of the travesties that has kept injury rates high for firefighters.

Establishing good habits takes nothing more than self-discipline and practice. The benefits of self-discipline applied to PPE completeness pay a dividend in the form of **acclimation**. PPE acclimation simply refers to "getting used to" the wearing of PPE. The restrictive, encompassing, and stifling heat initially associated with most firefighter PPE simply becomes less of an issue and physically less taxing with acclimation. No firefighter particularly enjoys donning full structural PPE in 100°F heat. The firefighter who is acclimated physically and mentally to the PPE will, however, accept the merits of the PPE and have less stress—and certainly less risk—than the firefighter who complains that the gear is "too hot" for the weather. The firefighter with good PPE habits and a positive PPE attitude knows that acclimation, rest, and on-scene **rehab** will beat the discomfort that some firefighters complain about.

> **Streetsmart Tip** Hydration prior to wearing PPE ensembles can help prevent stress-related injuries. Firefighters should have a bottle of water available near their gear or at their seat on the apparatus, and take a few drinks as they respond to the emergency.

When donning PPE, the firefighter should follow the manufacturer's instructions and ensure that shortcuts have been eliminated. Firefighters who have experienced a rash of false alarms (i.e., unintentional

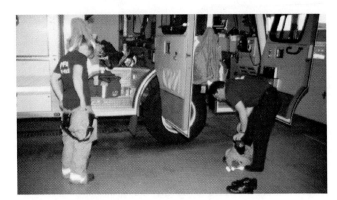

Figure 6-19 Duty personnel should set up their gear for rapid—and complete—donning. Establish good habits to help eliminate shortcuts.

fire alarm activations in a building) seem to be particularly guilty when it comes to shortcutting their PPE. The firefighter with an appropriate PPE attitude will view each one of these incidents as an opportunity to fine-tune donning procedures and achieve **mastery** in PPE donning. Further, it makes sense that a firefighter who has appropriately donned the PPE is in a strong readiness position to be immediately assigned duties in IDLH atmospheres. Firefighters must resist the temptation to get caught in the "cry wolf" syndrome. They must always be ready!

Streetsmart Suggestions

Daily, firefighters utilize PPE as their first and last defense against injury and illness. From these experiences come many "streetsmart" suggestions to help new firefighters:

- Keep PPE clean, inspected, and serviceable—an ownership attitude!
- Practice "team checks." Firefighters should check each others' PPE for readiness, **Figure 6-20**.
- When on duty in a station, position PPE for rapid, complete donning. A practical example is to place a structural hood across bunker pants and boots as a reminder to don the hood first so it will interface appropriately under the bunker coat.
- When doffing PPE, take the time to prepare it for the next response. Set the clothing up for rapid donning—preparation time spent now speeds the next donning.
- Do not let the urgency of the situation override prudent judgment.
- PPE also includes a flashlight, personal tool, radio, earplugs, safety glasses/goggles, accountability tag (see Chapter 23), and a partner, **Figure 6-21**.
- Practice doesn't make perfect—*perfect practice makes perfect!* Firefighters need to use PPE the right way every time to cement good habits. Can firefighters don their structural PPE ensemble in less than a minute (not including SCBA)? Can they add the SCBA in less than a minute? The most important question: Can they do both *correctly* in less than two minutes?
- Be the PPE success example—not an injury statistic.
- When wearing any PPE ensemble, the firefighter must increase water consumption to help stay hydrated.

Mastery of PPE skills allows the firefighter to concentrate on the important task at hand.

Figure 6-21 A full structural protective system includes NFPA-compliant gear, as well as a flashlight, forcible-entry tool, radio, SCBA, and a *partner!*

Figure 6-20 Firefighting teams should check each others' PPE for complete donning. This "team check" can help prevent a burn or other injury.

PERSONAL PROTECTIVE CLOTHING AND ENSEMBLES

Lessons Learned

Firefighter personal protective equipment and ensembles are the first and last defense against injuries and illness in a profession that is extremely hazardous. The firefighter who develops good PPE habits and a positive PPE attitude is likely to be quicker, more prepared, and better acclimated to the real-world confines of the PPE. Shortcutting the use of PPE is a dangerous practice that can contribute to injuries.

Through national consensus standards like the NFPA process, firefighter protective equipment has advanced significantly in the past two decades. NFPA standards address the performance requirements and manufacturing criteria for PPE as well as "use" standards for firefighters and fire departments. Federal government entities also influence PPE use and guidelines.

Different PPE ensembles exist for firefighters, including structural, proximity, and wildland ensembles. The key with any ensemble is the simple concept that an entire ensemble must be donned completely in order for the ensemble to protect the wearer. A shortcut can open the door to an injury. SCBA, harnesses, hearing protection, eye protection, and other PPE components must also be considered to protect the firefighter from hazards.

Care and maintenance of PPE are essential to its readiness. The firefighter must understand manufacturers' instructions and information regarding the fit, use, cleaning, and serviceability of assigned PPE. A sense of ownership is important here.

Finally, PPE effectiveness is dependent on the user's (individual firefighter's) good PPE habits and positive PPE attitude. Donning gear for the worst-case scenario prior to incident engagement is preferable to a "wait and see" approach. The firefighter must *be ready!*

KEY TERMS

Acclimation The act of becoming accustomed or used to something. Typically achieved through repeated practice within a given set of conditions.

Brush Gear Another term for a wildland personal protective ensemble.

Bunkers A slang term that is used mostly to describe the components of a structural firefighting ensemble. The original use of the term *bunkers* referred only to the pant/boot combination that firefighters wore at night and placed next to their "bunks" for rapid donning.

Fire Shelter A last-resort protective device for wildland firefighters caught or trapped in an environment where a firestorm or blowup is imminent.

Mastery The concept that an individual can achieve 90 percent of an objective 90 percent of the time.

Personal Alert Safety System (PASS) A device that emits a loud alert or warning that the wearer is motionless.

Rehab A shortened word meaning *rehabilitation*. Rehab typically consists of rest, medical evaluation, hydration, and nourishment.

Thermal Protective Performance (TPP) A rating level, expressed in seconds, used to characterize the protective qualities of a PPE component before serious injury is experienced by the wearer.

Web Gear The term given to a whole host of personal items carried on a belt/harness arrangement worn by wildland firefighters. Items include water bottles, a fire shelter, radio, and day sack.

Zoning A term given to the establishment of specific hazard zones; that is, hot zone, warm zone, cold zone. Also collapse zones.

REVIEW QUESTIONS

1. Explain what is meant by the phrase "PPE is the first and last resort."
2. List four government and ancillary agencies or associations that influence PPE design or use.
3. Name the agency responsible for developing consensus firefighter PPE standards. Name the agency that writes PPE use regulations.

4. List the components of a structural PPE ensemble.
5. Describe the importance of TPP and "sealing" a structural ensemble.
6. Describe how a proximity ensemble differs from other ensembles.
7. List the components of a wildland PPE ensemble.
8. Describe the operational concept of a PASS device.
9. Explain the relationship of a work uniform and undergarments to PPE ensembles.
10. List three important guidelines relative to the washing of PPE garments.
11. Define the concept of acclimation in regards to PPE.
12. List five "street smart" suggestions for improved effectiveness with PPE.

Additional Resources

Angle, James S., *Safety in the Emergency Services.* Delmar Learning, a part of the Thomson Corporation, Clifton Park, NY, 1999.

Dodson, David W., *Fire Department Incident Safety Officer.* Delmar Learning, a part of the Thomson Corporation, Clifton Park, NY, 1999.

Emergency Incident Rehabilitation, FA-114, United States Fire Administration, Emmitsburg, MD, 1992.

NFPA 1500: Standard on Fire Department Occupational Safety and Health Programs. National Fire Protection Association, Quincy, MA, 2002.

NFPA 1971: Standard on Protective Ensemble for Structural Firefighting. National Fire Protection Association, Quincy, MA, 2002.

NFPA 1976: Standard on Protective Clothing for Proximity Firefighting. National Fire Protection Association, Quincy, MA, 2000.

NFPA 1977: Standard on Protective Clothing and Equipment for Wildland Firefighting. National Fire Protection Association, Quincy, MA, 1998.

NFPA 1851: Standard on Selection, Care, and Maintenance of Structural Fire Fighting Protective Ensembles, National Fire Protection Association, Quincy, MA, 2001.

Risk Management Practices for the Fire Service, FA-166, United States Fire Administration, Emmitsburg, MD, 1996.

CHAPTER 7

SELF-CONTAINED BREATHING APPARATUS

Thomas J. Wutz, Midway Fire Department and New York State Office of Fire Prevention and Control

OUTLINE

- Objectives
- Introduction
- Conditions Requiring Respiratory Protection
- Effects of Toxic Gases and Toxic Environments
- Legal Requirements for Self-Contained Breathing Apparatus Use
- Limitations of Self-Contained Breathing Apparatus
- Types of Self-Contained Breathing Apparatus
- Donning and Doffing Self-Contained Breathing Apparatus
- Self-Contained Breathing Apparatus Operation and Emergency Procedures
- Inspection and Maintenance of Self-Contained Breathing Apparatus
- Lessons Learned
- Key Terms
- Review Questions
- Endnote
- Additional Resources

STREET STORY

It was early one morning, close to 8 a.m., when we got a call that a house in our area had a leak in the gas main outside. Apparently the leak was very close to the foundation of the house and had filtered through the cellar. Before we arrived, the house blew apart—it was totally lifted off its foundation and then settled. By the time we got there, you could see little flames all around the front lawn, which meant the ground was saturated with natural gas. So we put our SCBA masks on immediately. Then we went into the building to extinguish a fire in the cellar.

It was precarious because the destruction left the ceilings only 5 feet instead of 8 feet, and it was completely dark. It wasn't that big of a fire, but we were exerting a lot of energy because we were crawling and pulling the hose with us. Under normal conditions a tank of air will last you maybe half an hour, but the harder you work—if you're huffing and puffing—the more air you end up using. So in this case, we'd been down there for maybe twenty minutes, when the bell went off on my air supply, meaning I had about five minutes to get out of there. Sometimes it's hard to tell if it's your bell or someone else's, so I reached back and felt the bell on my tank—it was vibrating, which indicated it was mine.

Our hand lights stopped operating properly. And my partner and I didn't know how we'd get out since we couldn't see anything—then we had an idea. We grabbed onto the hoseline and followed it back, crawling out. If I would have run out of air, I would have had to put my face as close to the floor as possible so I could get just a little oxygen (gas rises). It would have been dangerous, because there would be the possibility I still could have been overcome if the air had been toxic with gas fumes. That's why it's important to stick with your partner and wear your SCBA in any situation—you never know whether what's burning is toxic.

—Street Story by Bernard Lach, Retired Chief, Torrington Fire Department, Torrington, Connecticut

OBJECTIVES

After completing this chapter, the reader should be able to:

- List two conditions requiring respiratory protection.
- List and explain the effects of oxygen deficiency and toxic gases on the human body.
- List one legal requirement for use of self-contained breathing apparatus (SCBA).
- List two types of SCBA.
- List four components of the SCBA used by the authority having jurisdiction.
- Demonstrate two different SCBA donning procedures at 100 percent accuracy in the time limit established by the authority having jurisdiction.
- Demonstrate routine inspection procedures of SCBA in accordance with manufacturers' instructions.
- Demonstrate after-use maintenance and servicing of SCBA in accordance with manufacturers' instructions.
- Demonstrate the servicing of an SCBA cylinder with the air-filling system used by the authority having jurisdiction.

INTRODUCTION

Self-contained breathing apparatus (SCBA) is one of the most important items of personal protective equipment used by firefighters and rescue personnel. SCBA allows firefighters to enter hazardous atmospheres to perform critical interior operations including offensive fire attack; search, rescue, and removal of victims; ventilation; and overhaul. In addition, SCBA is used at non-fire incidents, such as hazardous material or confined space rescue situations involving toxic fumes or oxygen-deficient atmospheres.

The **respiratory system** of the human body is most vulnerable to injury, especially from toxic conditions and gases encountered during firefighting operations. To protect firefighters in this environment, fire departments must have a respiratory protection policy or "mask rule." This policy should require all personnel to not only wear, but use their SCBA during operations where an **immediate danger to life and health (IDLH)** atmosphere may be encountered. Because it is impossible to predetermine all IDLH atmospheres, this policy must include operations during interior or exterior fire attack, such as structure, vehicle, and dumpster fires, below-grade or confined space rescue, and hazardous materials incidents. Respiratory protection provided by SCBA is necessary, even during exterior defensive operations as shown in **Figure 7-1**.

Figure 7-1 Large volumes of smoke require the use of SCBA, even for exterior operations as shown here at a tire storage facility.

> **Safety** Hydrogen cyanide is a deadly gas and has a distinct odor of almonds. Any inhaled toxic gas can directly cause disease of the lung tissue. SCBA, keen senses, incident command, and a thorough safety program are vital to keep firefighters safe from these deadly toxins and other safety hazards.

Failure to understand and know how to use SCBA properly could result in injury or death to a firefighter, failed rescue attempts of victims, or deterioration of the emergency incident. In addition to the short-term effects experienced during the emergency incident, firefighters may suffer long-term health problems due to repeated exposure to toxic environments. The tasks of firefighters can only be accomplished effectively through the proper use of personal protective equipment (PPE), **Figure 7-2.**

Syracuse, New York, and Lubbock, Texas, are two cities separated by 1,700 miles. Although they are separated geographically, they have a common connection: line-of-duty deaths of firefighters. Within a year's time from April 1978 to March 1979, these two cities lost a total of seven firefighters in structure fire incidents.

At 0046 hours on April 9, 1978, the Syracuse Fire Department responded to a structure fire. The

Figure 7-2 These firefighters in full protective equipment including SCBA are ready to begin interior firefighting operations.

fire involved a three-story, wood frame, balloon construction apartment building that had been converted from a single-family residence prior to 1957. On arrival, the first engine company reported light smoke showing from the second and third floors. As firefighters searched those floors, fire in concealed spaces progressed. Light smoke conditions deteriorated. An extensive fire developed on the third floor and in the attic, and the atmosphere above the second floor became untenable. At approximately 0059 hours, firefighters were ordered to evacuate the third floor. For unknown reasons four firefighters did not leave. All four became trapped and died.

Less than a year later, on March 25, 1979, the Lubbock Fire Department responded to a structure fire at 0432 hours. This incident was in a one-story building of ordinary construction, occupied by a restaurant undergoing renovations. Firefighters attacked a fire in the kitchen area, bringing it under control in about twenty minutes. During the overhaul phase, three firefighters entered the building, became disoriented, and were overcome by carbon monoxide gas and died.

During the investigation of these incidents, questions concerning the design, duration, and use of the SCBA were reviewed, along with training the firefighters had received for its use. Since these deaths, many changes have occurred in SCBA including design and use. The most notable of these changes follow:

- *SCBA weight.* The weight of a standard thirty-minute SCBA unit used in the fire service has been reduced by 15 to 17 pounds, approximately 40 percent since 1980. This reduces the physical stress on the firefighter.
- *Positive pressure.* A constant supply of air is delivered, pressurizing the face piece, keeping toxic gases from entering. This pressure (1½ to 2 psi, depending on the manufacturer), which is slightly above atmospheric pressure, also helps maintain face piece seals.
- *Improved design.* Breathing tube, regulator, and harness designs have been improved to limit catastrophic failure during firefighting operations.
- *Improved regulator design.* New regulator designs provide increased airflow and redundancy, providing a backup in the event of a regulator/pressure-reducing failure.
- *SCBA maintenance.* Field- and factory-level maintenance programs have been implemented to ensure proper operation of SCBA units.
- *PASS devices.* Personal Alert Safety Systems (PASS) are audible warning devices that incorporate a motion detector to sense movement. The PASS will automatically sound an alarm signal (and usually a small flashing strobe) if movement is undetected for thirty seconds. This alarm alerts other firefighters to an unconscious or incapacitated firefighter. The device can also be activated manually to signal others that a firefighter is in trouble and needs help. Newer SCBA units have the PASS device incorporated in the SCBA design, while older removable PASS devices are attached to the SCBA unit, usually somewhere on the harness. These devices are in common use and are required by NFPA 1500.
- *Training.* **Mask confidence or "smoke divers" training** programs develop the firefighter's knowledge of and confidence in using an SCBA unit.
- *Increased regulation.* Implementation of **respiratory protection programs** is required by OSHA 29 CFR 1910.134 or NFPA 1500 and NFPA 1404.

These improvements are only as effective as the training firefighters receive and the proficiency they develop using SCBA. Technology, regulations, and mandates are only as good as an individual firefighter's commitment to use SCBA.

Note All SCBA used or manufactured for fire service use must be positive pressure.

CONDITIONS REQUIRING RESPIRATORY PROTECTION

Four conditions that present respiratory hazards are commonly found at fire or other emergency incidents:

- Oxygen deficiency
- High temperatures
- Smoke or other by-products of combustion
- Toxic environments.

These conditions can be found separately, such as an oxygen-deficient atmosphere in a confined space situation, or combined in a fire incident. Regardless of the incident, any time the potential for these environments exists or develops, firefighters must use their SCBA.

Oxygen-Deficient Environments

The human body and fire are similar as both require oxygen to survive. Fire (combustion) consumes oxygen and produces toxic gases that may displace or dilute the available oxygen. Atmospheres with oxygen concentrations below 19.5 percent are classified as **oxygen-deficient atmospheres**. As shown in **Table 7-1**, decreased oxygen affects the human body by causing muscular impairment, mental confusion, and eventually death. This in combination with toxic gases is the cause for most fire deaths for individuals lacking the protection of SCBA.

Streetsmart Tip Installed fire-extinguishing systems such as total flooding carbon dioxide or halon systems create an oxygen-deficient atmosphere. When entering a structure or area where this type of system has activated, even without a fire, the firefighter must have the protection of SCBA.

Elevated Temperatures

The respiratory system of the human body is extremely delicate and sensitive to elevated temperatures. Even air temperatures related to a recreational activity such as a sauna, commonly 120° to 130°F, taken into the lungs may cause a serious decrease in blood pressure and circulatory system problems. Inhaling of heated gases will cause an accumulation of fluid in the lungs, **pulmonary edema**, which can cause death from **asphyxiation**. In addition, the damage caused by inhaling heated air or gases is long term and not reversible by treatment of fresh, cool air.

Firefighter Fact Temperatures in a structure fire reach 1,000° to 2,400°F. One unprotected breath at this temperature level will cause death or severe damage to a firefighter's respiratory system.

Smoke

Smoke is the combination of unburned products of combustion, particles of carbon, tar, and associated gases such as carbon monoxide, carbon dioxide, sulfur dioxide, and hydrogen cyanide. This

Effects of Hypoxia (Reduced Oxygen)

OXYGEN PRESENT/AVAILABLE (%)	SYMPTOMS
21	Normal conditions, no effect
19.5	OSHA definition as oxygen deficient
17	Some muscular impairment, increased respiratory rate
12	Dizziness, headache, rapid fatigue
9	Unconsciousness
7 to 6	Death within a few minutes

Source: Browne & Crist, *Confined Space Rescue*, page 8, Delmar Learning, a part of the Thomson Corporation, Clifton Park, NY, 1999.

TABLE 7-1

combination of materials is an irritant to the respiratory system and, in many cases, small inhaled quantities may be fatal. In addition, the temperature of smoke may cause burns to the respiratory system.

EFFECTS OF TOXIC GASES AND TOXIC ENVIRONMENTS

The combustion process produces toxic gases and irritants that affect both the short- and long-term health of firefighters operating in hazardous environments. In addition, when products of combustion combine, they may form toxins that are lethal. A swimming pool at a residential occupancy, **Figure 7-3,** indicates storage of chemicals such as chlorine, which will produce a poisonous gas. Firefighters must understand that there is no "routine" fire. Even a light smoke condition at this type of structure could be deadly.

Some of the common gases produced in a fire affect not only the respiratory system, but the circulatory system as well, **Table 7-2.** In addition to the toxic products of combustion pesticides and herbicides present at an agricultural occupancy will produce many additional toxins, thus requiring a higher level of protection for firefighters, **Figure 7-4.**

Figure 7-3 This swimming pool is a warning to firefighters of the possible presence of chemicals in storage in this residence.

Carbon Monoxide

Carbon monoxide (CO) is produced in great quantities during the combustion process and is one of the most lethal gases found in a fire. This colorless and odorless gas is always present, especially during incomplete combustion or in areas of poor ventilation. Its presence is of great concern, especially during the overhaul stage when respiratory protection habits may be lax.

Toxic Gases Formed as Products of Combustion

GAS	TOXICOLOGICAL EFFECT	PRODUCED BY	IDLH (PPM)*
Carbon dioxide	Displaces oxygen	Free burning	40,000
Carbon monoxide	Displaces oxygen	Incomplete combustion	1,200 to 1,500
Hydrogen cyanide	Chemical asphyxiant	Wool, silk, nylon, polyurethane	50
Hydrogen chloride	Respiratory irritant	Polyvinyl chloride (PVC), building materials, furnishings	50
Nitrogen dioxide	Pulmonary irritant	Small quantities from fabrics, large quantities from cellulose nitrate	20
Phosgene	Poison	Burning refrigerants; produces hydrochloric acid when inhaled	25

*ppm = parts per million, ratio of volume of gas compared to volume of air.

TABLE 7-2

SELF-CONTAINED BREATHING APPARATUS 147

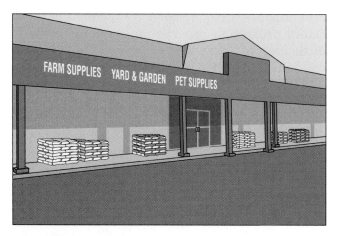

Figure 7-4 Agricultural occupancies will usually have a large supply of chemicals that are extremely hazardous during firefighting operations.

> **Caution** Firefighters should always use their SCBA, even when the fire is thought to be extinguished. Lethal amounts of carbon monoxide may be present.

Ordinarily, 98 percent of the oxygen carried by the blood is bound to the hemoglobin, which is contained in the red blood cells. CO attaches itself to the red blood cells and prevents oxygen from bonding with the hemoglobin. In fact, CO combines with blood almost 218 times easier than oxygen. It does not act directly on the body, but prevents oxygen from being distributed to the body, causing **hypoxia**, followed by death. **Table 7-3** shows the stages and symptoms of this process.

Intense physical activities during firefighting operations, **Figure 7-5**, require the respiratory system to increase the respiratory rate to deliver

Figure 7-5 The light smoke condition present during overhaul will contain large amounts of carbon monoxide requiring SCBA protection.

Symptoms of Carbon Monoxide Poisoning

SYMPTOMS	CARBON MONOXIDE CONCENTRATION (PPM)*
Mild headache	1,000
Headache, nausea	1,300
Unconsciousness after 1 hour[†]	1,500
10-minute exposure, dizziness, nausea	3,200
Fatal, less than 1-hour exposure	4,000
Danger of death in 1 to 3 minutes	10,000

*ppm = parts per million, ratio of volume of gas compared to volume of air.
[†]IDLH level.

TABLE 7-3

increased oxygen to the muscles and organs of the body. If a firefighter is working unprotected (without the protection of SCBA) in an atmosphere of CO, the increased respiration augments the ability of CO to incapacitate the firefighter. CO also does not metabolize out of the blood very quickly. During repeated exposures to low levels of CO during a shift, or during a number of fires, the blood level of CO is compounded. CO symptoms range from a mild headache to death, **Table 7-3.**

LEGAL REQUIREMENTS FOR SELF-CONTAINED BREATHING APPARATUS USE

Toxins are always present in the by-products of combustion. Common sense dictates that firefighters should use SCBA on every fire scene—from start to finish. For the firefighter's safety and health, a number of regulations have been developed for SCBA use. A number of organizations have also established regulations and standards concerning the design and use of SCBA, **Table 7-4.**

> **Firefighter Fact** Depending on the laws, rules, and regulations of a particular state or local government, OSHA standards may or may not be legal requirements. In general, the U.S. Department of Labor OSHA regulations do not apply to municipal employees, including firefighters. However, a number of states, **Figure 7-6**, have adopted their own plans, which must be as stringent as the federal requirements. In these states, the requirements do apply to municipal employees, including firefighters. Generally, volunteer firefighters are considered employees for the purposes of health and safety regulation. Simply said, firefighters must know the rules and regulations that apply to their organization.

Title 29 Code of Federal Regulations, Section 1910.134

OSHA 29 CFR 1910.134 establishes the standard for all entries into IDLH atmospheres. The April 1998 revision contains requirements related to interior structural firefighting and defines interior structural firefighting as an IDLH atmosphere. In addition to requiring the use of SCBA, the standard also establishes requirements for a complete respiratory protection program and regular medical evaluation of employees designated to wear and use SCBA.

Although employers (fire departments) are responsible for providing a safe and healthy work environment, all firefighters have a duty to understand and follow these regulations.

NFPA 1500: Standard on Fire Department Occupational Safety and Health Program

The National Fire Protection Association has established **NFPA 1500**, *Standard on Fire Department Occupational Safety and Health Program.* This standard covers many safety and health-related issues for firefighters, including respiratory protection. The difference between this standard and the OSHA regulations is that a government authority (city, town, county, or state), called the **authority having jurisdiction (AHJ)**, must adopt the standard as policy for the fire department.

The NFPA has two additional standards dealing with SCBA: **NFPA 1404**, *Standard for a Fire Department Self-Contained Breathing Apparatus Program,* and **NFPA 1981**, *Standard on Open-Circuit Self-Contained Breathing Apparatus for Fire Service.* The NFPA 1404 standard establishes minimum requirements for fire department respiratory protection programs including the use of, training, safety, emergency procedures, maintenance, and breathing air for SCBA used in the fire service. The NFPA 1981 standard establishes design and performance criteria, test methods, and certification for open-circuit SCBA intended for fire service use.

Both 29 CFR 1910.134 and NFPA 1500 contain similar requirements for respiratory protection. These are nationally recognized standards, and fire departments in jurisdictions without mandatory respiratory protection requirements should adopt or reference these for operations requiring the use of SCBA.

LIMITATIONS OF SELF-CONTAINED BREATHING APPARATUS

As with any other tool used by the fire service, SCBA has a number of limitations that firefighters must understand if they are to use the unit effectively and safely. These limitations are the SCBA

Organizations Concerned with SCBA Design and Use

ORGANIZATION	STANDARD	APPLICATION
National Institute for Occupational Safety & Health (NIOSH)	42 CFR Part 84	Requirements for design, testing, and certifying SCBA
Occupational Safety & Health Administration	29 CFR 1910.134	Respiratory protection programs for SCBA use
Occupational Safety & Health Administration	29 CFR 1910.156	Fire Brigade Standard, references 1910.134
National Fire Protection Association	NFPA 1404	Standard for a Fire Department SCBA Program
National Fire Protection Association	NFPA 1500	Standard on Fire Department Occupational Safety and Health Program
National Fire Protection Association	NFPA 1981	Standard on Open-Circuit SCBA for the Fire Service

TABLE 7-4

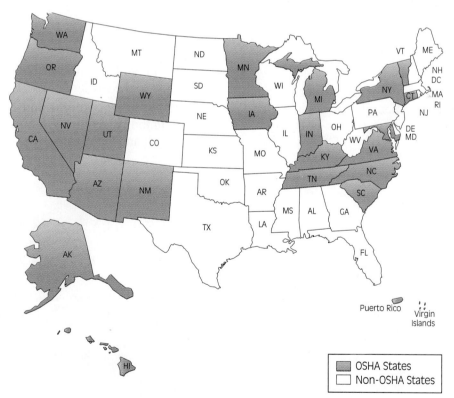

Figure 7-6 States requiring respiratory protection plans in accordance with OSHA's 29 CFR 1910.134 regulations.

unit itself (i.e., its size, weight, and limited air supply) and the physical and physiological condition of the user. Firefighters must be well trained to successfully complete all tasks and assignments requiring the use of SCBA, **Figure 7-7**.

SCBA Design and Size

The design and size of SCBA units vary greatly by manufacturer and by the age of the unit. Older units placed in service as little as ten years ago may be heavier and bulkier than newer ones. Firefighters must be conditioned to the added weight and bulk of the SCBA and the personal protective equipment they must wear. Depending on the activity level required and the physical condition of the user, *an SCBA cylinder with a rating of thirty minutes of air supply will usually be consumed more rapidly.* This factor limits the distance a firefighter may advance into a building or fire and requires frequent crew rotations at large incidents. Other concerns include:

- Visibility is restricted, peripheral vision is reduced, and fogging of the face piece can occur.
- Depending on the manufacturer's style, the age of the unit and the type of cylinder used, SCBA units will add 23 to 35 pounds of weight and 9 to 15 inches to the profile of the firefighter. This will limit the mobility of the firefighter.
- Unless the SCBA unit is equipped with a voice amplification accessory, the firefighter's voice will be muffled and difficult for others to understand, especially if the firefighter is trying to communicate by radio.
- The quantity of air is limited. It is vital that a firefighter know the status of the cylinder air supply prior to its use.

Streetsmart Tip The Philadelphia Fire Department conducted extensive testing in a firefighting skills proficiency course with 750 firefighters using SCBA. The average air consumption for SCBA rated for thirty minutes was less than fifteen minutes—from full cylinder to low air warning. Forcible entry and ventilation are two tasks requiring increased physical activity, **Figure 7-8**.

Limitations of the SCBA User

Firefighters themselves can be limited in their use of SCBA. A firefighter's physical, mental, and emotional condition can cause usage problems. When these limitations are coupled with the confines of the SCBA unit there can be serious problems.

Physical Limitations

- The total protective envelope including SCBA and PPE adds approximately 40 to 50 pounds of weight to the firefighter.
- The additional weight and bulk affect agility and mobility, requiring a high level of physical strength and endurance.

Figure 7-7 Continuous training with SCBA is one of the keys to effective firefighting operations.

SELF-CONTAINED BREATHING APPARATUS 151

(A)

(B)

Figure 7-8 (A) Ventilation and (B) forcible entry are physically demanding. They produce increased respiration rates and air consumption.

- Even though face pieces have been fitted and tested, weight loss or a twenty-four-hour growth of facial hair may affect the ability to obtain a good seal.
- In addition to strength and endurance, the stress of firefighting in elevated temperatures requires additional cardiovascular and respiratory conditioning. Lack of conditioning will increase stress on the body.

Firefighter Fact According to the U.S. Fire Administration, in the year 2000, seventy-two deaths occurred either at an emergency scene, or responding to or from the scene, with fourteen of the deaths occurring on wildland fires. There were also thirty nonemergency scene deaths involving training, administrative, or other duties. Forty-five (62.5 percent) of these fatalities were heart attack or stroke related. It is important for all fire departments to institute a program of physical fitness and nutrition training and regular medical examinations to combat firefighter stress and overexertion. Firefighters should be taught how to recognize their own physical and mental limitations and how to react when other company members reach their limitations.

Physiologic Limitations

- Lack of confidence in the SCBA unit and its ability to protect the firefighter may cause anxiety, increasing breathing rate.
- The degree of training or experience users have with SCBA affects their self-confidence and ability to function.
- Increased physical stress may cause anxiety.
- Emotional conditions, such as fear of being confined, excitement, or claustrophobia, may increase the user's breathing rate and air consumption.

These problems are addressed with proper training and education. Training in SCBA use is not a one-time deal; it must be continuous so firefighters maintain their proficiency and confidence with SCBA.

TYPES OF SELF-CONTAINED BREATHING APPARATUS

Two different types of SCBA are in general use in today's fire service: **open-circuit** and **closed-circuit** systems. In an open-circuit SCBA, exhaled air is vented to the outside atmosphere, whereas in a closed-circuit SCBA, the exhaled air stays in the system for filtering, cleaning, and circulation. The open-circuit SCBA, **Figure 7-9**, is the most commonly used for firefighting operations. The closed-circuit type is sometimes used for specialized rescue incidents.

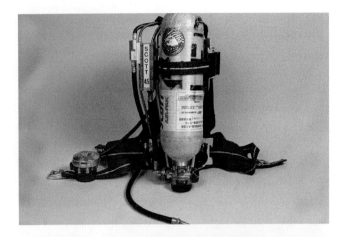

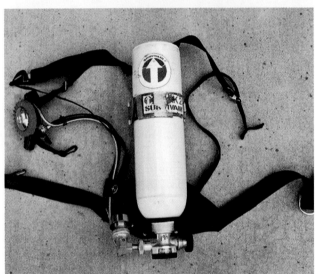

Figure 7-9 A number of manufacturers produce open-circuit SCBA for fire service use. (*Bottom right photo courtesy of Mine Safety Appliance Company*)

Open-Circuit Self-Contained Breathing Apparatus

A number of manufacturers supply SCBA for fire service use. Regardless of the manufacturer, SCBAs for fire service use are designed and built in accordance with **NIOSH** and NFPA standards. Certain parts of an SCBA unit may be interchangeable with parts from another manufacturer; however, this practice will void the NIOSH and manufacturers' certifications.

Generally, as shown in **Figure 7-10**, SCBA has four basic assembly components:

- *Backpack and harness:* The backpack holds the air cylinder and the harness allows the unit to be worn.
- *Cylinder:* Includes cylinder, valve, and cylinder pressure gauge.
- *Regulator:* Includes the high-pressure hose from the cylinder, the regulator, and end of service time indicator (EOSTI).
- *Face piece assembly:* Includes face piece, exhalation valve, and head harness.

Backpack and Harness Assembly

The backpack assembly, **Figure 7-11**, is designed to hold the air cylinder and provide the straps for securing the SCBA unit to the firefighter. It will consist of a metal or a high-temperature-resistant plastic frame, a mechanism for securing the air cylinder, and shoulder and waist straps. Depending on the manufacturer's design and style, the regulator may be attached to the waist straps. The straps are adjustable to the size of the user and designed to distribute the weight of the unit. Most models in use today are designed to have the greatest portion of the unit's weight carried on the hips by the waist straps.

SELF-CONTAINED BREATHING APPARATUS 153

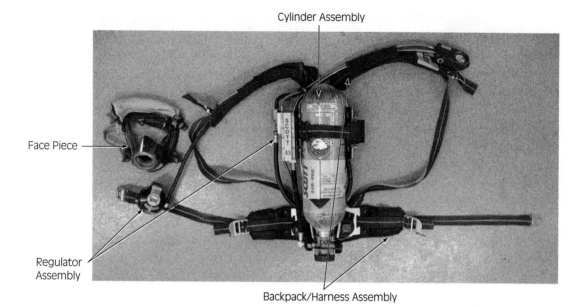

Figure 7-10 The four components of the open-circuit SCBA are the backpack/harness, cylinder, regulator, and face piece assemblies.

Cylinder Assembly

SCBA cylinders, **Figure 7-12**, contain the compressed air for breathing by the user. For this reason, the cylinder must be strong and durable and it constitutes most of the weight of the unit. Most cylinders in use today are aluminum, fiberglass/aluminum composite, and Kevlar or carbon composite materials. The change from steel to aluminum and then composite cylinders has reduced the weight of an SCBA cylinder by almost 50 percent. Typical cylinder capacities are listed in **Table 7-5**.

Types of Cylinders. The SCBA cylinders used in the fire service vary in material and type of manufacture. Listed here are general descriptions of the most common types of cylinders used in the fire service. Refer to manufacturers' information for exact specifications.

- *Steel.* Because of improvements in design and weight reduction using other materials, steel cylinders are generally not used for SCBA service. Most of the steel cylinders still in fire service use are used to provide air supply for various rescue tools. They have an unlimited service life as long as they pass a hydrostatic test every five years.

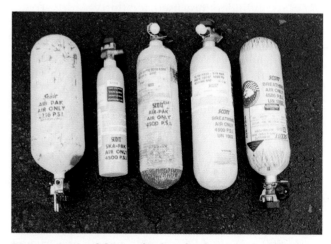

Figure 7-11 A typical SCBA backpack/harness assembly.

Figure 7-12 SCBA cylinders for fire service use, from left to right, are steel, aluminum, hoop-wrapped fiberglass, Kevlar composite, and carbon fiber composite.

SCBA Air Cylinder Capacities

RATED DURATION (MINUTES)	MATERIAL	CYLINDER PRESSURE (PSI)	CAPACITY-COMPRESSED AIR (FT³)
30	Aluminum or composite	2,216	44
30	Aluminum or composite	4,500	44
45	Carbon fiber	3,000	65
45	Aluminum or composite	2,216	65
45	Aluminum or composite	4,500	65
60	Aluminum or composite	4,500	88

TABLE 7-5

- *Aluminum.* Aluminum cylinders are used for thirty-minute (2,216-psi) rated SCBA units. This type of cylinder contains 44 cubic feet of compressed air and weighs about 22 pounds when charged. These cylinders have an unlimited service life as long as they pass a hydrostatic test every five years.
- *Fiberglass (hoop-wrapped).* Fiberglass cylinders are used for thirty-minute (2,216-psi) rated SCBA units and are constructed with an aluminum inner shell. The cylinder sides are wrapped with fiberglass for strength. This type of cylinder contains 44 cubic feet of compressed air and weighs about 16 pounds when charged. These cylinders have a service life limited to fifteen years and must have a hydrostatic test every three years. In accordance with current U.S. Department of Transportation (USDOT) regulations, these cylinders are destroyed at the end of their service life.
- *Fiberglass (full-wrapped).* Full-wrapped fiberglass cylinders are used for thirty-, forty-five-, and sixty-minute (4,500-psi) rated SCBA units. They are constructed with an aluminum inner shell and fully wrapped with fiberglass for strength. These cylinders contain 44 to 88 cubic feet of compressed air and weigh 16 (44 ft^3) to 26 (88 ft^3) pounds when charged. These cylinders have a service life limited to fifteen years and must have a hydrostatic test every three years. In accordance with current USDOT regulations, these cylinders are destroyed at the end of their service life.
- *Kevlar/carbon composites.* Composite cylinders are used for thirty-, forty-five-, and sixty-minute (4,500-psi) rated SCBA units and are constructed with an aluminum inner shell and fully wrapped with Kevlar or carbon fibers for strength. These cylinders contain 44 to 88 cubic feet of compressed air and weigh 12 (44 ft^3) to 22 (88 ft^3) pounds when charged. These cylinders have a service life limited to fifteen years and must have a hydrostatic test every three to five years—depending on the manufacturer of the cylinder. In accordance with current USDOT regulations, these cylinders are destroyed at the end of their service life.

Caution In the past few years the failure of composite-type cylinders has occurred with increasing frequency. In three instances, cylinders failed while stored on fire apparatus. Investigations concluded that the failures were caused by stress-corrosion cracking of the fiberglass wraps resulting from exposure to a strong corrosive agent. Fiberglass composite cylinders are particularly at risk for stress-corrosion cracking because the fibers are under constant tension due to the internal pressure. When the structural integrity of the overwrap is weakened, a catastrophic failure of a cylinder can occur that may result in serious injury or death. Persons responsible for maintenance of composite cylinders should take measures to ensure that they do not come in contact with strong corrosive agents. In addition, cylinders should be washed only with a mild soap and water solution, and all recommendations of the cylinder manufacturer or distributor with regard to maintenance, requalification, and use must be carefully followed. In addition, during the 1980s a number of aluminum cylinders developed catastrophic failures at the neck area, near the connection of the cylinder valve. Even with regulations and testing, failures do occur and firefighters must be aware of the correct maintenance procedures for SCBA cylinders.

SCBA Cylinder Failures. The rated duration of all SCBA cylinders is based on laboratory testing; actual duration will vary significantly with each individual user. One of the most important factors affecting air supply duration is the physical and physiologic condition of the user. Firefighting is physically demanding work, in a very harsh environment, and when combined with the stress of searching for victims usually leads to increased breathing and consumption of air. For this reason even firefighters in top physical condition may only be able to operate for fifteen to twenty minutes with a unit rated for thirty minutes.

Cylinder Testing. The USDOT regulates compressed gas cylinders, including those used for SCBA, and requires hydrostatic testing. This test is done to ensure that the cylinder is capable of withstanding its rated pressure and capacity and the stress created when the cylinder is being filled. Cylinder testing is usually accomplished by an outside vendor; however, some large fire departments may have their own service unit that conducts these tests.

Once the test is complete, each cylinder is labeled or stamped as shown in **Figure 7-13**. This label must show the licenses of the testing organization and the date of the most recent test. One should never attempt to fill a cylinder with an out-of-date test label.

> **Caution** In 1993 a firefighter filled a composite-type SCBA cylinder that was beyond its useful service life of fifteen years. After filling the cylinder, he placed it back on the apparatus and then a catastrophic failure occurred. The cylinder valve exploded out of the cylinder and killed the firefighter instantly. One should never service an SCBA cylinder that is out of date for hydrostatic testing or a composite cylinder that has passed its fifteen-year service life.

Air Quality for SCBA Use

The quality of the compressed breathing gas used in open-circuit SCBA has a direct effect on the performance of this equipment. Many standards address the quality or purity of compressed air used for breathing purposes. The most generally accepted are those established by the Compressed Gas Association in its pamphlet G-7.1-1989, which are incorporated by reference in OSHA 29 CFR 1910.134 and NFPA 1500. Briefly, the minimum quality of air used for SCBA is Grade D, as established in the pamphlet. This classification establishes maximum allowable quantities of impurities allowed in breathing air. In accordance with this specification the

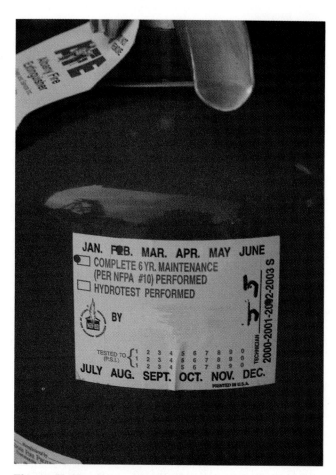

Figure 7-13 Depending on the type of material, SCBA cylinders are required to be hydrostatically tested every three to five years.

supply source (purifier/compressor) must be tested at least every three months and certified as shown on the certificate in **Figure 7-14**. NFPA 1404 has one exception to the Grade D breathing air requirement and that is a lower maximum moisture content of 24 ppm or drier.

Figure 7-14 All air systems for SCBA use must be tested every three months and certified as shown by this air test quality certificate.

Figure 7-15 SCBA designed for the fire service will use either (A) a face piece–mounted regulator or (B) a waist-mounted regulator.

This air is obtained from an air cylinder refill system owned by a fire department or from a commercial vendor. These systems, commonly called *cascade systems,* consist of a compressor, purifier, air quality monitor, storage system, booster compressor, control panel, and cylinder fill containers.

Regulator Assembly

Depending on the manufacture and style of the SCBA unit, the regulator assembly, **Figures 7-15A** and **B**, is attached to the face piece or waist straps. The regulator reduces the high-pressure air from the cylinder to a low, slightly above atmospheric pressure and controls the flow of air to the face piece. The regulator contains a diaphragm that is activated by the breathing action of the user. This action creates a pressure differential, which opens the diaphragm and allows air to flow to the face piece. In addition, all SCBA units designed for fire service use also maintain a constant positive pressure airflow to the face piece. The exhalation action moves the diaphragm to the closed position and opens exhalation valves, and the exhaled air is vented outside the face piece.

Again depending on the manufacture and style of the SCBA unit, the regulator will have color-coded valves to control both normal and emergency operation. One is the main line valve, usually colored yellow for normal operation; the second is the bypass valve, usually colored red for emergency operation.

During normal operation the main line valve is fully open, allowing maximum air to flow, and the bypass valve is in the fully closed position. The valves should remain in these positions for proper operation of the unit. In the event of a malfunction of the regulator or high-pressure reducer, the bypass valve must be manually operated.

The regulator may have a pressure gauge that is part of the regulator body or attached to the shoulder straps. This gauge should provide the same readings as those shown on the cylinder valve. There may be some difference, but the two gauges must read within 100 psi of each other or the unit should be taken out of service. Depending on the manufacturer, the increments shown on the gauge may be in percentages of a full cylinder or will display the actual psi remaining in the cylinder.

SCBA units designed for fire service use have a low air supply warning alarm, which is usually part of the regulator assembly. This alarm sounds when the cylinder pressure decreases to approximately one-fourth of the rated capacity of the cylinder. Depending on the manufacturer, some alarms on face piece–mounted regulators will also vibrate the face piece assembly to warn the firefighter of the low air supply. The 1997 edition of NFPA 1981 requires SCBA units to be equipped with two different types of low air alarms. The two alarms must function independently, failure of one alarm must not affect the operation of the second, and each must alert different senses, for instance, an audible alarm accompanied by a visual or tactile (vibration) alarm.

SELF-CONTAINED BREATHING APPARATUS   157

Figure 7-16 Two types of SCBA face pieces. The regulator connects to (A) the face piece, and (B) the hose connects to the regulator.

> **Streetsmart Tip** When using SCBA, firefighters must work in teams of a minimum of two, and *all* team members must exit the hazardous environment when a low air supply alarm sounds. A firefighter must never leave a partner or allow a partner to leave alone.

Face Piece Assembly

The face piece assembly, **Figures 7-16A** and **B**, provides fresh air to the firefighter wearing an SCBA unit. In addition, protection from the hazardous environment is provided to the face and eyes. The assembly consists of a flexible rubber or silicon mask with a lens, exhalation valves, and a harness with adjustable straps. Most manufacturers provide a number of different sizes for proper fit without leakage. As mentioned later in this chapter, 29 CFR 1910.134 and NFPA 1500 require an annual fit test to ensure proper sizing of the face piece. Depending on manufacturer and style, the face piece will have a low-pressure air hose to connect to the regulator or the regulator will connect directly to the face piece.

The exhalation valve is the outlet for exhaled breath and prevents toxic gases from entering the face piece. These valves must be inspected regularly because dirt or moisture in a cold climate may keep it partially open, allowing toxic gases to enter the face piece. The following section on donning procedures explains how to check the operation of the exhalation valves.

The face piece assembly may have a number of options depending on manufacturer and style. These include a "nose cup" to prevent fogging and "voice amplifiers" to facilitate communication. The NIOSH regulation requires the nose-cup option in cold climates.

The last part of the face piece assembly is the harness and straps. This provides a tight fit to the head and prevents the face piece from loosening during firefighting operations. Again there are variations with different manufacturers, but the types in use today are a web style or headnet style. When the face piece is stored, the harness straps must be in the full out position to reduce wear on the straps and facilitate quick donning. Also, regardless of manufacturer, when donning the face piece, the straps must be pulled straight back from the head to ensure proper fit.

Each SCBA manufacturer has specific instructions for cleaning individual face pieces. Failure to follow these instructions may result in damage to the face piece and potential failure during use. An SCBA face piece must be cleaned after each use or regularly to remove dust and particles and to prevent the spread of communicable diseases. To minimize this problem, many fire departments issue each individual an SCBA face piece.

Closed-Circuit Self-Contained Breathing Apparatus

Generally, closed-circuit SCBA is not used for firefighting operations. For services that do utilize this system, the most common use is for hazardous materials incidents. Below-grade rescue use requires an extended air supply. These units are available with air supplies that range from thirty minutes to approximately four hours.

Closed-circuit SCBAs contain a cylinder of oxygen, a filter system, a regulator, and valves. They work on the principle of cleaning and filtering exhaled breath and adding pure oxygen to continue operation. The duration of the air supply is based on the filtering/cleaning and oxygen capacity of the unit.

Open-Circuit Supplied Air Respirators

Open-circuit **supplied air respirators (SARs)**, also called airline respirators, **Figure 7-17**, are similar to SCBA units, except that the air supply cylinder is remote from the user. Air is supplied in the same manner as for a regular SCBA, but the hose connecting the cylinder and the SCBA unit may be 100 to 200 feet long, **Figure 17-7A.** These types of units are not used for firefighting operations, but are commonly used for hazardous materials incidents or confined space rescues. They provide the user with a long-duration air supply with mobility and agility. This type of unit must be equipped with an SCBA escape unit with duration of approximately five to ten minutes, **Figure 7-17B**.

DONNING AND DOFFING SELF-CONTAINED BREATHING APPARATUS

Depending on fire department apparatus and procedures, SCBA may be placed in service in a number of ways. The most common donning procedures are from a seat-mounted position in the apparatus, from a side compartment on the apparatus, or from a storage case. Regardless of how the SCBA unit is stored or mounted, firefighters should always refer to the manufacturers' instructions for specific procedures.

(A)

(B)

Figure 7-17 (A) A supplied air respirator with face piece, hose, and air supply. (B) Note the emergency escape cylinder.

Standard Daily Checks for SCBA

SCBA COMPONENT	CHECK FOR OPERATION
Cylinder gauge	Cylinder is 100% full; usually a green shaded area on gauge.
Cylinder valve (slowly open two to three full turns)	Listen for audible low alarm to activate. If it does not activate or continues to sound, place unit out of service.
Regulator or remote gauge and cylinder gauge	Compare gauges. They should read within 100 psi of each other.
PASS device	If unit has an integral PASS device, check for operation. If PASS device is a separate unit, check operation.
Valves	Check all valves for operation and proper position.
Valves and regulator	Close cylinder valve; bleed down system.
Face piece	Check that face piece is clean, free from cracks or deterioration; check condition of hose.
Backpack and harness	Check harness assembly for deterioration and that straps are fully extended. Check condition of straps and buckles. Make sure cylinder is secured.

TABLE 7-6

General Considerations

Before using any SCBA unit, regardless of manufacturer, model, or donning method, operational safety checks must be performed. As shown in **Table 7-6** and **Figure 7-18**, these checks must be conducted on a daily or regular basis, or immediately prior to using the SCBA unit.

Once these checks are completed and the unit operates properly, the SCBA is ready for use. If any of the components listed in **Table 7-6** does not operate properly, is broken, or is damaged, the unit is taken out of service immediately.

Storage Case

Generally, two methods are used to don SCBA units that are stored in their cases. These are the "over the head" and the "coat" methods, and the method used is usually a matter of personnel preference and training.

Firefighters should be proficient in donning SCBA in a number of different ways. Some departments have SCBA units mounted in the seat backs of the apparatus. Other services mount their SCBA units in apparatus compartments. Still other services store the SCBA units in hard-sided cases that are stored in compartments. Firefighters must be familiar and proficient with the proper donning method of their service. Firefighters must also be familiar with the "over the head" of "coat" donning method for quick response to an emergency or for donning after rest and rehabilitation during an emergency incident. For safety, the SCBA unit should only be donned once the apparatus has arrived on scene.

Donning Self-Contained Breathing Apparatus, Over the Head Method

These instructions and figures show the use of a face piece–mounted regulator. The procedure for a waist strap–mounted regulator is generally the

Figure 7-18 Firefighters must perform regular checks of SCBA to ensure the unit's ability to operate.

same, with the exception of donning the face piece. Refer to the face piece instructions for specific details. In addition, the hand position referred to may vary for left-handed users.

1. Check the air supply of the SCBA unit. The cylinder gauge must indicate the cylinder is full (100 percent), **JPR 7-1A**. Full reading will depend on the size of the unit, but on all units, the indicator needle will be in the green shaded area.
2. Slowly open the cylinder valve two to three full turns and listen for the low air alarm to activate as the regulator pressurizes. After the alarm sounds, *open the cylinder valve fully.* If it does not sound, or does not stop sounding, place the unit out of service (follow local SOP) and use another unit, again completing the first two steps.

> **Caution** On some units, if the high-pressure line was not bled down after the last use, the low air alarm may not activate. If the low air alarm does not activate, the firefighter should close the cylinder valve, bleed all air from the system, close all valves, again open the cylinder valve two to three turns, listen for the low air alarm to activate, and follow the rest of step 2 as described.

3. Compare the cylinder gauge to the regulator gauge, **JPR 7-1B**. These gauges must be within 100 psi of each other.
4. Spread and fully extend the harness assembly straps (shoulder and waist straps).

> **Streetsmart Tip** It is good practice to form the habit of readying the SCBA unit for use as it is put into storage on the apparatus. Fully extended and dressing all straps on the unit can save time on an emergency scene. Once the firefighter is tangled in an SCBA it might take precious time and possibly the assistance of another firefighter to untangle the mess.

5. With the cylinder valve pointing away from you, grasp the back plate and/or cylinder with both hands, **JPR 7-1C**. All straps should be outside of your hands.
6. With the regulator hanging freely, lower your head and lift the back plate/cylinder overhead, **JPR 7-1D**. Using your elbows to extend through the loops formed by the shoulder straps, pull the shoulder straps into the body and grasp them with the hands. Let the shoulder straps slide through the hands as the unit slides into place.
7. Balance the unit on your back by leaning forward, and tighten the shoulder straps by pulling out and down, **JPR 7-1E**. Grasp the waist strap and buckle the belt, adjusting the belt for a firm fit on the hips.

> **Caution** Firefighters should always refer to the specific donning instructions for the particular SCBA unit used by a fire department. Some units have a "chest" or "cross" strap, which should be fastened before the shoulder straps are tightened. In addition, some manufacturers recommend loosening the shoulder straps slightly so the weight of the SCBA is carried on the hips.

Once the SCBA unit is in place, refer to the section on the SCBA face piece for detailed donning instructions.

Donning Self-Contained Breathing Apparatus, Coat Method

These instructions and figures show the use of a face piece–mounted regulator. The procedure for a waist strap–mounted regulator is generally the same, with the exception of donning the face piece. Refer to the face piece instructions for specific details. In addition, the hand position referred to may vary for left-handed users.

1. Check the air supply of the SCBA unit. The cylinder gauge must indicate the cylinder is full (100 percent). Full reading will depend on the size of the unit, but on all units, the indicator needle will be in the green shaded area.
2. Slowly open the cylinder valve two to three full turns and listen for the low air alarm to activate as the regulator pressurizes. After the alarm sounds, *open the cylinder valve fully.* If it does not sound, or does not stop sounding, place the unit out of service (follow local SOP) and use another unit, again completing the first two steps.

On some units, if the high-pressure line was not bled down after the last use, the low air alarm may not activate. If the low air alarm does not activate, close the cylinder valve, bleed all air from the system, close all valves, again open the cylinder valve two to

JOB PERFORMANCE REQUIREMENT 7-1
Over the Head Method

A Checking the air supply is the first step before using SCBA.

B Always compare the cylinder and regulator gauges. They must be within 100 psi of each other.

C Grasp the back plate with both hands. Harness straps should be outside of your hands.

D Lift the back plate/cylinder overhead, using your elbows to extend through the loops formed by the shoulder straps.

E Balance the unit on your back by leaning forward, and tighten the shoulder straps by pulling down and out.

three turns, listen for the low air alarm to activate, and follow the rest of step 2 as described.

3. Compare the cylinder gauge to the regulator gauge. These gauges must be within 100 psi of each other.
4. Spread and fully extend the harness assembly straps (shoulder and waist straps).
5. Position the SCBA unit with the cylinder valve toward you. Using your left hand, grasp the left shoulder strap at the top of the back plate/harness assembly and grasp the lower portion of the same strap with your right hand, **JPR 7-2A**. When the unit is positioned correctly, the left shoulder strap will be on your right as you face the SCBA unit.
6. Lift the SCBA unit, swing it around the left shoulder and onto your back, **JPR 7-2B**. Both hands are still grasping the shoulder strap.
7. Holding the left shoulder strap with your left hand, release your right hand and put your right arm through the right shoulder strap.
8. Balance the unit on your back by leaning forward. Tighten the shoulder straps by pulling out and down on the straps, **JPR 7-2C**. Grasp the waist strap and buckle the belt, adjusting the belt for a firm fit on the hips.

Once the SCBA unit is in place, refer to the section on the SCBA face piece for detailed donning instructions.

Remember, always refer to the specific donning instructions for the particular SCBA unit used by a fire department.

Seat-Mounted Apparatus

With the use of enclosed firefighter riding positions becoming more common, many fire departments are mounting SCBA units at the seat position, as shown in **Figure 7-19**. This method allows for quick donning and the unit is readily available for regular inspection. Various types of mounting brackets and hardware are used for this type of mount, and firefighters must be familiar with the style used in their department. The specific steps for donning the SCBA unit from this position are detailed in the skill given later.

Three important safety requirements must be observed if SCBA units are mounted at the seat position. These are storing of the face piece, donning the unit while the vehicle is moving, and checking the cylinder gauge. The SCBA face piece should never be left connected to the regulator for storage. Ideally, the face piece should be stored in a bag to maintain cleanliness and keep the unit free of dust particles, which could affect operation or injure the user's eyes.

Figure 7-19 Many fire departments are mounting SCBA units at the seat position, allowing for easy access for inspection and quick donning.

In addition, a firefighter should never attempt to stand or don the SCBA unit while the apparatus is in motion. Donning gear while the vehicle is in motion may require loosening of the seat belt, and the firefighter may be thrown around the enclosed cab in the event of a sudden stop or turn. *Firefighters must always remain seated with seat belt fastened when the vehicle is moving.*

From the seated position, it is extremely difficult, if not impossible, to check the cylinder gauge and compare it to the regulator gauge. If this is not accomplished prior to response, then after dismounting the apparatus firefighters should use the buddy system to check each other's gauges, **Figure 7-20**.

> **Streetsmart Tip** It is a good rule of thumb for all firefighters to remain seated with seat belts fastened until the driver/operator of the apparatus has stopped the vehicle and set the parking brake.

JOB PERFORMANCE REQUIREMENT 7-2
Coat Method

A With your left hand, grasp the left shoulder strap at the top of the back plate/harness assembly and grasp the lower portion of the same strap with your right hand. When the unit is positioned correctly, the left shoulder strap will be on your right as you face the SCBA unit.

B Lift the SCBA unit, then swing it around the left shoulder and onto your back. Both hands are still grasping the shoulder strap.

C Balance the unit on your back by leaning forward. Tighten the shoulder straps by pulling out and down. Grasp the waist strap and buckle the belt, adjusting the belt for a firm fit on the hips.

Figure 7-20 After dismounting the apparatus, firefighters use the buddy system to check each other's cylinder gauges.

Donning Self-Contained Breathing Apparatus, Seat-Mounted Apparatus

These instructions and figures show the use of a face piece–mounted regulator. The procedure for a waist strap–mounted regulator is generally the same, with the exception of donning the face piece. Refer to the face piece instructions for specific details. In addition, the hand position referred to may vary for left-handed users.

1. Check the cylinder valve prior to response, if possible. *Do not attempt to check the unit when the vehicle is in motion.* Using the buddy system, turn the unit on and check the cylinder gauge when you dismount the apparatus.
2. With straps fully extended, place the right arm between the right shoulder strap and back plate assembly, **JPR 7-3A**. Repeat this action for the left arm.
3. Tighten the shoulder straps by pulling out and down on the straps, **JPR 7-3B**. Grasp the waist strap and buckle the belt, adjusting the belt for a firm fit on the hips. *Do not entangle the waist belt with the vehicle seat belt.*
4. Dismount the apparatus, **JPR 7-3C**, and recheck shoulder and waist straps.

Always refer to the specific donning instructions for the particular SCBA unit used by a fire department. Some units have a "chest" or "cross" strap, which should be fastened before the shoulder straps are tightened. In addition, some manufacturers recommend loosening the shoulder straps slightly so the weight of the SCBA is carried on the hips.

5. With a partner, **JPR 7-3D** and **E**, slowly open the cylinder valve two to three full turns and listen for the low air alarm to activate as the regulator pressurizes. After the alarm sounds, *open the cylinder valve fully.* If the alarm does not sound, or does not stop sounding, place the unit out of service (follow local SOP) and use another unit, again completing the first two steps. Compare the cylinder gauge to the regulator gauge. These gauges must read within 100 psi of each other.

On some units, if the high-pressure line was not bled down after the last use, the low air alarm may not activate. If the low air alarm does not activate, close the cylinder valve, bleed all air from the system, close all valves, again open the cylinder valve two to three turns, listen for the low air alarm to activate, and follow the rest of step 2 as described.

Once the SCBA unit is in place, refer to the section on the SCBA face piece for detailed donning instructions.

Compartment or Side-Mounted Apparatus

Some fire departments carry their SCBA units mounted on the side of apparatus or in compartments. Depending on the height of the apparatus compartment and the bracket used to mount the SCBA, the donning method may be similar to that for the seat-mounted position except that the firefighter is standing rather than sitting. If the cabinet height or mounting bracket position of the SCBA does not allow for ease of donning while standing, the firefighter should remove the SCBA unit and use the "coat" or "over the head" method of donning.

To don the SCBA unit from a compartment or side-mounted storage position, it is recommended to follow the donning instructions for the method best suited for the particular mounting style.

Donning the SCBA Face Piece

Most SCBA face pieces are donned in a similar manner, with the difference being in the style of head straps and the location of the regulator. Proper donning of the SCBA face piece is essential to

JOB PERFORMANCE REQUIREMENT 7-3
Seat-Mounted Apparatus

A With straps fully extended, place the right arm between the right shoulder strap and back plate assembly. Repeat this action for the left arm.

B Tighten the shoulder straps by pulling out and down on the straps. Grasp the waist strap and buckle the belt, adjusting the belt for a firm fit on the hips.

C Dismount the apparatus and recheck shoulder and waist straps.

D With a partner, slowly open the cylinder valve two to three full turns and listen for the low air alarm to activate as the regulator pressurizes.

E *Open the cylinder valve fully and compare the cylinder gauge to the regulator gauge.*

protect the firefighter from the effects of toxic gases and hazardous atmospheres. Each firefighter must be fitted for the face piece to be used with a particular manufacturer's SCBA as required by OSHA 29 CFR 1910.134:

29 CFR 1910.134 (f)(1)
The employer shall ensure that employees using a tight-fitting facepiece respirator pass an appropriate qualitative fit test (QLFT) or quantitative fit test (QNFT) as stated in this paragraph. 1910.134(f)(2)
The employer shall ensure that an employee using a tight-fitting facepiece respirator is fit tested prior to initial use of the respirator, whenever a different respirator facepiece (size, style, model or make) is used, and at least annually thereafter.[1]

In addition, 29 CFR 1910.134, NFPA 1404, and SCBA manufacturers' recommendations prohibit any facial hair, which may interfere with proper fit and seal of the face piece. Again, all firefighters must be familiar with the specific equipment used by their fire department.

Caution No glasses or facial hair are allowed between the skin and the sealing surface of the SCBA.

Donning Self-Contained Breathing Apparatus Face Piece

1. With the head straps/harness in the full out position, hold the head harness with one hand while placing the face piece on the face, with the chin properly located in the chin pocket, **JPR 7-4A**.
2. Fit the face piece to the face, pull the head harness over the head, and make sure straps are lying flat against the head with no twists.
3. Tighten the neck or lower straps by simultaneously pulling both straps to the rear, **JPR 7-4B**.
4. Check the fit of the head harness by stroking the harness down the back of the head using one or both hands. Retighten the neck straps at this time, **Figure 7-4C**.
5. Adjust the temple straps by simultaneously pulling both straps to the rear.
6. Check for proper seal by attaching the regulator and inhaling a breath, activating the regulator, **JPR 7-4D**. Hold that breath for about five seconds and listen and feel for any air leaks. If leaks occur, repeat steps 2 through 5. In addition, once the regulator or breathing tube is connected, listen for air leaks from the positive pressure mode. If leaks occur, repeat steps 2 through 5.

Streetsmart Tip It is common to see firefighters testing the seal of a face piece by placing a hand over the opening and inhaling. The vacuum will draw the mask to the face. This method is not accurate, as pressure can be inadvertently applied to the face piece, pushing it toward the face.

7. Pull the protective hood on over the entire head to cover exposed skin, **JPR 7-4E**. Ensure all long hair is inside the protective hood. Also ensure that the protective hood is tucked completely down inside the bunker coat. Pull up and fasten the bunker coat collar. Place the helmet on, and adjust it for the increased size of the head, due to the face piece and protective hood. Fasten the helmet strap. Make sure the helmet shroud is pulled over the neck and coat collar. Place protective gloves on.
8. As you approach the hazardous atmosphere, install regulator on face piece or connect the hose to the regulator for a waist strap–mounted regulator. Airflow to the face piece will begin with the breathing action. Again, listen for air leaks from the positive pressure mode. If leaks occur repeat steps 2 to 5.

Streetsmart Tip Firefighters should always work in pairs, and each firefighter should check the other for any situations that are unsafe. One firefighter should ensure all protective equipment is properly donned and that all areas of the skin are covered. All layers of PPE should interface with the additional layers. The roles are then reversed and the second firefighter is checked for safety.

Caution It is important to tighten either neck or lower straps first, pulling straight back. Pulling straps outward may damage the straps and prevent proper seal. The temple straps are adjusted by pulling the strap ends toward the rear of the head. Regardless of the type of regulator used, one should always check for proper face piece seal. A firefighter who does not get a leak-free seal should not enter the hazardous environment.

Removing/Doffing the SCBA Unit

Upon exiting the hazardous atmosphere, firefighters should remove the SCBA unit and rest. Generally, to remove the unit, the donning process is reversed

SELF-CONTAINED BREATHING APPARATUS

JOB PERFORMANCE REQUIREMENT 7-4
Face Piece

A With the head straps/harness in the full out position, hold the head harness with one hand while placing the face piece on the face, with the chin properly located in the chin pocket.

B Tighten the neck or lower straps by simultaneously pulling both straps to the rear.

C Check the fit of the head harness by stroking the harness down the back of the head. Retighten the neck straps as necessary.

D Check for proper seal by attaching the regulator and inhaling a breath, activating the regulator. Hold that breath for about five seconds and listen and feel for any air leaks.

E Pull on protective hood to cover exposed skin and head. Place helmet on head with earflaps down. Pull up coat collar and buckle. Snug or buckle helmet straps. Place gloves on hands.

following manufacturers' instructions for the type or model of unit used. If awaiting another assignment, the face piece should be removed to allow normal breathing and to conserve air. The regulator or face piece must not be contaminated by laying it on the ground or other dirty environment. This may cause injury if firefighters need to use the unit again.

Depending on local SOP, once an assignment with SCBA is complete, firefighters should report to rehabilitation for rest, fluids, and monitoring of vital statistics. Many respiratory protection programs limit firefighters to consumption of two SCBA cylinders, after which mandatory rehabilitation is required. Again, this rest period for fluids and monitoring of vital statistics reduces the physical and mental stress encountered during firefighting operations.

SELF-CONTAINED BREATHING APPARATUS OPERATION AND EMERGENCY PROCEDURES

In accordance with 29 CFR 1910.134, NFPA 1404, and NFPA 1500, fire departments must establish respiratory protection programs for firefighters using SCBA. In addition, the firefighters must be proficient in the safe use of SCBA, donning and doffing procedures, individual limitations, and limitations of the SCBA unit.

Safe Use of SCBA

Safe use of SCBA is essential to firefighter survival during operations requiring its use. Firefighting tasks are both mentally and physically demanding and firefighters must be in excellent physical condition. The SCBA unit and protective equipment add weight and bulk to the firefighter, causing increased exertion with loss of body fluids through perspiration. These actions increase during firefighting operations and firefighters must be aware of them and of the symptoms of heat stress and their own limitations and abilities.

The following items are essential for maximum safety while using SCBA:

- All firefighters using SCBA must be certified physically fit for respirator use in accordance with local policy or 29 CFR 1910.134.
- Fire departments must establish an accountability system, **Figure 7-21**, to track personnel entering an IDLH atmosphere.
- Firefighters must work in teams of two as a minimum. If one member of the team must leave for any reason, all members of the team must exit the IDLH.
- In accordance with OSHA CFR 29 1910.134, the "two in/two out" rule states that a rescue team of two firefighters must be available and ready to rescue or assist the firefighters operating in the IDLH.
- PASS devices, **Figures 7-22A** and **B**, must be activated in order to function. If the PASS is an integral part of the SCBA unit, it must be activated before entering a hazardous environment.
- Fire departments should establish policies for firefighter rehabilitation during operations

> **Firefighter Fact** New SCBA designs incorporate a variety of innovative technologies to provide maximum safety for the user. Some SCBAs incorporate the PASS device or distress alert as an integral component of the unit. These systems offer redundant visual and audible warning alarms, with the audible alarm located on the harness assembly and the visual warning, a flashing LED, located on the regulator or remote gauge.
>
> Other new designs include a technology that projects a display of the level of air in the cylinder to the face piece, straight ahead in the field of vision of the firefighter. This "Head's Up Display (HUD)" allows a firefighter to keep the head up, look straight, and continue to perform firefighting or other duties, while still being able to see the level of air remaining in the SCBA. The HUD uses an LED indication for air levels of full three-quarter, one-half, and one-quarter. Further, the HUD will flash visual notification of one-half and one-quarter remaining air supply. The HUD may also include a visual indicator of temperature. The HUD may be retrofitted into older SCBA units.
>
> Another new design feature is the Rapid Intervention Crew Universal Air Connection (RIC/UAC.) The RIC/UAC is a common style connector for use in emergency situations, such as rescuing a trapped firefighter. Rapid intervention teams (RIT) can carry an extra SCBA cylinder and recharge the trapped firefighter's air, as well as their own, during the rescue mission. The common connection is mounted on all compliant SCBA within 4 inches of the cylinder valve outlet and has a relief valve to prevent overfilling of lower pressure cylinders, such as 2,216-psi cylinder if a 4,500-psi cylinder is used as a filling source.

SELF-CONTAINED BREATHING APPARATUS 169

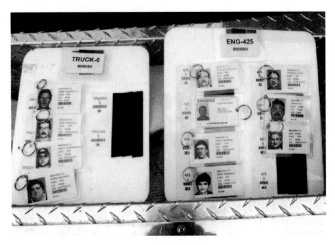

Figure 7-21 A typical accountability system to track personnel entering an IDLH atmosphere.

Figure 7-23 EMS personnel should monitor vital signs, and firefighters must hydrate to replace body fluids during firefighting operations.

requiring SCBA use. This policy should include the maximum number of cylinders a firefighter can use before mandatory rehabilitation.

- During rehabilitation, EMS personnel should monitor vital signs, **Figure 7-23**, and firefighters must hydrate to replace body fluids.
- Air consumption will vary with each individual's physical condition, the level of training, the task performed, and environment.
- The SCBA face piece should never be removed in a contaminated environment.
- Depending on an individual's air consumption and the amount of time required to exit a hostile environment, the low air alarm may not provide adequate time to exit. The individual should not panic, but move deliberately to a safe environment. All team members must exit together, even if only one low air alarm is sounding.

> **Streetsmart Tip** Air consumption: All firefighters must be aware of the amount of air an individual consumes using SCBA. This is accomplished by performing an air consumption activity. Firefighters in full protective equipment and SCBA perform simulated firefighting activities, such as advancing hoselines, search and rescue, and ventilation. The time is noted when the firefighter begins the activity, when the low air alarm sounds, and when the air supply is depleted. The air consumed divided by the time of the activity will provide an *approximate rate* (this is a controlled environment and actual fireground operations will result in higher air consumption rates) of air consumption and the amount of time/air supply available to the firefighter to exit after the low air supply alarm sounds. Each firefighter should perform the test a minimum of two times to provide an average rate. In addition, the test should be conducted on an annual basis to observe changes that may affect the air consumption rate.

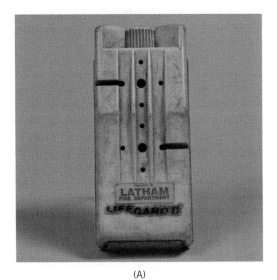

(A)

(B)

Figure 7-22 PASS devices used with SCBA include (A) the independent self-contained unit that attaches to the harness and (B) the integral type that is part of the SCBA unit.

Figure 7-24 Firefighters should always remain low. Heat and smoke from the fire will rise; floor level provides the best visibility.

Operating in a Hostile Environment

Firefighters will use SCBA in hostile environments and toxic atmospheres that will have limited visibility. During firefighting operations requiring the use of SCBA, teams conducting search or fire attack must be able to function effectively. Firefighters should follow these general rules:

- Always check in with the accountability/company officer when entering or exiting a building or when an assignment is complete.
- Always remain low, **Figure 7-24**. Heat and smoke from the fire will rise, so floor level provides the best visibility. This will also reduce the risk of injury to a firefighter from walking into a hole in the floor or falling into a pit or shaft.
- Check the environment; extend a gloved hand above the head to determine the heat conditions above. This can also be done with a tool, extending it above the head and checking the end for heat.
- Never remove the face piece.
- Maintain an awareness of location, floor level, front or rear of building.
- Ventilate as you advance, as long as it will not spread the fire. Ventilation allows the products of combustion to escape and provides a better environment.
- Check for outside openings such as windows and doors. This will provide a means of escape in an emergency and provide the firefighters' location to outside personnel.
- *Always* maintain direct contact with other team members. It is also acceptable for both firefighters to remain in contact by holding on to the same tool or a short piece of rope. There are also times where firefighters will be following a rope or hoseline into or out of a hostile environment. In this case, or in cases where direct contact is broken, firefighters should remain in constant voice communication until direct contact can be reestablished or until they are safely out of the IDLH atmosphere.
- Never enter a hostile environment alone.

Restricted Openings

As a general rule, areas with restricted openings that a firefighter cannot fit into should be probed with a tool. SCBA removal, or the loosening of straps and repositioning of the SCBA unit, can reduce the profile of the firefighter. These are last-resort measures and only used for escaping through a wall or a small opening under emergency conditions. Firefighters should remove or loosen only those parts of the SCBA unit necessary to negotiate the restricted opening. The exact procedures for accomplishing a reduced profile depend on the type of SCBA unit used. The following rules must be observed:

- Never remove the face piece to reduce the profile.
- Loosen straps and rotate the SCBA unit under the arm along the rib cage as a first step, **Figure 7-25.**
- As a last resort, perform a "full escape" by removing the harness assembly and holding the SCBA unit in front of you.
- Always maintain contact with the SCBA unit, **Figure 7-26**. Hold onto the shoulder straps and regulator assembly or regulator on a waist-mounted unit.
- Maintain control of the unit. Do not move it away or allow the face piece to be pulled away.
- Practice this procedure with an obscured face piece to accomplish it by feel.

> **Caution** The full escape procedure should only be used as a last resort for an emergency escape from a hostile environment. It may void the NIOSH certification for the SCBA; however, it is a tactic that can save lives.

SELF-CONTAINED BREATHING APPARATUS 171

Figure 7-25 To reduce profile with an SCBA unit, the firefighter loosens straps and rotates the SCBA unit under the arm along the rib cage.

Figure 7-26 Firefighters should only remove the SCBA unit as a last resort to escape from a life-threatening situation.

Emergency Procedures

SCBA units can be damaged and malfunction or the firefighter may become tangled in debris while operating. If this happens, a number of emergency procedures exist that will assist in the safe escape from the hazardous environment. Above all, firefighters must remain calm and rely on their training and knowledge.

Firefighters must be familiar with and practice the emergency procedure used by their fire department for failure of the unit. In the event of an SCBA unit malfunction, firefighters should follow these guidelines:

- Remain calm; rely on your training.
- Immediately exit the hazardous environment. All team members must exit—never leave a firefighter alone.
- Manually activate the PASS device.
- If equipped with a portable radio, announce a "Mayday" message, giving current location if possible.
- In case of a regulator failure, breathe using the bypass valve. Close the main line valve, open the bypass valve, breathe, and close the bypass valve to conserve air. This procedure may vary by manufacturer.
- For a face piece failure, use the bypass valve to increase airflow to keep the face piece free from toxic gases.

- If necessary, use the protective hood as a filter. *Use the "buddy breathing" attachment on an SCBA as a last resort because this will deplete the air supply at least twice as fast.* Many SCBA units are equipped with an emergency escape breathing support system, or a "buddy breathing" attachment. Current NIOSH and NFPA standards do not allow the use of this option. The use of these devices voids the NIOSH certification and use cannot be recommended at this time. Fire departments should review SOPs concerning "buddy breathing" and determine the best option for their needs.
- Practice these procedures during training. During an emergency is not the time to try something new.

If the firefighter becomes entangled in debris, other team members will assist in freeing the trapped firefighter. This is accomplished by cutting away the debris or removing the SCBA unit as a last resort.

> **Streetsmart Tip** Many firefighters carry a small tool or wire cutters to free themselves from mattress springs or suspended ceiling tie wires, some of the most common debris causing entanglement.

INSPECTION AND MAINTENANCE OF SELF-CONTAINED BREATHING APPARATUS

As with any piece of equipment used in the fire service, SCBA units must be ready to go on a moment's notice. Considering the importance of SCBA to the firefighter, inspection must be completed on a daily or regular basis.

There are a number of variations of maintenance procedures for inspection and servicing of SCBAs. Always follow the manufacturers' instructions and recommendations provided with the unit. The step-by-step instructions given in this chapter are intended for training purposes and may differ from the procedures recommended for the SCBA used by a particular organization.

Daily Maintenance

SCBA units should be checked daily to ensure they are secured and ready for operation. When an SCBA is used during an emergency scene or a training exercise, the unit should be serviced and checked in the same manner.

Daily Inspection of SCBA

1. Check to ensure the cylinder is full, **JPR 7-5A.**
2. Open the cylinder valve slowly, two to three full turns, until the low air alarm is activated. When the alarm stops sounding, open the cylinder valve completely. If the alarm does not activate or continues to sound, place the unit out of service following department procedures.

Depending on the manufacturer, some units—if the high-pressure line was not bled down after use—will not have activation of the low air alarm. In this case, close the cylinder valve, bleed all air from the system, close all valves and open the cylinder valve as described.

3. Compare the cylinder and regulator gauges, **JPR 7-5B.** Gauge readings should be within 100 psi.
4. Check to see that all hose connections are tight and free from leaks.
5. Check the bypass and the main line valves for operations.
6. Close the cylinder valve and (depending on manufacturer's instructions) drain all air from the system. Note that the low air warning alarm activates when the regulator gauge reads about 20 to 25 percent.
7. Ensure the main line valve (if the unit has one) is open and that the bypass valve is fully closed, **JPR 7-5C** and **D.**
8. Check the condition of the face piece, if one is assigned to the unit. Ensure cleanliness and check for cracks, for missing or broken straps, and that the diaphragm is in place. Clean face piece per local department policy, normally using a 1 percent bleach water solution.
9. Check the harness assembly, condition of belts, and cylinder fastening device.
10. Check the operation of the PASS device.

Monthly Maintenance

The monthly SCBA check contains all elements of the daily check but adds several checks of the mechanics of the system. Firefighters should be trained by their department to conduct the monthly maintenance. An example of a monthly check sheet is shown in **Figure 7-27.** Any irregularities should be noted and repaired, or the SCBA should be pulled from service until a department technician can repair the unit.

Annual and Biannual Maintenance

NIOSH and SCBA manufacturers require a number of different functional tests of SCBA units. Only a manufacturer's authorized or trained service personnel shall conduct these tests. Firefighters should refer to the instructions for the SCBA units used.

Changing SCBA Cylinders

SCBA cylinders must be changed after use, following local SOPs. Depending on the size and air consumption rate of a firefighter, allowing a cylinder at 90 percent full to remain in service could mean a loss of two to five minutes of air supply. This time may be the difference in being able to successfully exit or escape from a hazardous situation.

The firefighter should follow these procedures for cylinder replacement:

Cylinder Replacement Procedure

1. Have full air cylinder ready for use.
2. Place the SCBA unit on the ground with the cylinder valve toward you, **JPR 7-6A.**

JOB PERFORMANCE REQUIREMENT 7-5
Daily Inspection of SCBA

A Check to ensure that the cylinder is full.

B Compare cylinder gauge and regulator gauge. Gauge readings should be within 100 psi of each other.

C and **D** After checking for proper operation, ensure air is purged from the system, the main line valve (if the unit has one) is open, and the bypass valve is fully closed.

3. Push in the knob on the cylinder valve and close the cylinder valve.
4. Bleed residual air in the high-pressure line by slowly opening the bypass valve. When the flow of air from the face piece stops, close the valve. On some units, the pressure must be released by "breathing down" the regulator or opening the main line valve. Refer to the manufacturer's instructions for the unit in use.
5. Disconnect the high-pressure coupling from the cylinder, **JPR 7-6B**. If the knob is difficult to turn, the high-pressure line may still be pressurized; repeat step 5 and attempt to disconnect the coupling again. Lay the coupling on the back plate to prevent dirt or grit from contaminating the threads or seat.
6. Release the cylinder clamp or locking mechanism used on the unit and slide the empty cylinder down out of the harness back plate assembly, **JPR 7-6C**.
7. Inspect the O-ring on the seat of the high-pressure hose for any damage, scratches, or foreign matter, **JPR 7-6D**. If the O-ring is damaged, replace it in accordance with the manufacturer's instructions.

SCBA FIELD MAINTENANCE SHEET

DATE: _____ REDUCER #: _____ LOCATION: _____

Air-Pak shows no use; visual and function checked only _____

2.2 / 4.5   30 DAY / REPAIR

(REGULATOR ASSY.)

REG. COVER:	OK	Y / N	RETAINING RING:	OK	Y / N
DIAPHRAGM:	OK	Y / N	REG. GASKET:	OK	Y / N
PURGE VALVE:	OK	Y / N	DONNING SWITCH:	OK	Y / N
THUMB LATCH:	OK	Y / N	REG. HOSE:	OK	Y / N

(BACKFRAME ASSY.)

SHOULDER HARNESS:	OK	Y / N	WAIST BELTS:	OK	Y / N
REMOTE GAUGE:	OK	Y / N	ALL HOSES:	OK	Y / N
EBSS EQUIPMENT:	OK	Y / N	PACK ALERT:	OK	Y / N

(FUNCTION TEST & LEAK TEST)

BREATHING:	OK	Y / N	PURGE VALVE:	OK	Y / N
LOW AIR ALARM:	OK	Y / N	REMOTE GAUGE:	OK	Y / N
DONNING SWITCH:	OK	Y / N	ANY LEAKAGE:	OK	Y / N

(BOTTLE CONDITION)

Note any excessive wear or damage to bottle and/or valve. Bottle # _____

*LIST ANY PROBLEMS NEEDING REPAIRS: _____
_____
_____

*LIST ANY PARTS REPLACED: _____
_____
_____

Rev: 07-04-02   EQUIP. # _____

Figure 7-27 SCBA Field Maintenance Sheet.

8. Replace with a fully charged cylinder. Reverse the process in step 6 by sliding the cylinder into the back plate assembly. Do not secure the cylinder clamp until the high-pressure line is aligned and connected.
9. Align the high-pressure hose with the cylinder valve assembly and connect the hose. This connection should be hand tightened.
10. Lock the cylinder into place on the back plate assembly.

SELF-CONTAINED BREATHING APPARATUS

JOB PERFORMANCE REQUIREMENT 7-6
Cylinder Replacement Procedure

A Place the SCBA unit on the ground with the cylinder valve toward you.

B Disconnect the high-pressure coupling from the cylinder. If the knob is difficult to turn, the high-pressure line may still be pressurized.

C Release the cylinder clamp or locking mechanism used on the unit and slide the empty cylinder down out of the harness back plate assembly.

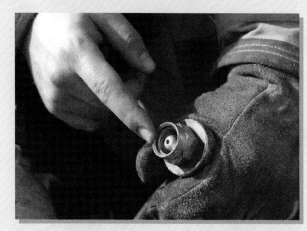

D Inspect the O-ring on the seat of the high-pressure hose for any damage, scratches, or foreign matter.

11. Open the cylinder valve, check for air leaks, and compare the cylinder valve reading with the regulator reading. These should be within 100 psi of each other.
12. Close the cylinder valve, bleed all air from the system, ensure that the main line and bypass valves are in the proper position, and place the unit in its mounting bracket or storage case.

Other units may have an integrated PASS device; once the pressure is off the system, the PASS device should be shut off.

Additional Steps for Two-person SCBA Cylinder Replacement

There are many times on a fire scene where firefighters will require other firefighters to change out a cylinder while they are wearing the SCBA. This can be done safely using the same sequence of events, with the following exceptions:

1. While wearing the SCBA, stop and lean forward, **JPR 7-7A** or kneel on the ground, **JPR 7-7B.** Ensure that your head is pointed down and out of the way of the cylinder as it is changed.

JOB PERFORMANCE REQUIREMENT 7-7
Additional Steps for Two-person SCBA Cylinder Replacement

A Stop and lean forward in a stable position.

B In some cases, it will be easier to kneel on the ground.

C Always know the amount of air in your cylinder.

2. The firefighter changing out the cylinder should follow the steps for removing the cylinder as noted in steps 1 through 7 above, taking care not to hit the head of the firefighter wearing the pack while removing the cylinder.
3. Before replacing a full cylinder, the firefighter changing the cylinder should ensure that it is full. It is good practice to quickly show the bottle gauge to the firefighter wearing the SCBA unit; this way both firefighters have seen the gauge and know the cylinder is full, **JPR 7-7C**.
4. The firefighter changing the cylinder should replace the cylinder as noted in steps 8 through 11 of JPR 7-6.

Servicing SCBA Cylinders

When the cylinder capacity is below full, the cylinder must be serviced. As noted earlier in this chapter, filling SCBA cylinders is completed using a

SELF-CONTAINED BREATHING APPARATUS

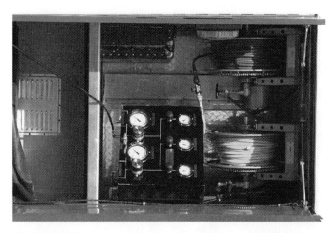

Figure 7-28 A cascade system is one of the systems available to service SCBA cylinders. These may be fixed or mobile units.

cascade system, **Figure 7-28**, or compressor/purifier system, **Figure 7-29**. Regardless of how this is accomplished, the following safety precautions must be followed:

- The air source being used must be tested and certified meeting the requirements of Compressed Gas Association Pamphlet G-7.1-1989, OSHA 29 CFR 1910.134, and NFPA 1500.
- All cylinders must have a current hydrostatic test date: no older than five years for aluminum/steel cylinders, or three years and less than fifteen years from manufacture date for composite cylinders.
- All fill stations must have fragmentation containment devices in case of cylinder failure.
- All manufacturers' recommendations should be followed, especially for recalls or safety notices concerning cylinder capacity.

Figure 7-29 A compressor/purifier system is the second type of system used to service SCBA cylinders. These are usually located at stations or other fixed facilities; however, larger departments may have mobile units.

- Fill rate may vary; 300 to 600 psi per minute is considered an acceptable range.

Servicing an SCBA Cylinder Using a Cascade System

Many organizations require certification before using a cascade system or air compressor/purifier system. Do not attempt this skill without training on the specific system used by a particular fire department and follow the procedures recommended by the system manufacturer. Never attempt to fill a cylinder above its rated capacity and never attempt to fill bottles with different rated capacities at the same time.

1. Check the hydrostatic test date of the cylinder, **JPR 7-8A**. Remove from service if test is out of date or usable service life for composite cylinders has passed.
2. Inspect the cylinder for any signs of physical damage, discoloration from heat, gouges, or nicks in the cylinder surface. *Never fill an SCBA cylinder that has visible signs of damage or an out-of-date hydrostatic test.*
3. Place the cylinder in the fragmentation containment device, **JPR 7-8B**.
4. Connect the fill hose to the cylinder, and if it is equipped with a bleed valve, close it.
5. Open the SCBA cylinder valve.

Caution A firefighter should never attempt to fill an SCBA cylinder without fragmentation protection. It is important never to fill a composite SCBA cylinder in a water-filled fragmentation protection device. Water may infiltrate between the composite layers, causing a weakness and failure.

JOB PERFORMANCE REQUIREMENT 7-8
Servicing an SCBA Cylinder Using a Cascade System

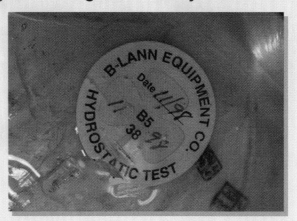

A Check the hydrostatic test date of the cylinder and inspect the cylinder for physical damage, discoloration from heat, gouges, or nicks in the cylinder surface.

B Place the cylinder in the fragmentation containment device. Connect the fill hose to the cylinder, and if it is equipped with a bleed valve, close it. Open the SCBA cylinder valve.

C Observe the cylinder gauge and watch that the cylinder fills at approximately 300 to 600 psi per minute. Close the cascade valve when pressure equalizes. If cylinder is not full—open cascade valve on cylinder with next highest pressure.

6. Open the valve at the cascade system manifold or the fill hose valve, depending on how the system is designed. Open both valves if the system has both.
7. Open the valve of the cascade cylinder with the lowest pressure. This pressure must be higher than the pressure in the cylinder being filled. *Control the fill rate of air to avoid excessive heating or chatter in the cylinder. If the cylinder heats or chatters, reduce the fill rate.*
8. Observe the cylinder gauge and watch that the cylinder fills at approximately 300 to 600 psi per minute, **JPR 7-8C**.
9. Close the cascade cylinder valve when the pressure equalizes with the pressure in the SCBA cylinder. If the SCBA cylinder is not full, open the valve on the cascade system

JOB PERFORMANCE REQUIREMENT 7-9
Servicing an SCBA Cylinder Using a Compressor/Purifier System

A Check the hydrostatic test date of the cylinder. Remove from service if the cylinder test date is not current or, if a composite cylinder, it is beyond the fifteen-year service life.

B Place the cylinder in the fragmentation containment device.

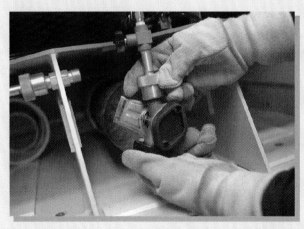

C Connect the fill hose to the cylinder and close the bleed valve.

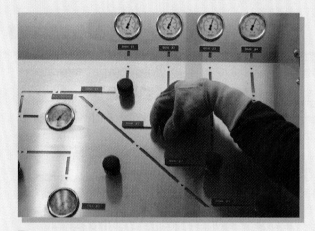

D Operate the compressor/purifier system in accordance with the manufacturer's instructions.

cylinder with next highest pressure. Repeat this step until the SCBA cylinder is full.
10. Close all valves on the cascade system, then close the SCBA cylinder valve.
11. Open the cascade system fill hose bleeder valve and bleed off excess line pressure. *Failure to follow this step could result in injury from the fill line as it discharges high-pressure air or blows off the O-ring seal on the end of the fill hose.*
12. Disconnect the cascade system fill hose from the cylinder valve, remove the cylinder from the fragmentation container, and place in storage.

Servicing an SCBA Cylinder Using a Compressor/Purifier System

Many organizations require certification before using either a cascade system or air compressor/purifier system. Do not attempt this skill without training on the specific system used by a particular fire department and follow the procedures recommended by the system manufacturer. Never attempt to fill a cylinder above its rated capacity.

1. Check the hydrostatic test date of the cylinder, **JPR 7-9A**. Remove from service if the cylinder test date is not current

or, if a composite cylinder, it is beyond the fifteen-year service life.
2. Inspect the cylinder for any signs of physical damage, discoloration from heat, gouges, or nicks in the cylinder surface. *Never fill an SCBA cylinder that has visible signs of damage or an out-of-date hydrostatic test.*
3. Place the cylinder in the fragmentation containment device, **JPR 7-9B**.
4. Connect the fill hose to the cylinder, and if equipped with a bleed valve, close valve, **JPR 7-9C**.
5. Open the SCBA cylinder valve.
6. Operate the compressor/purifier system in accordance with the manufacturer's instructions and open the outlet valve, **JPR 7-9D**.
7. Set the cylinder pressure adjustment on the compressor/purifier system to the correct pressure to fill the cylinder. Never attempt to fill a cylinder above its rated capacity.
8. Open the manifold valve and again check the fill pressure.
9. Open the fill station valve to begin filling the cylinder.
10. Observe the cylinder gauge and watch that the cylinder fills at approximately 300 to 600 psi per minute. Control the fill rate of air to avoid excessive heating or chatter in the cylinder. If the cylinder heats or chatters, reduce the fill rate.
11. Close the fill station valve when the cylinder is full.
12. Close the SCBA cylinder valve.
13. Open the fill hose bleeder valve and bleed off excess line pressure. *Failure to follow this step could result in injury from the fill line as it discharges excess air pressure or damages the O-ring seal.*
14. Disconnect the fill hose from the cylinder valve, remove the cylinder from the fragmentation container, and place in storage.

Lessons Learned

The SCBA unit is to a firefighter what a weapon is to a soldier. There are many events in military history where soldiers have lost their lives and battles because of lack of knowledge of their weapons. The same holds true for firefighting.

Streetsmart Tip Firefighters should be encouraged to perform a "buddy check" with their partner or team members before entering a hazardous environment. Firefighting is teamwork and this teamwork can be a lifesaver.

As with any firefighter skill, there is no substitute for proper training with SCBA. In addition to initial training and certification in SCBA use, continued practice and advanced training are necessary to ensure that firefighters maintain proficiency. The same is true for maintenance because some firefighters have a tendency to treat SCBA as a tool, such as a ladder or ax, rather than as the vital piece of protective equipment that it is. To prevent failures from occurring and endangering lives, it is imperative that firefighters thoroughly inspect and test the function of SCBA as often as possible.

Each year a number of line-of-duty deaths are documented where firefighters for various reasons have died because they did not rely on their training with self-contained breathing apparatus. Regardless of the nature of the alarm, firefighters must always be prepared to go in harm's way; to do so safely, they must be knowledgeable and proficient in the use of SCBA.

KEY TERMS

Asphyxiation Condition that causes death due to lack of oxygen and an excessive amount of carbon monoxide or other gases in the blood.

Authority Having Jurisdiction (AHJ) The responsible governing organization or body having legal jurisdiction.

Carbon Monoxide (CO) Colorless, odorless, poisonous gas that when inhaled combines with the red blood cells excluding oxygen.

Closed-Circuit SCBA A type of SCBA unit in which the exhaled air remains in the system to be filtered and mixed with oxygen for reuse.

Hypoxia A deficiency of oxygen.

Immediately Dangerous to Life and Health (IDLH) The maximum level of danger one could be exposed to and still escape without experiencing any effects that may impair escape or cause irreversible health effects.

Mask Confidence or "Smoke Divers" Training Training courses designed to develop a firefighter's skills and confidence for using SCBA.

NFPA 1404 National Fire Protection Association standard created by the Fire Service Training Committee detailing the requirements for fire service SCBA programs, including training and maintenance procedures.

NFPA 1500 National Fire Protection Association standard created by the Technical Committee on Fire Service Occupational Safety and Health that addresses a number of issues concerning protective equipment.

NFPA 1981 National Fire Protection Association standard specific to open-circuit SCBA for fire service use that contains additional requirements above the NIOSH certification.

NIOSH National Institute for Occupational Safety and Health, 42 CFR Part 84, sole responsibility for testing and certification of respiratory protection including fire service SCBA.

Open-Circuit SCBA A type of SCBA unit in which the exhaled air is vented to the outside atmosphere.

OSHA 29 CFR 1910.134 Standard establishing minimum medical, training, and equipment levels for respiratory protection programs.

Oxygen-Deficient Atmosphere An atmosphere with an oxygen content below 19.5 percent by volume.

Positive Pressure A feature of SCBA providing a continuous supply of air, delivered by the regulator to the face piece, keeping toxic gases from entering. This pressure (1½ to 2 psi, depending on the manufacturer) is slightly above atmospheric pressure.

Pulmonary Edema Fluid filling the lungs causing death by drowning.

Respiratory Protection Programs Management programs designed to ensure employee respiratory protection as required by OSHA 29 CFR 1910.134 and NFPA 1500.

Respiratory System The system of the human body that exchanges oxygen and waste gases to and from the circulatory system.

Self-Contained Breathing Apparatus (SCBA) A type of respiratory protection in which a self-contained air supply and related equipment are worn or attached to the user. Fire service SCBA is required to be of the positive pressure type.

Supplied Air Respirator (SAR) A type of SCBA in which the self-contained air supply is remote from the user, and the air is supplied by means of air hoses.

REVIEW QUESTIONS

1. List two conditions requiring respiratory protection and explain the effects of these conditions on the human body.
2. List and explain the effects of oxygen deficiency on the human body.
3. List and explain one legal requirement for use of SCBA.
4. List two different types of operation of self-contained breathing apparatus and describe each.
5. List the four basic assemblies of an SCBA unit.
6. Explain and demonstrate two different SCBA donning procedures at 100 percent accuracy in the time limit established by the authority having jurisdiction.
7. Explain and demonstrate routine inspection procedures for SCBA used by the authority having jurisdiction in accordance with the manufacturer's instructions.

8. Explain and demonstrate maintenance and servicing procedures for the SCBA used by the authority having jurisdiction in accordance with the manufacturer's instructions.

9. Explain and demonstrate the SCBA cylinder servicing equipment and procedures used by the authority having jurisdiction in accordance with the manufacturer's instructions.

Endnote

1. U.S. Department of Labor, Washington, DC. 29 CFR 1910.134 (f)(1).

Additional Resources

Because of the special nature of SCBA, readers should see specific manufacturers' instructions.

PORTABLE FIRE EXTINGUISHERS

Ric Koonce, J. Sargeant Reynolds Community College

 OUTLINE

- Objectives
- Introduction
- Fire Classification and Risk
- Types of Fire Extinguishers
- Rating Systems for Portable Extinguishers
- Limitations of Portable Extinguishers
- Portable Extinguisher Operation
- Care and Maintenance of Portable Extinguishers
- Inspection Requirements
- Lessons Learned
- Key Terms
- Review Questions
- Endnotes
- Additional Resources

STREET STORY

I remember going to a fire several years ago as a firefighter assigned to a ladder company. I was the can man for the tour, which meant my tool assignment was a 6-foot wooden hook and a 2½-gallon water extinguisher. We were out on building inspection duty, and an alarm came in that we were not assigned to; however, because we were only six blocks away the dispatcher told us to respond. As we were responding we were informed of children possibly trapped in a rear bedroom.

We were the first company to arrive, and heavy smoke was pushing from the second-floor front window of a four-story apartment house. We proceeded to the fire apartment, and as we forced entry, heavy smoke was pushing out of the apartment. Without the luxury of a hoseline, we proceeded to crawl down a narrow smoke- and heat-filled hallway. The heat was tremendous, and the lieutenant kept pushing me down the hall, telling me we were getting closer to the fire.

At this time, we were able to see fire lapping out of a bedroom door. I was instructed to use my water extinguisher to hit the fire both around the top of the door and the ceiling, driving the fire back into the bedroom. The lieutenant then took my 6-foot hook, grabbed the bottom of the bedroom door with it, and pulled it closed. Fire was now lapping from around the door, and I was able to give quick short blasts of the extinguisher and knock it down. The forcible entry man climbed over our backs and with the lieutenant was able to search the rear bedrooms, finding no one. My job now was to stay at the door to the fire and protect lives until a hoseline was stretched and operating. It only took about two minutes to get the hoseline operating and for me to join my company in the search effort, but it felt like a lifetime. I have been to many fires since that day, both as a firefighter and an officer, and I have seen much accomplished with a little 2½-gallon water extinguisher. It is an underrated piece of equipment that when properly used is worth its weight in gold.

—Street Story by Mike Gala, Jr., Lieutenant, Ladder 148, FDNY

OBJECTIVES

After completing this chapter, the reader should be able to:

- Explain the five classes of fire and the risks associated with each class.
- Identify the kinds of fire extinguishers used for each class.
- Explain the rating systems of portable extinguishers for Classes A, B, and C.
- Identify the limitations of portable extinguishers.
- Demonstrate the operation of portable fire extinguishers.
- Explain the care and maintenance of portable fire extinguishers.
- Discuss the inspection requirements of portable fire extinguishers.
- Given a fire scenario or an actual fire, choose the fire extinguisher of correct size, agent, and rating to extinguish the fire.

INTRODUCTION

Portable fire extinguishers are designed to fight small fires or unusual fires that are not easily extinguished with water. There are also some small fires that firefighters cannot reach quickly with hoselines. The use of a portable fire extinguisher can knock down a small fire or control a larger one until firefighters can stretch, or advance, a hoseline to the fire area. Citizens have also used fire extinguishers to control a small fire until the fire department arrives. When used by untrained persons, fire extinguishers have created deadly delays in alerting the fire department. Therefore, it is very important that firefighters be proficient in using fire extinguishers and in teaching proper extinguisher use to any interested citizens.

Firefighters usually use hose streams to fight fires; extinguishers are used only occasionally. However, this should not prevent firefighters from learning how to effectively use them. Firefighters must practice extinguisher usage to retain the knowledge and skill required to use the unit effectively. There are four basic steps in extinguisher use: Pull the pin, aim the nozzle, squeeze the handle, and sweep the base of the fire with the extinguishing agent. These steps can be practiced using a pressurized water extinguisher, which is easily refilled and very cost effective to use. These steps carry over to other types of extinguishers. It is important for firefighters to be knowledgeable about fire extinguisher types, extinguishing agents, and the five classes of fire.

Throughout the fire service, the phrase "Use the right tool for the right job" is prevalent.

Fire extinguishers come in a variety of types and sizes. Firefighters should know the extinguishers carried on each apparatus, as well as those inside the various buildings of the response district. During building inspections, firefighters should preplan special hazards and locations where an extinguisher would be a valuable tool to control a fire. Fire prevention will dictate the correct extinguishers for the occupancy. Firefighting skill will dictate the use of the right extinguisher for the conditions present during a fire. The general public will often ask firefighters questions about firefighting and fire extinguisher use. Firefighters must be prepared to answer questions and train citizens to use portable fire extinguishers, when requested.

FIRE CLASSIFICATION AND RISK

The type or nature of the material burning, that is, its fuel, defines the fire. Fuel is the key ingredient because as it varies so does the fire. The different types of fuels classify types of fires, and these classes of fire are used to identify the extinguishers and extinguishing agents used to put them out. An understanding of the fire classes leads to selection of the proper unit and agent. There are four traditional fire classes and one new one; however, the first three are the most common and the ones primarily covered by this chapter. Firefighters should be aware of places in their community with these types of fuels. This is the beginning of the process of planning for and dealing with potential emergencies and creating a means to deal with them. (For more information on fire classes, see Chapter 4.)

Class A

Class A fires involve ordinary combustibles such as wood, paper, cloth, plastics, and rubber. These fuels can be extinguished with water, water-based agents or foam, and multipurpose **dry chemicals**. Because of its availability, water is usually used by the fire department.

Class B

Class B fires involve flammable and combustible liquids, gases, and greases. Common products are gasoline, oils, alcohol, propane, and cooking oils. Pressurized flammable liquids and gases are special fire hazards that should not be extinguished unless the fuel can be immediately shut off. Flammable liquids that are flowing horizontally plus dripping or overflowing their container and spilling vertically (creating a three-dimensional flow), such as an overflowing tank, are also considered special hazards. *Special hazards*[1] refers to situations for which fire extinguishers have not been tested and therefore may be inadequate; it is important to carefully evaluate the situation prior to attacking these types of fires. Some solids under fire conditions may melt and act like flammable liquids. Common extinguishing agents for Class B fires are carbon dioxide (CO_2), regular and multipurpose dry chemical, and foam.

> **Safety** Pressurized flammable liquids and gases should be extinguished by stopping the flow of the fuel. In some cases, firefighters need to extinguish the fire to shut off control valves. These situations are extremely dangerous because firefighters have to work in a flammable atmosphere and any source of ignition may reignite that atmosphere with catastrophic results. This entry into a flammable atmosphere should be the last available option.

Class C

Class C fires involve energized electrical equipment, which eliminates the use of water-based agents to extinguish them. The recommended method of fighting these fires is to turn off or disconnect the electrical power and then use an appropriate extinguisher, depending on the remaining fuel source. Class C extinguishers have extinguishing agents and hoses with nozzles that will not conduct electricity. Class C only extinguishers are not made. Class C agents include carbon dioxide (CO_2) and regular and multipurpose dry chemicals.

Class D

Class D fires involve combustible metals and alloys such as magnesium, sodium, lithium, and potassium. These metals can be found in some lightweight motor vehicle engine components or lawn mower bodies. Great care must be used when attempting to extinguish a fire in these types of fuels. Water and other extinguishing agents can react violently when applied to burning combustible metals and can endanger nearby firefighters. Also, because of the differences in the metal and alloy fuels, there is no universal Class D extinguishing agent that works on all Class D materials; what works well on one metal may be relatively ineffective on a different alloy. Personnel must use the correct and uncontaminated (clean, dry, and without any other foreign materials in it—basically as it comes from the factory) extinguishing agent for each different Class D material. Facilities that use or store these materials are required to maintain adequate amounts of the proper extinguishing agents to combat any potential fire situation. Class D agents are called **dry powders** and should not be confused with *dry chemicals,* which, although dry and powdery, are not the same. (Some of these agents are dry sand, phosphate salts, or silica.) Other special agents such as Lith-X and Met-Ex are not commonly used and are only mentioned here. Firefighters using these special agents locally should seek additional information and training on their use.

> **Streetsmart Tip** Many facilities that use or process Class D fuels have frequent fires involving these materials. The local fire department may or may not be called to deal with these routine fires, especially if a fire brigade is on-site. Firefighters who do respond to these types of fires on a regular basis must ensure that they follow proper safety precautions with *each* fire incident and ensure they do not become complacent or relaxed about dealing with this dangerous class of fire.
>
> One fire department that regularly responds to a facility with a Class D material had successfully extinguished several fires over a period of a couple of weeks. When returning to the facility for another fire, they began their operations as they had previously done. While applying the extinguishing agent to the surface of the fuel, a violent explosion occurred that expelled some of the molten material onto several firefighters, causing injuries. The resulting investigation revealed that water had been used near the tank and had contaminated both the extinguishing agent and the shovels used to apply it.
>
> Class D materials are always considered hazardous materials. Proper hazardous materials procedures must be followed at *each* incident. Familiarity is the breeding ground of complacency and can result in injuries or worse.

Class K

Class K is a new classification of fire as of 1998 and involves fires in combustible cooking fuels such as vegetable or animal oils and fats. Its fuels are similar to Class B fuels but involve high-temperature cooking oils and therefore have special

Figure 8-1 Class K equipment.

Figure 8-2 Various types of fire extinguishers.

characteristics. Typically, firefighters have used Class B extinguishers on these types of fires, but they have been less effective on deep layers of cooking oils. Class K agents are usually **wet chemicals**, water-based solutions of potassium carbonate–based chemical, potassium acetate–based chemical, potassium citrate–based chemical, or a combination.[2] These agents are usually used in fixed systems, **Figure 8-1,** but some extinguishers are available.

TYPES OF FIRE EXTINGUISHERS

Many types of fire extinguishers are available for purchase and use, **Figure 8-2.** The best type of extinguisher depends on many factors and each should be considered prior to placing an extinguisher in use. *The wrong extinguisher can be worse than no extinguisher.*

Factors for selecting an extinguisher are the type of fuel, the person using the extinguisher, and the building or place where it will be used. The first factor to consider when choosing an extinguisher is the type and amount of fuel present. This will provide clues as to the type of fire to anticipate. The amount of fuel determines fire size, and the wrong size extinguisher will not completely extinguish the fire. The user of the extinguisher and the occupancy in which it will be used represent another factor. Are potential users trained to combat a fire effectively or will they use improper technique and exhaust the extinguisher before the fire is knocked down? What will the people in the building do with or to the extinguisher while it is stored? Will the extinguisher be tampered with, stolen, or otherwise ineffective when needed?

Another factor is the type of building construction and occupancy. What hazards or other conditions exist? What building areas need to be protected to avoid further problems? These potential hazards set the fire code requirements for selecting and placing the extinguishers. Environmental conditions must be examined for effects on the fire and the extinguishing agent. Temperature may eliminate using water-based agents that may freeze; corrosive atmospheres may require special protection for the user; the wind may blow away the extinguishing agent before it reaches the fire; and a confined space may create an unsafe or nonsurvivable atmosphere for the user.

A final factor would be the type of equipment protected. Often this last factor is given too much consideration. Delicate equipment and high-value items may require special considerations as they may be irreparably damaged in an extinguishing operation. It should be stressed, however, that the main objective is to extinguish the fire completely and effectively. Most extinguishing agents can be cleaned from equipment, whereas damage from a fire can render equipment permanently destroyed. It is important to remember that certain extinguishing agents are corrosive to certain substances. This should be a factor when choosing an extinguisher for the occupancy.

Types of Extinguishing Agents

Water is the basic fire-extinguishing agent for Class A materials. Its ability to absorb heat is one of the prime reasons for its effectiveness. The disadvantages are that it is subject to freezing and is ineffective—and sometimes dangerous—on other classes of fuels. **Loaded stream** combats the freezing problem by adding an alkali salt as an antifreezing agent. Water-based foam extinguishers for use on Class B fires have either **Aqueous Film-Forming Foam (AFFF)** or **Film-Forming Fluoroprotein Foam (FFFP)** of both the regular and **polar solvent type**. These agents are effective in cooling and smothering the fire and creating a vapor barrier. (See Chapter 11 for more foam information.)

Carbon dioxide (CO_2) is an inert gas stored under pressure as a liquid capable of being self-expelled. The colorless and odorless gas is effective in smothering a Class B or C fire. CO_2 works best in enclosed or semienclosed areas as the agent can be easily blown away by the wind. Personnel should take caution to avoid oxygen deprivation when CO_2 is used in small enclosed areas. This clean agent is still popular with those concerned about property damage, although in well-ventilated areas it can be ineffective. Some sensitive electrical equipment can be thermoshocked by CO_2.

Dry chemical extinguishing agents are particles propelled by a gaseous medium for distribution. There are three general categories: sodium bicarbonate–based, potassium–based, and multipurpose dry chemicals. The first two categories work on Class B and C fires, and are called regular dry chemical agents. Sodium bicarbonate, similar to baking soda, is the agent in the first category and is highly effective for cooking grease fires. The second category includes potassium bicarbonate, potassium chloride, and urea-based potassium bicarbonate—all of which are usually more effective than sodium bicarbonate. The third category is known as a multipurpose dry chemical agent—effective on Class A, B, and C fires. These multipurpose dry chemicals are monoammonium phosphate.[3] Dry chemicals are very effective due to their coating action, which reduces the chances of reignition. The coating action is a drawback, however, when protecting sensitive items such as computer mainframes.

Wet chemical agents are water-based solutions of potassium carbonate–based chemical, potassium acetate–based chemical, potassium citrate–based chemical, or a combination of those chemicals.[4] They are used for special applications, particularly Class K fires.

Clean agents are the agents used to replace halon or halogenated hydrocarbon extinguishers. These agents are thought to cause damage to the Earth's ozone layer and have been banned by an international treaty. These agents may still be in limited use, but must be phased out and replaced with clean agent systems. Clean agents are gases that do not conduct electricity or leave a residue, and are nonvolatile. They are divided into two classes—halocarbon agents and inerting gases—and are currently rated as somewhat ineffective for local application. Research on their application continues.[5]

Kinds of Extinguishers

Many types of portable fire extinguishers are in use today. Some are small and handheld, while others are so large that they require a wheeled cart to move them. Most extinguishers operate on the same basic principle of storing and expelling an extinguishing agent. Fire extinguishers are labeled to make their firefighting rating quick and easy to identify. The older versions of fire extinguishers are labeled with colored geometrical shapes with letter designations, **Figure 8-3A**. Newer fire extinguishers are labeled with a picture label system, **Figure 8-3B**. Many fire extinguishers are multiuse, that is, a single extinguisher may be used to fight more than one class of fire, **Figure 8-3C**. Class A and B fire extinguishers have a numerical rating discussed later in this chapter. Listed here are current kinds of fire extinguishers for each type of agent:[6]

1. *Water type* for Class A fires:
   a. Pump type
   b. Pressurized water
   c. Pressurized loaded stream
2. *Foam extinguishers* for Class A and B fires; stored pressure
3. *Carbon dioxide* for Class B and C fires; self-expelling gas, stored pressure
4. *Halon and clean agents* for Class B and C, or Class A, B, and C
5. *Dry chemical* for Class B and C or Class A, B, and C fires (also some dry powder, Class D extinguishers)
   a. Stored pressure
   b. Cartridge-operated type
6. *Wet chemical* for Class K fires; stored pressure

Pump-type extinguishers are hand-pumped devices of two designs, depending on whether the pump is internal or external to the tank. The pump tank extinguisher has the pump housed inside the water tank area and each stroke pumps water out of the tank. This type of extinguisher is simple to operate and maintain, **Figure 8-3D**. The second type is designed for wildfires and is a backpack pump that has the tank carried on the back with the hose and pump in front of the operator, **Figure 8-4** and **Figure 8-5**. Both usually come in a 2½-gallon size.

Pressurized water, pressurized loaded stream, and stored pressure extinguishers operate by means of

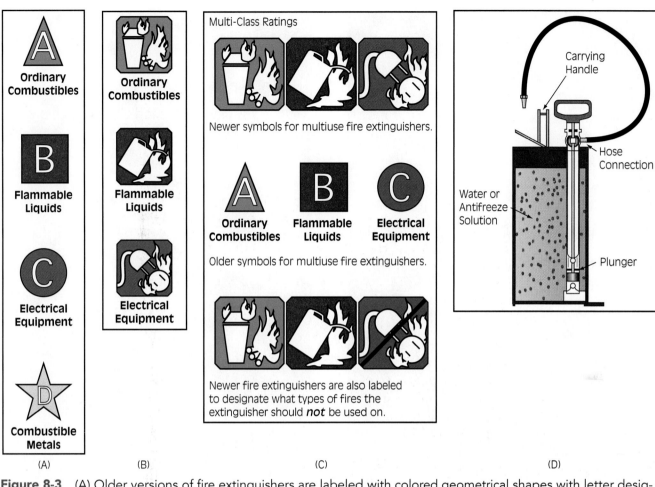

Figure 8-3 (A) Older versions of fire extinguishers are labeled with colored geometrical shapes with letter designations. (B) Newer fire extinguishers are labeled with a picture label system. (C) Many fire extinguishers can be used to fight more than one type of fire. (D) Pump tank extinguisher.

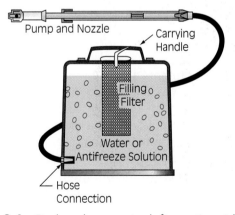

Figure 8-4 Backpack pump tank fire extinguisher.

an expelling gas that propels the agent out of the container. The differences are seen in the type of container and hose and nozzle assemblies. The container differences result from the various pressures required to store the agent. The hose and nozzle differences relate to the agent and its special need for application. Some agents may be dispensed over a wide area, others over a smaller area; some foam needs an air-aspirating nozzle or may require the nozzle to be nonconductive.

The basic principle is that the agent is inside the container with a pressurizing gas above it under constant pressure. Some use air or nitrogen as an expelling gas that is added to the container with the agent. Other agents, like carbon dioxide, are their own expelling gases. Most extinguishers have a gauge to measure the pressure of the gas, but CO_2 extinguishers do not. **Figures 8-6** through **8-15** show the various extinguishers and how they store their product.

Cartridge-operated extinguishers are used for some regular and multipurpose dry chemical and most dry powder Class D extinguishers. They are similar to stored pressure extinguishers except that instead of being under constant pressure, the expelling gas is stored in a cartridge on the side of the container. When the puncturing lever is

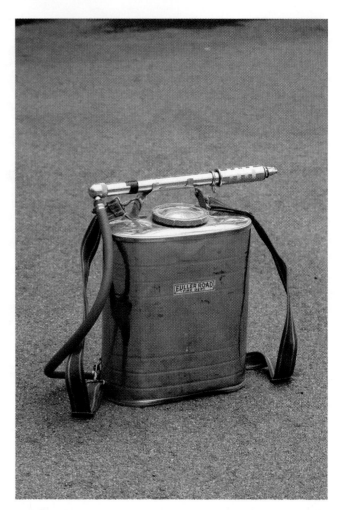

Figure 8-5 A backpack pump tank fire extinguisher. *(Courtesy of Fred Schall)*

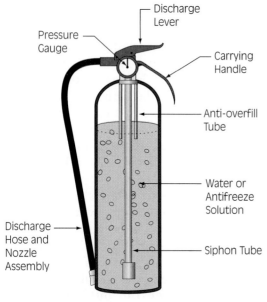

Figure 8-6 Stored pressure water extinguisher.

depressed, it ruptures a disk and the gas is expelled into the tank, which pressurizes it. The operator controls the flow of agent by a nozzle mounted at the end of the hose. Cartridge-type extinguishers are used when the agent may cake excessively and needs to be stirred. The agent is accessible without charging the pressure or dumping the agent. This type is also easy to refill by adding the agent into the tank and a new cylinder to the charging mechanism, **Figure 8-16** and **Figure 8-17**.

Note Inverting extinguishers were popular many years ago. These extinguishers are no longer manufactured and are seldom seen today.

PORTABLE FIRE EXTINGUISHERS 191

Figure 8-7 Stored pressure water extinguisher. *(Courtesy of Fred Schall)*

Figure 8-9 Stored pressure foam extinguisher.

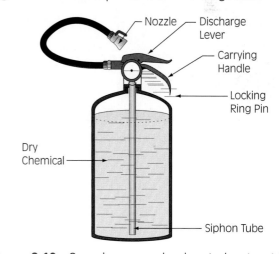

Figure 8-10 Stored pressure dry chemical extinguisher.

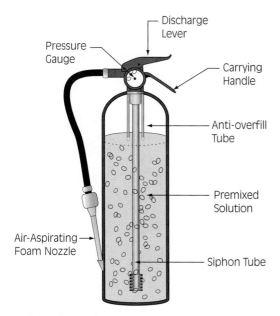

Figure 8-8 Stored pressure AFFF or FFFP extinguisher with air-aspirating nozzle.

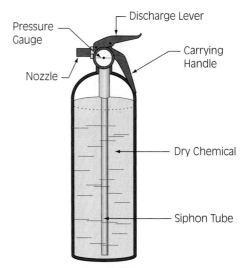

Figure 8-11 Stored pressure dry chemical extinguisher with fixed nozzle.

Figure 8-12 Stored pressure dry chemical extinguisher.

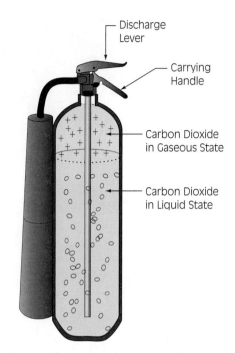

Figure 8-14 Carbon dioxide extinguisher.

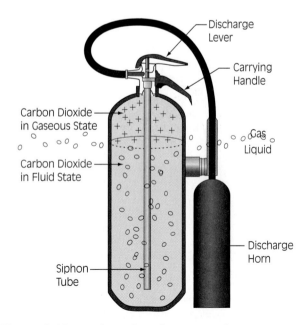

Figure 8-13 Carbon dioxide extinguisher.

Figure 8-15 Carbon dioxide extinguishers. *(Courtesy of Fred Schall)*

PORTABLE FIRE EXTINGUISHERS 193

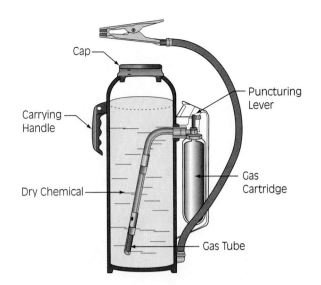

Figure 8-16 Cartridge-operated dry chemical extinguisher.

RATING SYSTEMS FOR PORTABLE EXTINGUISHERS

Each class of fuel is subjected to a separate type of extinguisher test for its class. The testing of extinguishers is usually conducted by an independent testing agency such as Underwriters Laboratories or Factory Mutual Research Corporation, or a government agency such as the Coast Guard. This section covers the tests for Classes A, B, and C. Classes D and K have special tests that are more specific to their fuels or application. The ratings are noted on the label of the extinguisher as is a symbol to show the class(es) of fire on which it works. (See Chapter 4 for these symbols.)

Class A

The testing of Class A extinguishers and agents uses a wood cribbing test. The cribbing is set on fire, allowed a pre-burn period, and then attacked with the extinguisher. For a 1-A rating, the extinguisher should extinguish a wood crib of about one cubic foot (0.03 cubic meters), **Figure 8-18**. The

Figure 8-17 Cartridge-operated dry chemical extinguisher.

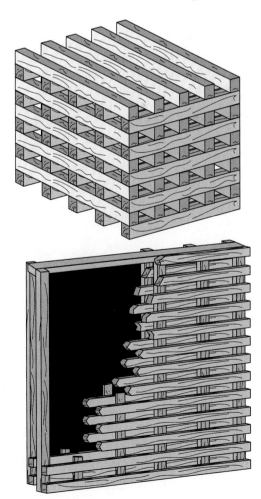

Figure 8-18 Wood cribbing for Class A extinguisher test.

ratings increase as the amount of fire suppressed increases; for instance, a 2-A extinguisher will put out twice the fire of a 1-A. In a wood paneling and excelsior test, the excelsior is ignited and allowed to spread to wood paneling before it is extinguished. Larger scale tests use only the wood cribbing and are conducted outside, while smaller tests are done indoors.

Class B

The test for Class B extinguishers involves igniting a pan of a flammable liquid (heptane), allowing a pre-burn period, and attacking the fire. The size of the pan determines the rating, **Figure 8-19**. For example, a 4-square-foot (0.37-square-meter) pan that is extinguished should yield a rating of 4-B. The rating assumes that an inexperienced operator using a 20-B rate extinguisher on a 20-square-foot fire should find the results adequate. The number rating compares approximately to the square footage to be extinguished, although larger units will not offer a direct relationship, as larger fires require more agent per area than the smaller ones.

Class C

The testing of Class C extinguishers and agents tests only the conductivity of the agent and the nozzle or hose and nozzle combination, **Figure 8-20**. There is no actual fire test for Class C agents or extinguishers. No numbers are assigned with the Class C rating.

Figure 8-19 Flammable pan for Class B extinguisher test.

LIMITATIONS OF PORTABLE EXTINGUISHERS

Fire extinguishers have limited capabilities, and attempting to exceed those capabilities can increase damage and cause injuries. They are designed for

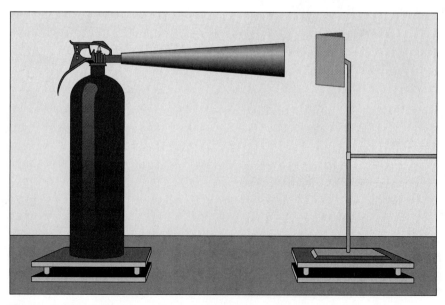

Figure 8-20 Class C test for conductivity.

specific purposes and are usually a first-aid method for fire extinguishment. Fire extinguishers are designed and rated with certain types and sizes of fires in mind; using the wrong class or the wrong size extinguisher may cause problems. When thinking of size, it is usually best to pick the larger size. However, picking the largest fire extinguisher available to put out a small fire can make it an expensive fire. The wrong class extinguisher will not do the job, will waste the agent, and can cause a reaction or electrical shock.

> **Note** Using the wrong class extinguisher will not do the job, wastes the agent, and can cause a reaction or electrical shock. *It is important to pick the right extinguisher for the job.*

PORTABLE EXTINGUISHER OPERATION

Suppose a wastebasket is on fire and the firefighter has chosen the correct extinguisher for putting the fire out. What is the right way to use it? The operation of most fire extinguishers is covered by the acronym *PASS* for the four simple steps in using an extinguisher: *P* is for pull the pin, *A* is for aim the nozzle, *S* is for squeeze the operating handle, and the second *S* stands for sweep the nozzle across the base of the fire.

> **Note** PASS, the four steps for using a fire extinguisher:
> *P*ull the pin.
> *A*im the nozzle.
> *S*queeze the handle.
> *S*weep the base of the fire.

Step one is *pull the pin* that prevents the handle from operating. Most extinguishers have the pin mounted at the operating handle to prevent it from being accidentally squeezed, **JPR 8-1A**. Cartridge-operated extinguishers have a pin at the activation lever for the cartridge, not on the operating handle at the nozzle. The pin is held in place by a wire or plastic tie; a simple tug removes this wire and the pin.

The second step is to *aim the extinguisher* toward the base of the fire closest to the firefighter. Firefighters must remember to have a path of escape behind them and *not allow the fire to get between them and the exit,* **JPR 8-1B**. The object of aiming at the base of the fire is to sweep the fire away from the firefighter, confining it and driving it back to its origin. Aiming at other points may cause the fire to spread beyond the capability of the extinguisher. The extinguisher must be aimed from the proper distance, which often is set by the room size and fire size. Various extinguishers can reach a variety of distances up to 50 feet (15 meters), but reaching and being effective are two different subjects. The practical effective range of a pressurized water or foam extinguisher is about 20 feet (6 meters), dry chemical units from 15 to 20 feet (4.5 to 6 meters), and a CO_2 type from 10 to 15 feet (3 to 4.5 meters). A final aiming caution when using a CO_2 extinguisher is to make sure that the aiming hand is on the handle, not on the horn where the cold can freeze the skin.

The third step is to *squeeze the handle* together to apply the agent, **JPR 8-1C**. The firefighter should continue to squeeze the handle for enough time to cover the whole fire area. Some short blasts are okay but too many will empty the extinguisher without extinguishing the fire.

Last, *the firefighter sweeps the area of the fire* by carefully keeping the nozzle aimed at the base of the fire, continuing to push the fire back, **JPR 8-1D**. The sweeping motions help extinguish the entire fire instead of creating a pocket that may allow the fire to circle around the firefighter. Foam extinguishers can also use the methods of regular foam application that are covered in Chapter 11. Bending over or sweeping the extinguishing agent at a low angle allows the agent to spread faster across the fire.

> **Safety** Firefighers must remember to have a path of escape behind them; *they cannot allow the fire to get between them and the exit.*

If the extinguisher does not put out the fire, the firefighter should not deploy a second extinguisher. Multiple extinguishers used at the same time, however, are more effective than using one at a time. Many buildings have been lost due to improper "one at a time" use of multiple extinguishers, and the untrained individuals are quickly forced to exit the building due to the fire's increasing intensity. In this situation, the fire department is not notified until the fire has increased greatly in size. Extinguishers are meant to combat small fires in the growth stage. Large fires are beyond the capabilities of portable extinguishers.

JOB PERFORMANCE REQUIREMENT 8-1
Portable Extinguisher Operation

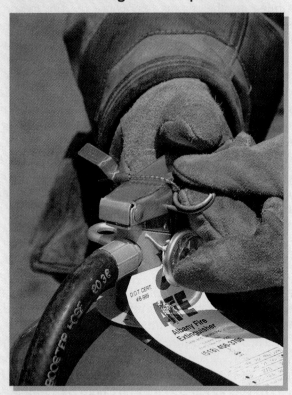

A Pull the pin.

B Aim the nozzle.

C Squeeze the handle.

D Sweep the base of the fire.

> **Note** Stored-pressure water extinguishers are designed to be carried in an upright position when approaching a fire. The firefighter is cautioned to keep the hose in hand at all times.

> **Streetsmart Tip** Civilians should be taught to call the fire department before attempting to use an extinguisher.[7] Often they plan to use just one extinguisher and then call the fire department. They almost get the fire out and run for another extinguisher, which again fails to put out the fire. One group of maintenance personnel used more than twenty extinguishers on an apartment fire, but when the first arriving officer, on a 1,000-gpm pumper, found fire coming through the roof, he called for a third-alarm assignment. The civilians got caught up in fighting the fire, rather than calling for help.

Firefighters must be adept at using all types of firefighting equipment. This includes portable fire extinguishers. However, in many services this tool is not used as often as a hoseline. It is, therefore, extremely important that firefighters train using portable fire extinguishers to stay proficient with the use of this tool and to effectively teach firefighting techniques to the general public.

> **Streetsmart Tip** In an emergency, a firefighter may use a CO_2 extinguisher on scenes where dangerous dogs or other animals are present. Normally the noise of the CO_2 extinguisher discharging will scare the animal. The cold cloud of agent will usually further frighten the animal into retreat. *Note:* Discharging the CO_2 extinguisher directly onto the animal can render permanent harm to the animal. The extinguisher should be thought of as a reasonable force tool. Only the amount of force (agent) to get the animal to retreat should be used and no more. This choice of protection should be a last resort for firefighters or other persons in danger.

CARE AND MAINTENANCE OF PORTABLE EXTINGUISHERS

Care and maintenance of fire extinguishers is fairly straightforward. Simple inspections (see next section) and careful storage prevent most problems. The extinguisher should be inspected carefully when first placed in service and properly stored in a cabinet or on a bracket. Keeping apparatus-mounted extinguishers in their brackets will prevent damage from falling or from striking other equipment. New apparatus standards require this and it is a good practice on older vehicles. Routine apparatus equipment checks should require the extinguisher to be visually examined for proper pressure and visible damage and to ensure the pin seal is in the proper place. If any problems are noted, the unit should be removed and sent for repairs. Vehicle operators should periodically remove the unit from its bracket to hand test its weight and to do a better visual check. Dry chemical or dry powder extinguishers carried on apparatus should occasionally be rotated upside down and shaken to keep powders from packing at the bottom as a result of vehicle vibration. Cleaning of any dirt or grit will help keep the unit in good working order. Finally, any extinguisher that has been discharged, that has low pressure on the gauge, or that has a broken seal should be removed from the vehicle until thoroughly examined and/or recharged.

INSPECTION REQUIREMENTS

Some very popular fire extinguishers of the past have now been declared obsolete and should be removed from service. These obsolete extinguishers include soda acid, chemical foam (except film-forming foam), vaporizing liquids such as carbon tetrachloride, cartridge-operated water or loaded stream, and any copper or brass extinguisher with solder or rivets (except pump tanks). Some of these have been removed because of the hazards of the extinguishing agents. Most of these extinguishers have the potential to explode when pressure is applied during use. Firefighters should never attempt to operate an obsolete extinguisher and should instead have them decommissioned by trained technicians.

The inspection of fire extinguishers by firefighters is usually a visual inspection, **Figure 8-21**. If something does not look right, the unit should be removed and replaced. Extinguishers in buildings should be checked every thirty days, and extinguishers on apparatus should be inspected each time the vehicle is inspected. Extinguishers that are new or old ones being returned to service should be examined prior to placing back on apparatus. The visual inspection should first check that the unit has a tag indicating its last service and/or inspection. Annual inspection and maintenance by a service technician is recommended. The pin seal should be firmly in place. If the unit has a gauge, it should register the proper pressure. If a gauge is not present, lifting the unit or weighing it will determine if it is properly filled. Some extinguishers such as carbon dioxide extinguishers can only be truly checked by weighing; the filled weight is stamped on the cylinder. A careful look at the unit should detect any damage, dents, or corrosion to the shell, hose, nozzle, and other parts. If any defects are present, the unit should be removed and replaced. A last check should be to see if the unit is within the time frame for its hydrostatic testing, which is a pressure test to ensure that the unit will not explode when operated. The test period ranges from five to twelve years; firefighters should consult their service center or the NFPA standard to verify extinguishers are compliant.[8]

Figure 8-21 Conducting visual inspections. *(Courtesy of Fred Schall)*

Lessons Learned

Fire extinguishers can be used as initial response tools or to fight fires in special situations. Firefighters must know how to use fire extinguishers and to teach the public how to use them. Part of that knowledge includes classifying fires by their fuels, especially ordinary combustibles, flammable liquids and gases, and energized electrical equipment. Knowing the classes of fire is the first step toward choosing the right extinguisher and the correct size. The four-step process for using an extinguisher is *PASS:* pull, aim, squeeze, and sweep.

KEY TERMS

Aqueous Film-Forming Foam (AFFF) A synthetic foam that as it breaks down forms an aqueous layer or film over a flammable liquid.

Carbon Dioxide (CO_2) An inert colorless and odorless gas that is stored under pressure as a liquid that is capable of being self-expelled and is effective in smothering Class B and C fires.

Class A Classification of fire involving ordinary combustibles such as wood, paper, cloth, plastics, and rubber.

Class B Classification of fire involving flammable and combustible liquids, gases, and greases. Common products are gasoline, oils, alcohol, propane, and cooking oils.

Class C   Classification of fire involving energized electrical equipment, which eliminates using water-based agents.

Class D   Classification of fire involving combustible metals and alloys such as magnesium, sodium, lithium, and potassium.

Class K   A new classification of fire as of 1998 that involves fires in combustible cooking fuels such as vegetable or animal oils and fats.

Dry Chemicals   Dry extinguishing agents divided into two categories. Regular dry chemicals work on Class B and C fires; multi-purpose dry chemicals work on Class A, B, and C fires.

Dry Powders   Extinguishing agents for Class D fires.

Film-Forming Fluoroprotein Foam (FFFP)   Foam that incorporates the features of AFFF and fluoroprotein foams with good resistance and the film-forming barrier.

Loaded Stream   Combats the water freezing problem by adding an alkali salt as an antifreezing agent.

Polar Solvent Type of Foam or Alcohol-Resistant Foam   Foam that is compatible with alcohol and/or polar solvents by creating a polymeric barrier between the water in the foam and the polar solvent.

Wet Chemicals   Extinguishing agents that are water-based solutions of potassium carbonate–based chemicals, potassium acetate–based chemicals, potassium citrate–based chemicals, or a combination.

REVIEW QUESTIONS

1. Name the four traditional classes of fire, give an example of each and an extinguishing agent for each class.
2. Name the new class of fire, describe its materials, and name an extinguishing agent.
3. What are the three categories of agents of dry chemicals?
4. What is a wet chemical?
5. How does carbon dioxide extinguish a fire?
6. Why are halons no longer used for fire extinguishers?
7. How does a stored pressure extinguisher work?
8. What are the two types of pump-type extinguishers?
9. How are Class A extinguishers rated?
10. What does a 10-B rating mean?
11. What is a Class C extinguisher tested for?
12. Describe the four steps for operating an extinguisher.
13. Describe a visual inspection process on a fire extinguisher.

Endnotes

1. "Special hazard" is a term that has several meanings. In this case, it denotes special fire situations that are so unique that it would be impossible to test and rate the effectiveness of a fire extinguisher for each special case.
2. *NFPA 17A: Standard for Wet Chemical Extinguishing Systems, 1998 Edition*, p. 17A-5. National Fire Protection Association, Quincy, MA.
3. Gagnon, Robert, *Design of Special Hazard and Fire Alarm Systems*, pp. 149–150. Delmar Learning, a part of the Thomson Corporation, Clifton Park, NY, 1998.
4. *NFPA 17A: Standard for Wet Chemical Extinguishing Systems, 1998 Edition*, p. 17A-5. National Fire Protection Association, Quincy, MA.
5. Gagnon, Robert, *Design of Special Hazard and Fire Alarm Systems*, pp. 101–102. Delmar Learning, a part of the Thomson Corporation, Clifton Park, NY, 1998.
6. Gagnon, Robert, *Design of Special Hazard and Fire Alarm Systems*, p. 30. Delmar Learning, a part of the Thomson Corporation, Clifton Park, NY, 1998.
7. OSHA regulations limit untrained or limited trained persons to fight only incipient fires.
8. *NFPA 10: Standard for Portable Fire Extinguishers, 1998 Edition*, Chaps. 4 and 5. National Fire Protection Association, Quincy, MA.

Additional Resources

From Delmar Learning, a part of the Thomson Corporation:

Diamantes, David, *Fire Prevention, Inspection and Code Enforcement*. Delmar Learning, a part of the Thomson Corporation, Clifton Park, NY, 1997.

Gagnon, Robert, *Design of Water-Based Fire Protection Systems*. Delmar Learning, a part of the Thomson Corporation, Clifton Park, NY, 1996.

Gagnon, Robert, *Design of Special Hazard and Fire Alarm Systems*. Delmar Learning, a part of the Thomson Corporation, Clifton Park, NY, 1998.

Klinoff, Robert, *Introduction to Fire Protection*. Delmar Learning, a part of the Thomson Corporation, Clifton Park, NY, 1997.

Sturtevant, Thomas, *Introduction to Fire Pump Operations*. Delmar Learning, a part of the Thomson Corporation, Clifton Park, NY, 1997.

From NFPA:

NFPA 10: Standard for Portable Fire Extinguishers, 1998 Edition. National Fire Protection Association, Quincy, MA.

NFPA 17: Standard for Dry Chemical Extinguishing Systems, 1998 Edition. National Fire Protection Association, Quincy, MA.

NFPA 17A: Standard for Wet Chemical Extinguishing Systems, 1998 Edition. National Fire Protection Association, Quincy, MA.

CHAPTER 9

WATER SUPPLY

Ric Koonce, J. Sargeant Reynolds Community College

OUTLINE

- Objectives
- Introduction
- Sources of Water Supply
- Water Distribution Systems
- Fire Hydrants
- Valves Associated with Water Distribution Systems
- Rural Water Supply
- Pressure Associated with Water Distribution Systems
- Testing Operability and Flow of Hydrants
- Determining Static, Residual, and Flow Pressures
- Obstructions and Damage to Fire Hydrants and Mains
- Lessons Learned
- Key Terms
- Review Questions
- Endnotes
- Additional Resources

Courtesy of Fred Schall

STREET STORY

Several years ago, I was a brand new assistant fire chief in a rural area of Florida. Having come from the New York metro area, "rural water supply" was new to me. In New York, if you needed water, you just used a hydrant! But this was not the case in the new department.

We were dispatched to a reported building fire. As I approached the scene, I found a very large single-family dwelling with heavy fire involvement in approximately 30 percent of the structure. I knew that there were no hydrants in that part of the district, so I started to radio the Emergency Communications Center for multiple mobile water supply apparatus. I wanted every mobile water supply apparatus for miles around. And I got them. Eventually.

As the incident progressed, I was interrupted two or three times by some of our firefighters who were mumbling "something" about water supply. I was busy and made it clear that I had it covered. As one might expect, the fire progressed faster than the arrival of the mobile water supply apparatus, and for all intents and purposes, the building burned to the ground. Not a good night.

In the mop-up stages of the fire, I went up to some of the firefighters who were speaking to me earlier and I asked them what they were trying to tell me. What they'd been trying to tell me was that a major river was less than 100 feet away from the building—and they were going to suggest we draft out of that river for our water supply. Hmmm. Great idea.

I learned a few lessons from that fire. The first was that if you're a Yankee, and you burn down a 150-year-old "pride of the South" building, you will hear about it for 20+ years! But more importantly, I learned there's no excuse not to be prepared when you know your district has areas without municipal water supply. You know what areas you protect right now—and certainly know where you have and don't have municipal water supply. Plan ahead for rural water supply before the fire comes in. Know what mobile water supply apparatus will be due to respond and have them respond on the first alarm. And drill with them before a fire to make sure that your radios and hose connections are all compatible. At the fire scene is too late to find out. Also, a few seconds of listening to the firefighters may have made a big difference in our tactical operation. This fire incident now prompts me to do a walk-around on almost any incident I'm involved with. The walk-around gives me a view of the big picture and allows me to determine our strategy—had I done it that night, it would have been clear to me that water, and lots of it, was flowing behind the building.

—Street Story by Battalion Chief Billy Goldfeder, E.F.O., Loveland-Symmes Fire Department, Loveland, Ohio

OBJECTIVES

After completing this chapter, the reader should be able to:

- Explain the value of proper water supply to other goals of firefighting.
- Identify sources of water supply for drinking and firefighting.
- Explain the difference between groundwater and surface water.
- Explain the purpose of mobile water supply apparatus.
- Explain the features of water distribution systems.
- Identify types of fire hydrants and their uses.
- Identify valves associated with water distribution systems.
- Explain how to operate a rural water supply.
- Explain a portable water tank operation.
- Explain tender operations.
- Identify the proper pressures associated with water distribution systems.
- Conduct a test of the operability and flow of fire hydrants.
- Determine the static, residual, and flow pressures of water sources.
- Identify cause of obstructions and damage to fire hydrants and mains.

INTRODUCTION

Water supply is one of the most critical elements of firefighting *because water is the most common extinguishing agent.* In considering the expression "the first five minutes of fighting a fire sets the stage for the entire battle," remember that without water it is not even a fight. Water supply is important in areas with a water distribution system and even more important in areas without a system because one must be created. Firefighters should understand the basics of water supply and how to establish one because without a water supply, hoses, appliances, and fire streams are useless.

Where the water comes from, how much is available, and how it is delivered are key questions. Water supply dictates the **fire flow capacity** or amount of water that can be flowed. The **fire flow requirement** is the amount of water required for putting out the fire.[1] Knowing the amount of water available and how much is needed lets firefighters select the right strategy, tactics, hose appliances, and fire streams. All are interrelated, and facing a large fire without sufficient amounts of water leads to a situation in which a fire will continue to burn.

Water is the most common extinguishing agent. It is economical, readily available, and abundant. It has the ability to absorb large quantities of heat. It is inexpensive when getting it directly from a river or pond. From a municipal water supply, the additional needs for fire protection make it more expensive but not as much as other extinguishing agents. Its availability and abundance often depend on the fire's location. Desert or rural areas have limited or no water. A city can have areas where the availability of water is limited. In such situations, firefighters must haul in the necessary water. Finally, water's ability to absorb large quantities of heat makes it extremely effective in fighting fire. Water absorbs such large amounts of heat from the fire that it cools the fuel below its ignition temperature. (More information on this can be found in Chapter 4.)

SOURCES OF WATER SUPPLY

An understanding of the water supply situation for firefighting requires first knowing where the water comes from and how it gets from that point to the fire scene. When using a lake as the water source, it is pretty obvious, but other cases are not as clear. How does the water get to a fire hydrant and what factors are involved in making sure that enough water is available for the needs? A wide range of natural and man-made factors affects water sources. The weather is probably the greatest of these factors. The Earth's water is in a constant cycle of change. The sun evaporates water into the atmosphere. It condenses into clouds that carry it to other places where it eventually falls as rain. Some areas get so much rain they are called rain forests, whereas other places get none and are deserts. Some areas alternate between rainy and dry seasons; others have plenty of water but it is frozen some or all of the time. These are just some problems encountered with water supply.

Groundwater

Most of the Earth's freshwater supply is groundwater. Groundwater is water that seeps into the ground from rain and other surface sources. It collects in pockets of the Earth or permeates into layers of water-bearing soil. These pockets are called **aquifers** and some are large underground rivers. The level of the water under the Earth's surface is the **water table**, which can rise and fall.

Springs are groundwater sources that naturally flow to the surface; others are under great pressure and when drilled may force their way to the surface. This pressure may decrease over the years such that pumps will be needed to draw the water to the surface, as is the case with most wells, **Figure 9-1**. Shallow wells are closer to the Earth's surface and are cheaper to drill but are more prone to changes in the water table and are subject to contamination and saltwater infiltration. Deep wells may penetrate through several layers of water before finding an aquifer. They are a more predictable water source with less chance of contamination. Wells used for municipal water supply and fire protection must be of sufficient volume; often multiple wells are employed. Domestic and farm wells are insufficient in volume and pressure for firefighting.

Surface Water

Surface water is the world's most common source of water—almost 75 percent of the Earth is covered by water. Unfortunately, most is found in the oceans and seas with limited availability. Other natural sources of surface water are rivers, lakes, and ponds, **Figure 9-2**. Man-made surface sources include lakes, ponds, reservoirs, swimming pools, and water tanks. Surface water may be fresh, salt, or brackish water, and many surface water sources are influenced by tidal changes. **Tidal changes** are the rising and falling of the surface water levels due to the gravitational effects between the Earth and the moon. In some areas, these changes are insignificant but in others a more than 40-foot difference exists between high and low tide. When using a tidal water source, it is important to understand the effects of the tide and know when the changes will occur or the water source may not be available when it is needed.

Figure 9-2 A surface water source. *(Photo courtesy of Donald Fischer)*

Mobile Water Supply Apparatus

For many small fires and in those areas without a water distribution system, the water tank on fire apparatus supplies the water, having gotten its water from another source. The tank sizes of fire apparatus vary, but most engines today have 500-gallon or larger tanks. If the vehicle has a tank of over 1,000 gallons and is primarily used to transport water, it is considered a mobile water supply apparatus, **Figure 9-3**. The term **water tender** is used to describe mobile water supply apparatus, although some jurisdictions use the term **tanker**. The fire service is still divided on this language to some degree. The National Fire Service Incident Management System Consortium uses the term *water tender* to describe land-based water supply apparatus in its *Model Procedures Guide for Structural Firefighting*. This guide has been endorsed by the International Association of Firefighters (IAF), International Association of Fire Chiefs (IAFC), National Wildfire

Figure 9-1 Well pump with storage tanks.

Figure 9-3 Mobile water supply apparatus.

Coordinating Group (NWCG), and Fire Resources of Southern California Organized for Potential Emergencies (FIRESCOPE) groups. The guide refers to a tanker as an aircraft capable of carrying and dropping water for firefighting operations. Although regional, some fire departments use *tanker* to describe the land-based water supply apparatus and **tender** to describe hose-carrying apparatus. This chapter will use the term *tender*, because it is becoming the preferred standard term for ground vehicle units, while tankers are understood to be air units, such as helicopters and airplanes. Tenders might or might not have pumps and hose. Mobile water supply apparatus range from 1,000 to over 8,000 gallons with some units being tractor-trailer units. Tenders combined with **portable water tanks** can efficiently provide large volumes of water to a fireground operation. Tenders may have a small booster or attack pump of at least 250 gpm, a fire pump of at least 750 gpm, or a transfer pump of at least 250 gpm. Transfer pumps use a vacuum-operated pump to draw and expel their load.

> **Safety** Mobile water supply apparatus have been involved in a large percentage of fire department vehicle accidents and firefighter injuries and fatalities. The rural environment that requires these large, heavy vehicles has older, narrower, crowned roads that are often driven by persons with less experience, training, and supervision. Some fire departments cannot afford a new vehicle so they modify an old delivery tank truck into a water supply unit without the necessary safety modifications such as extra baffles and heavier brakes. Another problem is that a gasoline tank truck with 7,600 gallons of water has a 20 percent heavier load than if carrying lighter weight gasoline. Fire departments using this type of apparatus need to examine all aspects of this problem and work to provide safe and effective fire protection.[2]

Tanks, Ponds, and Cisterns

An additional source of water can come from a developed source of water such as water tanks, ponds, and cisterns. Water tanks may be underground, ground level, or elevated and may be connected to a piping system, have a **dry hydrant** or other connection, or have just a drafting point. Ponds may be developed for fire protection reasons and may be lined or unlined with or without dry hydrants. A **cistern** is an underground water tank made from natural rock or concrete. Cisterns can store large quantities of water—30,000 gallons or more—in areas without other water supplies or as backup. Connections are dry hydrants, drafting pit, or other type. Firefighters should also be aware of other sources of water available to them such as swimming pools and any other body of water that can be accessed for firefighting.

WATER DISTRIBUTION SYSTEMS

How does the water get to the fire? This is a good question in areas with fire hydrants because the source may be many miles away. Water distribution systems have several components, including a method of getting the water, filtration or treatment processes, and a method of storage, supply, and distribution. The important parts for firefighters are the supply and distribution system, including storage facilities. The water can be obtained from a surface or groundwater source. Water supply and distribution systems should be designed to meet the community's domestic, industrial, and fire protection needs at peak periods.

Small groundwater systems need a well with a pumping station that can also treat and store the water and that connects to a local supply system. In larger well or surface systems, multiple supply, processing, and storage units with massive feeder and distribution lines are often used.

Water is supplied in three ways. The first is *gravity fed*, **Figure 9-4A**, in which the water source is at a higher elevation than where the water is used, such as in a mountain reservoir feeding a valley city. The next is a *pumped* system, **Figure 9-4B**, in which pumps are used to draw water from a well, river, or reservoir and pump it through the system. The third type is a *combination gravity-pumped system*, **Figure 9-4C**, where part of the system relies on pumps and the remainder on gravity. This is the most common type and even good gravity systems find areas where distance or elevation requires additional pumping and storage for maintaining pressure. It is considered to be a combined system because elevated water tanks that are filled by pumps in turn supply the piping system by gravity.

After treatment for drinking, water goes into the distribution system or water mains. The mains are divided into feeders and distribution lines. Primary feeders supply secondary feeders and then distribution lines. Primary and secondary feeder lines in large cities can be measured in feet; some of New York City's aqueducts are 20 feet in diameter. Some older and larger cities have separate water systems

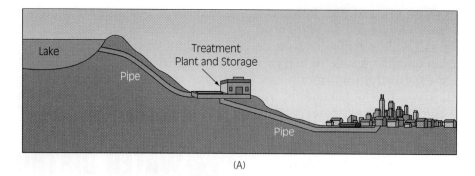

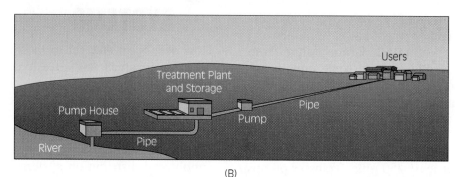

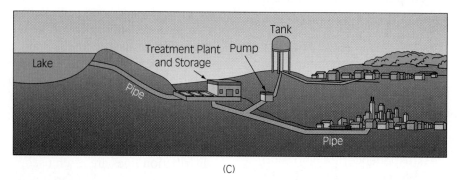

Figure 9-4 (A) Gravity-fed water distribution system. (B) A direct pump water distribution system. (C) A combination gravity-pumped water distribution system.

just for fire protection. Distributor lines are the water mains with the building connections and fire hydrants, and for fire protection; they range in size from 6 to 16 inches. Good water distribution systems are interconnected into *loops* and *grids* that allow multidirectional supply and reduce dead-end mains that cannot provide adequate pressure or volume, **Figure 9-5**.

Streetsmart Tip Firefighters should know the location of dead-end water mains and, where possible, avoid using them. Connecting to a dead-end water main may not provide adequate water, or if another unit connects to another hydrant on the same dead-end main, it may rob the original unit of what was a good water supply.

FIRE HYDRANTS

Fire hydrants allow the fire department to access water supply systems. A fire hydrant is also known as the "plug." The term *plug* is a throwback to days when firefighters drilled a hole in wooden water mains, then capped the hole with a plug when they were finished. The two major hydrant types are wet and dry barrel hydrants. Another hydrant type is a dry hydrant that is neither dry nor a true hydrant but a pipe system for drafting from a static water source. Some specialty hydrants are also used.

Wet Barrel

Wet barrel hydrants, **Figure 9-6** and **Figure 9-7**, have water in the barrel up to the valves of each outlet. They are used in areas that are not subject to

WATER SUPPLY ■ 207

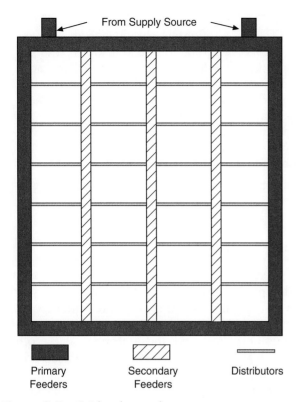

Figure 9-5 Grid or looped system.

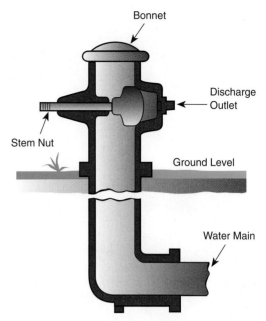

Figure 9-7 Typical schematic of wet barrel hydrant.

Figure 9-6 Multivalved wet barrel hydrant.

freezing temperatures, primarily California and Florida. The wet barrel hydrant allows each outlet to be controlled by a separate valve and can have additional lines taken off or supplied if an outlet is available. This additional connection does not require the flow through other outlets to be stopped. Some also have a main control valve that can control the flow to all outlets.

Dry Barrel

Dry barrel hydrants, **Figure 9-8** and **Figure 9-9**, are used in areas where freezing temperatures could damage the hydrant. It uses a valve at the base of the hydrant to control water flow to all outlets. The base and this valve are below ground at the level of the water main. Connecting additional lines to a flowing dry barrel hydrant requires the entire hydrant to be shut down until the new connect is made and then reopening the valve or adding a gate on one of the outlets prior to charging the hydrant. When the valve is fully closed or partially opened, a drain allows the water to drain from the hydrant preventing damage from freezing. Firefighters should ensure that the hydrant is operated and shut down with the valve in either the fully opened or closed position. When this has not been done, it has caused undermining of the hydrant and ground with damage to nearby roads, buildings, and fire apparatus.

Figure 9-8 Typical dry barrel hydrant.

Caution Firefighters should ensure that fire hydrants are operated and shut down with the valve in either the fully opened or closed position. When the valve is not fully opened or closed, undermining of the hydrant and ground with damage to nearby roads, buildings, and fire apparatus has occurred, plus it may restrict volume. Also, anytime a valve is operated, it should be done slowly to prevent a water hammer.

Dry Hydrant

A dry hydrant is not really a fire hydrant but a connection point for **drafting** from a static water source such as a pond or stream, not a pressurized one. A dry hydrant is a pipe system with a pumper suction connection at one end and a strainer at the other, **Figure 9-10** and **Figure 9-11**. They are used primarily in rural areas with no water system but may be found in urban and suburban areas as a backup or where certain buildings may be of great distance from other water sources but a pond or stream is located nearby. Placing dry hydrants allows fire departments to have better access to the water source; some even have graveled areas for parking and a turnaround point for the apparatus. The landowner benefits by having a ready source of water in case of fire and may even get a discount on insurance.

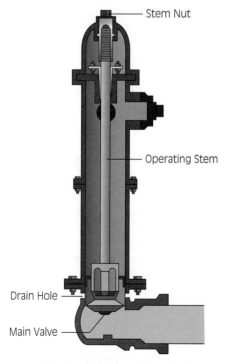

Figure 9-9 Typical schematic of dry barrel hydrant.

Figure 9-10 Dry hydrant.

WATER SUPPLY 209

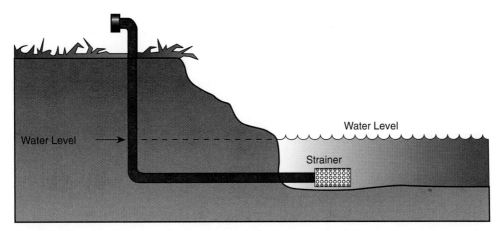

Figure 9-11 Schematic of dry hydrant installation.

Specialty Hydrants

Wall hydrants are hydrants mounted on the wall of a building after the water line has been run into the building, **Figure 9-12**. Firefighters should make sure that they are connecting to a wall hydrant and not the fire pump test connection outlets. A wall hydrant has a direct connection to the water supply system, whereas a fire pump test connection is an inspector's way to test the flow of the fire pump. The test connection will not supply any additional water; it will just rob the water from the fire pump.

Figure 9-12 Wall fire hydrant.

A flush-type hydrant is mounted below grade level and is usually found in a pit, vaults, or valve box. The purpose of mounting the hydrant below grade allows access to a water source in areas where an above-grade hydrant would interfere with operation of the facility. Airports, shipping terminals, and many European cities utilize flush-mounted hydrants.

High-pressure hydrants are hydrants that are connected to a separate high-pressure water system used only for fire protection purposes. Some larger cities, San Francisco is one, have these types of systems installed in the downtown commercial and industrial areas to provide high volume and pressure for large fires. Some of these systems draw their water directly from the city harbor, and the pumps are often maintained by the fire department. These hydrants are either of the wet or dry barrel design.

VALVES ASSOCIATED WITH WATER DISTRIBUTION SYSTEMS

The valves in public water systems are usually nonindicating type **gate valves** and **check valves**. The gate valves are butterfly valves that are opened and closed to control water flow, **Figure 9-13**. Nonindicating gate valves are installed at interconnections of water mains, at intermediate points on long sections of water mains, and before each hydrant and major building connection. Check valves are installed to control water flow in one direction, typically when different systems are interconnected. **Backflow preventers** are a check valve or pair of

Figure 9-13 Hydrant with plumbing. Note the location of a gate valve between the water main and hydrant.

check valves that prevent a backflow of water from one system into another and are mostly required where a building water or fire protection system connects with the public water system. Backflow preventers are being required for environmental and health reasons. The backflow preventer is installed on the building's private water piping after the building's main water shutoff. Private water valves are of the indicating type such as a post indicator valve (PIV), wall indicator valve (WIV), or an outside stem and yoke valve (OS&Y). These valves are commonly associated with standpipe and sprinkler systems.

RURAL WATER SUPPLY

Rural water supply or tender-based water supply operations can occur anywhere, not just in rural areas. Urban and suburban areas often have undeveloped areas or other places where the hydrants are just too far away to use effectively. Limited access highways and large bridges can be found to have water supply problems. Even redundant water supply systems can suffer a catastrophic event that knocks out or limits the water supply. The basics of rural water supply should be understood by all firefighters. Rural firefighters should find it easier to adjust to water supply operations in an urban area than urban firefighters could adjust to nonhydrant areas.

Rural water supply operations require careful coordination and control. A water supply officer should be part of the incident command system with full authority over tender operations. Firefighters assigned to water supply operations may find the work less glamorous than actual firefighting, but the nozzle person cannot put out the fire without this vital support.

Portable Water Tanks

Since tenders are designed to transport water, they need to be able to quickly drop off their load of water and return to the **fill site** as soon as possible. To speed the operation, each tender should have a portable water tank with a capacity equal to or greater than its tank size, **Figure 9-14**. The collapsible or inflatable tanks are filled by unloading or dumping the tender, allowing it to return to the fill site and bringing back an additional load of water. The portable tank is either set up next to the attack engine or, if space is not available there, it is assigned an engine that drafts from the tank and supplies the attack engines. Depending on the size of the fire and availability of tanks and tenders, several tanks may be used to increase the fire flow. Many large modern tenders use a **jet dump** or **jet siphon**

WATER SUPPLY 211

Figure 9-14 Portable water tanks are an essential piece of equipment for shuttle operations.

Figure 9-16 Tender at dump site dropping water directly into portable tank. *(Photo courtesy of Donald Fischer)*

to speed the unloading of tanks, **Figure 9-15**. The jet dump or siphon draws the water out as opposed to just using gravity in a nonassisted system.

The portable tank is erected or inflated at the **dump site** in a location where the loaded tenders can offload their water and an engine can place its suction hose into the tank to draft and supply the water to attack engines. Placing a heavy tarp on the ground prior to opening the portable tank can help protect the portable tank liner before water is dumped. In larger operations, multiple tanks may be set up together. Departments that use portable tanks and ones that may support those companies should regularly practice tender shuttle operations.

Tender Operation

A tender operation is a **shuttle operation** that involves tenders moving large quantities of water between a dump site and a fill site. The dump site is where the water is delivered for quick unloading, and because the tenders arrive already full, it is set up first. Using portable tanks and a supply engine to draft from the tanks speeds the process and allows the tender to return for another load. Personnel at the dump site connect hoselines or operate dump valves to fill the portable tanks and help set up the drafting operation for supplying the attack pumpers, **Figure 9-16**. Dump sites should be selected for availability to unload multiple tenders, turnaround area for the tenders, operational area, continued access to the fireground, and safety of personnel. The fill site is the location of the water source where the tenders are filled, **Figure 9-17**. The fill site should be staffed with enough personnel to quickly connect hoselines to fill the tender and may require a pumper to draft from a static source, **Figure 9-18**. The fill site should be picked for an adequate water source including filling multiple units, the ability to either turn around the vehicles or, better yet, have a drive-through site, and the safe operation of the site.

Figure 9-15 Jet siphon valve. *(Photo courtesy of Donald Fischer)*

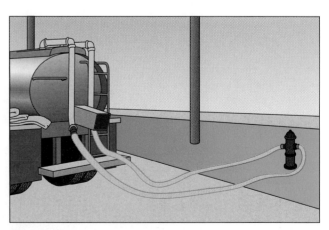

Figure 9-17 Tender at fill site with pressurized water source.

Figure 9-18 A pumper at fill site drafting water.
(Photo courtesy of Donald Fischer)

Shuttle operations control the fire flow capacity of the incident, **Figure 9-19**. Fire flow rates can be calculated by adding the time to fill a tender, turn the unit around and return to the dump site, drop the water, and turn around and return to the fill site divided by the quantity of water carried. If a 1,000-gallon (3,785-L) tender[3] has a 10-minute shuttle time, the flow rate cannot exceed 100 gpm (378 L/min) [1,000 gallons (3,785 L) ÷ 10 minutes]. Reducing the time or adding more tenders or larger tanks would increase the flow rate. With five 2,000-gallon (7,570-L) tenders operating on 10-minute cycles, the flow rate would be 1,000 gpm (3,785 L/min) [10,000 gallons (37,850 L) ÷ 10 minutes]. Tender operations cannot be sped up by increasing vehicle speed on the highway. Time should only be gained at dump and fill sites by increased efficiency of personnel and equipment.

Safety It is important to note that attempts to lower shuttle time cannot and should not be done by increasing vehicle speed. Time should only be gained at dump and fill sites by increased efficiency of personnel and equipment.

PRESSURE ASSOCIATED WITH WATER DISTRIBUTION SYSTEMS

All of the Earth's water is under **pressure**. (See Chapter 11 for more discussion on pressure.) Even seawater is under **atmospheric pressure**, which is 14.7 pounds per square inch (psi) (101 kPa) at sea level, **Figure 9-20**. Drafting water from a lake or the sea is accomplished by taking advantage of this atmospheric pressure. Creating a partial vacuum or low atmospheric pressure area inside a pump causes the atmospheric pressure on the water's surface to force the water up the suction hose and into the pump, which adds pressure and pumps it out. Pressure is the force, or weight, of water measured over an area, **Figure 9-21**. A 1-foot (0.305-meter) column of water 1 inch square weighs 0.433 pounds (0.19 kg), creating a pressure at the bottom of 0.433 psi (3 kPa). Pressure can also be expressed as feet of head with the 1-foot column having a head of 1 foot (0.305 meters) and 2.31 feet (0.7 meter) of head equaling 1 psi (6.895 kPa). An atmospheric pressure of 14.7 psi (101 kPa) would create a head of almost 34 feet (10.4 meters) [14.7 psi × 2.31 ft/lbs = 33.9 feet] if a perfect vacuum was created or the ability to lift a column of water 33 feet (10.1

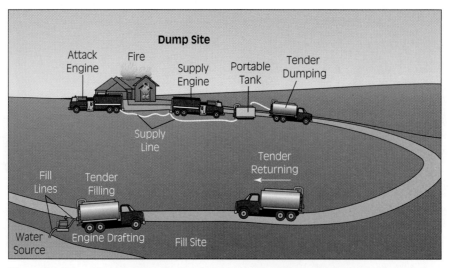

Figure 9-19 Tender shuttle operation.

WATER SUPPLY 213

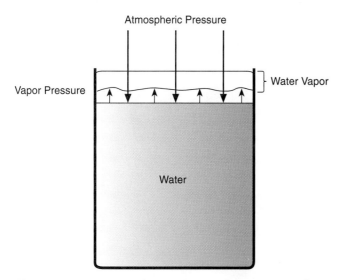

Figure 9-20 Atmospheric pressure being exerted on container of water.

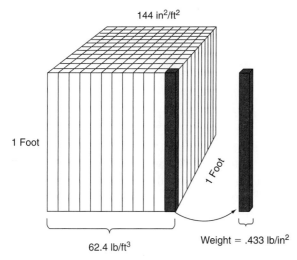

Figure 9-21 Pressure is expressed in pounds per unit area (psi).

meters). Practicality limits this ability to about 30 feet (9 meters). A tank in a distribution system that stores water 100 feet (30 meters) high would create 100 feet (30 meters) of head or 43 psi (296 kPa) [100 feet ÷ 2.31 ft/lbs = 43.3 psi or 100 feet × 0.433 lbs/ft = 43.3 psi]. Pressure in a nonflowing closed system is equal at all points, but while flowing, it is reduced by friction created by its movement and the loss of pressure at the opening. Pressure in an open system is subject to atmospheric pressure.[4]

Water distribution systems are supplied under pressure to meet domestic needs and fire protection. This dual use means the pressure is never as high as firefighters would prefer because it could cause leaks in private plumbing. The average pressures in the United States are between 65 and 80 psi (448 to 551 kPa) with a high of 150 psi (1,034 kPa) before damaging plumbing and a usual minimum of 20 psi (137 kPa).[5] The 20-psi (137-kPa) figure is the recommended low **residual pressure** when pumping from a hydrant; pressures below this may create a vacuum in part of the supply system that could cause damage to the system. Fire departments should know the normal and flowing pressures and capacity of the water distribution system including areas that are high or low in either category. High-flow and/or high-pressure areas are an advantage, whereas low-pressure and/or low-flow areas should be avoided if possible.

TESTING OPERABILITY AND FLOW OF HYDRANTS

Testing should be conducted on fire hydrants periodically to ensure that they are operable and to determine the flow rate of the hydrants, **Figure 9-22**. Testing prevents the fire department from finding out that the hydrants do not work or are insufficient while operating at a fire. Testing of the hydrant may be the responsibility of the water department, but the fire

Figure 9-22 Firefighters inspecting and servicing a hydrant.

Figure 9-23 Typical pump panel gauges are Bourdon gauges.

department should at least have access to the results of the testing. Some fire departments do the testing and maintenance of the hydrants, whereas others have joint programs with the water department. All testing should be coordinated between the two agencies.

The first set of tests involves the operability of the hydrants. On wet and dry barrel hydrants, the test starts with a visual inspection for damage. Then after ensuring the hydrant is off, inspectors remove all of the caps and check the threads and gaskets for damage. The next task would be to check that all valves on the hydrant allow water to flow. A dry barrel hydrant should be checked to ensure the drain valve is working properly by determining that the water level is dropping inside the barrel after the hydrant has been shut off. For a dry hydrant, a visual inspection of the piping, caps, and gaskets should be conducted. A flow test would be the next step to ensure water is available, followed by backflushing of the hydrant to clear any debris from the strainer. Backflushing involves pumping into the outlet of the hydrant so that water flows out of its intake. *This should not be done on other types of hydrants, because it will contaminate the drinking water supply.* Replacing any caps, oiling any moving parts, and painting may complete this inspection process.

The next tests are flow tests that should be conducted during the normal operations of the supply system. There are two ways to do a flow test in a pressurized system; one is a fireground method and the other is more extensive and is done for fire department planning purposes and insurance ratings. The fireground method uses the gauges on the pumper using the hydrant while the other requires a 2½ inch (63-mm) cap with a **Bourdon gauge**, **Figure 9-23**, hydrant wrenches, **Pitot gauges**, **Figure 9-24**, a ruler, and paper to record the data.[6] The tests are conducted in the next section.

Figure 9-24 Hydrant testing with a Pitot gauge.

DETERMINING STATIC, RESIDUAL, AND FLOW PRESSURES

The fireground method of flow testing involves connecting a pumper to a hydrant and turning it on. Prior to charging any lines, the **static pressure** is read on the main intake compound gauge. Static pressure is the pressure in the system with no hydrants or water flowing. The pump operator then charges the first line with the desired volume, noting

Percentage Drop Measurements

PERCENTAGE DROP 1	PERCENTAGE DROP 2	AMOUNT OF ADDITIONAL WATER
0–10% drop	0–5% drop	Three times the amount
11–15% drop	6–10% drop	Two times the amount
16–25% drop	11–20% drop	One times the amount

TABLE 9-1

the pressure first. With this flow going, the operator again reads the intake gauge and gets the residual pressure or the remaining pressure left in the system after the flow and friction loss from the flow. The pump operator then compares the percentage of pressure drop from static to residual and determines the amount of additional volumes that may be pumped from that hydrant. Two comparisons are used for percent drop as shown in **Table 9-1**.

Consider this example using the first column of drop: a static pressure of 50 psi (345 kPa) with a flow of 500 gpm (1,893 L/min) at a residual of 45 psi (310 kPa). Percent drop is 50 − 45 = 5 ÷ 50 = 10 percent. With 10 percent drop, the first column would allow an additional three times or 1,500 gpm (56,775 L/min) to come from the hydrant. The second column is more conservative and would allow only an additional 1,000 gpm (3,785 L/min) to be taken. Department SOPs should define which chart is to be used. In any case, the residual pressure should not drop below 20 psi (137 kPa).

The second test involves testing multiple hydrants and is not conducted during fire operations. The first hydrant selected is used with the cap gauge to take a static pressure with no hydrants flowing, and then use continues while measuring the system's residual pressure. The second hydrant is opened and the flow measured, preferably at a 2½-inch (63-mm) outlet, with a Pitot gauge while the residual reading at the first hydrant is taken. The Pitot gauge is inserted into the stream's midpoint and out about half the distance of the diameter, **Figure 9-25** and **Figure 9-26**. This process would continue if additional hydrants are to be tested, gathering each flowing hydrant pressure on the Pitot gauge and at the same time gathering the residual on the first hydrant, **Figure 9-27**. The next step involves the calculations of discharge for the second hydrant and beyond. Using a chart is easiest or the following formula can be used:

$$Q = 29.83\, c\, d^2 \sqrt{p}$$

where Q is quantity in gallons, c is the coefficient of the outlet, d^2 is the diameter of the outlet in inches squared, and $\sqrt{p}$ is the square root of the pressure. Each type of outlet or nozzle has a coefficient with most hydrants having a number of 0.90, 0.80, or 0.70.[7] The way to determine the correct coefficient for hydrants is simply to feel the inside edge of the discharge orifice. A smooth or rounded edge uses the .9 coefficient. A square edge requires a .8, and a rough lip or edge uses .7. For example, for a hydrant with a 2½-inch outlet with a coefficient of 0.90 and a Pitot reading of 49 psi, we would get:

$$Q = 29.83\, c\, d^2 \sqrt{p}$$
$$= 29.83 \times (0.9) \times (2.5^2) \times (\sqrt{49})$$
$$= 29.83 \times (0.9) \times (6.25) \times (7)$$
$$= 1{,}175 \text{ gpm}$$

OBSTRUCTIONS AND DAMAGE TO FIRE HYDRANTS AND MAINS

Obstructions and damage can occur to fire hydrants and water mains as they can to any other type of system. Nature, vandals, accidents, and

Figure 9-25 Pitot gauge and cap gauge with Bourdon gauge.

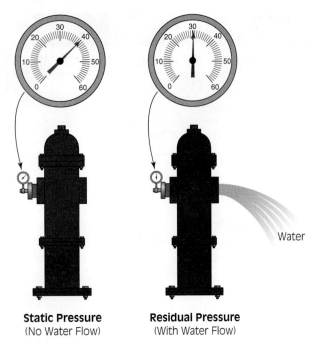

Figure 9-26 Static and residual pressures.

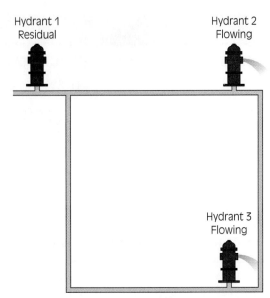

Figure 9-27 Flow testing diagram showing flowing of two hydrants with another used for residual pressures.

improper actions by members of the fire department can cause problems. The natural causes can be sudden events or an event occurring over a long term. Sudden events such as an earthquake or flood can undermine or break mains or hydrants. Other natural events are long term such as tree roots working into the piping to block it or crack the piping. Encrustation and blockage of the piping by minerals or organisms that survive the treatment process can slowly build up and restrict or completely block the flow. Vandals can do damage by opening or closing valves that can shut off the flow of water or undermine a hydrant. Opening a hydrant in freezing weather without allowing water to flow can cause the hydrant to freeze and crack. Vandals can also put debris into the system damaging the hydrant or the pump; the firefighter should check the outlet prior to connecting the hose and flush it if necessary. Accidents can crack pipes or break off a hydrant. For example, a bank building began to lose its front wall after a broken hydrant and main undermined the street, sidewalk, and then the foundation of the bank. All this damage could have been prevented if the engine company had closed off the gate valve on the water main at the hydrant instead of calling for the water company. Damage from improper actions by firefighters includes opening and closing valves or hydrants too quickly, thus creating a **water hammer**, failing to open or close a hydrant or other valve fully, or cross-threading the threads on the connections.

Lessons Learned

Water is the most common fire extinguishing agent used by firefighters and it must be supplied in sufficient quantity to accomplish extinguishment. Firefighters must understand the relationship between the amount of water that can be supplied and the amount needed. If they do not have enough water, their efforts will not be effective.

Supplying the water needed requires an understanding of where water comes from, how it gets to the firefighter, or where they need to go to get it. This leads to learning about water sources and distribution systems. Distribution systems have valves and hydrants that must be operated. The more experienced firefighters must know about pressure in the system and determining flow. They must also know about problems with the system.

Firefighters in areas without a distribution system must create one using mobile supply apparatus, portable water tanks, and shuttle operations. They quickly learn the value of water because if they run out, they must wait until the tender gets back before they can continue to fight the fire.

KEY TERMS

Aquifer A formation of permeable rock, gravel, or sand holding water or allowing water to flow through it.

Atmospheric Pressure The pressure exerted by the atmosphere, which for Earth is 14.7 pounds per square inch at sea level.

Backflow Preventers A check valve or set of valves used to prevent a backflow of water from one system into another. Required where a building water or fire protection system connects with the public water system. Backflow preventers are being required for environmental and health reasons.

Bourdon Gauge The type of gauge found on most fire apparatus that operates by pressure in a curved tube moving an indicating needle.

Check Valves Valves installed to control water flow in one direction, typically when different systems are interconnected.

Cistern An underground water tank made from natural rock or concrete. Cisterns store large quantities of water—30,000 gallons or more—in areas without other water supplies or as a backup supply.

Drafting The pumping of water from a static source by taking advantage of atmospheric pressure to force water from the source into the pump.

Dry Hydrant A piping system for drafting from a static water source with a fire department connection at one end and a strainer at the water end.

Dump Site The area where tenders are unloaded or their load dumped.

Fill Site The area where tenders are filled or get their water.

Fire Flow Capacity The amount of water available or amount that the water distribution system is capable of flowing.

Fire Flow Requirement A measure comparing the amount of heat the fire is capable of generating versus the amount of water required for cooling the fuels below their ignition temperature.

Gate Valves Indicating and nonindicating valves that are opened and closed to control water flow.

Jet Dump A device that speeds the process of dumping a load of water from a tanker/tender.

Jet Siphon A device that speeds the process of transferring water from one tank to another.

Pitot Gauge A device with an opening in its blade-shaped section that allows water to flow to a Bourdon gauge and registers the flowing discharge pressure of an orifice.

Portable Water Tanks Collapsible or inflatable temporary tanks for the storage of water that is dumped from tankers or tenders. Usually carried by the tender to set up a dump site.

Pressure The force, or weight, of a substance, usually water, measured over an area.

Residual Pressure The pressure in a system after water has begun flowing.

Shuttle Operation The cycle in which mobile water supply apparatus is dumped, moves to a fill site for refilling, and is returned to the dump site.

Static Pressure The pressure in the system with no hydrants or water flowing.

Tanker The term given to aircraft capable of carrying and dropping water or fire retardant. Some departments still use the term to describe land-based water apparatus.

Tender The abbreviated term for *water tender*. A water tender is defined as a land-based mobile water supply apparatus. Some departments still use the term *tender* to describe a hose-carrying support apparatus.

Tidal Changes The rising and falling of the surface water levels due to the gravitational effects between the Earth and the moon. In some areas, these changes are insignificant, but in others there is more than 40 feet of difference between high and low tide.

Water Hammer A sudden surge of pressure created by the quick opening or closing of valves in a water system. The surge is capable of damaging piping and valves.

Water Table The level of groundwater under the surface.

Water Tender The term given to land-based water supply apparatus.

REVIEW QUESTIONS

1. How does the fire flow capacity relate to the fire flow requirements?
2. What are the four reasons for using water for firefighting?
3. Where is most freshwater found?
4. Name three surface water sources.
5. What are the two names for mobile water supply apparatus?
6. Water is distributed or supplied through a supply system in what three ways?
7. What is a dead-end main and why should it be avoided?
8. Explain the two main types of fire hydrants and where they are found.
9. Why is a dry hydrant different from other hydrants?
10. What purpose does a backflow preventer serve?
11. What does a gate valve do?
12. Describe a portable water tank.
13. Explain a tender shuttle.
14. What is pressure?
15. What is the difference between static and residual pressure?
16. If a hydrant's water pressure drops when flowing a line by 9 percent, how many more lines can be taken from that hydrant?

Endnotes

1. See Sturtevant, Thomas, *Introduction to Fire Pump Operations,* Chap. 11. Delmar Learning, a part of the Thomson Corporation, Clifton Park, NY, 1997; *NFPA 1231: Standard on Suburban and Rural Water Supply;* or the NFPA's *Fire Protection Handbook,* 18th ed., Chap. 6-5 for additional information.
2. See many of the fire service publications about this safety issue. To examine closely the accident, injury, and fatality statistics, see the *NFPA Fire Journal* issues containing annual firefighter injuries and fatalities surveys.
3. Tank size is the total capacity of the tank, but often the tender cannot actually offload its capacity or the time to do so would not be productive. Departments operating tenders should conduct tests in which the tender is loaded and offloaded as if under fireground conditions and weigh the vehicle and divide the offloaded amount by 8.33 lb/gallon (or 1 kg/L) to determine its operating capacity.
4. See Sturtevant, Thomas, *Introduction to Fire Pump Operations,* Chap. 11. Delmar Learning, a part of the Thomson Corporation, Clifton Park, NY, 1997, for more information.
5. *Fire Protection Handbook,* 18th ed., pp. 6–72.
6. *NFPA 291: Recommended Practices for Fire Flow Testing and Marking of Fire Hydrants,* 2002 Edition.
7. *NFPA 291: Recommended Practices for Fire Flow Testing and Marking of Fire Hydrants,* 2002 Edition.

Additional Resources

Crapo, William F., *Hydraulics for Firefighting.* Delmar Learning, a part of the Thomson Corporation, Clifton Park, NY, 2002.

From Delmar Learning, a part of the Thomson Corporation:

Gagnon, Robert, *Design of Special Hazard and Fire Alarm Systems.* Delmar Learning, a part of the Thomson Corporation, Clifton Park, NY, 1998.

Klinoff, Robert, *Introduction to Fire Protection,* 2nd ed. Delmar Learning, a part of the Thomson Corporation, Clifton Park, NY, 1997.

Sturtevant, Thomas, *Introduction to Fire Pump Operations.* Delmar Learning, a part of the Thomson Corporation, Clifton Park, NY, 1997.

Other Publishers:

Cote, Arthur, and John Linville, *Fire Protection Handbook,* 18th ed. National Fire Protection Association, Quincy, MA, 1997.

NFPA 22: Standard for Water Tanks for Private Fire Protection, 1998 Edition. National Fire Protection Association, Quincy, MA.

NFPA 24: Standard for the Installation of Private Fire Service Mains and Their Appurtenances, 2002 Edition. National Fire Protection Association, Quincy, MA.

NFPA 1410: Standard on Training for Initial Emergency Scene Operations, 2000 Edition. National Fire Protection Association, Quincy, MA.

NFPA 291: Recommended Practices for Fire Flow Testing and Marking of Fire Hydrants, 2002 Edition. National Fire Protection Association, Quincy, MA.

NFPA 1142: Standard on Water Supplies for Suburban and Rural Firefighting, 2001 Edition. National Fire Protection Association, Quincy, MA.

NFPA 1901: Standard on Automotive Fire Apparatus, 1999 Edition. National Fire Protection Association, Quincy, MA.

CHAPTER 10

FIRE HOSE AND APPLIANCES

Ric Koonce, J. Sargeant Reynolds Community College

OUTLINE

- Objectives
- Introduction
- Construction of Fire Hose
- Care and Maintenance of Fire Hose
- Types of Hose Coupling
- Care and Maintenance of Couplings
- Hose Tools and Appliances
- Coupling and Uncoupling Hose
- Hose Rolls
- Hose Carries
- Hose Loads
- Advancing Hoselines—Charged/Uncharged
- Establishing a Water Supply Connection
- Extending Hoselines
- Replacing Sections of Burst Hose
- Hose Lay Procedures
- Deploying Master Stream Devices
- Service Testing of Fire Hose
- Lessons Learned
- Key Terms
- Review Questions
- Endnotes
- Additional Resources

STREET STORY

My station covers an area that has lots of old abandoned homes with wood construction. One evening a call came in as a house fire, and as we approached from several blocks away we could see the smoke. When we arrived, the house was fully involved. We were told that a homeless person had taken residency in this shack—it had no power. As we attempted to go in, we found out that he was a major pack rat. There were several appliances, including washing machines and refrigerators, and parts throughout the front of the house, and that made it nearly impossible to gain access. We had to squeeze through the door sideways and walk over things. Inside there were more piles: candles, cans, papers, etc.

It was an incredible obstacle just for the the rescue crew to get to the victim. (Eventually they found him behind a locked door. He wasn't breathing but he did have a pulse. He also had a gun in one hand and a knife in the other!) In terms of advancing the hose, we decided to do it as a team. We needed an action plan. Normally it's a three-person job, but in this case it took two entire crews. Instead of one person at the front, we had somebody every 4 or 5 feet, because otherwise as soon as you pulled you'd get stuck on something.

That day was really long and by the end of the incident more than half of the house was destroyed. But we really learned to adapt to a difficult situation. Most important is remembering and applying the basic skills you know to that specific incident. I remember learning hoseline rolling in training, but it's not something you use much in application. In this case, though, it was the best way to get the job accomplished. I also learned the importance of teamwork. You may come in on a fire assigned to one job, but when there's an issue like this, you all have to work together on one part.

—Street Story by Michelle Steele, Lieutenant, Miami-Dade Fire and Rescue, Miami, Florida

OBJECTIVES

After completing this chapter, the reader should be able to:

- Identify and explain the construction of fire hose.
- Demonstrate the care and maintenance of fire hose.
- Identify the types of hose couplings and threads.
- Demonstrate the care and maintenance of hose couplings.
- Identify and explain the use of hose tools and appliances.
- Demonstrate the coupling and uncoupling of fire hose.
- Demonstrate the rolling, carrying, and loading of fire hose.
- Demonstrate the advancing of fire hoselines, both charged and uncharged.
- Demonstrate the establishment of a water supply connection.
- Demonstrate the extending of hoselines.
- Demonstrate the replacement of burst hose sections.
- Demonstrate the procedures for laying hoselines for water supply.
- Demonstrate the deployment of master stream devices.
- Demonstrate the service testing of hose.

INTRODUCTION

This chapter introduces the firefighter to one of the most important tools for fighting fires: hose. Hose is the tool used to move water from one place to the fire. The firefighter should know about the hose, how it is made, and how to care for it. More importantly, the firefighter must know how to properly store it on the apparatus and how to quickly move it from that apparatus to the location where the firefighting will take place. The techniques learned in this chapter and the next one on fire streams and nozzles are the basic techniques of fire control. These chapters, along with many others in this book, are the foundation of firefighting in a safe, efficient, and effective manner. Firefighters should learn them well and practice them often. Many experienced fire officers often comment when things do not go well on the fireground of the need to "get back to the basics." These are those basics.

Fire hose is a flexible conduit used to convey water or other agent from a source to the fire. Early firefighters used bucket brigades to supply the water to the engines. Top-mounted nozzles sprayed the water on the fire. Leather hose was invented to first allow firefighters to advance nozzles from the engine to the fire and then to supply the engine. Today, many different materials have replaced leather hose but the basic tasks performed for supply and attack remain the same. Firefighters, fire brigade members, wildland firefighters, and some building occupants use hose today.

Couplings, adapters, and appliances are used in modern firefighting to connect hose and adapt it to different sizes or ways of operating. Hose couplings were a problem at several major fires at the beginning of the twentieth century. The different fire departments fighting these fires discovered that their hose couplings did not match and the hoses could not be connected. Today, departments that do not use National Standard Hose Threads use adapters to make connections between their threads and the standard ones. Adapters and appliances have been created or made lighter weight to make the job of firefighting easier or better. This chapter covers many of these, but others may be used locally to solve special problems.

> **Firefighter Fact** Large conflagrations in Boston in 1872 and Baltimore in 1904 resulted in many large cities sending fire units to help fight the fire only to discover that their hose couplings could not be connected.

CONSTRUCTION OF FIRE HOSE

Fire hose has two components, the hose itself and the couplings that connect the hose sections or appliances used with them. Fire hose is made in three types of construction: wrapped, braided, and woven, **Figure 10-1**. In wrapped construction, a fabric or mesh material is then impregnated with rubber or a plastic and then wrapped around a rubber **liner** and covered with rubber or plastic. Braided construction uses a yarn braided over the rubber liner, and a rubber or plastic layer separates it from the next layer. A rubber or plastic cover encases the entire hose. Woven hose has a rubber liner and one or more outer layers called a **jacket** that is woven of cotton, synthetic, or blended materials. Some hoses have jackets of rubber or plastic-impregnated woven construction. The jacket is designed to protect the

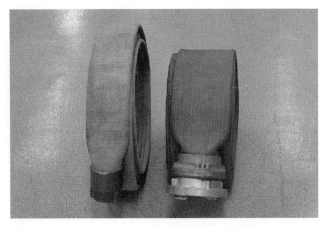

Figure 10-1 Woven and rubber-coated fire hose.

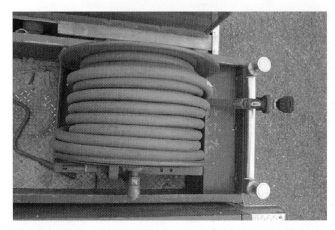

Figure 10-2 Booster hose on reel.

hoselining from heat and mechanical damage. In addition, **hard suction hose** has a plastic or wire **helix** to provide addition support and prevent collapse under a vacuum when drafting. Some newer synthetic types of hose combine the jacket and liner. Fire hose is divided into attack, supply, occupant use, forestry, and hard suction hose. Each type of hose has certain requirements for construction, size, and pressure. Most hose comes in 50- or 100-foot (15- or 30-m) lengths except as noted.

Firefighter Fact Cotton and other natural fibers replaced leather hose. Rubber liners were added in the late 1800s. Prior to the rubber liner, firefighters had to wait for the fibers to absorb enough water to cause them to swell and slow the leakage of water. Some fire departments and building occupant standpipe hoses used unlined single-jacket hose. Current standards now recommend lined hose.

Booster hose is smaller diameter, rubber-coated hose of ¾- or 1-inch (19- to 25-mm) size usually mounted on reel that can be used for outside fires or overhaul operations after the fire is out, **Figure 10-2**. It is not to be used for structural firefighting. Booster lines have a limited flow rate of up to 30 gpm (114 L/min) and, although easy to maneuver and control, they are unsafe for structural firefighting. Many departments have stopped purchasing this type of hose.

Attack hose is used by trained personnel to fight fires. This hose is connected to nozzles and distributors, master stream appliances, standpipes and sprinkler systems, hydrants, and fire pumps. Attack hose is a minimum of 1½ inches (38 mm) and can include large-diameter hose when supplying master stream devices. NFPA requires pumpers to carry 400 feet of attack hose. Attack hose is usually service tested annually at 250 psi (1720 kPa), and its use above this pressure may result in hose damage or personnel injury. Attack hose can be either single- or double-jacketed hose with the most common being double-jacketed for additional protection of the hose, **Figure 10-3**. Standpipe packs often use single-jacket hose to reduce the weight of the pack. **Medium-diameter hose (MDH)** is either 2½- or 3-inch (63- or 75-mm) hose, and sizes smaller than that are called **small lines** or **small-diameter hose**.

Supply hose or **large-diameter hose (LDH)** is larger hose [3½ inches (90 mm) or bigger] used to

Figure 10-3 Various hoselines in a hose bed.

move water from one engine to another or from the water source to a portable hydrant for distribution. Common sizes are 4 and 5 inches (100 to 125 mm) with some 6 inch (150 mm). Supply hose uses its larger diameter to move water at lower pressure. NFPA requires pumpers to carry at least 800 feet of 2½-inch or larger supply line.

Hard suction hose is rubber or plastic-coated hose with helix bands that resist collapse from the vacuum created by drafting. It comes in sizes from 2½ to 6 inches (65 to 150 mm) in diameter and is usually matched to the size of the main intake of the pump for which it will be used. Because of its more rigid construction, it is less flexible and movable than other hose. Hard suction hose is tested on its ability to withstand a vacuum. Hard suction hose is standard in 10-foot (3-m) lengths.

Soft suction hose is large-diameter woven hose used to connect a pumper to a hydrant and is also known as a soft sleeve, **Figure 10-4**. It is also sized according to the size of the intake of the pumper and usually comes in sizes from 4½ to 6 inches (112 to 150 mm). The lengths of a section vary: 20, 25, and 30 feet (6, 8, and 9 m). NFPA requires pumpers to carry at least 15 feet of soft sleeve intake hose.

Occupant use hose is hose used in standpipe systems for building occupants to fight incipient fires. It is usually 1½-inch (38-mm)-diameter single-jacket hose similar to attack hose. Older types of this hose may be unlined. Most fire departments do not use the occupant use hose, and if they use the standpipe system, they replace it with their own hoselines.

Forestry hose is specially designed for use in forestry and wildland firefighting, **Figure 10-5**. It comes in 1- and 1½-inch (25- and 38-mm) sizes and should meet U.S. Forestry Service specifications. Forestry hose is woven hose and comes in 50-, 75-, and 100-foot (15-, 23-, and 30-m) lengths.

Figure 10-4 A soft suction hose.

CARE AND MAINTENANCE OF FIRE HOSE

The care and maintenance of fire hose includes some basic measures. The care of hose begins with the careful placement and folding of the dry hose on the apparatus to prevent sharp turns or bends that can damage the liner (see Hose Loads section later). Previously used hose should be folded at different places to prevent overstressing one area of the hose. Hose standards still recommend that the hose be changed every thirty days with a visual inspection at the time of change but many departments no longer do so. The **hose bed** should be designed to allow the circulation of air to allow drying. This prevents mildew of natural fiber hose and vehicle rusting. The use of a hose bed cover will prevent rain and other road contaminants from getting into hose. The cover also blocks sunlight, which can affect rubber and plastic-coated hose. Wet hose should be thoroughly dried prior to returning it to apparatus but the newer synthetic hoses are often loaded wet.

When used on the fireground or for training, several steps can be taken to reduce damage to hose. Avoid laying hose over sharp or rough corners; use a hose roller or a salvage cover to prevent damage. Remove any kinks in hose prior to charging or soon afterward. Do not allow traffic to run over the hose, which could damage it, especially the couplings, which could become misshapen. Vehicles cause more damage to uncharged hose than charged hose because the inner liner is not protected when uncharged. Clean dirt and grit from hose with a brush or plain water to prevent these materials from creating abrasions on the hose. Firefighting exposes hose to heat and burning embers plus broken glass, nails, and other sharp objects that can quickly damage it. Avoid heat, burning embers, and chemicals, especially gasoline and oil. Mild chemicals can be removed using detergent, but large amounts or strong chemicals should be removed using hazardous materials decontamination procedures. Care should be taken to prevent hose from freezing. If it does freeze, it should be carefully moved to a place to thaw prior to any folding or bending. Any hose that has been subjected to any damage should be service tested prior to returning to service.

> **Note** The most common damage to hose is wear and tear due to dragging hose—this is especially damaging to couplings. Excessive pressure and water hammers should be avoided, as this can cause sudden rupture of the hose or separation of the hose from the coupling.

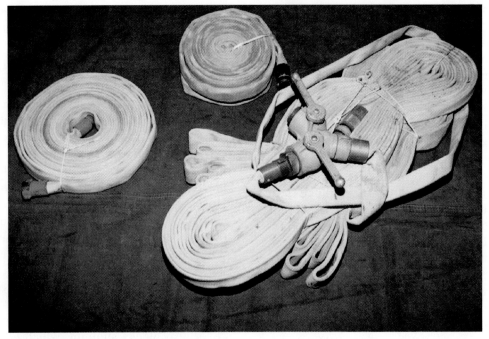

Figure 10-5 Hose packs and rolls for wildland firefighting.

Using a long, small-diameter hose and overpumping a hoseline at the incorrect pressure are common causes of excessive pressure. A water hammer is the pressure surge that occurs when water valves are rapidly opened or closed. Water that is not flowing has its pressure equalized throughout the whole system. Slamming a valve shut or opening it quickly allows this pressure to rapidly act on the parts of the system and damage hose, valves, pumps, and the water supply system including water mains, hydrants, and valves. Water hammers result from mistakes made by people operating valves and nozzles too quickly. The correct action is to operate all valves slowly and let the pressure either bleed off or build up slowly without the surge.

Dirty or contaminated hose can be rolled up and returned to the station for cleaning, or a portable washer can be used on scene or in a nearby parking lot. Cleaning may be done by placing the hose on a flat surface, rinsing with water, and scrubbing with mild detergent. A thorough rinsing and hanging in a hose tower or storage rack until dried will get the hose ready for reuse. Commercial hose washers are also available, but are not capable of adequately cleaning the threads of a coupling. Some work by connecting the hose washer to a hydrant and as the hose is thread through it is rinsed, **Figure 10-6**. This type still requires hand scrubbing by personnel. Another type has the ability to scrub, wash, rinse, and reroll the hose, **Figure 10-7**. Hose drying cabinets are also available that have heaters to dry the hose. Care must be taken not to overheat and damage the hose. Stored hose also requires maintenance. Occasionally water should be flowed through stored sections to prevent the liner from drying and cracking, and then it should be redried and stored.

The final part of the care of hose involves a regular visual inspection by firefighters after each use. Firefighters should carefully inspect the outer cover and couplings for any damage. Coupling care is discussed later but the outer cover should be checked as hose is reloaded for discoloration and abrasions. Further care is conducted during the annual service test.

Figure 10-6 Hose washer on hydrant.

FIRE HOSE AND APPLIANCES

Figure 10-7 An automatic hose washer.

TYPES OF HOSE COUPLING

Couplings allow hose and appliances to be joined or connected using a set of connection devices. Couplings are divided into two types: threaded and nonthreaded. Threaded couplings use a screw thread that secures the two sections of hose together, while nonthreaded couplings use locks or cams. Couplings are made of brass, aluminum, or an alloy called pyrolite, which is lighter than brass but more resistant to bending.

Threaded couplings are further separated into two different types: one with external threads, the male, and the other with internal threads, the female, with a matching thread type. A threaded hose coupling set is typically of three-piece construction with a hose bowl on each piece and a swivel attached to the internal coupling. The bowls are attached to the hose using an expansion ring with a tail gasket to prevent leaks. The internal swivel also has a swivel or thread gasket to stop leakage where the threads connect, **Figure 10-8**.

(A)

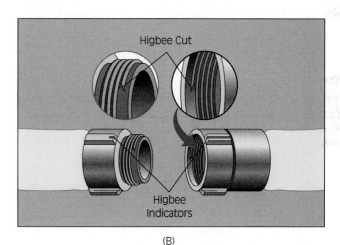
(B)

(C)

(D)

Figure 10-8 (A) Male and female threaded couplings. (B) A Higbee or blunt end cut on the thread of a coupling. (C) Storz coupling. (D) Quarter-turn thread. *(Photos A, C, and D by Fred Schall)*

Threaded couplings are of common design but with many nonstandard types of threads. Today, although a National Standard Thread (NST) (also called National Hose Thread, NHT) exists, there are other "standard" hose threads. Some include National Pipe Thread, Eastern Hose Thread, Pacific Coast Thread, Quebec Standard Thread, and Canadian Standards Association. New York and Chicago each have both a fire department thread and a city hose thread. Personnel checking new hose and connection equipment should ensure the correct threads are used on all hose, appliances, connections, and so on. Finding out the wrong threads are present while trying to make a connection during a fire is not the right time. Hose threads allow the couplings to screw or fasten together. Regular pipe threads normally screw grooves that go to the end of the fitting. National Standard and other fire hose type threads have blunt end threads. The blunt end threads can be started at only one point, thus reducing the possibility of cross-threading the couplings. These blunt ends are called a **Higbee cut**. A groove is notched into coupling lugs to help firefighters align the Higbee cuts. This notch is called the Higbee indicator.

> **Caution** Personnel checking new hose connection equipment should ensure that the correct threads are used on all hose, appliances, connections, and so on. During a fire is not the right time to find out the wrong threads are present while trying to make a connection.

Nonthreaded couplings use locks or cams to secure the connection. Both couplings are identical and either end can be used to connect to another coupling. The couplings are aligned and twisted one-quarter turn, locking them together. **Storz couplings** are the most popular of these types but there are other kinds. Some nonthreaded couplings use expansion rings and others are attached by wire or straps on the exterior of the hose that bind the hose to the couplings.

To assist firefighters in making, tightening, or breaking the connection of a coupling set, lugs or handles are placed on the couplings. The most common type of lug is the rocker lug, but pins and recessed pins are also used. Handles are used mostly on hard and soft sleeves. Spanner wrenches are used with the lugs to assist in tightening and loosening the couplings.

CARE AND MAINTENANCE OF COUPLINGS

Care and maintenance of hose couplings requires keeping them clean and preventing mechanical damage. Proper storage, rolling, and simply paying attention to where the coupling is placed can solve most problems. Mechanical damage prevention starts with caution being used to prevent dropping the couplings, which can damage the threads or the coupling itself. Couplings should not be dragged. Whenever possible, personnel should carry hose near the couplings to avoid damaging them. They should avoid the placement of hose couplings in vehicle traffic areas where the couplings may be struck or rolled over. Vehicles can easily crush couplings or misshape them, preventing the free operation of the couplings or breaking them entirely. Good storage and rolling methods can prevent damage to exposed threads, **Figure 10-9**.

The care of couplings should include a visual inspection each time the hose is reloaded. The cou-

Figure 10-9 Hose storage rack.

plings should be checked for any damage to the coupling and threads, the proper operation of the swivel, and the presence and condition of the gasket. The gasket should be pulled out and checked for cracks and pliability. One last check would be for any movement or slippage of the hose from the coupling. If dirt or grit is detected in the coupling, rinse it and take a brush to the coupling and threads to clean it. Remember, dirt and grit may prevent tightening or loosening of a connection.

HOSE TOOLS AND APPLIANCES

Hose tools are accessories that help firefighters move or operate hoselines. Appliances are devices that water flows through, including adapters and connectors. There are many types and varieties of such items; this chapter covers the more common ones. Many of these devices must be the correct size to work with the size(s) of hose being used.

Tools include rope hose tools, wrenches, rollers, clamps, and other items, **Figure 10-10**. A **rope hose tool** is about 6 feet (2 m) of ½-inch (13-mm) rope spliced into a loop with a large metal hook at one end and a 2-inch (50-mm) ring at the other. It is used to tie in hose and ladders, carrying hose, and many other tasks requiring a short piece of rope. Some departments have an engine or spanner belt to accomplish the same tasks. A **hose strap** is a variation of the rope hose tool, although shorter in length. The hose strap is perhaps the most useful tool in handling a charged hoseline due to its effective design of a strap, cinch clip, and forged handle. **Spanner wrenches** come in several sizes and are used to tighten or loosen couplings. They may also be useful as a pry bar, door chock, gas valve control, and for many other uses. A **hydrant wrench** is the tool used to operate the valves on a hydrant and may also be used as a spanner wrench. Some are plain wrenches and others have a ratchet feature to speed the operation of the valve.

A **hose roller** or **hoist** has a metal frame, with a securing rope, shaped to fit over a windowsill or edge of a roof with two rollers to allow the hose to roll over the edge, preventing chafe. A **hose clamp** is a device used to control the flow of water by squeezing or clamping the hose shut. Some work by pushing a lever, which closes the jaws of the device, and others have a screw mechanism or hydraulic pump that closes the jaws. Hose clamps create a pressure buildup in the hose. It is important to place the clamp at least 5 feet from any coupling (toward the supply side) and at least 20 feet from an apparatus. Another device used for stopping leaks without shutting down the line is a **hose jacket**. Hose jackets may be a metal or leather device that is fitted over the leaking area and either clamped or strapped together to control the leak. A **hose bridge** is a device that allows vehicles to pass over a section of hose without damaging it. A **hose cart** is a handcart or flat cart modified to be able to carry hose and other equipment around large buildings. Some departments use them for high-rise situations.

Figure 10-10 Various hose tools.

Appliances allow firefighters to connect various hoses and nozzles, **Figure 10-11**. Nozzles and foam equipment are covered in Chapter 11, and master stream devices are covered later in this chapter. A **double female** is a device that allows two male ends of hose to be connected, and a **double male** does the same for two female ends. A double male is used to connect the two female thread couplings when doing a split lay. An **increaser** is used to connect a smaller hose to a larger one, and a **reducer** connects a larger hose to a smaller one. An **adapter** is a device that adapts or changes one type of hose thread to another, allowing connection of two different lines. Adapters have a male end on one side and a female on the other with each side being a different thread type, for example, an iron pipe to national standard adapter.

Gate valves have an inlet and outlet of the same size with a gate to control water flow; they can be on apparatus or in hoselines. Large-diameter hose requires an **intake relief valve** at the receiving engine. These valves may function as a combined overpressurization relief valve, a gate valve, and an air bleed-off. A **wye** is a device that divides one hoseline into two or more. The wye lines may be the same size or smaller size, and the wye may or may not have gate control valves to control the water flow. A variation of the wye is the **water thief**, which has one inlet and one outlet of the same size plus two smaller outlets with all of the outlets being gated. The standard water thief usually has a 2½-inch (65-mm) inlet with one 2½-inch (65-mm) and two 1½-inch (38-mm) outlets. A **portable hydrant** or **manifold** is like a large water thief and may have one or more intakes and numerous outlets to allow multiple hoselines to be utilized with or without a pumper being directly at the fire location. A **siamese** is a device that connects two or more hoselines into one line. Siamese valves either have clapper valves or gate valves to prevent loss of water if only one line is connected. Clapper valves are spring-loaded and operate automatically, whereas the gate valve requires that a firefighter operate the valve to the open position. **Hydrant valves** or **switch valves** are used on a hydrant to allow an engine to connect and charge its supply line immediately but also allow an additional engine to connect to the same hydrant without shutting down the hydrant. A **strainer** is placed over the end of a suction hose to prevent debris from being sucked into the pump. Some strainers have a float attached to keep them at or near the water's surface. A strainer designed to draw water from a shallow, flat surface is typically called a "duck's foot." The duck's foot strainer is most often used for drafting from portable water tanks. A different style of strainer or screen is located on each intake of a pump. **Distributor pipe** or **extension pipe** allows a nozzle or other device to be directed into holes to reach basements, attic, and floors when access by personnel is not possible. The distributor pipe has self-supporting brackets that help hold it in place when in use. A **hose cap** does not really let water flow through it, but rather caps the end of a hoseline or appliance to prevent water flow.

COUPLING AND UNCOUPLING HOSE

Connecting hose couplings can be accomplished in several ways, depending on the number of personnel available. The use of spanner wrenches can assist in tightening and loosening hose couplings. Most threaded couplings should be tightened until they are "hand tight." Arbitrarily tightening threaded couplings with a spanner wrench causes stress and premature wear on gaskets. Spanners should be used to tighten leaking couplings or to assist the uncoupling of hose.

The first coupling technique is a foot-tilt method for one person. With the male coupling lying flat on the ground, place the left foot directly behind the coupling, stabilizing it and raising it upward, **JPR 10-1A**. With the female coupling in the hands, bend down and bring the two couplings together. When the couplings are aligned, turn the swivels first left to align the Higbee cut and then to the right to tighten, **JPR 10-1B**.

Figure 10-11 Various hose appliances.

FIRE HOSE AND APPLIANCES

JOB PERFORMANCE REQUIREMENT 10-1
One-Person Foot-Tilt Method

A With the male coupling lying flat on the ground, place the left foot directly behind the coupling, stabilizing it and raising it upward.

B With the female coupling in the right hand, bend down and bring the two couplings together. When aligned, turn the swivels to the right.

> **Note** Remember when making a connection, "righty tighty-lefty loosey."

A second one-person method is the over-the-hip method. Hold the hose with the female coupling slightly below the waist, **JPR 10-2A**. Pick up the male coupling, align the two couplings, and swivel the female coupling to make the connection, **JPR 10-2B**.

In the two-person method, the two firefighters face each other with each holding a coupling. The couplings are brought together, aligned, and swiveled, **JPR 10-3**. The firefighter with the female coupling watches and adjusts the alignment of the couplings.

Uncoupling hose is usually done in the opposite manner in which it was connected, but sometimes a connection is too tight to easily break the connection. If the hose has been charged, it should be

JOB PERFORMANCE REQUIREMENT 10-2
One-Person Over-the-Hip Method

A Hold the hose with the female coupling slightly below the waist.

B Pick up the male coupling, align the two couplings, and swivel the female coupling to make the connection.

JOB PERFORMANCE REQUIREMENT 10-3
Two-Person Coupling Method

Two firefighters face each other, each holding a coupling. The couplings are brought together, aligned, and swiveled.

uncharged and the pressure bled off. The first recommendation is to get a pair of spanner wrenches. With the hose lying flat on the ground, place the spanners on the lugs going in opposite directions, **JPR 10-4A.** Push the spanners downward to break the connection, **JPR 10-4B.** If spanners are unavailable, one of the following methods can be used.

The one-person uncoupling method is known as the knee-press method. Fold the hose coupling over on itself on the ground and press your knee down on the coupling, **JPR 10-5A.** Try to twist the coupling to the left, breaking the connection, **JPR 10-5B.**

The two-person method is called the stiff-arm method and is similar to the two-person coupling, but the firefighters brace themselves and twist the coupling while holding their arms stiff with elbows slightly bent, **Figure 10-12.**

HOSE ROLLS

Hose is rolled either to store it or to have it ready for use with hoselines or appliances.

Straight/Storage

The straight or storage hose roll is the easiest to work with. Start with the hose flat on the ground. From the male end, to protect the threads, roll it straight to the opposite end, **JPR 10-6.** This is called a storage roll because of its common use, and it is often used when picking up after a fire. An exception used by some departments is to store damaged hose with the male end out and/or tying a knot at the end.

JOB PERFORMANCE REQUIREMENT 10-4
Uncoupling Hose with Spanners

A With the hose lying flat on the ground, place the spanners on the lugs going in opposite directions.

B Turn the spanners downward to break the connection.

FIRE HOSE AND APPLIANCES 231

JOB PERFORMANCE REQUIREMENT 10-5
Knee-Press One-Person Uncoupling Method

A Fold the hose coupling over on itself on the ground and press your knee down on the coupling.

B Twist the coupling to the left, breaking the connection.

Single Donut

The single-donut hose roll is used when access to either or both couplings may be needed. There are several ways to do a donut roll. The first method has the hose lying flat. Fold the hose on top of itself with the male coupling on top about 3 feet (1 m) short of the female coupling, **JPR 10-7A.** Starting at the fold, roll the hose toward the couplings with the extra hose on the female end protecting the male coupling, **JPR 10-7B.** Leave a small space at the center of the roll to provide a handhold, **JPR 10-7C.** A second firefighter can assist in guiding the hose and adjusting the slack as it is rolled.

The second method used has the hose laid out flat. Starting at a point off-center about 6 feet (2 m) toward the male coupling, roll the hose toward the female coupling, **JPR 10-8A.** Once again the extra hose will protect the male coupling and the female coupling will be exposed. The handhold space and the extra firefighter are also useful with this method, **JPR 10-8B.**

Twin or Double Donut

The twin or double-donut roll is used for special applications and works best with 1½- to 2-inch (38- to 50-mm) hose. First the hose is laid flat with both couplings at one end and each half lying parallel to the other, **JPR 10-9A.** At the center, the loop is folded over the top of both halves. The roll is started toward the couplings at the same time, **JPR 10-9B** and **C.** At the end, the roll may be tied together for carrying, **JPR 10-9D.** The twin donut can be secured by using the hose itself. This is called a self-locking roll. To accomplish this, extend the amount of hose that is used for the starting fold and loop. Allow this excessive hose to "flop" as the twin donuts are rolled. When finished, use the extra hose at the center to form a bight around the two end couplings.

Figure 10-12 Two-person stiff-arm method. The firefighters brace themselves and twist the coupling while holding their arms stiff with elbows slightly bent.

JOB PERFORMANCE REQUIREMENT 10-6
Straight or Storage Hose Roll

A Lay the hose flat on the ground. Starting at one end, usually the male end to protect the threads, roll it straight to the opposite end.

B The finished roll.

JOB PERFORMANCE REQUIREMENT 10-7
Single-Donut Hose Roll

A The first method has the hose lying flat. Fold the hose on top of itself with the male coupling on top and about 3 feet (1 m) short of the female coupling.

B Starting at the fold, roll the hose toward the couplings with the extra hose on the female end protecting the male coupling. A second firefighter can assist in guiding the hose and adjusting the slack as it is rolled.

C Leaving a small space at the center of the roll allows a handhold.

JOB PERFORMANCE REQUIREMENT 10-8
Single-Donut Hose Roll

A Lay the hose out flat. Starting at an off-centered point about 6 feet (2 m) toward the male coupling, roll the hose toward the female coupling.

B The handhold space and the extra firefighter are also useful with this method.

JOB PERFORMANCE REQUIREMENT 10-9
Twin Donut Roll

A First lay the hose flat with both couplings at one end and each half laid parallel to the other.

B At the center, fold the loop over the top of both halves.

C Start the roll toward the couplings at the same time.

D At the end, you may tie the roll together for carrying.

HOSE CARRIES

Drain and Carry

The drain and carry method is used to combine the two steps of draining and carrying a section of hose into one operation. This is usually done with one section of hose. The firefighter starts at one end of the hose and with the coupling held waist height feeds the hose over the shoulder and back down to the waist, **JPR 10-10A**. A fold is created and the hose is laid on itself back to the front, **JPR 10-10B**. The firefighter continues to walk forward folding and refolding the hose at the waist until finished. The hose can then be carried to the new location, **JPR 10-10C**.

Shoulder Loop Carry

In the shoulder loop carry, the firefighter places the nozzle or end of hose over the shoulder resting against the back **JPR 10-11A**. The firefighter walks forward about 3 feet (1 m), picks up the hose, and forms a bight to bring the hose back up

JOB PERFORMANCE REQUIREMENT 10-10
Drain and Carry

A Start at one end of the hose. With the coupling held waist height, feed the hose over the shoulder and back down to the waist.

B Create a fold and lay the hose on itself back to the front.

C Continue to walk forward folding the hose at the waist until finished so it can be carried to the new location.

FIRE HOSE AND APPLIANCES ■ 235

and over the shoulder, creating a loop, **JPR 10-11B.** (This is similar to rolling an electrical cord around one's arm, but with bigger loops.) The action is continued as each section is picked up and carried forward, **JPR 10-11C.** If the firefighter needs to move in the opposite direction, the loops are collected and raised by the hands and the firefighter rotates to the opposite direction and returns the hose to the opposite shoulder moving forward in the new direction, **JPR 10-11D** and **E.**

JOB PERFORMANCE REQUIREMENT 10-11
Shoulder Loop Carry

A Place the nozzle or end of the hose over the shoulder, where it rests against the back.

B Walk forward about 3 feet (1 m), pick the hose up, and form a bight to bring the hose back up and over the shoulder, creating a loop.

C Continue as each section is picked up and carried forward.

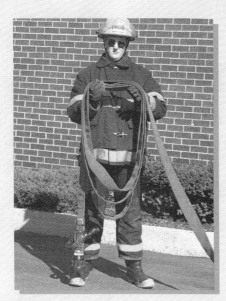

D If you need to move in the opposite direction, collect and raise the loops with your hands and rotate to the opposite direction.

E Return the hose to the opposite shoulder as you move forward in the new direction.

Single-Section Street Drag

The simple hose drag technique can move one or two hoselines. The firefighter puts a section of hose over a shoulder with the coupling in front at waist height. The firefighter then walks away dragging the line behind, **JPR 10-12A.** Placing a line over each shoulder allows the firefighter to pull two lines, **JPR 10-12B.** If additional sections are needed, additional firefighters can do the same with the following sections until the desired amount of hose is stretched, **JPR 10-12C.** Note that this technique can cause damage to the coupling being dragged.

HOSE LOADS

Apparatus hose loads are designed so that the engine can carry out any of the required fire attack or water supply evolutions. These include water supply, small and medium attack lines, and supply for protective systems and master stream devices. Some of these lines are made up in advance and are preconnected to the pump, while others are made up depending on the incident. A well-trained engine crew should be able to perform any of these required tasks quickly and efficiently. Part of this efficiency is how the crew trains and performs while the design and layout of the hose bed, the type of hose, and its couplings determine other parts. Hose loads should consider the need to have the right hose coupling or adapter at the end that will be used farthest from the engine.

> **Note** Time and care taken during the loading of hose will be repaid many times over when pulling the hose loads off during an incident.

JOB PERFORMANCE REQUIREMENT 10-12
Hose Drag

A Put the end of a section of hose over your shoulder with the coupling in front at waist height and walk away, dragging the line.

B Place a line over each shoulder to pull two lines.

C If additional sections are needed, additional firefighters can do the same to the following sections until the desired amount of hose is stretched.

FIRE HOSE AND APPLIANCES 237

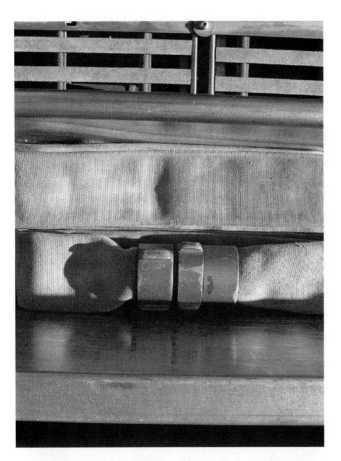

Figure 10-13 A dutchman is a short fold of hose or a reverse fold that is used when loading hose and a coupling comes at a point where a fold should take place or when two couplings end up on top of each other. The dutchman moves the coupling to another point in the load.

A **dutchman** is a short fold of hose or a reverse fold that is used when loading hose and a coupling comes at a point where a fold should take place or when two sets of couplings end up on top of or next to each other. The dutchman moves the coupling to another point in the load, **Figure 10-13**.

Accordion Load

Accordion hose loads can be used for preconnected, made-up hoselines or for providing additional supply line. The accordion load is ideal for making up **shoulder loads**. Care should be used with the accordion load because overpacking the hose bed can stress and crack the liner of the hose. For use with preconnected lines, the accordion load should be started with the female coupling at the pump outlet or the front of the bed, **JPR 10-13A**. The hose is placed in the bed on its side rather than lying flat.

The hose is brought forward in the bed until it reaches the rear, where it is folded and returned to the front where the process is repeated, **JPR 10-13B**. When the layer of the hose fills the bed, the hose is brought up at the rear and folded back on itself starting the next layer, again on the left side, **JPR 10-13C**. Watch the placement of hose couplings near the wall of the hose bed; they should be on their way out or have enough room to flip around when pulled from the bed. Continue until the bed is filled or the desired amount is loaded. Add a nozzle if it is a preconnected line. If a supply line, it may be connected to another bed or left with the male end exposed. The male end can be loaded into the bed first if the hose is being used for a forward lay. A blind cap, double male, or double female can be added to protect threads or to allow connection to a hydrant, **JPR 10-13D**.

Making up a hoseline from an accordion hose load is an easy task. A nozzle or other appliance is attached to the end of the hose and the desired amount is selected, **JPR 10-14A**. The firefighter pulls the amount of hose about halfway out of the bed and places it on one shoulder with the nozzle at the bottom. Use care not to have too much of the hose in front, which could cause tripping; it is better to drag the excess behind, **JPR 10-14B**. Stepping away from the bed removes the rest of the load. Pulling another number of folds and dropping them to the ground will allow some hose to be dragged prior to using the shoulder load, **JPR 10-14C**. When the point is reached where the shoulder load is to be deployed, the firefighter allows the hose to flake off one fold at a time while moving forward, **JPR 10-14D**. If additional hose is needed, another firefighter may repeat this process with the attached hose instead of the nozzle.

Flat Load

The flat load is used for supply lines and some attack lines. It involves simply laying the hose flat by starting at either end of the hose depending on use. An example of a flat load for water supply would be to start with the male end of the hose on the right side of the bed at the rear, **JPR 10-15A**. This will allow this hose bed to be connected to another one for long hose lays. Lay the hose flat until the front is reached where the hose is folded over and brought to the rear. When back at the rear, turn the hose off the first layer and lay it flat next to the first row, **JPR 10-15B** and **C**. When the edge of

JOB PERFORMANCE REQUIREMENT 10-13
Accordion Load

A For use with preconnected lines or for making up lines, start the accordion load with the female coupling at the pump outlet or the front of the bed.

B Place the hose in the bed on its side rather than lying flat. Bring the hose forward in the bed until it reaches the rear, where it is folded and returned to the front. Repeat the process.

C When the layer of hose fills the bed, bring the hose up at the rear and fold it back on itself to start the next layer.

D Continue until the bed is filled or the desired amount is loaded. Add a nozzle if it is a preconnected line. If a supply line, add a blind cap and adapters.

the hose bed is reached, a second layer is started by doing a double layer and then moving toward the right again with the rows over the top of the first layer's rows, **JPR 10-15D** and **E**.

Making a shoulder load from a flat load is slightly more difficult than the accordion load because an accordion load must be created. The nozzle or appliance is again attached. Next the firefighter gathers the amount of hose to be placed on the shoulder, **JPR 10-16A**. Sometimes this can be done by simply twisting the folds of the hose to the side, creating a layer of accordion-style hose and following the accordion load directions, **JPR 10-16B**. When this cannot be done, the firefighter must begin loading the folds one at a time on the shoulder. Place the fold with the nozzle side down over the shoulder and pull the hose until it is properly positioned. Grab the next fold of the hose and place it on top of the first until the desired amount is reached. Step away from the bed and again pull some extra folds. The hose should now deploy like an accordion load, **JPR 10-16C** and **D**.

JOB PERFORMANCE REQUIREMENT 10-14
Advancing an Accordion Load

A Attach a nozzle or other appliance to the end of the hose and select the desired amount.

B Pull the amount of hose about halfway out of the bed and place it on the shoulder with the nozzle at the bottom. Use care not to have too much of the hose in front, which could cause tripping; it is better to drag the excess behind.

C Pull another number of folds and drop them to the ground to allow some hose to be dragged prior to using the shoulder load.

D When you reach the point where the shoulder load is to be deployed, allow the hose to flake off one fold at a time while moving forward.

Horseshoe Load

The horseshoe load also has several applications. The load starts at a rear corner of the hose bed with the hose on its side, **JPR 10-17A**. The hose is laid toward the front wall of the hose bed and then along the wall to the other side and back toward the rear. At the rear, the hose is folded alongside itself and heads back to where it started. It is folded again and another "U" is formed, **JPR 10-17B** and **C**. When the bed is filled a new layer may be started by laying the hose flat to the front of the bed where it is turned again on its side and moved to the rear on the same side as originally started, **JPR 10-17D** and **E**.

Making a shoulder load from the horseshoe bed is somewhat similar to the accordion load. Care should be taken when reaching the outer edges of the hose bed as the increasing length of hose between folds may cause problems advancing the line. Again the nozzle is placed on the hose and the desired amount selected, **JPR 10-18A**. The firefighter pulls the hose and twists it into place on the shoulder and stepping away pulls the hose out of the bed. The remaining steps are as given earlier, **JPR 10-18B** and **C**.

JOB PERFORMANCE REQUIREMENT 10-15
Flat Load

A To use the flat load for water supply, start with the male end of the hose on the right side of the bed at the rear.

B Lay the hose flat until the front is reached where the hose is folded over and brought back to the rear.

C When back at the rear, turn the hose off the first layer and lay it flat next to the first row.

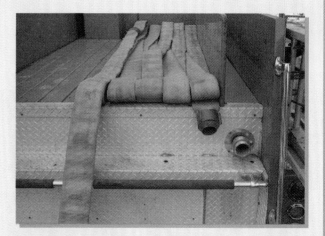

D When the edge of the hose bed is reached, do a double layer and then move toward the right again with the rows over the top of the first layer's rows to start a second layer.

E The finished load.

JOB PERFORMANCE REQUIREMENT 10-16
Advancing a Flat Load from a Supply Bed

A Attach the nozzle or appliance and begin to gather the amount of hose needed to be placed on the shoulder.

B Sometimes this can be done by simply twisting the folds of the hose to their sides, creating a layer of accordion-style hose and following the accordion load directions.

C When this cannot be done, you must begin loading the folds one at a time on the shoulder. Place the fold with the nozzle side down over the shoulder and pull the hose until it is properly positioned. Grab the next fold of the hose and place it on top of the first until the desired amount is reached.

D Step away from the bed and again pull some extra folds. The hose should now deploy like an accordion load.

JOB PERFORMANCE REQUIREMENT 10-17
Horseshoe Load

A Start the load at a rear corner of the hose bed with the hose on its side.

B Lay the hose toward the front wall of the hose bed and then along the wall to the other side and back toward the rear.

C At the rear, fold the hose alongside itself and head back to where it started. Fold it again and another "U" is formed.

D When the bed is filled, start a new layer by laying the hose flat to the front of the bed where it is turned again on its side and moved to the rear on the same side as originally started.

E The finished bed.

FIRE HOSE AND APPLIANCES   243

JOB PERFORMANCE REQUIREMENT 10-18
Advancing a Horseshoe Load

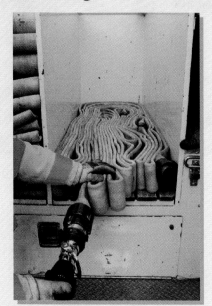

A Place the nozzle on the hose and select the desired amount.

B Pull the hose and twist it into place on the shoulder.

C Step away to pull the hose out of the bed.

Finish Loads and Preconnected Loads

Finish loads and preconnected loads can utilize the three methods of loading hose just discussed. Some use combinations of these loads in different layers to assist firefighters in advancing hose quickly from an engine. Finish loads can be used for assisting in laying supply line or attack lines. Preconnected loads are usually used for attack lines. Preconnected loads are typically preattached to a pump discharge. Therefore, the hose must be loaded in such a way that the entire hose load can be pulled off and deployed easily. Failure to clear all the hose from the bed will cause kinking of the hose and jamming of the hose in the bed when the pump operator charges the line. Apparatus damage may also occur if hose loaded in a bed is charged prematurely.

A straight finish load is used usually with a straight hose lay and simply involves taking the final length or two of a load and laying it flat across the top of the load. A rope with adapters, a spanner wrench, and a hydrant wrench attached allows the layout person to quickly have all the necessary tools and enough hose to make the hydrant connection, **Figure 10-14**. A reverse horseshoe load can be used

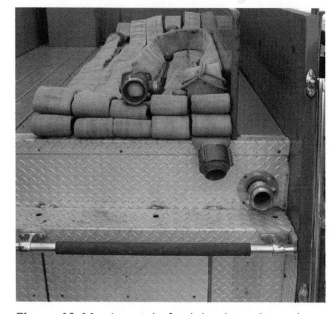

Figure 10-14 A straight finish load simply involves taking the final length or two of a load and laying it flat across the top of the load. A rope with adapters, a spanner wrench, and a hydrant wrench attached allows the layout person to quickly have all the necessary tools and enough hose to make the hydrant connection.

for laying out or making an attack line. For laying out, a horseshoe load is made on top of the hose load but in the reverse direction (front to back), and at the center point of the "U" of the horseshoe, the rope with adapters and wrenches is attached. The first portion of the hose may have to have a twist in it to get it to change direction, **Figure 10-15**.

Similarly an attack line can be attached to the end of a hose load. This works when a **backstretch** or **flying stretch** will be utilized. A wye or gate valve can be installed at the end of the hose load before the horseshoe. (One will have to be used if changing hose size.) The reverse horseshoe, including any needed twist, is created again on top of the hose bed. When the amount of hose is loaded, a nozzle is attached and laid in the center of the horseshoe with the nozzle at the rear of the hose bed. The firefighter grabs the nozzle and a few lengths of hose over the shoulder. The horseshoe load is then grabbed at the bottom of the "U," pulling it out of the hose bed and allowing it to flake out while moving to the fire, **JPR 10-19A** and **B**.

Figure 10-15 A reverse horseshoe load for laying out is made on top of the hose load but in the reverse direction (front to back), and at the center point of the "U" of the horseshoe the rope with adapters and wrenches is attached. The first portion of the hose may have to have a twist in it to get it to change direction.

JOB PERFORMANCE REQUIREMENT 10-19
Reverse Horseshoe Load for an Attack Line

A For an attack line, create the reverse horseshoe again on top of the hose bed. When the amount of hose is loaded, attach a nozzle and lay it in the center of the horseshoe with the nozzle at the rear of the hose bed.

B Grab the nozzle and place a few of the hose folds over the shoulder. Grab the horseshoe load at the bottom of the "U," pulling it out of the hose bed and allowing it to flake out while moving to the fire.

FIRE HOSE AND APPLIANCES   245

Preconnected lines can be made up using the accordion or horseshoe loads, and some use a combination. Since they are preconnected to a discharge they will always start with the female end where the discharge is located and may require some twists to start. It is possible to use preconnected loads with all accordion or horseshoe layers. An example of a combination load would be a horseshoe bottom layer, with the top layer accordioned. Other combinations include horseshoe, accordion, accordion layers, or alternating horseshoe and accordion layers, **Figure 10-16A** and **B**. The combinations are often designed based on a specific need at a certain property within the response district. When using multiple layers or various hose loads, it is common to place **ears** on the hose to assist in pulling the layer, **Figure 10-17**.

Flat Load, Minuteman Load, and Triple-Layer Load

Preconnected loads must rapidly remove the hose from a slot or bed. Many preconnected loads exist, and there are many innovative ways to accomplish the rapid deployment of fire attack lines. The following loads are among the more popular. The flat load, as a preconnect, is also based on the flat load described earlier and starts at the discharge. The hose is laid flat as described earlier. At a point from one-third to one-half the length of the line an ear or row of ears should be added to assist in pulling the line, **JPR 10-20A**. The line is advanced by grabbing the nozzle and placing it over the shoulder with the other hand reaching around and pulling the ear(s). The firefighter then walks away pulling the line behind, **JPR 10-20B** and **C**.

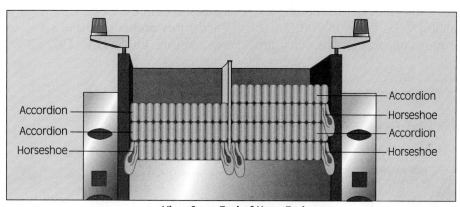

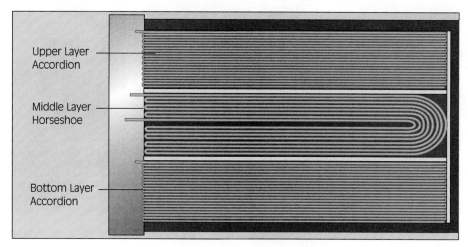

Figure 10-16 Preconnected combination loads include horseshoe, accordion, accordion layers or alternating horseshoe and accordion layers. (A) Horseshoe, accordion, accordion layers. (B) Alternating horseshoe and accordion layers.

246 ■ CHAPTER 10

Figure 10-17 Ears on hose load.

The minuteman load is a preconnected load using a narrower section of the hose bed. This narrower section is called a **slot load** and can allow more hoselines to be placed in the same space. Some longer length versions may use more than one slot, and all require attention when being loaded. These loads have become more popular because they are quicker and more efficient than some of the previous loads described, **Figure 10-18**.

The minuteman load has the female end connected to the discharge, and several sections of hose are flat loaded into the slot with the last few feet (about a meter) placed aside out of the bed. An ear may be placed one or two folds from the bottom to assist in pulling, **JPR 10-21A.** Next a section of hose with the nozzle is started and the hose flat loaded above that. The nozzle will therefore be in the middle of the slot at the point where it will be pulled, **JPR 10-21B.** When the designated number of sections is finished, the set-aside piece is connected to the pieces with the nozzle attached and the connected sections are on the top layer, **JPR 10-21C.**

To advance the minuteman load, the firefighter lifts the nozzle and several layers of hose while pulling them out and placing them midway on the shoulder, **JPR 10-22A.** The firefighter then steps away to remove the remainder of the top layers and turns around and pulls the ear to remove the remaining hose, **JPR 10-22B** and **C.** When the

---

JOB PERFORMANCE REQUIREMENT 10-20
Advancing the Flat Load from a Preconnected Bed

A Start the flat load at the discharge with the hose laid. At a point from one-third to one-half the length of the line, an ear or row of ears should be added to assist in pulling the line.

B To advance the line, grab the nozzle and place it over the shoulder with the other hand reaching around and pulling the ear(s).

C Walk away, pulling the line behind.

FIRE HOSE AND APPLIANCES 247

Figure 10-18 The minuteman load or slot load is a preconnected load using a narrower section of the hose bed, which allow more hoselines to be placed in the same space. Some longer length versions may use more than one slot.

bottom sections are fully stretched out, the shoulder load is allowed to flake out toward the fire, **JPR 10-22D**.

The triple-layer load is another modified flat load where the hose is folded over on itself three times. All the hose is connected to the discharge, and the nozzle is attached at the opposite end with the hose stretched out its full length, **JPR 10-23A.** At a point two-thirds of the way from the hose bed, a fold is made and the hose is doubled on itself back toward the engine. This fold creates three layers of hose. One is the original bottom layer, one the folded back middle layer, and the top third layer has the hose with the nozzle attached. These three layers are now treated as one for loading on the engine, **JPR 10-23B**. The triple-layered hose is loaded into the slot as a single layer. A fold is made at the edge of the hose bed and the hose doubled over creating two other layers. This continues until all of the hose is loaded. The nozzle's layer should be at the edge of the bed for pulling, **JPR 10-23C**.

JOB PERFORMANCE REQUIREMENT 10-21
The Minuteman or Slot Load Being Loaded

A This is a narrower version of the flat load with a slight modification. Connect the female end to the discharge and flat load several sections of hose into the slot with the last few feet (about a meter) placed aside out of the bed. An ear may be placed one or two folds from the bottom to assist in pulling.

B Next, start a section of hose with the nozzle, and load the hose flat above that. The nozzle will therefore be in the middle of the slot at the point where it will be pulled.

C When the designated number of sections is finished, connect the set-aside piece to the pieces with the nozzle attached. The connected sections are on the top layer.

JOB PERFORMANCE REQUIREMENT 10-22
Advancing the Minuteman Load

A Lift up the nozzle and layers above it while pulling them out and placing them midway on the shoulder.

B Step away to remove the remainder of the top layers.

C Turn around and pull the ear to remove the remaining hose.

D When the bottom sections are fully stretched out, allow the shoulder load to flake out toward the fire.

Advancing the line has the firefighter grab the layer with the nozzle and place it on the shoulder, **JPR 10-24A**. Either another firefighter grabs the next layer or the nozzleperson pulls them out of the slot and the hose is stretched to the fire, **JPR 10-24B and C**.

Stored Hose Loads/Packs

Apparatus typically carry stored hose rolls and special application hose packs. Hose rolls are just that—extra sections of rolled hose for replacing damaged hose or short sections to help assemble an evolution. These hose rolls can be stored as a straight roll, donut roll, or double donut. Hose packs can be numerous in design and makeup. High-rise (or standpipe) packs are hose loads that are pre-assembled and bundled to be easily carried into a building, **Figure 10-19**. High-rise packs may also include a tool and appliance bag containing adapters, spanners, and hose straps. Wildland packs are covered in the next section.

Wildland Firefighting Hose Loads

Wildland firefighting often requires firefighters to stretch hoselines a great distance from the engine while fighting the fire. To accomplish this task, the hose is rolled and bundled together to allow it to be

JOB PERFORMANCE REQUIREMENT 10-23
The Triple-Layer Load

Another modified flat load where the hose is folded over on itself three times.

A Connect all the hose to the discharge and attach the nozzle at the other end with the hose stretched out its full length.

B At a point two-thirds of the way from the hose bed, make a fold and double the hose on itself back toward the engine. Make another fold, the last part with the nozzle, to create the third layer. These three layers are now treated as one for loading on the engine.

C Load the hose in the slot in a layer and make a fold, creating another layer, until all hose is loaded. The nozzle's layer should be at the edge of the bed for pulling.

JOB PERFORMANCE REQUIREMENT 10-24
Advancing the Triple-Layer Load

A Grab the layer with the nozzle and place it on the shoulder.

B Pull the layers out of the slot, or another firefighter can grab the next layer.

C Stretch the hose to the fire.

---

carried on the firefighters' backs while carrying tools or stretching and advancing the line. Placing two bundles together allows each firefighter to carry 200 feet (60 m) of 1-inch (25-mm) or 1½-inch (38-mm) hose.

A standard wildland hose load is the modified Gasner bar pack, which provides ease of rolling and stretching the line, convenience of carrying with hands free, and protection for the couplings. The first step starts at the male end of the hose [100 feet (30 m)] and the firefighter forms a roll 30 inches (0.8 m) in diameter on flat ground, **JPR 10-25A**. Continue to roll the hose in the direction of the rows and make the rolls as tight as possible, **JPR 10-25B**. The completed field roll, with the female coupling on the outside, is ready for packing. Grab the outside of the hose roll near the female coupling and pull toward the center of the roll, **JPR 10-25C**. The fold should now protect the female end. The top layer of hose can now be tied to the bottom layer using a piece of rope 32 inches long (0.9 m). Two rolls of hose may be placed together and tied off with rope on the outside of the rolls with a shoulder strap to allow carrying, **JPR 10-25D**.

A modification would be to add a gated wye or tee at the end of each roll of hose. To do this merely requires that when the roll is closed together, the female end is left to one side, on the outside, and the gated wye or tee is attached after tying the roll off. When placing the two rolls together, they should be placed so that the gated wyes are on opposite sides of the pack, **Figure 10-20**.

FIRE HOSE AND APPLIANCES 251

Figure 10-19 High-rise (standpipe) hose packs are preassembled in a bundle for deployment.

ADVANCING HOSELINES— CHARGED/UNCHARGED

The engine company's purpose is to advance hoselines to the fire's seat to apply water for extinguishment. These tasks should be accomplished in the most efficient manner, whether around, onto, or into buildings. The following evolutions are based on a four-person crew with one person being at the layout position and connecting to the water source. As crew situations change, the crew positions may also change. Remember that two in/two out firefighter safety rules apply, and conducting any fireground operation without sufficient personnel for a firefighter rescue team can be highly dangerous and unlawful. (See Chapters 2 and 5 for more information.)

When advancing hoselines in any of the evolutions described, the nozzleperson will advance the first shoulder load with the nozzle. The officer takes the second position on the line if it is a two- or three-person line. The engine driver will take the third position on a three-person line. If additional personnel are available, they will be substituted for the officer and driver. The driver is responsible for clearing the hose bed prior to charging any of the lines.

The officer determines the length of the hoseline by figuring the distance from the engine to the location of the fire. Upon reaching the fire area, there should be about 50 feet (15 m) of hose left for movement around the fire area. When taking a hoseline up the outside of a building a 50-foot (15-m)

Top Layer

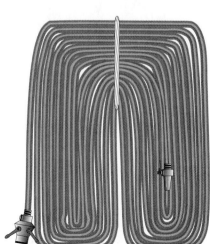

Add Gated Wye (Closed) and Remove Hose Clamp

Bottom Layer

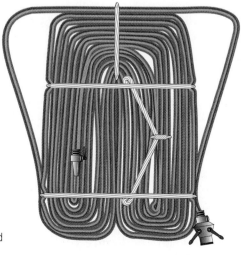

Figure 10-20 Modified Gasner bar pack with added gated wyes.

JOB PERFORMANCE REQUIREMENT 10-25
Modified Gasner Bar Pack

A Start at the male end of the hose [100 feet (30 m)] and form a roll 30 inches (0.8 m) in diameter on flat ground.

B Continue to roll the hose in the direction of the rows and make the rolls as tight as possible.

C The completed field roll, with the female coupling on the outside, is ready for packing. Grab the outside of the hose roll near the female coupling and pull toward the center of the roll.

D The top layer of hose can now be tied to the bottom layer using a piece of rope 32 inches long (0.9 m). Two rolls of hose may be placed together and tied off with rope on the outside of the rolls with a shoulder strap to allow carrying.

---

section will reach four floors, while taking the same section inside the building via a stairwell will reach only one individual floor, **Figure 10-21**.

When operating a charged hoseline in one location, if the line is then ordered moved to another position, such as to another floor, it is best for time and efficiency to shut down, drain, and advance the line as a dry line if fire conditions allow. When the new location is reached, the line is then recharged.

FIRE HOSE AND APPLIANCES

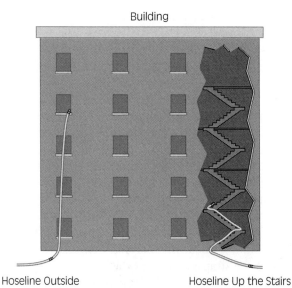

Figure 10-21 When taking a hoseline up the outside of a building a 50-foot (15-m) section will reach four floors, while taking the same section inside the building via a stairwell will reach only one individual floor.

Into Structures

Advancing a hoseline into a structure requires careful placement of the pumper and hoseline, proper selection of the correct size and length hoseline, and skillful execution by the hose crew. Park the engine at a safe location that allows good access for hoselines without hindering any other apparatus placement, especially a ladder company, and allows for any possible changes in fire conditions and equipment usage, **Figure 10-22**. The hoseline must be placed to attack the fire while also being at a position to prevent any fire spread. The hoseline should be large enough to suppress the fire, a minimum of 1½ inches (38 mm). The length has already been discussed. Finally, it is the skillful crew that carefully and safely advances the line, ensuring that they are ready to combat the fire.

To advance the line into a structure, the crew selects a hoseline and properly removes it from the engine, deploying it toward the entrance. A careful evaluation of the conditions is required prior to entering any structure or room. Structural firefighting involves the possibility of flashover, backdrafts, and building collapse. Entering a burning building without having a safe plan of attack and emergency escape can be fatal.

Caution A careful evaluation of the conditions is required prior to entering any structure or room. Structural firefighting involves the possibility of flashover, backdrafts, and building collapse. Entering a burning building without having a safe plan of attack and emergency escape can be fatal.

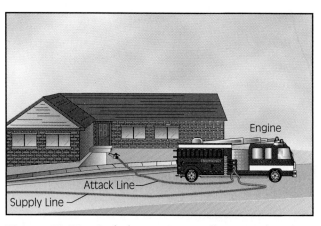

Figure 10-22 Park the engine to allow good access for hoselines without hindering any other apparatus placement.

At the entrance doorway, there should still be plenty of hose available. If the fire is not near this door, the line may be extended into the building as a shoulder load. However, when reaching the entrance or doorway to the area of the fire, the excess hose is removed from the shoulder and carefully laid in a manner to allow it to be easily pulled into the building or room. This would include straightening the hose and, if necessary, looping it back and forth so that it will pull without kinking. The nozzleperson and other firefighters should all be on the same side of the hoseline and properly spaced on the line for easier movement of the hose. The nozzleperson is at the nozzle, the officer or second firefighter a few feet (a meter or so) back, and the next firefighter, if extra ones are available, between the midpoint of the first section and the coupling, **Figure 10-23**. Prior to entering the structure, the door is checked for heat with the back of the hand with the glove off. With the line charged, the firefighter opens the nozzle to bleed off air from the hose and selects the proper pattern. (See nozzle operation in Chapter 11.) Crouching in a low or kneeling position while positioned to the side of the door, the firefighter opens the door and moves the line forward. When advancing the line past doorways, around corners, and furniture, remember that the extra time spent ensuring the line can smoothly move past will be saved when the line runs short and the crew has to wait while someone crawls back to free it.

Hoselines are usually advanced into buildings but to attack some fires the hose may have to be brought through an adjoining property. In these cases with the fire near the building exit, the doorway going outside should be treated as if it were the doorway going into the building and the steps just cited followed.

Figure 10-23 The nozzleperson is at the nozzle, the officer or second firefighter a few feet (a meter or so) back, and the next firefighter, if there is one, between the midpoint of the first section and the coupling.

Up and Down Stairs

When the fire is on a level other than the ground floor, firefighters will need to advance the line up or down stairs. The method used to advance the line to another floor depends on whether the fire is in an open stairwell or has taken control of the landing on the other floor. The first method discussed is for situations where the fire is not involving the stairs or landing, and the second is for situations where there is fire in the stairs.

If the fire does not involve the stairs, the best method is to advance an uncharged line to the fire floor. Make sure that the proper length and size of hoseline is chosen considering the length and size of the structure. The hoseline is advanced from the engine as described and the crew carries the shoulder loads into the building and up the stairs, allowing it to play out as needed, **Figure 10-24**. Use caution so that the hose does not slip back down the stairs. Do not allow it to get caught on stair handrails where it can become wedged between them. To help avoid this and the resulting kinking of the hose, lay the hose against the outside edge or wall of the stairs. If it is necessary to run the hoseline up between the handrails, it should be done in a vertical manner and tied off with a rope hose tool or strap at the upper level. The hose tool is passed around the hose and pulled snugly, near a coupling if possible. The rope or strap is wrapped around the railing several times and secured back on itself. When the desired floor is reached, the shoulder load is carefully deployed to avoid kinks. Do not just drop it into a pile; after it is charged the pressure will be reduced by the kinks and the "pile of spaghetti" will make movement on the landing nearly impossible. One ideal solution for the extra hose, if possible, is to run the line up the stairs toward the next level and back down to the fire floor. This creates a large loop that will be easy to pull, especially since gravity will help it move down, **Figure 10-25**. With the hoseline

Figure 10-24 A crew advancing an uncharged hoseline up stairs.

Figure 10-25 Extra hose on stairs creates a large loop that will be easy to pull.

FIRE HOSE AND APPLIANCES

in place, the nozzleperson has the line charged, bleeds the nozzle, and is ready to advance onto the floor.

If the stairwell or the landing on the desired floor is involved with fire, the crew will need to advance a charged line. The line is brought to the entrance or landing as described earlier, charged, and advanced at the point where the attack is to be made, **JPR 10-26A.** Care is taken to ensure that sufficient hose and personnel are available to move to the next floor without having to stop on the stairs, **JPR 10-26B.** The nozzle is operated to darken down the fire at the next landing, then shut down, and quickly advanced to that level, where it is opened again to secure the landing and moved forward to maintain the floor, **JPR 10-26C.** While advancing a hoseline on a stairwell is always a challenge, it is more so when going down the stairs. Firefighters going down the stairs will encounter the chimney effect of rising heat, hot gases, and possibly fire. "Making the floor" when the fire is on a lower level is a true test of a fire attack crew.

Using a Standpipe System

Advancing hoselines using a standpipe system involves two different hoseline evolutions. The first is the engine driver connecting to the fire department connection on the structure, and the second is

JOB PERFORMANCE REQUIREMENT 10-26
Advancing a Charged Hoseline Up a Stairwell

A Apply water from bottom of stairs to darken down fire.

B Shut down nozzle and advance quickly up stairs.

C At top of stairs, open nozzle and continue fire attack.

the hose crew connecting to the standpipe outlet and advancing the hoseline to attack the fire.

The pumper first establishes a water supply from the nearest hydrant whenever possible. Pre-emergency planning should have determined the water supply for each standpipe system, including water supply and any necessary hose evolutions if the fire department connection is not near the building entrance. The engine driver advances a supply line of medium- or large-diameter hose to the fire department connection. Local standard operating procedures (SOPs) should outline what size hose should be connected. Depending on the length of the lay and the style of hose load, the driver can drag the hose or make up a shoulder load. The hose load and type of coupling will determine the need for any adapters; the male end is needed at the siamese. Reaching the siamese, the driver removes all caps, usually two, and checks the operation of the clapper valve, **Figure 10-26**. If the clapper is inoperative but open, then advance a second line to the connection prior to charging the system. The driver connects the hose first to the left outlet, as the rotation of the coupling will be easier to do when a second line is connected, **Figure 10-27**. Returning to the engine, straightening any kinks, the driver disconnects the hose from the bed and connects to the pump's discharge outlet. The line(s) are charged to the proper pressure. If a second line has not already been run, it should now be done.

On occasion the fire department connection will be damaged or inoperable. Several steps can be taken to still use the system. Pick the one that gets the system into service the quickest. If the siamese swivel does not turn, place a double male into the siamese and add a double female to connect the hose, **Figure 10-28**. If the siamese cannot be used at all, the supply line must be extended into the building to a first floor outlet. To connect the hose, a double female will be required, **Figure 10-29**. After connecting the hose but before opening the outlet, the driver must return to the engine and break the supply line and connect to the discharge outlet and charge the line to the proper pressure. The driver then returns to the outlet and opens the valve, **Figure 10-30**. Opening the outlet valve prior to breaking, connecting, and charging the line will result in the standpipe draining into the hoseline and charging the hose bed. A second line should then be run to the next most convenient outlet following the steps outlined earlier.

The advancing of the interior hoseline to attack the fire should use a standpipe pack that includes the attack hose and nozzle, a gated wye, a short section of supply hose, and necessary wrenches and adapters. The crew proceeds first to the floor below the fire with the required equipment and then to the

Figure 10-26 Driver checking standpipe connection and clapper valve.

Figure 10-27 Driver attaching second supply line to FD connection. Note the first line is on left inlet.

FIRE HOSE AND APPLIANCES ■ **257**

Figure 10-28 If siamese swivel is inoperable, place a double male into the siamese and add a double female to connect the hose.

Figure 10-29 If the siamese cannot be used at all, the supply line must be extended into the building to a first floor outlet. To connect the hose a double female will be required.

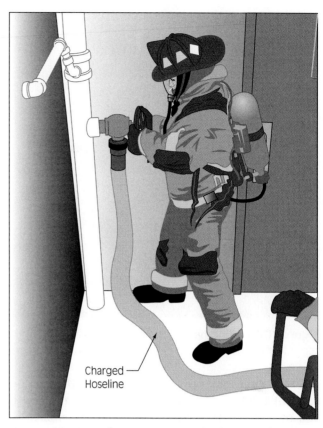

Figure 10-30 After connecting the hose at the outlet and the discharge outlet of the pump, the line is charged to the proper pressure. The driver then returns to the outlet and opens the valve.

fire floor. Depending on fire conditions and location of the standpipe outlets in relation to the fire, an outlet on a lower floor can be chosen. The crew disassembles the hose pack and connects to the outlet, **Figure 10-31** and **Figure 10-32**. The crew stretches the hose either in the stairwell or hallway depending on fire location and conditions. The crew advances the line similar to a line from a stairs or door.

Working Hose Off Ladders

Advancing a hoseline up a ladder can be done with the line charged or uncharged. The best and safest manner is to advance an uncharged hoseline up the ladder and into the building or onto a fire escape before it is charged. The other method advances a charged hoseline up a ladder and either into the building or for operation from a ladder. This is the least preferred option and should be done using great caution when conditions require this method.

When advancing an uncharged hoseline over a ladder, start with the nozzleperson and or firefighters removing and advancing the shoulder loads as described earlier to the base of the ladder. The shoulder load is laid on the left side of the ladder,

Figure 10-31 Engine crew with standpipe equipment.

JPR 10-27A. The nozzleperson takes the nozzle and pulls it under the left armpit, across the chest, and over the right shoulder, allowing the nozzle to rest in the small of the back, **JPR 10-27B.** The nozzleperson, with both hands free for climbing, begins climbing the ladder to a point about 20 feet (6 m) up and stops. The next firefighter places the hoseline over the left shoulder at the next coupling and begins to climb the ladder. The two firefighters coordinate and maintain their distance on the ladder and hose, **JPR 10-27C** and **D**. Additional lengths and firefighters are added as needed and space allows.

Upon reaching the entry point of the building—a window, balcony, or the roof—the firefighters stop. The nozzleperson removes the nozzle and places it over the tip of the ladder into the building, **JPR 10-28A.** The nozzleperson climbs into the building, using the top rung of the ladder as a hose roller, and pulls the hoseline up as the next firefighter climbs, **JPR 10-28B.** The nozzleperson must complete pulling up the section as the next firefighter arrives at the tip, **JPR 10-28C.** The second firefighter pulls the remainder of the hose into the building while the nozzleperson advances the line. To remove the line from the building after use, the line is drained and the process of advancing the hose is reversed.

Another method to advance the line over a ladder involves using rope hose tools or straps wrapped around the line just below the nozzle and at the next coupling and the hose tool looped over the shoulder of the firefighters carrying the line up the ladder, **Figure 10-33**. Distances on the hose and ladder are the same as for the previous method.

Figure 10-32 Engine crew deployed with hose connected to standpipe.

Safety The best and safest manner is to advance an uncharged hoseline up the ladder and into the building or onto a fire escape before it is charged.

Advancing a charged hoseline over a ladder requires more firefighters to advance the hoseline, because the line is heavier and climbing with the line is very dangerous. It is usually done when the line is to be operated from the ladder rather than advanced into the building. The hoseline is actually passed by the firefighters rather than advanced by one or more firefighters. The hoseline is brought to the base of the ladder and charged and bled. Firefighters then climb the ladder spacing themselves at the opening and then at the feet of the next firefighter. When they are in place, the hoseline is passed up the ladder from one firefighter to the next until it reaches the opening, **Figure 10-34**.

JOB PERFORMANCE REQUIREMENT 10-27
Advancing an Uncharged Hoseline Over a Ladder

A The shoulder load is laid on the left side of the ladder.

B The nozzleperson takes the nozzle and pulls it under the left armpit, across the chest, and over the right shoulder, allowing the nozzle to rest in the small of the back.

C The next firefighter places the hoseline over the left shoulder at the next coupling and begins to climb the ladder.

D The two firefighters coordinate and maintain their distance on the ladder and hose.

JOB PERFORMANCE REQUIREMENT 10-28
Advancing an Uncharged Hoseline at Entry Point of Building

A The nozzleperson removes the nozzle and places it over the tip of the ladder into the building.

B The nozzleperson in the building, using the top rung of the ladder as a hose roller, pulls the hoseline up as the next firefighter climbs.

C The nozzleperson must complete pulling up the section as the next firefighter arrives at the tip.

FIRE HOSE AND APPLIANCES   261

Figure 10-33 Another method to advance the line over a ladder involves using rope hose tools or straps wrapped around the line just below the nozzle and at the next coupling. The hose tool or strap is looped over the shoulder of the firefighters to carry the line up the ladder.

Caution Advancing a charged hoseline over a ladder requires more firefighters to advance the hoseline, because the line is heavier and climbing with the line is very dangerous.

If the line is to be used from the ladder, the nozzleperson puts the nozzle through the top two rungs of the ladder allowing 1 foot (0.3 m) to be extended beyond the rung for movement and the line secured with a rope hose tool. The rope hose tool is wrapped around the hoseline snugly and then two or three wraps are taken around the second rung and the hose tool secured on itself. The hoseline is also secured by rope hose tools every 20 feet (6 m) and at the base of the ladder, **Figure 10-35**. Operating a hoseline from a ground ladder requires the ladder to be securely tied in and heeled. (See Chapter 14.)

Figure 10-34 Firefighters passing a charged hoseline up a ladder from one firefighter to the next until it reaches the opening.

Safety Operating a hoseline from a ground ladder requires the ladder to be securely tied in and heeled.

An uncharged hoseline can also be advanced to the tip of a ladder, secured in the same manner as the charged line, and then operated. The same safety precautions apply plus the line should be charged in a very careful and slow manner.

ESTABLISHING A WATER SUPPLY CONNECTION

Several different methods exist for establishing a water supply depending on the type of water source (static or hydrant), style of hydrant, the hose lays

Figure 10-35 A charged hoseline on a ladder. The hoseline is also secured by rope hose tools every 20 feet (6 m) and at the base of the ladder.

used (discussed later), and whether a pumper will be used at the water source. Static water sources obviously require a pumper at the source, but a hydrant source may not, depending on the volume and pressure of the hydrant and its distance from the fire. A supported fire hydrant is one that has a pumper at the fire and another pumper at the hydrant.

Firefighter Fact Early fire department pumpers needed one unit to do the pumping at the water source and another to carry the hose to the fire. The pumper was called the pumper and the hose cart was called a wagon. Even after modern engines were capable of supplying adequate pressure and volume, some fire departments continued to operate a two-piece engine company using a wagon–pumper combination until staffing requirements limited the concept.

Firefighters need to be able to connect directly to a fire hydrant or assist the engine driver in making the connections to hydrant and static sources.

From Hydrants

Using an unsupported hydrant requires a hoseline to be connected to a fire hydrant without an engine at the hydrant. This can be done using a single hoseline, multiple lines, or with adapters or valves to allow additional lines or a pumper to be connected later. Most departments use some type of finished hose load for a layout load. This load has a roll or several folds of hose to make any connections, a hydrant wrench and necessary adapters, and a rope to tie it all together and to allow the hydrant to be wrapped. "Wrapping the hydrant" allows the hose to be pulled far enough to reach the hydrant and helps prevent the hose from being dragged by the engine as it pulls away. The engine stops at the hydrant or layout point and the officer gives instruction for the hose lay. The layout person dismounts the vehicle and removes the layout load from the hose bed. Any additional adapters should also be removed from the engine. The layout person pulls sufficient hose and advances it to the hydrant and wraps either the hose or rope around the hydrant to secure it. The firefighter should be positioned to prevent being pinned between the hose and the hydrant. At this point the engine can move forward completing the hose lay, **Figure 10-36**.

Safety Caution should be used to ensure that the layout person is safely away from the vehicle prior to moving it. Firefighters have been struck by their own engine or other vehicles while laying out. Many departments now require the officer to dismount the vehicle to supervise this operation, and the engine does not move until the officer has remounted the vehicle.

After unwrapping the hydrant and removing the rope and tools, the hose is stretched out to remove any kinks. The layout person picks the proper hydrant outlet and removes the cap. This outlet depends on hydrant style and size and SOPs. On a wet barrel hydrant, some departments utilize a gate valve or a switch valve for additional lines or a pumper to be connected without shutting down the hydrant, **Figure 10-37**. To remove the cap, the hydrant wrench may be needed. The hydrant itself should then be briefly flushed prior to connecting the hose. The hose or adapter is leveled at the outlet and the coupling is screwed on, carefully ensuring that the threads are not crossed. The coupling should be hand tightened, although wrench tightening may be needed after turning the hydrant on, **Figure 10-38**. The hydrant valve is opened when water is called for and leaks and kinks are removed as the layout person returns to the engine.

FIRE HOSE AND APPLIANCES ■ **263**

Figure 10-36 The layout person pulls the layout section and enough hose to reach and wrap the hydrant.

Streetsmart Tip Care should be taken in removing kinks in large-diameter hose (LDH). For safety reasons, the firefighter should always try to "kick" the kink out of LDH. If it is necessary to lift the LDH to remove a kink, it should be done on the charged and pressurized side of the hose (toward the hydrant). It is important to use a strong lifting position (using the legs and a tight "core" posture).

An engine can be connected directly to a fire hydrant or to a switch valve already in service to supply attack lines or to support another engine. A soft sleeve hose is typically connected to the large

Figure 10-37 A supply line with a switch valve connected to hydrant.

hydrant outlet or the switch valve. The hard sleeve may also be used and other supply hose as well. Remember that the larger the hose, the more water that can be moved through it. The engine should be set up with gate valves at the inlet to speed up the process; in fact, some fire departments carry the soft sleeve preconnected. Care should be taken to make sure that the hose is connected to an inlet rather than a discharge valve of the pumper.

Connecting a soft sleeve to a hydrant starts with proper parking of the engine. The charged line should

Figure 10-38 The layout person making the hose connection to a hydrant.

be without kinks and undue stress on the hose or couplings. If the hose rests on the ground, it should have a chafing block between it and the ground. The hose is removed from its compartment or bed and stretched from the engine to the hydrant, **JPR 10-29A.** The outlet cap is removed, the hydrant flushed, and the coupling connected, **JPR 10-29B.** If the line is not preconnected, the hose must be attached to the engine's inlet. If a gate valve is on the inlet, the hose is connected directly to it. If not preconnected, the inlet cap must be removed. This will cause water in the pump to flow out; the tank-to-pump valve should be closed. The sleeve is then connected to the inlet, **JPR 10-29C.** The hydrant valve can now be opened slowly to avoid a water hammer. The gate valve on the engine or switch valve on the hydrant must also be opened if used.

Connecting a hard sleeve to a hydrant can also be done, but it is not recommended due to the chance of drawing a vacuum on the hydrant or piping and causing damage. The positioning of the engine is even more critical with the hard sleeve. The best way to do this is to stop the engine just prior to the hydrant. If the side intake is used, the engine should be at a slight angle. The hard sleeve is removed from its bed and attached to the inlet of the engine, **JPR 10-30A.** The engine is then moved forward toward the hydrant with several firefighters carrying and positioning the sleeve, **JPR 10-30B.** When the hard sleeve is aligned with the outlet, the engine is stopped and the sleeve connected, **JPR 10-30C.**

Medium- or large-diameter hose is used to connect an engine to a hydrant as a standard operation or used when the soft sleeve cannot reach the hydrant. The connection would be similar to connecting a layout line to the hydrant with a shorter distance.

From Static Water Supplies

The supply of water from a static source requires use of an engine and its hard sleeves to draft the water. The engine has to be able to position itself close enough to place the hard sleeve and strainer into the water or else reach the dry hydrant connection. Connecting a hard sleeve to a dry hydrant, not a dry barrel hydrant, is the same procedure as connecting to a regular hydrant without the normal concerns surrounding creation of a vacuum. A vacuum has to be created and it is important to make a tight connection to prevent leakage and loss of the vacuum.

The placement of a hard sleeve and strainer directly into a water source is somewhat similar to using it on a hydrant. The engine is placed near the water source but slightly back and at an angle, **JPR 10-31A.** The hard sleeve and strainer are attached to the pumper's inlet and the engine moves forward as

JOB PERFORMANCE REQUIREMENT 10-29
Soft Sleeve Hydrant Connection

A Stretch a non-preconnected soft sleeve for making a connection.

B Connect to the steamer connection on the hydrant.

C Connect to the steamer connection on the pumper.

FIRE HOSE AND APPLIANCES ■ 265

JOB PERFORMANCE REQUIREMENT 10-30
Hard Sleeve Hydrant Connection

A Connect hard sleeve to engine inlet.

B Move engine and sleeve toward hydrant.

C Attach sleeve to hydrant outlet.

the crew positions the sleeve into the water. A rope is attached to the strainer to assist in its placement, **JPR 10-31B** and **C.** If a nonfloating strainer is used, the rope also keeps the strainer from resting on the bottom. Nonfloating strainers should be well under the water's surface to prevent drawing in of air. When in place the engine is stopped, the line is secured, and drafting operations begin, **JPR 10-31D** and **E.**

EXTENDING HOSELINES

Despite the best efforts of judging the needed length of a hoseline, there will be occasions when the line comes up short and will need to be extended. Wildland firefighters encounter this often, not because of misjudgment, but because they have successfully driven the fire back beyond the length of their line.

Whatever the case, all firefighters should be familiar with techniques used to extend their hoselines. While the pump operator can shut down the line when the additional hose is needed, it is easier to extend the line using either a break-apart nozzle or a hose clamp.

The preferred method of extending a hoseline is to use a break-apart nozzle. The additional hose is brought to the nozzle end of the hoseline and the line shut down with the control handle. The nozzle tip is removed and the additional hose added if it has 1½-inch (38-mm) couplings, **JPR 10-32A** and **B.** If extending a line with hose couplings larger than 1½ inch (38 mm), then a 1½-inch (38-mm) to 2½-inch (65-mm) increaser will be required at the control handle, **JPR 10-32C.** The additional line with the nozzle tip attached is advanced and charged from the control valve, **JPR 10-32D.**

JOB PERFORMANCE REQUIREMENT 10-31
Setting up for Drafting Operation

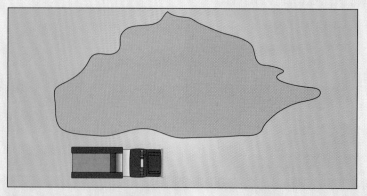

A Pre-position the engine at an angle to the water source.

B Connect the hard sleeve to the engine inlet.

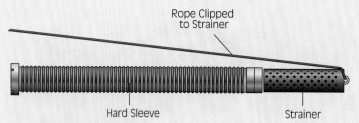

C Attach a strainer and guide rope to the other end of the sleeve.

D Move the engine and sleeve toward the water source.

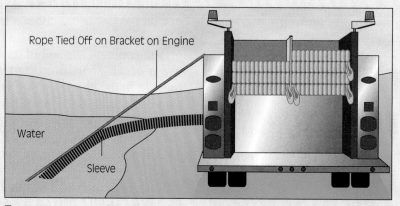

E Ensure that sleeve is in position and strainer is tied off for operations.

JOB PERFORMANCE REQUIREMENT 10-32
Extending a Hoseline with a Break-Apart Nozzle

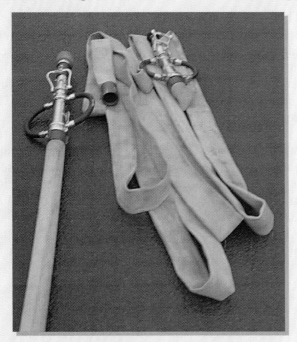

A Bring additional hose to the end of the hoseline and shut the line down. Remove the nozzle tip and add the additional hose if it has 1½-inch (38-mm) couplings.

B Remove the nozzle from the break-apart handle and connect hose.

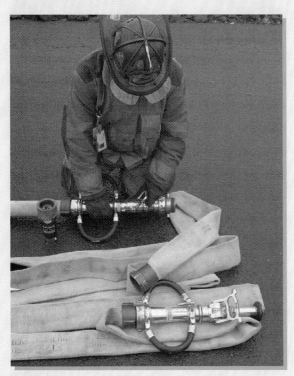

C If extending a line with hose couplings larger than 1½ inches (38 mm), then a 1½-inch (38-mm) to 2½-inch (65-mm) increaser will be required at the control handle.

D The additional line with the nozzle tip attached is advanced and charged from the control valve.

The other method is to use a hose clamp. The clamp and additional hose are brought up to the nozzle end of the hose. Clamping a hose can be dangerous, so firefighters need to ensure that the clamp is safely operated and is fully locked before releasing the clamp. The firefighter is positioned to the side of the clamp to prevent injury if it slips. The hoseline is clamped just before the nozzle and the nozzle removed. The additional line is attached as is the nozzle and the line is then advanced. When water is called for, a firefighter removes the hose clamp, **JPR 10-33A** and **B**.

Wildland firefighters often extend their hoselines as they advance on a fire. Some of their hose loads plan for this with the addition of a gated wye at each layer of hose. To extend these lines, the firefighters can simply go back to the last connection and close the gate until the additional hose is added, the nozzle reattached, and the gate reopened. They also use the hose clamp method, **JPR 10-34A–D**.

REPLACING SECTIONS OF BURST HOSE

The bursting of a section of hose is a dangerous situation that stops the flow of water to the fire and can injure firefighters and bystanders while causing property or water damage. It is a situation that requires immediate attention, especially when firefighters are operating under fire conditions. The hoseline must immediately be shut down.

The other firefighter should then remove the burst section and replace the section or if enough hose is available, reconnect the two adjacent sections. If enough hose is not available, a new section must quickly be brought and connected. The line may then be recharged.

Typically, the pump operator will shut down the line. If this is not possible, a hose clamp can be used

JOB PERFORMANCE REQUIREMENT 10-33
Extending a Hoseline Using a Hose Clamp

A Place and clamp hose clamp on hose behind nozzle.

B Remove nozzle, add additional hose, and release hose clamp.

FIRE HOSE AND APPLIANCES

JOB PERFORMANCE REQUIREMENT 10-34
Wildland Hose Advancing and Extension

A A preconnected line is pulled to begin the hose lay. As the line is moved forward, it is operated for suppression and crew protection.

B When the line is almost fully extended, a second line is rolled back in preparation for the next length. A working area is created with the water before the line is clamped off.

C With the line clamped, the nozzle is removed and placed on the new length, and the hoses are connected.

D The clamp is released and the line advanced. When the line is again fully extended, the team repeats the clamping and extending process.

Photos by Brentt Sporn, California Fire Photo

---

to stop the water flow. If no clamp is available, a firefighter can fold the hose twice over itself and kneel down to hold pressure buildup in the kinks. This will stop the flow sufficiently to have another firefighter replace the burst section.

HOSE LAY PROCEDURES

Hose lay procedures bring the water to the fire location by placing supply hose between the water source and the attack engine. SOPs should cover preferred hose lays and water supply operations, and apparatus is then designed for these preferences. The direction of the engine's travel to or from the water source gives each procedure its name. Supply lines and the hose beds on apparatus are designed to use at least one of the following three hose lays techniques; many are designed to allow all three. Any needed adapters are either attached to the supply line or readily accessible to the layout person. Each of these may work with either a hydrant or static source of water with the static source needing a supply engine at the source.

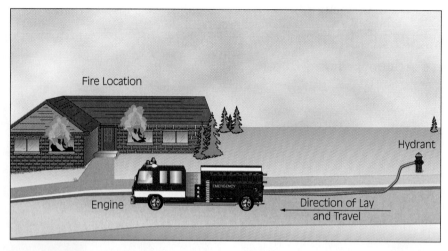

Figure 10-39 The forward or straight hose lay.

Forward Lay

Forward lay refers to the engine stopping first at the water source to drop off a supply line(s) and then advancing to the location of the fire, **Figure 10-39**. The forward lay is a preferred lay because a water supply can be established by connecting and charging the hydrant and may need no additional engines for the water supply. The attack engine is closer to the fire with access to additional hoselines and equipment. The forward lay works best when a water source is located in the approach path of the engine. An additional engine may still be required at the water source if needed to support the pressure or volume and will be required if the source is a static water source. The forward lay can also be used with a portable water tank operation where the fire is off the main road and the tank dump site is on the main road.

Reverse Lay

A reverse lay is the opposite of the forward lay with the supply line being dropped off at the fire location and the engine laying the hose toward the water source, **Figure 10-40**. Typically, the first arriving engine is used as the attack engine with the second engine doing the reverse lay and being the support or supply engine. Reverse lays can be used where a manifold, attack lines, and equipment can be dropped off at the fire with the engine going to the water source. This is less preferred but may be necessary in areas with few responding units and poor or static water sources.

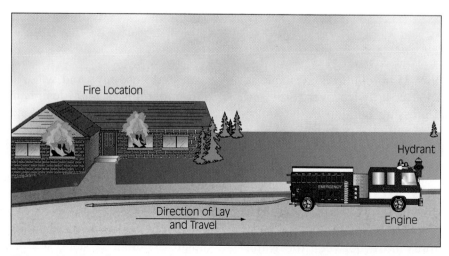

Figure 10-40 The reverse hose lay.

FIRE HOSE AND APPLIANCES 271

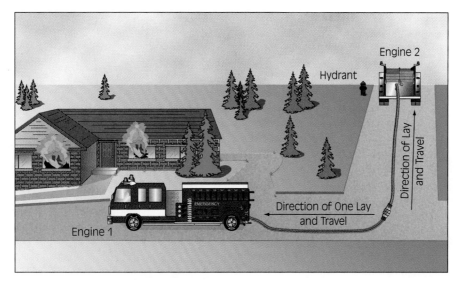

Figure 10-41 The split hose lay.

Split Lay

The split lay is used where the fire and the water source are in two different directions, such as on two different streets, thus needing to split the lay between two engines. The split lay has the first engine laying its line from a point or intersection to the fire location and the second engine laying its line from the point to the water source, **Figure 10-41**.

DEPLOYING MASTER STREAM DEVICES

Master streams or **heavy appliances** are non-handheld water applicators capable of flowing over 350 gallons of water per minute (1325 L/min). Four basic types of master stream devices are used, although the names are sometimes interchanged. The wagon pipe is a permanently mounted master stream device on an engine that has either prepiped water connection or needs a short section of hose to connect it to the pump, **Figure 10-42**. A similar looking device called a deluge set is not permanently mounted so it can also be removed and operated from the ground. Different models of deluge sets have one base that can be used both mounted on the apparatus or on the ground while others have separate bases for each mounting. A monitor pipe is a permanently attached master stream device with a prepiped waterway on an aerial device such as an aerial ladder or platform. A ladder pipe is a non-permanently mounted device that needs a hoseline for its waterway on an aerial ladder or platform. Some ladder trucks have both a monitor pipe and a ladder pipe, **Figure 10-43**.

> **Safety** Master stream appliances are often called the "artillery" of the fire service. Because of the power of these streams, great care should be used in the setup and operation of these appliances. One long-standing rule states that no master stream shall be introduced into a structure where interior firefighters are operating. Individual companies should be well versed in setting up a master stream to protect exterior exposures or to make a quick knock on a fire before firefighters enter a structure. Orders to charge a master stream after fire attack operations have begun should come only from the incident commander or operations section chief as part of a well-communicated and coordinated incident action plan (IAP).

When operating master streams with solid stream tips, the following rules apply for figuring reach, vertically and horizontally. For every foot (0.3 m) of vertical reach needed, move the device 1 foot (0.3 m) away. For horizontal reach, each pound of pressure (6.89 kPa) equals 1 foot (0.3 m) of reach. An angle of 45 degrees is about the maximum angle that can produce an efficient stream. The reach is generally limited to three floors above the nozzle's height.

The wagon pipe is a permanently mounted device that usually has a prepipe water connection needing no hoselines to be attached to it. Some departments refer to wagon pipes as a "deck gun." If it is without permanent piping, then either short or regular sections of supply hose can be run from the discharge to the inlets of the device. It is about 9 feet (2.8 m) in the air and can reach about four stories with a solid stream nozzle. A wagon pipe can also be directed at lower angles than a ground-mounted deluge set.

Figure 10-42 (A) A wagon pipe. (B) A deluge set and a wagon pipe operating.

Figure 10-43 A ladder truck with both a monitor pipe and ladder pipe.

The monitor pipe has a prepiped waterway with either a direct discharge valve if the unit also has a pump or needs a supply hoseline(s) if not equipped with a pump. If a hoseline from an engine is needed, the monitor pipe has a siamese or manifold that can be supplied similarly to a standpipe system, as described earlier.

Deluge sets when operated from the top of an engine may have prepiped water connections to a pump and need no supply lines attached. Some newer ones come with an extension device or are able to be adjusted in several positions that extend their reach several feet. Supply lines can be run up from the discharge to the inlet if no prepipe water connection is available. When securely attached to the engine, they can also be operated at low angles or can reach four stories. Portable deluge sets operated from the ground may not be operated at an angle lower than 25 degrees. Most deluge sets have a pin or other device that prevents them from going below this angle, and this pin should only be removed when the set is bolted down to the apparatus. Going below this angle can cause the set to become unstable.

Safety Portable deluge sets operated from the ground may not be operated at an angle lower than 25 degrees. Most deluge sets have a pin that prevents them from going below this angle, and this pin should only be removed when the set is bolted down to the apparatus. Going below this angle can cause the set to become very unstable.

FIRE HOSE AND APPLIANCES 273

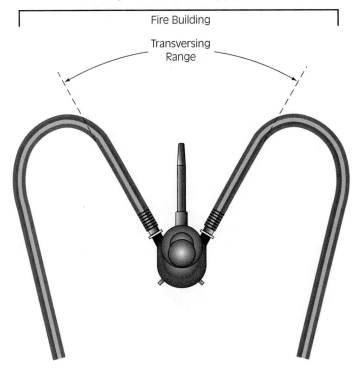

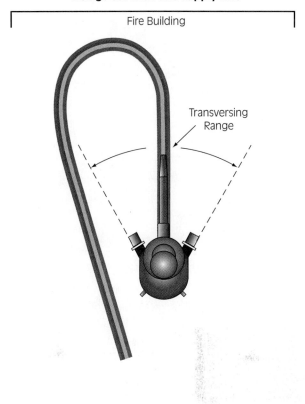

Figure 10-44 Proper operations of portable deluge sets.

When operating a portable deluge set on the ground, the intakes normally should be facing the fire building to counteract the nozzle reaction. Some newer balanced flow models do not need to have their intakes facing the building. Hoselines connected to the intakes should be balanced among the intakes and if only one hose is run, the nozzle shall be operated directly over that intake. When operating the set at the minimum 25-degree angle, the nozzle should not pass beyond the two outboard intakes, **Figure 10-44**. The advancing of medium- or large-diameter hoselines to the deluge set is also similar to the standpipe siamese described earlier, **Figure 10-45**.

The ladder pipe is not permanently mounted and therefore needs a hoseline for rigging it up the ladder and another for supply. The ladder pipe is often assembled, completely or partially, with its siamese, hose for the ladder, guide ropes, and the pipe itself. The ladder pipe is removed from its bracket and mounted on the aerial ladder in the center. The nozzle and **stream straighter** are adjusted to the operating position and guide ropes attached. If the hose is also attached, it is deployed down the aerial to the turntable and the ground. If the hose is not attached, it is often run up the ladder from the turntable and attached to the pipe. In either case the hose is run in the center of the ladder and secured at the top with a rope hose tool. The bottom of the hose and the

Figure 10-45 A portable deluge set in operation.

(A) (B)

Figure 10-46 Ladder pipe operations. (A) Ladder pipe on ladder. (B) Ladder pipe hose with rope hose tools secured to the hose. Notice the hose is run in the center of the ladder.

siamese are placed on the ground and the aerial is elevated to its operating position. Another rope hose tool is attached at the base of the aerial, **Figure 10-46**. A supply line(s) is run from an engine to the siamese and attached. When water is called for, the lines are charged slowly and carefully to ensure everything is properly connected.

SERVICE TESTING OF FIRE HOSE

Fire hose needs to be tested prior to being placed in use and then retested annually during its lifetime. Hose also should be tested after being damaged and after repairs have been made. To ensure accuracy of the testing program and for personnel safety, a record-keeping system must be utilized. The record system should have identification numbers, dates of testing, repairs, and other important notations about each section.[1]

The testing of hose begins with a visual inspection of the hose coupling. This inspection should be done with the annual test and during routine reloading and reconnection of hose sections. The inspection should check thread damage. The coupling should not be misshapened and it should swivel freely. Missing, worn, or damaged gaskets should be replaced and the inspector should note any missing lugs, slippage of the hose, or loose collars.[2] As the hose is loaded, it should also be visually inspected for any type of damage. Any damaged or suspect sections should be removed from service until tested or repaired.

The service test consists of testing the hose under pressure or, in the case of a hard sleeve, under a vacuum. These tests can be done with a hose testing machine, a fire apparatus pump, or a stationary pump. Pressure testing is designed to check for hose failure, and proper safety precautions should always be taken. Firefighters should follow manufacturers' instructions or NFPA standards. The hose is laid out in a straight line and a visual inspection conducted. The hose is marked, using a new color each year, at the point where the couplings are connected to check for any slippage of the coupling under pressure, **Figure 10-47**. A hose test valve, a gate valve with a ¼-inch (6.4-mm) opening drilled into it, is placed on the discharge of the pump. This hose test valve is designed to limit the flow rate of the water into the hose to reduce chances of injury if the hose fails during the test.

The hoselines are attached to the pumping device with no more than 300 feet (91 m) of length per line and a nozzle or test cap with a bleeder valve at the other end. The hoseline is charged to about 45 psi (310 kPa) with the nozzle open to bleed off any air

FIRE HOSE AND APPLIANCES 275

Figure 10-47 The hose is marked at the point where the couplings are connected to check for any slippage of the coupling under pressure.

and then closed. The hoseline and couplings are checked for leaks with couplings tightened with a spanner if necessary. The hose is now ready for the pressure testing.

Each new length of hose made after 1987 has its service test pressure rating stamped on it, **Figure 10-48**. In most cases, this is about 110 percent above its maximum operating pressure and maintained for at least three minutes.[3] The highest pressure is usually about 250 psi (1720 kPa) as the fire apparatus pumps are usually tested with this as their maximum pressure. SOPs should define maximum operating pressures for hoselines and pumps for each department. After the test is completed the hose has the current date written on it, and it is cleaned, dried, and returned to service or storage.

> **Safety** Hose testing is a destructive process that identifies weak hose by causing it to burst. Firefighters should secure the nozzle end of the hose and operate hose valves from a distance to prevent personal injury if the hoseline breaks and begins to whip about.

Hard sleeves are tested by being connected to a suction source and capped at the opposite end with a transparent cap. A vacuum of 22 inches of mercury (74.5 kPa) must be reached and maintained for 10 minutes while the inner lining is viewed for any collapse. If the hard sleeve is used under positive pressure, it must pass a service test of 165 psi (1138 kPa).[4]

Figure 10-48 Each new length of hose made after 1987 has its pressure testing rating stamped on it.

Lessons Learned

Fire hose, adapters, and appliances allow firefighters to move water at a distance from its source and from the apparatus's pumps. Without these valuable tools, firefighters would be extremely limited in the ability to move water and to conduct most fire suppression operations, especially interior fire attack. Firefighters must understand the proper use and care of these tools and how to connect, advance, and operate these tools. These are the basics of firefighting, and the best method of learning these basics is practical application.

KEY TERMS

Adapter   Device that adapts or changes one type of hose thread to another, allowing connection of two different lines. Adapters have a male end on one side and a female on the other with each side being a different thread type, for example, an iron pipe to national standard adapter.

Attack Hose   Small- to large-diameter hose used to supply nozzles and other applicators or protective system for fire attack. Attack hose commonly means handheld hoselines from 1½ to 2½ inches (38 or 63 mm) in diameter.

Backstretch or **Flying Stretch**   An attack line lay where the engine is at the hydrant and the line is stretched back from the engine to the fire. The flying stretch is a version of the backstretch where the engine stops in front of the fire, the attack portion is removed, and the engine proceeds to the hydrant.

Booster Hose   Smaller diameter, flexible hard-rubber-coated hose of ¾- or 1-inch (19- to 25-mm) size usually mounted on reel that can be used for small trash and grass fires or overhaul operations after the fire is out.

Distributor Pipe or **Extension Pipe**   Devices that allow a nozzle or other device to be directed into holes to reach basements, attic, and floors that cannot be accessed by personnel. The distributor pipe has self-supporting brackets that help hold it into place when in use.

Double Female   Allows the two male ends of hose to be connected.

Double Male   Used to connect two female thread couplings.

Dutchman   A short fold of hose or a reverse fold that is used when loading hose and a coupling comes at a point where a fold should take place or when two sets of couplings end up on top of or next to each other. The dutchman moves the coupling to another point in the load.

Ears   Elongated folds or flaps at the ends of a layer of hose to assist in pulling that layer.

Fire Hose   A flexible conduit used to convey water or other agent from a water source to the fire.

Forestry Hose   Specially designed hose for use in forestry and wildland firefighting. It comes in 1- and 1½-inch (25- and 38-mm) sizes and should meet U.S. Forestry Service specifications.

Hard Suction Hose   A special type of hose that does not collapse when used for drafting.

Helix   The metal or plastic bands or rings used in hard suction hose to prevent its collapse under drafting conditions.

Higbee Cut   The blunt ending of the threads of fire hose couplings that allows the threads to be properly matched, avoiding cross-threading.

Hose Bed   The portion or compartment of fire apparatus that carries the hose.

Hose Bridges   Devices that allow vehicles to pass over a section of hose without damaging it.

Hose Cap   Does not allow water to flow through it. Instead, it caps the end of a hoseline or appliance to prevent water flow.

Hose Cart   A handcart or flat cart modified to be able to carry hose and other equipment around large buildings. Some departments use them for high-rise situations.

Hose Clamp   A device to control the flow of water by squeezing or clamping the hose shut. Some work by pushing a lever that closes the jaws of the device, and others have a screw mechanism or hydraulic pump that closes the jaws.

Hose Jackets   Metal or leather devices used for stopping leaks without shutting down the line that is fitted over the leaking area and either clamped or strapped together to control the leak.

Hose Roller or **Hoist**   A metal frame, with a securing rope, shaped to fit over a windowsill or edge of a roof with two rollers to allow the hose to roll over the edge, preventing chafe.

Hose Strap   A short strap with a forged handle and cinch clip attached. Used to help maneuver hose and attach hose to ladders and stair rails.

Hydrant Valves or **Switch Valves**   Valve used on a hydrant that allows an engine to connect and charge its supply line immediately but also allows an additional engine to connect to the same hydrant without shutting down the hydrant, and increases the flow of the hydrant.

Hydrant Wrenches   Tools used to operate the valves on a hydrant. May also be used as a spanner wrench. Some are plain wrenches and others have a ratchet feature to speed the operation of the valve.

Increaser   Used to connect a smaller hose to a larger one.

Intake Relief Valve   Required on large-diameter hose at the receiving engine that functions as a combined overpressurization relief valve, a gate valve, and an air bleed-off.

Jacket   The outer part of the hose, often a woven cloth or rubberized material, which protects the hose from mechanical and other damage.

Liner   The inner layer of fire hose, usually made of rubber or a plastic material, that keeps the water in the tubing of the hose.

Master Stream or **Heavy Appliances**   Non-handheld water applicator capable of flowing over 350 gallons of water per minute (1325 L/min).

Medium-Diameter Hose (MDH)   Either 2½- or 3-inch (63- or 75-mm) hose.

Occupant Use Hose   Hose that is used in standpipe systems for building occupants to fight incipient fires. It is usually 1½-inch (38-mm) single-jacket hose similar to attack hose.

Portable Hydrant or **Manifold**   Like a large water thief and may have one or more intakes and numerous outlets to allow multiple hoselines to be utilized with or without a pumper at the fire location.

Reducers   Used to connect a larger hose to a smaller one.

Rope Hose Tool   About 6 feet (2 m) of ½-inch (13-mm) rope spliced into a loop with a large metal hook at one end and a 2-inch (50-mm) ring at the other. Used to tie in hose and ladders, carry hose, and perform many other tasks requiring a short piece of rope.

Shoulder Load   Hose load designed to be carried on the shoulders of firefighters.

Siamese   A device that connects two or more hoselines into one line with either a clapper valve or gate valve to prevent loss of water if only one line is connected.

Slot Loads   Narrow section of a hose bed where hose is flat loaded in the slot.

Small Lines or **Small-Diameter Hose**   Hose less than 2½ inches (63 mm) in diameter.

Soft Suction Hose   Large-diameter woven hose used to connect a pumper to a hydrant. Also known as a soft sleeve.

Spanner Wrenches   Used to tighten or loosen couplings. They may also be useful as a pry bar, door chock, gas valve control, and so on.

Storz Couplings   The most popular of the non-threaded hose couplings.

Strainers   Placed over the end of a suction hose to prevent debris from being sucked into the pump. Some strainers have a float attached to keep them at or near the water's surface. A different style of strainer or screen is located on each intake of a pump.

Stream Straighter   A metal tube, commonly with metal vanes inside it, between a master stream appliance and its solid nozzle tip. The purpose is to reduce any turbulence in the stream, allowing it to flow straighter.

Supply Hose or **Large-Diameter Hose (LDH)**   Larger hose [3½ inches (90 mm) or bigger] used to move water from the water source to attack units. Common sizes are 4 and 5 inches (100 to 125 mm).

Water Thief   A variation of the wye that has one inlet and one outlet of the same size plus two smaller outlets with all of the outlets being gated. The standard water thief usually has a 2½-inch (65-mm) inlet with one 2½-inch (65-mm) and two 1½-inch (38-mm) outlets.

Wye   A device that divides one hoseline into two or more. The wye lines may be the same size or smaller size, and the wye may or may not have gate control valves to control the water flow.

REVIEW QUESTIONS

1. Describe the various types of hose construction and how each type is used by firefighters.
2. What is the difference between single- and double-jacketed hose?
3. What are the different types of hose couplings important to firefighters?
4. What is a hose clamp?
5. What types of wrenches are used with hose?
6. What are double males and double females used for?
7. Describe the foot-tilt method of coupling hose.
8. What is a dutchman?
9. Describe flat, accordion, and horseshoe loads.
10. How is a hoseline advanced up a stairwell?
11. Describe connecting a supply line to a hydrant.
12. How can a hoseline be extended without shutting down the line at the pump?
13. What is a forward lay?
14. What is a deluge set?
15. What is a triple-layer load?
16. Describe the testing of fire hose.

Endnotes

1. *NFPA 1962: Standard for Care, Use and Testing of Fire Hose Including Couplings and Nozzles,* pp. 1962-7. National Fire Protection Association, Quincy, MA, 1998.
2. *NFPA 1962: Standard for Care, Use and Testing of Fire Hose Including Couplings and Nozzles,* pp. 1962-7. National Fire Protection Association, Quincy, MA, 1998.
3. *NFPA 1962: Standard for Care, Use and Testing of Fire Hose Including Couplings and Nozzles,* pp. 1962-9. National Fire Protection Association, Quincy, MA, 1998.
4. *NFPA 1962: Standard for Care, Use and Testing of Fire Hose Including Couplings and Nozzles,* pp. 1962-9. National Fire Protection Association, Quincy, MA, 1998.

Additional Resources

Richman, Harold, *Engine Company Fireground Operations,* 2nd ed. National Fire Protection Association, Quincy, MA, 1986.

Smoke, Clinton H., *Company Officer,* Appendix B: Suggested Training Evolutions. Delmar Learning, a part of the Thomson Corporation, Clifton Park, NY, 1999.

Sturtevant, Thomas, *Introduction to Fire Pump Operations,* Chapter 4: Hose, Nozzles, and Appliances. Delmar Learning, a part of the Thomson Corporation, Clifton Park, NY, 1997.

CHAPTER 11

NOZZLES, FIRE STREAMS, AND FOAM

Ric Koonce, J. Sargeant Reynolds Community College

OUTLINE

- Objectives
- Introduction
- Definition of Fire Stream
- Nozzles
- Operating Hoselines
- Stream Application, Hydraulics, and Adverse Conditions
- Types of Foam and Foam Systems
- Foam Characteristics
- Classification of Fuels
- Application of Foam
- Lessons Learned
- Key Terms
- Review Questions
- Endnotes
- Additional Resources

Photo by Fred Schall

STREET STORY

A few of the senior firefighters in my engine company had told me that when I was the nozzle man to always keep the nozzle pointed up and in the ready position when waiting to move into a burning room. In this way it will be ready to quickly knock down any fire that might roll out the top of the door when the door was forced open by the forcible entry team. If the nozzle was pointed downward, toward the floor, I might not get it pointing up and open it in time to protect the firefighters in the hall and, if the hallway was crowded with firefighters, it might not be possible to raise the nozzle quickly if it was blocked by people. Also, anytime I was in nozzle position, I was told never to put the nozzle down, even when things seemed under control, because you never know. As nozzle man, I was expected to be ready to open the line immediately, should the need arise.

This lesson served me well in a cellar fire in an apartment building that we were called to in the early 1970s. The fire involved the gas heating unit and had spread to most of the cellar. I had extinguished the fire, and the truck company reported that the gas to the heating unit was turned off. There were three firefighters overhauling in the room containing the heating unit. I was kneeling at the door to the room with the nozzle in my hands pointed up and forward toward the ceiling when suddenly the entire room lit up—leaking gas had ignited, filling the room with flame. The sudden burst of flame knocked me on my back. Instinctively, I pulled back on the nozzle's handle, opening the nozzle and directing the water from my hose stream into the room and onto the ceiling of the room, quickly extinguishing the flames and allowing the firefighters in the room to scramble to safety. Luckily, I had the nozzle in ready position even though the fire was out and everything had seemed to be under control. While extinguishing a gas flame is not the proper way to handle a gas fire, in this case it was necessary to allow firefighters to escape the flaming room. Proper training and attention to seemingly minor details can save lives. Never let your guard down. If you are the nozzle man, stay alert, hold onto the nozzle, and always be ready to open it.

—Street Story by Frank Montagna, Batallion Chief, FDNY, New York City

OBJECTIVES

After completing this chapter, the reader should be able to:

- Define a fire stream.
- Identify the purposes of a fire stream.
- Identify the various types of fire streams.
- Identify the types of nozzles.
- Explain the pattern and use of each type of nozzle.
- Demonstrate the operation of the various types of nozzles.
- Explain the operation and characteristics of various sizes (diameters) of fire streams.
- Explain the reach and application of various sizes of fire streams.
- Identify the three types of fire attack.
- Explain the factors in choosing the type of fire attack.
- Identify and explain the principles of hydraulics relating to fire streams.
- Define and explain friction loss.
- Define and explain nozzle pressures and reactions.
- Define and explain elevation as a factor in fire streams.
- Explain adverse factors in operations of fire streams.
- Explain the selection factors for fire streams in overall fire operations.
- Define foam.
- Identify the types of foam.
- Explain the principles of foam for fire suppression.
- Explain the operation of foam-making equipment.

INTRODUCTION

Fires are usually extinguished using water to cool the heat produced. Foam is added to improve water's extinguishment ability or on fuels where plain water is ineffective. Water and foam are delivered using nozzles and fire streams to reach the seat of the fire at the proper quantity. This chapter examines fire streams of water and foam, various nozzles and appliances, attack applications, and basic hydraulic principles. Proper selection of the right nozzle gives the attack crew the tool needed to fight the fire successfully, and no one nozzle or extinguishing agent is perfect for every fire situation.

DEFINITION OF FIRE STREAM

A **fire stream** is the water or other agent that leaves the nozzle toward its target, usually the fire. The four elements of a fire stream are the pump, water, hose, and nozzle. The fire stream is essential to the fire suppression effort because it targets the enemy, the fire. Properly developed and aimed fire streams are successful in extinguishing fires, while poorly developed or improperly aimed ones fail to reach the target, allowing the fire to continue to burn. Proper fire streams are the result of the knowledge, skills, and abilities of the firefighter on the nozzle, the company officer, and the **pump operator**. A proper fire stream is one that has sufficient volume, pressure, and direction and reaches the target in the desired shape or form and pattern. It is important that firefighters understand fire streams and how they are applied to various firefighting situations. (See Chapter 19 for situations.)

> **Streetsmart Tip** Which type of nozzle to use has been a fire service controversy ever since the first fog nozzle was invented and challenged the solid stream. Today, solid stream, fog, or automatic fog nozzles of various designs are all available to fire departments. Some only use one type, others use two, and still others all three. The type used by a particular department has been chosen after careful consideration has been given to local fires and conditions and based on testing and experimenting with the various kinds of nozzles available, and the training and experience of firefighters using the nozzles. Equipment changes and improves regularly, so fire departments should constantly reevaluate their equipment and tactics to ensure that they are using the best equipment for the conditions encountered. Nozzles are seldom used at their maximum performance capabilities, but are often used to maximize the performance of their operators. Regardless of which type of nozzle is used, firefighters should learn how to use it in the best manner possible.

NOZZLES

Nozzles are the appliances that apply the water or extinguishing agent. *Webster's Eleventh Collegiate Dictionary* states a nozzle "a short tube with a taper or constriction used (as on a hose) to speed up or direct a flow of fluid." The two basic types of nozzles are **solid stream** (also called a smooth bore, straight bore, or solid tip) and **fog nozzles** with different styles available for each kind, especially fog nozzles. **Combination nozzles** are capable of providing straight stream and spray patterns, which can be varied or adjusted by the operator. Some fire departments primarily use either solid stream or combination fog nozzles; however, firefighters should note that each type of nozzle has its advantages and disadvantages.

> **Safety** Selecting the proper type of nozzle to match the fire situation improves fire operations and personnel safety.

Important factors in nozzle selection are the nozzle pressure, nozzle flow, nozzle reach, stream shape, and nozzle reaction. **Nozzle pressure** is the pressure required for effective nozzle operation and relates to flow and reach. Nozzles are designed to operate at a certain pressure, usually 50, 75, 80, or 100 psi (345, 517, 552, or 690 kPa). Pressure is measured in pounds per square inch (psi) or kilopascals (kPa). **Nozzle flow** is the amount or volume of water that a nozzle will provide at a given pressure. Flow is critical because the amount of water provided determines the amount of heat absorbed or cooled. Some nozzles only flow a set volume at a set pressure, while others can be adjusted manually or automatically adjust their flow. Flow is measured in gallons per minute (gpm) or liters per minute (L/min).

Nozzle reach is the distance the water will travel after leaving the nozzle. Greater reach is important in large rooms or during exterior fire operations. Reach is a function of the pressure, which is converted to velocity or speed, of the water leaving the nozzle. Reach is measured in feet or meters. Reach is affected by the other factors, especially shape, pressure, wind direction, gravity, and friction of the air. The angle of the nozzle can affect the reach: Maximum horizontal reach is optimum at 32 degrees, while maximum vertical reach is obtained at 65 to 70 degrees. The objective is to reach the fire with maximum effect, not to be at the maximum distance from the fire.

Stream shape, also called stream pattern, is the arrangement or configuration of the droplets of water or foam as they leave the nozzle, **Figure 11-1**. The shape of the pattern helps determine the reach of the fire stream. **Nozzle reaction** is the force of nature that makes the nozzle move in the opposite direction of the water flow, **Figure 11-2**. The nozzle operator must counteract or fight the backward thrust exerted by the nozzle to maintain control of the nozzle and to direct it to the correct location. The nozzle pressure and stream shape affect nozzle reaction.

Figure 11-1 Nozzles showing the stream shape for straight, solid, and wide pattern streams.

Solid Tip or Stream

Solid tip, solid stream, or smooth bore nozzles deliver an unbroken or solid stream of water at the tip and toward the fire, **Figure 11-3** and **Figure 11-4**. The solid stream nozzle can deliver its water as a solid mass or cone of water or, when bounced off a ceiling, wall, or other object, as large water droplets. This solid mass breaks or shears apart the farther the

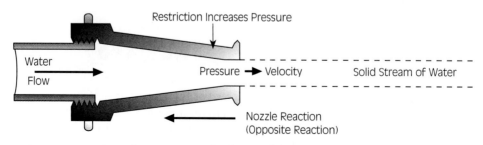

Figure 11-2 Nozzle reaction is the force of the water as it escapes and pushes back toward the nozzle and nozzleperson.

NOZZLES, FIRE STREAMS, AND FOAM

Figure 11-3 Various solid tips (stacked).

water travels. The solid stream nozzle's flow is a factor of the tip size at a certain nozzle pressure. Excessive or reduced nozzle pressures have adverse effects on stream performance. Handlines use tips from ¾ to 1¼ inches (19 to 32 mm) at 50 psi (345 kPa), and master streams use tip sizes of 1¼ inches (32 mm) and larger at 80 psi (552 kPa).

Solid stream handlines can reach over 70 feet (21 m) and master streams about 100 feet (33 m). They have the ability to penetrate through the fire's heat without absorbing that heat before reaching the target. A solid stream has less effect on a room's thermal balance, that is, the layers of heated gases in a burning room (see Chapter 4 on fire behavior), and produces minimal amounts of steam compared to narrow and wide fog patterns. Disruption of the thermal balance of a room can cause steam burns to firefighters as heavier water vapor descends onto firefighters after discharging the water. Fog streams are more apt to do this. Many fire departments have tried to minimize this chance of burns through the utilization of solid stream nozzles for aggressive, interior fire attack. It has good penetration into piles of materials to quench the fire. Smooth bore nozzles are more durable and easier to maintain than fog nozzles because they have fewer parts. The disadvantages of a solid stream are that there is no volume control other than changing tip sizes or adjusting the shutoff, the

lack of fog protection when working close to the fire, and a higher nozzle reaction at the same pressure than a fog nozzle.

Fog

Fog nozzles deliver either a fixed spray pattern or a variable combination pattern with both **straight stream** and spray patterns that can be adjusted by the nozzleperson. Fixed spray pattern nozzles are of the impinging design in which the nozzle has a series of holes at the end that creates a water spray, **Figure 11-5**. The variable fog patterns vary from the straight stream pattern, which is similar to a solid stream pattern, to a wide-angle fog pattern of at least 100 degrees, **Figure 11-6** and **Figure 11-7**. The different types of combination fog nozzles depend on the variations allowed and include constant volume or set gallonage nozzles; variable, adjustable, or selectable gallonage nozzles; and automatic or constant pressure nozzles, **Figure 11-8**. The **constant** or **set volume nozzle** has one set volume at a set pressure, for example, 60 gpm at 100 psi (227 L/min at 690 kPa), and the only adjustment is the pattern. **Variable, adjustable,** or **selectable gallonage nozzles** allow the nozzleperson to select from two or three flow choices and the pattern. This allows the flow choice to be made at the nozzle closest to the fire. The nozzleperson or officer should ensure that the pump operator is aware of this flow choice to maintain the correct pressure.

The **automatic** or **constant pressure nozzle** has a flow that can be adjusted by the pump operator, who increases the pressure, which in turn increases the gallons flowing. The pattern can be adjusted by the nozzleperson. Some newer automatic nozzles have a dual-pressure option for normal and low-pressure situations that is controlled by the nozzleperson. The automatic feature has a spring mechanism built around the baffle that reacts to

Solid Stream

Figure 11-4 The flow pattern of a solid tip nozzle.

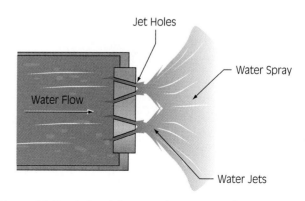

Figure 11-5 A fixed fog nozzle pattern of impinging design.

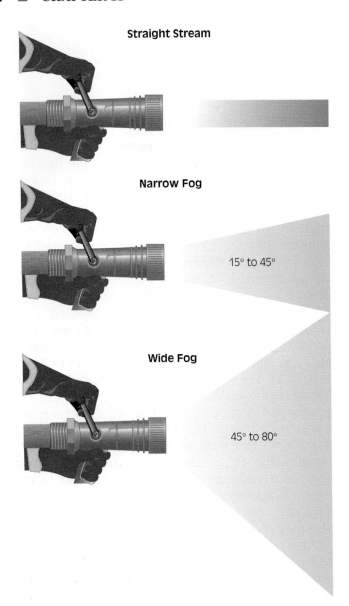

Figure 11-6 Variable combination fog nozzle patterns. From top to bottom: straight stream, narrow fog, and wide fog.

increased pressure, hence adjusting the flow and resultant reach of the nozzle, **Figure 11-9**. The disadvantage is that the pump operator, who is farther away from the fire, controls the maximum flow, while gating down at the nozzle will reduce the flow. The officer or nozzleperson should be in contact with the pump operator to control this flow. Fog nozzles have traditionally operated at 100 psi (690 kPa) but new low-pressure nozzles operating at 50 or 75 psi (345 or 517 kPa) have been approved. The lower nozzle pressure gives the same volume, but nozzle reaction is reduced and greater lengths of hose can be used when at the maximum pump pressure. On the low-pressure settings, additional flow is available at higher pressures.

Fog nozzles provide good reach that varies with the pattern from 25 to over 100 feet (7.5 to +33 m). They also provide good penetration. Fog nozzles that are adjustable provide personnel protection because of the screening effect between them and the fire (convected and radiated heat). Fog provides better heat absorption, but this can cause the water to change to steam before reaching the seat of the fire. Fog streams can produce more steam, which can extinguish hidden fire and is good for indirect attack. Fog streams can be used as a fan due to their ability to move large volumes of air. This advantage is seen when used at a window for horizontal ventilation, but can create problems for firefighters by drawing the fire's heat back toward them, **Figures 11-10 A** and **B**. The fog nozzle can be used to assist horizontal ventilation. Air movement is created to swirl and mix conditions including affecting the thermal balance, which can pull heat down to the firefight-

Figure 11-7 Parts of a fog nozzle.

Figure 11-8 Various styles of fog nozzles.

NOZZLES, FIRE STREAMS, AND FOAM 285

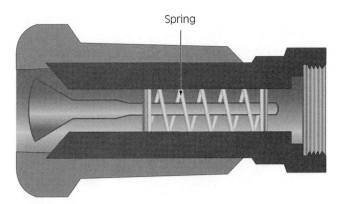

Figure 11-9 An automatic or constant pressure fog nozzle.

pass around the baffle of the nozzle. This creates an opening in the pattern, which may allow air into the stream and reduce its reach, **Figure 11-11**. Newer fog nozzle designs, especially the automatic nozzles, only have this hollow effect from the tip, and it is a short distance to refocus the stream to create a solid stream with good reach and penetration abilities. In fact, some are better than solid stream nozzles.

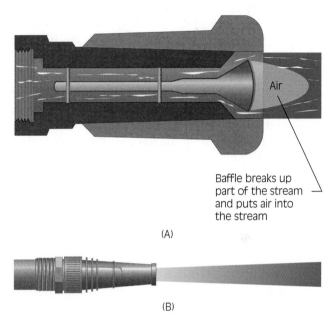

Figure 11-11 Comparison of (A) straight and (B) solid streams at tip.

ers at the floor level. For this reason, firefighters should use straight streams when attacking fires in high-heat environments or when trying to cool interior compartments while advancing toward the fire. A 1½-inch (38-mm) fog nozzle moves more air than a 14-inch (0.36-m) smoke ejector, and a 2½-inch (63-mm) fog nozzle moves more air than a 24-inch (0.6-m) smoke ejector.

Straight Stream

The straight stream nozzle pattern creates a hollow type stream that is similar in shape to the solid stream pattern, but the straight stream pattern must

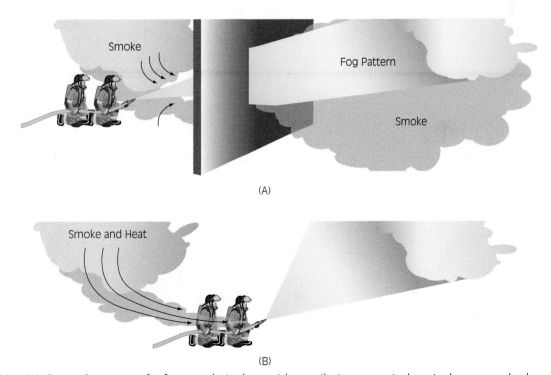

Figure 11-10 (A) One advantage of a fog nozzle is that, with ventilation at a window, it draws out the heat and smoke. (B) The disadvantage is that heat and gases are drawn onto the firefighters at the nozzle.

Special Purpose

Special-purpose nozzles were developed for use in limited types of situations by some fire departments. Special-purpose nozzles are not often used but firefighters should know when and how to use them. **Cellar nozzles** and **Bresnan distributors** can be used to fight localized fires in basements or cellars when firefighters cannot make a direct attack on the fire, **Figures 11-12A** and **B**. The cellar nozzle has four spray nozzles, and the Bresnan distributor has six or nine solid tips or broken stream openings that are designed to rotate in a circular spray pattern.

Piercing nozzles were originally designed to penetrate the skin of aircraft and now have been modified to pierce through building walls and floors, **Figure 11-13**. Some have striking points that allow them to be driven through the material. Other nozzles include high-pressure fog that can operate at 1,000 psi (6,895 kPa) and industrial and forestry nozzles that are combination fog nozzles with a shutoff built into the design.

Another special nozzle is a **water curtain nozzle**, which is designed to spray water to protect against exposure to heat, **Figure 11-14**. When using a water curtain, its spray should be directed on the object being protected to absorb the heat, not just up into the air. This is because radiant heat passes through clear materials, such as water, without being absorbed.

Playpipes and Shutoffs

Originally nozzles did not have a valve other than that at the pump. Today, Underwriters' and Factory Mutual playpipes, which are used for testing pur-

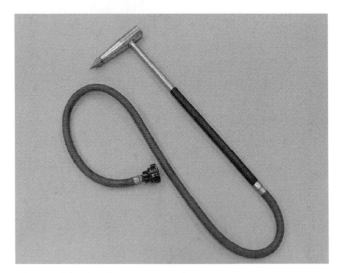

Figure 11-13 Piercing nozzle.

poses, not firefighting, are still made without a shutoff mechanism, **Figure 11-15**. Some nozzles or just nozzle tips are placed on hoselines for occupants' use, not for firefighters or fire brigades. The shutoff at the nozzle places the water flow control with the nozzleperson. The most commonly used is the lever type, bale, or handle, which operates in a line with the waterway, usually by moving a ball valve. The shutoff is opened by pulling back on the lever and closed by pushing the handle toward the nozzle. It can come built into the nozzle or as a break-apart type as a shutoff, pistol grip, or playpipe. The rotating type operates by either rotating a gate a quarter turn or rotating from a seat to open the waterway. These come either built in or as a separate shutoff, **Figure 11-16**. The nozzle is opened by rotating it counterclockwise or to the left, and it is closed in the opposite direction.

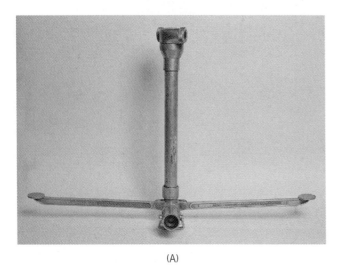

(A)

(B)

Figure 11-12 (A) Cellar nozzle and (B) Bresnan distributor.

NOZZLES, FIRE STREAMS, AND FOAM ■ **287**

Figure 11-14 Water curtain nozzle.

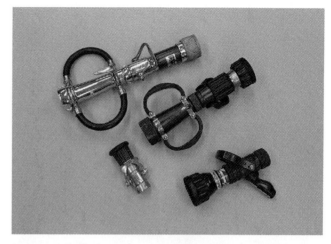

Figure 11-16 Various shutoff devices for nozzles. From bottom right to top left: built-in lever with pistol grip, built-in lever, rotating with playpipe, and break-apart playpipe.

Nozzle Operations

> **Note** Handline nozzles usually have shutoffs, but many other appliances do not provide shutoffs at the nozzle and the only control is the ability to aim it.

Some older appliance nozzles without shutoffs have adjustable fog patterns, and they are operated similarly to nozzles described in the following paragraphs. Appliances with shutoffs are also operated as described next.

Solid stream nozzles are simple to operate with the nozzleperson controlling the shutoff, tip size, and aim. The typical shutoff for a solid tip nozzle is a lever and can be fully opened or closed or sometimes partially opened. The other control is selecting the tip or nozzle size to match the desired flow, coordinated with the pump operator for proper pressure. Screwing them on or off is one way to change tips; many are designed to stack onto each other. Even with this stacking effect, the firefighter needs to be able to carry the nozzle tips when switching to a different size.

Fog nozzles have either the lever-type open/close shutoff, which is the most common, or the rotating type. In addition, the fog pattern can be adjusted by rotating the nozzle barrel counterclockwise to move from straight stream to narrow fog to wide-angle fog, **Figure 11-17**. Rotating clockwise adjusts the pattern the opposite way. Fog nozzles with variable gallonage have an addition ring on the collar that rotates

Figure 11-15 Underwriters' playpipe.

Figure 11-17 Close-up of rotating type of nozzle shutoff.

Figure 11-18 Close-up of combination fog nozzle showing the adjustments for selecting patterns and for selecting gallonage.

Figure 11-19 Firefighter operating nozzle in the crouching or kneeling position.

from one gallonage to the next, **Figure 11-18**. Both gallonage and pattern adjustments can be detected in the dark because the nozzle clicks at each position. Fog nozzles have more applications and many firefighters consider them more effective than smooth bore nozzles.

OPERATING HOSELINES

Advancing hoselines was covered in Chapter 10 including some of the initial operation of the nozzle. Included was straightening the hose, having firefighters properly spaced on the same side of the line, and with the line charged, bleeding off air from the hose and nozzle, and selecting the proper pattern. Most hoselines are operated from a crouching or kneeling position, but lying, standing, or sitting positions are also used, **Figure 11-19**, **Figure 11-20**, and **Figure 11-21**. Chapter 19, on fire suppression, has a table that highlights some of the characteristics of various size hose streams and covers hoseline operations in fire situations.

Small-Diameter Handlines

Small-diameter handlines are typically 1½, 1¾, or 2 inches (38, 44, or 50 mm) in diameter and flow from 100 to over 250 gpm (378 to over 946 L/min).[1] When flowing at the lower volumes, these lines can be operated by one person; larger volumes require two persons.[2] Both fog and solid tip nozzles can be used for small lines. Small lines are popular because of their ease of mobility, the number of personnel required to operate them, and their ability to extinguish one to three typical rooms of fire with their flow.

Figure 11-20 Firefighter operating nozzle in the lying position.

Figure 11-21 Firefighter operating nozzle in the sitting position. Notice that the hoseline is placed in a loop to counter nozzle reaction.

Medium-Diameter Handlines

Medium-diameter hose for handlines is 2½-inch (75-mm) hose and can be used with solid tip and fog nozzles, flowing from 165 to 325 gpm (625 to 1,230 L/min). This is the standard size hoseline for the fire service, used for calculating friction loss, pump capacity, and appliance discharge values. Many fire departments abandoned the 2½-inch hose as a fire attack tool, advocating 1¾-inch and 2-inch lines because of increased maneuverability. However, increased Btu development of fires due to plastics and low-mass fuels has caused many to reintroduce the 2½-inch hoseline to increase gpm flow to quench the higher heat. This is especially important in large, commercial structures or buildings with high fire loading (like an auto body repair shop or high rack storage).

Medium-size hoselines usually require two or more personnel to operate them because they are less mobile than small lines, which has been a major drawback. Some departments have gone to 3-inch (75-mm) hoses with 2½-inch (65-mm) couplings, while other departments do not use any handline over 2½ inch (65 mm).

Master Stream Devices

Master stream devices (see Chapter 10) are capable of flowing over 350 gpm (1,325 L/min), and some have capabilities of many thousands of gallons per minute. This is the artillery of the fire service and is used when large volumes are required. These devices must be mounted or secured properly and safety should be a major concern during their operation. Specific operating instructions are necessary for each type of device. They require only one person to operate them, but have either poor or no mobility.

STREAM APPLICATION, HYDRAULICS, AND ADVERSE CONDITIONS

Application of fire streams depends on the method of fire attack and conditions encountered, including environmental factors and water supply. The fire stream must have the proper pressure and flow, and an understanding of hydraulics is needed to provide those factors. Improper hydraulic calculations are the leading causes of poor fire streams.

Direct, Indirect, and Combination Attack

> **Note** The method of stream application or fire attack depends on the fire's fuel, its location, and the equipment of the fire department, especially the size and type of hoseline and nozzle.

Three methods of fire attack exist: direct, indirect, and combination. **Direct fire attack** is used to attack the fire by aiming the flow of water directly at the seat of the fire, **Figure 11-22**. Direct fire attack is used on most deep-seated fires that require penetration by the hose stream.

Indirect fire attack is used to attack interior fires by applying a fog stream into a closed room or compartment, converting the water into steam to extinguish the fire, **Figure 11-23**. Firefighters apply the water at the doorway and then close the door, allowing the steam to put the fire out. The closed compartment is needed to contain the steam, thereby smothering the fire. The estimated quantity of water applied is the amount needed for total conversion to

Figure 11-22 Firefighter directly attacking a fire.

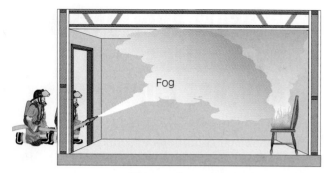

Figure 11-23 Firefighter using indirect attack by applying water into room and then closing the door.

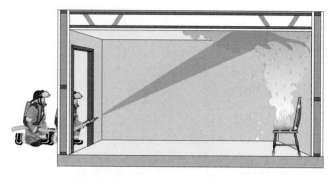

Figure 11-25 Firefighter using combination fire attack.

steam to fill the room. As water is converted to steam at 212°F (100°C), it expands 1,600 to 1,700 times its volume; 1 ft³ of water (0.028 m³ or 7.48 gallons or 28 L) would fill a 1,700-ft³ (48.1-m³) room, **Figure 11-24**. At 1,000°F (500°C), this expansion is over 4,000 times. Because the entire space is filled with steam, the indirect attack should not be used when people are in the space.

A **combination fire attack** uses a blend of the direct and indirect fire attacks, with firefighters applying water to both the fuel and the atmosphere of the room, **Figure 11-25**. To achieve this, the nozzleperson opens the nozzle and directs the stream toward the ceiling and then the fire with a circular, "Z," or "T" motion. This puts water on the seat of the fire and the atmosphere, while creating limited amounts of steam to extinguish the fire in any hidden areas. Ventilation with this combination attack controls the flow of fire gases and steam, improving the survival chances of victims, and makes this attack the type typically used by firefighters in structural firefighting.

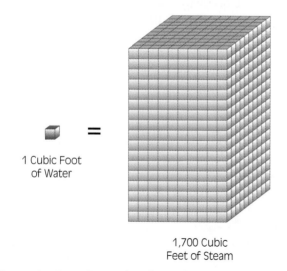

Figure 11-24 One cubic foot of water in liquid form expands 1,700 times when converted to steam at 212°F.

Basic Hydraulics, Friction Loss, and Pressure Losses in Hoselines

Hydraulics is the study of fluids at rest and in motion, which describes the flow pattern of water supply and fire streams. Water moving through a hoseline and out a nozzle is an example of water in motion, while use of the nozzle shutoff leads to water at rest. An effective fire stream must have sufficient volume and pressure and be delivered in the correct form. This relationship begins with the water supply into the fire pump and then out the lines to the nozzle tip and to the fire (see Chapter 9 for more information). Hydraulic principles are a whole field of study; this section examines only the basics.

Pressure is force divided over an area, usually expressed in pounds per square inch (psi) or kilopascals (kPa) where 1 psi = 6.895 kPa. Pressure is required to lift, push, or move water. Force is a measurement of weight and is measured in pounds (kilograms). The pull of gravity on a mass of water creates a force. Other external pushing and pulling, like the actions of a fire pump, applied to a substance (water) can also create motion. Water weighs 62.4 pounds per cubic foot (1,000 kg/m³), which creates a force of 62.4 pounds (28.3 kg). This would also create a pressure of 62.4 pounds per square foot (or 1,000 kg/m²) or 0.434 pounds per square inch or psi (3 kPa), **Figure 11-26**. Pressure can also be measured in feet (meters) of head or the height of a column of water. Thus, a column of water 100 feet (30 m) would create a head of 100 feet (30 m) or a pressure of 43.4 psi (300 kPa).

Firefighters should be familiar with the several types of pressure: Atmospheric pressure is the pressure exerted by the atmosphere or the weight of the air at the Earth's surface. At sea level this pressure is 14.7 psi (101 kPa) and gauges reading this pressure show **absolute pressure**, which is indicated as pounds per square inch absolute (psia). **Gauge pressure** normally measures pres-

NOZZLES, FIRE STREAMS, AND FOAM

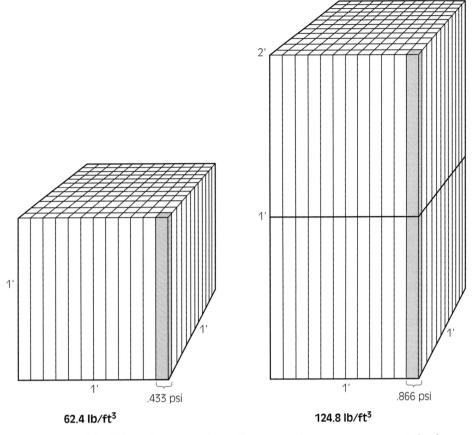

Figure 11-26 Water in a container (like a hose) exerts pressure at the lowest point.

sure minus atmospheric pressure and is measured in psi or psig (pounds per square inch gauge). Most fire department readings are from gauge pressure and begin at zero, **Figure 11-27**. **Vacuum (negative) pressure** is the measurement of pressure less than atmospheric pressure, which is usually read in inches of mercury (in. Hg or mm Hg). Fire apparatus capable of drafting have at least one gauge that measures vacuum pressure, called a compound gauge. Today, it is common for all gauges on apparatus to be of the compound type. **Head pressure** measures the pressure at the bottom of a column of water in feet (meters) (see example given earlier) and head pressure can be gained or lost when water is being pumped above or below the level of the pump. A head of 2.31 feet (0.7 m) would equal 1 psi (6.895 kPa). This is also called elevation pressure and is usually rounded to 5 psi (35 kPa) per floor when pumping water up or down in a building. **Velocity pressure** is the pressure in a hose being converted to velocity or speed as it leaves an opening. When leaving a nozzle, this becomes the nozzle pressure. Chapter 9 discussed static pressure, the pressure with no water flowing, and residual pressure, the pressure in a system after water has been flowing.

Flow is the rate and quantity of water delivered and is usually measured in gallons per minute or liters per minute (1 gpm = 3.785 L/min). The **needed** or **required flow** is the amount of water required to put out the fire and is determined by what and how much is burning. The **available flow** is the amount of water that can be moved to extinguish the fire. The water supply, pump(s) and

Figure 11-27 Typical pump pressure gauges.

their capability, the size and length of hose, and the type(s) of nozzles determine available flow. Next to be determined is the **discharge flow**, the amount of water flowing from the discharge side of the pump. The discharge flow is the flow of each hoseline and of all hoselines and outlets from the pump or the total gallons per minute. Water flow can be determined using a flowmeter or pressure gauge. The flowmeter measures the flow of water and is simpler to use than the pressure gauge, **Figure 11-28**. Flowmeters have only recently become effective for fire apparatus use. Pressure gauges have been primarily used for fire apparatus for determining pressure requirements to generate a certain flow. (Actual flow measurement calculations were discussed in Chapter 9; see Eq. 9-1.) The pressure gauge method requires more complex calculations using flow rate and pressures of the nozzle and any **friction** in the system, that is, the **friction loss**. Note that friction loss in hose and appliances, including nozzle pressures, and elevation losses or gains only occur when water is in motion or flowing. When all water flow stops, the system equalizes at the highest pressure in the system or hoseline. A sudden stop of water flow creates a water hammer or pressure surge that could damage equipment, piping, and hose. (See Chapter 10 for more on water hammer.)

After determining what flow is needed, the pump operator adds the pressures of the various parts of the pump to nozzle system and increases engine pressure to generate that flow. Errors made in this formula can create too much or too little pressure, which affects nozzle performance. Hose and appliances are constantly being improved. When new equipment is purchased, water flow and pressure evaluations should be conducted to revalidate these formulas. To calculate the discharge pressure of a pump, the following formula is used:

$$EP = NP + FL \pm E + SA \qquad (Eq.\ 11\text{-}1)$$

where:
- EP = engine pressure, also called pump discharge pressure.
- NP = nozzle pressure, the pressure required to operate the nozzle and the pressure as it discharges from the nozzle.
- FL = friction loss in the hose(s). (There can be multiple cases of friction losses, such as the hose between the pump and a standpipe connection and also the attack line hose.)
- E = elevation gain or loss from moving the weight of the water up or down in relation to the pump.
- SA = friction loss in any special appliance, not hose or elevation. It is also called appliance friction loss.

Each of the formula's components is either a given number or must be calculated. Engine pressure is the addition of the other components and is typically the highest value for any of the hoselines. If more than one hoseline is pumped, the lower pressured line is gated down. In reality, the engine pressure is figured for each line, the highest one being the main pump pressure. Nozzle pressure is usually a given value for each type of nozzle. **Table 11-1** gives the typical values.

Nozzle pressure pushes the water out of the nozzle, but it also pushes the nozzle in the opposite direction against the firefighter or whatever is holding it. A higher nozzle reaction means more effort is required by the nozzleperson and hose crew to control the nozzle and hoseline to keep them from getting away or when advancing an opened line. Now it is easier to understand why operating a nozzle in the sitting or lying position is less stressful than the standing position—the ground helps absorb this reaction. Nozzle reaction can make an unsecured hoseline unsafe and dance like a wild snake. It is important to keep a firm grip. Nozzle reaction is a relationship between the nozzle pressure and the type of nozzle. Nozzle reaction for a smooth bore nozzle is:

Figure 11-28 Flowmeters eliminate the need for friction loss calculations during pump operations.

(Photo courtesy of Class One)

Nozzle Types and Pressure

TYPE OF NOZZLE	NOZZLE PRESSURE, psi (kPa)
Smooth bore handline	50 (345)
Fog handline, normal	100 (690)
Fog handline, mid pressure	75 (517)
Fog handline, low pressure	50 (345)
Smooth bore master stream	80 (552)
Fog master stream	100 (690)

TABLE 11-1

$$NR = 1.57 \times d^2 \times NP \quad \text{(Eq. 11-2)}$$

where

NR = nozzle reaction
1.57 = a constant (some round to 1.5)
d = diameter of the nozzle tip in inches
NP = nozzle pressure

For example, a 1-inch nozzle with 50 psi of nozzle pressure flowing at 210 gpm would have a nozzle reaction of 78.5 lb.

$$\begin{aligned} NR &= 1.57 \times d^2 \times NP \\ &= 1.57 \times (1 \text{ in.})^2 \times 50 \text{ psi} \\ &= 1.57 \times 1 \text{ in.}^2 \times 50 \text{ psi or } 1.57 \times 50 \text{ psi} \\ &= 78.5 \text{ lb} \end{aligned}$$

The nozzle reaction for a fog nozzle is figured using the following formula:

$$NR = gpm \times \sqrt{NP} \times 0.0505 \quad \text{(Eq. 11-3)}$$

where NR = nozzle reaction
gpm = gallons per minute
$\sqrt{NP}$ = square root of the nozzle pressure
0.0505 = constant

A fog nozzle flowing 200 gpm at 50 psi of nozzle pressure would have a nozzle reaction as follows:

$$\begin{aligned} NR &= gpm \times \sqrt{NP} \times 0.0505 \\ &= 200 \text{ gpm} \times \sqrt{50} \text{ psi} \times 0.0505 \\ &= 200 \text{ gpm} \times 7.07 \times 0.0505 \\ &= 71.4 \text{ lb} \end{aligned}$$

A fireground formula for fog nozzles operating with a nozzle pressure of 100 psi is

$$NR = gpm \times 0.5 \quad \text{(Eq. 11-4)}$$

Note that, as previously stated, at the same pressure and gpm, the smooth bore nozzle has a higher nozzle reaction than the fog nozzle.

Friction loss is the loss of energy from the turbulence or rubbing of the moving water through the hose. The friction loss is compensated for by the pump operator increasing the pump pressure and providing the correct pressure to the nozzle. Friction loss is based on four principles. The first is that the friction loss varies directly with the length of the hose if all other variables are held constant. This means that if the length of the hose doubles, so does the friction loss. Second, with all other variables held constant, friction loss varies approximately with the square of the flow. Therefore, if the flow rate doubles, then the friction loss will increase four times or be squared ($2^2 = 2 \times 2 = 4$). Third, when the flow remains constant, friction loss varies inversely with the hose diameter. This means that at the same flow, the friction loss will decrease as the size of the hose diameter gets larger. The fourth and final principle states that for any given velocity, the friction loss will be about the same regardless of the water pressure. This says that the speed of the water flow rather than the pressure is what determines friction loss.[3] Friction loss can be calculated using Eq. 11-5, which factors these principles into the formula:

$$FL = Q^2 \times c \times L \quad \text{(Eq. 11-5)}$$

where
FL = friction loss of a hoseline
Q = quantity of water in hundreds of gallons per minute, that is, the flow:

$$Q = gpm/100 \quad \text{(Eq. 11-6)}$$

c = friction loss coefficient for the diameter of the hose with 2½-inch (65-mm) hose being used as the standard (see **Table 11-2**)
L = length of hose in hundreds of feet:

$$L = \text{hose length}/100 \quad \text{(Eq. 11-7)}$$

Friction Loss Coefficients

HOSE DIAMETER, INCHES (MM)	COEFFICIENT VALUE (c)
1½ (38)	24
1¾ (44)	15.5
2 (50)	8
2½ (65)	2
3 (76) with 2½-inch (65-mm) couplings	0.8
4 (101)	0.2
5 (127)	0.08
6 (152)	0.05

Note: Use caution because the hose diameter required may not be the same as the labeled hose diameter.

TABLE 11-2

Streetsmart Tip When discussing friction loss formulas and the values of the coefficient c, many firefighters will use different formulas. The formula $FL = Q^2 \times c \times L$ was once given as $FL = 2Q^2 \times c \times L$ but with different values for c. The different values depended on whether the coefficient had already been multiplied by 2. Many older firefighters remember the formula as $FL = (2Q^2 + Q) \times c \times L$, which has changed over time as hose construction has reduced the friction created in the line.

An example calculating the friction loss formula is given next, using a 2-inch hoseline that is 150 feet long with a flow of 200 gpm, **Figure 11-29**.

$$\begin{aligned} FL &= Q^2 \times c \times L \\ &= (200/100)^2 \times 8 \times (150/100) \\ &= (2)^2 \times 8 \times 1.5 \\ &= (4) \times 8 \times 1.5 \\ &= 48 \text{ psi} \end{aligned}$$

Elevation can be a positive number if the nozzle is above the level of the pump or a negative number if below the pump. If even, elevation is not a factor. Elevation is equal to head pressure of 0.434 psi per foot (9.81 kPa/m) of height. [Note: This is usually rounded to 0.5 psi per foot or 5 psi per floor (10 kPa/m) for practical fireground purposes.]

Special appliances or appliance friction loss is the friction loss created by movement of water through the valves and turns of the appliances. Special appliance friction losses vary with each device. Examples include wagon pipe or deluge set, 20 psi; ladder pipe, 50 psi; and 2½ to 2½ siamese or wye, 5 psi. Manufacturer's specifications or department SOPs should be consulted for guidance in determining these friction loss factors.

The next example uses the preceding example's 2-inch hoseline that is 150 feet long flowing at 200 gpm, with a fog nozzle and operating on the second floor. Using that friction loss equation, engine pressure (EP) would be:

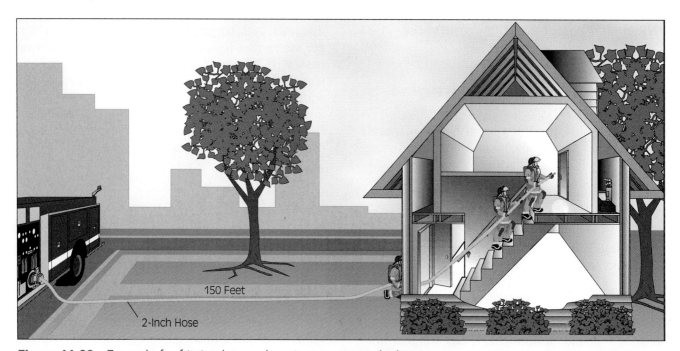

Figure 11-29 Example for friction loss and engine pressure calculations.

EP = NP + FL ± E + SA
 = 100 + 48 + 5 + 0
   or shorter EP = 100 + 48 + 5
 = 153 psi

Adverse Conditions That Affect Fire Streams

Adverse conditions that affect fire streams are of two types: natural and man-made. The major natural factor affecting a fire stream is the wind and wind direction. Anyone who has tried to wash a car on a windy day knows the wind will break up a stream and deflect it from its target unless the flow is being aimed downwind. This also occurs with fire streams. Rain, snow, hail, and objects such as tree branches and wires are solid or semisolid objects that deflect and break up hose streams. Gravity and air friction are also natural factors, especially the farther the travel distance of the stream. These natural causes cannot be removed but getting the stream closer to its target or in a better position allow these effects to be reduced.

> **Streetsmart Tip** Man-made causes are typically from improper operation of the pump or nozzle. Except on automatic nozzles, too much and too little pressure are both causes of poor fire streams as is failing to operate the nozzle properly, including fully opening the shutoff valve. Automatic nozzles use the shutoff to control the flow while still providing a proper stream.

These problems can be remedied by following the correct procedures for the nozzle being used.

TYPES OF FOAM AND FOAM SYSTEMS

Foam is an aggregate of gas-filled bubbles formed from aqueous solutions of specially formulated concentrated liquid foaming agents.[4] The bubbles are filled with a gas, usually air, creating a blanket over the surface of the fuel to cool and smother the fire, while sealing the escaping vapors. The mechanical action of mixing the foam concentrate in the water makes a foam solution to which air is added. Combining these three components makes the foam lighter than the flammable liquids and gives it the ability to float over their surface, **Figure 11-30**.

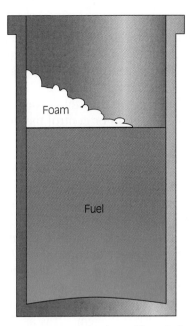

Figure 11-30 Foam applied on Class B fuel.

Class A foam is an aggregate of gas-filled bubbles formed from aqueous solutions of detergent or soap-based surfactants used to penetrate ordinary combustible materials and to keep the fuel wet that is unable to burn.

FOAM CHARACTERISTICS

Foam's ability to extinguish fires is based on several characteristics. The various foams available differ in their abilities to supply these characteristics. **Application rate** is the amount of foam or foam solution needed to extinguish a fire and is usually expressed in gallons per minute per square foot (gpm/ft^2) or liters per minute per square meter (L/min/m^2). Application rates vary depending on the fuel type and severity and fuel depth. See NFPA Standard 11 for recommended application rates.

Application rates must be maintained for a minimum amount of time, and additional foam will need to be reapplied to any remaining fuel to prevent reignition. The use of application rates allows firefighters to preplan for probable incidents or assess needs for foam prior to beginning a foam attack. For example, a shallow spill fire of 1,000 ft^2 (90 m^2) of regular gasoline would require an application rate of 0.1 gpm/ft^2 using a 3 percent foam concentrate with a minimum run time of 15 minutes. This would equal a need for 100 gpm of foam solution per minute or 3 gallons of foam concentrate for 15 minutes for a total of 45 gallons of foam concentrate required.

Heat resistance is the ability of the foam to stand up to the heat of the fire or to hot surfaces near the fire. **Knockdown speed** is how fast the foam spreads across the surface of a fuel. **Fuel resistance** is the ability to tolerate the fuel and to avoid being saturated or picking up the fuel. **Vapor suppression** is the ability to contain or control the production of fuel vapors.

Protein foam is made from chemically broken down natural protein materials, such as animal blood, that have metallic salts added for foaming. The foam created is stiff, elastic, and has a high water retention rate and heat resistance. It is highly effective in vapor suppression but can be blown away by the wind, does not spread easily, and has a limited shelf life. It cannot be saturated or dipped into the fuel (poor fuel resistance), has slow knockdown time, and breaks down when used with dry chemical extinguishing agents. It is applied in 3 and 6 percent concentrations.

Fluoroprotein foam was designed as improved protein foam and has a fluorinated surfactant added. This surfactant allows the foam to be dipped into the fuel. In fact, fluoroprotein can be subsurface injected into a fuel tank to rise through the fuel and to the surface. It also has the ability to work with dry chemical extinguishing agents. It has high heat resistance and vapor suppression capabilities and a better knockdown time than protein foam. Fluoroprotein foam is popular at fuel processing and storage facilities and is applied in 1, 3, and 6 percent concentrations.

Aqueous film forming foam (AFFF) is made from fluorochemical surfactants and synthetic foaming agents that have a quick drain-down time. This feature creates a liquid that forms a film or layer of water that spreads quickly over the surface of the flammable liquid, **Figure 11-31**. This film provides a faster knockdown time but gives up some heat resistance, fuel tolerance, and vapor suppression compared to fluoroprotein foam. AFFF can be applied with regular fog nozzles and comes in alcohol-type concentrates (ATC) (see later discussion). Because of its effectiveness, ease of use, and reduced need for special applicators, it is carried by most fire departments. AFFF can also be used with dry chemical extinguishing agents. The quick drain-down does require continued application with a poorer resistance to burnback than fluoroprotein foam. It is also applied in 1, 3, and 6 percent concentrations.

Fluoroprotein film-forming foam (FFFP) combines protein with the film-forming fluorosurfactants of AFFF to improve on the qualities of both types of foam. It has the fast film-forming capabilities of AFFF and the burnback and heat resistance of protein foams. It has some alcohol resistance.

Detergent-type foams use synthetic surfactants to break down the surface tension of water and create a foaming blanket. These foams originally were good at penetrating into Class A materials but lacked heat resistance and stability. Newer types have overcome these drawbacks, and these types are used both as an active firefighting agent and as a protective barrier against fire spread. A special type of detergent foam is used for high expansion foam, which is used to fill up entire areas such as mine shafts or buildings.

CLASSIFICATION OF FUELS

Foams are used for Class A and B fires and some specific considerations affect their use.

Class A

Class A fuels are ordinary combustibles that have traditionally been extinguished with water. Sometimes piles of Class A materials are extinguished using a wetting agent, often a detergent-like substance, to help soak through the piles. Today, a new class of foams is available for Class A materials that uses a detergent base that extinguishes by reducing the surface tension of the water, allowing greater penetration into the materials. The foamy water solution has the ability to cling to the sides of objects, thus enhancing protection. This clinging ability is now being used to protect homes in urban interface areas during wildland fires, **Figure 11-32**. Urban fire departments are using it for interior firefighting and have reported faster fire extinguishment with less water. Class A foams use application ranges of 0.03 to 1.0 percent, which require a separate and more accurate proportioning system than Class B foams. Some Class B foams like AFFF have a detergent action that can be used on Class A materials.

Disadvantages of Class A foams include cost of equipment and agent, additional possibilities of

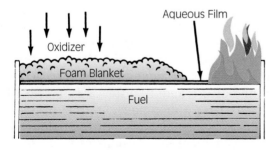

Figure 11-31 AFFF applied on Class B fuel. Note the film barrier on the surface.

NOZZLES, FIRE STREAMS, AND FOAM ■ 297

Figure 11-32 Firefighters applying Class A foam in wildland–urban interface. *(Photo courtesy of Ansul, Inc.)*

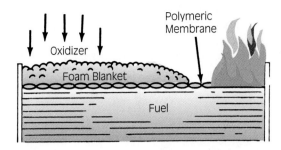

Figure 11-33 Polar solvent or alcohol-type foam applied on Class B polar solvent fuel. Note the polymeric film barrier on the surface.

equipment failure, possible effects on the environment and fire investigation laboratory tests.[5] Other potential problems are more difficult salvage operations and a customer service problem of explaining the milky residue on furniture and possessions to the homeowner.

Class B

Class B fuels include flammable liquids in two categories: hydrocarbons and polar solvents. Because gases should be extinguished by shutting off the flow of the fuel, firefighters do not use foam on them.

Hydrocarbons and Polar Solvents

Hydrocarbons cover a wide range of substances in forms from gaseous to liquid to semisolid and solid. Common examples are heating oil, diesel fuel, gasoline, kerosene, petroleum jelly, paraffin, and asphalt. These fuels do not mix with water; they are not **miscible** or water soluble and the best method to extinguish large quantities is foam. These foams work by cooling and smothering to extinguish the fire and provide a vapor barrier.

Polar solvents mix with water and this ability to mix with water causes either a breakdown of the foam or mixing of flammable vapors into the bubbles of ordinary foams. These liquids include alcohols, both methyl and ethyl alcohol, lacquer thinners, acetone, ketones, acrylonitrile, and acetates. Perhaps the most commonly used is the ethyl alcohol, which when added to water with some flavoring makes alcoholic beverages such as whiskey. Many gasoline blends today have polar solvents in solution and require alcohol-resistant foams for effective extinguishment. To prevent the breakdown of ordinary foams, special foams called polar solvent-type, alcohol-resistant concentrate, or alcohol-type foams have been developed. These polar solvent-type foams create a **polymeric barrier** that separates the polar solvent from the liquid of the foam, **Figure 11-33**.

APPLICATION OF FOAM

Foam is a mixture that requires a device to proportion, meter, or mix the foam concentrate into the water. Air must then be added to the foam solution to complete the process. There are several ways of adding air to the foam concentrate. Concentrations are usually expressed as a percentage of foam concentration to water in the solution. For example, a 3 percent concentration would have three parts of foam concentrate to every ninety-seven parts of water. In premixed systems, the foam concentrate is added directly to the tank, and the solution is pumped as the tank is drained. The problems are the resulting limited supply, the size of the tank, and the possible damage caused to the pump and valves by the foam solution.

One common proportioner is an **eductor, Figure 11-34,** which works on the **venturi principle**. In a venturi eductor, water enters its inlet which is then reduced down, causing the water to move faster through the smaller opening. As the speed increases, the pressure drops and creates suction. At the point of suction, the foam concentrate pick-up tube is attached and the concentrate is drawn up

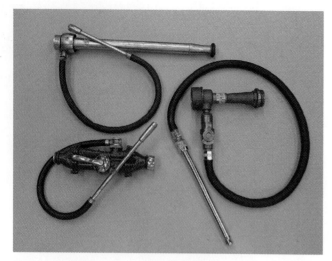

Figure 11-34 From right to top left clockwise: an in-line eductor, bypass eductor, and a foam eductor nozzle.

the tube to mix with the water, **Figure 11-35**. **Figure 11-36** shows how to connect and operate an in-line eductor.

Several types of eductors can be permanently piped into or added to a hoseline. One that is always piped through the venturi is an **in-line eductor,** and one that has a separate waterway and valve to allow plain water to pass by the venturi is called a **bypass eductor**, **Figure 11-34** and **Figure 11-37**.

Eductors must have the proper gpm flow, have the correct pressure, be kept clean, and not have any back-pressure situations such as hose kinks, too much hose, a partially opened nozzle, or too much elevation to work properly. Firefighters must ensure that the proper setup is used to match the eductor, nozzle, and hoseline to give this correct flow. The eductor and nozzle must be clean of dried foam concentrate and other debris. Things as simple as a few pebbles in the nozzle baffle or kinks in the hose can reduce the flow and prevent foam pickup. The eductor must be set at the correct percentage for the concentrate.

Practice and maintenance are often keys to good foam operations. An around-the-pump proportioner uses an eductor between the intake and discharge side of the pump. The discharged water pressure is pumped through the eductor to siphon the concentrate into the intake side of the pump and through the pump. This system allows the pump to operate at full flow, but may cause damage to the pump and valves without proper backflushing of the entire pump. Balanced pressure systems use one of two

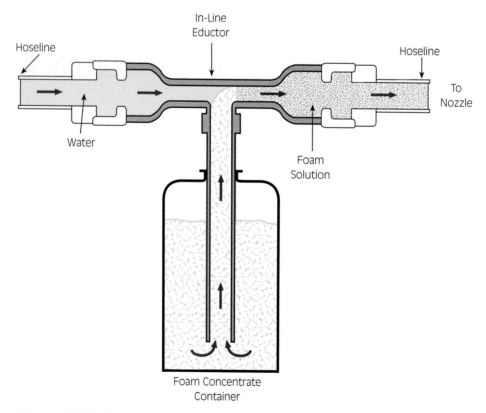

Figure 11-35 Foam eductor using the venturi principle.

Figure 11-36 The in-line eductor is connected either directly to the pump discharge or to a supply hoseline. Eductors require a specific amount of hose between the eductor and the nozzle, plus the gallonage of the nozzle must match the eductor. The eductor tube is placed into the can of concentrate, and any necessary metering adjustments are made. The eductor only operates when water is flowing.

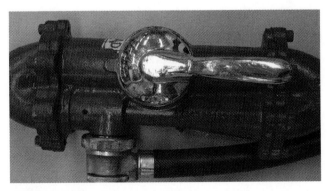

Figure 11-37 Close-up of bypass eductor bypass valve.

foam pumping methods, in which the foam is pumped under pressure into a metering chamber that balances the pressure of the concentrate and water controlling the flow of the foam solution, **Figure 11-38**.

In **compressed air foam systems (CAFS)** or dual-injection systems, the concentrate is in a separate foam tank and a foam pump pumps the concentrate directly into the hoseline, which is metered by a flow-metered microprocessor, **Figure 11-39**. After the foam solution is created, an air compressor line injects air into the hoseline to create a light and

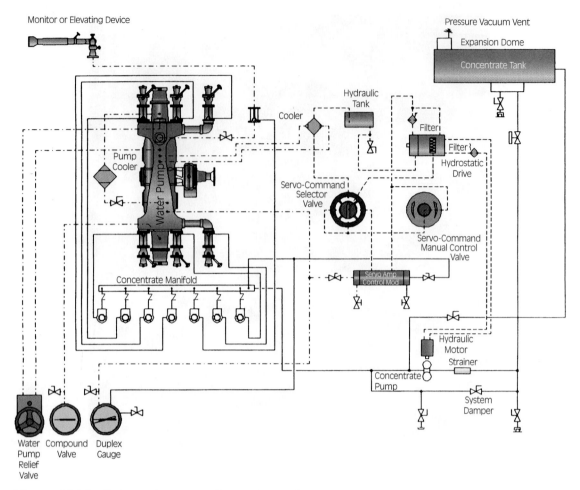

Figure 11-38 Balanced pressure demand–type foam proportioner system.

fluffy foam, **Figure 11-40**. These systems are used for Class A foams.

To finish making the foam, air must be added to the foam solution, normally at the nozzle. The various types of foam nozzles, regular fog nozzles, and foam generators have the ability to aspirate various quantities of air into the foam solution. The goal is to get the correct expansion ratio, that is, amount of air added to the solution. The expansion ratio is expressed as the volume of air added to the volume of solution. For example, 10:1 refers to 10 volumes of air added to 1 volume of solution. Foam's expansion ratios are in three ranges: low, medium, and high. Low expansion foam has an air to solution ratio of up to 20 to 1, or as much as 20 ft³ (0.56 m³) of finished foam for each cubic foot (0.028 m³) of foam solution. Medium expansion foam has ratios ranging from 20:1 to 200:1, and high expansion foam is rated from 200:1 and higher (the upper limit is about 1,000:1). The creation of medium and high expansion foams requires a special foam generator. This special unit has a nozzle that sprays the foam solution onto a screen or mesh and a fan that pushes large volumes of air to create millions of bubbles. These foam generators may be gas, electric, hydraulic, or water powered, **Figure 11-41** and **Figure 11-42**.

Fog Nozzles versus Foam Nozzles

Originally, foam making required a special foam nozzle to properly aspirate the air into the foam solution. Some of these foam nozzles incorporated the eductor into the nozzle, combining all of the foam-making steps. Some of these combined eductor nozzles are still in use but because they require having the foam concentrate containers and the personnel to move the containers at the nozzle, these devices are not very popular. Foam nozzles are designed today to aspirate the proper amounts of air and apply the foam to the fuel. They do this by having air vents or ports built into the nozzle, and the movement of the foam solution past these ports draws the air into the solution, and it mixes both

NOZZLES, FIRE STREAMS, AND FOAM 301

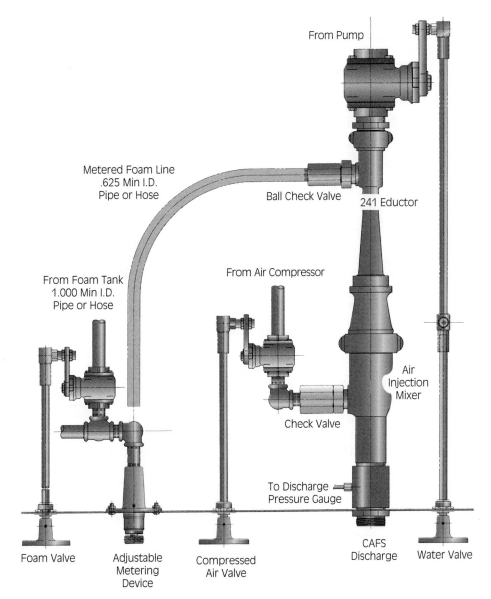

Figure 11-39 Typical compressed air foam system (CAFS).

Figure 11-40 Foam from CAFS being applied. *(Photo courtesy of Ansul, Inc.)*

Figure 11-41 Handline-operated medium expansion foam generator. *(Photo courtesy of Task Force Tips)*

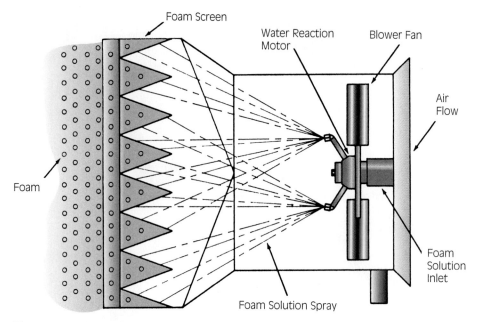

Figure 11-42 High expansion foam generator.

before and after the tip, **Figure 11-43**. Foam nozzles are designed for low and medium expansion foams usually having expansion ratios of 8:1 to 20:1 in the low range and up to 50:1 in the medium range. Foam nozzles are recommended with protein and fluoroprotein foams.

When AFFF was introduced, one of its additional advantages was that it did not require a special foam nozzle for application. Fog nozzles typically used by the fire department could be used. This helped reduce the costs of purchasing an effective foam system, and many departments began carrying foam concentrate and foam eductors on all pumpers. While the cost is lower, so is the expansion ratio, with 8:1 being at the high end. To increase this expansion efficiency while keeping costs lower, some manufacturers now have clip-on or snap-on foam nozzle adapters that attach to the fog nozzle and make it a foam nozzle, **Figure 11-44**. These are effective in increasing the expansion ratio and are almost as good as a specially designed foam nozzle. CAFSs, by preinduction of air, are effective with a fog nozzle and even with solid tip nozzles.

Foam from nozzles is applied using one of three techniques. The first is the *bank-in technique,* in which the foam strikes the ground before the fire and rolls into the fire, **Figure 11-45**. The second is the *bank-back* or *bounce-off technique,* in which the foam is banked off a wall or other object and rolls back into the fire, **Figure 11-46**. The third technique is the *raindown* or *snowflake technique,* in which the foam is sprayed high into the air over the fire and it floats down onto it, **Figure 11-47**.

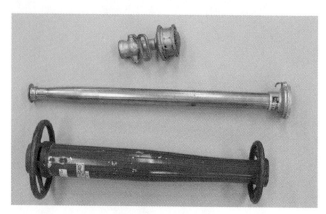

Figure 11-43 Typical foam nozzles.

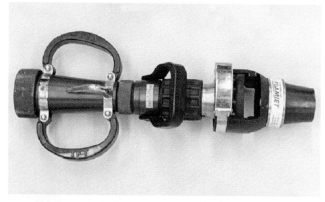

Figure 11-44 Clip-on foam attachment for fog nozzle.

NOZZLES, FIRE STREAMS, AND FOAM   303

Figure 11-45   The bank-in technique of foam application.

Figure 11-46   The bank-back technique of foam application.

Figure 11-47   The raindown technique of foam application.

Lessons Learned

Fire streams are made of water that leaves a nozzle and heads toward the target. The nozzle defines the characteristics of the fire stream. The two main types of nozzles are solid tip and fog, and they need to match the fire conditions and fire department resources. Each type of nozzle has its advantages and disadvantages; firefighters should use the one that will do the best job of suppressing the fire. Proper use of nozzles requires shutoffs and even some special types of nozzles. Fire streams and fire conditions determine how firefighters will attack a fire.

An understanding of fire streams is not possible without understanding the basic hydraulics of moving the water from a source to the fire.

Correct hydraulic calculations result in a good fire stream being delivered in the proper amount, with a good shape. To do this properly requires an understanding of the pressures and the amount of friction loss in the system. Calculations of the flow, friction loss, and pressure are combined as needed by the hoseline(s) and nozzle(s) being used.

When the fuels involved in a fire are not compatible with water, other agents must be used to fight the fire and, commonly, this other agent is foam. Firefighters should know the different kinds of fuels and types of foams available. Foam requires special equipment to create and apply it, and some special application techniques are used.

KEY TERMS

Absolute Pressure The measurement of pressure, including atmospheric pressure. Measured in pounds per square inch absolute.

Application Rate Amount of foam or foam solution needed to extinguish a fire. Usually expressed in gallons per minute per square foot or liters per minute per square meter.

Automatic or **Constant Pressure Nozzle** Nozzle with a spring mechanism built in that reacts to pressure changes and adjusts the flow and resultant reach of the nozzle.

Available Flow Amount of water that can be moved to extinguish the fire. Depends on the water supply, pump(s) and their capabilities, and the size and length of hose.

Bresnan Distributors Has six or nine solid tips or broken stream openings designed to rotate in a circular spray pattern. Used to fight fire in basements or cellars when firefighters cannot make a direct attack on the fire.

Bypass Eductor Eductor with two waterways and a valve that allows plain water to pass by the venturi or through the venturi to create foam solution.

Cellar Nozzles Has four spray nozzles designed to rotate in a circular spray pattern for fighting fires in basements or cellars when firefighters cannot make a direct attack on the fire.

Combination Fire Attack A blend of the direct and indirect fire attack methods, with firefighters applying water to both the fuel and the atmosphere of the room.

Combination Nozzle A spray nozzle that is capable of providing straight stream and spray patterns, which are adjustable or variable by the operator. Most fog nozzles used today are combination nozzles.

Compressed Air Foam Systems (CAFS) A foam system where compressed air is injected into the foam solution prior to entering any hoselines. The fluffy foam created needs no further aspiration of air by the nozzle.

Constant or **Set Volume Nozzle** Nozzle with one set volume at a set pressure. For example, 60 gpm at 100 psi (227 L/min 690 kPa). The only adjustment is the pattern.

Detergent-Type Foams Use synthetic surfactants to break down the surface tension of water to create a foaming blanket.

Direct Fire Attack An attack on the fire made by aiming the flow of water directly at the material on fire.

Discharge Flow Total amount of water flowing from the discharge side of the pump.

Eductor Device that siphons a liquid from a container into a moving stream.

Fire Stream The water or other agent as it leaves the hose and nozzle toward its objective, usually the fire.

Flow The rate or quantity of water delivered, usually measured in gallons per minute or liters per minute (1 gpm = 3.785 L/min).

Fluoroprotein Film-Forming Foam (FFFP) Combines protein with the film-forming fluorinated surfactants of AFFF to improve on the qualities of both types of foam.

Fluoroprotein Foam   Designed as an improved protein foam with a fluorinated surfactant added.

Foam   An aggregate of gas-filled bubbles formed from aqueous solutions of specially formulated concentrated liquid foaming agents.

Fog Nozzle   Delivers either a fixed spray pattern or variable combination of straight stream and spray patterns.

Friction   Caused by the rubbing of materials against each other while in movement and converts or robs some of the movement energy into heat energy.

Friction Loss   Measurement of friction in a system such as a hoseline.

Fuel Resistance   Ability to tolerate the fuel and to avoid being saturated by or picking up the fuel.

Gauge Pressure   Measures pressure without atmospheric pressure. Normally fire department gauges do not measure atmospheric pressure. Gauge pressure is measured in psi or psig.

Head Pressure   Measures the pressure of a column of water in feet (meters). Head pressure gain or loss results when water is being pumped above or below the level of the pump. A head of 2.31 feet (0.7 m) would equal 1 psi (6.895 kPa).

Heat Resistance   The ability of foam to stand up to the heat of the fire or to hot surfaces near the fire.

Hydraulics   The study of fluids at rest and in motion.

Indirect Fire Attack   An attack made on interior fires by applying a fog stream into a closed room or compartment, thus converting the water into steam to extinguish the fire.

In-Line Eductor   Eductor in which the waterway is always piped through a venturi.

Knockdown Speed   Speed with which foam spreads across the surface of a fuel.

Miscible   Having the ability to mix with water.

Needed or **Required Flow**   Estimate of the amount of water required to extinguish a fire in a certain time period. Based on the type and amount of fuel burning.

Nozzle   A tapered or constricted tube used to increase the speed or change the direction of water or other fluids.

Nozzle Flow   The amount or volume of water that a nozzle will provide. Flow is measured in gallons per minute or liters per minute.

Nozzle Pressure   The pressure required to effectively operate a nozzle. Pressure is measured in pounds per square inch or kilopascals.

Nozzle Reach   The distance the water will travel after leaving the nozzle. Reach is a function of the pressure, which is converted to velocity or speed of the water leaving the nozzle.

Nozzle Reaction   The force of nature that makes the nozzle move in the opposite direction of the water flow. The nozzle operator must counteract the thrust exerted by the nozzle to maintain control.

Piercing Nozzles   Originally designed to penetrate the skin of aircraft and now have been modified to pierce through building walls and floors.

Polymeric Barrier   A separation barrier made up of polymer or a chain of molecules linked in a series of long strands. This separates a polar solvent from an ATC foam blanket.

Protein Foam   Made from chemically broken down natural protein materials, such as animal blood, that have metallic salts added for foaming.

Pump Operator   A generic term to describe the person responsible for operating a fire apparatus pump. Other commonly used titles include motor pump operator, engineer, technician, chauffeur, and driver/operator.

Solid Stream Nozzles   Type of nozzle that delivers an unbroken or solid stream of water to the fire. Also called solid tip, straight bore, or smooth bore.

Straight Stream   A nozzle pattern that creates a hollow stream, similar in shape to the solid stream pattern, but the straight stream pattern must pass around the baffle of the nozzle. Newer fog nozzle designs, especially the automatic nozzles, only have this hollow effect from the tip on and, hence, create a solid stream with good reach and penetration abilities, some better than solid stream nozzles.

Stream Shape   The arrangement or configuration of the water or other agent droplets as they leave the nozzle.

Vacuum (Negative) Pressure   The measurement of the pressure less than atmospheric pressure, which is usually read in inches of mercury (in. Hg or mm Hg) on a compound gauge.

Vapor Suppression   Ability to contain or control the production of fuel vapors.

Variable, Adjustable, or **Selectable Gallonage Nozzle**   Nozzle that allows the nozzleperson to select the flow, with usually two or three choices, and the pattern.

Velocity Pressure   The forward pressure of water as it leaves an opening.

Venturi Principle A process that creates a low-pressure area in the induction chamber of the eductor and allows the foam concentrate to be drawn into and mixed with the water stream.

Water Curtain Nozzle Designed to spray water to protect exposures against heat by wetting the exposure's surface.

REVIEW QUESTIONS

1. Define a fire stream.
2. Explain what a nozzle is and how it works.
3. What is the amount or volume of water leaving the nozzle called?
4. Explain the term *nozzle reaction*.
5. Explain the difference between a solid tip nozzle pattern and a straight stream nozzle pattern.
6. A selectable or variable nozzle allows the operator to adjust what feature?
7. What is the purpose of a Bresnan distributor?
8. Why should the pattern of a water curtain be applied onto a solid object?
9. What is the most common type of shutoff device for a nozzle and how does it work?
10. Explain the difference between a direct and indirect attack.
11. What is the difference between atmospheric and gauge pressure?
12. Identify common nozzle pressures for solid tip and fog nozzles.
13. To reduce the friction loss in hose, what two things can be done?
14. Define foam and name the three major components.
15. How does AFFF work on a flammable liquid fire?
16. If a foam concentration was 6 percent, how many gallons of concentrate would be added to how many gallons of water to equal 100 gallons?

Endnotes

1. A 1½-inch (38-mm) hoseline is the minimum size hoseline for structural firefighting.
2. Although the hoseline can be operated by one person, safety regulations require operation with a second team member.
3. Sturtevant, Thomas, *Introduction to Fire Pump Operations,* pp. 228–232. Delmar Learning, a part of the Thomson Corporation, Clifton Park, NY, 1997.
4. Cote, Arthur and John Linville, *Fire Protection Handbook,* 18th ed., pp. 6-349. National Fire Protection Association, Quincy, MA, 1997.
5. Stern, Jeff and J. Gordon Routley, *Class A Foam for Structural Firefighting.* U.S. Fire Administration, Emmittsburg, MD, December 1996.

Additional Resources

Gagnon, Robert M., *Design of Special Hazard and Fire Alarm Systems.* Delmar Learning, a part of the Thomson Corporation, Clifton Park, NY, 1998.

Klinoff, Robert, *Introduction to Fire Protection.* Delmar Learning, a part of the Thomson Corporation, Clifton Park, NY, 1997.

Sturtevant, Thomas, *Introduction to Fire Pump Operations.* Delmar Learning, a part of the Thomson Corporation, Clifton Park, NY, 1997.

Crapo, William F., *Hydraulics for Firefighting,* Delmar Learning, a part of the Thomson Corporation, Clifton Park, NY, 2002.

NFPA 11: Standard for Low, Medium, and High Expansion Foam Systems. National Fire Protection Association, Quincy, MA, 2002.

NFPA 1961: Standard on Fire Hose. National Fire Protection Association, Quincy, MA, 2002.

NFPA 1962: Standard for Care, Use and Testing of Fire Hose Including Couplings and Nozzles. National Fire Protection Association, Quincy, MA, 1998.

NFPA 1964: Standard for Spray Nozzles (Shutoff and Tip). National Fire Protection Association, Quincy, MA, 1998.

Stern, Jeff and J. Gordon Routley, *Class A Foam for Structural Firefighting.* U.S. Fire Administration, Emmittsburg, MD, December 1996.

CHAPTER 12

PROTECTIVE SYSTEMS

Ric Koonce, J. Sargeant Reynolds Community College

OUTLINE

- Objectives
- Introduction
- Detection Systems
- Sprinkler Systems
- Sprinklers and Life Safety
- Sprinkler Head Design and Operation
- Types of Sprinkler Systems
- Sprinkler System Connections and Piping
- Control Devices for Sprinkler Systems
- Returning Sprinkler Systems to Service
- Standpipe Classifications
- Standpipe System Connections and Piping
- Alarms for Standpipes and Sprinklers
- Other Protective Systems
- Fire Department Operations with Protective Systems
- Lessons Learned
- Key Terms
- Review Questions
- Endnotes
- Additional Resources

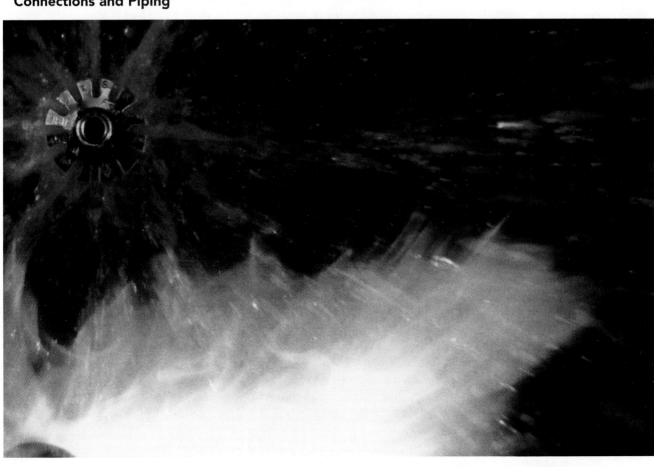

STREET STORY

One winter night in the heart of downtown Philadelphia an entire fire protection system failed; a trio of firefighters was lost, a thirty-eight-story skyscraper was destroyed, and the confidence of those dependent on fixed fire protection systems was shaken. It was the devastating No. 1 Meridian Plaza fire that changed my view on active and passive protective devices.

Upon arrival of the first-alarm companies, fire was venting from the twenty-second floor. At the onset it was a controllable fire. Then the fire protection systems began to quickly fail. Initially, it was the electrical systems. The primary and secondary wires "shorted" and placed the entire high-rise office building into total darkness. With the elevator stalled, access to the upper floors was limited to climbing the three fire towers. After firefighters hauled their equipment to the twentieth floor they discovered the pressure-reducing valves on the standpipe system were improperly set. Generating an effective fire streams system was impossible. Later, without an operable HVAC the core temperature in the building began to escalate. Pipe chases for the bathroom plumbing were never sealed. These "poke-holes" permitted the toxic smoke to spread in all directions. Smoke began to bank down, and the ventilation group became disoriented and lost and perished on a smoke-charged upper floor. With conditions rapidly deteriorating, a partially sprinklered structure, crews working with only hand lights, and practically no way to ventilate the building, the odds did not favor the men and women of the Philadelphia Fire Department.

After twelve alarms were struck, bringing more than 300 firefighters to the scene, and nineteen long hours, the blaze was contained to eight floors. In the end, ten sprinkler heads stopped the upward progress of the flames on the thirtieth floor. The lessons from No. 1 Meridian Plaza are plentiful, but one undisputed fact is that fire protection systems must be properly designed, installed, and maintained.

Recently, at the Canarsia Senior Citizen Center on Vandalla Avenue in New York City, a self-closing steel fire door remained ajar and a sprinkler system did not operate in a tenth-floor common hallway. When heavy winds pushed the flames like a blowtorch from the upper floor apartment into the hallway, a search and rescue group was engulfed in the fireball. Tragically, again three firefighters were killed because essential components of the fire protection system did not properly function.

Daily across this nation, emergency responders are witness to incidents that reinforce the need for knowledge of these basic systems. Somewhere in the nation this week a rate-of-rise alarm in a commercial building warned of an incipient fire, a sprinkler head activated in a factory and extinguished a minor fire before it spread to a manufacturing process, a closed fire-rated door in a hospital contained the smoke to a particular room before a massive evacuation was needed, a damper over a kitchen grill closed to prevent the involvement of a metal duct that crisscrossed vital structural supports in a restaurant, and the annunciator panel in an enclosed shopping mall pinpointed the exact location of a potential problem. Unfortunately, when these fire protection systems malfunction, the lives and safety of the occupants and emergency responders are at a much higher risk.

What is the price of success? In the fire rescue service, personnel safety and operational efficiency are surely two of the keys; often both hinge on fire protection systems. Firefighters often see firsthand the economic value of a properly maintained and tested fire protection system. I know both the economic and emotional value of these auxiliary appliances. I believe that an in-depth knowledge of sprinkler and standpipe systems will help to "save our own."

—Street Story by William Shouldis, Deputy Chief, Philadelphia Fire Department, Philadelphia, Pennsylvania

OBJECTIVES

After completing this chapter, the reader should be able to:

- Identify the value of protective systems in protecting life and property.
- Identify and explain the operation of the various types of detection devices.
- Explain and recognize the types of sprinkler heads and how they operate.
- Identify the various types of sprinkler systems and the components of each type.
- Identify the piping arrangements of sprinkler systems and connections.
- Demonstrate how to connect to a fire department connection.
- Identify control valves for systems and explain their operation.
- Explain the methods used to return a sprinkler system to service.
- Demonstrate techniques for stopping a flowing sprinkler head.
- Identify standpipe classes and types of systems.
- Identify piping and connections for standpipe systems.
- Demonstrate how to connect supply and attack lines to standpipe connections.
- Identify alarm systems for protective systems.
- Explain fire department procedures at protective properties.
- Identify other protective systems, their components, and their benefits and hazards.

INTRODUCTION

Protective systems help protect lives and property by detecting and/or suppressing fires. Detection systems detect the presence of a fire and alert building occupants and/or the fire department and may also activate a suppression system. Suppression systems are devices that help firefighters in controlling fires. Some systems are designed to automatically detect and suppress a fire, while others assist firefighters and occupants in putting out the fire. They are also called **auxiliary appliances**. Regardless of the term used, firefighters must know about these systems and how to use them. Great efforts have been expended to require these systems in building and fire prevention codes. To builders, owners, and occupants, they are expensive to purchase and maintain, and the expectation is that the fire department will use them correctly.

Note Remember that these devices can make the job of firefighting safer and easier.

Detection systems are varied. Some require people to detect the fire and manually activate an alarm. Others are highly complex systems that can detect a fire almost at its ignition and sound an alarm. They may also activate suppression systems.

The two main suppression systems are sprinklers and standpipes. These are devices that allow firefighting's favorite extinguishing agent, water, to be used. When the situation dictates that another agent may be more effective, other types of protective systems are required. Some use the agents found in portable fire extinguishers to cover room or building size areas. These systems include dry and wet chemicals, inerting gases, and halogenated agents.

Detection systems, suppression systems, and alarm systems (covered briefly in Chapter 3) can all be tied together to form a protective ensemble for building occupants. This chapter describes individual detection and suppression components as well as the interface of each. It also covers fire department operations using protective systems.

DETECTION SYSTEMS

Detection systems are designed to notify people of a fire or other problem. Simple detection systems can actually be viewed as a warning system requiring a person to recognize a danger. More complex systems use a series of devices to automatically detect a fire or event and initiate a warning alarm. Fire detection and warning can also come from the initiation of a suppression system. For example, a sprinkler activates, tripping a water flow alarm that can notify occupants. Regardless of how complex or simple, detection systems are designed to notify *people* that a potentially life-endangering event is happening. By understanding the basic principles of operation of various detection systems, firefighters more effectively utilize these systems to perform their job. Communities *expect* firefighters to understand these systems.

People or Manual Systems

People can alert other building occupants and can call the fire department after discovering a fire. Some buildings are equipped with a manual fire alarm system that requires a person to discover the fire and then pull a lever or push a button. A restaurant cooking area or a laboratory hood with a local application system that sounds an alarm and activates the suppression system can have a manual switch for people to operate. Street box systems are also manual fire alarm systems.

Manual systems suffer from two typical problems. The first is that if they are to work, a person must be present, awake, and alert to discover the fire and then activate the system. Unfortunately, sleeping people and pets make poor fire detectors, because all of their detection senses are also asleep. The second problem is that many manual fire detection systems are local only, meaning that they will alert any building occupants but will not summon any additional help. A public education program and proper signs at the pull station can help teach people about the need to call the fire department.

Heat Detectors

Heat detectors operate by detecting the heat of a fire at a fixed temperature or as the rising temperature builds at a rapid rate. These can be used as part of a suppression system such as a sprinkler system, can operate a fire protection device such as closing a door, or are merely detection devices that operate an alarm. Heat detectors are slow to detect fire and hence should not be used for life safety. They are, however, inexpensive and have a low rate of false alarms. Heat detectors can be of the spot type, which covers just one area, or of the line type, which covers a larger area.

The *rate-of-rise heat detector* measures temperature increases above a predetermined rate. The pneumatic or air-sensitive device uses a diaphragm with a small hole or orifice that acts as a relief valve for slow temperature increases. When a fast temperature increase occurs, the diaphragm is pushed toward a contact point and the alarm is activated, **Figure 12-1**.

The *fixed temperature heat detector* is usually electrically operated with a bimetallic strip of two metals that expand at different rates, eventually bending to touch a contact point and complete the alarm circuit, **Figure 12-2**. Another fixed temperature type uses a metal element similar to solder that melts at a fixed temperature and breaks a circuit or

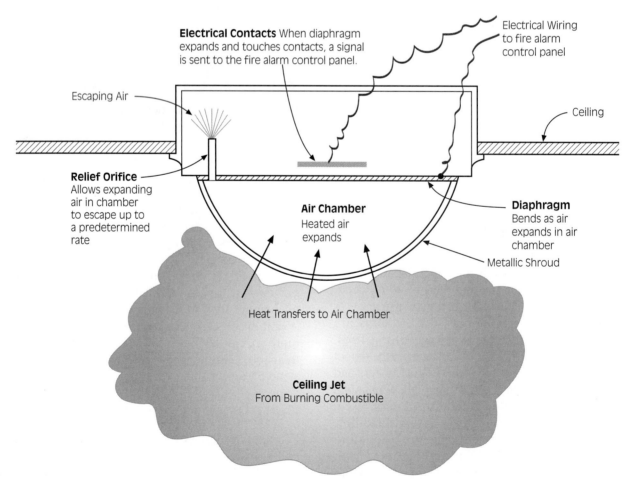

Figure 12-1 Rate-of-rise heat detector.

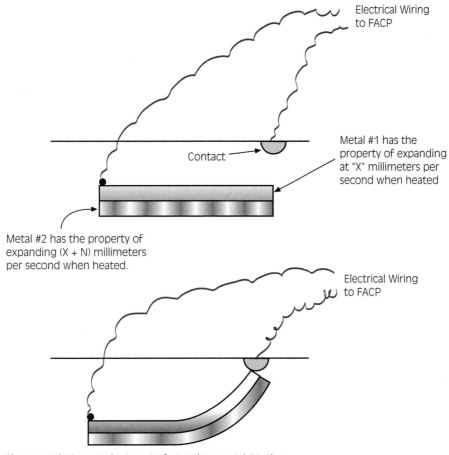

Figure 12-2 Fixed temperature heat detector.

opens a cap. Line-type detectors can be either pneumatic or electric, **Figure 12-3**.

Smoke Detectors

Smoke and toxic gases are the leading killers of people in the home. Smoke detectors have greatly improved residential life safety by quickly detecting and alerting building occupants to the harmful effects of fire. Smoke detectors are the most prevalent automatic detection system. Even with the abundance of smoke detectors installed in homes across North America, many occupants do not understand how they work—or how to fix problems. It is important that firefighters understand the operation of smoke detectors. It is *essential* that they can describe problems and suggest solutions to the building owner or occupant.

Smoke detectors are hard or permanently wired, battery operated, or a combination of hard wiring with a battery backup. Smoke detectors work primarily on two principles: ionization and photoelectric.

Ionization detectors are the most common type of smoke detector—the kind installed in most homes and apartments since the early 1980s. These detectors use a radioactive element that emits ions into a chamber. The positive and negative ions are measured on an electrically charged

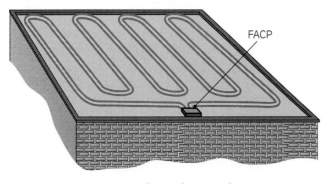

Figure 12-3 Line-type heat detector layout.

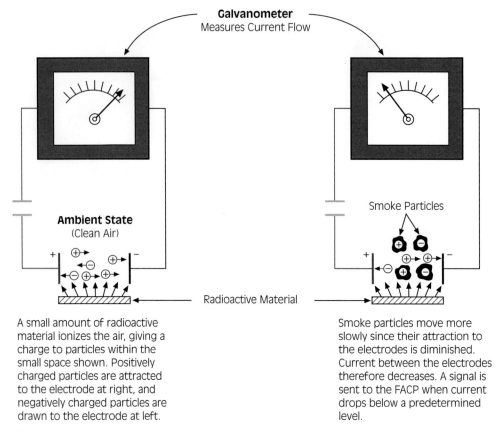

Figure 12-4 Ionization-type smoke detector.

electrode. When smoke enters the chamber, the flow of ions to the electrode changes and an alarm is activated, **Figure 12-4**. Ionization detectors are spot-type devices that are inexpensive and are very quick to react, especially to small smoke particles from flaming fires. Ionization smoke detectors can detect invisible smoke—certain fire gases that are colorless can disrupt ion flow within the chamber. Ionization smoke detectors are quite reliable but can be activated by cooking, steam (even from a hot shower), electrostatic discharge (from a thunderstorm), and dust buildup. Some people may classify these activations as false alarms. In actuality, these activations indicate that the detector is working as designed. Most ionization detectors will reset themselves once the smoke has cleared. Sometimes it is necessary to "blow out" the detector to clear the internal chamber of smoke products and dust. Ionization detectors that are activated by an electrostatic discharge may not reset; the detector will likely need to be replaced. Smoke detectors that "chirp" typically indicate that the battery is low.

Photoelectric detectors use two different methods to detect the smoke of a fire and can be spot or line detectors. The first is the *light obscuration detector*, which has a light beam aimed at a light sensor. When the smoke particles obscure or block the light beam, the light sensor notes the loss of light and activates the alarm, **Figure 12-5**. The *light-scattering detector* operates by having a light beam aimed at the end of the chamber with a light sensor in an angled-off chamber, **Figure 12-6**. When the smoke enters the chamber, the smoke particles scatter the light and the scattered light strikes the light sensor, activating the alarm. Photoelectric detectors are activated by visible smoke. They make a better smoke detector for areas where some smoke may be developed but is normally dissipated visually, such as in cooking areas and where fuel-powered equipment is working. Like ionization detectors, dust and steam can cause activation of the photoelectric detector. If the smoke causing activation of the photoelectric detector is especially oily (like from a grease fire), the detector may not reset once the smoke clears. A film has likely obscured the light sensor.

PROTECTIVE SYSTEMS ■ 313

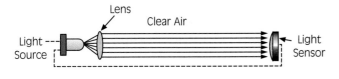

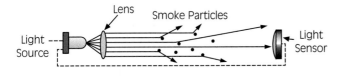

Figure 12-5 Light obscuration photoelectric smoke detector.

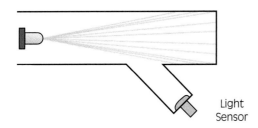

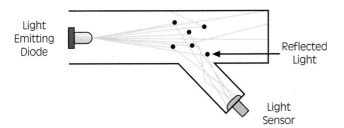

Figure 12-6 Light-scattering photoelectric smoke detector.

Gas Detectors

Gas-sensing detectors are designed to detect the presence of a certain gas or gases prior to the gases reaching a concentration that can cause danger. A flammable gas detector would detect the presence of the gas, such as a petroleum product, before it reached its ignitable concentration. Other gases would be detected prior to reaching a concentration that could cause death or injury, such as carbon monoxide. Gas detectors can be permanently mounted devices or portable units.

Carbon monoxide (CO) detectors for the home are becoming popular and are even required in some communities. This popularity has placed an additional service load on fire departments. Like smoke detectors, CO detectors have saved many lives. CO detectors function using several different methods. Consulting manufacturers' literature is perhaps the best approach to understand how each operates and how each needs to be reset. Most CO detectors, however, provide an early warning alarm (like an intermittent beep) when low-level CO is detected. Some CO detectors have an actual LED display that indicates the level (parts per million) of CO being detected. Often, persons reporting *smoke* detector activation may actually be hearing a CO detector activation—and vice versa. CO is odorless and colorless. Even though smoke is not visible, dangerous levels of CO may be present. A firefighter must be careful not to get caught investigating a smoke alarm activation only to find a CO detector has alarmed to dangerous CO levels. All detector activations should be treated as a worst-case scenario. Many fire departments carry gas detectors to check for flammable gases, carbon monoxide and other hazardous materials, and the oxygen content of the air.

Flame Detectors

Flame detection devices that detect the flames or lightwaves of the fire are mainly of three types: ultraviolet (UV), infrared (IR), or a combined ultraviolet-infrared (UV-IR) detector. These detectors are extremely rapid and are to protect petroleum and chemical facilities where fast and high temperature fire can occur.

Ultraviolet flame detectors detect the radiation of high-intensity flaming fires by detecting the lightwaves emitted in the UV spectrum (below 4,000 angstroms), which is below the visual light wavelengths. The UV detectors must have proper shielding from light sources, especially sunlight and welding. Dust or moisture can also fog the lens of the UV detector and cause detection problems.

Infrared flame detectors detect the infrared radiation (8,500 to 12,000 angstroms) of a fire using a photocell, which is above the visual light wavelengths. IR detectors are effective in rapidly spreading fires, but must be screened from sunlight and have a limited distance range. Beyond that range, unwanted signals will occur.

Combined *ultraviolet-infrared flame detectors* are used to rapidly detect the flames of a fire but without the false alarm problems of either a UV or IR detector. This is because the detector requires that both the UV and IR components be activated before an alarm is sent. Because the weakness of one does not affect the other, the double positive usually indicates a fire.

SPRINKLER SYSTEMS

Sprinkler systems are designed to automatically distribute water through sprinklers that are placed at set intervals on a system of piping, usually in the ceiling area, to extinguish or control the spread of fires. Most sprinkler heads detect the heat of a fire and begin to apply water directly over the source of the heat. Sprinkler heads, unless deluge-type heads, are heat-sensitive devices that react to a fixed temperature and disperse water in a specific pattern and quantity over a set area. Sprinkler systems are highly effective. In fact, some people describe the benefit of sprinklers as similar to having a firefighter constantly on duty in the protected building. The NFPA publishes standards for installation of sprinkler systems and their inspection and maintenance.[1]

SPRINKLERS AND LIFE SAFETY

Sprinkler systems were originally designed in the late 1800s to protect property, especially businesses and factories, from total loss from fires. They are almost 100 percent effective. The times when they do not work properly usually involve human action such as improper maintenance or turning off of the water supply. In the early 1900s, the idea that sprinklers might be able to save lives was beginning to take shape.

Firefighter Fact New York City's tragic Triangle Shirtwaist Factory fire in 1911 killed 146, mostly women, many of whom jumped to their deaths trying to avoid being burned. This fire was the beginning of the concept of adding sprinklers to buildings for **life safety** and providing occupational safety for workers.

America's fire loss statistics show that most structure fires, most fire damage, and, of greatest importance, most fire injuries and fatalities occur in residential properties. Yet the American home is one of the least protected properties. It is usually built of wood and other combustible materials. Until the mid-1970s most American homes did not have smoke detectors, and only a few non-highrise residential units had sprinklers. Recognizing the potential benefit of these types of systems, the U.S. Fire Administration, National Fire Protection Association, International Association of Fire Chiefs, and Factory Mutual, among many others, led the research efforts to develop faster responding and less expensive residential sprinkler systems. The first standard on residential sprinkler systems was developed in 1975.[2] Some cities and counties now have mandatory sprinkler requirements in all newly constructed residential structures, and many states require them in multifamily dwellings. There is still, however, a long way to go to place them in all new residential buildings. The American heritage of protecting personal property rights often allows citizens the right to die in their own homes, the least protected and inspected of any occupancy classes.

Note Firefighters must realize that protective systems are designed for specific purposes and will not completely defend a property unless they are designed with that goal in mind.

Residential sprinklers are designed for life safety and not necessarily to protect property. While a residential sprinkler system may completely extinguish a fire in a protected area, unprotected areas are allowed in the design of such systems. Fires that may occur in these unprotected areas can gain great headway or move to other unprotected areas and the system may not be able to prevent a total loss. Fires have occurred in apartment buildings, motels, and hotels protected by a life safety sprinkler system, yet resulting in a total loss of the property. When looking at these fires and rating the effectiveness of the protection, the question is, Were there any occupant fatalities or injuries? If the answer is no, the system did what it was designed to do.

Sprinkler systems are also not totally effective in life safety in some other types of fires. These include slow burning and smothering fires that produce fatal quantities of smoke without generating sufficient heat to activate the system, flash fires that rapidly engulf a person, or a fire in the immediate vicinity of the victim where the fire inflicts the injuries prior to system activation. Another situation is one in which the victims are disabled or unable to remove themselves from the fire area. Protecting against all situations is impossible, but residential sprinkler systems combined with an effective smoke detection system would allow maximum life safety in the home.

SPRINKLER HEAD DESIGN AND OPERATION

Sprinklers or sprinkler heads are the key components of the system, **Figure 12-7**. Most important are the heat-sensitive parts that usually detect heat and apply water to the fire. (Note the word *usually*. Some types of sprinkler systems such as a deluge-type system have separate detection devices but the head is still the applicator. See deluge system discussion later). Sprinkler heads must be appropriate in design and performance, orientation, application or environment, and temperature for the type of property to be protected.

Sprinkler heads come in many designs and this affects their performance. One main difference in sprinkler heads is the new and old-style sprinkler head, **Figure 12-8**. Sprinklers designed and manufactured up to the early 1950s are called *old-style sprinklers*. They deflected the majority of the water toward the ceiling to further break up its pattern. The effect was that the water was concentrated in a smaller area.

The new style or standard sprinklers have a much more even flow across the coverage area and do not bounce the water off the ceiling,

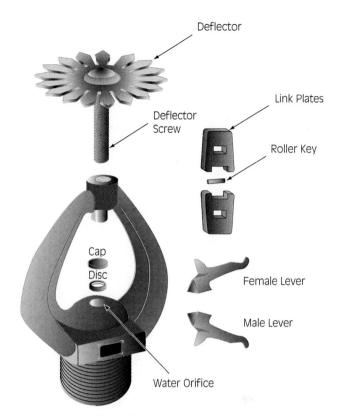

Figure 12-7 Sprinkler head parts.

Figure 12-8. Standard sprinklers are marked with *SSU* (Standard Sprinkler Upright) or *SSP* (Standard Sprinkler Pendent) on the deflector. Orientation means up, down, or sideways. Upright sprinklers have the head vertically above the piping with the deflector at the top, while pendents have the head suspended below the piping with the deflector at the lowest point, **Figure 12-9**. The design of sidewall sprinklers allows them to be placed near the wall and, although the deflectors look bent, they provide the correct pattern. Sidewall heads may be pendent, upright, or horizontal. Each must be properly positioned.

Application and environment may require corrosion-resistant heads, special dry head with extension piping, rack storage head, or decorative head. The orifice or the size of the water opening varies from ¼ to ¾ inch (6.4 to 19 mm) depending on the occupancy protected. Sizes other than ½ and $^{17}/_{32}$ inch (12.7 and 13.5 mm) are noted on the sprinkler frame.[3] Temperature ratings range from ordinary at 135°F (57°C) to ultra high at 625°F (343°C) with the most common temperature being 165°F (74°C),[4] **Table 12-1**.

The operation of a sprinkler head begins with the fire's heat raising the temperature of the sprinkler

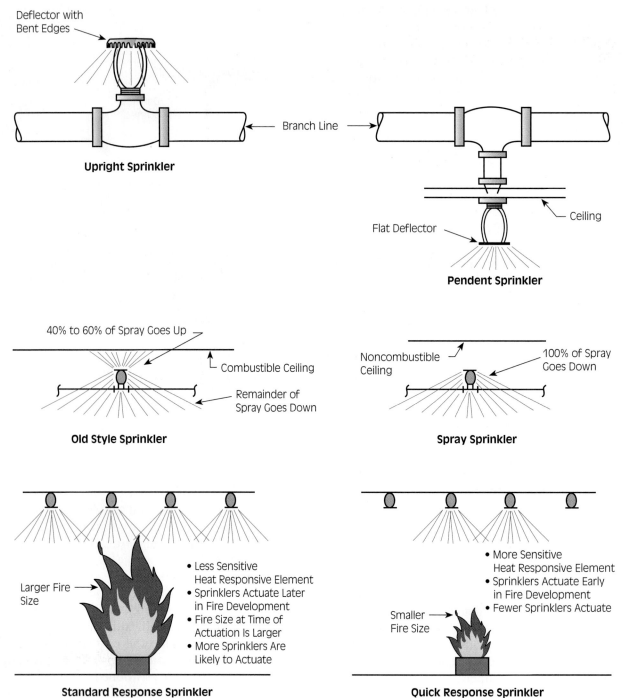

Figure 12-8 Sprinkler differences.

head and its fusible element to its fusing point. The fusible elements can be of three types. The first is a fusible link that has a metal or solder that melts at a fixed temperature. The second is a bulb filled with a liquid, leaving only a small air bubble, which expands and bursts at a fixed temperature. The third type uses a chemical pellet that melts at a fixed temperature. The fusible element melts or bursts and the levers holding the cap then fall out. The cap pops out and the water flows, striking the deflector and spraying into the designed pattern, **Figure 12-10**. Another newer innovation is an on/off head that operates at a fixed temperature, but when the fire's temperature drops, it operates a spring to close the waterway and can reopen if the temperature rises again.

PROTECTIVE SYSTEMS 317

(A)

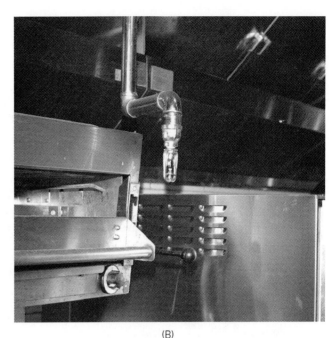

(B)

(C)

Figure 12-9 Types of sprinklers: (A) upright, (B) pendent (without guard), and (C) pendent sidewall (with guard). *(Photos by Fred Schall)*

Temperature Ratings, Classifications, and Color Coding

TEMPERATURE RATING °F	°C	TEMPERATURE CLASSIFICATION	COLOR CODING	GLASS BULB COLORS
135–170	57–77	Ordinary	Uncolored or black	Orange or red
175–225	79–107	Intermediate	White	Yellow or green
250–300	121–149	High	Blue	Blue
325–375	163–191	Extra high	Red	Purple
400–475	204–246	Very extra high	Green	Black
500–575	260–302	Ultra high	Orange	Black
625	343	Ultra high	Orange	Black

Note: A portion of Table 6-10c is reprinted with permission from *Fire Protection Handbook*, 18th edition. Copyright © 1997, National Fire Protection Association, Quincy, MA 02269. This reprinted material is not the complete and official position of the National Fire Protection Association on the referenced subject, which is represented only by the complete table in its entirety.

TABLE 12-1

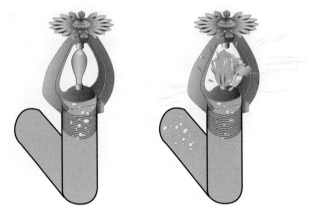

Figure 12-10 Sprinkler head closed and opened or operating.

TYPES OF SPRINKLER SYSTEMS

The four major types of sprinkler systems and the residential system are discussed in the following subsections. Specialty sprinkler systems include some combination-type sprinkler systems and systems that cannot meet the standards for some reasons. They may have an inadequate water source or supply or may be a partial or outside system. Even if a system does not completely meet a standard, it provides a higher measure of protection than if no protection were available. Fire departments using special systems should become familiar with their limitations.

Wet Pipe Systems

A **wet pipe sprinkler system** has automatic sprinklers attached to pipes with water under pressure all the time, **Figure 12-11**. This allows the quickest response when the head is opened. The wet pipe system is the simplest sprinkler system in design and operation. The main or alarm valve is a one-way check or clapper valve that prevents water from reentering the water supply and, when closed, shuts off the water flow to the alarm line. Both sides of the alarm valve have pressure gauges that register the water pressure of the supply and the system. The system side gauge should read a slightly higher pressure, because the reclosing of the clapper valve would trap any pressure surges. The alarm line piping usually has a **retard chamber** that acts to prevent false alarms from a sudden pressure surge in the water supply, **Figure 12-12**. The chamber collects a small volume of water before allowing a continued flow to the alarm device. The water from a surge is drained from a small hole in the bottom of the collection chamber. A waterflow indicator, a vane or paddle in the waterway, detects the water flow and activates an alarm signaling system.

A wet pipe system has three more valves. The first is the control valve, which is used to shut off the supply of water to the system and is usually an **outside stem and yoke (OS&Y) valve**, **Figure 12-13**. The second valve is the main drain, which allows the system to be drained for maintenance or to be restored from a fire, and the last valve is an inspector test valve. The test valve is at the farthest end of the system and is used to simulate the flow of a single head and to measure the response time of the system.

The operation of a wet pipe system starts with the fusing or bursting of a sprinkler head, which causes it to begin applying water to the fire. As the water pressure in the system begins to fall, the main check valve opens and water flows into the system and the alarm line, filling the retard chamber, and then activating the automatic alarm and water motor gong. The alarm signal may be used to notify the fire department or an alarm company. *After ensuring the fire is out or completely under control,* the control valve is closed. Sprinkler maintenance personnel replace the head and restore the system. When the system is shut down, a firefighter with a radio

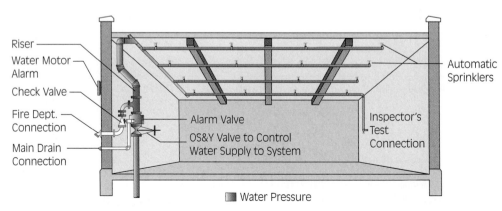

Figure 12-11 Wet pipe sprinkler system.

PROTECTIVE SYSTEMS 319

should be posted at the control valve and be ready to reopen the valve.

Note Some wet pipe systems are located in or have parts of the system located in areas subject to freezing temperatures. To protect these systems, an antifreeze solution can be added to the water in the system. These systems require special attention to restore and maintain.

Streetsmart Tip One of the reasons that sprinklers have not been 100 percent effective is that firefighters or building personnel shut off the control valve prior to the complete extinguishment of the fire. The fire returns and, before the system is turned back on, it gains enough headway to cause additional loss. Building personnel are often overly afraid of water damage to property. Most water-damaged materials can be salvaged but fire-damaged materials are usually destroyed. *Firefighters should not close down any sprinkler system until making sure the fire is out or completely under control. They should keep a hoseline in place to make sure it stays that way.*

Figure 12-12 Retard chamber.

Figure 12-13 OS&Y valves chained in the opened position.

Dry Pipe Systems

In **dry pipe systems**, air under pressure replaces the water in the system to protect against freezing temperatures. The system uses a dry pipe valve to keep pressurized air maintained above with the supply water under pressure below the valve, **Figure 12-14** and **Figure 12-15**. A small amount of water at the seat of the valve, called the priming water, maintains the seal at the valve and is filled to the priming level. The clapper valve has a locking mechanism that keeps the clapper open until it is manually reset to prevent **water columning**.

Dry pipe valves use an air differential system having a smaller air pressure maintained over the larger head surface of the clapper valve, which keeps back the higher water pressure exerted on the smaller water side of the clapper valve. If water were allowed to fill the riser above the clapper, the water column's weight would never allow the clapper to be forced open and make the system inoperative. When a sprinkler head is fused by heat, air is first discharged. As the air pressure drops below the pressure of the supply water, the clapper valve is opened and locked. Because air is in the system rather than water, dry pipe systems are slightly

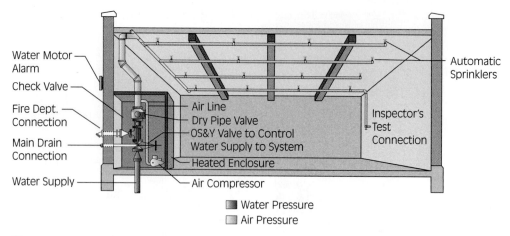

Figure 12-14 Dry pipe system schematic.

slower to activate than wet pipe systems. Most systems have either an **exhauster** or an **accelerator** to speed up the operation of the dry pipe valve. The exhauster detects the decrease in air pressure and helps bleed off air. The accelerator detects the decrease in air pressure and pipes air pressure below the clapper valve, speeding its opening. Drain and alarm valves are similar to wet pipe systems. Dry pipe systems are used in unheated buildings, in buildings that refrigerate or freeze materials, and in outdoor applications where freezing temperatures occur, but the valve room must be heated.

The dry pipe system is more complex in design than a wet pipe system and also harder to restore because it requires the dry pipe valve cover to be opened after draining the system and resetting the lock on the valve. The valve is primed, air pressure charged in the lines, and the control valve opened carefully to prevent retripping and creation of a water column.

Deluge Systems

Deluge systems are designed to protect areas that may have a fast-spreading fire that could engulf an entire area. Petroleum-handling facilities, aircraft hangars, some manufacturing facilities, and hazardous materials storage areas are all examples of occupancies that may have a deluge system. The essential difference between a standard wet system and a deluge system is the individual sprinkler head. Deluge systems utilize open heads, without any fusible elements. When a fire is detected, the deluge system delivers water (or foam solution) to all heads in the area—allowing total coverage of the area.

Operationally, the deluge system must interface with a detection system. Once the detection system activates, it sends a signal to a "deluge valve" which opens, sending water or foam solution to all the open heads, **Figure 12-16.** Most municipal water systems lack the pressure to effectively supply the numerous open heads in the deluge system. For this reason, deluge systems also incorporate a fire pump to boost pressure in the system. Likewise, deluge systems that deliver foam solution require a foam supply tank, foam pump, and foam mixing device.

As can be imagined, deluge system activation will cause tremendous quantities of water to flow. While this is essential to fire control of high hazards, activation of the system in absence of a fire will likely cause much water damage and a significant cleanup effort. For this reason, deluge system activation usually requires activation of several detectors. For example, several flame detectors must activate before a signal is sent to the deluge valve. Some sys-

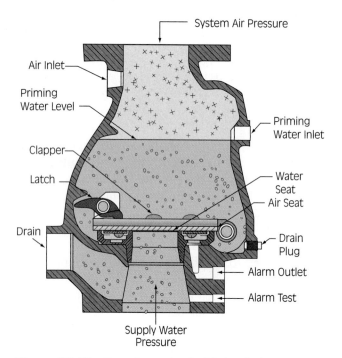

Figure 12-15 Dry pipe valve holds back the water with air pressure.

PROTECTIVE SYSTEMS 321

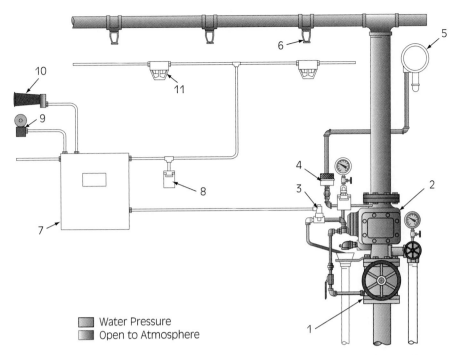

Figure 12-16 Deluge system schematic. 1, OS&Y valve; 2, deluge valve with basic trim; 3, solenoid valve and electric actuation trim; 4, pressure alarm switch; 5, water motor alarm; 6, spray nozzles or open sprinklers; 7, deluge releasing panel; 8, electric manual control stations; 9, fire alarm bell; 10, trouble horn; 11, heat detectors. *(© Copyright Simplex Grinnell. All rights reserved)*

tems require detection from different kinds of devices (i.e., activation of a flame detector and heat detector). To prevent unneeded damage from activation, the deluge system may utilize a manual override alarm and "deadman" switch. In these systems, an alarm is sent telling occupants that the deluge system is ready to fire. If someone does not push a button in a prescribed time, the system will deluge the area. A person who does push the hold button cannot release the button until other persons reset the detectors that have activated. If the person does release the button prior to system reset, the deluge valve will open—hence the deadman label.

Preaction Systems

Preaction systems are similar to the dry pipe and deluge system. The system has closed piping and heads with air under no or little pressure, but the water does not flow until signaled open from a separate fire detection system, **Figure 12-17**. The preaction valve then opens and allows water to flow through the system to the closed heads. When an individual head is heat activated, it opens and water attacks the fire. Preaction systems are used in areas where the materials protected are of high value and water damage would be expensive, such as computer rooms and historical items.

Residential Systems

Residential sprinkler systems are smaller and more affordable versions of wet or dry pipe sprinkler systems, **Figure 12-18**. They are designed to control the level of fire involvement such that residents can escape. The water supply is combined with the domestic water supply, and flow rates are designed for one or a few heads in operation. Residential sprinkler heads have a faster response time and use a lighter and smaller pipe than wet/dry pipe systems, and were the first to use plastic piping. Residential systems use a check valve, waterflow alarms, and drains similar to the bigger systems and are not required to have a fire department connection, although some do have one. Some residential systems use antifreeze to protect all or part of the system.

SPRINKLER SYSTEM CONNECTIONS AND PIPING

The connections and piping for a sprinkler system provide the water from its source to the heads, and they comprise most of the system. The main water supply can come from a public or private water

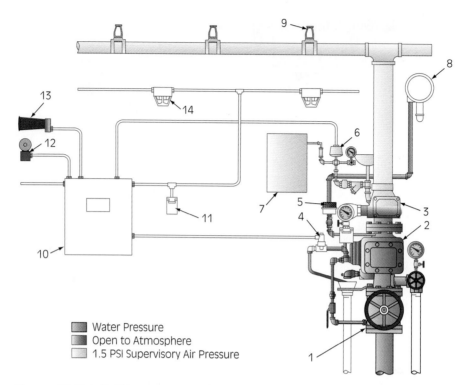

Figure 12-17 Preaction system schematic. 1, OS&Y valve; 2, deluge valve with basic trim; 3, check valve; 4, solenoid valve and electric actuation trim; 5, water pressure alarm switch; 6, 1.5-psi low air pressure alarm switch; 7, 1.5-psi supervisory air pressure control; 8, water motor alarm; 9, automatic sprinklers; 10, deluge releasing panel; 11, electric manual control stations; 12, fire alarm bell; 13, trouble horn; 14, heat detectors. (© Copyright Simplex Grinnell. All rights reserved)

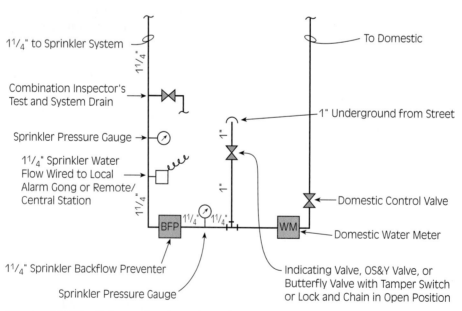

Figure 12-18 Schematic of residential sprinkler system.

company, the protected property's own supply system, a variety of tanks such as gravity or suction, or a cistern, pond, or stream. Depending on the requirements of the system and the water source, a fire pump may be included. A secondary water source is a fire department siamese connection, which allows pumpers to supplement the water supply, **Figure 12-19**.

PROTECTIVE SYSTEMS 323

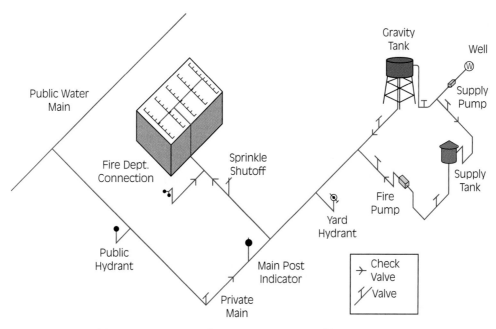

Figure 12-19 Various water supply systems to a sprinkler system.

> **Note** A serious consideration is where the fire department gets its water supply to determine if water is being robbed from the sprinkler system.

It is best to use a different water main or supply source than the one being used by the sprinkler system. The fire department connection is usually piped in past the main control valve and can supply the system even if the main control valve is closed.

Another required valve in most systems is a backflow preventer, a set of one-way check valves designed to keep water in a sprinkler or other system from reentering the public water supply. These are an environmental requirement, and most codes require and mandate them even on old systems. The system may have other control valves and check valves and will have a water control valve if connected to a public water supply.

Next in line is the main control valve, which is of the wet, dry, deluge, or preaction type as described earlier. Above that is the riser. The riser pipe feeds the mains that feed the cross mains to the branch lines, **Figure 12-20**. The branch lines have the sprinkler heads attached to them. Some larger systems may have sectional valves that divide the system into floors or other subareas of the system. The location of these valves can be important when shutting down the system or doing maintenance. They should be kept locked and supervised for tampering. Tamper alarms alert the alarm company whenever someone operates the valve, and unauthorized valve operations are quickly checked to prevent an arsonist from disabling the system prior to setting a fire.

Firefighters should be able to properly connect a supply line to a fire department connection for either a sprinkler or standpipe. To do this requires the correct amount of hose to reach from the pumper's discharge gate to the connection, a spanner wrench to tighten the couplings, and any adapters needed to complete the connection. The firefighter should check the connection for damage, remove the cap cover, and check inside the siamese for debris, damage, and operation of a clapper valve and an O-ring gasket. If no clapper valve is provided, all necessary connections must be made prior

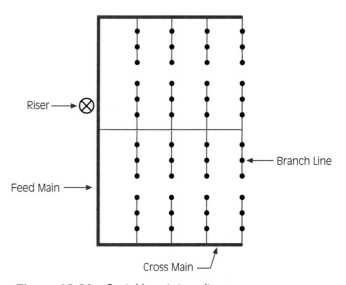

Figure 12-20 Sprinkler piping diagram.

to charging the system. When connecting to a siamese, the outlet on the far left should be chosen first, because this will allow better access for using the spanner wrench to tighten the coupling. After completing the connection, the pump operator should be advised that the line is ready for charging.

> **Streetsmart Tip** When connecting to a siamese, the firefighter should choose the outlet on the far left first, because this will allow better access for using the spanner wrench to tighten the coupling. Some connections only have clappers installed on the right side and must be connected on the left side first.

CONTROL DEVICES FOR SPRINKLER SYSTEMS

The three main control devices for sprinkler and standpipe systems are the outside stem and yoke valve (OS&Y), the **post indicator valve (PIV)**, and the **wall indicator valve (WIV)**, **Figure 12-21**, **Figure 12-22**, and **Figure 12-23**. The valves have either a butterfly or gate valve type. The names come from the appearance of the valves. The OS&Y valve has a wheel on a stem housed in a yoke or housing. When the stem is exposed or outside, the valve is open. These valves must have a chain lock on them to prevent tampering. If the system is

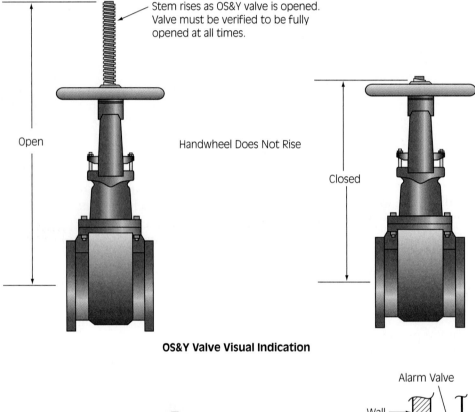

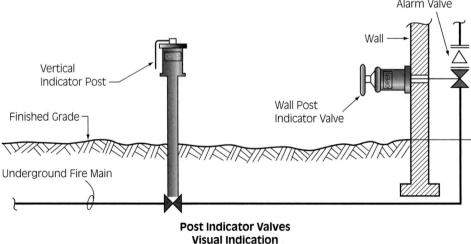

Figure 12-21 OS&Y valves and post indicator valves.

PROTECTIVE SYSTEMS 325

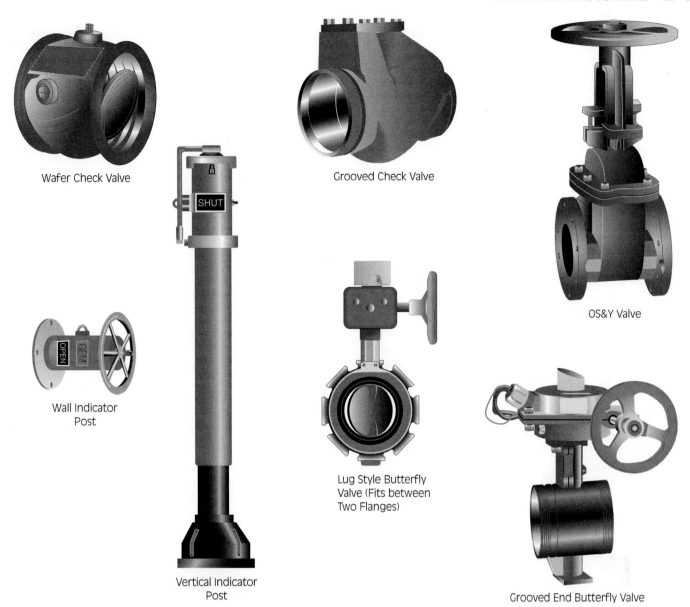

Figure 12-22 Fire protection valves.

Figure 12-23 Fire department connection and post indicator valve. The box on the side of the PIV is a tamper alarm switch box.

supervised, a tamper switch will also be found on the stem. PIVs and WIVs are very similar, but one is mounted on a post in the ground, and the other is mounted on a wall. Both valves are housed in a metal case with a small window, reading either "OPEN" or "SHUT." A wrench or a wheel controls these valves; a padlock and chain are used to lock them opened. Some water control valves may also be of the nonindicating type; public water system valves usually are of the nonindicating type.

RETURNING SPRINKLER SYSTEMS TO SERVICE

Fire departments used to regularly restore sprinkler and other protective systems to service for building owners. Many also regularly tested these systems.

Many engine and ladder companies carried a sprinkler kit with wrenches, extra heads, stops, and other equipment to do this. Today many departments no longer provide this service; instead they either stop the flowing head or shut down the system, letting the owner contact a fire protection service company. This has become necessary for liability issues. Fire departments should have local SOPs that address the response, operations, and return-to-service guidelines for protective systems. Firefighters must understand how to properly shut down either individual heads or the entire system to stop excessive water damage until the heads can be completely restored.

When the fire is completely out, the first step to restore the system is to shut down any pumper supplying the system. Firefighters should not disconnect the hoselines at this point; the fire may reignite so hoselines should stay connected until overhaul is complete. If the system is to be restored immediately or a large number of heads have opened, either the main sprinkler or a sectional valve should be shut down and drained. The sprinkler head is replaced with an identical one from the spare heads that are required to be kept in a cabinet in the sprinkler control room. The sprinkler valves must be reset or reopened.

The simplest—and often quickest—way to stop water flow from an individual sprinkler head is to

Figure 12-24 Sprinkler tongs and wedges.

insert a stop. Sprinkler stops may include manufactured sprinkler tongs or improvised wedges, dowels, or clamps, **Figure 12-24**. While it may seem easy, stopping water flow from a sprinkler head requires practice to effectively stop the flow and establish a leak-free seal, **Figure 12-25**. The firefighter assigned to stop sprinkler flow at the head will get seriously wet. It is important to use caution when climbing a wet ladder and to make sure eye protection is used to prevent injury. Following a step-by-step plan will minimize the exposure and speed the water stoppage, **JPR 12-1**.

Figure 12-25 Sprinkler tongs and wood wedges stopping sprinkler flow.

PROTECTIVE SYSTEMS 327

JOB PERFORMANCE REQUIREMENT 12-1
Using "Stops" to Stem the Flow of a Sprinkler Head

A Empty your PPE pockets of radios or other items that can be damaged from getting soaked.

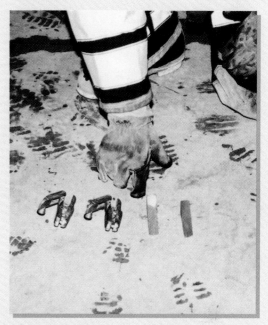

B Select the right stop for the head discharging water. Remember, simple wedges may not fit into recessed sprinkler heads found in some finished ceilings.

C Choose a stable ladder, desk, or other platform to reach the head.

Photos courtesy Captain Pete Evers, City of Auburn Fire Department, CA

D Put your face shield down or don eye protection to help you protect your eyes from the deluge when you climb the platform and secure the wedges into place.

(continued)

JOB PERFORMANCE REQUIREMENT 12-1
Using "Stops" to Stem the Flow of a Sprinkler Head (Continued)

E Most times you will be unable to "see" through the rain of water and be required to stop the water flow blind. Feel your way, and do not rely on sight.

F Carefully reach up to the head with the stop.

G Apply the stop and tighten it until the flow stops.

Photos courtesy Captain Pete Evers, City of Auburn Fire Department, CA

H Sometimes, using two opposing wedges is easier than trying to force in a single wedge. The sliding motion from two directions overcomes water pressure quicker and provides a better seal than a single wedge.

> **Streetsmart Tip** Firefighters will find that carrying a small assortment of tools and equipment in their turnout gear will be invaluable. A flashlight is essential. Tools should include straight and Phillips screwdrivers, a pair of pliers, an adjustable wrench, and a pocketknife. Other items would include several wooden wedges that can be used for sprinkler stops, door chocks, and other tools that come to mind. Fire departments should also provide search tags, utility shutoff tags, and for firefighters acting as officers, pocket-size versions of a command chart.

The other way to shut off the water flow from opened sprinklers is to shut down either the main sprinkler valve or a floor or sectional valve. This task will require a key or set of bolt cutters because main valves are locked as discussed earlier. If the main valve is an OS&Y valve, it is turned until the stem is no longer exposed. If it is a PIV or WIV valve, it is turned until it reads "SHUT." Personnel with a portable radio should stand by the valves until the system is recharged or until the fire is determined to be out. If the entire system is shut down or if there is a sectional drain, that valve can be opened. After draining the system, this valve should be closed. Once the system is shut down and drained, the heads may be replaced. Replacement of the heads is recommended as soon as possible because in the event of an additional fire, even with the system shut down, the fire department can pump through the fire department connection or turn the system back on and have it operational for firefighting. Refilling a system should be done slowly to prevent damaging the system.

To avoid liability of restoring sprinkler systems, some fire departments *do not* replace heads and restore the system. In these cases, the fire department should require that a **fire watch** be established until the system has been restored by a qualified technician. Some occupancies such as hotels, public assemblies, and high-rises *require* a sprinkler system for life safety. In these cases, the fire department should *not* allow the building to be occupied without an approved fire watch system or sprinkler restoration.

STANDPIPE CLASSIFICATIONS

Standpipe systems are designed to allow firefighters to fight fires in larger buildings by prepiping water lines for fire streams through the building. Some systems allow both occupants and firefighters to use the system, and the systems for building occupants have hoselines and nozzles attached. Standpipe systems are not an automatic; they must be manned by people, who use them to attack the fire. They are used in high-rise buildings; large commercial, retail, and industrial buildings; places of public assembly; and other areas where advancing hoselines from hydrants or fire apparatus would be difficult due to the size of the building. Tunnel systems, such as subway systems, and shopping malls have horizontal standpipe systems. Large does not always mean tall—providing access is the key goal.

Standpipe systems are classified according to the type of intended user and each class sets requirements for volume and size of outlets. *Class I systems* are designed for use by the fire department or trained personnel such as a fire brigade, **Figure 12-26**. The systems use 2½-inch (65-mm) outlets. A single standpipe should flow a minimum of 500 gpm (1,890 L/min) plus an additional 250 gpm (945 L/min) for each additional riser. No hose is provided for this class. Outlets are found in stairwells on each floor, near the exits, or in the hallways on each floor near the stairs.

A *Class II system* is designed for use by untrained building occupants and has a 1½-inch (38-mm) outlet with a minimum flow of 100 gpm (378 L/min), **Figure 12-27**. Hose is provided and is either 1 or 1½ inch (25 or 38 mm). Firefighters wishing to use a Class II system should not use the hose provided, because it is not tested and may be single-jacketed unlined hose with a nozzle that has no shutoff. Firefighters using this system should replace the hoseline with one of their own.

A *Class III system* is used by either users of the other two classes but meets the Class I requirements for flow. It has both 1½-inch (38-mm) and 2½-inch (65-mm) outlets or may have a 2½-inch (65-mm) outlet with a 2½- to 1½-inch (65- to 38-mm) reducer. All of the flows given are at the top outlet with a residual pressure of 65 psi (448 kPa).[5] Any of these systems may have a **pressure-regulating device**, which is designed to maintain or reduce the pressure to a set amount regardless of the height of a building. Standpipes have height limits on each section [about 280 feet (88 m)], but the pressure difference in an unregulated system could vary by almost 140 psi (966 kPa) from top to bottom.

> **Caution** Firefighters wishing to use a Class II or III standpipe system should not use the hose provided, because it is not tested and may be single-jacketed unlined hose with a nozzle that has no shutoff. Firefighters using this system should replace the hoseline with one of their own.

Figure 12-26 Class I standpipe system.

> **Caution** Pressure-regulating valves on standpipe systems may cause firefighters problems if improperly installed or reset. Some are preset at the factory and must be installed on the correct floor to properly regulate the pressure. Others are adjustable and can be reset after installation. Improperly installed pressure-regulating devices have contributed to some firefighter fatalities in high-rise firefighting. Careful inspection and testing of these devices is highly recommended on a regular basis due to maintenance problems. (See Street Story in this chapter.)

The types of standpipe systems are differentiated based on their water supply. An automatic wet pipe system has a water supply that is ready to conduct firefighting, whereas automatic dry and semiautomatic dry pipe systems use dry pipe until a hose station (an outlet) is opened to allow water to flow into the system. The semiautomatic system requires an activation valve to be opened manually. The manual dry pipe is completely dry and relies on the fire department connection for water supply, whereas a manual wet pipe system has the piping filled with water but needs the fire department connection for firefighting supply.[6]

STANDPIPE SYSTEM CONNECTIONS AND PIPING

Standpipe systems can range from very simple to highly complex ones with multiple connections, risers, sources of water supply, and many outlets on each floor or area. A manual dry system can be a single riser having a fire department connection and a drain at the bottom and an outlet on each floor that is capped at the top. The wet manual type may just add a small water line to maintain water in the piping. The other systems become more complex as a water supply is added and are found in larger buildings with more requirements.

Figure 12-27 Class II standpipe system.

Some standpipe systems are combined with the building sprinkler system with outlets coming off a branch line.

The components of a standpipe system are the piping, outlets with hose and other attachments, the valves, the fire department connection, and any monitoring devices. The piping includes the riser or risers and any attachments needed to plumb between the water sources and the uppermost or furthest outlet. The piping must be capable of discharging the required volume and pressure with minimal friction loss. The outlets are usually placed in the building stairwell or may be a wall-mounted cabinet. Attached to the outlet would be any pressure-regulating device required. If it were a Class II or III system, a hoseline and nozzle would be attached. The hose connection usually has a wheel-type handle to operate the valve and a cap covering the threads. The nozzles may be a combination fog-straight stream or a solid tip nozzle and quite often have no shutoff, which could confuse the untrained user.

Standpipe valves are similar to the types used on sprinkler systems, including gate valves of various types and check valves, with the addition of hose valves at the outlets. If a water supply is provided, an indicating gate valve is required and a check valve or backflow preventers. A check valve is also added with a drain on the fire department connection line to allow drainage and prevent freezing. Gate valves should also be installed on each section or separate riser to allow maintenance work to be performed without shutting down the entire system. Drain valves are installed on each section. The fire department connection(s) is an inlet or siamese device with protective caps and a raised letter sign marked "Standpipe." If there are multiple connections that are not interconnected, the sign should state the zone covered. Monitoring systems are covered in the next section.

ALARMS FOR STANDPIPES AND SPRINKLERS

Alarm and monitoring systems are found in most sprinkler systems and many standpipe systems, especially Class II and III systems. Sprinkler systems are both fire detection and suppression devices and are designed to notify at least the building occupants. Occupant-used standpipe systems should be monitored in case the occupants forget to notify the

Figure 12-28 Water motor gong.

fire department prior to fighting the fire. Most of these systems require monitoring to prevent tampering or deliberate attempts to disable the system. These may be electronically or mechanically activated by the movement of a gate valve. Waterflow alarms are also electrical or mechanical. The most familiar mechanical one is the water motor gong, which operates like a waterwheel with the flow of the water causing it to turn and strike a gong, **Figure 12-28**. The water motor gong, unless connected to another alarm, is a local alarm only. Other waterflow devices are hooked to an electronic sensor that is connected to a paddle or vane inside the riser. A monitoring alarm company notifies the fire department about fires, but its own personnel respond to tamper alarms.

OTHER PROTECTIVE SYSTEMS

Many other types of protective systems are used today. Some are rather simple, for instance, those designed to protect the grill, fryer, and ductwork of a local restaurant, whereas others are extremely complex and are designed to prevent or suppress an explosion in highly hazardous locations.

Figure 12-29 Local application system (notice nozzles) protecting cooking area and ductwork.

Figure 12-30 Dry chemical nozzles are located above the island and at ground level.

Firefighters need to understand the most common types of protective systems. Fire departments responding to the more complex systems should ensure that all firefighters are familiar with them. Firefighters should be aware of how these systems operate, what the hazards of the fires and extinguishing agents are to themselves and any occupants, and how to safely operate these systems to ensure that the fire is extinguished and the building is safe.

Local Application and Hood Systems

One of the most common types of protective systems is a **local application system**, especially the systems protecting the cooking areas of restaurants, **Figure 12-29**. Local application means that the system is designed to protect only a certain or local portion of the building, usually directly where the hazard will occur or spread. Local applications are also used in laboratory hoods, paint booths, and other small hazardous locations, **Figure 12-30**. Most of the local application systems use dry or wet chemical agents with the most popular being an ABC dry chemical. Higher temperature cooking units and oils have created the need for a new Class K classification, and these systems also use dry or wet chemicals. Hood systems use a heat-sensitive device or a manual switch for activation.

Total Flooding Systems

Total flooding systems are used to protect an entire area, room, or building. The total flooding system discharges an extinguishing agent that completely fills or floods the area with the extinguishing agent to smother or cool the fire or break the chain reaction. Total flooding systems can use carbon dioxide or other inert gases, halogenated or clean agents, dry chemicals, or foam as extinguishing agents (see Chapter 8 for a discussion of these agents). They are effective as long as the proper amount discharges, the area is contained to prevent loss of agent, or the fire goes out prior to ventilation of the agent. There are hazards associated with each type of application with which firefighters and building personnel should be familiar.

Caution Carbon dioxide (CO_2) and other inert gases such as nitrogen can be used to smother a fire. CO_2 is most often used for firefighting, while nitrogen is used to inert an atmosphere to allow welding or flammable activities to take place near combustible materials. CO_2 systems can be either high- or low-pressure storage systems and are designed to fill the area with a concentration of the gas that extinguishes the fire, **Figure 12-31**. Unfortunately, this smothering effect is also fatal to any living beings in the area, and an evacuation period is built into the system activation. Firefighters, using SCBA, may enter the area to ensure the fire is extinguished and all persons have been safely removed. This agent is very clean and is often used in computer and electronics areas. If the fire is out, ventilation of the area and restoration of the system are the only cleanup activities required.

Caution Dry chemicals and foam flooding systems are found in areas with highly hazardous processes and fast-moving fires. Dry chemical total flooding systems discharge enough agent to cover all surface areas and sufficiently extinguish the fire. A caution is in order when dealing with a deep-seated Class A fire, because the lasting effects of these agents may not completely extinguish these types of fires. Foam systems use high-expansion foam generators to completely fill the area with foam. These are effective in fighting flash-type fires and will also work in Class A materials. High-expansion foam was designed for fighting coal mine fires, and some fire departments have used it to fight large commercial basement fires when access for personnel was inadequate. If the system works properly, firefighters using SCBA and hoselines merely need to mop up any remaining hot spots.

Caution Halon and the newer clean agents are also used for total flooding applications. Depending on the toxicity and concentration of agent needed, some of these systems may even be used when people are still occupying the area. Firefighters should still use SCBA when entering areas where even safe agents have been discharged to ensure that the fire is out. As with CO_2, electronics and other high-cost or sensitive valuables are protected by these systems. The halons are being phased out of use, but many buildings will continue to have them for the foreseeable future. The replacement clean agents are in use now and will continue to take over these types of properties.

FIRE DEPARTMENT OPERATIONS WITH PROTECTIVE SYSTEMS

Standpipe and sprinkler systems plus any other protective system should be part of the fire department overall strategic plan to provide fire protection for its community. This strategic plan recognizes the community's hazards and tries to keep those hazards within certain limits. When properties or processes create hazards beyond those limits, protective systems are required. The fire department

Figure 12-31 Low-pressure carbon dioxide extinguishing system.

must survey its hazards and also its resources, both public and private. A program of maintenance, inspections, and pre-emergency planning keeps the department in compliance with the strategic plan. The property owner does maintenance, the fire prevention bureau does the inspections, and fire companies the preplanning. Preplanning identifies hazards and resources, and some key resources are protective systems. Fire companies should know where every protective system is and how it operates. For standpipe and sprinkler systems, fire company personnel should know the type of system, the location of all fire department connections, two closest water supply points, and the location of key valves. This information should be noted on area maps, pre-emergency plans, and dispatch information. Knowing where the systems are and how to use them is the first step in coordinating a proper action plan. The next step is having SOPs that give a recommended course of action in buildings with these systems.

From an operational point of view, protective systems can be separate components requiring the fire department to look in different areas to understand what has been activated. In newer systems, the protective systems are integrated into a single "smart" system that includes an annunciator panel, fire alarm control panel, and system override controls. Typically, annunciator panels are near the primary entrance to the building and simply "announce" what has been activated and where. The fire alarm control panel and other protective system controls are in a locked room and allow the fire department to reset, silence, or otherwise control the system. Large occupancies such as a high-rise may have a complete fire command center where the fire department can monitor and/or control all protective systems as well as HVAC systems and building intercom/communications systems. Some of the operational procedures for protective systems are addressed in this chapter.

Standpipe Operations

Standpipe operations start with establishing a water supply to the fire department connection with a minimum of one hoseline. Additional hoselines are added as needed. The pump operator should immediately charge the standpipe system for the pressure required at the reported fire level. The pressure should account for the required nozzle pressure, friction loss in the supply and attack lines and the standpipe piping, plus any elevation. The first arriving unit should have its personnel check the annunicator panel and then go to the reported location of the fire. Methods of going to this location will vary depending on the situation but may involve use of stairs or elevators and should be covered in SOPs. Firefighters using elevators should do so only if they are equipped with firefighter service and use this control system. They should *not* take the elevator directly to the reported fire floor because this can be hazardous. They should stop at least two floors below.

Personnel should have full personal protective equipment, standpipe pack, and forcible entry equipment. The suggested standpipe pack equipment is a minimum of 150 feet (46 m) of 1½-inch (38-mm) hose or larger [many departments are now using 1¾ or 2 inch (45 and 50 mm)], a 2½- to 1½-inch (65- to 38-mm) gated wye with 1½- to 2½-inch (38- to 65-mm) increaser, 5 to 10 feet (1½ to 3 m) of 2½-inch (65-mm) hose, spanner wrench, adjustable wrench, and valve wheel. Additional sections of hose are recommended if personnel allow, **Figure 12-32**. Personnel should connect their standpipe pack to the nearest and safest outlet in relation to the fire location. This may require using the floor below and advancing the hoseline toward the fire location. Depending on the fire's location, the line may be charged prior to entering the hallway or delayed until reaching the door of the room or unit on fire. Additional lines may be run off the gated wye, another standpipe outlet on that floor, or another floor.

> **Caution** Firefighters using elevators should do so only if they are equipped with firefighter service and use this control system. They should *not* take the elevator directly to the reported fire floor because this can be hazardous. They should stop at least two floors below.

Figure 12-32 Hose pack for standpipe use.

> **Streetsmart Tip** When advancing a hoseline from a standpipe system in a stairwell, or any hose in a stairwell, extra hose should be looped up the stairs and then down so that when pulled, gravity will assist in pulling the hose down instead of pulling it up the stairs.

Sprinkler System Operations

Sprinkler system operations start with an investigation of the building to determine if water is flowing or a fire is present. Typically, the first arriving officer checks the annunciator panel and then directs the crew to pull fire attack lines (or begin standpipe operations). Typically, SOPs require the first arriving crew to start fire attack while the pump operator sets up to support the sprinkler system. Establish a water supply to the fire department connection with a minimum of one medium- or large-diameter hoseline. Additional hoselines are added as needed; the second line can be a backup for the first. The pump operator should charge the sprinkler system when smoke or fire is showing, the water motor gong is operating, or when ordered by the officer-in-charge. The system should be charged immediately if it is combined with a standpipe system. Sprinkler systems are normally charged and maintained at 150 psi (1035 kPa). Personnel should advance hoselines into the fire area and conduct extinguishment and overhaul operations. If no fire is found, the officer may direct firefighters to stop sprinkler flow as discussed earlier in this chapter. It is important to note that all sprinkler system alarms should be treated as an actual fire until confirmed otherwise. Just because a building is fully sprinkled does not mean the fire department should prepare to fight fire differently. Being fully prepared with PPE, tools, radios, and so forth, should take precedence.

Detector Activation Operations

Operationally, the generic "fire alarm activation" response can range from investigating an alarm malfunction to a full-blown fire operation. Histories of false alarms have caused some firefighters to take a less than ready approach to their operations. Not only is this dangerous, it can be fatal. Most experienced firefighters can tell a story of arriving at a fire alarm activation, only to be surprised that an actual fire existed. These firefighters now respond to all alarms just as if an actual fire were awaiting them. It cannot be overemphasized: Firefighters must treat all fire alarm activations as an actual fire and prepare for the worst.

First arriving firefighters at an alarm activation should dismount the apparatus wearing full PPE and carrying tools and radio. Typically, the first step at an alarm activation is to investigate. At residential structures, firefighters should meet with the building occupant if no fire signs or conditions exist. In cases where nobody is awaiting fire department arrival, the firefighters should systematically walk around the structure and check for smoke or other indicators of an internal fire. Local SOPs should address further steps if no visible signs exist. These SOPs can range from notifying neighbors, to keeping an eye on the structure, to forcible entry and interior search. If access can be made to the structure, caution should be employed. Smoke detector and CO detector activation may sound similar, but the CO can incapacitate unprepared firefighters with little warning.

At commercial structures, the first arriving crew should first check the annunciator panel to ascertain which device has activated. If no annunciator exists, firefighters may have to access the fire alarm control panel if no obvious signs are found. After business hours, firefighters should perform a walk-around to look for signs. If the building is so equipped, the officer will then access the lockbox and enter the structure. Anytime a lockbox is utilized, the officer should radio dispatch that locked entry is being made. After checking the annunciator panel, the firefighter should proceed to the indicated zone and investigate. If any signs of smoke or fire are visible, the operation resorts to a fire attack mode.

Buildings equipped with strobes and loud audible warning devices can be annoying to occupants if they see no signs of fire. Firefighters must not get trapped in a "false alarm mentality" and arbitrarily silence these alarms to investigate the activated zone. Premature silencing may send a message to occupants that they can return to their business. Using simple foam earplugs (which should be in the pocket of firefighters' PPE) can reduce the potential of temporary hearing loss while investigating the alarm. Once the activated detector is found, the immediate area can be checked for potential causes. If no fire signs are found, the firefighters can then take steps to clear the detector (discussed previously) and try to reset the alarm. Systems that will not reset become the property owners' responsibility to correct. Often, the fire department needs to advise the building representatives to contact their alarm company or a repair technician. All fire department actions must be documented on the incident report.

Operations for Other Protective Systems

Local SOPs should address the operations for total flooding, foam, dry chemical, and other unique systems. Some general procedures, however, can be addressed here. Total flooding systems that have discharged present a hazard to occupants and firefighters. Firefighters must never enter a total flooding environment without first engaging their SCBA. It is essential to be prepared for the worst. The introduction of air into total flooding environments can reignite the fire. Hood systems that have discharged will require an overhaul effort to ensure that the fire is totally out. Firefighters should check ceiling spaces and the rooftop ventilator to make sure the fire has not extended out of the hood and ducts. Dry chemical systems are quite effective for fire control but require a large cleanup effort. This cleanup is not the responsibility of firefighters although it is important to notify occupants that breathing dry chemical can cause lung irritation. Activation of suppression systems may cause secondary damage to electrical circuits and other building infrastructure. It is important to be aware of this potential and protect firefighters and occupants accordingly.

Lessons Learned

Protective systems are devices designed to either automatically detect or suppress a fire or to assist people in extinguishing the fire. They can apply water or other extinguishing agents on a fire. Protective systems have been credited with saving both lives and property and are essential fire protection tools. Firefighters need to understand the value and operation of these systems to protect their communities.

Sprinkler systems are used for detection and suppression and can apply water or foam to extinguish the fire. Most sprinkler systems have a closed sprinkler head that thermally detects the fire, causing it to open and begin spraying water on the fire while sounding an alarm. The heads are designed to react to set temperatures and to spray water over a certain pattern. The four main types of systems are wet pipe with water in the pipes for immediate action, dry pipe for areas with freezing temperatures, deluge systems for fast-spreading fires, and preaction systems to prevent accidental discharge of water onto valuable property. Each type of system uses a different type of alarm valve but many of the other features are similar. Gate and check valves such as OS&Y or PIV valves are commonly used. Firefighters are still needed in sprinkler-protected property to finish putting out any fire that the system could not.

Standpipe systems facilitate manual fire suppression in which people do the firefighting. Standpipes supply water in large buildings where it would be hard to drag hoselines to and from fire apparatus, such as in high-rise and other large buildings. Building occupants can also use some standpipe systems, which are a system of pipes with a water supply connection at one end and a series of outlets, some with hose and nozzles, at the other. Standpipe operations require careful coordination of firefighters due to the size of the buildings involved.

The last group of protective systems detects fire and applies other extinguishing agents to fires in proximity to the hazard or fills an entire area. Extinguishing agents range from carbon dioxide to wet and dry chemicals to halogenated agents. Restaurants and kitchens usually have a hood system protecting the cooking areas, while other systems are used to protect high hazards or high value items. These protective systems are highly specialized but firefighters must understand their operation and any hazards of the agents.

Fire department operations at buildings with protective systems are typically outlined in local SOPs. Firefighters should treat all fire alarm activations as the worst-case scenario. This means investigating with full PPE, tools, and radios. Operations involving standpipe and sprinkler systems require the fire department to support the system with a water supply and fire attack hoselines. Stopping water flow from sprinkler systems should happen only after the fire has been extinguished. Hoselines should stay in place while the sprinkler is shut down. Firefighters need to practice using stops to stem the flow of activated sprinkler heads.

KEY TERMS

Accelerator A device to speed the operation of the dry pipe valve by detecting the decrease in air pressure. It pipes air pressure below the clapper valve, speeding its opening.

Auxiliary Appliances Another term for protective devices, particularly sprinkler and standpipe systems.

Deluge Systems Designed to protect areas that may have a fast-spreading fire engulfing the entire area. All of its sprinkler heads are already open, and the piping contains atmospheric air. When the system operates, water flows to all heads, allowing total coverage. The system uses a deluge valve that opens when a separate fire detection system senses the fire and signals to trip the valve open.

Dry Pipe System Air under pressure replaces the water in the system to protect against freezing temperatures. The sprinkler control valve uses a dry pipe valve to keep pressurized air maintained above with the supply water under pressure below the valve.

Exhauster A device to speed the operation of the dry pipe valve by detecting the decrease in air pressure. It helps bleed off air.

Fire Watch An organized patrol of a protected property when the sprinkler or other protection system is down for maintenance. Personnel from the property regularly check to make sure a fire has not started and assist in evacuation and prompt notification of the fire department.

Life Safety Term applied to the fire protection concept in which buildings are designed to allow for the escape of building occupants without injuries. Life safety usually makes the building more fire resistant, but this is not the main goal.

Local Application System Designed to protect only a certain or local portion of the building, usually directly where the hazard will occur or spread.

Outside Stem and Yoke (OS&Y) Valve Has a wheel on a stem housed in a yoke or housing. When the stem is exposed or outside, the valve is open. Also called an outside screw and yoke valve.

Post Indicator Valve (PIV) A control valve that is mounted on a post case with a small window, reading either "OPEN" or "SHUT."

Preaction System Similar to the dry pipe and deluge systems. The system has closed piping and heads with air under no or little pressure, but the water does not flow until signaled open from a separate fire detection system. The preaction valve then opens and allows water to flow through the system to the closed heads. When an individual head is heat activated, it opens and water attacks the fire. Usually used when water can cause a large dollar loss.

Pressure-Regulating Device Designed to control the head pressure at the outlet of a standpipe system to prevent excessive nozzle pressures in hoselines.

Residential Sprinkler System Smaller and more affordable version of a wet or dry pipe sprinkler system designed to control the level of fire involvement such that residents can escape.

Retard Chamber Acts to prevent false alarms from a sudden pressure surge in the water supply by collecting a small volume of water before allowing a continued flow to the alarm device. The water from a surge is drained from a small hole in the bottom of the collection chamber.

Sprinkler Systems Designed to automatically distribute water through sprinklers placed at set intervals on a system of piping, usually in the ceiling area, to extinguish or control the spread of fires.

Standpipe Systems Piping systems that allow for the manual application of water in large buildings.

Total Flooding System Used to protect an entire area, room, or building by discharging an extinguishing agent that completely fills or floods the area with the extinguishing agent to smother or cool the fire or break the chain reaction.

Wall Indicator Valve (WIV) A control valve that is mounted on a wall in a metal case with a small window, reading either "OPEN" or "SHUT."

Water Columning A condition in a dry pipe sprinkler system in which the weight of the water column in the riser prevents the operation of the dry pipe valve.

Wet Pipe Sprinkler System Has automatic sprinklers attached to pipes with water under pressure all the time.

REVIEW QUESTIONS

1. What are the two most common styles of smoke detectors?
2. How does an ionization detector operate?
3. Why should firefighters treat a smoke detector activation like a CO detector activation?
4. What is an upright and pendent sprinkler?
5. What happens when a fire occurs in a sprinkled room?
6. What is an OS&Y valve and how does it work?
7. What is a fire department connection and how is hose connected to it?
8. What are the steps to restoring a wet pipe sprinkler system?
9. Explain how to stop a sprinkler head from flowing.
10. Name the three classes of standpipes and who can use them.
11. What is a waterflow alarm?
12. What equipment should be carried by firefighters in a high-rise fire?
13. What is the hazard with a carbon dioxide total flooding extinguishing system?
14. What is the difference between an annunciator panel and a fire alarm control panel?
15. What precautions should firefighters take when using an elevator to access upper floors?
16. When should firefighters silence the audible fire alarm devices in a building?

Endnotes

1. See NFPA Standards 13, 13D, 13R, and 25 for more information.
2. *NFPA 13D: Standard for the Installation of Sprinkler Systems in One- and Two-Family Dwellings and Manufactured Homes, 2002 Edition.* National Fire Protection Association, Quincy, MA.
3. Cote, Arthur and John Linville, *Fire Protection Handbook,* 18th ed., pp. 6-145. National Fire Protection Association, Quincy, MA, 1997.
4. *NFPA 13: Standard for the Installation of Sprinkler Systems, 2002 Edition,* 13-11. National Fire Protection Association, Quincy, MA.
5. Cote et al., *Fire Protection Handbook,* pp. 6-249–6-250.
6. Ibid., pp. 6-250.

Additional Resources

Diamantes, David, *Fire Prevention, Inspection and Code Enforcement.* Delmar Learning, a part of the Thomson Corporation, Clifton Park, NY, 1997.

Gagnon, Robert, *Design of Special Hazard and Fire Alarm Systems.* Delmar Learning, a part of the Thomson Corporation, Clifton Park, NY, 1998.

Gagnon, Robert, *Design of Water Bases Fire Protection Systems.* Delmar Learning, a part of the Thomson Corporation, Clifton Park, NY, 1998.

Klinoff, Robert, *Introduction to Fire Protection.* Delmar Learning, a part of the Thomson Corporation, Clifton Park, NY, 1997.

NFPA 13: Standard for the Installation of Sprinkler Systems, 2002 Edition. National Fire Protection Association, Quincy, MA.

NFPA 13D: Standard for the Installation of Sprinkler Systems in One- and Two-Family Dwellings and Manufactured Homes, 2002 Edition. National Fire Protection Association, Quincy, MA.

NFPA 13E: Guide for Fire Department Operations in Properties Protected by Sprinkler and Standpipe Systems, 2000 Edition. National Fire Protection Association, Quincy, MA.

NFPA 13R: Standard for the Installation of Sprinkler Systems in Residential Occupancies up to and Including Four Stories in Height, 2002 Edition. National Fire Protection Association, Quincy, MA.

NFPA 14: Standard for the Installation of Standpipe, Private Hydrants, and Hose Systems, 2003 Edition. National Fire Protection Association, Quincy, MA.

NFPA 25: Standard for the Inspection, Testing, and Maintenance of Water-Based Fire Protection Systems, 1998 Edition. National Fire Protection Association, Quincy, MA.

Sturtevant, Thomas, *Introduction to Fire Pump Operations.* Delmar Learning, a part of the Thomson Corporation, Clifton Park, NY, 1997.

CHAPTER 13

BUILDING CONSTRUCTION

David Dodson, Lead Instructor, Response Solutions, Colorado

OUTLINE

- Objectives
- Introduction
- Building Construction Terms and Mechanics
- Structural Elements
- Fire Effects on Common Building Construction Materials
- Types of Building Construction
- Collapse Hazards At Structure Fires
- Lessons Learned
- Key Terms
- Review Questions
- Additional Resources

STREET STORY

A firefighter's worst fear is being trapped in a collapsing building and burning to death. Real-life encounters become implanted in our memory. It is one thing to read or hear about certain occurrences, it is another thing to have lived through one.

As a chief officer arriving at a fire scene, I encountered a large body of fire on the top floor of a three-story building. The first-due companies were initiating an offensive attack. My initial reaction was that the tactics being employed were correct. I surveyed the scene by doing a 360-degree walk-around of the building. I realized that conditions were not improving. I contacted the interior sector officer, and he informed me that interior conditions were deteriorating. I had noticed the following collapse indicators in my size-up: a large volume of unabated fire despite the aggressive interior attack; heavy smoke conditions indicating that the fire was probably attacking the building's structural supports; the ordinary constructed building contained a corbelled brick cornice, which added an eccentric load to the bearing wall supporting it; and lack of progress in the interior attack.

I had witnessed similar indicators in previous collapses. I ordered the units to withdraw from the building, set up a collapse zone, and initiated a defensive attack. The building's masonry wall collapsed shortly after the units' removal. The ensuing investigation found that the collapse occurred due to the combination of the collapse indicators.

Knowledge of building construction is vital before firefighters can attempt to fight a fire within a structure. Each type of construction contains inherent problems or positive features that can impede or assist in our ability to control a building fire. Examine the building. If it contains collapse indicators, decide their potential impact on the building's stability. If in doubt, it is better to err on the side of safety. Remember that once all civilians are removed from a structure, the life safety of firefighters is our utmost consideration.

—Street Story by James P. Smith, Deputy Chief, Philadelphia Fire Department, Philadelphia, Pennsylvania

OBJECTIVES

After completing this chapter, the reader should be able to:

- Describe the relationship between loads, imposition of loads, and forces.
- List and define four structural elements.
- Identify the effects of fire on five common building materials.
- List and define the five general types of building construction.
- List and define hazards associated with alternative building construction types.
- List five building collapse hazards associated with fire suppression operations.
- List five indicators of collapse or structural failure that might be found during fire suppression operations.

INTRODUCTION

Many fire departments pride themselves in their ability to launch aggressive interior structural fire attacks. Unfortunately, many firefighters are injured and killed when that same structure collapses, **Figure 13-1.** Often, buildings collapse without a "visual" warning such as sagging floors and roofs, leaning walls, and cracks. To keep from getting trapped in a collapse, firefighters must understand the types of structures they enter from the perspective of how the buildings are assembled, what materials are used, and how buildings react to fire. Additionally, firefighters must understand how fire travels through a building and choose appropriate tactics to stop the fire before key structural elements are attacked by the fire. Many firefighter fatality investigations conclude that fire departments need more training and education on building construction and the effects of fire on buildings. This chapter introduces several key topics regarding building construction and how fire affects buildings. It is important to note, however, that this chapter is merely an introduction. Firefighters must bridge the information in this chapter with a long-term commitment to study and research building construction and, more importantly, to explore the buildings within their jurisdiction.

Figure 13-1 This collapse happened seconds after firefighters were repositioned.

Streetsmart Tip Firefighters must realize that most of the buildings they will work in are already in place, and it is very difficult to determine the type of construction and fire-resistive rating by driving by or standing in front of a structure. Conducting in-service inspections, and securing the building owner's permission to walk through a building and get an "inside view" are ways of learning more about the structures in a particular jurisdiction.

If new construction is being conducted in a firefighter's response area or jurisdiction, contact the building department and secure a set of the plans. The fire department should be involved in the plan review process; however, this is not always the case. Also, take photographs of how the building is built and what materials are used for future training references. Many new materials and construction methods are introduced regularly. Firefighters should find out what materials are in the buildings in their jurisdictions.

This chapter begins by exploring some basic terms and mechanics of building construction, and then examines structural hierarchy and fire effects on materials. That information is then applied to classic and new construction types. The chapter concludes with a look at collapse hazards associated with structural fires.

BUILDING CONSTRUCTION TERMS AND MECHANICS

Firefighters need a basic understanding of certain terms and concepts associated with building construction. Obviously, buildings are constructed to provide a protected space to shield occupants from elements. The building must be built to resist wind, snow, rain, and still resist the force of gravity. Additionally, the intended use of the building can add a tremendous amount of weight, placing more stress on the building's ability to resist gravity. In building terms, these elements create building **loading**. Loads are then *imposed* on building materials. This imposition causes stress on the materials, called *force*. Forces must be delivered to the earth in order for the building to be structurally sound. With this basic understanding, we can start to define terms and mechanics.

Types of Loads

Loads can be divided into two broad categories as it relates to building construction: **dead loads** and **live loads**. Dead loads include the weight of all materials and equipment that are permanently attached to the building. Live loads include equipment, people, movement, and materials not attached to the structure. Dead loads and live loads can be more specifically described using the following terms:

- **Concentrated load**: A concentrated load is a load that is applied to a small area, **Figure 13-2.** An example of a concentrated load is a heating, ventilation, and air-conditioning (**HVAC**) unit on a roof.
- **Distributed load**: A distributed load is a load applied equally over a broad area, **Figure 13-3.** Examples of this include snow on a roof or a hoist attached to numerous roof supports.
- **Impact load**: Impact load is a load that is in motion when applied, **Figure 13-4.** Crowds of people, fire streams, and wind gusts are examples of impact loads.

Figure 13-2 The steel stairs and air-conditioning unit apply a concentrated load on this roof structure. Also note the potential instability of the air-conditioning unit placed on cement blocks.

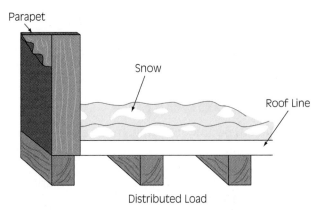

Figure 13-3 The weight of snow is a distributed load on this roof structure.

Figure 13-4 This ladder pipe operation is applying an impact load to the wall of this structure. As the wall weakens, it will eventually collapse. *(Photo courtesy of William H. Schmitt, Jr.)*

Historically Significant Building Collapses

Contributed by Dave Dodson

Many firefighters have been killed as a result of building collapse during firefighting operations. With each of these tragic losses, lessons can be learned. The following is a brief look at some of the more significant collapses. Each of these events should be researched to find all the contributing factors that led to the event. The italicized lessons are perceptions that are shared in the spirit of preventing firefighter injuries and death.

New York City, 1966

Firefighters responded to a commercial structure fire only to discover that the building shared a basement with another building. The concealed fire advanced rapidly and undetected. The ensuing firefight trapped and killed twelve firefighters. *It is vitally important to preplan buildings prior to a fire event. Older buildings may have access ways sealed from adjoining buildings. Shared utilities and other hidden voids can facilitate fire spread.*

The Boston Vendome Hotel Fire, 1972

Nine firefighters died during overhaul of a fire in an old, remodeled hotel. The investigation revealed that a masonry wall had been breached to make way for an air duct. Just above the breach, a column carried the load from floors above. A corner of the five-story building collapsed, trapping the firefighters. There were no obvious signs of impending collapse. *The Vendome building was brought back from disrepair in 1971, and many alterations were made that were unknown to the fire department. Firefighters should take an interest in the construction activities in and around buildings. Remodeling and restoration can compromise structural elements.*

Detroit, Abandoned Building, 1980

A fire was reported in a large, abandoned building that was scheduled for demolition. Responding firefighters found "light smoke" showing. The fire escalated rapidly due to the poor interior conditions and wide-open spaces. One firefighter died while trying to escape. Two other firefighters died when a firewall collapsed. *Abandoned buildings are much like a building under construction. Firefighters cannot take anything for granted. Rapid fire spread and suspect integrity should be the order of the day. Defensive operations must respect collapse zones.*

Hackensack, New Jersey, Ford Dealer, 1988

A fire was discovered in the attic space above an automobile repair garage. Responding firefighters launched an aggressive attack. The bowstring truss roof space was being used as a parts storage area, placing additional load on the structure. During firefighting operations, the roof collapsed, trapping and killing five firefighters. *Fighting fires in truss spaces is like playing Russian roulette. Trusses help form a wide, open space beneath. Clear spans are a warning sign of quick collapse should the truss space be involved in fire. Where there are no occupants to rescue, firefighters should reduce their risk and fight fire from safe attack points.*

Orange County, Florida, 1989

Firefighters responded to a fire in a single-story commercial structure. Interior conditions were described as light smoke and no heat. The fire had gained headway in a truss space above the ceiling. The tile-covered roof collapsed twelve minutes after firefighters arrived and killed two firefighters. *A fire can be roaring over firefighters' heads without them being aware. Firefighters should routinely inspect the ceiling space above their heads for fire. Once fire or heavy, dark smoke conditions are found in truss spaces, tactics should change. Tile roof coverings are quite attractive but show very little signs that the roof supporting them is about to collapse.*

Brackenridge, Pennsylvania, 1991

Firefighters were attempting to attack a fire in the basement of a large commercial building with concrete floors supported by steel columns. During the attack, the floor collapsed, trapping and killing four firefighters. *Basement fires present many difficult challenges to firefighters. Limited access, trapped heat and smoke, and the storage nature of basements must be factored in fire attack. Unfinished basements allow the fire to attack the floor above and floor supports rather quickly. Unprotected steel exposed to fire will soften quickly, leading to collapse.*

(continued)

Historically Significant Building Collapses (Continued)

One Meridian Plaza, Philadelphia, Pennsylvania, 1991

Three firefighters died when they became disoriented and ran out of air while fighting fire in the high-rise building. The fire started on the twentieth floor and ran up to the thirtieth floor where a sprinkler system extinguished the fire. Although the firefighters did not die from a collapse, this event is significant in that fire officers feared a catastrophic collapse due to stress cracks found in the concrete stair towers and withdrew their firefighters. *There are many lessons learned from this event. The fatality and fire investigation details many building construction issues associated with high-rise firefighting. It is available from the U.S. Fire Administration (http://www.fema.gov/usfa).*

Mary Pang Fire, Seattle, Washington, 1995

An arson fire in a multiuse commercial building caused the deaths of four firefighters when a portion of the floor collapsed. The building had been altered several times, and a lightweight "pony wall" had been used to replace a portion of a load-bearing wall. The building had a confusing layout including entry points at two different elevations (like a walkout basement). The advanced fire was not apparent from the "front" side of the building. An aggressive interior attack was under way when the floor collapsed. *Buildings that have gone through several owners and occupancy changes should always be suspect. Prefire planning helps uncover hazards that could change firefighting tactics. Firefighters should make a habit of reporting fire and building conditions and observations.*

Stockton, California, 1997

Two firefighters died when a home addition collapsed during interior firefighting operations. The homeowners had built a large, two-story clear-span addition on the back of the home. Firefighters entered the front of the building and found a heavy fire and dense smoke conditions. *Homeowners do not necessarily follow established building practices and codes when making additions. A "360" prior to fire attack can uncover significant hazards when "reading" the building. Firefighters cannot preplan every home, so they must rely on their ability to read buildings and read smoke conditions.*

World Trade Center, New York City, 2001

Terrorists hijacked two large airliners and hit the twin towers of the World Trade Center. The Fire Department of New York (FDNY) responded and began the biggest rescue effort in fire service history. The high-rise towers collapsed, killing 343 of FDNY's bravest. Thousands of civilians also died in the collapse. *Steel high-rise construction relies on fire-resistive coatings to protect the steel. The combination of burning jet fuel and the trauma of the aircraft strikes rendered the steel unprotected. Failure came much quicker than the four-hour time limit prescribed for Type I fire-resistive construction found in high-rises. Firefighters should never rely on fire protection time ratings when making decisions for fire attack. History will always remember that the FDNY firefighters died trying to rescue trapped civilians. They have the undying respect of people throughout the world.*

Design load: Design loads are loads that an engineer has planned for or anticipated in the structural design.

Undesigned load: Undesigned load is a load that was not planned for or anticipated. Buildings that are altered or are being used for occupancy other than original intent create an undesigned load. One common example is a residential structure that is converted to a print shop or legal office. These buildings were not designed to hold the additional live loads caused by the change in occupancy.

Fire load: A fire load is the number of British thermal units (Btus) generated when the building and its contents burn. It is important to note that the construction industry does not recognize *fire load* in its vocabulary—it is a fire engineering term.

Imposition of Loads

Loads must be transmitted to structural elements. This is called *imposition of loads*. Terms associated with imposition include axial load, eccentric load, and torsion load, **Figure 13-5.**

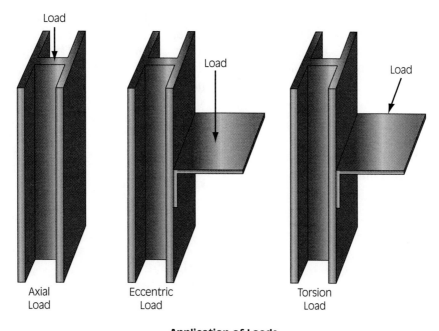

Application of Loads

Figure 13-5 There are three types of loads that can be transmitted through a structural member: axial, eccentric, and torsion.

An **axial load** is a load that is transmitted through the center of an element and runs perpendicular to the element. **Eccentric load** is applied perpendicular to an element and, subsequently, does not pass through the center of the element. **Torsion load** is a load that is applied offset to an element, causing a twisting stress to the material.

Forces

Loads imposed on materials create stress and strain on the materials used to make the element. Stress and strain are defined as forces applied to materials. These forces are defined as compression, tension, and shear, **Figure 13-6**. In **compression**, forces tend to push materials together. **Tension** occurs when forces tend to pull a material apart. **Shear** occurs when a force tends to "tear" a material apart—the molecules of the material are sliding past each other.

All loads—and the forces they create—must eventually pass through the structure and be delivered to the earth through the foundation of the building. Under normal conditions, structures will resist failure. Under fire conditions, the materials used to resist forces start breaking down. Eventually, gravity takes over and pushes the building to the earth.

Streetsmart Tip As a building burns, the structural elements decompose and lose their strength. This causes a change in the forces and the way the design loads are applied, leading to structural failure and collapse.

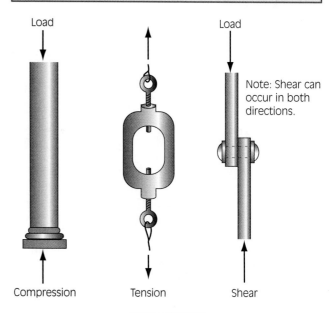

Types of Loads

Figure 13-6 Loads are applied to a structural member as compression, tension, and shear forces.

The time it takes for gravity to overcome the structure during a fire is not predictable. A number of variables determine the amount of time a material can resist gravity and fire degradation. These include:

- Material mass
- Surface-to-mass ratio
- Overall load being imposed
- Btu development (fire load)
- Type of construction (assembly method)
- Alterations (undesigned loading)
- Age deterioration/care and maintenance of the structure
- Firefighting impact loads
- Condition of fire-resistive barriers

> **Streetsmart Tip** Surface-to-Mass Ratio: **Surface-to-mass ratio** is defined as the exposed exterior surface area of a material divided by its weight. In simple terms, smaller, lighter structural members will have a large surface with small mass when compared to larger structural members capable of carrying an equal load. A 3- × 14-inch solid wood beam may carry the same design load as six 2- × 4-inch parallel chord trusses. The trusses have much more wood surface exposed and are more likely to ignite and burn rapidly.
>
> The larger the surface area, the smaller the mass, the quicker it will burn or fail. Also, in combustible construction, the large surface area provides more fuel for the fire. Surface-to-mass ratio may also be applied to lightweight steel or pre-engineered buildings. These lightweight structural steel members will absorb heat quickly, and the steel elements will lose their strength and fail, causing collapse.

STRUCTURAL ELEMENTS

Buildings are an assembly of structural elements designed to transfer loads to the earth. Structural elements can be defined simply as beams, columns, and walls. Each of these elements must be connected in some fashion in order to effectively make the load transfers to the building foundation, which delivers the building live and dead loads to earth.

Beams

A **beam** is a structural element that delivers loads perpendicular to its length. Obviously, something must support the beam—usually a wall or column.

It stands to reason that beams are used to create a covered space. In doing so, the beam is subjected to an eccentric load. This load causes the beam to deflect. The top of the beam is subjected to a compressive force whereas the bottom of the beam is subjected to tension, **Figure 13-7.** The distance between the top of the beam and the bottom of the beam dictates the amount of load the beam can carry. I beams are very typical and usually refer to the use of steel to form a beam. The top of the I is known as the top **chord**; the bottom of the I is called the bottom chord. The material in between is known as the **web.** There are numerous types of beams although the principal method of load transfer remains the same. A few types of beams include:

- **Simple beam:** A beam supported at the two points near its ends.
- **Continuous beam:** A beam that is supported in three or more places.
- **Cantilever beam:** A beam that is supported at only one end—or a beam that extends over a support in such a way that the unsupported overhang places the top of the beam in tension and the bottom in compression.
- **Lintel:** A beam that spans an opening in a load-bearing masonry wall—such as over a garage door opening. In wood construction, the same beam is often called a *header*.
- **Girder:** A beam that supports other beams.
- **Joist:** A wood framing member used to support floors or roof decking. A **rafter** is a joist that is attached to a ridge board to help form a peak.
- **Truss:** A series of triangles used to form a structural element that in many ways is really a "fake" beam. That is, a truss uses geometric shapes, lightweight materials, and connections to transfer loads just like a beam. Trusses will be covered in detail later in this chapter.
- **Purlin:** A series of wood beams placed perpendicular to steel trusses to help support roof decking.

Columns

A **column** is any structural component that transmits a compressive force parallel through its center. Columns typically support beams and other columns, **Figure 13-8.** Columns are typically viewed as the vertical supports of a building; however, columns can be diagonal or even horizontal. The guiding principle is that a column is totally in compression.

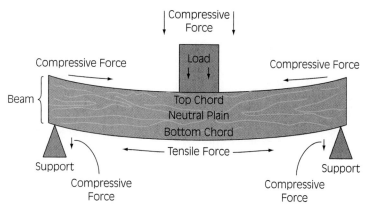

Figure 13-7 A beam transfers a load perpendicular to the load—creating compressive and tensile forces within the beam.

Walls

A wall is also a component that transmits compressive force through its center. Simply put, a wall is a really long, but slender, column. Walls are subdivided into two categories: **load-bearing** and non-load-bearing. A load-bearing wall carries the weight of beams, other walls, floors, roofs, or other structural elements as well as the weight of the wall itself. A non-load-bearing wall need only support its own weight. A partition wall is an example of a non-load-bearing wall.

Connections

As mentioned previously, beams, columns, and walls must be connected in some fashion in order to effectively transfer loads. Often, the connection is the weak link as it relates to structural failure during fires. The connection point is often a small, low-mass material that lacks the capacity to absorb much heat, thereby failing quicker than an element that has more mass such as a column or wall. Connections fall into three categories: pinned, rigid, and gravity. Pinned connections use bolts, screws, nails, rivets, and similar devices to transfer load. Rigid connections refer to a system where the elements are bonded together such that all the columns (or load-bearing walls) are bonded to all the beams. Typically, failure of one element will cause the loads to be transferred to other elements. Gravity connections are just that—the load from an element is held in place by gravity alone.

Together, structural elements defy gravity and make a building sound. A series of columns and beams used to hold up a building are often referred to as the *skeletal frame*. *Post and beam* describes the same concept. Beams resting on walls are simply called wall-bearing buildings. One factor that has not been discussed is the suitability of the materials used to form structural elements. The next section covers these materials and how they act during fires.

Figure 13-8 This column is supporting a beam, flooring, and another column. Columns are subjected to compressive forces.

FIRE EFFECTS ON COMMON BUILDING CONSTRUCTION MATERIALS

Many factors determine which material is used to form structural elements. Quality, cost, application, engineering capabilities, and adaptability all play in the suitability of a material. In some cases, the material chosen for a structural application

Tribute to the Ol' Professor
Contributed by Dave Dodson

Francis L. Brannigan is a true friend to the fire service. For over thirty-five years Mr. Brannigan has shared his knowledge of fire and, more specifically, effects of fire on buildings. His book *Building Construction for the Fire Service* (see the Additional Resources section at the end of this chapter) is a must-read for any firefighter and critical reading for anyone wanting to promote into fireground decision-making positions.

The fire service has affectionately called Mr. Brannigan the "Ol' Professor." His teachings have saved untold numbers of firefighters. I will never forget my first exposure to Mr. Brannigan. It was at a national conference in Cincinnati. The Ol' Professor was teaching a daylong class on steel buildings. As a young and inexperienced firefighter, I was all ears. The class taught me two things. First, I had much to learn about fire effects on building construction. I was way behind in my knowledge, and the Ol' Professor motivated me to make a never-ending knowledge quest to understand buildings. Second, I realized that reading buildings and reading smoke were the keys to rescuing people and putting the "wet stuff on the red stuff."

Over the years, Mr. Brannigan has coined many powerful—and lifesaving—phrases. These bits of advice have remained part of the teachings of many fire instructors. Among my favorites:

Trusses

"BEWARE the TRUSS!"

"A truss is a truss, is a truss . . ."

—in reply to the notion that a bowstring truss is more dangerous than other trusses.

"The bottom chord of a truss is under tension—It's like you hanging on a rope. If the rope gets cut, you will fall. So it is with a truss."

"Failure of one element of a truss may cause the entire truss to fail—failure of one truss can cause other trusses to fail."

Columns

"The failure of a column is likely to be more sudden than failure of a beam."

"The slightest indication of column failure should cause the building to be cleared immediately."

On Collapses

"There is a tendency among those concerned about building stability to make light of partial collapse . . . a partial collapse is very important to at least two groups—those under it and those on top of it!"

"From an engineering point of view, (lightweight, trussed) buildings are made to be disposable . . . we don't make disposable firefighters!"

In response to those who claim a building collapsed without warning during a fire:

"The warning is the brain—in your ability to understand buildings and anticipate how they will react to a fire."

The fire service is indebted to the Ol' Professor. Thanks for all you have done!

---

needs to meet fire resilience criteria. Regardless, the firefighter needs to understand how these materials react to fire. In the past, the fire service looked at the characteristics of four basic material types: wood, steel, concrete, and masonry. Each of these materials can be found together or separately. Each material reacts to fire in a different way, **Table 13-1**. Now, advanced material technology has found its way into structural elements. Buildings are being assembled using plastics, graphites, wood derivatives, and other composites. This section covers the four basic building materials as well as some of the new composites.

Wood

Wood is perhaps the most common building material. It is used in millions of residential and commercial buildings. Wood is relatively inexpensive, easy to manipulate, and a replenishable natural resource (although that can be argued). Wood has marginal resistance to forces compared to its weight, but it does the job for most residential and small commercial buildings. Wood also burns—and in doing so gives away its mass. The more mass a section of wood has, the more material it must burn away before strength is lost. This is true of native wood—that is, wood that

Performance of Common Building Materials under Stress and Fire

MATERIAL	COMPRESSION	TENSION	SHEAR	FIRE EXPOSURE
Brick	Good	Poor	Poor	Fractures, spalls, crumbles
Masonry block	Good	Poor	Poor	Fractures, spalls
Concrete	Good	Poor	Poor	Spalls
Reinforced concrete	Good	Fair	Fair	Spalls
Stone	Good	Poor	Fair	Fractures, spalls
Wood	Good w/grain; poor across grain	Marginal	Poor	Burns, loss of material
Structural steel	Good	Good	Good	Softens, bends, loses strength
Cast iron*	Good	Poor	Poor	Fractures

*Some cast iron may be ornamental in nature and not part of the structure or load bearing.

TABLE 13-1

has been cut from a tree. Engineered wood can react differently when exposed to heat from a fire. Engineered wood includes a host of products that take many pieces of native wood and glue them together to make a sheet, longer beam (trees only grow so tall!), or stronger column. Plywood delaminates when exposed to fire. Some newer wood products such as composites, which are discussed later in this chapter, present safety concerns for all firefighters.

Steel

Steel is a mixture of carbon and iron ore heated and rolled into structural shapes to form elements for a building. Steel has excellent tensile, shear, and compressive strength. For this reason, steel is a popular choice for girders, lintels, cantilevered beams, and columns. Additionally, steel has high factory control. It is easy to change its shape, increase its strength, and otherwise manipulate it during production.

As it relates to fires, steel loses strength as temperatures increase. The specific range of temperatures depends on how the steel was manufactured. Cold drawn steel, like cables, bolts, rebar, and lightweight fasteners, loses 55 percent of its strength at 800°F. Extruded structural steel used for beams and columns loses 50 percent of its strength at 1,100°F. Structural steel will also elongate or expand as temperatures rise. At 1,000°F, a 100-foot-long beam will elongate 10 inches. Imagine what that could do to a building. If a beam is fixed at two ends, it will try to expand—and likely deform, buckle, and collapse. If the beam sits in a pocket of a masonry wall, it will stretch outward and place a shear force on the wall—which was designed only for a compressive force. *This could knock down the whole wall!*

Because steel is an excellent conductor of heat, it will carry heat of a fire to other combustibles. This can cause additional fire spread, sometimes a considerable distance from the original fire.

> **Caution** Steel softens, elongates, and sags when heated, leading to collapse. Cooling structural steel with fire streams is just as important as attacking the fire.

Concrete

Concrete is a mixture of portland cement, sand, gravel, and water. It has excellent compressive strength when cured. The curing process creates a chemical reaction that bonds the mixture to achieve strength. The final strength of concrete depends on the ratio of these materials, especially the ratio of water to portland cement. Because concrete has poor tensile and shear strength, steel is added as reinforcement. Steel can be added to concrete in many ways. Concrete can be poured over steel rebar and become part of the concrete mass when cured. Cables can be placed through the plane of concrete and be tensioned, compressing the concrete to give it required strength. Cables can be pretensioned (at a factory) or posttensioned (at the job site). *Precast concrete* refers to slabs of concrete that are poured at a factory and then shipped to a job site. Precast slabs are "tilted up" to form load-bearing walls—thus the term *tilt-up construction*.

All concrete contains some moisture and continues to absorb moisture as it ages. When heated, this moisture content will expand, causing the concrete to crack or spall. **Spalling** refers to a large pocket of concrete that has basically crumbled into fine particles, taking away the mass of the concrete. Reinforcing steel that becomes exposed to a fire can transmit heat within the concrete, causing catastrophic spalling and failure of the structure. Unlike steel, concrete is a heat sink and tends to absorb and retain heat rather than conduct it. This heat is not easily reduced. Concrete can stay hot long after the fire is out, causing additional thermal stress to firefighters performing overhaul.

Masonry

Masonry is a common term that refers to brick, concrete block, and stone. Masonry is used to form load-bearing walls because of its compressive strength. Masonry can also be used to build a **veneer** wall. A veneer wall supports only its own weight and is most commonly used as a decorative finish. Masonry units (blocks, bricks, and stone) are held together using **mortar**. Mortar mixes are varied but usually contain a mixture of lime, portland cement, water, and sand. These mixes have little to no tensile or shear strength. They rely on compressive forces to give a masonry wall strength. A lateral force that exceeds the compressive forces within a masonry wall will cause quick collapse of the wall.

> **Streetsmart Tip** **Masonry Walls Collapse:** Masonry has very little lateral stability, and in many cases the roof or floor structure of a building holds the walls in place. Steel beams or joists will expand during a fire, creating lateral loads that the walls were not designed to withstand. In addition, wood roof structures will burn away or collapse during a fire, leaving little lateral support. The effects of the fire, pulling forces of the collapsing wood structure, or the force of a hose stream may cause the wall to collapse. Remember, the designer and contractor did not plan for the building to burn down.

Figure 13-9 Prior to a fire, the effects of age will take their toll on masonry walls. How stable is this wall with joint deterioration and lack of full mortar bond?

Brick, concrete block, and stone have excellent fire-resistive qualities when taken individually. Many masonry walls are typically still standing after a fire has ravaged the interior of the building. Unfortunately, the mortar used to bond the masonry is subject to spalling, age deterioration, and washout. Whether from age, water, or fire, the loss of bond will cause a masonry wall to be very unstable, **Figure 13-9.**

Composites

New material technologies have introduced some interesting challenges for the firefighting community. Composites are a combination of the four basic materials listed above as well as various plastics, glues, and assembly techniques. Of particular interest are the many wood products that are widely used for structural elements.

Lightweight wooden I beams (joists) are nothing more than wood chips that are press-glued together into the shape of an I beam, **Figure 13-10.** While structurally strong (stronger than a comparable solid wood joist), the wooden I beam fails quickly when heated. Actually, no fire contact is required. Ambient heating causes the binding glue to fail, leading to a quick collapse. The bottom of a beam is under tensile forces. If the bottom of the beam falls off, due to glue failure, the beam will immediately snap and collapse.

New products, known as FiRP (fiber-reinforced products) are becoming common in the construction industry. FiRP can be plastic fibers mixed with wood to give the wood increased tensile strength. As with most plastics, fire exposure can cause quick failure as the plastic melts.

The mixture of steel and wood as a structural element can cause rapid collapse because steel expands

BUILDING CONSTRUCTION ■ **351**

Figure 13-10 To save on materials and cost, the use of composites or engineered wood structural members is becoming popular. Shown here is a typical wood I beam with 2- × 3-in. flanges and a ⅜-in. structure board web. In addition to providing a large surface-to-mass ratio, the flanges and web are fastened with glue, which may deteriorate quickly under fire conditions.

Figure 13-11 A composite truss. Rapid heating will cause the steel to separate from the wood chords.

(A)

(B)

Figure 13-12 These are wall panels that are load-bearing (A). Expanded polystyrene is sandwiched between OSB sheets. Fire can easily enter the wall space. (B) Failure of the wall panel will cause instability in the roof structure.

faster than wood. This causes stress at the intersection of the two materials, **Figure 13-11.**

Structural insulated panels (SIP) are another interesting composite. This technique is characterized by large wall panels made of expanded polystyrene sandwiched between two sheets of oriented strand board (OSB). OSB is a sheet of wood chips bonded by glue, **Figure 13-12A** and **B.** The OSB is covered with a typical wall finish. It is anticipated that a fire will cause rapid deterioration of the load-bearing panel.

To review, this chapter has so far explored the basic terminology, mechanics, elements, and materials used in the construction of buildings. It has also discussed a bit about how fire attacks materials and causes failure. The next section explores the various methods used to assemble buildings.

TYPES OF BUILDING CONSTRUCTION

Over time, five broad categories of building construction types have been developed to help classify structures. These categories give firefighters a basic

Types of Construction from NFPA 220

	TYPE I		TYPE II			TYPE III		TYPE IV	TYPE V	
	443	332	222	111	000	211	200	2HH	111	000
Exterior Bearing Walls—										
Supporting more than one floor,										
columns, or other bearing walls	4	3	2	1	0	2	2	2	1	0
Supporting one floor only	4	3	2	1	0	2	2	2	1	0
Supporting a roof only	4	3	1	1	0	2	2	2	1	0
Interior Bearing Walls—										
Supporting more than one floor,										
columns, or other bearing walls	4	3	2	1	0	1	0	2	1	0
Supporting one floor only	3	2	2	1	0	1	0	1	1	0
Supporting roofs only	3	2	1	1	0	1	0	1	1	0
Columns—										
Supporting more than one floor,										
columns, or other bearing walls	4	3	2	1	0	1	0	H[1]	1	0
Supporting one floor only	3	2	2	1	0	1	0	H[1]	1	0
Supporting roofs only	3	2	1	1	0	1	0	H[1]	1	0
Beams, Girders, Trusses										
& Arches—										
Supporting more than one floor,										
columns, or other bearing walls	4	3	2	1	0	1	0	H[1]	1	0
Supporting one floor only	3	2	2	1	0	1	0	H[1]	1	0
Supporting roofs only	3	2	1	1	0	1	0	H[1]	1	0
Floor Construction	3	2	2	1	0	1	0	H[1]	1	0
Roof Construction	2	1½	1	1	0	1	0	H[1]	1	0
Exterior Nonbearing Walls	0	0	0	0	0	0	0	0	0	0

☐ Those members that shall be permitted to be of approved combustible material.

[1]"H" indicates heavy timber members; see text for requirements.

Source: Reprinted with permission from NFPA 220, *Types of Building Construction*, copyright © 1995, National Fire Protection Association, Quincy, MA 02269. This reprinted material is not the complete and official position of the National Fire Protection Association on the referenced subject which is represented only by the standard in its entirety.

TABLE 13-2

understanding of the arrangement of structural elements and the materials used to construct the building. Unfortunately, these broad classifications are dangerously incomplete for firefighters and may lead to deadly assumptions about the makeup of a building. As stated before, firefighters need to explore the buildings within their jurisdiction to determine how buildings are assembled.

It is important to note that buildings are built to meet certain codes. These codes are designed to give occupants time to escape during a fire. Concrete is **fire resistive**—meaning it has some capacity to withstand the effects of fire. Other materials, like steel and wood, need fire-resistive assistance to give occupants a chance to escape. Building codes outline **fire-resistive ratings**, occupancy classifications, and means of egress based on five general types of buildings. The features of each type of construction will be discussed shortly, but first it is important to understand fire resistance for structural elements. **Table 13-2** outlines the number of hours that a structural element needs to be protected for the five types of construction. Simply put, firefighting time is not part of the fire-resistive and building construction equation. Fire-resistive ratings are established in a laboratory. In the real world, fire resistance ratings could "underperform" due to many factors. For example, a structural element with a two-hour fire rating may fail in thirty minutes if it was not assembled correctly or if improperly inspected. Fire resistance for structural members can be achieved by various methods including drywall (gypsum wallboard), spray-on coatings, and concrete, **Figure 13-13.** Aging, alterations, and wear can damage fire-resistive methods to the point that structural elements have no fire resistance protection.

The following paragraphs outline the basic definition of each building type, its general configuration, and some historical fire spread problems associated with each. Also included are some construction methods that do not fit into the five common types.

Type I: Fire-Resistive

Type I fire-resistive construction is a type in which structural elements are of an approved noncombustible or limited combustible material with sufficient fire-resistive rating to withstand the effects of fire and prevent its spread from story to story. Concrete-encased steel, **Figure 13-14,** monolithic-poured cement, and steel with spray-on fire protection coatings are typical of Type I, **Figure 13-15.** Generally, the fire-resistive rating must be three to four hours depending on the specific structural element. Fire-resistive construction is used for high-rises, large sporting arenas, and other buildings where a high volume of people are expected to occupy the building.

Most Type I buildings are typically large, multi-storied structures with multiple exit points. Fires are difficult to fight due to the large size of the building and the subsequent high fire load. Type I buildings rely on protective systems to rapidly detect and extinguish fires. If these systems do not contain the fire, a difficult firefight will be required. Fire can spread from floor to floor on high-rises as windows break and the next floor windows fail, allowing the fire to jump. Fire can also make vertical runs through utility and elevator shafts. Regardless, firefighters are relying on the fire-resistive methods to protect the structure from collapsing. The collapse

Figure 13-13 This parking garage is of Type II construction. The protective coating applied to the structural steel may increase the fire-resistive rating, but note that the unprotected corrugated metal flooring and interior steel structure are not protected. These unprotected structural members may fail early in a fire. *(Photo courtesy of William H. Schmitt, Jr.)*

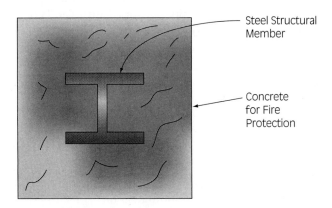

Figure 13-14 To achieve a Type I fire-resistive rating, structural steel members are encased with concrete to prevent failure from the effects of a fire.

Figure 13-15 A typical Type I building, with structural members designed to resist the effects of fire for three to four hours. This building is of reinforced concrete construction.

Figure 13-16 Buildings of Type II construction will have structural elements with little or no protection from the effects of fire. Remember, in the event of a fire, these unprotected steel structural members may fail and collapse with little warning.

of fire-resistive structures can be massive, as we are reminded from the World Trade Center collapse in New York City.

Type II: Noncombustible

Type II noncombustible construction is a type in which structural elements do not qualify for Type I construction and are of an approved noncombustible or limited combustible material with sufficient fire-resistive rating to withstand the effects of fire and prevent its spread from story to story. More often than not, Type II buildings are steel, **Figure 13-16.** Modern warehouses, small arenas, and newer churches and schools are built as noncombustible. Because the steel is not required to have significant fire-resistive coatings, Type II buildings are susceptible to steel deformation and resulting collapse. Fire spread in Type II buildings is influenced by the contents. While the structure itself will not burn, rapid collapse is possible from the content Btu release stressing the steel.

Suburban strip malls with concrete block load-bearing walls and steel roof structures can be classified as Type II. Fires can spread from store to store through wall openings and shared ceiling and roof support spaces. The roof structure is often of lightweight steel that fails rapidly. More often than not, the fire-resistive device used to protect the roof structure is a dropped-in ceiling. Missing ceiling tiles, damaged drywall, and utility penetrations can render the steel unprotected. These buildings may have combustible attachments such as facades and signs as well as significant content fire loading.

Type III: Ordinary

The term **Type III ordinary construction** is often misapplied to wood frame buildings. By definition, ordinary construction includes buildings where the load-bearing walls are noncombustible, and the roof and floor assemblies are wood. Most commonly, this is load-bearing brick or concrete block with wood roofs and floors. Ordinary con-

Figure 13-17 Buildings of Type III, ordinary construction, are common throughout North America. These typical "Downtown USA" buildings provide many challenges to firefighters, such as void spaces and common walls allowing rapid fire extension and little structural protection, with early collapse during firefighting operations.

> **Firefighter Fact** Sagging or bowing load-bearing walls are often pulled back in alignment by tightening a steel rod that runs through the building from wall to wall. A small interior fire can elongate this steel and cause catastrophic wall failure. These buildings can be spotted by decorative stars or ornaments (called *spreaders*) on the outside brick wall.

struction is prevalent in most downtown or "main street" areas of established towns and villages, **Figure 13-17**. Firefighters have long called ordinary construction "taxpayers." This slang is derived from landlords who built buildings with shops and/or restaurants on the first floor with apartments above in order to maximize income to help pay property taxes. Newer Type III buildings include strip malls with block walls and wood truss roofs, **Figure 13-18**.

Ordinary construction presents many challenges to firefighters. In older buildings, numerous remodels, restorations, and repairs have created suspect wall stability and hidden dangers.

Ordinary construction has many void spaces where fire can spread undetected. Common hallways, utilities, and attic spaces can communicate fire rapidly. Masonry walls hold heat inside, making for difficult firefighting. Wood floors and roof beams are often gravity fit within the masonry walls. These can release quickly and cause a general collapse, leaving an unsupported masonry wall. Older Type III buildings have structural mass; therefore, they burn for a long time.

Type IV: Heavy Timber

Type IV heavy timber construction can be defined as those buildings that have block or brick exterior load-bearing walls and interior structural members, roofs, floors, and arches of solid or laminated wood without concealed spaces. The minimum dimensions for structural wood must meet the criteria in **Table 13-3**. Heavy timber buildings, as the name suggests, are quite stout and are used for warehouses, manufacturing buildings, and some older churches, **Figure 13-19**. In many ways, a Type IV building is like a Type III—just larger dimension lumber instead of common wood beams and trusses.

Figure 13-18 One of the most common uses of Type III, ordinary construction, is the "strip mall" with masonry walls and lightweight steel or wood trusses. Common problems associated with this type of construction are void spaces allowing for rapid fire extension and collapse of lightweight structural elements.

Heavy Timber Dimensions

TYPE OF ELEMENT	USE	SIZE
Column	Supporting floor load	8- × 8-in. minimum any dimension
Column	Supporting roof load	6-in. smallest dimension, 8-in. depth minimum
Beams and girders	Supporting floor load	6-in. width and 10-in. depth minimum
Beams, girders, and roof framing	Supporting roof loads only	4-in. width minimum, 6-in. depth minimum
Framed or laminated arches	As designed	8-in. minimum dimension
Tongued and grooved planks	Floor systems	3-in. minimum thickness with additional 1-in. boards at right angles
Tongued and grooved planks	Roof decking	2-in. minimum thickness

TABLE 13-3

Figure 13-19 Type IV buildings, heavy timber construction, have large wood structural elements with great mass. The mass of these structural members requires a long burn time for failure. The connections, usually steel, are the weak points in this type of construction.

Some firefighters mistakenly call Type IV buildings "mill construction." Mill construction is a much more stout, collapse-resistive building that may or may not have block walls. A new Type IV building is hard to find. The cost of large-dimension lumber and laminated wood beams makes this type of construction rare.

Fire spread in a heavy timber building can be fast due to wide-open areas and content exposure. The exposed timbers contribute Btus to the fire. Because of the mass and large quantity of exposed structural wood, fires burn a long time. If the building housed machinery at one time, oil-soaked floors will add more heat to the fire and accelerate collapse. Once floors and roofs start to sag, heavy timber beams may release from the walls. This is accomplished by making a fire-cut on the beam, and the beam is gravity fit into a pocket within the exterior load-bearing masonry wall, **Figure 13-20**. As the floor sags, it loses its contact point with the wall and simply slides out of its pocket without damage to the wall. It is important to recall that a free-standing masonry wall has little lateral support and requires compressive weight from floors and roofs to make it sound.

Type V: Wood Frame

Type V wood frame construction is perhaps the most common construction type. Homes, newer small businesses, and even chain hotels are built primarily with wood, **Figure 13-21**. Older wood frame buildings were built as **balloon frame**—

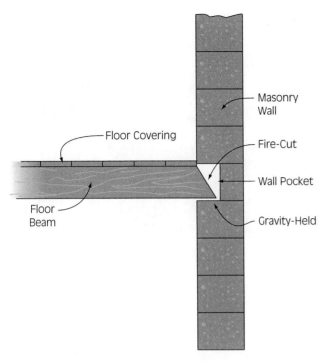

Figure 13-20 Wood and heavy timber beams were often "fire-cut" so that a fire-damaged, sagging floor would simply slide out of the wall pocket in order to preserve the wall.

Figure 13-21 The wood frame structure, Type V construction, is the most common type of construction in North America.

that is, wood studs ran from the foundation to the roof and floors were "hung" on the studs. As can be envisioned, fire could enter the wall space and run straight to the attic. In the early 1950s, builders started using a **platform framing** arrangement where one floor was built as a platform for the next floor. This created **fire stopping** to help minimize fire spread. Newer wood frame buildings utilize lightweight wood trusses for roofs and floors. This is akin to a "horizontal balloon frame" that can allow quicker lateral fire spread. Coupled with high surface-to-mass wood exposure, collapse becomes a real possibility, **Figure 13-22.** Some codes require truss spaces to have fire stopping every 500 square feet. Even with this fire stopping, it remains dangerous to step onto the 500 square feet where the fire is. Wood frame structures may appear more like a Type III ordinary building because of a brick-wall appearance. Remember, brickwork may be a simple veneer to add aesthetics.

To protect structural members from a fire, wood frame construction typically uses gypsum board (drywall or the brand name Sheetrock). Once finished, wood frame buildings typically have many rooms that can help compartmentalize content fires. Fire that penetrates wall, floor, or attic spaces becomes a significant collapse threat, especially in newer buildings. Often, the only warning that fire has penetrated these spaces is the issuance of smoke from crawl space vents, gable end vents, and eaves.

Other Construction Types

As mentioned earlier, the five broad building types can actually lead to dangerous assumptions. Newer construction and alternative building methods may not fit cleanly into one of the above types. Some buildings are actually two types of construction. For example, a particular restaurant located in Colorado is built as a Type II noncombustible yet is topped with a large wood frame structure to hide rooftop HVACs and cooking vent hoods, **Figure 13-23A** and **B.** The square feet space of the false dormers and wood frame structure exceeds that of most homes.

New lightweight steel homes resemble wood frame homes. These buildings are actually a "post and beam" steel building with lightweight steel studs to help partition the home. OSB is added to the studs to help make the house more "stiff" and increase wind-load strength, **Figure 13-24.** Another interesting construction type uses foam blocks to make a form for a lightweight concrete mud mixture. The concrete is not contiguous—there are many voids, utility runs, and foam block spacers (made of plastic or galvanized steel), **Figure 13-25.** These structures are called "ICF" or insulated concrete formed." However, it is important not to be fooled by any claims that these buildings are concrete or are less combustible. In reality, these composite buildings are assembled with plastics, polystyrene, lightweight steel, and lightweight concrete. When finished, these buildings may resemble wood frame or even ordinary construction.

(A)

(B)

(C)

Figure 13-22 (A, B) Note the void spaces at the first and second floor levels and in the attic area that are created by the use of truss systems. (C) The building after it was completed. Firefighters should survey the buildings in their area before they are completed.

Extended window and door jambs are clues that indicate the wall is thicker than that of typical wood or masonry built buildings.

The fire service has very little research information on the stability of these new types of buildings during fires. One thing is certain: Firefighters should expect rapid collapse due to the low-mass, high surface-to-mass exposure of structural elements.

Manufactured buildings can be defined as those structures that are built at a factory and then trucked to a job site. These building are quite light with little mass. Where a stick-built home uses 2 × 4 or 2 × 6 lumber, the manufactured home uses 1 × 2 and 2 × 2 lumber. These buildings use galvanized strapping to give required strength. In any case, these buildings burn quickly and collapse equally fast.

BUILDING CONSTRUCTION

Figure 13-23 (A) The decorative roof assembly is a Type V wood frame structure while the occupancy space is Type II noncombustible. (B) This building uses two types of construction.

Figure 13-24 This lightweight steel home is built similar to a Type V. OSB sheeting gives the steel rigidity to torsional loads like wind.

Relationship of Construction Type to Occupancy Use

Before considering the basic types of construction, many officials and builders first look at the anticipated use of the building—its occupancy type. **Occupancy classifications** are called many different names around the country, but they are usually broken down into five basic arenas: residential, commercial, business, industrial, and educational. Each of these general occupancies has a number of hazards that firefighters must understand, **Table 13-4.** Remember, a building may have been built for one type of occupancy only to be sold and converted to another occupancy type for which it may not have been designed. Firefighters should go out and explore the buildings in their community.

 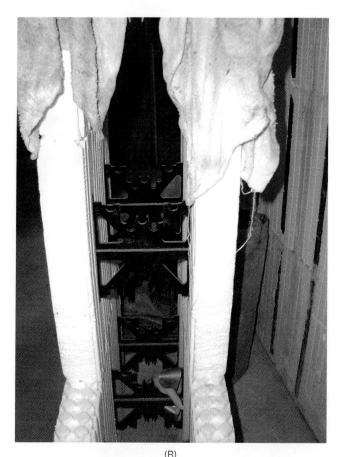

Figure 13-25 (A) This wall is a load-bearing foam block unit filled with a lightweight concrete mud mix. (B) Note the black plastic spacers that will fail early in a fire.

Typical Hazards Associated with Occupancies

OCCUPANCY	TYPE OF CONSTRUCTION	HAZARDS
Residential	Type V, most common	Fire loading, truss construction, owner alterations, rapid fire extension in void spaces
Commercial	Type III, most common	Fire loading, truss construction, rapid fire extension in void spaces, unknown occupancy change
Educational	Type II, most common	Unprotected structural steel, collapse, high fire load in some areas
Business	Types II and III, most common	Unknown change in occupancy, high fire load, difficult to ventilate
Industrial	Types I and II, most common	Hazardous materials, difficult to ventilate

TABLE 13-4

COLLAPSE HAZARDS AT STRUCTURE FIRES

It cannot be overstated that firefighters have to understand the buildings in their jurisdiction. Constant reading, study, and site visits will help them "read" buildings. Reading buildings is essential to anticipating collapse proactively. This section addresses some specific collapse threats that the fire service has experienced throughout history and the importance of understanding buildings and how they react at structure fires.

Figure 13-26 Wood trusses provide a large surface-to-mass ratio, fuel load, and void spaces—three of the worst conditions a firefighter will encounter during structural firefighting operations.

Figure 13-27 A typical parallel chord truss. The gusset plates on this truss are pressed into the wood. In addition to the decomposing of the wood element, the light-gauge steel plates will deform and pull away from the wood under fire conditions.

Trusses

Truss roof collapses have killed many firefighters. As stated previously, a truss is actually a fake beam. A truss uses geometric shapes (the triangle) to create a structural element similar to a beam. A wood truss can actually be stronger than a like-sized solid wood beam, and does so with less material, **Figure 13-26.** It is this loss of material and subsequent increase in exposed surface area that make them so vulnerable during fires. Trusses rely on each and every part of the truss to carry a portion of the imposed load. Like a beam, the top of the truss (called the top chord) is typically under a compressive force. The bottom chord is under tension. In between the two chords, connecting members (the web) transfer the two forces creating stress and strain. Failure of one part of the truss will likely cause the whole truss to fail. This distributes the weight of the failed truss to other trusses—which may not have the capacity to take that weight—thereby starting a domino effect collapse. Trusses come in many styles and shapes. Bowstring truss, parallel chord truss, and open web joists are some of the more common names. Trusses are also classified by the type of material used in assembly.

Wood Trusses

Wood trusses are an assembly of many pieces of wood. Some may even be press-glued particles. These pieces are connected using **gusset plates**. A gusset plate is a simple galvanized steel plate (very thin) with perforations punched into the plate. The perforations are used to pierce into wood fibers to hold pieces together. These perforations only penetrate the wood a fraction of an inch (3/8 inch is typical), **Figure 13-27.** During fires, the steel gusset heats up and transfers heat into the very wood fibers that are being held. If the heating is slow (like a smoke-filled attic), the wood decomposes, allowing the gusset plate to fall out. If the heating is fast, like sudden exposure to flame, the steel expands too quickly for the wood and the gusset simply pops out. Either way, truss failure is imminent. Sometimes the truss gingerly stays together because of the weight of roofing or flooring materials, yet a sudden force, like a walking firefighter, will cause the truss to disassemble and suddenly collapse.

Wood trusses are mass-produced at a factory where quality control may not be adequate. Further, the truss gusset plates may vibrate or be damaged while being delivered to a job site. Once on the job site, contractors may use shortcuts to lift the truss into position, furthering the damage to gusset plates.

Steel Trusses

Steel trusses are no less susceptible to collapse than wood trusses. Like wood, steel trusses are an assembly of pieces—typically angle iron for the chords and cold-drawn round stock for the web. The pieces are tack-welded together to form the truss unit. While not a true joist by definition, many call the common steel truss an open web steel joist, **Figure 13-28.** The term *bar joist* is also used to describe an open web steel joist. These trusses expose a large surface area to heat during fires. Given the lack of mass, the truss heats quickly and will soften and expand. The expansion can cause wall movement. (Remember, masonry walls must be loaded axial with compressive force.) Lateral movement can cause wall collapse. If the wall does not move, the steel truss will twist and buckle to allow expansion. It is very important to keep steel trusses cool.

Figure 13-28 Unprotected open web steel joists present a large surface area to absorb the heat of a fire, expand, and collapse. Structural steel will lose 50 percent of its strength at temperatures of 1,000°F.

Figure 13-29 Lightweight floor truss systems have many void spaces. This could be called "horizontal balloon frame."

Void Spaces

Trusses create large void spaces. The area between the chords of trusses will allow fires to spread horizontally, **Figure 13-29.** Some codes require fire stopping in floor truss spaces but may allow wide-open attic spaces. Fires can start in void spaces due to electrical and other utility problems. In Type III ordinary construction, voids are numerous. Some voids may pass through masonry walls, causing fire spread from one store to the next in a row of buildings. The obvious collapse danger with void spaces is that the fire may be undetected with simultaneous destruction of structural elements.

Roof Structures

The roof of a building can be flat, pitched, or inverted. Many factors help determine how and why the roof is built the way it is. Sometimes the roof is designed just to hide rooftop HVACs. Other times the roof shape is designed to shed snow, accommodate a vaulted ceiling, or merely give the building character, **Figure 13-30.** As it relates to

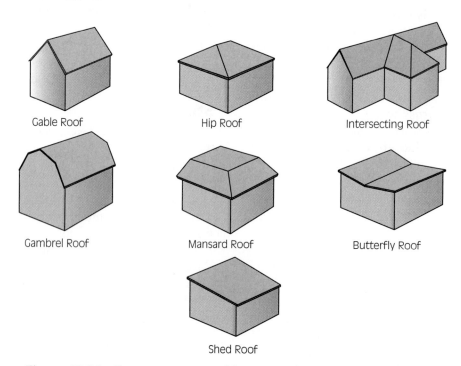

Figure 13-30 Some common roof framing styles used in wood frame or ordinary construction.

BUILDING CONSTRUCTION 363

structural collapse, the roof style may allow a large volume of fire to develop. Other roofs, like the mansard, have many concealed spaces. Dormers are protrusions from a roof structure. Dormers can be used to introduce daylight into a roof space that is converted into a living space, **Figure 13-31.** Other dormers are actually aesthetic (false) and can fool ventilation crews attempting to relieve heat from a roof space.

Stairs

First arriving firefighting crews rely on internal stairways to help gain access for rescue and fire attack. For years, firefighters have found stairways to be durable and a bit stronger than other interior components. This is a dangerous assumption in newer wood frame buildings. Stairs are now being built offsite and simply hung in place using light metal strapping, **Figure 13-32.** Additionally, stairs are being made using lightweight engineered wood products that fail quickly when heated. Remember, press-glued wood chip products can fail from the heat of smoke—no flame is required.

Parapet Walls

A **parapet** wall is the extension of a wall past the top of the roof. Parapets are used to help hide unsightly roof equipment and HVACs and give a building a finished look. Typically masonry, these walls are free-standing with little stability. Collapse may be caused by the failure of the roof structure, **Figure 13-33.** Business owners hang signs, utility connections, and other loads on the parapet. During a fire, the steel cables and bolts holding these will weaken and subsequently pull down the parapet, **Figure 13-34.**

Figure 13-32 This prefab stair assembly is hung in place by thin metal strapping. Note the staples and plastic shims that can quickly fail under fire conditions.

Figure 13-31 An internal view of dormers in a mansard roof. Fire can run through the many voids between the exterior and the roof.

Figure 13-33 This is the scene of a typical "parapet" wall failure common in Type III ordinary construction.

Figure 13-34 This electrical service entrance and weather head may be the eccentric load causing an early failure of this parapet wall.

Collapse Signs

Firefighters must rely on building material knowledge, building construction principles, and an understanding of fire effects on buildings in order to predict or anticipate collapse. Waiting for a visual sign that a building will collapse is dangerous, especially in newer buildings. There are, however, some factors and observations that can be used to help anticipate collapse. These include:

- Overall age and condition of the building
- Deterioration of mortar joints and masonry
- Cracks, in anything
- Signs of building repair including reinforcing cables and tie-rods
- Large open spans
- Bulges and bowing of walls
- Sagging floors
- Abandoned buildings
- Large volume of fire
- Long firefighting operations—remember gravity?
- Smoke coming from cracks in walls
- Dark smoke coming from truss roof or floor spaces (Brown smoke indicates that wood is being heated significantly; black smoke means combustibles have ignited or are near ignition.)
- Multiple fires in the same building or damage from previous fires

Buildings under Construction

Buildings are especially unsafe during construction, remodeling, and restoration. The word *unsafe* applies not only to fire operations but also to rescues, odor investigations, and on-site inspections. Buildings need only meet fire and life safety codes when they are completed. During construction, many of the protective features and fire-resistive components are incomplete. Additionally, stacked construction material may overload other structural components. This is not to say contractors are using unsafe practices, but to underscore that exposed structural elements, incomplete assemblies, and material stacks will contribute to a rapid collapse if a fire were to develop.

> **Streetsmart Tip** **Beware of Buildings under Construction:** A building as a complete unit has a number of interdependent parts. During construction, these parts may not be fully connected (steel), may not be at their full design strength (concrete), and may lack any type of fire protection (gypsum board or concrete). Also, many structural elements may be held in place by scaffoldings, false work, or forms. Because of this, a building under construction is exposed to early collapse due to the effects of fire or other elements such as high winds.

Historical building restoration and general remodel projects in buildings are similar to buildings under construction. Firefighters may find temporary shoring up of walls, floors, and roofs while other structural components are being updated, replaced, or strengthened. Contractors may use simple 2 × 4s to temporarily shore up heavy timber, leading to disastrous results during fire conditions. The best approach for firefighters to take when responding to fires in buildings under construction is to be defensive. They should make sure everyone is out and accounted for and then attack the fire from a safe location. A building under construction can be replaced—a firefighter's life cannot.

Preparing for Collapse

There are no time limits for firefighting operations within a building. Tests have shown that the age-old "twenty-minute rule" used by previous fire officers is no longer accurate. Roofs and walls can collapse within minutes of fire involvement given certain conditions. An overloaded (due to improper storage or other factors) truss can collapse immediately when heated.

BUILDING CONSTRUCTION ■ **365**

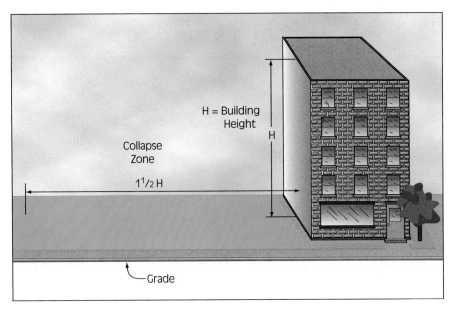

Figure 13-35 A minimum collapse zone should be 1½ times the height of the building.

Streetsmart Tip **Collapse:** Every firefighter must understand two rules about structural collapse during fire operations. The first is that the potential for structural failure during a fire always exists. Do not set artificial time limits based on experience. The second rule is to establish a collapse zone, as shown in **Figure 13-35,** which is an area around and away from a building where debris will land if the building fails. As an absolute minimum this distance must be at least 1½ times the height of the building. The walls may crumble into a pile or they may tip out the full height of the building. Also you need to provide extra room for cascading debris.

Once a building has been searched for occupants, the risks firefighters take to control the fire should be reduced—after all, it is now a property issue. Many firefighters have been killed fighting interior fires, only to have the building torn down after the investigation. Outside (defensive) firefighting operations can be equally dangerous if firefighters wander into the **collapse zone, Figure 13-36.**

Figure 13-36 These photos show the effects of fire on a masonry wall. Note the debris and distance the bricks fell away from the building. Firefighters should always establish a collapse zone, as shown in Figure 13-35.

Lessons Learned

Many firefighters have been killed as a result of building collapse from structural fires. To prevent future deaths, firefighters must understand the buildings in which they fight fires. This understanding comes from a long-term commitment to read and study building construction information. Additionally, firefighters must get into buildings within their jurisdictions to survey and explore the way buildings are assembled, remodeled, and used in the real world. Knowledge of building construction starts with an understanding of loads, forces, and materials found in the structural makeup of buildings. Firefighters also study the effects of fires on materials and construction types. The five classic types of construction are being challenged by new construction methods. Trusses are used in virtually all new buildings. Trusses have high surface-to-mass characteristics that rapidly absorb heat and subsequently fail quickly. Failure of one truss can cause failure of other trusses. There are no rules for how long a building will last while on fire. Many factors determine when materials and construction design fail and gravity pushes down the building. Buildings under construction are losers from a firefighting point of view—they collapse quickly.

KEY TERMS

Axial Load A load passing through the center of the mass of the supporting element, perpendicular to its cross section.

Balloon Frame A style of wood frame construction in which studs are continuous the full height of a building.

Beam A structural member subjected to loads perpendicular to its length.

Cantilever Beam A beam that is supported at only one end.

Chord The top and bottom components of a beam or truss. The top chord is subjected to compressive force; the bottom chord is subjected to tensile force.

Collapse Zone The area around a building where debris will land when it falls. As an absolute minimum this distance must be at least 1½ times the height of the building.

Column A structural element that is subjected to compressive forces—typically a vertical member.

Compression A force that tends to push materials together.

Concentrated Load A load applied to a small area.

Continuous Beam A beam that is supported in three or more places.

Dead Load The weight of the building materials and any part of the building permanently attached or built in.

Design Load A load the engineer planned for or anticipated in the structural design.

Distributed Load A load applied equally over a broad area.

Eccentric Load A load perpendicular to the cross section of the supporting element that does not pass through the center of mass.

Fire Load The amount of heat energy released when combustibles burn in a given area or building—expressed in British thermal units (Btus).

Fire Resistive The capacity of a material to withstand the effects of fire.

Fire-Resistive Rating The time in hours that a material or assembly can withstand fire exposure. Fire-resistive ratings are usually provided for testing organizations. The ratings are expressed in a time frame, usually hours or portions thereof.

Fire Stopping Pieces of material, usually wood or masonry, placed in stud or joist channels to slow the extension of fire.

Girder A large structural member used to support beams or joists—that is, a beam that supports beams.

Gusset Plate A connecting plate used in truss construction. In steel trusses, these plates are flat steel stock. In wood trusses, the plates are either light-gauge metal or plywood.

HVAC Acronym for heating, ventilation, and air-conditioning unit. HVACs are typically a rooftop unit on commercial buildings. Buildings may have one or dozens of these units.

Impact Load A load that is in motion when it is applied.

Joist A wood framing member that supports floor or roof decking.

Lintel A beam that spans an opening in a load-bearing masonry wall.

Live Load The weight of all materials and people associated with but not part of a structure.

Load-Bearing Wall Any wall that supports other walls, floors, or roofs.

Loading The weight of building materials or objects in a building.

Mortar Mixture of sand, lime, and portland cement used as a bonding material in masonry construction.

Occupancy Classifications The use for which a building or structure is designed.

Parapet The projection of a wall above the roofline of a building.

Platform Framing A style of wood frame construction in which each story is built on a platform, providing fire stopping at each level.

Purlins A series of wood beams placed perpendicular to steel trusses to help support roof decking.

Rafter A wood joist that is attached to a ridge board to help form a peak.

Shear A force that tends to tear a material by causing its molecules to slide past each other.

Simple Beam A beam supported at the two points near its end.

Spalling Deterioration of concrete by the loss of surface material due to the expansion of moisture when exposed to heat.

Surface-to-Mass Ratio Exposed exterior surface area of a material divided by its weight.

Tension A force that pulls materials apart.

Torsion Load A load parallel to the cross section of the supporting member that does not pass through the long axis. A torsion load tries to "twist" a structural element.

Truss A rigid framework using the triangle as its basic shape to emulate a beam.

Type I, Fire-Resistive Construction Type in which the structural members, including walls, columns, beams, girders, trusses, arches, floors, and roofs, are of approved noncombustible or limited combustible materials with sufficient fire-resistive rating to withstand the effects of fire and prevent its spread from story to story.

Type II, Noncombustible Construction Type not qualifying as Type I construction, in which the structural members, including walls, columns, beams, girders, trusses, arches, floors, and roofs, are of approved noncombustible or limited combustible materials with sufficient fire-resistive rating to withstand the effects of fire and prevent its spread from story to story.

Type III, Ordinary Construction Type in which the exterior walls and structural members that are portions of exterior walls are of approved noncombustible or limited combustible materials, and interior structural members, including wall, columns, beams, girders, trusses, arches, floors, and roofs, are entirely or partially of wood of smaller dimension than required for Type IV construction or of approved noncombustible or limited combustible materials.

Type IV, Heavy Timber Construction Type in which exterior and interior walls and structural members that are portions of such walls are of approved noncombustible or limited combustible materials. Other interior structural members, including columns, beams, girders, trusses, arches, floors, and roofs, shall be of solid or laminated wood without concealed spaces.

Type V, Wood Frame Construction Type in which the exterior walls, bearing walls, columns, beams, girders, trusses, arches, floors, and roofs are entirely or partially of wood or other approved combustible material smaller than the material required for Type IV construction.

Undesigned Load A load not planned for or anticipated.

Veneer A covering or facing, not a load-bearing wall, usually with brick or stone.

Web The portion of a truss or I beam that connects the top chord with the bottom chord.

REVIEW QUESTIONS

1. What are three ways loads are imposed on materials?
2. List the three types of forces created when loads are imposed on materials.
3. Name three kinds of beams.
4. Explain the effects of fire on steel structural elements.
5. How does a masonry wall achieve strength?
6. List and define the five common types of building construction. Give an example of each type that is located in your district or response area.
7. List the three parts of a truss and explain what forces are being applied to each.
8. List three buildings in your district or response area that have truss construction.
9. List how fire affects the four more common building materials in use today.
10. Diagram and label four different roof shapes.
11. List and describe eight conditions or observations that might indicate potential structural collapse.

Additional Resources

Brannigan, Francis, *Building Construction for the Fire Service,* 3rd ed. National Fire Protection Association, Quincy, MA, 1992.

"High Rise Office Building Fire, One Meridian Plaza, Philadelphia, PA," Technical Report Series, Report #049, U.S. Fire Administration, Washington, DC.

"Without Warning, A Report on the Hotel Vendome Fire," Boston Sparks Association.

Building Construction for Fire Suppression Forces—Wood & Ordinary. National Fire Academy, National Emergency Training Center, Emmittsburg, MD, 1986.

Angle, James; Harlow, David; Gala, Michael; Maciuba, Craig; and Lombardo, Williams; *Firefighting Strategies and Tactics.* Delmar Learning, a part of the Thomson Corporation, Clifton Park, NY, 2001.

Diamantes, David, *Fire Prevention: Inspection & Code Enforcement,* 2nd ed. Delmar Learning, a part of the Thomson Corporation, Clifton Park, NY, 2003.

CHAPTER 14

LADDERS

Frank J. Miale, Retired Battalion Chief, New York City Fire Department and Lake Carmel Volunteer Fire Department

OUTLINE

- Objectives
- Introduction
- Ladder Terminology
- Ladder Companies
- Types of Truck-Mounted Ladders
- Types of Ground or Portable Ladders
- Use and Care of Portable or Ground Ladders
- Maintenance, Cleaning, and Inspection
- Ladder Safety
- Ladder Uses
- Ladder Selection
- Special Uses
- Safety
- Miscellaneous Ladder Information
- Ladder Skills
- Raising Skills
- Lessons Learned
- Key Terms
- Review Questions
- Additional Resources

STREET STORY

With newer and more technical equipment being introduced to the fire service almost on a daily basis, it is hard to stay focused on basic firefighting. Ground ladders are as basic as you can get; however, they are still as important today as they ever were. When you place a ladder to a window, you have an instant exit for both occupants and firefighters. Place a roof ladder on a roof and you have a safe place to start ventilation. Aside from the obvious gravity-defying uses for ladders, others are breaking out windows, and ice rescue. Of course, do not forget about the occasional cat stuck in a tree. As long as rescuing cats does not interfere with more serious firefighting duties, this type of rescue provides training and good public relations. The more you use ladders, the more uses you will find for them.

New firefighters often wonder why they must practice raising ground ladders over and over, almost to the point where they no longer have to consciously think about what they are doing. That is exactly the point. Anytime, day or night, snow or rain, heat or cold, firefighters must use ground ladders with practiced skill and also must understand the features and limitations of them.

This concept became apparent to me several years ago at the scene of a 2½-story wood frame occupied structure fire. As my crew approached the fireground, we noticed a man visible through the lower sash of a double-hung window on the second floor. His outline was obscured by thick black smoke with an orange tint to it, which meant that his seconds were numbered. From a different vantage point, another crew observed the same scene. In an instant, both groups bolted for their respective ladder trucks to retrieve the appropriate ground ladders needed to effect the rescue. Both crews returned with exactly the same sizes and types of ground ladders for the job. This was no coincidence. The members of both crews had received their training at different times and by different instructors, but the proper use of ground ladders had been embedded in the minds of both crews, so they responded to the task in the same way.

Keep in mind that the rescue was successful because of ground ladders, not a thermal imaging camera, not an articulated telescoping tower ladder, not any other specialized expensive equipment—just ground ladders and firefighters who knew what they were doing. Of course, there are situations that call for special equipment, but the proper knowledge of and use of ground ladders will save lives and reduce property damage at almost every fire.

—Street Story by Peter F. Kertzie, Captain, Buffalo Fire Department, Buffalo, New York

OBJECTIVES

After completing this chapter, the reader should be able to:

- Name the parts of a ladder.
- Describe the many functions for which a ladder can be used.
- Name the different types of mounted ladder apparatus.
- Describe the function of the different types of ground ladders.
- Describe the care of ladders.
- Cite maintenance, cleaning, and inspection functions for ladders.
- Exhibit ladder operation safety.
- Name different types of ladder uses.
- Describe the ladder selection process.
- Describe the concepts behind different ladder-raising techniques.
- Cite safety concerns of ladders and their use.
- Demonstrate skills associated with ladders, such as raising, leg locks, rope handling, mounting and dismounting of ladders, and use of roof ladders.
- Describe fundamentals of ladder placement.
- Determine how far away from a building a ladder should be placed.

INTRODUCTION

Ladders are used in the fire service for many purposes besides providing access to elevated locations. Used with ropes, they can assist in shoring a wall as an emergency support. When bound and tied together and placed on their side with tarps draped over them, they can be used to build emergency walls for liquid pools. Covered with tarpaulins, they can be used to channel water to drains. Placed over openings, they can act as barriers to prevent accidental falls into holes. When used with handlines, they can serve to provide elevated stream penetration. Some types of doors can even be forced open with ladders, a technique that was used as a standard form of forcible entry in the past. When lashed together in an A-frame setup, they can be used as a hoist point for pulley placement. These applications are described in this chapter.

Ladders were originally constructed of wood. As the need for greater height became evident, the supporting solid wood beams were replaced with newer technology truss-type beams. This truss construction is a design that removes the nonsupport portions of a solid structural member without reducing strength but significantly reducing weight. While this design permitted longer ladders to be constructed, there were still limits. Introduction of lightweight high-strength aluminum replaced the wood, first in solid beam construction, then in truss construction. Later, higher strength aluminum alloys were developed and later still, fiberglass. Along with the change in the ladder's material makeup, new design technology continued to meet the needs. For instance, several smaller ladders were arranged to be adjustable by sliding them against one another through connecting channels. These extension ladders were designed to reach a range of heights. As ladders got longer and longer, additional features such as tormentor poles were designed into the ladder to assist in raising.

Ladders can be used for many purposes besides climbing. This chapter describes many innovations that have evolved from hands-on practical use.

LADDER TERMINOLOGY

A ladder is defined as "a structure consisting of two long sides crossed by parallel rungs, used to climb up and down" and as "a means of ascent and descent." Today, fire departments boast many different ladder types and lengths.

There are many parts to a ladder. It is important for every firefighter to know each part by its name, **Figure 14-1**.

In different departments, the same ladder part might have a different name.

> **Safety** Common terminology usage will reduce miscommunication when passing along information or requesting assistance.

Parts of a Ladder

- *Beam:* The side of the ladder. It is the rail that runs the full length of the ladder from top to bottom from which rungs span, **Figure 14-1A**.
- *Bed section:* The part of the ladder that is the foundation, usually the part of the ladder that is in touch with the ground or attached to the body of an aerial ladder truck. It is the section from which all other sections (see Fly section entry) are raised in extension ladders. This is also called the *bed ladder.*

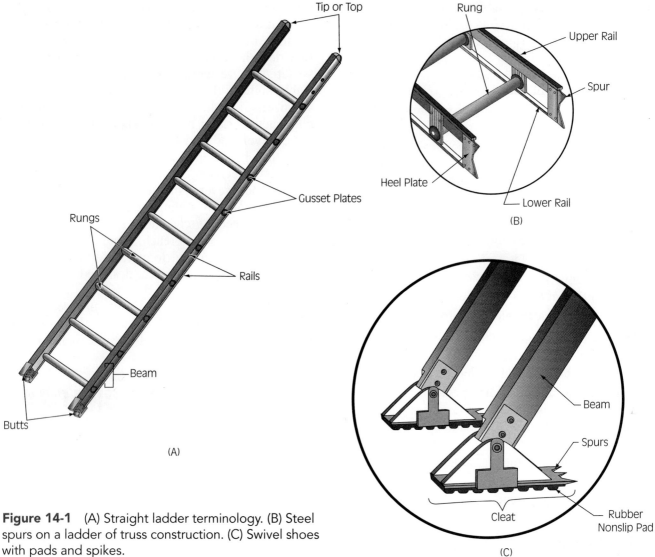

Figure 14-1 (A) Straight ladder terminology. (B) Steel spurs on a ladder of truss construction. (C) Swivel shoes with pads and spikes.

- *Heel (also called foot, base, or butt):* The bottommost part of a ladder. It is usually a reinforced section with points, spurs, or rubber pads to reduce slipping on various surfaces.
- *Spurs, spikes, cleats, shoes, and butt plates:* The pointed shoes that are attached to the base of a ladder to dig into the surface and prevent slippage during use, **Figure 14-1B**.
- *Dogs, pawls, rung locks, or ladder locks:* The mechanisms in an extension ladder that ride up along with a fly section and engage a rung of the section from which they extend to prevent retraction through the use of a spring-loaded lock.
- *Fly section:* The section (or sections) of an extension ladder that is (are) raised. Also called the *fly ladder*.
- *Pads:* Nonslip pieces of rubber or plastic that attach to the bottom of a ladder, usually in a swivel-type foot designed to lie flat against the ground or floor to prevent slippage on smooth surfaces, **Figure 14-1C**.
- *Guides/channels:* Channels of the bed ladder that permit the fly sections to ride up and maintain stability.
- *Halyard:* The cable or rope, made of nylon, hemp, or steel, that is used to raise or lower the fly section out of or into the bed section through the use of pulleys.
- *Sensor label:* A heat-sensitive label affixed to the ladder to alert firefighters that the ladder has been exposed to a potentially damaging heat level and that testing should be performed before it is used again.
- *Hooks:* Retractable hooks that permit certain types of ladders to be placed on a slanted roof surface and used for footing stability.
- *Rung locks:* See Dogs.
- *Pawls:* See Dogs.

- *Protection plates:* Reinforced metal that is built up at chafing points to avoid weakening created by rubbing and friction wear.
- *Pulley:* A wheel with a groove through which the halyard passes. It is used for raising or lowering the fly sections of extension ladders.
- *Rails:* Running lengthwise, the upper and lower surfaces of the beams.
- *Rungs:* The "steps" of a ladder that connect one beam to the other. Generally round, they can also be flattened on the upper side for more secure footing.
- *Stops:* Limiters built into the bed section to prevent the extension fly sections from being overextended. They are found in the shape of solid blocks or angled metal.
- *Tie-rods:* Found on wooden ladders, these metal rods secure the two beams and prevent them from spreading apart when the rungs are doweled into the beams. These rods prevent the rungs from pulling out, which would result in a ladder collapse.
- *Tip:* The top of the ladder. In an extension ladder, it would be the top of the fly section that attains the greatest height.

LADDER COMPANIES

In the past, most fire departments employed a single apparatus and it was the "fire department." Eventually as the number of fire companies grew, they became specialized to perform specific duties. The tasks required were logically associated with the apparatus. Ladder companies are the apparatus that carry ladders and other devices, tools, and personnel to upper levels. **Tower ladders** and **articulating boom ladders** are included in this category, although technically, they are not ladders. These units, however, carry ladders that are more traditional in construction and use.

As more and more fire departments created special units to serve particular functions, the personnel assigned to those apparatus were trained to operate the equipment on that vehicle. With specialized tools designed to complete a specific function, a certain "division of labor" developed in which firefighters on certain types of apparatus performed designated functions. Engine companies carried hose, nozzles, and appliances that served the water delivery function. Ladder companies, originally developed to gain access to upper floors, became responsible for tasks associated with entry. That included forcible entry of the fire area, access to the roof for ventilation operations, and access to upper floors for entry, search, and rescue. Eventually, the need for access to higher levels of the building became evident. Higher reaching ladders, better designed tools, and a wider measure of knowledge became the trademark of the ladder company and its personnel.

TYPES OF TRUCK-MOUNTED LADDERS

Many types of ladder trucks are used by the fire service today. Each was invented and designed to serve a particular function and has been named by its function.

Aerial Ladder

An aerial ladder is an apparatus-mounted ladder capable of reaching heights of generally up to 100 feet with some that even go beyond 110 feet, **Figure 14-2**. An aerial ladder with a reach capability of 144 feet was once tested in field service. It was taken out of service after a short time because of limited use, safety concerns, and maintenance.

Because of the heavy-duty use and reach, most of these ladders are very heavy and are constructed of some combination of steel, aluminum, and other metallic alloys designed for strength using a truss-style construction for maximum strength.

Aerial ladders are designed so that various sections slide out from one another to produce greater reach, **Figure 14-3**. Each **fly ladder** section is designed to overlap the section below sufficiently to maintain stability. It is usually made of a steel/aluminum alloy for lighter weight. The lowest section of the ladder is called the **bed ladder**. It is usually made of steel and is attached to the main body of the apparatus by a combination of pins and pistons that allow it to be raised

Figure 14-2 Rearmount aerial ladder.

Figure 14-3 Extended aerial ladder. Note the sliding sections.

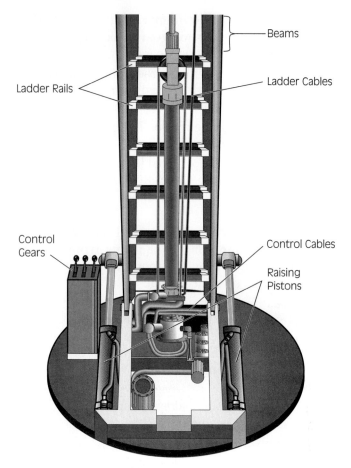

Figure 14-4 Aerial ladder raising mechanisms as seen from under a raised bed ladder.

from the horizontal to the vertical positions. The bed ladder is attached to a **turntable**, a 360-degree rotatable platform that is attached to the framework of the apparatus. The fly sections of an aerial ladder are the sections that extend out from the bed ladder and one another to reach whatever height is desired.

The bed ladder, including the **nested** fly sections, is raised out of the bed of the apparatus through the use of **hydraulic pistons**, **Figure 14-4**. The turntable rotates the ladder to align with the target, and the fly sections are extended to reach the target.

Proper terminology for use of these ladders is very important, **Figure 14-5**. The same word must mean the same thing to all people. **Table 14-1** describes the terms that should be used to identify the operation desired.

> **Safety** To avoid confusion firefighters should use standard terminology when working with ladders.

The word *raise* should refer only to the raising of the bed ladder out of the bed and not to making the ladder longer. The order to make the ladder longer would be to *extend* the ladder.

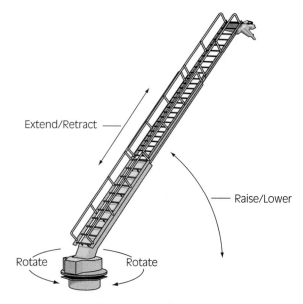

Figure 14-5 Ladder positioning terminology.

Terminology for Ladder Raising

Raise:	Lifts the ladder from horizontal to vertical angles.
Lower:	Lowers the ladder from vertical angles to horizontal.
Rotate:	Turns ladder right and left.
Extend:	Increases the length and reach by extending fly sections.
Retract:	Decreases length and reach by nesting the fly sections.

TABLE 14-1

> **Note** In many cases, the operator of the ladder might not be in visual contact with the target and is being directed by a forward observer. In cases such as this, common terminology becomes essential.

The ladder's reach is achieved through the use of cables and pulleys. Using the physics of mechanical advantage associated with block and tackle, a continuous cable is spun off a drum and woven through the various fly sections of the ladder, **Figure 14-6**. As the first fly section moves out from the bed ladder, the cable and pulley configuration pulls out the second and third sections at the same time. When in operation, each fly section moves faster than its parent section. This permits quick placement of the ladder at the target. Depending on the pulley and cable arrangement, the upper fly section can be moving three times faster than the lowest fly section.

Once the mainstay of a fire department's high elevation needs, the aerial ladder has become only one of the fire service's high elevation tools. Today's fire service has other apparatus such as tower ladders and articulating boom ladders that are equally capable, and, in some cases, better designed for increased versatility. Newer designs incorporate features of both the tower and the articulating boom.

Tower Ladder

The tower ladder is often found as a standard piece of equipment in any moderate to large fire department. It is still referred to as a ladder company because it carries a complement of ground and portable ladders. Performing many of the same functions as the aerial ladder, in many departments it has replaced older aerial ladders as they were retired from service.

The tower ladder, **Figure 14-7**, has a telescopic boom with a mounted basket capable of holding from 750 to more than 1,000 pounds. Like the aerial ladder, it is capable of rotation, extension, and retraction. The fly sections of the boom telescope out of the bed boom section. It has an advantage over the aerial ladder in that it can hold many people at the same time and is capable of many uses including use as a work platform, an observation vantage point, and an elevated water stream appliance. The tower ladder can remove many trapped victims at the same time with limited commitment of resources. One firefighter at the turntable as a safety person and one in the basket can safely remove many victims, including children and small animals from an elevated location such as a window or a

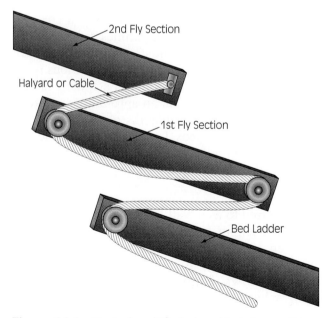

Figure 14-6 Typical multifly halyard hoisting pulley arrangement. As the halyard is pulled, the fly sections are hoisted in unison with all the rungs aligning at each locking position.

Figure 14-7 Tower ladder.

rooftop. A ground ladder would generally require the use of a separate firefighter for each victim removed to assist in climbing down the ladder.

In addition, the tower ladder affords people who suffer from a fear of height, psychologically paralyzed victims, and persons with physical and mental disabilities a safe escape.

The tower ladder is a very versatile piece of equipment that achieves many fireground elevation needs. It has a tendency, however, to take slightly longer to place into operation and position than the aerial ladder.

Articulating Boom Ladder

The articulating boom (sometimes known as a snorkel ladder) truck was among the first designs for elevated platform use in the fire service. First appearing in the early 1960s, it broke new ground that eventually led to changes in fire service tactics and equipment design. Some of the original designs had a ladder attached to the booms for escape, but its use proved very impractical because of the angles and accessibility shortcomings. Although not a true ladder, it has been used to replace aging aerial ladders. Newer designs have incorporated elements of the telescopic and articulating boom designs, **Figure 14-8**.

Through the use of several articulating booms, a snorkel ladder uses balance and individual extension and retraction capability to place the bucket into places that are not reachable by tower ladders due to obstructions. By positioning the booms at different angles, the basket can be lowered behind an obstruction. Tower and aerial ladders would be unable to reach the same objective.

An articulating boom ladder can also be used as an elevated water application platform or observation point and can simultaneously remove several victims. The disadvantage of a snorkel-type ladder is that it

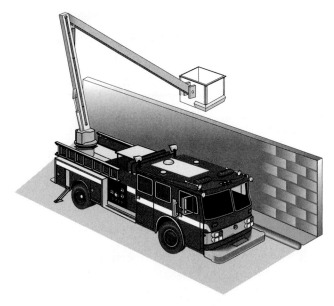

Figure 14-8 A snorkel basket can reach places not possible with other types of apparatus.

requires observation of many possible points of contact. Each articulation joint is capable of striking an object, building, electric wires, or other obstacle.

TYPES OF GROUND OR PORTABLE LADDERS

Carried on all ladder truck apparatus is a complement of ground ladders, sometimes referred to as portable ladders. The most common types are straight ladders, extension ladders, and various types of specialized ladders. Although standard ladder lengths are often found on an apparatus, each truck might contain different lengths depending on the department and its needs. Ladders are also carried on engine company apparatus for limited applications.

Straight Ladder

Also referred to as a wall ladder, the straight ladder is a fixed length ladder. Usually found in lengths between 12 and 20 feet, they are generally long enough to gain access into first-floor and second-floor windows, **Figure 14-9**. The surrounding topography of the ground will affect this capability. Beyond the height of the fixed ladders, **extension ladders** may be employed. The straight ladder is generally light, can be carried by one person, and can often be raised by one person using special techniques designed to perform that function. Straight ladders can be used for access, ventilation of upper floor windows, and escape.

LADDERS

Figure 14-9 Straight wall ladder.

Figure 14-10 Ground ladder nested on the side of a ladder truck.

Extension Ladder

An extension ladder consists of two or more ladders that operate as a unit. The bed ladder acts as the nest for the movable fly ladder(s). In the two-piece extension ladder, the fly ladder slides in channels built into the bed ladder, **Figure 14-11**. Some ladders that reach beyond 25 feet can have two or more fly ladders with each ladder running through channels of the ladder beneath it. Extended by use of a rope **halyard**, these ladders are designed to maximize mechanical advantage through the use of pulleys. These pulleys are arranged so that as the fly ladders extend, the rungs of the various sections come to rest in alignment at each stop. Each level that the ladder can reach is locked into place on the bed ladder by the use of rung locks, also called dogs or pawls. These locks secure the unit into place. The value of an extension ladder is that one ladder can be used to reach various heights by moving the fly ladders up and out from the bed ladder. One adjustable ladder can serve the function of several fixed length ladders. In addition, if the need for different heights is required at a remote location from the apparatus, the same ladder can be used. Access to a second-story window and then to a third or even a roof can be accomplished without the need for a second ladder. Primarily carried on the ladder truck, they are also found on pumpers for use when a ladder company is not on the scene.

When an extension ladder exceeds 40 feet, it is required to be equipped with **staypoles,** also called tormentor poles. Called Bangor ladders or pole ladders, these ground ladders generally do not exceed 50 feet. The staypoles are used only for raising, and, once the ladder is in a raised position, they are not

In departments without ladder companies, several ladders chosen to serve the needs of the response area are carried on the pumping engines or tankers. Several lengths of ladders can be carried nested into one another and secured to the apparatus by restraining devices. This arrangement maximizes carry capability with minimal storage space use, **Figure 14-10**.

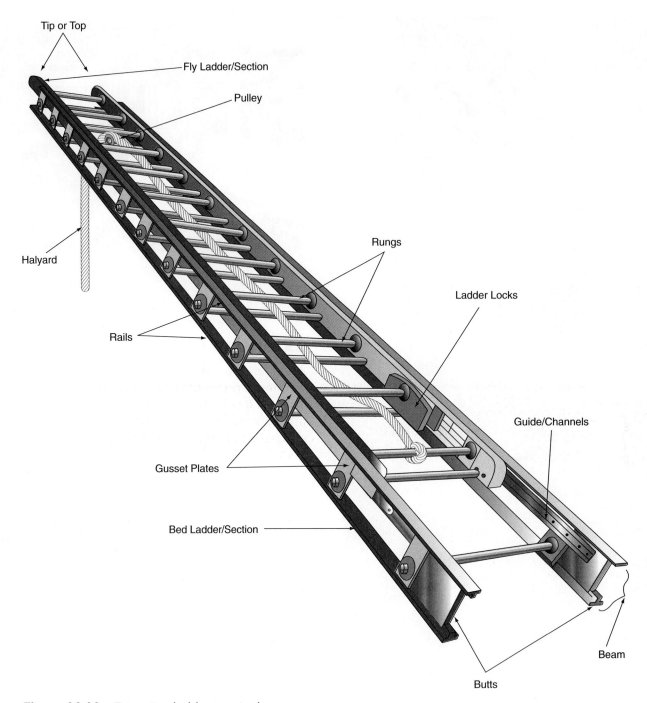

Figure 14-11 Extension ladder terminology.

used to support any weight on the ladder. Because of the weight and size of these ladders, the staypoles are needed to push up the tip of the ladder in the initial stages of raising. Used primarily for elevated access in remote locations where aerial apparatus cannot be placed, they require up to six firefighters and are generally used when no other means of access is available, **Figure 14-12**.

Roof or Hook Ladder

The roof or hook ladder is basically a straight wall ladder that possesses a set of retractable hooks at the tip end, **Figure 14-13**. Used when operating on a sloped roof, it enables a firefighter to work with more secure footing. When extended, the hooks are placed over the ridge of a peaked roof

LADDERS

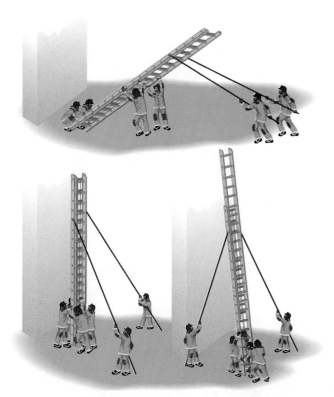

Figure 14-12 Bangor ladder raise (also called a pole ladder). The staypoles are used only for raising.

Figure 14-13 Hook ladder with retracted hooks being used as a straight ladder.

while the ladder rests on the sloped roof. The rungs are then used by the firefighter to gain a foothold where it would otherwise be impossible. In the retracted position, the hooks remain out of the way, enabling this ladder to be used as a standard straight wall ladder. A roof or hook ladder is usually found in lengths of from 12 to 24 feet. A hook ladder is not designed to be used as a hanging ladder.

Folding Ladder

The folding ladder is known by many names. Also called a suitcase ladder, an attic ladder, or a closet ladder, its function is to enable transport into narrow and confined places where a ladder is required. Although usually carried in a 10-foot model, it is available in lengths from 8 to 16 feet. It is a straight ladder that can provide access to attic spaces through hatches from the top floor of a structure. Its collapsible feature is very useful when attic access hatches are found in closets. When bedded (or folded), it is very portable through tight corners and into narrow spaces, **Figure 14-14A**. It can be used to remove trapped occupants through an elevator car's roof hatch. It is an ideal ladder to use to reach a sprinkler shutoff located high off the floor in a large open area of a warehouse. Footpads are provided as a safety feature to prevent slipping because this ladder is very often used indoors on structure floors, **Figure 14-14B**.

A-Frame Combination Ladder

An A-frame ladder is a combination ladder that can be used in various configurations. When nested, it is easily stored and acts as a mini-extension ladder, **Figure 14-15A**. Without the use of halyards, the fly ladder can be manually raised to the desired level and locked into place, **Figure 14-15B**. When fully articulated and extended, it can be locked into place as a full fixed straight ladder. When articulated into an A shape, it becomes a stepladder, **Figure 14-15C**. Each position is provided with locking mechanisms to secure the ladder into that configuration. Like the folding ladder, its greatest asset is its mobility in tight places because it can

(A)   (B)

Figure 14-14 Folding ladder (also called a suitcase ladder and attic ladder). (A) Folded. (B) Opened.

(A)   (B)   (C)

Figure 14-15 Combination A-frame ladder. (A) Used as a short extension ladder. (B) Being converted from an extension ladder to an A-frame stepladder. (C) In the A-frame stepladder mode.

reach locations that would otherwise be unsafe or difficult with conventional length ladders.

Pompier Ladder

The Pompier ladder, also called the scaling ladder, is no longer an approved ladder according to **NFPA Standard 1931**, *Design and Verification Test for Fire Department Ground Ladders*. Because it is still used for training in some places, it is described here, **Figure 14-16**. The ladder is a single-beam configuration with rungs emanating from the central beam axis. On the top is a large hook that is designed to actually break through the window of an upper floor and hook onto the windowsill or roof parapet. The firefighter then uses this hanging ladder to climb up to the floor where the hook is attached. Then, while straddling the windowsill, the firefighter lifts up the ladder and, in a hand-over-hand motion, raises it up to the next window level, hooks it in, and repeats the climbing process. Using this procedure, a firefighter can literally climb up an entire building one floor at a time. In years past this ladder was used successfully in many rescues where the victim was out of reach of standard ladders.

Safety Because of the danger in removing a victim with a Pompier ladder and the provision of other safer methods of reaching victims through the interior with breathing apparatus and from the exterior with ropes, this ladder was removed from front-line service.

Figure 14-16 Pompier (or scaling) ladder.

USE AND CARE OF PORTABLE OR GROUND LADDERS

Because of the harsh environment in which ladders will be used on a regular basis, as much care as possible must be employed to prevent the avoidable causes of ladder damage. See **Table 14-2** for a list of some tips that should be second nature to all firefighters.

Ground Ladder Tips

- Store ladders in clean, dry places.
- Store ladders with multiple support points to prevent bowing or sagging.
- Avoid moisture and keep ladders as dry as possible.
- Do not place ladders in a location where they are subjected to heat or exhaust.
- Do not lean ladders on movable objects such as apparatus.
- Do not place ladders in out-of-sight areas where they can become tripping hazards.
- When left unattended, secure ladders at their tips to prevent them from toppling.
- Paint will tend to hide defects. Paint only 14 inches of the tips or heels for identification.

TABLE 14-2

NFPA Standard 1932, *Use, Maintenance and Service Testing of Fire Department Ground Ladders,* covers the use, maintenance, and service testing of ground ladders.

MAINTENANCE, CLEANING, AND INSPECTION

Because of the nature of firefighting, ladders will be subjected to a great deal of misuse, for instance, in emergency situations where on-site adaptations are employed when time is short. As a result, extra vigilance must be exercised when inspecting ladders, **Table 14-3**. With high levels of heat, overloading, and general rough handling, fire service ladders experience abuse that warrants special inspection attention. Ladders should be inspected at regular intervals and after each use—before they are stowed on the apparatus. NFPA Standard 1931 sets out the standards to which fire service ladders must conform. Also required is a certification label affixed to the ladder that provides evidence that the ladder is in conformance with the NFPA standard. Any ladder that needs repair should be immediately taken out of service.

Defects are uncovered through regular inspection and maintenance, **Table 14-4**. When a ladder is put through regular maintenance routines, it is generally

General Inspection Guidelines

- Check halyard ropes for undue wear. Replace if rope is kinked or frayed.
- Check rivets for tightness.
- Make sure that rub plates are secure, without burrs, and not worn out.
- Check support plates for tightness.
- Check the heat sensor label for presence and condition.
- Check for splitting or cracks, which are cause for repair or replacement.
- Make sure that bolts are secure and not overtightened.
- Make sure that rungs have no play or movement.
- Examine any discoloration closely for possibility of heat damage.
- Check for cracks in welds, which are cause for concern.
- Check for any wavy or deformed areas on the surface of the ladder that might indicate damage.

TABLE 14-3

Extension Ladder Inspection Guidelines

- Ensure that dogs are freely operating and springs are operational and in place.
- Make sure rope halyard is not knotted, kinked, or worn.
- Make sure there is no excessive play in the halyard.
- Ensure that pulley is not out of round and operates freely.
- Ensure that fly ladders operate and slide freely in channels.
- Ensure that any operating latches on Bangor ladders do not bind.
- Ensure that the halyard cable of multiple-fly extension ladders are snug when in bedded position. Inspect and replace cable if it is kinked or frayed.
- In the bedded and extended positions, ensure that all rungs line up when pawls engage.
- With any of the special folding type or articulating ladders, check that the hinges are secure.

TABLE 14-4

verified that it is in a state of readiness. Any work beyond general maintenance should be performed by trained ladder repair technicians. In addition, detailed maintenance records need to be kept to confirm that required inspections are being performed regularly.

Cleaning Ladders

The act of cleaning a ladder is not just a function of improving its appearance. Dirt can cake on moving parts and prevent proper operation. Dirt and caustic substances can also act as an abrasive, causing accelerated wear of moving parts. Generally, warm, soapy detergent and a scrub brush will remove most dirt. During the cleaning process, closer inspection of the ladder will occur as opposed to a general overview inspection. The nooks and crannies of the ladder will be observed, and defects are more likely to be uncovered. If a buildup of tar or grease is discovered, proper solvents and thorough cleaning are indicated with renewed lubrication of moving parts if warranted. Manufacturer's recommendations should be consulted for the proper choice of a cleaning agent for use with ladder material.

LADDER SAFETY

In most cases, ladder safety is equated with common sense.

> **Safety** Something that would not be done on the ground should not be done on a ladder.

Overreaching, overbalancing, or overloading ladder limits are the most common causes of ladder-related injuries in general use. At fires, a host of new dangers is added to the list. Ladder placement is a critical element. The obvious is generally recognized, but placement of a ladder where it might be a hazard later in the operation is one that is easily and often overlooked. Placing a ladder in front of a lower floor window or door is not a good practice at any time. It becomes a critical danger if and when fire erupts out that lower window or door, exposing the ladder and cutting off the escape route of firefighters using the ladder.

The use of gloves reduces the possibility of pinching fingers or skin, which can occur with articulating, folding, or extension ladders. When climbing, placing, or positioning a ladder with debris falling, the firefighter must be looking toward the target. Eye protection is essential.

The use of the correct ladder is important. With an extension ladder, the length is adjustable. With a fixed straight ladder, a target that is beyond the ladder length should not be reached by decreasing the angle to the building or by standing on a set of rungs closer to the tip.

All ladders are electrical conductors despite their construction material. Moisture on the surface negates any composite insulation. Overhead wires must always be suspected of being live and improperly insulated. The quality of the insulation should never be trusted. Age, weather, or heat from the fire can reduce the insulation value to nothing.

When moving on a ladder, firefighters should always keep at least three limbs in contact with the ladder. They should not move to the next rung (hand or foot) until the previous limb is in place. Some firefighters choose to slide one hand along the underside of the beam while using the other hand to move from rung to rung. This movement makes the three-limb contact rule easier for them. Others choose to move their hands alternately from rung to rung. No matter which method the firefighters use, movement up or down a ladder should be rhythmic, safe, and smooth. Firefighters carrying a tool up or down a ladder should slide their free hand along the underside of the beam while ascending or descending, as described above. A ladder should always be butted (also referred to as *footed, heeled,* or *tended*) at the bottom by another firefighter. In the event a ladder must be left unattended, it should be secured by a short rope or cord at the tip to prevent toppling. Finally, whenever the job requires reaching and working off a ladder, the firefighter should be secured to the ladder with a ladder belt or the firefighter should utilize a leg lock. (The leg lock is described later in this chapter in the Ladder Skills section.)

> **Safety** When ascending ladders, firefighters should keep their eyes focused level and ahead, occasionally looking up the climbing path for hazards and to view the objective.

LADDER USES

Ladders are used for many purposes in the fire service. Used primarily for climbing, they can also be used as a shoring tool, a fence, a means to hold back loose debris, or a chute to channel water with a tarp—and these are just some of the other available applications. Some of the more exotic uses are covered later in the Special Uses section.

Access

The most obvious use of a ladder is for access. A ladder can provide a path to an otherwise inaccessible opening or height. The first image often pictured is an elevated climb, but a ladder can be used to descend into an opening or as a bridge between two points at the same level.

Rescue

The use of a ladder for rescue is probably the most recognizable use at a fire scene. The drama in extracting a victim from the "jaws of death" is played out in most daily newspapers and television news programs. Not to diminish the importance of this role, it is probably the use of ladders that is employed *least often!*

Stability

The use of a special ladder called a roof or hook ladder will provide footing stability to a firefighter working on a sloped roof.

Ventilation

Using a ladder for ventilation can take place in one of two ways. A firefighter can use the ladder to remove glass with a tool from an elevated position. Also, the ladder itself can be used as the tool that takes out the glass. As always, safety must be paramount in either use.

Bridging

A ladder can be an effective bridge between two points. It can be used to support weight over a weakened floor or afford stable footing over a collapsed stairway. It has been successfully used to reach victims through windows from across a shaft, such as an alley. When bridging, a bedded extension ladder is the safest to use. By using it in this fashion, the firefighter is afforded the benefit of two ladders that support one another. When bridging, the portion of the ladder that rests on the support points should be at least 5 feet on each end of the ladder for each 10 feet spanned. In other words, the portion of the ladder that rests should be one-half of the distance spanned.

Elevated Streams

The use of a ground ladder for an elevated stream has lost its place in the fire service over the years with the introduction of the tower ladder and **ladder pipe**. It is still, however, an option if needed. The application of water from an exterior location off a ground ladder may be employed when no other approach to the fire is available. This certainly does not advocate the practice of exterior firefighting over the interior approach, but in some cases there might be an advantage to such a practice. For example, in a structure that has had its interior stairs burned away with no other access, firefighters might find the elevated handline stream an asset. Apparatus-mounted appliances in the form of tower ladders, ladder pipes, or deck guns might be unable to be positioned for remote position water application.

Elevated Work Position

A ladder can serve as an exterior work platform. Just as a painter uses a ladder for reach, so can a firefighter. There may be a need to remove something or check for heat during overhauling.

LADDER SELECTION

> **Note** One of the first considerations that will have to be made is the location where the ladder will be placed. The second will be the length of the ladder needed to reach the objective. A general rule of thumb for ladder selection involves the normal heights of the floors of residential and commercial structures. Residences measure approximately 8 to 10 feet from floor to floor. The average commercial story will average approximately 10 to 12 feet from floor to floor. This information is essential for choosing the correct length ladder to use. When in doubt, firefighters should use a longer ladder.

Once a general tactical attack plan is formulated, ladder selection can take place. Among the many ladder choices will be straight, extension, aerial, and tower. Once the target is identified, a secondary set of considerations should be entertained. What length ladder is necessary, and what will be done with the ladder after the initial use? When being used as an access path to a fire or search area it must be left in place. Or will the firefighter be using it at several locations, as would be the case for ventilating a series of upper level locations? Whereas a straight ladder would be ideal in the first application, an extension ladder might be more appropriate if different heights will be necessary after the first goal is attained. The ease of portability and use of the straight ladder might have to be sacrificed in order to meet the needs of the second and third goals. Some of the considerations that generally need to be entertained are as follows:

- *Ground condition.* Soft, hard, muddy, gravel, concrete, slippery.
- *Height needed.* Single height, several heights, same level but different ground slope.
- *Purpose.* Access will require stationary placement. Rescue might entail a number of people and also ventilation.
- *Slope of ground.* Might prevent setting up at some locations. Different heights from front to rear.
- *Accessibility of location.* Ladder portability through passageways to rear or sides, over the roof, and down the rear of the structure, or over fences.
- *Available personnel.* The number of personnel available for assistance might affect the choice of ladders.
- *Overhead considerations.* Electrical wires, porch overhangs, tree limbs, clotheslines, signs.
- *Raising space considerations.* **Tip arc** clearance will be necessary for raising, or the ladder might need to be raised first and then carried to the location in vertical mode.
- *Stability.* Stability of the structure where the ladder's tip will rest.

Because ladder work usually requires two people and not always the same two people, specific guidelines must be established in the raising of a department's ladders. There are many ways to raise a ladder, and the individual past history of each firefighter who may have worked for a utility company, construction, or in general household duties will present an array of different raising techniques. One standard technique should be established so that each firefighter will be able to perform the same function, same foot location, same hand placement, and so on. At any time, the firefighter can be interchangeable but the routine will remain the same. This eliminates wasted effort and unnecessary discussions on how to perform the skill at the scene.

The skill should be practiced and become second nature to all firefighters. This in no way implies that discussion and communication should be curtailed, but, on operations such as this, discussion should not be necessary beyond the identification of the target and a general "thinking out loud" so each firefighter knows what the other is going to do.

Butt Section

If the butt or heel of the ladder will be placed on flat ground, there generally will not be a problem deploying a ladder. However, if the ground slopes, raising a ladder might be impossible. The practice of chocking one beam with fillers such as wood or chocks is dangerous and should not be employed under any circumstances.

> **Caution** Ladder checking is dangerous! One shift of the ladder as weight is being distributed can cause the entire ladder to topple. The operation is too dangerous and should not be permitted.

The point where the butt is positioned should be directly under the target with an appropriate distance from the wall. It is easier to adjust the ladder's angle to the building than to move the raised ladder laterally. When transporting a ladder, the heel or butt should be carried in the direction of the target. Once the target is reached, the butt can be planted and the ladder raised almost without hesitation. If the butt were carried in reverse, the lead firefighter (carrying the tip end) would have to pass the target before the butt can be planted. This practice wastes valuable time and requires additional effort.

Fly Section

Just as the butt of the ladder dictates the ladder placement, the tip of the fly dictates how the ladder will be used. There are several specific locations where the placement of the tip will be important and contribute to the success of the operation and achievement of the goals.

Windows

When the tip is placed at a window, two general guidelines should be followed for placement:

1. If the ladder will be used for access or escape, the tip should not extend into the window frame, **Figure 14-17A.** For every inch the ladder penetrates the opening, the size of the opening is reduced. Furthermore, the higher the extended ladder is into the opening, the more the victims or rescuers will be exposed to the heat escaping the window at the upper level as they attempt to mount the ladder. The ideal location for access is for the ladder tip to be level or slightly below (no more than a few inches) the windowsill. This permits the firefighter to climb in face first over the windowsill below any heat. In

Figure 14-17 (A) Ladder placed with the tip below the windowsill. (B) Ladder placed with the tip at the top of the windowsill to either side.

addition, this ladder placement will provide some measure of safety to a victim or firefighter if immediate escape is required. The ladder location will at least provide a place on which to land if immediate escape is absolutely required.

2. Placement of the ladder with the tip at the top of the window frame to either side of the window, **Figure 14-17B,** has several positive and negative aspects. On the positive side, it affords the firefighter a relatively safe location from which to perform ventilation. The firefighter can remove windows with a tool while remaining below the venting heat. However, to make access once the window is opened, the firefighter must negotiate the distance from the window to the ladder that is created by the angle of the window to the wall. As the distance from the tip increases, the distance of the beam from the building increases. At the sill level, depending on the height of the ladder, the distance could be from 18 to 24 inches. This gap can create difficulty in victim removal or access to a window that is spewing heat and smoke.

Roof Level

When placed for access to a roof, the tip should extend above the roof level approximately five rungs. That will provide about 5 feet of visible ladder over the roof. There are several reasons for this overextension. For one, the firefighter should climb onto the roof without overreaching and causing imbalance. The extended tip affords a handhold that can be used while dismounting the ladder. Second, the tip will be visible to any roof occupants that need to find it, especially if it is an escape route.

> **Note** Some ladder companies have installed mini-strobe lights on the tips of aerial ladders that only illuminate in the raised position for visibility at night or in smoke.

Third, the upper rungs of the ladder might be needed as a tie-off point for a rope where one is needed and the roof structures offer no alternatives. This practice should be employed only when absolutely necessary, such as when a lifesaving effort is involved. It should not be used merely as a tie-off point for rope in any other application because it would immobilize the ladder, which might be needed for a rescue effort elsewhere.

Fire Escapes

When raised to a fire escape, the position of the ladder will depend on the intended use for the given situation:

1. If the ladder tip is on the fire escape itself, the tip should be level with or below the upper rail so it can be mounted by straddling the rail and stepping onto the ladder.

2. If the ladder tip can be placed against a wall adjoining the fire escape, it should be several rungs above the fire escape rail so a person can swing a leg over the rail and step onto a rung while holding on to an upper rung. This is the preferred method. It will provide a victim with a stronger sense of stability.

When placing a ladder on a fire escape for victim removal, the priority and sequence should be that the first ladder is placed opposite the drop ladder. The second ladder is placed on the same side as the drop ladder on the floor above, **Figure 14-18**. In this manner, the evacuees can mount ladders at three points without having to wait for the person in front to clear the area. Additionally, if all the escape ladders can only be served by a single file access, removal will only occur as fast as one person can climb on a ladder. That speed is usually slow.

 SPECIAL USES

Ladders can be used as tools or as a portable stairs as long as their integrity is not compromised. Whether for venting upper floor windows with the tip or as a fence, the firefighter should be alert to visualizing adaptive uses for a ladder to achieve the firefighting goal.

Removal of Numerous Victims

The usual method of removing a victim is to raise the ladder, ascend and secure the victim onto the ladder, and then descend escorting the victim. When multiple victims need removal, taking one person at a time can be a very time-consuming process, even if several ladders and firefighters are available. A method to "keep the flow going" is to place two or more ladders at the escape point. This will usually be a large area where victims might be collecting such as a roof of a lower building, an area of escape refuge, or perhaps a school classroom where children have been brought. One ladder is used by the firefighters strictly to ascend, and the others are used to descend escorting a victim. The one "supply" ladder can service many escape ladders, **Figure 14-19**.

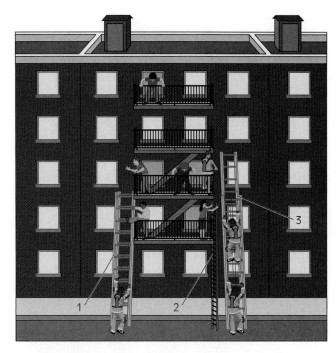

Figure 14-18 Ladder placement at fire escapes: (1) Drop ladder or stairway. (2) Opposite drop ladder or stairway. (3) Balcony above drop ladder on the same side of balcony. This provides for maximum escape flow.

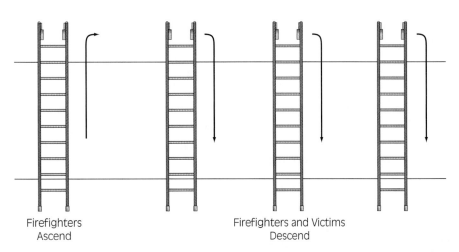

Firefighters Ascend   Firefighters and Victims Descend

Figure 14-19 Multiple ladder removal technique.

Figure 14-20 Ladder with a salvage cover, plastic sheet, or tarpaulin used as a chute to divert and discharge water.

Figure 14-21 Ladders can be used to climb over a high fence.

Figure 14-22 A handline can be used off a ground ladder for difficult-to-reach areas. Note the use of a ladder belt.

Chute with a Tarp

If water from a leak or as part of an upper floor firefighting operation is threatening lower floors with damage, a ladder with a tarp draped over it can be used as a makeshift water chute to direct water coming through the ceiling out a window, **Figure 14-20**.

Over a Fence

Two short ladders tied together at the tip in an A-frame format can be used to climb over a fence that might impede operations and for some reason cannot be cut away, **Figure 14-21**.

Elevated Hose Streams

In a location that could not be approached by conventional apparatus with elevated stream capability, a handline off a portable ground ladder was a standard firefighting technique in the days before mechanized master streams. It can still be effective when no other approach is available, **Figure 14-22**. When using this technique, basic safety practices must be employed. The firefighter and hose must be secured to the ladder, and the ladder must be stabilized at the base or at the tip. Each department must develop this operation in a form that uses its individual inventory of tools without compromising safety.

Portable Pool

If an emergency pool is needed, three or four ladders tied together to form a crib can be lined with a tarp and filled with water, either for firefighting use or to capture runoff, **Figure 14-23**.

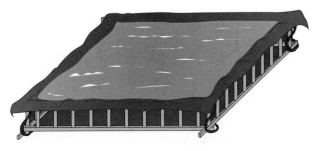

Figure 14-23 Ladders can be turned into an emergency water pool or collection area.

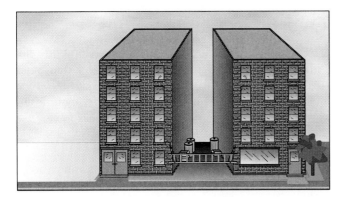

Figure 14-24 A ladder can be used as a barrier.

Figure 14-25 Ladder used as a shoring tool. A ladder secured to substantial objects by ropes can assist in stabilizing a structural defect as an emergency measure.

Barrier

A ladder mounted and tied off to secure objects can be used as a makeshift fence in cases where a barrier tape will not suffice. This would be especially useful where, for example, a large hole needs to be barricaded and an individual not paying close attention might accidentally compromise the presence of flimsy tape. The ladder barrier will offer a positive visual as well as physical mechanism to prevent a pedestrian from passing, **Figure 14-24**.

Support

A dangling sign or cornice that needs to be supported across a span and is beyond the capability of a mere rope can be supported by a ladder that spans the unsafe structural component. Using ropes, each end can be tied off to secure objects only as an emergency structural stabilizer. This temporary emergency measure should be replaced as soon as possible with tools designed for the situation. The ladder must be thoroughly inspected before being placed back into service, **Figure 14-25**.

Hoist Point

A set of ladders tied off at the tip and at the base into an A-frame configuration can be used as an emergency hoist point if a pulley and rope are

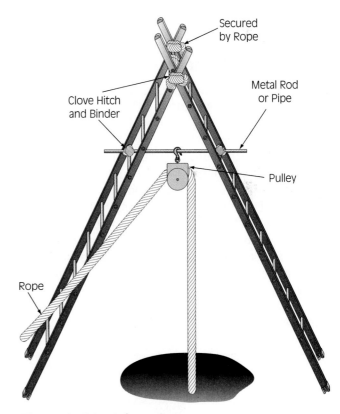

Figure 14-26 A-frame hoist.

attached to the apex of the ladder triangle. Ropes are lashed at the beam intersections and where the rungs run parallel. In addition, the support for the pulley is lashed to the rungs, creating a stable base, **Figure 14-26**. Special care should be used to make sure the ladder weight limits are not exceeded.

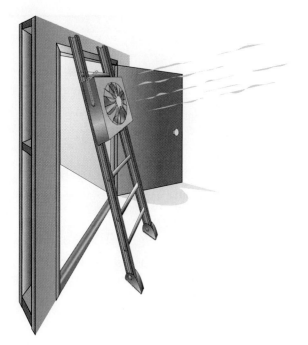

Figure 14-27 A ladder can be used to support a fan in a doorway.

Ventilation Fan Supports

A short ladder spanning an opening—such as a cellar opening or a hole made in a floor—can support a ventilation fan or blower. It can also be used to hang a fan at an upper level opening if such fan placement is desired, **Figure 14-27**. The fan can be placed on the ladder, in front of the opening, and operated from an exterior location.

These are just some of the examples where a ladder can be used as other than a climbing tool. As is always the case, innovation must be tempered with safety, and the weight limits of a ladder should never be exceeded. *In addition, whenever a ladder is used in a fashion other than that for which it was designed, a thorough inspection is essential before the ladder can be placed back into service.*

 **SAFETY**

Even though ladder use is common in the fire services, firefighters must always be conscious of safety concerns and potentially dangerous situations. Some common safety concerns are listed in this section. Dangerous situations change from one fire scene to another. Firefighters must learn to quickly scan the area in which a ladder will be deployed. Anything considered a hazard should be removed. If removal is not possible, the ladder should be deployed elsewhere.

Overhead Obstructions

Tree limbs, structural overhangs, television and telephone wires, and, of course, overhead electrical lines are all potential injury producers. The path in which the ladder will be raised must be visualized before setting the ladder base and before the ladder is lowered.

> **Safety** All overhead obstructions must be considered when placing a ladder.

Overhead electrical wires are always a hazard. Although insulated when installed, weathering can create cracks, which in turn can provide total exposure to a bare current-carrying cable underneath. There are many other dangers to be alert for when raising a ladder. If a fire has caused the insulation of a wire to be burned away and the wire is in contact with electrically conductive material (aluminum siding, a fire escape railing, television cable wires, etc.), this material could be electrically charged, **Figure 14-28**. If a ladder is placed against an energized material, the ladder could become an immediate path to ground. If the firefighter is in contact with the ladder, the ground path may course electricity through the firefighter's body.

Figure 14-28 A ladder can make an electrical connection to ground.

Safety Overhead wires have always been a safety consideration when raising ladders. There are situations where a ladder must be placed in a certain region in the presence of overhead wires. This presence will require an alternate raising tactic or site selection. A safety officer should always be on scene when ladders are raised in the vicinity of overhead wire hazards.

Climbing Path

The climbing path is the imaginary passageway a firefighter climbs through while ascending a ladder. The path is not only the straight line created by the ladder butt to the ladder tip, but also the pass-through "tunnel" area that is necessary for a person to climb, **Figure 14-29.** If a firefighter is required to alter the normal climbing angle or squeeze through a tight space created by wires, tree limbs, or structural components, the climbing path is obstructed. The space a breathing apparatus cylinder will occupy must be accounted for when estimating the space needed for the climbing path.

Ground Considerations

An assurance that the ladder will be stable must be confirmed before raising. The ground under the ladder must be level and not create a dangerous lateral lean. The longer the ladder, the more a lateral lean will be exaggerated, **Figure 14-30**. What might be an acceptable lateral lean at a very low

Figure 14-29 "Climbing path" pass-through area.

level might be dangerously beyond a safe position at an extended height.

Ladder Load

The number of people permitted on a ladder at one time will vary. The load capacity is based on weight, not the number of people. The load limits

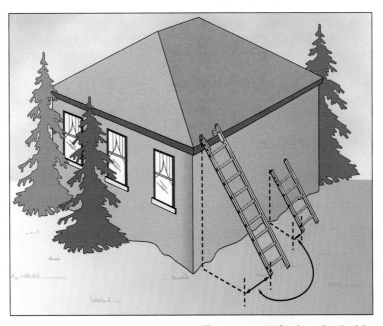

Figure 14-30 Uneven ground effect is magnified as the ladder increases in height.

Figure 14-31 Ladders must be positioned properly and not overloaded.

of a ladder should always be understood, known, and not exceeded, **Figure 14-31**. Because each type of ladder can have different construction and weight-carrying capacity or be built differently by various manufacturers, firefighters should be aware of their department's equipment or at least know where to look for the capacity statement. The recommended maximum load will be found on the manufacturer's label affixed to the ladder, **Figure 14-32**.

Working Off a Ladder

When working off a ladder, it is imperative that the firefighter be secured to the ladder. Ladder belts or safety harnesses have the ability to hook into a rung to secure the firefighter should balance be lost. Another method of locking into a ladder is by use of a leg lock. The firefighter inserts a leg through the rungs and articulates it back through the next lower rung opening. Then the instep of the foot is locked onto the beam, **Figure 14-33**. This maneu-

Figure 14-32 Ladder load limit is found on the label attached to the ladder.

Figure 14-33 Leg lock.

ver enables the firefighter to operate with both hands free of the ladder.

MISCELLANEOUS LADDER INFORMATION

Ladder Storage

When ladders are stored for extended periods it is important that they be supported by more than just two support points. Ideally, ladders should be stored on a flat surface with the ladder on its beam or flat on both beams. By avoiding specific contact points, the tendency for a ladder to sag is reduced. For ladders that are often removed and replaced, a storage technique not using the same contact points will reduce any wear and point weakening. A ladder that is continually hung on its upper beam and constantly slid on and off along a hanger could develop wear spots and weak points at those locations on the beam.

Apparatus Ladder Storage

Under ideal conditions, ladders should be stored under cover in compartments, **Figure 14-34**. With protection, ladders will not be subjected to weather that could accelerate deterioration of halyards or permit caking of grime or debris on moving parts. It will also prevent accumulation of ice, snow, or water on the ladders during inclement weather. This is especially relevant during below-freezing conditions when ice can accumulate on ladder rungs or extension ladder operating parts.

Ladder Apparatus Parking

Most new ladder apparatus have ladder storage compartments that unload at the rear of the apparatus. Very often, units responding to the emergency

Figure 14-34 Various ground ladder storage practices.

place their apparatus directly behind a parked ladder apparatus, impeding ground ladder access, and, in some cases, preventing use of ladders all together, **Figure 14-35A**. The best way to prevent the actions of another from affecting the operation is to be proactive. When parking the ladder apparatus, the ladder apparatus driver should place the apparatus at an angle to the fire building. The ground ladders will still be able to be removed, and the turntable ladder position will remain unaffected, **Figure 14-35B**. In addition, all department apparatus drivers should be made aware of this so that, in their haste to place their own company into operation, they do not block the activities of another.

Ladder Painting

Ladders should never be painted as a means of maintenance, especially wooden ladders. With good intent, ladders are painted in their entirety with a specific color for unit identification or for general appearance, but paint obscures imperfections and hides defects that would otherwise be discovered during inspections. However, small areas of ladders should be painted for identification, visibility, and quick reference.

- *Identification.* Some departments use a color code for individual companies to quickly identify equipment during the postoperation fireground take-up (breakdown) period. It is acceptable to paint the base of a ladder up to the first rung for this purpose, but no part of any structural connecting plates, operating parts, or rivets should be painted.
- *Visibility.* The ladder may be painted from the tip down to the top rung for visibility. Yellow, white, and/or reflective paint helps make the tip visible in dark conditions and can be a genuine aid in raising. It is also a valuable feature for the

(A)

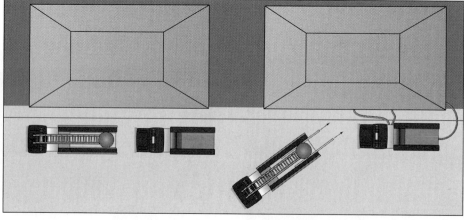

(B)

Figure 14-35 (A) It is important to leave room to remove portable ground ladders. (B) Parking apparatus can impede ladder access, but parking on an angle can be a simple solution.

firefighters using a raised ladder as an escape path from a roof or upper level location. Aerial ladder tips can also be painted to serve this safety function.

- *Quick reference.* The ladder size should be affixed on places that would be instantly visible. In addition to enabling the size to be ascertained immediately when obtaining a ladder, it would also be visible from the ground when the ladder is in a raised position. The use of ladder length identification numbers on the butt (or base) end of the ladder will assist in choosing the correct ladder from the ladder storage bed.
- *Hoist points.* One type of rope hoist method employs using a bowline loop inserted through a specific set of rungs. Small painted identifier strips may encircle the rungs involved for quick recognition (see Ladder Skills section).

Certification and Testing Procedures

NFPA Standard 1931, *Design and Verification Tests for Fire Department Ground Ladders,* outlines specific procedures for ladder testing and certification. This standard requires that a ladder manufacturer attach a certification label to each ladder verifying that the ladder was manufactured according to the NFPA standard and to OSHA guidelines.

When a ladder model is designed, it is subjected to rigid testing to ensure it will withstand the rigorous tasks required of it on the fireground. Ladder weight loads are tested while the ladder is in a horizontal position and resting on one beam, supported only at the tip and the butt, and again while the ladder is in the climbing position of 75 degrees. The ladder is tested to ensure that both butt spurs remain on the ground under simulated climbing conditions and that the rungs do not fail under simulated fireground operations. The ladder is tested to ensure that swaying and twisting will not result and that the butt spurs will not move across a surface. Special-use ladders (roof, pole combination ladders, etc.) are tested to ensure they are capable of accomplishing their specialized functions. It is important to note that a *design model ladder* is tested and certified, not each individual ladder. After the testing process, the ladder is discarded. The ladder label will attest that the model of ladder is in compliance with the NFPA standard and OSHA guidelines. For more detailed information on ladder certification and testing, please consult NFPA Standard 1931.

LADDER SKILLS

Moving a ladder from point A to point B and setting it in position may seem to be a simple task. However, without common terminology and technique, the moving and positioning of a ladder can be chaotic. It is important that firefighters understand commands given to carry, position, and raise a ladder.

Carrying Ladders

Several techniques are used for carrying ladders. Each technique has its own advantages and disadvantages. For short distances, the suitcase carry is sufficient. If longer distances are involved, the shoulder carry is more suitable to reduce fatigue. The butt (or base) firefighter should be the lead and the tip firefighter should be in the rear. Once the raising point is reached, the firefighters plant the base and raise the ladder in almost one motion.

Carrying Ladders on the Ground: The Suitcase Carry

The suitcase carry is used primarily as a short carry maneuver. Although it is very effective and a quick method of moving a ladder, carrying a ladder in this manner for long distances can be fatiguing. With the butt of the ladder facing the direction of travel the firefighters position themselves on the same side of the ladder facing in the direction of travel; one firefighter is at the tip and one is at the butt, **JPR 14-1A.** The tip firefighter controls the operation and gives the commands. On command, both firefighters raise the ladder to stand on one beam, **JPR 14-1B.** Grasping the upper ladder beam with the hand closest to the ladder, in unison, the firefighters lift and carry the ladder to the objective as if it were a suitcase, **JPR 14-1C.**

The firefighter at the butt is leading and can use the other hand to push away minor obstructions, open gates, and so on. The firefighter at the tip is the guide. It is up to this guide firefighter to ensure that turns are not too sharp or that angles are not created that cannot be negotiated by the tip firefighter. Communication is key, and the guide should dictate an alternate route if the travel path is unacceptable or if another carry method is necessary. If the ladder is heavy or the distance traveled is lengthy, a shoulder carry should be used. However, if the suitcase carry is selected a third firefighter can assist, **JPR 14-1D.**

The Shoulder Carry

The shoulder carry is useful for operations where firefighters will have to carry a ladder a good distance. The tip firefighter controls the operation and gives the commands. With the ladder flat on the ground, the two firefighters position themselves, facing opposite the direction of travel (facing the tip) on one side of the ladder beam. One is at the tip and one is at the butt, **JPR 14-2A.** On command, both firefighters squat down and grasp a rung and raise the ladder to stand on one beam, **JPR14-2B.** In unison, the firefighters lift the ladder using their legs for the power maneuver, **JPR 14-2C.** Both firefighters pivot in unison toward the direction of travel (facing the butt), place their free arm through the space between two rungs, and place the underside of the upper beam to rest on their shoulders, **JPR 14-2D.** The hand of the arm that was inserted through the rungs reaches to the next rung forward and grasps it, **JPR 14-2E.** The ladder is then carried to the objective. The tip firefighter is the guide. The free hand of either firefighter may be used to push away obstructions. In the case of large ladders, three or more firefighters might be needed to perform the carry. The firefighters must use on-the-spot adjustments to account for different firefighter body heights when positioning along the ladder.

Flat Carry

The three-person flat carry is a particularly useful technique when great disparity exists in the height of the firefighters. It also helps to evenly distribute the weight of the ladder. With the ladder flat on the ground, the two firefighters who are about the same height will position themselves, facing the direction of travel, on one side of the ladder beam. One is at the tip and one is at the butt. The firefighter who is taller or shorter locates midway on the other beam,

---

JOB PERFORMANCE REQUIREMENT 14-1
The Suitcase Carry

A Position of members before lifting.

B Raise the ladder to stand on one beam.

C Suitcase carry.

D Three-person suitcase carry.

JOB PERFORMANCE REQUIREMENT 14-2
The Shoulder Carry

A Two firefighters position on the same side of the ladder. One firefighter is at the tip of the ladder, and one firefighter is at the butt. Both firefighters are facing opposite the direction of travel, and both face the tip.

B On command, both firefightrs squat down and grasp a rung and raise the ladder to stand on one beam.

C In unison, the firefighters lift the ladder using their legs for the power maneuver.

D Both firefighters pivot in unison toward the direction of travel, rest the beam on their shoulders.

E The firefighters place an arm through the space between two rungs and grasp the next rung forward.

JOB PERFORMANCE REQUIREMENT 14-3
The Flat Carry

A Three-person flat carry.

B Four-person flat carry.

JPR 14-3A. The tip firefighter controls the operation and gives the commands. On command, all three firefighters squat down and grasp a beam with the hand nearest the ladder. Using their legs for the power maneuver the firefighters stand with the ladder. The ladder will slant up or down toward the odd-sized person, but the weight will be evenly distributed. As with the other carries, the tip firefighter is the guide. The butt firefighter leads the way to the objective.

The sequence is basically the same for the four-person flat carry, except that two firefighters are at the tip and two firefighters are at the butt. Either the beam or the rung can be grasped during the carry, depending on personal choice, **JPR 14-3B**.

Single Carry

A single firefighter can carry a small ladder individually by hoisting it up onto the shoulder and inserting an arm through the ladder rungs and grasping the next forward rung set or over the upper beam for weight balance, **Figure 14-36**. When initially placed on the shoulder, the tip should be slightly elevated with the greater weight on the butt (or base) of the ladder. The ladder should be balanced so that it can be adjusted by moving the hand on the rung or the

LADDERS

Figure 14-36 Single-person ladder carry.

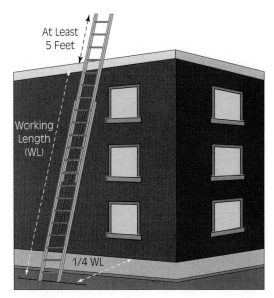

Figure 14-37 Working length is the distance from the ground to the point of contact with the building. Ladder is placed at a point approximately one-quarter of the working length.

beam to make a final weight equalization adjustment. When being carried, the butt (or base) of the ladder should be slightly dipped for maximum visibility and control.

Carrying Victims

Many types of carry techniques are used when dealing with a victim on a ladder. The level of consciousness, fear on the part of the victim, and trust in the firefighter are factors that will aid in removing a victim down a ladder. This is covered in greater detail in Chapter 16, Rescue Procedures.

RAISING SKILLS

Several considerations must be weighed when raising a ladder. The heel or foot of the ladder must be a calculated distance from the building for stability. The distance will depend on the target height. With the correct distance from the building for the given height, the ladder will be at a safe climbing angle, ideally about 75 degrees. For most people, this angle will be naturally comfortable and not require excessive bending or reaching to accomplish tasks performed while on the ladder. In addition, the ladder is designed to carry its rated weight at this angle. At lower angles, the ladder will actually bend due to the weight it is carrying. At higher angles, climbing and attempting any tasks while on the ladder may be unsafe. The distance between the ground and the point of contact with the structure is called the **working length, Figure 14-37.** If the ladder tip goes beyond the building when it is placed against the roof, the working length is from the ground to the point where the ladder contacts the roof level.

Raising Ladders

The number of firefighters needed to raise a ladder will vary. Depending on its size, as many as six firefighters may be needed to raise one ladder. Usually one, two, or three firefighters will be able to raise almost any of the ladders carried by a specific department. When conducting ladder operations, common terminology and practice is the key. There is no specific set of commands or one correct way to raise a ladder. Ladder operations depend on factors such as the specific goal to be attained, the size of the ladder, the manpower available, the topography of the area, and so on. However, it is very important that every firefighter from the same department use like methods and common ladder terminology. All firefighters must attempt to accomplish the same objective in the same way.

Most firefighting situations call for two firefighters to raise a ladder. The average ladder used at a fire is in the 18- to 35-foot range. The two people required are referred to by the functions they perform. The pivot firefighter is referred to as the *heel, butt,* or *footer,* while the firefighter raising the ladder is called the *raiser, tip,* or *fly.*

Note There are many different ladder commands used. The actual verbiage is not important. What is paramount is that common terminology be used. Ladder commands must be consistent for each agency, and interagency operations should be considered when deciding what actual commands the department will choose. When a ladder command is given, usually by the tip firefighter, every member of the operation must be thinking of the same potential actions required to accomplish the goal.

Streetsmart Tip The standard formula for determining the proper distance the foot or butt should stand out from the building is *one-quarter of the working distance* of the ladder from the base of the wall. A good way to estimate this is to stand with the toes against the foot of the beams and then, with outstretched arms, reach for a rung at about shoulder height. If the firefighter can grasp the rung, the angle is within a range that is acceptable. Remember that this is a rule of thumb for the *average* firefighter. An exceptionally tall or short firefighter might have to make some adjustments. The time for this adjustment calculation is during drills, not during the fireground operations.

Rung and Beam Raises

There are two methods of raising a ladder to the vertical position, **Figures 14-38A** and **B.** In one method, called the beam raise, the raiser brings the ladder up using the beam in a hand-over-hand motion. In the rung raise, the raiser uses the same approach but turns the ladder 90 degrees and utilizes the rungs for raising in a hand over hand motion. The beam raise is a good technique when the approach to the objective is parallel to the wall to be scaled. The beam raise is especially helpful when buildings are in close proximity with limited access. This would also be true where the presence of overhead obstructions would impede the operation. The rung raise is a good technique to use when the approach to the objective is perpendicular to the wall to be scaled.

Two-Person Rung Raise

With the rung raise, the ladder is brought to the raising point and positioned perpendicular and out from the wall approximately one-quarter of the working length from the wall or objective. The footer posi-

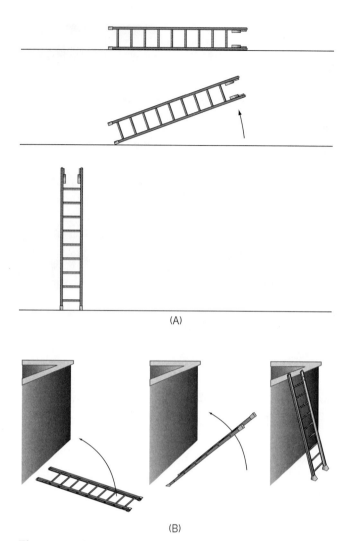

Figure 14-38 (A) Beam raise. (B) Rung raise.

tions the ladder on the ground and places a foot at each beam to plant the pivot point of the raise. The toes of the boots are snubbed up against the heel of the ladder for stability. The raiser is positioned approximately two-thirds of the length of the ladder away from the base and toward the tip, **JPR 14-4A.** This distance can be adjusted depending on the length of the ladder and the comfort level of the firefighter.

When ready, the raiser grasps the beams or the rungs and raises the ladder to hip level using the strength of the legs as lifting mechanisms, **JPR 14-4B.** Using the upper body with the back straight and legs slightly bent, the raiser lifts the ladder overhead while stepping under the ladder and pivoting to face the footer in one fluid motion, **JPR 14-4C.** Using a hand-over-hand motion, the raiser with outstretched arms, moves from rung to rung while walking toward the footer until the ladder is in the vertical position, **JPR 14-4D.** The footer constantly surveys the area around the

JOB PERFORMANCE REQUIREMENT 14-4
Two-Person Rung Raise

A Rung raise—ready.

B Tip lifts to hip level.

C Swings underneath and lifts to full arm length.

D Tip walks to butt until ladder is in vertical position.

ladder for hazards that could impede the operation. The ladder is then lowered onto the building.

> **Safety** The hands may slide along the beam rather than from rung to rung to prevent a missed grip on a rung that would permit the ladder to fall onto the firefighter underneath. The sliding hands move and maintain contact with the ladder at all times, thus minimizing the possibility of losing control of the ladder.

Two-Person Beam Raise

The ladder is placed down on the beam parallel to the building approximately one-quarter the working distance of the ladder. The footer is positioned in such a way that the firefighter's back is to the structure. The footer places a foot on the beam of the ladder to plant the pivot point of the raise. The toe of the boot is snubbed up against the heel of the ladder for stability and to secure the pivot point. The raiser is positioned approximately two-thirds of the length of the ladder away from the base and toward the tip, **JPR 14-5A,** adjusting this position as necessary. When ready, the raiser grasps the beams or the rungs and raises the ladder to hip level using the strength of the legs as lifting mechanisms, **JPR 14-5B.** Using the upper body with the back straight and legs slightly bent, the raiser then lifts the ladder overhead while stepping under the ladder and pivoting to face the footer in one fluid motion, **JPR 14-5C.** Using a hand-over-hand motion, the raiser moves along the beam with outstretched arms while walking toward the footer, **JPR 14-5D,** until the ladder is in the vertical position. Simultaneously, the footer uses both hands to steady the ladder by moving up the beam in a hand-over-hand method as the ladder ascends to the vertical position, **JPR 14-5E.** The ladder is then lowered into the building.

Fly Extension Raise

If the ladder is an extension ladder, there are additional factors and steps to consider before lowering the ladder into the building. Any extension ladder over 25 feet will usually require three persons to carry it to the objective because of the sheer weight of the ladder and its contribution to instability. The third firefighter can assume a raiser position if the ladder is too heavy for one firefighter to raise. When positioning an extension ladder, the fly section is positioned away from the building, **Figure 14-39.** Once the ladder is positioned and vertical, the footer unties the halyard while the raiser accepts the weight and stabilizes

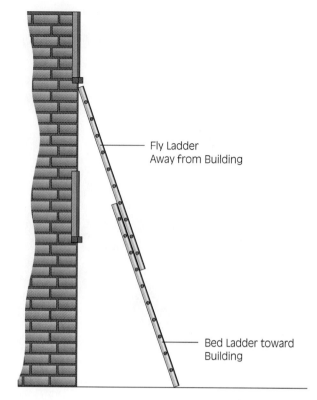

Figure 14-39 Extension ladder. The fly ladder is positioned away from the building, the bed ladder toward the building.

the ladder, **JPR 14-6A.** The footer extends the ladder to the desired height by pulling down on the rope, **JPR 14-6B.** The footer must watch to ensure the fly section is locked and secured in place once the desired height is reached, **JPR 14-6C.** The footer then secures the rope around the rungs and ties a bowline knot in the halyard as a safety precaution and slides the rope to the side against the beam so as not to interfere with the climbing, **JPR 14-6D.** For personal safety, it is important that the raiser hold the weight of the ladder by grasping the bed section of the ladder on the beams where the sliding fly section can operate free of any hand contact, **JPR 14-6E.**

> **Safety** When raising the fly ladder with the halyard, the firefighter should grasp the rope so that the heel of the hand is the uppermost part of the hand on the rope. The hand is twisted so that it creates a 90-degree angle with the rope with the balance of the rope passing through the fist and out between the firefighter and the thumb, **Figure 14-40.**

JOB PERFORMANCE REQUIREMENT 14-5
Two-Person Beam Raise

A Firefighters position the ladder on the beam, parallel to the building and approximately one-quarter the working distance of the ladder. The footer's back is to the structure. The footer snubs the heel of the ladder for stability and to secure the pivot point. The raiser positions approximately two-thirds of the length of the ladder away from the base and toward the tip.

B The raiser grasps the beams or rungs and rises the ladder to hip level.

C The raiser lifts the ladder overhead while stepping under the ladder and pivoting to face the footer in one fluid motion.

(continued)

404 CHAPTER 14

JOB PERFORMANCE REQUIREMENT 14-5
Two-Person Beam Raise (Continued)

D Using a hand-over-hand motion, the raiser, with outstretched arms, moves along the beam while walking toward the footer until the ladder is in the vertical position.

E The ladder is then lowered into the building.

Figure 14-40 Hand grasps rope with heel of hand at 90 degrees to rope for better grip.

Lower Ladder into Building

From the raising procedure, the footer is left standing between the ladder and the structure, while the raiser is standing on the outside of the ladder facing the building. The raiser grasps a set of rungs at shoulder height and places one foot on the lowest rung. The footer grasps both beams at shoulder height, **JPR 14-7A.** As the raiser lowers the ladder into the structure, the footer accepts the weight of the ladder in a controlled maneuver until the ladder is resting against the building, **JPR 14-7B.** If any angle adjustment is necessary, the firefighters can lift the ladder off the ground by each grasping a rung while positioned on either side of the ladder beams. The ladder can then be safely adjusted by lifting and pulling it away from the building until

JOB PERFORMANCE REQUIREMENT 14-6
Fly Extension Raise

A Footer unties the halyard while the raiser stabilizes the ladder.

B Footer extends the ladder while the raiser stabilizes the ladder.

C Rung locks should be secured.

D Footer ties off the halyard and moves it to the side.

E Raiser's hands are on the bed section rails.

JOB PERFORMANCE REQUIREMENT 14-7
Lower Ladder onto Building

A The raiser grasps a set of rungs at shoulder height and places one foot on the lowest rung. The footer grasps both beams at shoulder height.

B As the raiser lowers the ladder into the structure, the footer accepts the weight of the ladder in a controlled maneuver until the ladder is resting against the building.

the desired angle is attained. Fine adjustments can be made by moving the ladder out and letting the ladder tip ride down the wall. In like manner, the tip can be pushed up the wall if the angle needs to be decreased.

One-Person Raise

In some cases the situation might call for only one person to raise the ladder. The condition in which a firefighter would be working alone is rare and generally prohibited by NFPA standards. However, it is possible for a ladder to be raised by one person provided the ladder is of a length and weight that is not beyond the capability of that person. Under normal two-person raising conditions, one firefighter will secure the pivot of the ladder foot or butt while the other raises the tip section. With only one person, the base pivot firefighter is absent. A substitute can be employed through the use of a wall, fence, bush, or anything that is substantial enough to act as a foot for the ladder. The ladder is placed on the ground in a rung raise position with the ladder heel against the substitute foot, **JPR 14-8A.** The firefighter stands at the butt of the ladder facing the building and raises the butt of the ladder with a beam in each hand to the hip level, **JPR 14-8B.** With an upward swing the firefighter raises the ladder to an overhead position and walks the ladder into the building using a sliding hand motion along the beams to support the weight of the ladder, **JPR 14-8C,** as the ladder ascends into a vertical position. When the ladder is in a vertical position it will be against the building from base to tip, **JPR 14-8D.** The firefighter must continue to support the weight of the ladder. The firefighter then sets the ladder angle by lifting the ladder off the ground and backing away while the tip rests against the building until the desired angle is attained. The

JOB PERFORMANCE REQUIREMENT 14-8
One-Person Raise

A Single-person raise—ready.

B Raise ladder to hip level.

C Raise ladder overhead and walk toward wall using a sliding hand motion on beams.

D Bring ladder to vertical position.

E Rule of thumb for ladder angle: toes at base, hands on beam at shoulder level.

ladder butt is then placed on the ground. The ladder will be at approximately the correct angle when the firefighter's toes are at the base and the outstretched arms are able to reach the beams at shoulder length, **JPR 14-8E**. As normal fireground take-down operations are not urgent, there should never be a reason for a firefighter to bring down a ladder alone. Takedown should be accomplished as described in the previous ladder raise sections.

Leg Lock

The leg lock is used to secure the firefighter to the ladder when both hands must be used to perform a task and a ladder belt is not available. The firefighter first chooses the desired height, **JPR 14-9A**, and climbs up one additional rung. Then the locking leg is inserted into the space between the rungs, **JPR 14-9B**, and over the rung that was at knee height.

Using the supporting leg as an assist, the firefighter lowers the entire body until the leg inserted between the rungs can be inserted through the next lower set of rungs such that the foot projects beyond the ladder beam, **JPR 14-9C**. Twisting the ankle slightly, the toes are projected beyond the ladder beam. Then the instep of the foot becomes the locking mechanism to the ladder or, alternatively, just hooking a lower rung of the ladder. Then, the desired operation may be performed with the stability of the firefighter in place. To remove the leg lock, perform the leg lock steps in reverse.

Carrying Tools

When carrying tools on a ladder, a certain amount of security is sacrificed. The positive grip that would be afforded by the hand holding the tool is negated. When possible, tools should be passed up to another firefighter first. Another technique is to hang the tool on an upper rung, **JPR 14-10A**, and then climb up to it, **JPR 14-10B**, reposition the tool to another upper rung and again climb up to it, **JPR 14-10C**. One last—but not the best—technique is to ride the tool up the rails while climbing, **JPR 14-10D**. This last technique is suited for climbing an aerial ladder where high side rails are available. A ground ladder is not a good candidate for this method, and special care must be exercised.

JOB PERFORMANCE REQUIREMENT 14-9
The Leg Lock

A Leg lock—ready.

B Step up one rung higher and insert other leg into rung space.

C Weave leg back through next rung opening and hook instep on beam or lower rung.

JOB PERFORMANCE REQUIREMENT 14-10
Carrying Tools

A Hook tool as high as comfortable.

B Climb up to top of tool so tip of tool is at hip level.

C Lift tool and push it to a higher rung level.

D On ladder with rails, slide tool up to top of rail to help maintain balance.

Mounting and Dismounting

Getting on and off a ladder is the most difficult action for the uninitiated. It requires releasing a grip on the very thing that is supporting the climber off the ground. As height increases, so does the fear of falling. A person who has a fear of heights dreads the very act of moving from one rung to another.

Before a firefighter ascends a ladder it should be secured. If a ladder is not secured with a rope, it should be heeled by another firefighter. There are two methods of heeling a ladder. Each method uses the objective to assist in steadying the ladder by applying pressure on the ladder toward the objective. In these operations, the heeling firefighter should be wearing full protective gear including gloves and a helmet with eye protection. A firefighter may stand under (or inside) the ladder, grasping a beam with each hand at about eye level, **JPR 14-11A.** The firefighter then leans away from the ladder, pulling it toward the objective. This firefighter must be disciplined to look forward and *never* look up as falling objects could strike the firefighter, and neck, facial, or other serious injuries could result.

A firefighter may also heel a ladder from the outside of the ladder, facing the building or objective. After another firefighter begins to ascend the ladder, the heeler moves into position, grasps a beam with each hand at about eye level, and chocks the ladder with a foot by snubbing the toe of the boot against the butt, **JPR 14-11B,** or by placing a foot on the lower rung, **JPR 14-11C.** The heeler pushes the ladder toward the objective. This firefighter must be prepared to move out of the way as a descending firefighter nears the bottom of the ladder.

Again, to be as safe as possible, at least three limbs should be in contact with either the ladder or the objective at all times if practical. Only one limb should move at a time, and the next limb to move should not do so until the previous one is secure. Whether mounting a ladder from an elevated location or dismounting the ladder to a structural item such as a fire escape or a window, a few simple commonsense actions must be followed. First, the firefighter should make sure to step onto a secure and stable objective. Using the hand-move first will permit the climber to get an immediate "feel" for whether the target is unstable, shaky, or questionable. This action will occur while the full weight of the firefighter is still on the ladder. The next foot movement permits the firefighter to gradually apply some weight onto the target while one hand and foot are still on the ladder. If at any time the objective becomes unstable as more weight is shifted, the firefighter can still pull back and retreat to the safety of the stable ladder. Once the second foot is on the objective and secure, the second hand can release its grip from the ladder.

> **Caution** Carelessness on a ladder can result in severe injury or death. Simple rules that should become the standard must be followed and be second nature to the ladder climber.

If the dismount will be made onto a balcony, a parapet, or a roof where there is no structural element to grasp, the firefighter should first check the structural stability with a tool. If stability seems to be intact, the firefighter can move one foot onto the objective and, while still grasping the ladder with both hands, shift weight to that foot. If any unstableness develops, immediate retreat is possible without a loss of balance. If stable, the weight shift can continue until the majority of the firefighter's weight is on the objective. When ensured that stability is present, the firefighter can move the other foot to the objective and let go of the ladder.

Window Dismount

When climbing into a window from a ladder, two methods are used. The method chosen should be used based on the conditions. When the ladder is placed on the side of the window, a step-in method is satisfactory provided the ladder is close enough to the building. It is always better to step down onto the windowsill so that if there is structural instability, weight can be withdrawn. If the step-up method is used, the firefighter's weight is shifting onto the structure, and, if failure occurs, the firefighter might already be committed and fall from the ladder. Mounting the ladder is essentially the same process in reverse. It is best to step up to and on the ladder rung rather than step down to the rung. This permits a more positive grip and less chance of loss of balance.

If entry into the window must be made under heat and smoke conditions, the ladder's tip is placed under the windowsill and the firefighter actually climbs right over the tip of the ladder and into the window. This method keeps the firefighter low. Exit from the window would require a

JOB PERFORMANCE REQUIREMENT 14-11
Heeling A Ladder

A Heeler assuming the inside position to secure the ladder.

B Heeler assuming the outside position to secure the ladder.

C Alternate foot position for outside heeling.

somewhat different approach. The firefighter would sit on the sill and roll over onto the ladder and descend facing the ladder.

Ornamental Works

When mounting or dismounting a ladder from an ornamental works such as a fire escape or a raised parapet wall at the front of the building, special care should always be observed to account for structural weakening from fire, weather, or just plain deterioration from lack of maintenance. The weight shift from the ladder should be slow and calculated to permit quick withdrawal should structural stability become questionable.

Engaging the Hook on a Hook Ladder

The hook (or roof) ladder is held hip high, **JPR 14-12A**. Using the palm of the hand, **JPR 14-12B**, the hook is depressed against its spring-loaded resistance. Once at the limit of the hook's travel, it is rotated 90 degrees to the perpendicular position from the ladder, **JPR 14-12C**. The hook is able to rotate in either direction.

When fully rotated, the pressure on the hook is gradually released, and it is permitted to return to its limit in the up/open position, **JPR 14-12D**. The same procedure is used for the other hook.

Roof Ladder Deployment

A ladder of any size or type is first raised to the eave of the roof directly under the desired access point. The hook ladder (with hooks retracted) is raised alongside the prepositioned ladder using the beam-raise method. One firefighter climbs the non-hook ladder. While secured into the ladder by use of a life belt or a leg lock, the firefighter at the tip of the ladder engages the hooks while the heel of the hook ladder is still in contact with the ground. The footer maintains the stability of the climbing ladder and assists with the hoisting of the hook ladder if a third firefighter is not available or if the climbing ladder is secured. The hook ladder is slid up the climbing ladder on its beam until it reaches its balance point on the edge of the roof. At that point it can be gently tilted until it rests firmly on the roof on one beam. The ladder is pushed up the slope of the roof on its beam until the hooks clear the peak. The hook ladder is then rotated so the hooks are down. Once over the peak, a firm tug is employed to secure the hook ladder into the roof.

Hoisting Ladders by Rope

On occasion, the need to use a ladder from an elevated location might arise. For example, there might be a window that can be reached from the flat roof of a one-story building, or the only way to get to the roof of a four-story building is to climb up a ladder from the roof of a three-story building. In any case, the need to get a ladder to an elevated location might be present. All ladders are treated the same, and the same technique can be used for any ladder that has two beams and rungs. When lifting an extension ladder, the fly ladder should be nested or bedded before raising. A utility rope of sufficient strength is used for this operation. A bowline knot is tied at one end at a point that will create a loop that is 9 feet in circumference or at least sufficient for the bowline knot to rest above the next set of rungs, **JPR 14-13A**. The middle of the ladder is ascertained by counting rungs. Once the middle is established, the firefighter counts up two rungs and passes the rope through the hole that the rungs create. Strips of electrical tape or painted bands can be used to semipermanently identify these rungs.

With the ladder on the ground, the butt (or base) of the ladder is gently lifted and the rope slid up to the marked rungs while the slack rope is taken up, **JPR 14-13B**. The knot should be approximately in the center of the ladder between the beams. The ladder is then rolled over so that the rope runs under the ladder rungs, **JPR 14-13C**. The rope is arranged to be between the ladder and the building. Then the command to hoist may be given.

The tip will tend to lean away from the building and prevent the ladder tip from snagging on any projections caused by windowsills, protruding brickwork, or any other structural element that might cause difficulty, **JPR 14-13D**. Once the rope knot is at the point of raising, the ladder is pivoted on the edge of the structure and gently pulled down on the roof. Then the remaining portion of the ladder is pulled onto the roof. When lowering the ladder, the procedure is reversed. However, during lowering, the ladder is placed between the rope and the building to avoid any snagging on building protrusions. The tip will drag along the building, and the base or heel of the ladder will tend to stay out from the wall a few inches and ride over any obstructions.

JOB PERFORMANCE REQUIREMENT 14-12
Engaging the Hook on a Hook Ladder

A Raise hook ladder to hip level on ground.

B Depress hook with the palm of the hand.

C Push hook in and turn 90 degrees to right or left.

D When fully rotated, release hook into extended position.

JOB PERFORMANCE REQUIREMENT 14-13
Hoisting Ladders by Rope

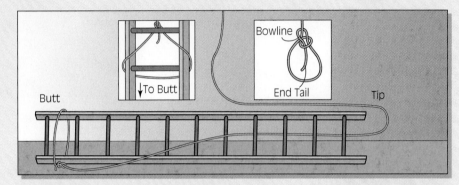

A A bowline knot is tied at one end at a point that will create a loop. The rope is passed between the rungs at a point a little more than half the distance of the ladder.

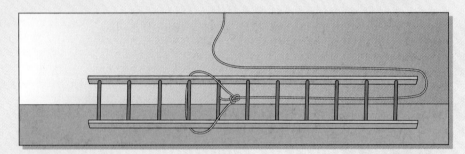

B With the ladder on the ground, the butt (or base) of the ladder is lifted and the rope slid up the beams while the slack rope is taken up.

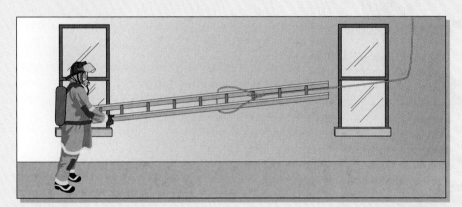

C The ladder is rolled over so that the rope runs under the ladder rungs.

D The rope is arranged to be between the ladder and the building and the ladder may be hoisted.

Lessons Learned

Ladders have many applications for use in the fire service beyond just climbing. With safety always in the forefront, innovative techniques can provide greater uses for ladders. However, it is important to remember what use a ladder was designed for and where the strengths and weaknesses lie. Any use other than such requires a thorough inspection of the ladder before returning it to service on the apparatus.

The different types of ladders, from truck-mounted aerial and tower ladders to portable ground ladders, are designed for a specific use. Roof ladders are used to stabilize footing on sloped roofs, and folding ladders are used for tight space application. New designs are always being developed. Some new designs in truck-mounted ladders include combination aerial ladder/bucket combinations and telescopic boom/articulating boom combinations. Ground ladders are also undergoing constant redesign. Collapsible telescopic ground ladders that save storage space and can be hand-carried into tight places and then extended are currently on the drawing board and might be introduced in the new generation of ground ladders. From climbing to bridging and from hoisting to shoring, ladders have many uses that can provide a superior application when used with judicious common sense.

Ladder use is packed with additional dangers that each firefighter must respect. Maintenance duties, inspection, and recording results all have a place where safety is the underlying motive and the overlying concern.

KEY TERMS

Articulating Boom Ladder An apparatus with a series of booms and a platform on the end. It is maneuvered into position by adjusting the various boom sections into place to position the platform at the desired location.

Bed Ladder The nonextending part of an extension ladder.

Extension Ladder A ladder consisting of two or more sections that has the ability to be extended to a desired height through the use of a halyard.

Fly Ladder That portion of a ladder that extends out from the bed ladder. Also called *fly section*.

Halyard A rope or cable that is used to raise the fly ladders of an extension ladder.

Hydraulic Pistons Mechanical rams that operate by pressure exerted through the use of a liquid, usually some form of oil.

Ladder Pipe An appliance that is attached to the underside of an aerial ladder for an elevated water application.

Nested The state when all the ladders of an extension ladder are unextended.

NFPA Standard 1931 The standard issued by the National Fire Protection Association that governs fire service ladder testing and certification.

Staypoles The stabilizer poles attached to the sides of Bangor ladders that are used to assist in the raising of this type of ladder. Once raised, they are not used to support the extended ladder.

Tip Arc The path that a ladder's tip will take while being raised.

Tower Ladder An apparatus with a telescopic boom that has a platform on the end of the boom or ladder. It can be extended or retracted and rotated like an aerial ladder.

Turntable The rotating platform of a ladder that affords an elevating ladder device the ability to turn to any target from a fixed position.

Working Length The length of the ladder that spans the distance from the ground to the point of contact with the structure. This does not include any distance the ladder might go beyond the point of contact as would be the case when the tip extends beyond the roof.

REVIEW QUESTIONS

1. What is the main difference between a wall ladder and an extension ladder?
2. What is the function of the hooks on a roof ladder?
3. What is the main function of cleaning and maintenance duties?
4. What is the main use of ladders at fire operations?
5. Under what conditions is a ground ladder used as an elevated water stream application device?
6. Describe the rules for mounting and dismounting a ladder.
7. When using a ladder for access, why should it be left in position?
8. What is the "climbing path"?
9. How far away from the building should the ladder be placed?
10. Describe the two techniques a firefighter can use to heel or foot a ladder.
11. When stabilizing an extension ladder, the raiser who is stabilizing the ladder must take care to ensure proper hand placement. Why?
12. What are the hazards a firefighter should look for when raising a ladder? What additional resource is advisable when a ladder must be raised in the area of charged power lines or other overhead hazards?

Additional Resources

Casey, James F., *Fire Chief's Handbook.* Dun-Donnelley Publishing Corporation, New York, 1978.

NFPA 1931: Standard on Design and Verification Tests for Fire Department Ground Ladders, National Fire Protection Association, Quincy, MA.

Training Bulletins of the Fire Department of the City of New York, *Firefighting Procedures* and *Ladder Company Tactics* (vols. 1–6), New York, 1997.

Walsh, Charles V., *Firefighting Strategy and Leadership.* McGraw-Hill Book Company, New York, 1963.

CHAPTER 15

ROPES AND KNOTS

Robert F. Hancock, Hillsborough County Fire Rescue

OUTLINE

- Objectives
- Introduction
- Rope Materials and Their Characteristics
- Construction Methods and Their Characteristics
- Primary Uses
- Fire Service Knots
- Inspection
- Maintenance
- Rigging for Hoisting
- Lessons Learned
- Key Terms
- Review Questions
- Additional Resources

STREET STORY

For many rookie firefighters, the prospect of memorizing the array of knots taught in recruit school seems daunting and relatively unimportant. Advancing hoselines into burning buildings and performing other fireground operations are much more exciting and seemingly more important skills to master.

However, a number of experiences in my twenty-five-year career as a firefighter have convinced me of the necessity of not just learning the knots but practicing them over and over until their tying becomes almost instinctual.

As a young chief I once had the experience of ordering a vent saw to the roof of a three-story building. I watched as a young firefighter started and briefly warmed up a K12 saw according to department protocols and then shut off the saw and tied it into the center of the rope hanging down from the roof so that the tail of the rope could be used as a tag line. Confident that the saw would soon arrive at the roof to cut the vent hole that would allow the hose team to advance to the top floor, I turned to answer a question. The startled expression etched on this captain's face was matched only by the obvious anguish of the firefighter whose knot had slipped loose, sending the saw crashing to the ground from the eaves of the roof. Luckily no one was in the saw's path, and that rookie survived his lesson to become a department chief and a master of fire service knots!

My first high-angle rescue also taught a similar lesson. The call was for a teenager who had fallen to the bottom of a 110-foot gorge. After reaching the unconscious but still breathing patient and stabilizing his multiple fractures and bleeding, a decision was made to secure him to a backboard and haul him back up the cliff face in a Stokes basket.

The necessary ropes and webbing slings were lowered to our position, and I proceeded to diamond lash the young man into the litter and attach haul lines to the litter bridle in the way that I had been trained. As the litter was being hauled up the cliff face, my partner and I used tag lines affixed to each end of the litter to pull the basket away from the ledges and snags. Everything was going according to plan until the litter was approximately halfway up the cliff. The patient suddenly regained consciousness. The combination of his being disoriented by his surroundings coupled with multiple facial injuries that further compromised his airway caused him to begin thrashing around quite violently against his restraints. Visions of this panicked young man clawing his way out of the litter and plunging to his death on rocks at the bottom of the falls made the few minutes that it took to complete the haul to the top of the cliff an eternity. I replayed the tying of the knots that was used to secure the patient over and over again in my mind as the litter inched its way up the cliff face.

Fortunately, the knots had all been tied correctly and safetied in just the way I had been taught, and the patient survived both the fall and the rescue, eventually making a full recovery. Since that day, a rope short has hung from the top drawer of the desk in my office, and I will often throw knots when some phone caller has me on hold.

Mastery of the unique basic skills associated with an occupation is one of the defining characteristics of any profession. Like suturing for physicians or writing legal briefs for lawyers, knot tying is one of the most basic and perhaps most critical of the core skills associated with our profession. Take pride in your ability to tie knots correctly, safely, and efficiently, for you might be asked to on your next call.

—Street Story by Joseph De Francesco, Coordinator, Madison County Fire and Rescue Services

OBJECTIVES

After completing this chapter, the reader should be able to:

- Identify the different materials that fire service rope is constructed from and their characteristics.
- Describe the differences between life safety and utility ropes.
- Define the basic terminology used when discussing ropes and knots.
- Identify the basic knots used by the fire service, how to tie each of them, and their uses.
- Describe the proper methods of inspection, maintenance, and storage of ropes.
- Describe the method of rigging basic firefighting equipment to be hoisted.
- Explain reasons for placing rope out of service.

INTRODUCTION

Rope is one of the most important and routinely used tools in the fire service. In this chapter the firefighter will learn how to select the proper rope for a given application based on the material that it is constructed of and the type of construction method. Also the firefighter will learn the primary uses of rope in the fire service, how to tie the knots utilized in the fire service, and how to properly inspect, maintain, and store rope.

ROPE MATERIALS AND THEIR CHARACTERISTICS

Rope is constructed of a wide variety of natural and synthetic materials. Each material has a different set of characteristics that impact its appropriateness or inappropriateness for utilization in the fire/rescue service. Some materials present characteristics that make them perfect for specific applications while rendering them useless for other applications.

The earliest ropes made by man and used in the fire service were made of natural materials. The use of natural material ropes by the vast majority of departments continued until the 1980s. As happens historically in the fire service (unfortunately), a series of incidents occurred that resulted in firefighters being killed or seriously injured while using natural fiber ropes. These incidents forced the whole fire service to reexamine the type of rope materials being utilized as **life safety lines**. This review ultimately resulted in the drafting of NFPA 1983, **Figure 15-1**, which deals with life safety ropes, harnesses, and hardware. NFPA 1983 established, for the first time (effective date June 6, 1985), minimum standards for this type of equipment if it is to be used by firefighters during the performance of their duties.

The following section presents the types of materials and characteristics of the different ropes used in the fire service.

Natural Materials

The materials that fall into the natural materials category include manila, sisal, and cotton. Because they are all natural materials, they share some of the same poor characteristics with regard to rot, mildew, abrasion resistance, natural deterioration/degradation due to age, a very low strength-to-weight ratio (when compared to synthetic materials), and a low **shock load** absorption capability.

> **Safety** Any ropes manufactured from natural materials should be used only as utility lines (ladder halyards, hoisting lines, etc.) and never in a life safety situation.

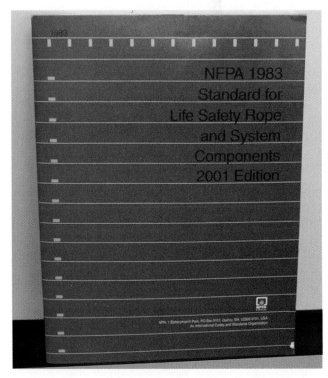

Figure 15-1 2001 Edition of NFPA 1983 Standard.

Manila

Manila rope, **Figure 15-2**, is made from the fibers that grow in the leafstalk of the abaca plant, which grows predominantly in the Philippines (Manila), hence the name. These are relatively short fibers that are twisted tightly together during the manufacturing process. This twisting creates the bond, through friction, that gives the rope its strength. Although the rope appears to be one continuous length, it is actually innumerable short fibers twisted together.

Manila rope is available in different types, with Type 1 being the higher quality and the one that was used most often by fire departments. Type 1 manila rope is manufactured from the higher quality inner fiber of the abaca plant leafstalk and can be identified by the colored string or ribbon that is twisted into the strands of the rope.

For years it was a common practice to soak new manila rope in water to make it limber. In actuality, this soaking reduced its strength by approximately 50 percent (which it never regained). This same reduction in strength occurs when a manila rope is stored in a compartment or area that has a high degree of humidity.

Sisal

Sisal is another fiber obtained from plant leaves, in this case the agave plant, native to southern Mexico. Ropes made of sisal fiber have approximately 25 percent less tensile strength than manila fiber ropes of similar diameter.

Cotton

Cotton fiber rope is made from the seed hairs obtained from the cotton boll. As can be expected, this rope tends to be very soft and pliable. It is also much lower in tensile strength (approximately 50 percent) when compared to manila, due in part to the shortness of the fibers relative to manila or sisal. It is very susceptible to damage from abrasion.

Figure 15-2 Type 1 manila rope.

Synthetic Materials

The primary synthetic materials utilized in the manufacture of ropes are nylon, polypropylene, polyethylene, and polyester. In the fire service today ropes made of these materials are the rule rather than the exception. These materials have excellent properties resisting rot, mildew, and natural degradation due to age. They are also much more resistant to physical damage and damage from abrasion. One of the major advantages/differences between these materials and the natural materials is that the fibers making up the rope are continuous from end to end, as opposed to the short fibers inherent to natural material ropes. This difference is readily identifiable in **Figure 15-3**.

Some of the other significant advantages that ropes manufactured from synthetic materials have over those manufactured from natural materials are a high strength-to-weight ratio, much greater shock load absorption, higher resistance to acids (polypropylene and polyethylene—not nylon), and no permanent loss of significant strength when they become wet. The temperature at which the various rope materials begin to lose strength and either melt or char is shown in **Table 15-1**.

Nylon

The material nylon was introduced in 1938 by E. I. Du Pont de Nemours and Company. It began to be heavily used to make ropes during World War II due to the shortage of natural fibers. The simplest example of nylon line is monofilament (single filament) fishing line, **Figure 15-4**. Nylon ropes are constructed of this same type of filament configured in multifilament bundles.

Figure 15-3 The manila fibers can be seen sticking out, while the nylon kernmantle rope is smooth and unbroken.

Rope Strengths

	TEMPERATURE AT WHICH THERE IS A LOSS OF STRENGTH (°F)	CHAR OR MELTING TEMPERATURE (°F)
Manila	180	375
Nylon	300	400–500
Polypropylene	200	275–300
Polyethylene	230	285
Polyester	300	450–650

TABLE 15-1

> **Note** Nylon has the following positive properties: high melting point (400–500°F), excellent abrasion resistance, can be bent sharply, high tensile strength (3 to 3½ times that of equal size manila), and a resistance to most chemicals (not acids).

Negative properties of nylon include susceptibility to damage by acids (*particularly battery acid*), loss of up to 25 percent strength when wet and/or frozen, stretches (elongates) when under load (use caution when utilizing to stabilize objects), and will not float on water.

Polyester

The most significant differences between nylon and polyester materials are that polyester has a good resistance to both acids and alkalis (however, an 80%+ acid and 10%+ base will affect it), it has low elongation under load, and its strength is not negatively affected by being wet or frozen. However, polyester does not handle shock loading very well. A correlation can be drawn between its low elongation characteristic and its ability to handle shock loading.

Polypropylene

Ropes constructed from polypropylene, **Figure 15-5**, are primarily used for water rescue operations by the fire service. This is due to the fact that water has no effect on their strength, and even more beneficial is the fact that they will float. Ropes of this material may be utilized in industrial operations where there is a high likelihood of exposure to chemicals and/or acids due to their high resistance to these materials. Polypropylene also has good resistance to rot and mildew.

Polypropylene ropes have a low melting point, have low resistance to abrasion, are susceptible to damage from sunlight, and have a relatively low breaking strength.

Polyethylene

Polyethylene ropes, **Figure 15-6**, have properties that are very similar to those of polypropylene.

Figure 15-4 Fishing line is the simplest example of nylon line.

Figure 15-5 Polypropylene rope floating on water.

Figure 15-6 Brightly colored polyethylene rope floating on water.

used combines to determine a given rope's properties. The two broad categories that rope falls into are **static** or **dynamic**. Static ropes have very little (less than 2 percent) elongation at normal safe working loads. This characteristic is very beneficial in the rescue field and the primary reason why most departments use static rope for rescue operations. Dynamic rope on the other hand has a much higher degree of elongation (10 to 15 percent) at normal safe working loads. This is the primary reason why this is the type of rope most preferred by mountaineers and others where fall protection is the primary purpose of a rope.

Laid (Twisted)

As previously mentioned, the laid method is the most common type of construction for natural fiber ropes. It is also utilized with synthetic fibers, but is by no means the construction method of choice for rescue rope. Laid ropes are formed by twisting individual fibers together to form strands or bundles. These strands (three or more for larger ropes) are then laid (twisted) together to form the finished rope. One of the major drawbacks to this type of construction method is that with all of the twisting to form the strands every fiber is exposed on the outside of the rope. This means that every fiber has potential to be damaged by abrasion, sunlight, chemicals, and so on.

However, they will float indefinitely and can be purchased in bright, highly visible colors if desired.

CONSTRUCTION METHODS AND THEIR CHARACTERISTICS

When natural fiber ropes dominated the fire service, they were basically constructed utilizing the laid (twisted) method, **Figure 15-7**. When synthetic materials were introduced, a variety of other construction methods were also introduced.

Modern ropes utilize a number of different construction techniques (braided, braid-on-braid, and kernmantle), each with its own particular set of characteristics, which in conjunction with the type of fiber

Another type of laid rope is used primarily for mountaineering and called *hard laid*. It is generally much stiffer than other types and tends to get stiffer the more it is used. It is also difficult to form knots in, and one has to be very careful to use safety knots, especially when using a bowline knot. This stiffness also makes it difficult to handle and store.

A characteristic that is common to both types of laid ropes is that it tends to impart and/or accentuate spinning and subsequent twisting and possibly knotting at the bottom end of the rope when used for a rappel rope.

The one real advantage of laid rope is that, because all fibers are exposed, it is easy to inspect. Again, it is important to remember that this also makes all of the exposed fibers susceptible to damage.

Braided

The braided method of rope construction, **Figure 15-8**, is utilized predominantly with synthetic fibers, although there are some natural fiber braided ropes. Braided ropes are formed by weaving small bundles (not twisted) of fibers together, much the same as braiding hair, uniformly and systematically. These ropes are generally smooth to the touch and have a good degree of flexibility.

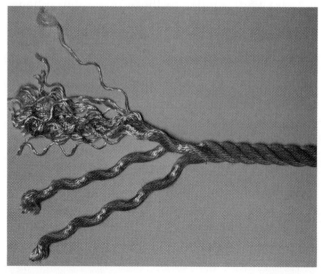

Figure 15-7 Example of laid construction method.

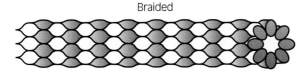

Figure 15-8 Example of braided construction method.

While braided rope does not induce or accentuate spinning during rappelling, the braiding process exposes every fiber to abrasion, sunlight, and other forms of physical damage.

Braid-on-Braid

As can be assumed from the name, this type of rope is formed by braiding a sheath over a smaller braided core, **Figure 15-9**. This type of construction often results in an approximately 50/50 split between the core and the sheath in the total strength of the rope, although there is at least one manufacturer that has designed braid-on-braid rope so that the core maintains 80 percent of the strength. Ropes of this construction method tend to be quite dynamic (stretchy) and most have a soft sheath that is more susceptible to damage from abrasion.

Kernmantle

The term *kernmantle,* when separated, very accurately describes this construction method, **Figure 15-10**. The first part, **kern**, is a derivative of the term *kernel,* which is defined as "the central, most important part of something; core; essence." The second part of the word, **mantle**, is defined as "anything that cloaks, envelops, covers, or conceals." In a kernmantle rope, the kern generally carries the vast majority of the load, accounting for approximately 75 percent of the strength of the rope, with the mantle making up the remaining 25 percent. The mantle's main purpose is to protect the kern from physical damage.

Kernmantle ropes can be either dynamic or static depending on the configuration of the fibers of the kern. When the kern fibers are twisted together and the mantle woven over them, the result is a dynamic kernmantle rope. When the kern fibers run parallel continuously and the mantle is woven over them, the result is a static kernmantle rope.

Kernmantle ropes tend to be quite resistant to abrasion and other forms of physical damage since

Figure 15-9 Example of braid-on-braid construction method.

Figure 15-10 Example of kernmantle construction method with the outer kern cut and pulled back at the end showing the inner mantle section.

the main strength of the rope, the kern, is protected by the mantle. However, since the vast majority of the kernmantle ropes used in the fire service are manufactured from nylon, they are subject to the possible sources of damage listed previously in the synthetic materials section.

> **Caution** Care must be taken that lifelines are not stored in an area that has the potential of exposing them to acids, such as near the batteries or battery box vent of an apparatus.

PRIMARY USES

Although ropes have many uses, the fire service tends to utilize them on a regular basis for a few tasks and operations that can be divided into two classifications: utility and life safety. These classifications are discussed in detail in the following section. If a department/area has a particular evolution not shown here as a part of their standard operating procedures, this is by no means intended to intimate that they are wrong or improper.

A firefighter should be familiar with department standard rope use, as well as a broad range of *possible* uses. Firefighters are faced with new and different challenges and unusual situations every day. Experience and a broad knowledge base are vital assets of firefighting.

Utility

Rope used only for utility purposes has no standard governing materials, required strength, safety factors, number of uses, associated hardware, and so on, so it is the responsibility of the department to determine what type of material and construction method is desired for its utility ropes. The department should use good judgment and look at the tasks for which the ropes will be used.

> **Safety** It is not only inappropriate, it is dangerous and contrary to NFPA 1983, *Standard on Fire Service Life Safety Rope and System Components*, to use the same rope for both utility and life safety operations.

Some of the tasks for which **utility ropes** are used are shown in **Figure 15-11**. There are many uses too numerous to list, but firefighters should become intimately familiar with the common uses within their own department.

Firefighting and Rescue Uses

Ropes used for structural search and rescue guide ropes and other non-life-supporting operations do not fall into the category of life safety ropes according to NFPA 1983. However, it would be prudent to utilize either a life safety rope or a rope used specifically for these purposes, as opposed to utilizing everyday utility rope.

Ropes, harnesses, and hardware utilized anywhere there will be a life supported, as shown in **Figure 15-12**, must comply with NFPA 1983, which sets minimum standards for equipment used in these types of situations. NFPA 1983 categorizes life safety ropes as **one- or two-person ropes** and sets minimum **tensile strength** requirements for each.

> **Safety** A one-person rope requires a minimum tensile strength of 4,500 pounds, and a two-person rope requires a minimum tensile strength of 9,000 pounds.

While this chapter does not go into depth on life safety rope operations, firefighters need to know the basics in order not to endanger themselves, other firefighters, or the persons they are attempting to rescue.

FIRE SERVICE KNOTS

While some of the knots covered here have been around for as long as people have been tying vines together, some were introduced along with the switch to synthetic fibers and modern construction techniques; some of the knots that have been used for years in the fire service do not hold well in these

(A)

(B)

(C)

Figure 15-11 (A) Utility rope hoisting a pike pole. (B) Hoisting a ladder. (C) Hoisting a charged hoseline.

new ropes. The bowline is an excellent example. For years, the bowline was the standard knot for fire service application. However, this knot was found not to work as well with modern synthetic rope. Cavers and mountaineers used synthetic ropes for many years. Because of synthetic rope characteristics and life safety applications, these mountaineers began to utilize and perfect their technique. With the fire service switch to synthetic ropes, these knot-tying techniques were incorporated into department applications.

Nomenclature of Rope and Knots

To understand knot tying instructions, firefighters need to know the basic terms used to describe the parts of a rope and knot.

Parts of a Rope

A rope can be divided into three separate and distinct parts, **Figure 15-13.**

> **Note** It helps to think of the rope as having two ends and a part or portion between the ends.
> 1. The **working end** is the end of the rope utilized to tie the knot.
> 2. The **standing part** is between the working end and the running end.
> 3. The **running end** is used for work such as hoisting a tool.

Figure 15-12 Life safety rope.

Figure 15-13 The three parts of a rope: working end, standing part, and running end.

Figure 15-14 Left to right: a round turn, a bight, and a loop.

Elements of a Knot

Just as firefighters need to know the basic terms used to describe the parts of a rope, they also need to know the terms used to describe the elements that are combined to form a knot, **Figure 15-14**. A **round turn** is formed by continuing the loop on around until the sections of the standing part on either side of the round turn are parallel to one another. A **bight** is a doubled section of rope, usually made along the standing part. A bight forms a U-turn in the rope that does not cross itself. A **loop** is a turn in the standing part that crosses itself and results in the standing part continuing on in the original direction of travel.

> **Streetsmart Tip** Studying the photos in the Job Performance Requirements (JPRs) and reading the associated directions will assist firefighters with the skills necessary to master knot-tying techniques. It is helpful to have another firefighter read the instructions step by step and cross-check the steps with the photographs in the JPRs.

Knots

The following pages provide step-by-step instructions for tying the knots that are utilized most often by the fire service. They are tried and true and straightforward. Fellow firefighters should not have any problem recognizing the knot when it needs to be untied. Later, this chapter shows how each of these knots can be used for various tasks that may be required at any given incident.

It is important to point out that the step-by-step instructions are intended to teach firefighters the steps necessary to tie the various knots. Very seldom are firefighters called on to tie any of these knots while standing in a brightly lighted, air-conditioned classroom or apparatus bay.

> **Streetsmart Tip** For the purposes of consistency and safety, it is the firefighter's responsibility to learn and practice tying knots in all of the different types of situations: in the open, around an object (both vertical and horizontal), in low visibility, when wet, with gloves on, and so forth.

All knots should be neat and tight. The terms used to describe this practice are to *dress* and *set* a knot. **Dressing** a knot is the practice of making sure that all parts of the knot are lying in the proper orientation to the other parts and look exactly as the pictures indicated. **Setting** a knot is the finishing step of making sure the knot is snug in all directions of pull. This is an important step because a perfect knot, if not set, will slip when a load is applied. This is due to the lack of appropriate friction and cinching effects that cause knots to hold. **Figure 15-15** shows knots incorrectly and correctly dressed and set.

(A)

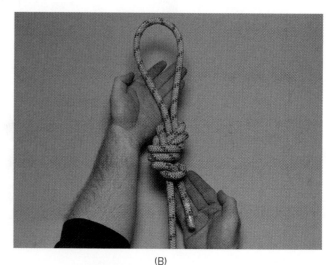
(B)

Figure 15-15 (A) A loose and sloppy knot. (B) A knot properly dressed and set.

JOB PERFORMANCE REQUIREMENT 15-1
Making a Half Hitch Around an Object

Make a round turn in the standing portion of the rope, and slide the round turn down over the object being hoisted.

Half Hitch

A half hitch, **JPR 15-1**, is almost always utilized in conjunction with some other knot and is used to maintain the proper orientation of the object being hoisted.

1. Make a round turn in the standing portion of the rope.
2. Slide the round turn down over the object being hoisted, making sure that the running end passes under the working end. If necessary, on particularly long objects such as pike poles more than one half hitch can be used.

Overhand (Safety) Knot

An overhand knot, **JPR 15-2**, is generally used to secure the loose end of the working end after tying a knot. All of the knots described in this chapter should be secure in and of themselves, but the addition of this simple knot provides an extra measure of security.

> **Note** A knot should not be considered complete until a safety knot is tied to back it up.

1. Take the loose end of the working end after tying your primary knot and secure it by making a round turn around the standing part and bringing the loose end through between this round turn and the primary knot.

Clove Hitch

The clove hitch is used to attach a rope to an object, such as a pole, tree, or fence post, or to a tool, such as a pike pole or hoseline. A clove hitch can be tied anywhere along the rope if needed. It will hold

JOB PERFORMANCE REQUIREMENT 15-2
Tying an Overhand Safety Knot Around an Object

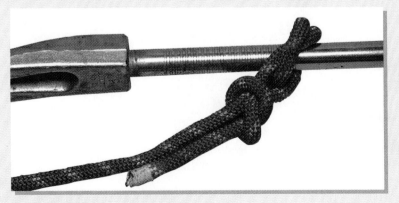

Take the loose end of the working end after tying the primary knot, and secure it by making a round turn around the standing knot and bringing the loose end through.

equally well when tension is applied from either direction—if tied correctly.

There are basically two ways of tying a clove hitch. Step-by-step instructions and figures are provided for each method. The first way, which is fastest, is to tie a clove hitch in hand (sometimes referred to as in the open) and then slip it over the object. In many cases, however, this is not possible, so it is important that firefighters also know how to tie the clove hitch around an object.

Clove Hitch Tied in the Open (JPR 15-3)

1. Hold the rope in your hands with the working end to your right, **JPR 15-3A**.
2. Form a loop with the working end passing in front of the standing part (between you and the rope).
3. Move the rope down to the right and form another loop exactly like the one formed in step 2, **JPR 15-3B**.

JOB PERFORMANCE REQUIREMENT 15-3
Clove Hitch Tied in the Open

A Form a loop with the working end passing in front of the standing part (between you and the rope).

B Move the rope down to the right and form another loop exactly like the one formed in step A.

C Place the loop formed in step B behind the loop formed in step A. You now have a clove hitch.

D Keeping the clove hitch loops in the proper orientation to one another, slide them over the object to be secured.

E Finish with a safety knot.

4. Place the loop formed in step 3 behind the loop formed in step 2. You now have a clove hitch, **JPR 15-3C**.
5. Keep the clove hitch loops in the proper orientation to one another and slide them over the object to be secured, **JPR 15-3D**.
6. Finish with a safety knot, **JPR 15-3E**.

Clove Hitch Tied around an Object (JPR 15-4)

1. With the working end, make a round turn around the object, **JPR 15-4A**.
2. Cross the working end over the standing part and make another round turn around the object.
3. Bring the working end under at the point you crossed with the first round turn. At this point your round turns should be side by side with the standing part and the loose end of the working end opposing one another, **JPR 15-4B**.
4. Finish with a safety knot, **JPR 15-4C**.

Becket Bend and Double Becket Bend

Also known as the sheet bend and double sheet bend, the becket bend and the double becket bend are very useful knots for tying ropes together. The becket bend is utilized to tie ropes of equal diameter together, while the double becket bend is used most often when tying ropes of unequal diameter.

Tying a Becket Bend (JPR 15-5)

1. Form a bight in the running end of the line that needs to be extended, **JPR 15-5A**.
2. Utilizing the working end of the rope being added, bring it up through the bight, **JPR 15-B**.

JOB PERFORMANCE REQUIREMENT 15-4
Clove Hitch Tied Around an Object

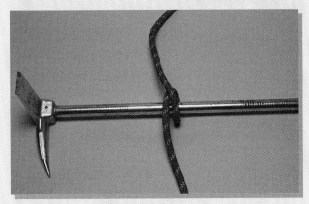

A With the working end, make a round turn around the object. Cross the working end over the standing part and make another round turn around the object.

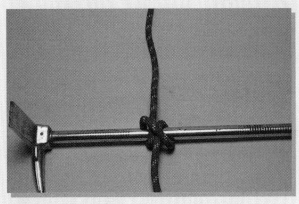

B Bring the working end under at the point you crossed with the first round turn. At this point your round turns should be side by side with the standing part and the loose end of the working end opposing one another.

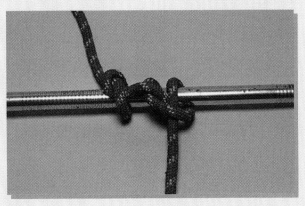

C Finish with a safety knot.

3. Take this same end around both sides of the bight, forming a loop around the bight, **JPR 15-5C**.
4. Still utilizing this same end, pass it between itself and the rope in which you formed the bight.
5. When dressed and set, the end you were working with should be at approximately 90 degrees to the rope, **JPR 15-D**.
6. Finish with a safety knot, **JPR 15-5E**.

Tying a Double Becket Bend (JPR 15-6)

1. Form a bight in the running end of the larger line, **(JPR 15-6A)**.
2. Utilizing the working end of the smaller rope, bring it up through the bight, **JPR 15-6B**.
3. Take this same end all the way around both sides of the bight forming a round turn around the bight, **JPR 15-6C**.
4. Continue around the bight a second time, **JPR 15-6D**.

JOB PERFORMANCE REQUIREMENT 15-5
Tying a Becket Bend

A Form a bight in the running end of the line needing to be extended.

B Utilizing the working end of the rope being added, bring it up through the bight.

C Take this same end around both sides of the bight, forming a loop around the bight. Still utilizing this same end, pass it between itself and the rope in which you formed the bight.

D When dressed and set, the end you were working with should be at approximately 90 degrees to the rope.

E Finish with a safety knot.

5. Still utilizing this same end, pass it between itself and the rope in which you formed the bight. At this point the rope you have been working with should have formed what looks like the number eight (8), **JPR 15-6E.**
6. When dressed and set, the end you were working with should be at approximately 90 degrees to the rope, **JPR 15-6F.**
7. Finish with a safety knot.

Bowline Knot

Although it was the mainstay of fire service knots for years, the advent of synthetic fiber ropes has greatly reduced the utilization of the bowline knot. This is due in large part to the fact that the bowline is an inherently loose knot, which is beneficial when it needs to be untied but has a dangerous tendency to slip when used on synthetic fiber ropes.

JOB PERFORMANCE REQUIREMENT 15-6
Tying a Double Becket Bend

A Form a bight in the running end of the larger line.

B Utilizing the working end of the smaller rope, bring it up through the bight.

C Take this same end all the way around both sides of the bight, forming a round turn around the bight.

D Continue around the bight a second time.

E Still utilizing this same end, pass it between itself and the rope in which you formed the bight.

F When dressed and set, the end you were working with should be at approximately 90 degrees to the rope. Finish with a safety knot.

> **Caution** If a bowline is not properly dressed, set, and secured with a safety knot there is a possibility of it inverting and becoming a slip knot.

Tying a Bowline (JPR 15-7)

1. Holding the standing part in your hand with the working end coming toward you, form a loop in the rope with the working end on your side when it passes the standing part, **JPR 15-7A.**
2. Still utilizing the working end, bring it up through the loop formed in step 1, **JPR 15-7B.**
3. Pass this same end around behind the standing part and then bring it back down through the loop you brought it up through.
4. Be sure that the working end is on the inside of the loop that you have just formed with the completed knot. If left on the outside there is a much greater chance of it getting caught and the knot inverting to a slip knot, **JPR 15-7C.**
5. Finish with a safety knot, **JPR 15-7D.**

Figure Eight Knots

The figure eight knots came into use in the fire service during the same time line as synthetic fiber rope. This was a natural transition because cavers and

JOB PERFORMANCE REQUIREMENT 15-7
Tying a Bowline

A Holding the standing part in your hand with the working end coming toward you, form a loop in the rope with the working end on your side when it passes the standing part.

B Utilizing the working end, bring it up through the loop formed in step 1.

C Pass this same end around behind the standing part and then bring it back down through the loop you brought it up through. Be sure that the working end is on the inside of the loop that you have just formed with the completed knot. If left on the outside there is a much greater chance of it getting caught and the knot inverting to a slip knot.

D Finish with a safety knot.

mountaineers, from whom the fire service adopted synthetic fiber ropes, had been utilizing figure eight knots for years. Figure eight knots have several good features. They are very simple to tie; they form a knot that will not slip or pull from the rope (as a bowline knot will); and they place less stress on the rope, as the turns in the figure eight are not as sharp as in the other knots. Figure eight knots are the preferred knots when working with synthetic rope.

The basic name *figure eight knot* is used to describe many different knots with a modifier added to the basic name to delineate one from the other. This section covers the basic figure eight, the follow-through figure eight, and the figure eight on a bight.

Basic Figure Eight Knot

The basic figure eight knot shown in **JPR 15-8** is useful when an "end of the line" knot is needed, such as when a rappel rope is not long enough to reach the ground or solid landing. This knot is also useful when joining two ropes of equal diameter together.

Tying a Figure Eight (JPR 15-8)

1. Form a bight in the working end of the rope, holding the bight out in front of you by the apex so that both sides of the bight are hanging parallel, **JPR 15-8A.**
2. Take the short end (working end) of the rope and wrap it around the standing part from front to rear, **JPR 15-8B.**
3. After the working end comes around the standing part, you will have formed an eye. Now take the working end through the eye from front to rear, **JPR 15-8C.**
4. If you are joining two ropes, at this point you will take the working end of the second rope and, coming from the opposite direction, trace the path of the original rope as it flows through the knot. You should end up with both working ends lying parallel to the opposite rope's standing part, **JPR 15-8D.**
5. Finish with a safety knot with each working end around the opposite rope's standing part, **JPR 15-8E.**

Follow-Through Figure Eight Knot

The follow-through figure eight knot shown in **JPR 15-9** is very useful when attaching a utility or life safety line rope to an object that does not have a free end available.

Tying a Follow-Through Figure Eight (JPR 15-9)

1. Follow the directions for tying a basic figure eight; however, you need to make sure that you have an adequate amount of rope left on the working end side of the basic knot to go around the object you are securing the rope to, **JPR 15-9A.**
2. If you are securing the rope to an object, wrap the working end around the object at this point, **JPR 15-9B.** (From this point on the procedure is very similar to the procedure for joining two ropes together.)
3. As the name implies, you now follow the other rope through the knot, coming from the opposite direction. In effect, you have formed a figure eight on a bight (the next knot described) around an object, **JPR 15-9C.**
4. You should conclude with the end of the working end parallel to the standing part, **JPR 15-9D.**
5. Finish with a safety knot, **JPR 15-9E.**

Figure Eight on a Bight

The figure eight on a bight shown in **JPR 15-10** is utilized when the object being attached to has a free end available to place the rope over and is very useful in forming a loop in the line that will not slip or cinch up. This knot can also be used when a loop is needed in the middle of a rope.

Tying a Figure Eight on a Bight (JPR 15-10)

1. Start by forming a bight. The size of the bight needs to be relative to the object the finished loop is going to be placed over or around, **JPR 15-10A.**
2. Grab both ropes so that the bight is hanging down from one side of your hand while the standing part and the end of the working part are hanging out of the opposite side of your hand. Basically you are forming a second bight of doubled ropes, **JPR 15-10B.**
3. Hold this bight out in front of you by the apex so that both sides (four parts of the same rope) of the bight are hanging parallel, **JPR 15-10C.**
4. Take the first bight and wrap it around the standing part from front to rear, **JPR 15-10D.**
5. After the bight comes around the standing part, you will have formed an eye. Now take the bight through the eye from rear to front, **JPR 15-10E.**
6. You have now formed an easy, reliable loop that will not slip. Do not forget to dress and set your knot and apply a safety with the end of the working end, **JPR 15-10F.**

JOB PERFORMANCE REQUIREMENT 15-8
Tying a Figure Eight

A Form a bight in the working end of the rope, holding the bight out in front of you by the apex so that both sides of the bight are hanging parallel.

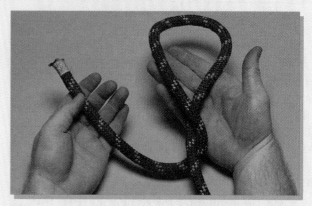

B Take the short end (working end) of the rope and wrap it around the standing part from front to rear.

C After the working end comes around the standing part you will have formed an eye. Now take the working end through the eye from front to rear.

D If you are joining two ropes, at this point you will take the working end of the second rope and, coming from the opposite direction, trace the path of the original rope as it flows through the knot. You should end up with both working ends lying parallel to the opposite rope's standing part.

E Finish with a safety knot with each working end around the opposite rope's standing part.

JOB PERFORMANCE REQUIREMENT 15-9
Tying a Follow-Through Figure Eight

A Follow the directions for tying a basic figure eight; however, you need to make sure that you have an adequate amount of rope left on the working end side of the basic knot to go around the object you are securing the rope to.

B If you are securing the rope to an object, wrap the working end around the object at this point.

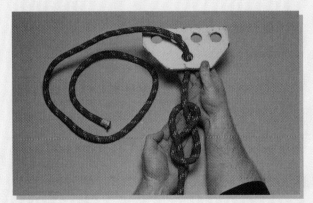

C As the name implies, you now follow the other rope through the knot, coming from the opposite direction. In effect, you have formed a figure eight on a bight around an object.

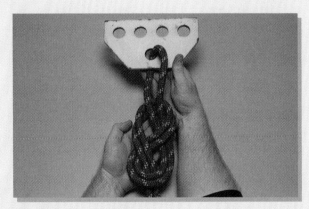

D You should conclude with the end of the working end parallel to the standing part.

E Finish with a safety knot.

JOB PERFORMANCE REQUIREMENT 15-10
Tying a Figure Eight on a Bight

A Start by forming a bight. The size of the bight needs to be relative to the object the finished loop is going to be placed over or around.

B Grab both ropes so that the bight is hanging down from one side of your hand while the standing part and the end of the working part are hanging out of the opposite side of your hand. Basically, you are forming a second bight of doubled ropes.

C Hold this bight out in front of you by the apex so that both sides (four parts of the same rope) of the bight are hanging parallel.

D Take the first bight and wrap it around the standing part from front to rear.

E After the bight comes around the standing part, you will have formed an eye. Now take the bight through the eye from front to rear.

F You have now formed an easy, reliable loop that will not slip. Do not forget to dress and set your knot and apply a safety with the end of the working end.

ROPES AND KNOTS

Rescue Knot

Almost every department and/or recruit school has a knot or combination of knots that they recognize as the best knot to be used as a rescue knot. The rescue knot described in this section is by no means intended to imply that their choice is incorrect or to disparage their choice in any way. The knot presented here and shown in **JPR 15-11** is one of the easiest, fastest, and surest among the many that have been tried over the years. It should come as no surprise that the primary knot utilized is a member of the figure eight family, a figure eight on a bight.

This rescue knot can be used on oneself, a conscious patient, or an unconscious patient. The basic method of applying this rescue knot does not change with the patient; however, the difficulty factor does go up if the patient is unconscious. The following instructions are based on applying the rescue knot to oneself. However, it is also important to practice the rescue knot on other people to simulate victims needing rescue. An emergency scene is not the proper venue for learning skills.

Tying a Rescue Knot (JPR 15-11)

1. Tie a figure eight on a bight. Your initial bight should be approximately 6 feet long. This will allow you to tie the figure eight on a bight, including a safety at the working end, and still have a finished loop approximately 4 to 5 feet long. This loop is going to need to be long enough to go around the victim's waist and back up between the legs. The loop does not need a precise measurement, but it is important to remember that the length cannot be adjusted once the knot is tied. Time should be spent sizing and applying this knot on different size victims in practice scenarios. Firefighters should rehearse this skill until they feel comfortable and can be consistently successful with this maneuver on the first try, **JPR 15-11A**.

2. Place the loop on the floor or step into it with the actual figure eight knot behind you. Raise the loop to waist level, being careful not to get above the top of the hip bones, **JPR 15-11B**.

JOB PERFORMANCE REQUIREMENT 15-11
Tying a Rescue Knot

A Tie a figure eight on a bight. Your initial bight should be approximately 6 feet long. This will allow you to tie the figure eight on a bight, including a safety at the working end, and still have a finished loop approximately 4 to 5 feet long.

B Place the loop on the floor or step into it with the actual figure eight knot behind you. Raise the loop to waist level being careful not to get above the top of the hip bones.

C Reach down between your legs pulling the standing part of the rope up in front of you. The standing part should not be between your body and the loop but on the outside of the loop.

(continued)

JOB PERFORMANCE REQUIREMENT 15-11
Tying a Rescue Knot (Continued)

D Continue to hold the rope in front of you and raise your hand, holding the rope above your head. Reach out with the other hand and grasp both sections of rope.

E Make three full twists with the hand over your head and form a loop. Place the loop over your head and shoulders so that it comes to rest under your arms.

F Keep the standing part pointed down and adjust the slack out of the rope between the figure eight knot coming from your legs and the twists forming the loop around your chest. Raise the standing part up over your head and allow the twists to wrap around the rope. Cinch down on the rope, forming the loop around your chest. Arrange the ropes between your legs as necessary.

3. Reach down between your legs, pulling the standing part of the rope up in front of you. The standing part should not be between your body and the loop but on the outside of the loop, **JPR 15-11C**.
4. Continuing to hold the rope in front of you, raise your hand holding the rope above your head. Reach out with the other hand and grasp the ropes, both the one coming from between your legs and the one hanging down from the hand over your head, **JPR 15-11D**.
5. With the hand over your head make three full twists, forming a loop. Place the loop over your head and shoulders so that it comes to rest in your underarms, **JPR 15-11E**.
6. Keeping the standing part pointed generally down, adjust what slack you can out of the rope between the figure eight knot coming from between your legs and the twists forming the loop that is now around your chest, **JPR 15-11F**.
7. Raise the standing part up over your head, allowing the twists to wrap around themselves and cinching down on the rope forming the loop around your chest. This is important to keep the loop from cinching down on your chest.
8. Arranging the ropes that come up between your legs as your weight is lifted by the rope will help ease the discomfort somewhat. Remember this tip when lifting a conscious or unconscious patient.

Water Knot

Webbing has become very popular and is carried by many firefighters in their personal protective clothing for use as an emergency harness or hose holding strap.

> **Safety** The water knot, **JPR 15-12**, is the only knot that is recommended for use when tying webbing.

This is one knot that absolutely must be kept neat and be tightly set.

Tying a Water Knot (JPR 15-12)

1. Tie a simple overhand knot in one end of the webbing, making sure the webbing lies flat as it crosses back over itself. Leave enough tail beyond the overhand knot to be able to place a safety knot on that side of the water knot when finished, **JPR 15-12A**.
2. Take the other end of the webbing and thread it through in the opposite direction, again making sure that the webbing lies flat as it crosses itself. Make sure you feed enough webbing through so that you are able to place a safety knot on this side of the water knot, **JPR 15-12B**.

 INSPECTION

As with any emergency service tool, all ropes must be inspected and properly maintained to ensure they are in good shape for use during an emergency incident. These inspections should be a matter of department policy and done on a regular basis; many departments conduct monthly inspections. It is also recommended that all inspections of life safety ropes be logged on a form specifically designed for this purpose. These forms can be purchased from most rescue supply houses or the department can develop its own. An example is shown in **Figure 15-16**. In order for an inspection log to be useful, the ropes themselves need to be individually identified.

If a life safety rope is found to be damaged or suspect it should be immediately removed from service. According to NFPA 1983 a life safety rope must also be removed from emergency operation service once it has been used during an actual

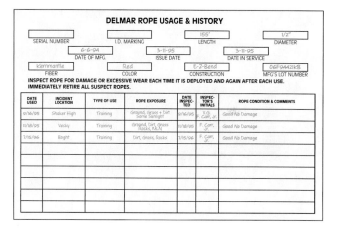

Figure 15-16 A standard life safety rope inspection form.

JOB PERFORMANCE REQUIREMENT 15-12
Tying a Water Knot

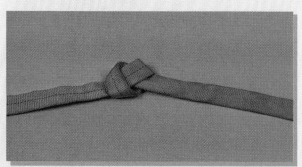

A Tie a simple overhand knot in one end of the webbing, making sure the webbing lies flat as it crosses back over itself. Leave enough tail beyond the overhand knot to be able to place a safety knot on that side of the water knot when finished.

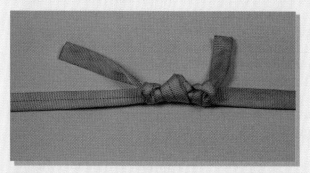

B Take the other end of the webbing and thread it through in the opposite direction, again making sure that the webbing lies flat as it crosses itself. Make sure you feed enough webbing through so that you are able to place a safety knot on this side of the water knot.

Figure 15-17 It is very important that rope is inspected as it is being put back into a rope bag.

Figure 15-18 Damaged rope.

emergency operation. This is mainly due to the unavoidable rigors that can be placed on a life safety rope during emergency operations, such as shock loading, overloading, chafing, abrasion, and exposure to chemicals/acids. These ropes, although removed from actual emergency operations, can be used for training as long as they are inspected and determined to be undamaged. *Training ropes should be inspected after every use.*

Ropes should be inspected along their entire length. This can be done when they are being placed back in storage, coiled, or bagged, **Figure 15-17**. It is best to inspect ropes after they have been cleaned. Generally, all ropes should be visually inspected for abrasion, laceration, chemical exposure, melting, and excessive fuzzing, many of which can be seen in **Figure 15-18**. They should also be inspected **tactilely** by running the rope through the hands. For this reason, firefighters should not wear gloves when inspecting and storing a rope, because they are feeling for foreign material embedded in the rope, slippery spots (possible chemical damage), voids in the center fibers (kernmantle and braid-on-braid), stiff or hard spots, and soft spots.

Laid (Twisted)

When inspecting laid ropes, firefighters should look and feel for all of the items just mentioned. In addition, laid ropes should be untwisted at random intervals to inspect the inside between the strands, **Figure 15-19**. When untwisting the strands of a natural fiber rope, it is important to watch for small particles such as sand or gravel that may damage the rope fibers. If a rope is stored wet or stored in a very humid location there may be signs of mold, mildew, or rot. The presence of mildew is not necessarily a terminal problem if found early enough. If these signs are present, the rope should be

Figure 15-19 It is important to twist apart a laid rope to inspect between strands.

removed from service and the problem should be addressed as soon as possible.

Braided

Braided rope should be inspected visually and tactilely as described previously. It cannot be twisted apart, so inspecting the inside is not possible. With braided rope, all strands will appear on the surface somewhere along the rope.

Braid-on-Braid

It is very important to remember, when inspecting braid-on-braid rope, that there is no way to see the inside braided rope. The person doing the inspection needs to pay particular attention to the tactile part of the inspection, feeling for possible types of damage previously listed. Another problem with this type of rope is that the outside braid will sometimes slip over the inner braid, causing the rope to invert on itself. A rope with this problem should be immediately removed from service. The outside braid typically represents 50 percent of the strength, so any damage to the outside braid will have a significant impact on the remaining strength.

Kernmantle

As with braid-on-braid rope, there is no way to see the kern portion, which typically represents 75 percent of the total strength of a kernmantle rope. The tactile inspection is the best and only way to discover damage to the kern. It is possible (but unusual) for the kern to be damaged without damaging the mantle. One should be alert for any change in the regular uniform feel of the rope as it is run through ungloved hands. Irregularities that can be discovered in a thorough tactile inspection include flat spots, voids, bunches, stiffness, and limpness. The key is to search for a different feel. If a minor imperfection seems to have been located, it may not necessitate the removal of the rope, but would certainly indicate the need for another thorough inspection.

> **Streetsmart Tip** Another way to inspect the rope, especially if a minor imperfection seems to have been located, is to tie it off to an object and do a thorough tactile inspection while placing slight tension on it. This may cause the minor imperfections to be more pronounced and therefore easier to feel.

MAINTENANCE

As with all firefighting and life safety tools and equipment, proper maintenance of utility and life safety ropes is required in order to ensure that they are available, safe, and ready for immediate use at an emergency scene. The maintenance of ropes is not difficult or complicated; it is generally limited to cleaning, inspecting, and storing.

Occasionally the firefighter may be called on to assist in placing new rope in service. *If the rope was purchased in bulk, it is important that the manufacturer's instructions with regard to cutting, sealing, whipping, and so on, be carefully adhered to.* The same goes for cleaning and storage of ropes. Most new ropes (especially life safety ropes) purchased from a rescue equipment dealer come with instructions and warnings, either attached to or packaged with the rope. An example is shown in **Figure 15-20**.

Cleaning

As mentioned in the previous section, the best policy for cleaning is to follow the manufacturer's instructions; however, since not all rope comes with instructions, the following are some general guidelines for the cleaning of rope.

Natural Materials

The cleaning of natural fiber rope is very difficult because water cannot be utilized. Remember from the earlier discussion that natural fiber ropes lose approximately 50 percent of their strength when wet and do not regain the lost strength when they dry. Firefighters are limited to brushing off the loose dirt and foreign materials with a stiff broom or brush,

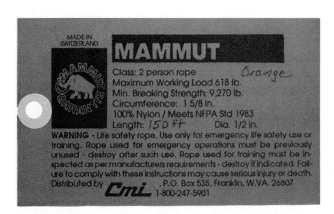

Figure 15-20 A standard manufacturer's warning/instruction label.

Figure 15-21 Manila rope is best cleaned by laying it out and brushing it off.

Figure 15-21. This is another reason why the fire service is moving away from the use of natural fiber rope, even as utility rope.

Synthetic Materials

Ropes manufactured from synthetic materials can be cleaned in a number of ways. Once again it is always the best policy to follow the manufacturer's instructions. In the absence of these, however, the following procedures can be utilized:

1. Use tap/cold water (high-temperature hot water may damage rope).
2. If detergent use is necessary, use only mild detergent that has been well diluted. (Use absolutely no bleach or any cleaning product containing bleach.)

Hand Washing

The most basic method of washing rope is to wash it by hand, using a large utility sink, bucket, or hose and scrub brush. The first two methods (sink or bucket) require immersing the rope to be washed in the sink or bucket, letting it soak for a few minutes, and then agitating and scrubbing it with the hands or small brushes, **Figure 15-22**. The third method (hose and scrub brush) described is to lay the rope out on the apparatus bay floor or apron, wet it with a garden hose, and scrub it with a brush or broom. If detergent is needed, it can be mixed in a bucket and the brush or broom can be dipped into it.

If detergent is used with any of these methods, all residue must be removed. If the sink or bucket

Figure 15-22 A kernmantle (nylon) rope being washed in a bucket.

method is used, this can be accomplished by changing the water or using other sinks or buckets full of fresh water to thoroughly rinse the rope. With the hose method, the rope must be rinsed thoroughly by occasionally moving it once or twice so that the soap trapped underneath it can be rinsed away.

Rope Washer

A number of different rope washers are available for purchase from rescue equipment suppliers. These are generally small PVC devices that fasten directly to a hose bib or to the male end of a garden hose, **Figure 15-23**. They are designed so that the water is directed at the rope from all different directions at the same time. The rope is inserted in one end (the direction of travel is usually marked) of the washer and slowly pulled out the other end. Without some method of injecting detergent into the hose stream, detergent cannot be used with rope washers.

Washing Machine

If using a clothes washing machine to wash rope, it is very important to make sure that the machine itself cannot damage the rope. Only a front-loading machine (with a glass window) should be used. These machines do not have an agitator for the rope to get caught on or wrapped around. Placing the rope in a large mesh bag before placing it in a machine, **Figure 15-24**, is the best way to keep it from getting fouled in the machinery and terribly knotted. If a large mesh bag is not available, the rope can be "chained," **Figure 15-25**, to try to keep it from getting fouled and/or knotted. Once again the previous directions regarding use of detergents should be followed and all residue rinsed out of the rope.

Drying Ropes

No matter what cleaning method is utilized, the rope must be completely dry prior to being stored. A few different methods of drying a rope are discussed next.

Laying Flat to Dry. The rope can be laid out on the apparatus bay floor or any other area that is clean, dry, and out of direct sunlight. Rope should be turned as it dries so it dries uniformly throughout.

Hanging to Dry. This is one of the easier methods of drying ropes. They can be hung in a hose tower, from the bar joists/trusses of the apparatus bay, or from any other place that is clean, dry, and out of direct sunlight. The rope must not come in contact with roofing tar, paint, bug spray, or other contaminants that may have been sprayed on bar joists or trusses. It is important to use care when removing

Figure 15-24 A kernmantle rope can be placed in a mesh bag and washed in a front-loading washing machine.

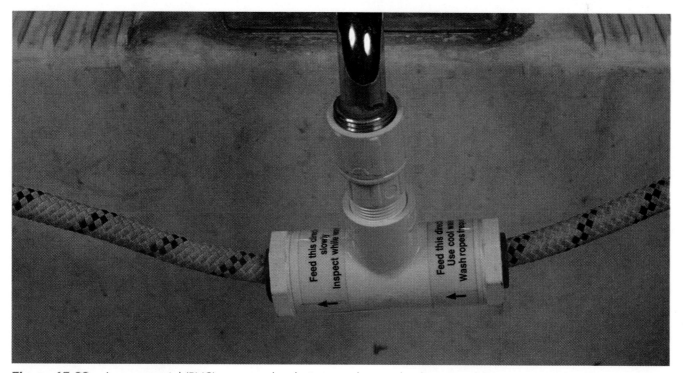

Figure 15-23 A commercial (PVC) rope washer being used to wash a kernmantle rope.

Figure 15-25 A kernmantle rope can also be "chained" and washed in a front-loading washing machine.

Figure 15-26 Properly identified, bagged, and stored life safety ropes.

the rope so that it does not get snagged or otherwise damaged by the truss plates on wood trusses or slag from welds on steel bar joists.

Machine Drying. Some departments utilize and some texts list clothes dryers as a viable alternative for drying rope.

> **Caution** Extreme caution is necessary if a clothes dryer is to be used.

Not only is there the same issue as with the washer (i.e., the possibility of the rope becoming entangled in the machinery), but with a dryer (gas or electric) it is very difficult to tell what the interior temperature is, even if there is a "low-temperature" setting. The temperature could surpass the manufacturer's recommendations.

Storage

> **Note** It is very important that rope be stored in a manner that allows for quick identification, access, and deployment at emergency operations, **Figure 15-26**.

Quick identification is very important so that a firefighter can rapidly select the proper type (utility or life safety) and length required for a given application. There are many ways to accomplish quick identification, **Figure 15-27**. Many departments use different material types, sizes, or colors of bags or tags to differentiate between utility ropes, life safety ropes, and ropes of different lengths. While there is no "right" way, to comply with NFPA 1983, each department needs to establish a policy.

ROPES AND KNOTS 445

Figure 15-27 Different types, sizes, and colors of bags are useful for quickly identifying different sizes and types of ropes.

Figure 15-28 The two most common methods of storing life safety ropes (bagged or coiled).

Quick access is usually not a problem on fire or rescue apparatus, especially if there is a routine inspection policy in place that keeps the ropes from being buried behind other equipment. Remember that ropes must be stored away from areas where they might be exposed to battery acid, fuel (including fumes from either), or sunlight.

The issue of quick deployment has been addressed for the most part by the storage of ropes in specially designed bags, although some departments still utilize the coiling method to store their ropes. While either method, **Figure 15-28**, will work to keep the rope neatly stored, the bags have some significant advantages. Depending on its construction, a bag offers protection from dirt, liquid contaminants, sunlight, abrasion, and so on. If a rope is properly bagged, it is easier to deploy than a rope that is coiled.

Coiling

The standard coil used by fire departments for years will work with either natural or synthetic ropes equally well. If the coiled method of storage is selected by a department, it is imperative that all members of the department be familiar with the process of both coiling and deploying from the coil. The proper process for coiling a rope is shown in **JPR 15-13**.

Some departments have built special apparatus to assist with the coiling of their ropes, but this is not really necessary. It is possible to use improvised standards and common items to coil a rope. The heel or tip of a ground ladder (do not use the tip of a roof ladder), a four-legged kitchen chair, or any item that has parallel posts (legs) approximately 18 to 24 inches apart will work for rope coiling. The distance between the standards and the number of wraps per layer are dependent on the size and length of the rope being coiled.

Coiling a Rope (JPR 15-13)

1. Measure off an amount of rope equal to approximately three times the distance between the standards to be used in tying the finished coil, **JPR 15-13A**.
2. Starting with this point measured at one of the standards, wrap the rope around the standards, forming the loops and moving toward you until you have nine to twelve loops on the standards. Then form a second layer on top of the first, moving away from you. Continue in this manner until approximately twice the amount measured off in step 1 remains, **JPR 15-13B**.

JOB PERFORMANCE REQUIREMENT 15-13
Coiling a Rope

A Measure off an amount of rope equal to approximately three times the distance between the standards to be used in tying the finished coil.

B Starting with this point at one of the standards, wrap the rope around the standards until you have nine to twelve loops on the standards. Then form a second layer on top of the first, going back the other way. Continue in this manner until approximately twice the amount measured off in step 1 remains.

C Begin wrapping this remaining rope around the coils that you formed in step 2. Finish these wraps off with a clove hitch.

D Utilizing the rope measured off in step 1, form a bight, leaving the loose end of the rope coming out toward you. Feed the bight from the back side to the front side on the wraps. You should now have the loose end and the bight sticking out toward you.

E Place the loose end through the bight and adjust the bight until it is snug where it passes through the bight.

3. Begin wrapping this remaining rope around the coils that you formed in step 2. Finish these wraps off with a clove hitch, **JPR 15-13C.**
4. Utilizing the rope measured off in step 1, form a bight leaving the loose end of the rope coming out toward you. Feed the bight from the back side to the front side on the wraps. You should now have the loose end and the bight sticking out toward you, **JPR 15-13D.**
5. Place the loose end through the bight and adjust the bight until it is snug where it passes through the bight, **JPR 15-13E.**

To deploy a rope that has been coiled in this manner, the firefighter pulls the loose end out of the bight and pulls the bight out from between the loops. Next, the firefighter pulls a couple of loops out of the center and drops the coil if in an elevated position or feeds the rope out, pulling it from the center, as needed. A firefighter must be aware that if the coil is dropped, it may hang up before it reaches the ground, either on itself or on some structural member. If extending the rope horizontally, **Figure 15-29**, it is usually easier if one person holds the coil on the ground while another person pulls the rope out of the coil.

Figure 15-29 A coiled rope in the process of being deployed.

Bagging

The utilization of special bags for the storage of rope is another of the things that came about with the switch to synthetic fiber ropes. Ideally rope bags should have attached watertight storage for the rope log. If they do not, the rope and bag need to be clearly identified and associated with a rope log on file in the apparatus or at the station.

The process for placing a rope in a rope bag, **JPR 15-14**, is very simple (especially when compared to coiling a rope) and can be easily accomplished by one or two persons. Many bags have holes in the bottom for drainage and/or to feed the rope out of. Department policy should determine whether the rope is fed out of the bottom hole.

Bagging a Rope (JPR 15-14)

1. If policy is for the rope to be fed out the bottom hole, start by feeding enough rope out the bottom to tie a basic figure eight knot. If not, tie a basic figure eight knot in the end of the rope and place it inside at the bottom of the bag, **JPR 15-14A.**
2. Begin placing the rope in the bag by sliding your hand approximately 12 to 18 inches up the rope at a time and "stuffing" it in the bag. *Do not coil the rope in the bag; it will hang up almost every time if you do,* **JPR 15-14B.**
3. Continue this until you are at the end of the rope. Tie a basic figure eight knot in the end and place it on top of the rope in the bag. Secure the bag closed and you are finished, **JPR 15-14C.**

The process for deploying a rope stored in a bag is very straightforward. If the policy is to deploy through the bottom hole, the firefighter grasps the figure eight tied just outside the hole and either drops the bag or feeds the rope out of the bag as needed. As with a coil, if the bag is dropped, the rope may hang up on either itself or the structure. Feeding the rope out allows the firefighter to keep the bag in case a problem develops, **Figure 15-30**. If deploying the rope horizontally, the firefighter can either have someone take the end of the rope or the bag and walk in the desired direction of deployment. The rope will feed out as the person walks.

JOB PERFORMANCE REQUIREMENT 15-14
Bagging a Rope

A If policy is for the rope to be fed out the bottom hole, start by feeding enough rope out the bottom to tie a basic figure eight knot. If not, tie a basic figure eight knot in the end of the rope and place it in the bottom of the bag.

B Begin placing the rope in the bag by sliding your hand approximately 12 to 18 inches up the rope at a time and "stuffing" it in the bag. *Do not coil the rope in the bag; it will hang up almost every time if you do.*

C Continue doing this until you are at the end of the rope. Tie a basic figure eight knot in the end and place it on top of the rope in the bag. Close the bag and you are finished.

ROPES AND KNOTS 449

Figure 15-30 A bagged lifeline in the process of being deployed.

Figure 15-31 A firefighter deploying a rope from an elevated position in order to hoist a tool.

RIGGING FOR HOISTING

One of the primary uses of rope on an emergency scene is for hoisting of tools and equipment to the needed elevation/location. This is not the type of job that requires a life safety rope. Instead, a much smaller diameter rope can be utilized, which translates to a lighter, easier-to-carry rope. Ropes used for hoisting can be stored either coiled or bagged. It is usually much easier if the rope is fed out by the person who will be hoisting the item, **Figure 15-31**, as opposed to dropping the whole bag or trying to pass the end of the rope up to the person at the higher elevation.

This section presents examples of how to hoist a few specific tools and equipment and the proper knots to utilize when doing so.

> **Note** If a firefighter has good knowledge and is skilled in tying the knots presented in this chapter, there should be no problem with hoisting almost any type of tool or equipment.

In general, anything that has a closed handle can be hoisted with a figure eight or bowline, while longer cylindrical tools (i.e., ax and pike pole) can be hoisted using a clove hitch and half hitches.

Some departments have policies requiring the use of **tag/guide lines**, which are guide ropes held and controlled by firefighters on the ground, **Figure 15-32**. There are some circumstances in which tag lines should be used whether required or not. Examples of these circumstances are when an overhang(s) exists that the item is likely to get caught on, when the item rubbing against the side of the structure may be damaged, when there is a strong wind that may cause the item to blow out of control, or any time firefighters think a tag/guide line is necessary. All knots must be dressed, be set, and have safeties.

Specific Tools and Equipment

Ax

A small figure eight on a bight with a half hitch up the handle is the easiest and quickest way to hoist an ax.

450 ■ CHAPTER 15

Figure 15-32 Use of a tag/guide line.

Hoisting an Ax (JPR 15-15)

1. Tie a figure eight on a bight forming a small loop. Drop the loop over the ax handle, **JPR 15-15A**.
2. Take the loop around the head of the ax, bringing it back up, paralleling the handle. Place a half hitch approximately 6 to 8 inches below the handle, **JPR 15-15B** and **C**.

Pike Pole

Pike poles should be hoisted point up.

Hoisting a Pike Pole (JPR 15-16)

1. To hoist a pike pole point up, place a clove hitch near the end of the handle, **JPR 15-16A**.
2. Place two half hitches around the handle between the clove and the point, with the last one being located immediately below the head, **JPR 15-16B**.
3. Raise the pole with the head at the highest point, **JPR 15-16C**.

JOB PERFORMANCE REQUIREMENT 15-15
Hoisting an Ax

A Tie a figure eight knot and slip the knot over the ax handle.

B Wrap the rope over the head and back toward the handle.

C Place a half hitch near the end of the handle.

JOB PERFORMANCE REQUIREMENT 15-16
Hoisting a Pike Pole

A Tie a clove hitch near the end of the handle.

B Place half hitches on the pole. Put one immediately below the pike pole head.

C Hoist the pole up the building.

Hoselines

Hoselines can be hoisted either charged or uncharged. A charged hoseline is going to be dramatically heavier than an uncharged one.

Hoisting a Charged Hoseline (JPR 15-17)

1. With the nozzle bale in the closed (forward) position, tie a clove hitch around the hoseline 18 to 24 inches behind the nozzle, **JPR 15-17A**.
2. Form a bight in the rope and feed it through the bale from the coupling side, flipping it over before slipping it over the nozzle tip. You will have formed a half hitch by doing so, **JPR 15-17B** and **C**. Hoisted by this method, the rope will actually hold the nozzle closed should the bale be caught on something while being hoisted, **JPR 15-17D** and **E**.

Hoisting an Uncharged Hoseline (JPR 15-18)

1. Fold the nozzle back on the hose approximately 3 to 4 feet. Tie a clove hitch around the hose and nozzle to hold them together, **JPR 15-18A**.
2. Place a half hitch around the end of the hose approximately 6 inches from where it is folded back, **JPR 15-18B** and **C**.

Smoke Ejector, Chain Saw, Rotary Saw

All of these items and a host of others used on emergency scenes have closed handles or support pieces that can have rope tied around them for hoisting.

Hoisting Small Equipment (JPR 15-19)

1. Tie a follow-through figure eight through the closed handle.
2. The use of a tag line is highly recommended for items of this type, which tend to be heavy and hard to control with only the hoisting line from above.

Ladders

Both ground ladders and roof ladders are hoisted on a regular basis at emergency scenes. The hoisting procedure is the same for both types. Once

JOB PERFORMANCE REQUIREMENT 15-17
Hoisting a Charged Hoseline

A Tie a clove hitch around the charged hoseline 18 to 24 inches behind the nozzle.

B Form a bight in the rope and feed it through the bale.

C Form a half hitch.

again the use of a tag line is recommended, especially in those cases where the firefighter on the ground will not be able to guide the bottom of the ladder as it is hoisted.

Hoisting a Ladder (JPR 15-20)

1. Tie a large figure eight on a bight forming a loop approximately 3 to 4 feet long, **JPR 15-20A.**
2. Go approximately one-third the length down the ladder and put the loop through the rungs, **JPR 15-20B.**
3. Pull the loop up and slip it around the top of the ladder allowing it to slide back down, securing the ladder, **JPR 15-20C** and **D.**

Securing a Rope between Two Objects

While the use of rope(s) to cordon off an area is not a common practice in many departments today, the need to secure a rope between two objects may arise at any emergency scene. If this need arises, a rope may be used as a barrier using one of two

JOB PERFORMANCE REQUIREMENT 15-17
Hoisting a Charged Hoseline (Continued)

D Slip the half hitch over the nozzle.

E The charged hoseline is rigged for hoisting.

JOB PERFORMANCE REQUIREMENT 15-18
Hoisting an Uncharged Hoseline

A Fold the nozzle back on the hose approximately 3 to 4 feet. Tie a clove hitch around the hose and nozzle together.

B Place a half hitch around the end of the hose, approximately 6 inches below the bend.

(continued)

JOB PERFORMANCE REQUIREMENT 15-18
Hoisting an Uncharged Hoseline (Continued)

C The uncharged hoseline is rigged for hoisting.

JOB PERFORMANCE REQUIREMENT 15-19
Hoisting Small Equipment

A rotary saw (closed handle) being hoisted with a tag line.

methods. A figure eight on a bight may be used to secure an anchor point.

Tying a Rope between Two Objects

1. Starting at one end, secure the rope to a solid object (anchor) by forming a figure eight on a bight and sliding the rope over the anchor, **JPR 15-21A**. If the rope cannot be placed over the anchor, utilize a follow-through figure eight to secure the anchor point, **JPR 15-21B**.
2. Lay the rope out, keeping it as straight as possible to minimize slack, until reaching the other objective to be tied, **JPR 15-21C**. Measure approximately one-third of the standing part of the rope toward the anchor and tie a figure eight on a bight resulting in a loop 6 to 12 inches long, **JPR 15-21D**.
3. Wrap the running end of the rope around the objective and bring it to the figure eight on a bight tied in step 2, **JPR 15-21E**.
4. Thread the running end through the loop and pull it back toward the objective, tightening the rope as necessary, **JPR 15-21F**.
5. Using the running end, tie three or four consecutive half hitches around both sections of rope, **JPR 15-21G**.

JOB PERFORMANCE TASK 15-20
Hoisting a Ladder

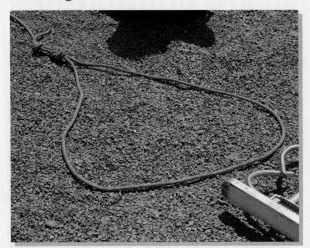

A Tie a large figure eight on a bight forming a loop approximately 3 to 4 feet long.

B Go approximately one-third the length down the ladder and pull the loop through the rungs.

C Pull the loop up and slip it around the top of the ladder.

D Allow the loop to slip down the ladder. Attach a tag line. Ladder is ready to be hoisted.

JOB PERFORMANCE REQUIREMENT 15-21
Tying a Rope Between Two Objects

A Secure the rope to an anchor using a figure eight on a bight.

B Secure the rope to an anchor using a follow-through figure eight.

C Lay the rope out to the objective.

D Measure approximately one-third of the standing part and tie a figure eight on a bight resulting in a loop 6 to 12 inches long.

E Wrap the running end of the rope around the objective and bring it to the figure eight on a bight tied in step 2.

F Thread the running end through the loop and pull it back toward the objective, tightening the rope as necessary.

JOB PERFORMANCE REQUIREMENT 15-21
Tying a Rope Between Two Objects (Continued)

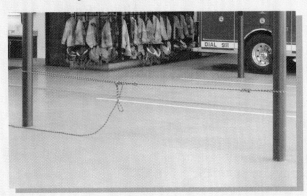

G Using the running end, tie a minimum of one half hitch around both sections of rope. Multiple half hitches will secure the rope and ensure that the knot is not accidentally untied.

Lessons Learned

This chapter presented basic information and introduced skills to accomplish basic tasks required at emergency scenes. It has not attempted to present all of the information about skills that firefighters may need as they progress in their careers and become involved in more technical and complex emergency operations. Firefighters are encouraged to regularly practice the basics learned in this chapter and pursue further information and training to expand their knowledge, skills, and abilities in this very interesting and exciting subject.

KEY TERMS

Bight A doubled section of rope, usually made along the standing part, that forms a U-turn in the rope that does not cross itself.

Dressing The practice of making sure that all parts of a knot are lying in the proper orientation to the other parts and look exactly as the pictures herein indicate.

Dynamic A rope having a high degree of elongation (10 to 15 percent) at normal safe working loads.

Kern A derivative of the term *kernel,* which is defined as "the central, most important part of something; core; essence."

Life Safety Line According to NFPA 1983, rope dedicated solely to the purpose of constructing lines for supporting people during rescue, firefighting, or other emergency operations, or during training evolutions.

Loop A turn in the standing part that crosses itself and results in the standing part continuing on in the original direction of travel.

Mantle Anything that cloaks, envelops, covers, or conceals.

One- or Two-Person Rope According to NFPA 1983, a one-person rope requires a minimum tensile strength of 4,500 pounds, and a two-person rope requires a minimum tensile strength of 9,000 pounds.

Round Turn Formed by continuing the loop on around until the sections of the standing part on either side of the round turn are parallel to one another.

Running End End of the rope that is not rigged or tied off.

Setting The finishing step, making sure that the knot is snug in all directions of pull.

Shock Load A load or impact being transferred to a rope suddenly and all at one time.

Standing Part The part of a rope that is not used to tie off.

Static A rope having very little (less than 2 percent) elongation at normal safe working loads.

Tactilely Using the sense of touch to feel for any differences or abnormality.

Tag/Guide Lines Tag lines are ropes held and controlled by firefighters on the ground or lower elevations in order to keep items being hoisted from banging against or getting caught on the structure as they are being hoisted.

Tensile Strength Breaking strength of a rope when a load is applied along the direction of the length, generally measured in pounds per square inch.

Utility Rope Rope used for utility purposes only. Some of the tasks utility ropes are used for in most every fire department are hoisting tools and equipment, cordoning off areas, and stabilizing objects. Also used as ladder halyards.

Webbing Nylon strapping, available in tubular and flat construction methods.

Working End The end of the rope that is utilized to secure/tie off the rope.

REVIEW QUESTIONS

1. What NFPA standard addresses life safety ropes and system components?
2. Polypropylene and polyethylene are two of the four synthetic materials discussed in this chapter. What is the major distinguishing feature between these two materials and nylon and polyester, the other two discussed?
3. Modern ropes manufactured from synthetic materials fall into two broad categories, static and dynamic. What is the major difference between the categories?
4. In a kernmantle rope the kern carries the vast majority of the load. What is the role of the mantle?
5. According to NFPA 1983, there are two classifications of life safety ropes. What are they and what are the minimum requirements for each?
6. Three terms are commonly utilized to describe parts of a rope. What are they and what does each mean?
7. To "dress" and to "set" a knot means to do what to it?
8. What is the best feature of figure eight knots?
9. What is the only knot recognized for tying webbing?
10. According to NFPA 1983, how many times can a life safety rope be utilized for emergency service after having been used during an emergency?
11. When storing ropes, it is important that the rope be stored in a manner that allows for what three things at emergency scenes?
12. If a rope, coiled or bagged, is dropped from an elevated position, what are some of the negative things that could occur?

Additional Resources

Frank, James and Jerrold Smith, *Rope Rescue Manual.* California Mountain Company Limited, Santa Barbara, CA, 1987.

Padgett, Alan and Bruce Smith, *On Rope.* National Speleological Society, Huntsville, AL, 1987.

Setnicka, Tim, *Wilderness Search and Rescue.* Appalachian Mountain Club, Boston, MA, 1980.

CHAPTER 16

RESCUE PROCEDURES

Robert F. Hancock, Hillsborough County Fire Rescue

OUTLINE

- Objectives
- Introduction
- Hazards Associated with Rescue Operations
- Search of Burning Structures
- Victim Removal, Drags, and Carries
- Extrication from Motor Vehicles
- Specialized Rescue Situations and Tools
- Lessons Learned
- Key Terms
- Review Questions
- Additional Resources

STREET STORY

We had just sat down for a steak dinner at 6 p.m. when the alarm came in. Over the enhanced 9-1-1, we got a dispatch that there was a person trapped in an elevator. We do two or three elevator rescues per week, but we knew this one was unusual because we heard the communications center dispatch a medic unit on the EMS channel. When we arrived at the scene—a historical home downtown that was being renovated—one person met us and said that the elevator was stuck between floors and someone's head was caught. Apparently, four kids, one of them a fourteen-year-old boy, had been starting the elevator, holding the door open, and trying to jump in and catch the car as it was moving down between flights. The final time, this boy got his legs and torso in but he'd gotten his head pinned between the floor landing and the top of the elevator car.

There was a stained-glass transom window over the elevator doorway on the first floor so we could see into the shaft that there were three hysterical kids in there, and the boy's legs were hanging into the car. The kids were holding his body up.

We had a three-person engine company, so we split up to do the size-up from a few different angles. We shut the power off first. In a rescue, the first thing to consider is not only the safety of the victim, but the safety of all the firefighters involved, and the big threat was that this elevator could move and we could get hurt. At first, it seemed the elevator was inaccessible, but then I realized I could make it in by breaking the transom window and climbing in the car. (If you had ever met me, you would never believe that at 6 feet 4 inches and 300 pounds, I made it through a 2-foot by 2-foot window!) The other two firefighters proceeded upstairs to evaluate the situation from there. They could only see the top of the kid's head; the area in which his head was compressed was only about 2 inches wide. I could see his neck and chin; it was obvious he was unconscious and not breathing. I concentrated on calming the hysterical kids, telling them everything would be all right. This was traumatic for them.

The team above determined that they would get something under the roof of the elevator car to pry it up. The other firefighters grabbed a hydraulic rescue tool, called a ladder and rescue for backup, and proceeded to pry enough to create an opening. Then they used some cribbing as a wedge to maintain the space, so it wouldn't spring back. I was able to pull the boy down into the elevator with me. I gave him a couple of rescue breaths before I passed him out the same window I had come in to the EMS crew that was waiting outside. Obviously you want to immobilize people as much as possible to prevent spinal injuries, but in this situation we couldn't get a backboard in that space, and the priority was to get him breathing, so we wanted to get him out right away. So I supported him as much as possible, passing him out feet first. As the EMS crew grabbed his legs, I held his shoulders, back, and head. Once they got him out, he was fully immobilized, intubated, and rushed to the hospital.

He survived the event—within ten days he walked into our fire station.

—Street Story by Mike Wisko, Acting Captain, Galveston Fire Department, Engine 1, B Shift, Galveston, Texas

OBJECTIVES

After completing this chapter, the reader should be able to:

- Recognize the hazards associated with various rescue operations.
- Describe the differences between primary and secondary searches.
- Demonstrate the proper procedures for victim drags and carries.
- Define the proper terminology utilized during motor vehicle extrication operations.
- Demonstrate proper and safe operation of vehicle extrication tools and equipment.
- Explain the various types of specialized rescue situations presented and the specific hazards associated with each of them.

INTRODUCTION

The term **rescue** in the emergency services has many meanings. In this chapter, *rescue* describes those actions that trained firefighters perform at emergency scenes to remove someone from imminent danger or to **extricate** them if they are already entrapped. Rescue is a very broad subject, and this chapter is going to touch on various rescue situations firefighters may find themselves confronted with. It is not the intent to make the reader an "expert" in all rescue situations discussed, but rather to bring the reader to an awareness level in order to recognize a situation for what it is, be aware of the dangers associated with it, and apply safe procedures to any potential rescue situation that may be encountered. The chapter does, however, cover building search, victim removal, and vehicle extrication in more detail since these are the areas that are most common to daily firefighting operations.

Streetsmart Tip Teamwork and safety are the key points to remember in any rescue operation.

HAZARDS ASSOCIATED WITH RESCUE OPERATIONS

Hazards are associated with every type of rescue operation. The most common hazards are presented along with discussions of the various rescue topics later in this chapter.

When firefighters are involved in a rescue operation, one of the biggest dangers that they must be aware of is the focusing of attention on a particular problem without proper regard for possible consequences or alternative approaches. This is generally referred to as **tunnel vision**. It is very easy to develop tunnel vision when a rescuer is involved in an unusually complex and/or lengthy rescue. Tunnel vision can keep, and in many cases has kept, the rescuer from seeing an obvious solution or more often an impending danger.

Note An injured rescuer does more than just add another patient to be cared for. As can be seen in **Figure 16-1**, it is not uncommon for the remaining rescue personnel to lose focus on the person originally injured and focus on the injured rescuer.

SEARCH OF BURNING STRUCTURES

Searching burning structures is one of the most dangerous rescue situations regularly faced by the majority of firefighters. As with other situations that firefighters respond to which are discussed in this chapter and elsewhere in this text, the best way to reduce the danger while searching involved structures is through training, practicing, and planning.

Safety Any time that firefighters enter a structure that is burning or in danger of becoming so, they must always wear appropriate personal protective equipment (PPE), including self-contained breathing apparatus (SCBA) and a personal alert safety system (PASS) device, for the known or potential threat/danger.

Figure 16-1 When a member of the rescue crew is injured, it is hard to maintain focus on the person originally injured.

Firefighters must always work in teams of two or more when entering an involved structure for any reason (i.e., interior firefighting, search and rescue, ventilation). In addition to this search team, a minimum of two firefighters must be standing by immediately outside in full protective clothing and SCBA with a charged hoseline (although the charged hoseline is not required, it is a good logical practice) ready to come in and assist the search team should a problem develop, **Figure 16-2**. This is commonly referred to as the **two in/two out** rule. Even if the search team also has a hoseline with them, this recommendation remains in effect.

As firefighters approach a structure that is going to be searched, they should perform a size-up to determine a "rescue profile." The rescue profile helps the firefighters prioritize the probability, location, and status of potential victims as well as the avenues of access and egress. While life, safety, and rescue always remain a priority, there are structural fire environments that are a "recovery" environment. Simply stated, a recovery environment means there is little to no chance of saving a victim due to fire and smoke conditions and/or building collapse potential. Factors that should be evaluated to determine the rescue profile include:

Occupancy type/time of day: Residential structures can be occupied at any time, although typical sleeping hours present the greatest indication that victims may need rescue. Commercial occupancies seldom have sleeping occupants and, therefore, a greater chance exists that the occupants have self-evacuated. In some high-cost resort areas, employees may sleep on the premises due to the high cost of domicile, therefore increasing the rescue profile. Typically, victims in a commercial structure are found in the path of exits, whereas residential victims can be anywhere. Children are likely to hide from the smoke and fire—and may be in closets, in cabinets, under beds, and even in their parents' bed.

Fire/smoke conditions: Turbulent smoke (pressurized with high heat) is an unsurvivable environment for occupants due to pain threshold and toxicity—a recovery environment versus a rescue environment. Other areas with less dense and lighter colored smoke indicate a higher rescue profile. Post-flashover compartments (rooms) are a recovery environment. Rooms adjacent and above an involved room have a higher rescue priority, assuming the smoke in these areas is not turbulent. Remember, flashover of one room is likely to cause rapid and rolling ignition of dense smoke.

Activity clues: Cars in the driveway or garage, toys strewn about, shoveled snow and footprints, and open windows provide clues that a home is occupied, **Figure 16-3.** If home

Figure 16-2 A rapid intervention team should stand by for immediate action should an interior team need rescue.

RESCUE PROCEDURES

Figure 16-3 A street side view of a typical residential occupancy. Note the clues that can be spotted in this photo: cars in driveway, toys, bicycles, and so on.

Figure 16-4 Well-equipped interior structural firefighting/search and rescue crews need a minimum of full PPE, SCBA, PASS, forcible entry tool, flashlight, portable radio, and thermal imaging camera.

occupants are not out waiting for the fire department, the rescue profile is high. Conversely, no cars, boarded-up windows, signs of neglect, and vacancy signs should indicate a lower rescue profile.

In addition to the protective clothing and equipment listed earlier, firefighters should carry with them a forcible entry tool (ax, pry bar, Halligan tool, etc.), flashlight, portable radio, and thermal imaging camera, **Figure 16-4**. The forcible entry tool can be useful in gaining access to locked or blocked rooms within the structure and is also useful in extending the searching firefighters' reach under and behind objects such as beds, dressers, and tables and can assist the firefighter in creating an emergency egress if necessary. The flashlight is useful in searching if the smoke is not too thick and can be useful to signal a rescue/backup crew should a firefighter get into trouble. The portable radio is helpful for keeping the incident commander informed of progress, fire/smoke conditions at a location, and the results of a search. It is very useful in communicating with a rescue/backup crew if a firefighter becomes disoriented or lost. Thermal imaging cameras and devices also allow the search team to see through the smoke, thereby increasing the speed with which searches can be accomplished. These devices are discussed later in this chapter.

In many single-family residential structures, it may be possible to conduct a search thoroughly and safely without a **guideline/lifeline** or hoseline by using the wall as a reference, **Figure 16-5**. Utilizing a wall in this manner is referred to as conducting a "right-hand" or a "left-hand" search, meaning that the firefighter maintains constant contact between the wall and that side of the body. By doing this, the firefighter will return to the point of entry while searching all the way around the room. However, this is not the case in larger mercantile, commercial, or industrial occupancies. In these types of occupancies, it is mandatory that a guideline/lifeline be

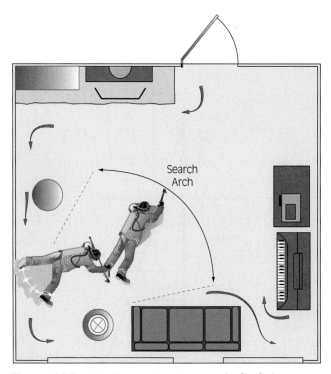

Figure 16-5 During an interior search, firefighters should stay in contact with a wall. If visibility is hampered, firefighters can reach into the center of the room using a tool and a "human chain" technique.

utilized. It is impossible to conduct a thorough search with an acceptable level of safety without one. It is very easy to become disoriented and completely turned around when in a large open area such as a department store, warehouse, or industrial plant. As shown in **Figure 16-6**, these types of occupancies generally not only have large open spaces, but often have obstructions (shelf units, machines, displays, etc.) at various and random locations within the open spaces.

Ropes are now available that are specifically designed to be utilized as guidelines. They have reflective material woven into the outer jacket of the rope to make them more visible. Ropes that are used as guidelines should not be used as lifelines since they are subjected to damage, both chemical and mechanical, while being dragged through structures.

> **Streetsmart Tip** Another way for firefighters to find their way back to where they entered a structure is to follow the hoseline back out. To be able to do this in firefighting conditions, it must be practiced during training until firefighters can tell which way the hose is going by feeling the couplings. The male coupling will be pointing toward the nozzle, away from the entry/exit.

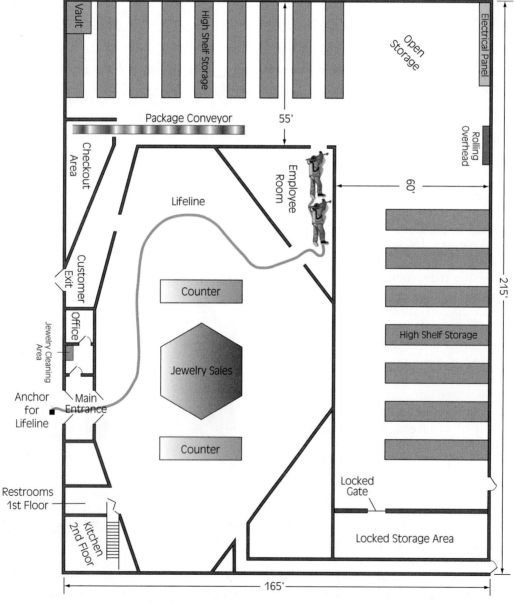

Figure 16-6 Commercial/industrial occupancies have large areas and various obstructions, machines, and storage. Firefighters must use a guideline/lifeline to help find their way out. The lifeline will also help an RIT find interior crews experiencing an emergency.

The biggest asset to conducting a safe and successful search is to have practiced and trained together prior to the actual emergency and to have a plan prior to entering the structure. "Having a plan" means that the search team members know who is in charge, in what direction they are going, on what side (left or right) to keep the wall, and any other pertinent information. Team members must stay together. If searching a small room or area with decent visibility, one team member may remain at the entrance but must be able to stay in constant voice communication with the other member so that a reference point can be maintained. Another way of covering more area is for one member to remain in contact with the wall while the other member holds onto the first member's arm, leg, or forcible entry tool, effectively doubling the distance that can be reached without leaving the wall. If a room is too hot to be entered, a firefighter can use a forcible entry tool to probe through the doorway or window.

Searching a building is completed in two different operations: the primary search and the secondary search. These operations are also two of the tactical benchmarks utilized by the incident commander.

Primary Search

The primary search is the first and most dangerous of the two. During the primary search, the team is often ahead of the attack lines and may be above the fire (the most dangerous place). They search the areas that are most likely to have victims in a rapid, but thorough, manner. In residential occupancies at night, these would be bedrooms, closets, near doorways, and bathrooms. In commercial occupancies during operating hours, victims would be expected near the exits or in offices or restrooms.

At times, the primary search is conducted from outside the structure, especially in residential structures. Called a *window search*, this primary search takes advantage of speed by opening windows of rooms uninvolved and doing a quick look into the room. This can be especially advantageous if the main path into the building is heavily charged with fire and/or turbulent, dense smoke. It is important to remember that "opening" the structure can increase fire intensity—although the accompanying ventilation can help relieve pressure and smoke conditions that are the main threat to those needing rescue. This approach is compatible with a *vent for life* strategy. While inside a structure, firefighters may "vent as they go" to help relieve smoke pressure, but only if such will not cause significant fire spread.

Often, it is quicker and safer for firefighters to ladder and break out an upstairs (bedroom) window and enter for a primary search versus a blind crawl through thick smoke and heat to find a stairway. A fire on the main floor will convect heat up open stairways and make it untenable for victims and quite dangerous for firefighters. Flashover of a room can lead to smoke cloud ignition that will literally blowtorch up the stairway. If firefighters are using stairways, they must stay low. When descending stairs, firefighters should stay low by crawling on their hands and knees and proceed feet first.

When conducting the primary search, visibility is often obscured by smoke conditions and darkness. The best visibility and portable light penetration will be closer to the floor. Additionally, the noise created by SCBA, radios, crawling, running into furniture and obstacles, as well as general fireground activities can mask the simple whimper of a child or the tapping of a trapped occupant who has little or no strength left. Firefighters should occasionally pause and listen for cries or signals for help.

Once the primary search is completed, crews should notify the incident commander that the assigned search is complete. Once all search crews report a complete search, the incident commander will broadcast an all clear. This is an incident tactical benchmark. Following an all clear, firefighters should reduce the degree of risk they take to help keep risk/benefit in balance.

> **Streetsmart Tip** Many people act in unpredictable ways when a fire occurs. Because they are scared, they will try to hide from the fire, which is why firefighters must check under beds and in closets and restrooms thoroughly.

Secondary Search

The secondary search is usually conducted once the fire is out or at least well under control. This search can be much more thorough since there is no immediate fire danger, and the visibility is much better once the smoke is markedly reduced. During the secondary search, the team can search through debris that has fallen or been knocked over. It is also possible during this search to locate areas that may still need to be extinguished. If the fire was extinguished before consuming the room contents, when firefighters return to do the secondary search, they may be able to see where the primary searchers missed areas. This an excellent learning opportunity.

Thermal Imaging Cameras

Mike West, Lieutenant, South Metro Fire and Rescue, Colorado, and Instructor for SAFE-IR

Introduction to Thermal Imaging Cameras

Firefighters have long fought the lack of visibility at structural fires. Many operations are conducted blindly due to dense smoke and darkness. Firefighters are beginning to overcome this issue with the use of technology. Specifically, thermal imaging cameras (TICs) help firefighters "see" in low-visibility environments. The TIC uses an electronic detector to receive heat energy and then turn that energy into a picture that is viewed by the firefighter on a display screen. TICs do not require visible light to operate and work well in most low-visibility environments. The military developed much of the thermal imaging technology used today, and some fire departments have used TICs for nearly twenty years. Its popularity with most departments, however, has increased only in the last few years.

Fundamental Operational Concept

All objects emit heat energy. Some things, like humans, animals, or fire, create their own energy. Humans, for instance, emit energy at about 98.6°F. Other objects simply absorb and reflect the heat around them. For example, furniture, walls, flooring, and just about anything that is not "living" inside a building will absorb and reflect heat. All TICs have detectors used to gather heat energy. Currently, there are several types of image detectors available, but all process the detected heat energy into a picture or image representing temperature differences. Images are displayed as black, white, and shades of gray. In most TICs, white represents the warmest object in the scene and black represents the coolest. Some TICs also display colors that are associated with specific temperatures. TICs are typically battery powered. The size, type, and display of the TIC will impact how long the battery can last before needing recharging. It is important that firefighters using the imager understand the specific camera and detector used in their department—all TICs have strengths and limitations and can vary in how they depict different scenes. Only through instruction and training will firefighters learn how to interpret the pictures of the specific camera their department uses. Thermal imagers have handheld or helmet-mounted housings. Some fire departments have fixed-mount thermal imagers on their apparatus. For instance, an airport rescue and firefighting apparatus may use fixed-mounted TICs to locate downed planes in dense fog. Technology already exists that allows some TICs to transmit the image to remote displays so outside fire officers can see what the interior crew is seeing on their TIC.

(Photo courtesy of SAFE-IR)

(Photo courtesy of SAFE-IR)

Thermal Imaging Cameras (Continued)

Uses

TICs allow the firefighter to see through smoke. Because of this, thermal imagers are excellent search and rescue tools. It is important to note that TICs *do not* take the place of standard search techniques like following a wall, rope, or hoseline. The camera, like any tool, can fail, leaving the firefighting crew in a position to be lost. Maintaining good orientation is essential with or without the use of TICs. Thermal imaging cameras can be used to direct hose streams toward the seat of the fire and are helpful in overhaul by locating areas that may have hidden fire. A TIC used on the roof can point crews toward the hottest point, making vertical ventilation more effective. At odor investigations, they can help locate overheated light ballasts and electrical motors. TICs are used at nonfire emergencies as well. At HAZMAT calls, imagers can monitor fluid levels inside single-wall tanks, and they can monitor foam blankets on hydrocarbon spills. TICs have also been used to search vegetation for victims that may have been ejected from a vehicle accident. In all operations, it is important to use standard safety rules and local departmental SOPs in the performance of the task.

Training

Training is vital to use the TIC effectively and safely. The training should be specific to the camera used by the fire department and should include classroom orientation and hands-on usage. Only through training and repeated use can the firefighter become proficient. Each fire department should use the manufacturer's literature and guidelines as a starting point for their training.

VICTIM REMOVAL, DRAGS, AND CARRIES

Victims must be removed as carefully and expeditiously as possible. The goal should be not to cause any further injury or aggravation of existing injuries during the rescue process. Many times it is not possible to utilize all of the patient handling and immobilization skills that firefighters have learned due to the imminent danger presented by the heat, smoke, and gases in a structural fire. Other types of rescue situations also sometimes prevent the rescuer from using all the care that the person would like to due to a continuing hazard, or being in a confining area or other hostile environment.

All carries and drags place additional stress on the rescuer's musculoskeletal system. Training and work hardening prepare the firefighter to perform these victim removal techniques. Of particular importance with all drags and carries is the need to keep a tight core. Simply put, firefighters performing drags and carries need to do so by tightening the core muscles around the hips, back, and torso. This creates a "power center" that aids in balance, strength, and injury prevention. Ideally, the firefighter should keep the spine (back) in a neutral position (straight and tight) and use the legs and buttocks for leverage and lifting power, **Figure 16-7**.

Figure 16-7 To avoid injury, firefighters should lift heavy objects using a "tight core" and leverage from the legs and buttocks.

Carries

Firefighter's Carry

The firefighter's carry can be utilized on both conscious and unconscious patients. It is a one-rescuer operation. While this is not the preferred method, it is very effective when one rescuer must carry an unconscious patient. As a rule, the firefighter's carry can be used to carry someone who weighs *less* than the rescuer. Obviously, very strong (muscular) firefighters may be able to carry someone who is heavier than they are, although safety should be paramount. By following the steps given next, it is possible for a single rescuer to pick up and carry the patient without any assistance.

Firefighter's Carry (JPR 16-1)

1. Lay the patient on the back, with arms laid alongside the torso, knees bent, and

JOB PERFORMANCE REQUIREMENT 16-1
Firefighter's Carry

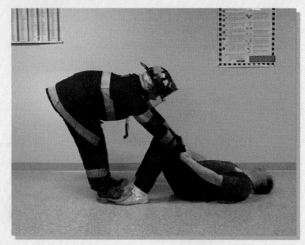

A Lay the patient on the back, with arms laid alongside torso, knees bent, and feet pushed back close to buttocks. Stand in front of the patient with your feet holding the patient's feet in place.

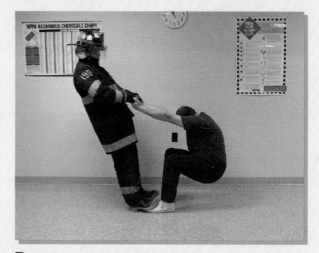

B Reach down and grasp the patient's hands and lean back as you stand, using your weight to help lift the patient. (It may be necessary to rock back and forth a couple of times in order to build sufficient momentum.)

C When you are ready, pull hard on the patient. As the patient comes up, duck your shoulder into the midsection and stand up. As you stand, wrap one arm around the patient's legs and maintain your grasp on the arm that will now be across your back/shoulder.

feet pushed back close to the buttocks, **JPR 16-1A**.
2. Stand in front of the patient with your feet holding the patient's feet in place.
3. Reach down and grasp the patient's hands and lean back as you stand using your weight to help lift the patient, **JPR 16-1B**. (It may be necessary to rock back and forth a couple of times in order to build sufficient momentum.)
4. When you are ready, pull hard on the patient. As the patient comes up, duck your shoulder into the midsection and stand up. Remember to keep a tight core and use the legs and buttocks for power. As you stand, wrap one arm around the patient's legs and maintain your grasp on the arm that will now be across your back/shoulder, **JPR 16-1C**.

Extremity Carry

The extremity carry (sometimes referred to as the cross-arm carry) can be utilized on both conscious and unconscious patients and requires two rescuers. However, it is much easier on the rescuers and patient than the firefighter's carry.

Extremity Carry (JPR 16-2)

1. Lay the patient on the back, with arms laid across the torso, knees bent, and feet pushed back approximately halfway to the buttocks, **JPR 16-2A**.
2. For an unconscious patient, it may be easiest to have the rescuer at the patient's feet reach down, grasp the patient's hands, and pull them into a sitting position using the rescuer's weight. Have a conscious patient assume a sitting posture with feet pulled back, lifting the knees above the floor.
3. The rescuer at the patient's head squats behind the patient, sliding arms under the patient's armpits and grasping the wrist of the patient's opposite arm (if possible). At the same time the rescuer at the patient's feet squats between the patient's feet and grasps the patient's legs under the knees (if possible) or as close as possible below the knees, **JPR 16-2B**.
4. When ready to lift, the rescuers need to communicate clearly with each other so that they can stand as one and carry the patient to safety, **JPR 16-2C**.

Seat Carry

The seat carry can be utilized on conscious patients only and requires two rescuers.

Seat Carry (JPR 16-3)

1. The rescuers face each other. Each rescuer grasps his own right forearm just above the wrist. The rescuers then grasp each other's left forearm just above the wrist, forming a square "seat," **JPR 16-3A**.
2. The rescuers lower the "seat" that they have just formed, allowing the patient to sit on the seat with arms across the shoulders of each rescuer, **JPR 16-3B**.
3. When ready to lift, the rescuers need to communicate clearly with each other so that they can stand as one and carry the patient to safety.

Drags

Rescuers can move a patient by placing him or her on a blanket, bunker coat, salvage cover, and so on, or by using the patient's own clothing as a handhold or by the utilization of a **webbing sling**. Each of these various types of drags is described below. All of the drags shown here can be carried out by a single rescuer.

Blanket Drag

As mentioned, the blanket drag can also be accomplished utilizing a bunker coat, salvage cover, or any other type of material of adequate size. The steps and procedures are the same regardless of the material being used.

> **Streetsmart Tip** A rescuer should always drag a patient head first.

Blanket Drag (JPR 16-4)

1. With the patient lying face up, lay the material the patient is to be placed on along one side of the patient's body with just over one-half of the material gathered close to the patient's side, **JPR 16-4A**.
2. Kneel on the opposite side (from the blanket) of the patient, extend the patient's arm (on your side) above the head if injuries permit, reach across the patient with one hand just above the waist and one just below the hips, and roll the patient toward you, **JPR 16-4B**.
3. While supporting the patient, now on the side, tuck the gathered blanket material close to the body.
4. Roll the patient back onto the blanket material, return the arm to the side, and

JOB PERFORMANCE REQUIREMENT 16-2
The Extremity Carry

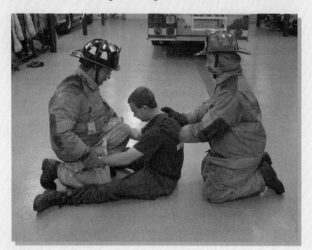

A Lay the patient on his back, with arms laid across torso, knees bent, and feet pushed back approximately halfway to buttocks. For an unconscious patient, it may be easiest to have the rescuer at the patient's feet reach down, grasp the patient's hands, and pull him into a sitting position using the rescuer's weight. Have a conscious patient assume a sitting posture with feet pulled back, lifting the knees above the floor.

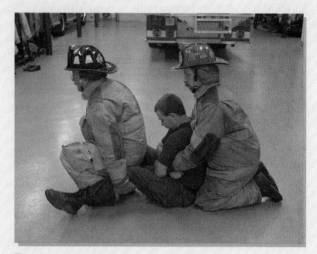

B The rescuer at the patient's head squats behind the patient, sliding arms under the patient's armpits and grasping the wrist of the patient's opposite arm (if possible). At the same time, the rescuer at the patient's feet squats between the patient's feet and grasps the patient's legs under the knees (if possible) or as close as possible below the knees.

C When ready to lift, the rescuers need to communicate clearly with each other so that they can stand as one and carry the patient to safety.

JOB PERFORMANCE REQUIREMENT 16-3
The Seat Carry

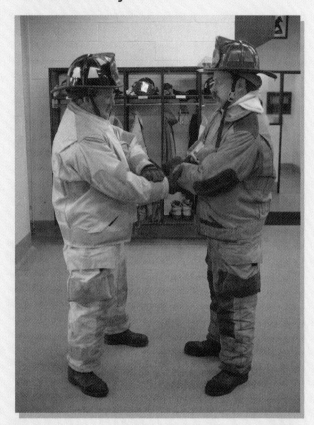

A The rescuers face each other. Each rescuer grasps his own right forearm just above the wrist. The rescuers then grasp each other's left forearm just above the wrist, forming a square "seat."

B The rescuers lower the "seat" that they have just formed, allowing the patient to sit on the seat with arms across the shoulders of each rescuer. When ready to lift, the rescuers need to communicate clearly with each other so that they can stand as one and carry the patient to safety.

wrap the blanket snugly around the patient, supporting the head and neck as much as possible, **JPR 16-4C.**

5. The patient is now ready to be dragged head first with the head and shoulders raised slightly off the floor, **JPR 16-4D.**

Clothing Drag

If a patient is wearing substantial clothing such as protective clothing or a heavy jacket, it may be possible to utilize this clothing to drag a patient to safety.

Clothing Drag (JPR 16-5)

1. Place the patient on the back arranging clothing to provide support to the head and neck. Be careful that the patient's ability to breathe is not compromised, **JPR 16-5A.**
2. Grasp the top of the patient's clothing on each side of the patient's head, supporting the head on your forearms. When using this method, it is important to keep the patient's head close to the floor to avoid causing the head to be pushed downward toward the chest, possibly causing difficulty breathing, **JPR 16-5B.**

JOB PERFORMANCE REQUIREMENT 16-4
The Blanket Drag

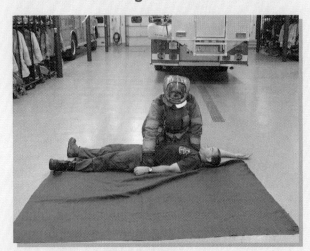

A With the patient lying face up, lay the material the patient is to be placed on along one side of the patient's body with just over one-half of the material gathered close to the patient's side.

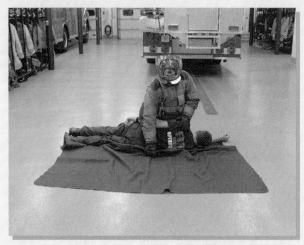

B Kneel on the opposite side (from the blanket) of the patient, extend the patient's arm (on your side) above the head if injuries permit, reach across the patient with one hand just above the waist and one just below the hips, and roll the patient toward you. While supporting the patient, now on the side, tuck the gathered blanket material close to the body.

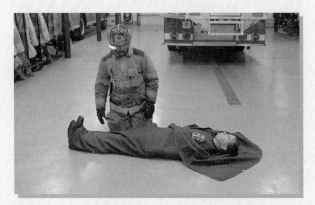

C Roll the patient back onto the blanket material, return the arm to the side, and wrap the blanket snugly around the patient, supporting the head and neck as much as possible.

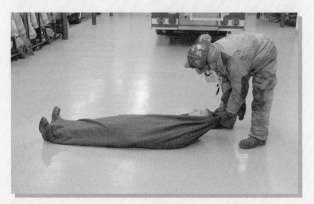

D The patient is now ready to be dragged head first with the head and shoulders raised slightly off the floor.

Webbing Sling Drag

For this drag the rescuer is going to need to have a piece of webbing or rope (although rope does not work nor carry as easily) approximately 12 to 15 feet long. The webbing drag enables a rescuer who is significantly smaller than the victim to perform a rescue.

Webbing Sling Drag (JPR 16-6)

1. Tie the webbing (using a water knot with safety) end to end forming a continuous loop, **JPR 16-6A**.
2. With the patient lying on the back, place the loop under each arm, coming up under the armpits.

JOB PERFORMANCE REQUIREMENT 16-5
The Clothing Drag

A Place the patient on the back, arranging the clothing to provide support to the head and neck. Be careful that the patient's ability to breathe is not compromised.

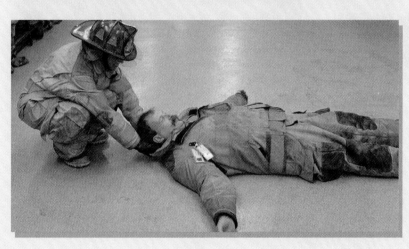

B Grasp the top of the patient's clothing on each side of the patient's head, supporting the head on your forearms. When using this method, it is important to keep the patient's head close to the floor to avoid causing the head to be pushed downward toward the chest, possibly causing difficulty breathing.

3. You will then have two loops stretched out above the patient's head. Adjust the loops so that the loop that comes out the underside of the armpits is snug against the patient's back. Feed the longer loop through between the body and the other webbing loop, **JPR 16-6B.**
4. When the patient is pulled by the long loop, the webbing should snug up under the armpits and provide some support to the patient's head, **JPR 16-6C.**

Sit and Drag Method

This method can be useful when the firefighter experiences trouble moving a patient due to size difference, sheer weight, or any other problem that may be encountered.

Sit and Drag Method (JPR 16-7)

1. Place the patient in a face-up position, **JPR 16-7A.**
2. Assume a sitting position at the head of the patient with legs to each side and hands on either side of the patient.
3. Grasp the patient under the arms. Pull the patient close to you and move under the patient so that the patient's head and back rest against your chest, and your thighs are under the patient's armpits with the patient's arms on the outside of your legs, **JPR 16-7B.**
4. Remove the patient from the area by sliding backward in the sitting position and using your legs to drag the patient along, **JPR 16-7C.**

JOB PERFORMANCE REQUIREMENT 16-6
Webbing Sling Drag

A Tie the webbing (using a water knot with a safety) end to end, forming a continuous loop. With the patient lying face up, place the loop under each arm, coming up under the armpits.

B You will then have two loops stretched out above the patient's head. Adjust the loops so that the loop that comes out the underside of the armpits is snug against the patient's back. Feed the longer loop through between the body and the other webbing loop.

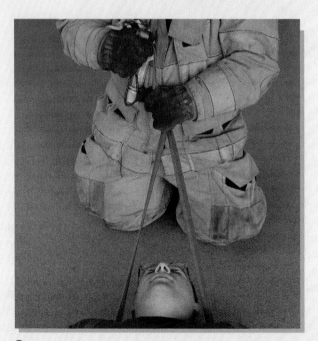

C When the patient is pulled by the long loop, the webbing should snug up under the armpits and provide some support to the patient's head.

JOB PERFORMANCE REQUIREMENT 16-7
Sit and Drag Method

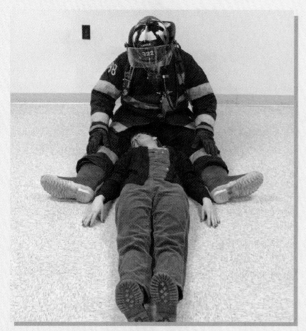

A Place the patient in a face-up position. Assume a sitting position at the head of the patient with legs to each side and hands on either side of the patient.

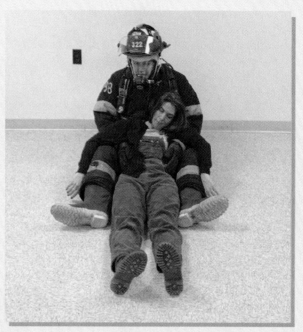

B Grasp the patient under the arms. Pull the patient close to you and move under the patient so that the patient's head and back rest against your chest, and your thighs are under the patient's armpits with the patient's arms on the outside of your legs.

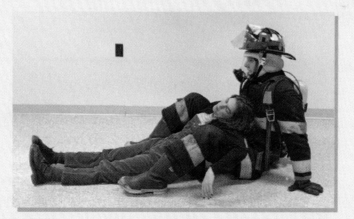

C Remove the patient from the area by sliding backward in the sitting position and using the legs to drag your patient along.

JOB PERFORMANCE REQUIREMENT 16-8
Firefighter's Drag

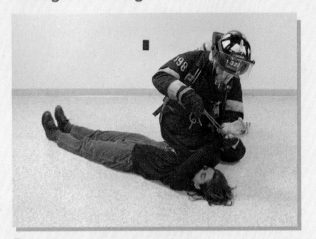

A Place the patient in a face-up position. Use a piece of rope, webbing, belt, handkerchief, or other available material to tie the patient's wrists together.

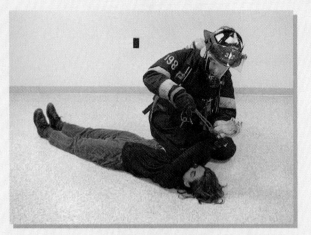

B Straddle the patient, facing the patient, and place the patient's bound wrists over your head and behind the neck.

C Crawl on hands and knees while dragging the patient out of the area.

Firefighter's Drag

This is another useful method of dragging a patient from a dangerous/hazardous area.

Firefighter's Drag (JPR 16-8)

1. Place the patient in a face-up position, **JPR 16-8A**.
2. Use a piece of rope, webbing, belt, handkerchief, or other available material to tie the patient's wrists together.
3. Straddle the patient, facing the patient, and place the patient's bound wrists over your head and behind the neck, **JPR 16-8B**.
4. Crawl on hands and knees while dragging the patient out of the area, **JPR 16-8C**.

Note For pregnant or large patients, the rescuer will need to drag the patient while crawling alongside rather than straddling the patient.

Rescue of a Firefighter Wearing a SCBA

This method works very well for removing an unconscious or incapacitated firefighter who is wearing a SCBA whether it is functioning or not.

JOB PERFORMANCE REQUIREMENT 16-9
Rescue of a Firefighter Wearing a SCBA

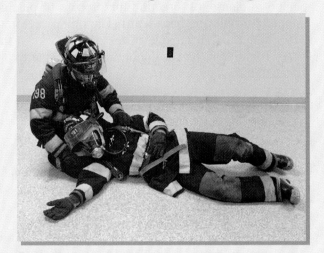

A Determine if the SCBA is functioning. If the SCBA is *not* functioning, disconnect the low-pressure tube and place inside the firefighter's coat or remove the regulator from the face piece, leaving the face piece in place.

B Roll the firefighter onto the side, ensuring that the air supply is not compromised.

C Verify that the SCBA is securely fastened on the firefighter.

D Grasp the shoulder straps of the firefighter's SCBA and drag the firefighter from the area.

Rescue of a Firefighter Wearing a SCBA (JPR 16-9)

1. Determine if the SCBA is functioning. If SCBA is *not* functioning, disconnect the low-pressure tube and place inside the firefighter's coat or remove the regulator from the face piece, leaving the face piece in place, **JPR 16-9A.**
2. Roll the firefighter onto the side, ensuring that the air supply is not compromised, **JPR 16-9B.**
3. Verify that the SCBA is securely fastened on the firefighter, **JPR 16-9C.**
4. Grasp the shoulder straps of the firefighter's SCBA and drag the firefighter from the area, **JPR 16-9D.**

Streetsmart Tip The rescuer must remember that none of these drags provides **spinal immobilization** and are intended to be utilized only in situations where greater harm will come to the patient if not immediately moved. If the patient is wearing a functioning SCBA, the rescuer needs to be careful not to break the face piece seal.

Backboard, Stretcher, and Litter Uses

While the carries and drags discussed are emergency methods of moving patients, it is much preferred to utilize a backboard, stretcher, or litter to transport patients who must be moved. These pieces of equipment are designed to provide the protection, immobilization, and safety needed by injured patients. Each of these items has specific characteristics that make a particular piece of equipment more appropriate for a given situation than the other.

Backboards

Backboards (also known as long spine boards) are designed to provide the maximum in spinal immobilization. They are manufactured (commercially and homemade) from many different types of materials, the most common of which is plywood. However, the more modern backboards are manufactured using materials that allow the patient to remain on the backboard while being x-rayed. This minimizes the amount of movement the patient is subjected to and therefore reduces the possibility of aggravation of injuries. These newer materials also simplify the cleanup and decontamination process.

Placing a patient who is suspected of having a spinal injury onto a backboard takes a coordinated team effort. It is best to have four personnel available for this task. The rescuer at the patient's head is responsible for maintaining traction (in line with spine) on the patient's cervical spine and is also the person in charge of directing the process of placing the patient onto the backboard. The other three rescuers kneel along the side of the patient, one at the upper torso area, one at the hip area, and one at the knee/shin area.

Placing a Patient on a Backboard (JPR 16-10)

1. While the rescuer at the patient's head is maintaining traction, the rescuer at the upper torso places a cervical collar (or some other immobilization device) on the patient. *Even with an immobilization device in place, traction must be maintained at all times during this process,* **JPR 16-10A.**
2. The backboard is laid alongside the patient on the opposite side from the three kneeling rescuers. When ready the rescuer at the patient's head directs the others to "prepare to roll." The three rescuers reach across and grasp the patient at the appropriate locations: The rescuer at the torso grasps the patient at the shoulder and upper arm area; the rescuer at the hips grasps the patient just above and below the hip area; the rescuer at the knee/shin

---

JOB PERFORMANCE REQUIREMENT 16-10
Placing a Patient on a Backboard

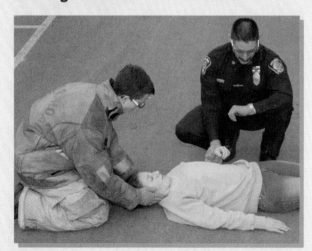

A A firefighter maintains manual stabilization while another first responder checks pulse, movement, and sensation.

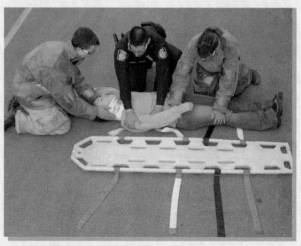

B A cervical collar is applied. While a firefighter maintains manual stabilization, two other firefighters take positions at the patient's shoulders and pelvis, reaching across the patient and grasping the patient's shoulders and pelvis, respectively.

JOB PERFORMANCE REQUIREMENT 16-10
Placing a Patient on a Backboard (Continued)

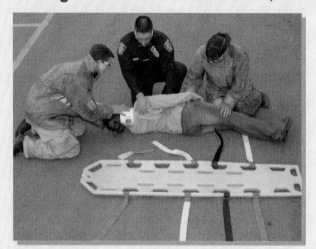

C On the command of the firefighter maintaining manual stabilization, the team rolls the patient onto the patient's side.

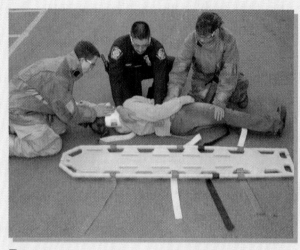

D One firefighter places the backboard under the patient with the bottom of the backboard at the patient's knees.

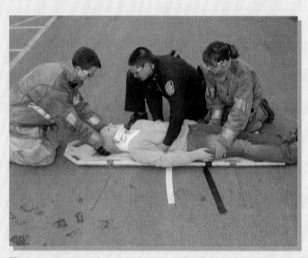

E On command, the team rolls the patient back onto the backboard, and the patient is pulled up to the center of the board using a long axis drag.

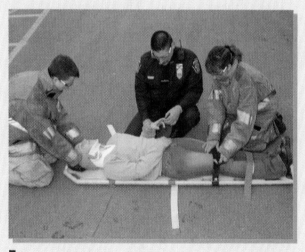

F Once the patient is centered on the backboard, the firefighter secures the patient to the backboard and reassesses distal pulses, movement, and sensation.

area grasps the patient just above and below the knee, **JPR 16-10B**.

3. When ready the rescuer at the head directs the others to "roll patient." At this time the three others roll the patient toward them on to the side. During all of these maneuvers it is essential for the rescuer at the head to rotate the patient's head along with the body, maintaining traction, **JPR 16-10C**.
4. With the patient on the side, one of the rescuers (usually the hip area rescuer) reaches over the patient and slides the backboard up tight against the patient's body. When this is completed, the rescuer at the head directs the others to "prepare to lower" and then "lower patient." At this command the three rescuers allow the patient to roll back down on top of the backboard, **JPR 16-10D**.
5. At this point the patient should be on the backboard; however, it is usually necessary to move the patient in order to

center the patient on the board. This needs to be done very carefully so as not to aggravate any injuries. The patient must be moved as a unit, meaning the patient's body is moved all at the same time. The safest way to accomplish moving the patient to the center of the backboard is to move the patient along the long axis of the body rather than try to move sideways, **JPR 16-10E.**

The rescuer at the head remains in charge. When ready to move the patient, the rescuer at the head directs the others to "prepare to slide." At this point the rescuers reach across or straddle the patient, grasping the upper arm/chest, waist, and lower legs. When ready to slide, the rescuer at the head directs the others to "slide." All of the rescuers then slide the patient down toward the foot of the board while moving slightly toward the center. When the patient has been moved enough, the rescuer at the head directs them to "stop." The patient now needs to be moved back up on the board to the proper location both top-to-bottom and side-to-side. This is accomplished by reversing the just completed movement, **JPR 16-10F.** The patient's distal pulses, movement, and sensation are then reassessed.

6. With the patient now in the proper location, the head is supported on both sides by sandbags, towels, commercial head blocks, or some other acceptable manner. The actual method of strapping the patient to the board varies widely from department to department, but whatever the particular method, the patient must be securely fastened to the board when complete. The strapping method shown in **JPR 16-10** is one that has been found to work very well and can be accomplished in an expeditious manner when practiced.

Stretchers

Stretcher is a universal term that can be applied to the patient-carrying device for ambulances as well as the common army litter. Other terms and slang words are used for the ambulance-type stretcher. Perhaps most used is the term cot, although pram and wheels are still prevalent. The venerable army litter is still used in some situations.

A variety of different types of stretchers are manufactured by various companies and are in use today. It would be very difficult and unnecessary to describe in detail the operation of all of the different units available. However, it is imperative that firefighters know the correct method of operating the type of stretcher(s) that is used in their community.

The most common methods of placing patients onto these types of stretchers are the extremity carry, by utilizing a backboard, or by having the patient lie directly onto the stretcher. Regardless of which of these methods is utilized, the rescuers need to assist and support the patient onto the stretcher. Once on the stretcher, the patient must be secured as soon as possible.

> **Streetsmart Tip** Here are some safety rules that apply to the operation of all types of ambulance stretchers, **JPR 16-11:**
>
> 1. Always make sure the patient is strapped securely prior to lifting, lowering, or moving, **JPR 16-11A.**
> 2. Make sure that your partner is ready and understands what movement is desired.
> 3. Any time that a stretcher is being moved a minimum of two rescuers should work together, **JPR 16-11B.**
> 4. Do not attempt to roll a stretcher across uneven or rough terrain. It should be carried by an adequate number of personnel to make it safe. If at all possible, no rescuers should be backing up as they transport it.
> 5. The stretcher should be placed in the transport unit by the personnel responsible for operating the unit. Many times when firefighters or others attempt to assist, the stretcher gets unbalanced or in worst cases turns over, **JPR 16-11C.**

The army stretcher is most commonly utilized in mass casualty incidents or at events where transport units are either not available or their use is not practical (large crowds, stadiums, evacuations, etc.). A patient may be placed on an army stretcher using the extremity carry, by having the patient lie directly on the stretcher, or by using the method just given for placing a patient on a backboard. Whichever method is utilized, the patient must be securely strapped to the stretcher as soon as possible. Rescuers should not back up when carrying a patient on an army stretcher. The handles allow room for both rescuers to face and walk in the same direction.

JOB PERFORMANCE REQUIREMENT 16-11
Placing a Patient on an Ambulance Stretcher

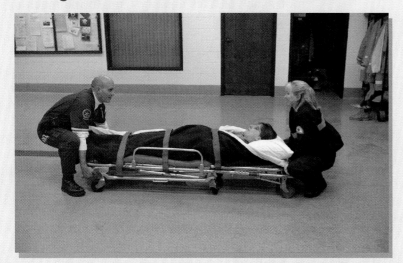

A Always make sure the patient is strapped securely prior to lifting, lowering, or moving. Make sure that your partner is ready and understands what movement is desired.

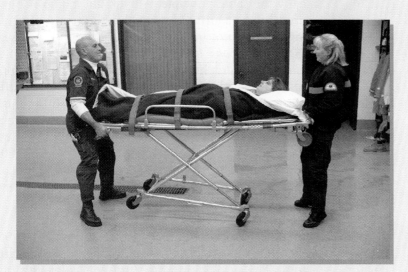

B Any time that a stretcher is being moved, a minimum of two rescuers should work together.

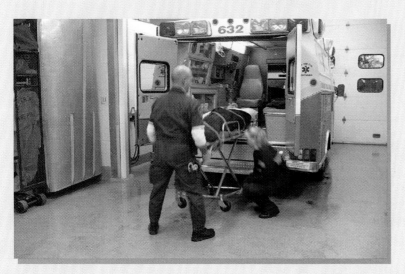

C The stretcher should be placed in the transport unit by the personnel responsible for operating the unit. Many times when firefighters or others attempt to assist, the stretcher gets unbalanced or in worst cases, turns over.

EXTRICATION FROM MOTOR VEHICLES

As was discussed in the introduction to this chapter, motor vehicle crashes are probably the most common rescue situation that today's firefighters respond to. The firefighters' most valuable tool at vehicle crash incidents is their knowledge, experience, and skill—not, as is often believed, more powerful tools. For the purpose of this section, a good working definition of extrication is "to set free, release or disentangle a patient from an entrapment."

Extrications can be as simple as removing a window, unlocking a door, or sliding a seat back, or as complicated as forcing and/or removing vehicle doors, roofs, dashes, seats, and so on. In general, as vehicles become larger and heavier (i.e., buses, tractor trailers, trains) extrication becomes more difficult due to the heavier structural components and severity of the crashes.

Operations at an extrication incident should follow a predetermined sequence of events. The following order of procedures works very well and ensures that none of the essential procedures or operations are overlooked:

1. Scene assessment (size-up)
2. Establishment of work areas
3. Vehicle stabilization
4. Patient access
5. Disentanglement
6. Patient removal
7. Scene stabilization

These procedures/operations are discussed in further detail later in this section.

Tools and Equipment

Tools and equipment utilized at vehicle crash incidents range from the most basic of firefighting tools, such as axes and pry bars, to the much more complex and specialized power hydraulic tools, air bags (low and high pressure), and battery-powered saws. A statement made earlier bears repeating here: The firefighters' most valuable tool at vehicle crash incidents is their knowledge, experience, and skill—not, as is often believed, more powerful tools. While power hydraulic tools, air bags, and the other advanced tools and equipment available today have certainly increased a firefighter's capabilities, they cannot replace the knowledge, skill, and ability developed through training and experience.

Power Hydraulic Tools

Many different companies manufacture power hydraulic tools today, but they all operate on basically the same principle. A hydraulic pump is powered by a gasoline engine, an electric motor, an air-driven motor, or the apparatus engine itself through a power take-off. Some manufacturers offer a manual hydraulic pump as a backup should the primary power hydraulic pump fail. The hydraulic pump, **Figure 16-8**, provides the required fluid and pressure to operate the variety of spreaders, cutters, and rams available. Because different companies' tools operate at different pressures (5,000 and 10,000 psi predominantly) and use different types of fluid, most of the manufacturers use hoses that will not connect to anything other than a compatible tool.

If a department has more than one type of power tool, personnel adding hydraulic fluid must be sure they have the right hydraulic fluid for the specific tool. The wrong type of fluid can cause serious damage.

Figure 16-8 Gasoline engine-powered hydraulic pumps for extrication equipment (left to right): Hurst, Genesis, Amkus. *(Photo courtesy of Rick Michalo)*

RESCUE PROCEDURES 483

Figure 16-9 Power hydraulic spreaders (left to right): Hurst, Genesis, Amkus. *(Photo courtesy of Rick Michalo)*

Figure 16-10 Power hydraulic cutters (left to right): Hurst, Genesis, Amkus. *(Photo courtesy of Rick Michalo)*

Spreaders were the first power hydraulic tool that became available to firefighters, and they are still widely used today. Spreaders can be used to both push and pull (if they are equipped with a chain attachment); they can be used to spread things apart or squeeze them together. Spreaders, **Figure 16-9**, come in various sizes of spread up to 32 inches with spreading forces up to 60,000 psi. Generally the pulling forces are less than the pushing forces for the same tool.

Cutters (also known as shears) were first introduced as a separate attachment to go on the spreader arms, either at the ends sticking out or coming back between the arms. *All cutters of this type should be removed from service since there is a possibility of the blades becoming crossed and shattering.* Modern cutters, **Figure 16-10**, are completely separate tools and come in a variety of sizes also. The smaller cutters are designed for working in close places to cut pedal supports or steel rods. Larger cutters have a tip opening in excess of 7 inches and a cutting force of greater than 80,000 psi.

Rams were the next hydraulic tool to appear with limited available sizes, but that has changed such that rams that measure as small as 10 inches extended up to more than 60 inches extended are available, **Figure 16-11**. Some rams are designed to both push and pull; however, the pulling strength is generally one-half that of the pushing strength.

Combination tools combine the functions of the spreader and the cutter. This type of tool incorporates the cutting blades on the inside of the spreader arms, **Figure 16-12**. These tools are generally less powerful than the individual stand-alone spreader or cutter.

Figure 16-11 Power hydraulic rams of different sizes (left to right): Hurst, Genesis, Amkus. *(Photo courtesy of Rick Michalo)*

Figure 16-12 Power hydraulic combination tools: Amkus (top); Hurst (bottom).

> **Safety** Firefighters should remember these important safety precautions when operating power hydraulic tools, **Figure 16-13**:
>
> - Wear full protective clothing, including eye protection.
> - Operating a tool is a one-person task.
> - Do not put hands or arms inside the arms or blades of an operating tool.
> - Make sure you are balanced.
> - Watch the movement of the operating handle, ensuring that hands or the hydraulic lines do not get pinched.

Air Bags

Air bags used in rescue operations come in high-pressure and low/medium-pressure styles, **Figures 16-14A** and **B**. Each style has operational properties that are peculiar to that style, and therefore, each style serves a specific purpose in the rescue field.

High-pressure bags operate at a maximum inflation pressure of approximately 130 psi. They come in sizes ranging from 6 by 6 inches up to the largest ones measuring upwards of 36 by 36 inches. The larger bags are capable of lifting up to approximately 80 tons. One of the major advantages of these bags is that when deflated they are only 1 inch thick, allowing them to be inserted into very small spaces when necessary, **Figure 16-15**. The biggest drawback is that even the largest bag will only lift a maximum of approximately 20 inches.

> **Caution** Although it is possible to stack high-pressure bags, extreme caution must be exercised to keep them from shifting under a load and being expelled at high velocity.

Figure 16-13 It is important to be properly dressed and balanced when operating a power hydraulic tool at a vehicle extrication scene. *(Photo courtesy of Rick Michalo)*

Low/medium-pressure bags operate at 7–10 psi/12–15 psi inflation pressure, respectively, depending on specific type and brand. These bags are bulky when deflated, not getting much less than 3 to 6 inches thick, but can lift heavy loads a considerable distance. Some of the larger bags can lift loads in excess of 6 feet, **Figure 16-16**.

> **Safety** Firefighters should remember these important safety precautions when operating air bags, **Figure 16-16**.
>
> - Bags must be on or against a solid base.
> - Never inflate bags against sharp or hot objects.
> - Make sure that air supply is adequate. (Do not use oxygen.)
> - You must crib as you lift.
> - Do not stack more than two bags, and if they are different sizes, the smaller bag goes on top.
> - Operate controls carefully, filling bags in a controlled manner.

(A)

(B)

Figure 16-14 (A) A typical high-pressure air bag set. (B) A typical low-pressure air bag set. *(Photo courtesy of Rick Michalo)*

Figure 16-15 A high-pressure bag can be placed in a very small space at a vehicle extrication scene. *(Photo courtesy of Rick Michalo)*

Figure 16-16 A low/medium-pressure bag can lift a load 3 to 4 feet. Load must be properly cribbed as it is lifted. *(Photo courtesy of Rick Michalo)*

Air Chisels and Reciprocating Electric Saws

Standard air chisels have been used in the rescue field for quite some time. These chisels are designed to be operated at between 100 and 150 psi. However, newer versions of the air chisel, **Figure 16-17**, have been designed specifically for rescue operations and operate at up to 300 psi. The performance of these new air chisels is markedly improved.

Reciprocating electric saws are one of the newer tools making their way into the extrication field. One of the drawbacks with the earlier versions was that it was necessary to have 120 volts to operate them. However, with the evolution of the high-capacity battery-powered saws on the market today, they are finding widespread acceptance and usage in the rescue field, **Figure 16-18**.

Figure 16-17 A high-pressure air chisel kit. *(Photo courtesy of Rick Michalo)*

Figure 16-18 A battery-powered reciprocating saw.

> **Safety** Many firefighters have been injured and killed by traffic and secondary crashes at vehicle incidents. For this reason, fire and rescue responders need to size up traffic conditions as a *priority*. Blind spots, poor visibility (weather/darkness), road surfaces, bridges, barriers, congestion, and passing traffic speeds are all factors that need to be assessed.

> **Safety** Firefighters should remember these important safety precautions when operating air chisels and reciprocating electric saws:
> - Do not use in hazardous or flammable atmospheres because they can cause sparks.
> - Make sure air supply is adequate. (Do not use oxygen.)
> - Make sure that no victims or rescuers are in the way when cutting.
> - Use caution cutting hardened steel because blades may chip and/or break.

The results of this assessment will help determine the need for additional resources or assistance that might include more ambulances, law enforcement, specialized equipment, power company response, or other ancillary needs. It is important to mention that the scene assessment is an ongoing process.

Establishment of Work Areas

Ideally, the fire department would like to shut down all traffic in and around the area of the vehicle incident. Unfortunately, the congestion this may cause can create secondary hazards and reduce the ability of additional responders to reach the scene. Many state and local law-enforcement agencies will work at all costs to keep traffic flowing. Because of this, it is essential that the fire department create protected work areas for rescuers and responders. Work areas are established using a combination of traffic barriers, traffic-calming strategies, and hazard zoning. Based on the scene assessment, the first arriving apparatus should be positioned to create a traffic barrier to help shield the greatest number of rescuers. Additional apparatus can increase the size of the barrier or be positioned *past* the scene so that they are also screened from traffic by the first arriving apparatus. While various positioning strategies exist, firefighters need to understand the logic and process of vehicle positioning so that they can work within the protection and limitations afforded by the barrier, **Figure 16-19.**

Scene Assessment (Size-Up)

Scene assessment should be a predetermined sequence of steps or actions that is used in evaluating a crash scene. This evaluation is usually carried out by the officer; however, a good firefighter who conducts an assessment will be able to foresee many of the needs and requests of the officer.

> **Note** While it is important that this assessment be done quickly, it is more important that it be done accurately and completely.

Scene Safety

Good scene assessment considers many facts and probabilities. Specifically, the following items need to be taken into consideration:

- Traffic (see safety box)
- Number and type of vehicles involved
- Potential number and apparent extent of patient injuries
- Hazardous conditions (fire, HazMat, electrical, water, structural integrity, weather, crowds)
- Degree of entanglement

Traffic Barrier

A large fire apparatus such as an engine or heavy rescue can form the initial barrier by stopping in a position that shields the area where firefighters need to work. The apparatus should park with a slight diagonal angle to help increase the work area and make the apparatus appear larger to approaching traffic. The diagonal park also helps approaching traffic recognize that the fire apparatus is *not* moving. Even with a barrier in place, firefighters should mentally preplan an escape route should the barrier apparatus get hit by traffic.

RESCUE PROCEDURES   487

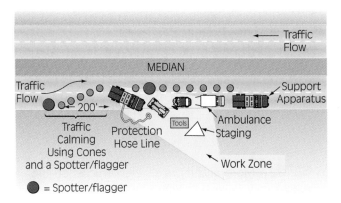

Figure 16-19 The first arriving large fire apparatus should position to create a traffic barrier and work zone. Cones and a spotter/flagger can help with traffic calming.

Traffic Calming

Efforts expended to warn approaching traffic of an upcoming hazard are known as traffic-calming strategies. Typically, law-enforcement agencies and the Department of Transportation will handle this task. However, in the early stages of the incident, the fire department can begin some simple traffic-calming strategies to slow traffic. Placing traffic cones on the street to help guide traffic is a popular strategy.

> **Note** Traffic cones should never be used to create a barrier. Cones should be placed well behind the traffic barrier (apparatus). Some departments still use flares (fusees) to help with traffic calming. These devices are incendiary. It is important to make sure they cannot roll and will not ignite flammable gases, liquids, or vegetation.

Other traffic-calming strategies include:

- Deploying portable Accident Ahead signs.
- Switching off white lights and strobe warning lights that cause night blindness to approaching traffic.
- Utilizing arrow sticks (flashing amber lights that signal in which direction traffic needs to move).
- Positioning a firefighter to wave down traffic speeds. This is done using a wand light or other highly visible device. The firefighter waves the wand up and down in a full arm length waving motion. This firefighter is exposed to traffic, but can also serve as a spotter for other rescuers. It is important to make sure an escape route is planned.

Hazards Zoning

Hazards that are identified in the scene assessment must be addressed. Initially, the fastest way to address the hazard is to create an exclusion zone (verbally or using flagging and/or traffic cones). Electrical wires will arc and move as automated systems try to reenergize the line. Pad-mounted transformers and electrical boxes can create a ground gradient of energy when they are damaged. Firefighters should treat them all as if they are energized and keep a 10-foot ring clear around the box. Damaged electrical poles can suddenly fall, dropping energized lines. Placing a lookout and preplanning an escape route can help zone these hazards. Hazardous materials should be treated as worst-case and zoned accordingly.

Most fire departments require that a fire attack line be pulled and charged as a standby measure whenever a vehicle extrication operation is under way. The firefighter assigned to this protective task can also serve as a lookout and remind other firefighters when they wander into an exclusion zone.

The vehicle(s) involved in the incident will likely present additional hazards. Fuel, electrical hazards, cargo, fire risk, and vehicle systems can be compromised in a motor vehicle incident. Understanding all the potential hazards a vehicle incident can present is well past the scope of this book; however, some hazards deserve a mention. Fuel systems on most vehicles are pressurized. If part of the fuel system is compromised, the fuel will escape under pressure initially.

Newer hybrid and alternative fuel systems present unique hazards ranging from asphyxiation to electrocution to toxicity. Alternative fuels can include compressed natural gas and propane. Hybrid systems include hydrogen and high-voltage battery systems combined with traditional fuels to increase engine efficiency. Each of these presents a unique set of hazards to responders. Firefighters should be encouraged to pursue understanding of new technologies through constant study of trade journals and extrication-specific training.

Vehicle Stabilization

Vehicle stabilization can be as simple as putting the transmission in gear or extremely complex, requiring cribbing, air bags, ropes, or other tools and equipment. The rescuer needs to use the information gathered during the assessment process to determine the amount of stabilization required. In any case, if a vehicle has an injured person inside, it must be stabilized by taking the weight off the vehicle's **suspension system**. (*Note:* Deflating the tires will not accomplish this.) This can be accomplished by using cribbing, **Figure 16-20**. In more complex situations, stabilization may require the use of a variety of tools and equipment, **Figure 16-21**. Stabilization struts, cables, "step chocks," winches, and booms are just a few examples of the many tools that can be used to help stabilize vehicles. Tow-vehicle and vehicle-recovery businesses may have equipment that can help with vehicle stabilization. Utilizing the expertise of vehicle-recovery technicians can be helpful in unusual stabilization situations. It is important to have established previous communication and training with these technicians so that safety and accountability issues are clear.

Patient Access

Accessing the patient refers to providing a pathway for the rescuer assigned to evaluate and care for the patient, **Figure 16-22**. Many times this can be through a doorway or window that can be quickly broken or removed. Once access is gained:

- The patient can be evaluated.
- Life support activities can be initiated.
- The patient's position as it relates to the extrication activities can be evaluated.
- The patient can be protected from further injuries.
- Patient **packaging** can be initiated.

> **Safety** More and more of the vehicles on the road today have supplemental restraint systems (SRS) or, as most of us know them, air bags. These systems were initially located in the steering wheel or dashboard on the passenger side, but they can now be found in many different areas of a vehicle. Either currently or soon to be in production are bags that deploy from the seat cushions, the headrest, the head liner above the window, the dashboard under the steering wheel, and other locations. Most of these systems are deployed by electrical sensors that react to sudden directional impacts of sufficient force. SRS devices that have *not* deployed should be considered "live." Firefighters can be severely injured should the device activate during patient disentanglement or removal. Although disconnecting the vehicle battery will in some cases render the system inoperative, manufacturers have various backup or stored energy systems in case the battery was destroyed during a crash. Some of the best ways to learn the different systems and how they operate is to read fire rescue periodicals, visit various dealerships as training exercises, and attend specialized training.

Figure 16-20 A properly cribbed vehicle using step and box cribbing styles. *(Photo courtesy of Rick Michalo)*

Figure 16-21 A properly cribbed vehicle using a combination of methods. *(Photo courtesy of Rick Michalo)*

Figure 16-22 Many times access to the patient can be made by removing the rear window. Note that the vehicle is properly cribbed, and the glass edges the patient attendant has to crawl over are covered. *(Photo courtesy of Rick Michalo)*

Disentanglement

The process of disentangling a patient begins with the rescuer who is tending the patient advising the incident commander as to the extent of injuries and mechanism of entrapment. The best pathway for patient removal is determined, keeping in mind that the pathway must provide ample working space and adequate space for removal of the packaged patient.

Once the pathway is selected the method or combination of methods for disentanglement must be selected. The options are:

> **Disassembly**: The actual taking apart of the vehicle components.
> **Distortion**: The bending of sheet metal or components, **Figure 16-23**.
> **Displacement**: The relocating of major parts (i.e., doors, roof, dash, steering column), **Figure 16-24**.
> **Severance**: The cutting off of components (i.e., brake pedal, steering wheel), **Figure 16-25**.

The method of disentanglement selected from those given here will provide information to the rescuer in selection of the proper tools to accomplish the desired goal. The safety of the rescuer(s) and the patient must always be at the forefront.

Entire texts have been written on effective disentanglement. Tried and tested extrication methods can be rendered ineffective with newer vehicle and material technologies. Training, tool capability understanding, practice, and inventiveness are the keys to effective disentanglement.

Patient Removal

Once the pathway has been created and made as safe as possible by covering sharp edges left by cut or torn metal, covering any broken glass that the patient could come in contact with, and removing all tools that may be in the way, the properly packaged patient should be carefully removed, **Figure 16-26**, again remembering that the goal is to minimize any aggravation of existing injuries.

Figure 16-24 The vehicle roof being displaced to allow better access to the patient. Note that the roof must be secured once it has been folded back to prevent it blowing back over on the patient and/or rescuers. *(Photo courtesy of Rick Michalo)*

Figure 16-23 A hydraulic ram is being used to distort (bend) the dashboard up and away from the patient. Note the long spine board being used for patient protection, victim cover in place, and attendant maintaining cervical support. *(Photo courtesy of Rick Michalo)*

Figure 16-25 A steering column being severed to allow for easier removal of patient. Note that protection must be provided between steering wheel and patient when performing this procedure. *(Photo courtesy of Rick Michalo)*

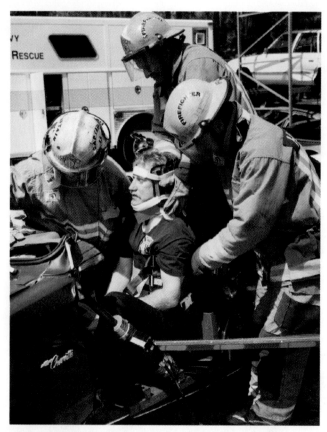

Figure 16-26 A properly packaged patient being carefully removed from a two-vehicle accident and placed on to a long spine board. *(Photo courtesy of Rick Michalo)*

Patient removal should be made only with the direct supervision of a certified EMS responder—typically a paramedic or experienced EMT. Appropriately, the supervising EMS provider should be focused on patient care. Often, the removal of the patient and subsequent movement toward an ambulance takes the rescuers and patient out of the work area protected by traffic barriers. Spotters should watch for traffic during patient movement, and escape routes must be planned.

Scene Stabilization

After patient removal, firefighters must refocus their energy toward securing the incident scene. Although this is not as exciting as the actual rescue, firefighters must continue to exercise safe practices. Extreme attention is required to release tools that are holding the vehicle—many components may "spring back" when the force of the rescue tool is relieved. Cribbing used to shore and support the vehicle might need to be retrieved, creating a potentially unstable vehicle again. At times, it may be best to allow tow vehicles to connect to the vehicle prior to cribbing removal. As tools are being removed from the wreckage, firefighters should inspect the cutting/contact surfaces, hydraulic lines, and safety features for damage and report any problems or observations to their officer. Additionally, the scene stabilization effort may include the following:

Vehicle Recovery

Fire resources are often asked to stand by for the tow vehicle to overturn or "hook" the damaged vehicle. The tow-vehicle operator must be advised of vehicle hazards such as fuel leaks and supplement restraint systems (SRS air bags) that have *not* deployed. Local SOPs may require that the battery of a damaged vehicle be disconnected. Only trained personnel should perform this disconnect. Tensioned tow cables introduce another hazard. It is a good practice to "envision" the tow cable breaking and stay clear of the path it will recoil. Likewise, it is important to stay clear of the vehicle as it is being moved and to predict its path should the tow cable break.

Fluid/Parts Cleanup

The use of a dry absorbent is typical for oils, engine coolant, and diesel fuel. Gasoline is best absorbed with specially designated "fuel pads" and disposed of in accordance with local SOPs. When handling fuel pads, it is best to use gloves that can be decontaminated. Battery acids are extremely caustic and should be handled with diligence. Vehicle parts that have been strewn about can be sharp. Law-enforcement officials may want strewn parts to be left in place to assist with accident reconstruction investigations. If not, carefully place all loose parts back into the wrecked vehicle.

Once the scene has been stabilized, the responding crews must secure all deployed equipment and pick up traffic-calming devices. This exposes rescuers to traffic hazards again. As a rule, they should not put their back to traffic—they should pick up traffic cones while facing traffic. Often, spotters are needed to help apparatus back out of position and into traffic. These spotters must also have a traffic lookout watching their back.

SPECIALIZED RESCUE SITUATIONS AND TOOLS

The previously discussed situations are the ones that firefighters are most commonly faced with. There are a variety of other rescue situations, however, that a firefighter may be called on to respond to. The following sections discuss a number of these, focusing on enabling firefighters to recognize the situation, call for the necessary assistance, not cause the situation to worsen for the victim, and not become victims themselves. Many of the techniques discussed earlier can and will be very beneficial to a firefighter responding to these special situations.

Vertical Rescue

Vertical rescue deals with a victim who is either above or below normal ground level and beyond the reach of available ladders or other practical means. Chapter 15 on rope discussed life safety rope; the requirements as far as size, usage, and care; and the appropriate knots and their uses. That chapter also referenced NFPA 1983, the standard that addresses life safety rope. In addition to rope, the standard also addresses **hardware** and **harnesses** utilized in rescue situations. Harnesses are classified into three categories:

Class 1: A single belt with an adjustable buckle that goes around the wearer's waist and some type of hook, **Figure 16-27**. This harness is commonly known as a ladder belt, and the hook is used to secure a firefighter to a ladder. It is intended to hold a firefighter in place, not provide fall protection. If a fall of more than 2 feet were to occur, serious injury can be expected to the abdominal and spine area. It is not acceptable as a life safety harness.

Class 2: A Class 2 harness consists of a waist strap and straps that go around each leg, creating a seat-type arrangement, **Figure 16-28**. Class 2 harnesses are required to be capable of supporting a rescuer and patient at the same time.

Class 3: A Class 3 harness consists of the same type of seat arrangement as the Class 2 harness, but has additional pieces that go over the shoulders and have additional attachment points, usually at the top of each shoulder and in the center of the upper back, **Figure 16-29**. They are designed to keep wearers in an upright position or to keep them from slipping out of the harness if they are using the waist attachment point and get in a head-down position. Class 3 harnesses are also required to be capable of supporting a rescuer and patient at the same time.

As with all rescue equipment, proper maintenance, cleaning, and inspection are required for life

Figure 16-27 A Class 1 harness.

Figure 16-28 A Class 2 harness.

492 CHAPTER 16

Figure 16-29 A Class 3 harness.

Figure 16-30 A firefighter using the reach method to rescue a victim in the water. Note the PFD and the use of a pole to extend the firefighter's reach.

safety harnesses. They should be inspected and cleaned following the same general criteria used to inspect life safety ropes with particular attention paid to the stitching.

If a patient is going to be suspended in a harness for any reason, it is highly recommended that a Class 3 harness be used. In emergency situations where waiting to get the proper harness could cause greater harm, a patient may be lowered or raised utilizing the rescue knot described in the previous chapter or the department-approved rescue knot.

If a patient is to be lowered using a life safety rope it is imperative that the personnel on scene have the training, skill, and equipment to do so safely. Many fire/rescue departments have developed special teams to respond to these types of incidents. The teams have the proper equipment, and the personnel receive specialized training and are required to maintain proficiency. However, firefighters should be familiar with the basic equipment and techniques utilized by their department at this type of incident.

Water Rescue

Water rescue operations can be very dangerous for the rescuers. In many cases, the victim(s) can be seen, which results in the emotions of the bystanders being very high. They may expect the firefighters to jump in the water immediately to rescue the person in trouble. That is the *last* thing a rescuer should do, and it should never be done without a personal flotation device (PFD).

> **Safety** Every firefighter at the scene of a water rescue incident should have a PFD on if at all possible. If not, those without PFDs should be kept well back from the water's edge. If they are allowed to be near the water's edge and something happens in front of them, they may go into the water without the proper equipment.

Following in priority order are the methods and procedures that firefighters should utilize to rescue a victim from static (lake) water and slow-moving rivers and streams:

1. *Reach.* If the victim can be reached without the rescuer entering the water, a pike pole, pool rescue pole, or anything that can be used to extend the rescuer's reach can and should be utilized, **Figure 16-30**.
2. *Throw.* If the victim cannot be reached directly but is within throwing range of a rescue throw bag or flotation device attached to a rope, then one should be thrown and the rope utilized to pull the victim to safety, **Figure 16-31**.
3. *Row.* If the victim cannot be reached by either of the preceding methods and there is a suitable boat available, the rescuer(s) should utilize it. Extreme caution should be exercised on approaching the victim. In a

RESCUE PROCEDURES 493

Figure 16-31 A firefighter using the throw method to rescue a victim in the water. Note the PFD and the underhand throwing technique.

Figure 16-32 A firefighter using the row method in a small boat to rescue a victim in the water. Note the extra PFD in the boat for the victim. Do not attempt to lift the victim into a small boat; instead have the victim hang on to the side.

panic, the victim could capsize the boat, resulting in the rescuers now becoming potential victims. The best course of action is *not* to try lifting the victim into the boat at all (unless it is designed for rescue operations), but to hold onto the person and return to shore, **Figure 16-32**.

4. *Go*. If none of the preceding options is possible, the absolute last method is for the rescuer to enter the water, **Figure 16-33**. Extreme caution must be exercised when approaching the victim. In a panic, the victim could injure the rescuer or make it difficult for the rescuer to assist him or her. If possible the rescuer should carry an additional PFD to the victim. In still or extremely slow-moving water it is a good idea for the rescuer to swim out holding a **tether line**. This will allow the rescuers on shore to pull the rescuer and victim back to shore.

Safety A rescuer should not use an attached tether line in moderate to fast moving water. It can pull the rescuer under.

Swift-water rescue is a specialized field requiring additional training, practice, and protective equipment. The force of a fast moving stream is considerable, even in shallow currents. As mentioned before, tether lines increase the danger to rescuers. Special rigging and procedures are required for a successful rescue. Local SOPs and training are required for swift-water rescues; however, some basic rescuer safety procedures can be discussed here. First, all rescuers must wear PFDs. They should try to encourage the victim to keep afloat until a swift-water team can arrive and set up. If an immediate rescue attempt must take place, rescuers can place themselves at the shore and attempt the throw method, making sure a float is attached to the end of the rope. A PFD is a great float and will help calm the victim. Rescuers should throw the rope well upstream of the victim and allow the current to drift the rope toward the victim. Once the victim has the rope, rescuers can encourage the victim

Figure 16-33 A firefighter using the go method to rescue a victim in the water. Note that the firefighter is carrying an extra PFD for the victim.

to don the PFD. The rope is *not* used as a tether; it acts merely as a guide rope. The victim holds the rope, and the rescuers allow plenty of slack in the line. As the victim drifts into the stream, the rescuers release just enough slack to let the victim swing toward the shore. Once again, this is a last-ditch method. Properly trained swift-water technicians are the key to a successful rescue.

Ice Rescue

Many of the safety precautions and rescue procedures just presented with regard to water rescue apply to ice rescue as well. The procedures and hazards are very similar with the added element of extreme cold. In place of or in addition to PFDs, rescue personnel working at an ice rescue scene should have thermal rescue suits, which provide the required personal flotation as well as protection from the extreme cold. The rescuers in **Figure 16-34** are wearing thermal suits.

If rescuers must go out onto the ice themselves, they need to use whatever is available to distribute their weight over as large an area as possible. Some items that work well for this are ladders, sheets of plywood, planks of wood, and backboards—even folded salvage covers would be better than nothing.

> **Caution** A rescuer should not approach too close to the hole itself. The ice is obviously not stable in that area.

The rescuer should use the reach (with extension) or throw method to keep from having to be right up next to the hole. Ice rescue victims may not have the strength or determination to grab a rope or pole. The effects of the cold water quickly rob the human body of energy. In these cases, firefighters with appropriate thermal flotation PPE should try to approach the victim in such a way as *not* to disturb the ice the victim is clinging to. The best approach is to come in behind the victim and use a rescue collar or other rope device to secure the victim. It is essential to coach the victim to hold tight to the rope but not to the rescuer. Again, these rescue skills are past the scope of this book. All firefighters should train and practice under the supervision of a competent instructor.

Structural Collapse Rescue

Structural collapse, while not a common incident, may occur for any number of reasons: weakening due to age or fire, environmental causes (earthquake, tornado, hurricane, flooding, rain, or snow buildup on roofs), or an explosion (accidental or intentional). Whatever the cause, the result can be any number of victims trapped, injured, or killed. One of the firefighter's first priorities at this type of incident is to determine the number of possible victims. This information along with the construction type, size, and occupancy of the collapsed structure will have a direct impact on the assistance required.

(A)

(B)

Figure 16-34 (A) A ladder, a sheet of plywood, or a specialized ice rescue boat can be used to spread a rescuer's weight over a larger area. (B) Using a throw bag or pike pole to reach out to a victim who has fallen through the ice prevents the rescuer from having to approach the weakened area too closely. *(Photo courtesy of Halfmoon-Waterford Fire District 1)*

RESCUE PROCEDURES   495

> **Safety** Safety of the rescuers needs to be the number one consideration as in all rescue operations. Some of the particular hazards to be aware of at collapse scenes are secondary collapse, live electrical wires, and gas leaks.

Rescue of victims who are not trapped or are lightly trapped should be done as soon as safely possible. It is very common for this process to already be under way by civilians from the immediate area when a rescue unit arrives. It is important to remember that as civilians and rescuers are climbing on and over debris there is a possibility that other victims may be trapped under that debris, **Figure 16-35**.

Structural collapses are generally classified into three different types, which are described next. However, it is not uncommon to encounter more than one type at any given collapse incident.

Pancake Collapse

A pancake collapse occurs when both sides of the supporting walls or the floor anchoring system fails. It is characterized by the roof or upper floor(s) falling parallel onto the one below, often causing a domino effect, which may result in several or all of the floors collapsing one upon the other until the lowest level is reached. This type of collapse often results in many small **voids** being created by debris between floors supporting the upper floor, **Figure 16-36**.

Lean-To Collapse

A lean-to collapse occurs when only one side of the supporting walls or floor anchoring system fails. It is characterized by one side of the collapsed roof or floor remaining attached or supported by the wall or anchoring system that did not fail. This type of collapse usually results in a significant void being created near the remaining wall, **Figure 16-37**.

V-Type Collapse

A V-type collapse occurs when the center of the floor or roof support system is overloaded (improper placement of stock, buildup of snow or rain, etc.) or becomes weakened for some reason (fire, rot, termites, improper removal of support beams, etc.). It is characterized by both sides remaining attached or supported by the walls or anchoring system but collapsing in the center. This type of collapse will

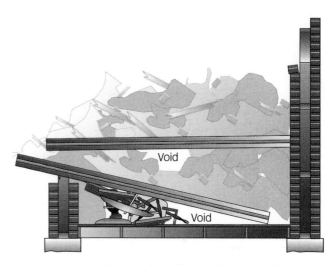

Figure 16-36 A pancake collapse. Note voids where survivors may be located that have been created by debris during structural collapse.

Figure 16-35 Firefighters need to remember that victims may be buried under the debris that they are climbing on and over at a collapse scene.

Figure 16-37 A lean-to collapse. Note voids where survivors may be located that have been created by debris during structural collapse.

usually result in voids being left on each side near the supporting walls, **Figure 16-38**.

It is important for the firefighter to know where these voids can be anticipated in each of the different types of collapses since this is where the greatest likelihood of finding survivors exists. Knowing where the survivors may be is just part of the problem. The firefighter must also know how to make the area as safe as possible prior to entering. Having dealt with the issues previously discussed (live wires, gas leaks, etc.) the firefighter needs to have a basic knowledge of cribbing, **shoring**, and **tunneling**: examples of each can be seen in **Figure 16-39**.

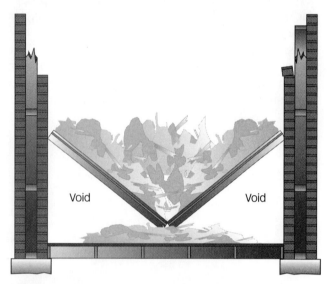

Figure 16-38 A V-type collapse. Note voids where survivors may be located that have been created by debris during structural collapse.

Safety It is important to remember in all of these operations that the firefighter is not lifting the fallen members, but merely supporting them where they are. To lift them may cause further collapse.

- *Cribbing.* As previously defined, cribbing is the use of various dimensions (2×4, 2×6, 4×4, 6×6, etc.) of lumber arranged in systematic stacks (pyramid, box, step, etc.) to support an unstable load. The same principle applies to the use of cribbing in collapse operations; however, it is common to use timbers of larger dimensions.
- *Shoring.* Shoring is the use of timbers to support and/or strengthen weakened structural members (roofs, floors, walls, etc.) in order to avoid a secondary collapse during the rescue operation. These can be very complex operations and

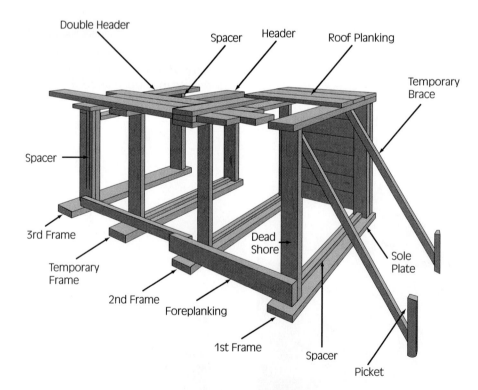

Figure 16-39 The proper usage and placement of cribbing to support fallen members, shoring to hold back the sides and roof debris, and tunneling supports are required prior to entering trench or collapse areas.

should not be undertaken without the input, advice, and guidance of experts in the engineering field, preferably someone with experience with collapsed structures.

- *Tunneling.* Tunneling may be required to reach survivors located in the voids created during the collapse. This is a very dangerous and time-consuming process and should not be undertaken unless all other avenues have been tried. Tunneling should not be utilized as a general search process; a specific destination is necessary along with a good indication that survivor(s) will be located when the destination is reached.

Trench and Below-Grade Rescue

Most trench or below-grade incidents occur at construction sites, utilities (gas, water, sewer, etc.), maintenance sites, or well digging sites, an example of which can be seen in **Figure 16-40**. However, every year a number of incidents are responded to that involve children digging a large hole for whatever reason they can think of. There are also a number of incidents that do not involve trenches or below-grade operations at all but are very similar in the challenges they present to rescuers and the life-threatening entrapment of the victim. These incidents occur in grain silos, cement hoppers, sawdust collectors, fertilizer hoppers, etc.—basically anywhere that a victim can be **engulfed** by a loose granular-type product. What makes the entrapment by dirt, sand, or any of the listed types of products so life threatening is not only that the engulfing material could cover the victim's mouth and nose, thereby compromising the airway, but second and possibly even more life threatening is that the engulfing material gets packed in around the victim's chest, preventing the ability to inhale. This is why a victim succumbs to **asphyxiation** even though the head is not buried.

The most immediate danger to the victim is usually a compromising of the ability to breathe. It follows then that the first priority on reaching a victim is to uncover the victim's head and as much of the chest as possible. Supplementing respiration with oxygen should also occur as soon as safely possible.

Even with the need for haste as described in the previous paragraph, rescuers should assume that, because the fire department has been called to the scene for a rescue, proper safety procedures were not being utilized to begin with. With this in mind, the rescue crew must approach the scene very carefully keeping their safety in the forefront. The rescuers should use **ground pads** or something else to distribute their weight over a larger area (the same as with ice rescue discussed earlier) when approaching the caved-in area, to prevent causing more material to fall in on the victim or others.

> **Safety** Even if coworkers or others are already attempting to assist the victim(s), the fire department must resist the urge to jump in and help until the area is made safe.

For a trench or below-grade incident to be resolved as safely as possible, the previously mentioned ground pads should be in place, all nonessential personnel and equipment should be kept away from the site, the cave-in area must be shored up, the air quality needs to be monitored, fresh air needs to be introduced into the area, and adequate egress needs to be provided for the rescuers, many of which can be seen in **Figure 16-41**.

Upon entering the caved-in area, the rescuer needs to have appropriate personal safety clothing and equipment and the correct tools to rescue the victim. In the case of a cave-in, power equipment and tools should not be utilized until the exact number, location, and position of the victim(s) are determined. Even then they must be used very cautiously, and the effect they are having on the possibility of further cave-in closely monitored. In most cases the safest (for victim and rescuers) and most effective method is to dig with small hand tools or the hands only.

While one rescue team is in the process of excavating the victim, there should be another team topside addressing the need to remove the victim once freed. It can be anticipated that the victim is going to have injuries such as trauma to the head, spine, torso, and extremities. The removal team needs to be factoring these possible injuries into the removal

Figure 16-40 An actual trench rescue operation. Note the amount and variety of equipment that can be seen.

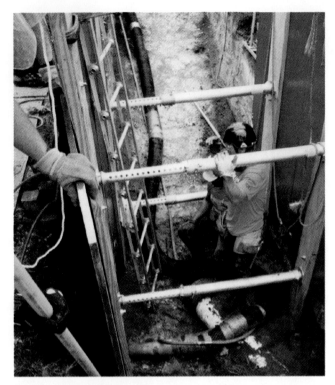

Figure 16-41 A trench rescue operation requires that all necessary safety equipment/precautions be in place prior to rescuers entering the trench.

Figure 16-42 The topside team needs to be prepared with any immobilization and hauling equipment that may be required.

plan and be prepared to provide adequate immobilization and hauling systems, **Figure 16-42**.

Confined Space Rescue

A **confined space** is defined as a space that is large enough to be entered but is not designed for continuous occupancy. Confined spaces come in many different forms; they can be found at grade level, below grade, above grade, and on board ships. Typical confined spaces are utility vaults, storm water or sewer culverts, septic tanks, industrial boilers, storage tanks, grain silos, cement hoppers, sawdust collectors, fertilizer hoppers, and areas aboard ships. As can be seen from this list, confined spaces can and do exist in every department's jurisdiction.

> **Note** It is estimated that approximately one-half of the people who die each year in confined space incidents do so while attempting to "rescue" a victim. Although the majority of these "rescuers" are civilians or coworkers, unfortunately some are firefighters.

The most common hazard encountered at a confined space incident is an oxygen-deficient atmosphere, followed closely by toxic and/or hazardous vapors. Entry into a confined space should never be attempted without having first sampled the atmosphere and confirmed it as tenable or identified the hazards. If the atmosphere is found to be oxygen deficient, this situation can be improved by the use of an air blower and flexible duct.

Whenever members of a rescue team enter a confined space, they must utilize appropriate personal protective clothing and equipment (for the known and possible hazards), including lifelines for each member and a radio or other communication system between the rescue team and the operations chief. The space must be constantly monitored, preferably by individual monitors on each entry team member, but at a minimum by one member of the entry team assigned this task. A backup crew must be fully equipped and dressed out to act as a rescue team for the entry team if a problem develops or the entry team needs assistance for any reason.

Rescue from Electrical Situations

Rescuers responding or arriving at an incident scene involving electrical wires, transformers, transfer stations, and so on, must assume these items are energized until advised otherwise by the utility company representative at the scene. Very often wires will appear to be de-energized because they are lying still on the ground and no arcing is visible. This is not necessarily the case; the wire may be well grounded at that time or there may be a time delay reset breaker on the line that will reset itself after a set amount of time and may do so a number of times before finally staying in the off position. Any victim who is in contact with an electrical wire must also be considered energized and cannot be touched. Firefighters need to remember that none of their protective clothing and/or equipment is designed to protect them from electrical current.

Vehicle accidents involving electrical equipment like pad transformers and downed wires are especially dangerous. It is best to leave victims in the vehicle and not approach. Often, the vehicle may be energized—yet the passengers are safe as long as they do not exit. A victim in an energized car is like a bird on an overhead power wire—that is, as long as the ground is not touched, no electrocution will take place. Anything or anyone who touches the vehicle and ground at the same time will cause an easier path for electricity to seek ground. Firefighters attempting to approach involved electrical equipment may feel a tingling sensation in their boots. This is a dangerous sign known as ground gradient. When the tingling is felt, firefighters must back away using a shuffle foot motion to keep their feet in contact with the ground. In all cases, the fire department shall verify that power company technicians have been dispatched.

If the electrical contact is within a building, it should be possible to shut the power off at the main disconnect. It is recommended that the main disconnect be utilized as opposed to individual breakers because sometimes circuits or equipment may be receiving electricity from more than one breaker. If this (shutting off main disconnect) method is used, a firefighter should be stationed at the disconnect, **Figure 16-43**, to ensure that no one turns the power back on either accidentally or because the person did not know a rescue was in progress.

Firefighters should never enter an enclosed electrical transfer station, substation, or vault without a company representative on scene to ensure that all electrical power is shut off. Every wire, bracket, frame, transformer, and so on, within the enclosed area should be assumed to be energized prior to the on-scene company representative confirming otherwise.

Some departments carry specialized equipment such as hot sticks, insulated cutters, and insulated lineman's gloves on their apparatus. If these items are to be utilized they must be stored, tested, and certified in accordance with the very strict standards that apply to such equipment. The personnel assigned to utilize this equipment must also receive specialized training in the use and care of this equipment.

Most fire departments have recognized that this is a highly specialized and dangerous operation and best left to the professionals with the utility company.

Any time large electrical wires are going to be cut (energized or not) one of the dangers to be aware of is what is known as **reel coil**. This is memory that the wire develops from being placed on the wooden spool as it is being manufactured. This reel coil will cause the wire to recoil (spring away) from the point at which it was cut, attempting to resume the coiled

Figure 16-43 Rescuers must take control of the electrical disconnect and remain in contact with the rest of the team via portable radio.

form again. This recoiling can be very quick and powerful in heavier wire and can cause severe injury from the impact even if it is de-energized.

Industrial Entrapment Rescue

Extricating a victim from an industrial entrapment is usually a very complex process. In most cases if the entrapment is relatively minor, the victims' coworkers will have already extricated them by the time the fire department arrives. Due to the number and complexity of machines used in industrial plants today, it is impossible to have a specific plan for each possible entrapment.

The operations at an industrial entrapment incident should follow a predetermined sequence of events. The process presented previously in the vehicle crash extrication section is applicable at an industrial entrapment with very minor changes.

The following order of procedures works very well and ensures that none of the essential procedures or operations is overlooked:

1. Assess the incident (size-up).
2. Identify on-scene or available experts.
3. Stabilize entrapping machinery.
4. Access the patient.
5. Disentangle the patient.
6. Remove the patient.

Incident assessment should be a predetermined sequence of steps or actions that is used in evaluating an entrapment. This evaluation is usually carried out by the officer. However, at an industrial entrapment incident, there may be a firefighter on the crew with knowledge and/or experience that could be invaluable. A good officer should know this ahead of time and never hesitate to utilize it. As previously stated, while it is important that this assessment be done quickly, it is more important that it be done accurately and completely.

Although identifying on-scene or available experts is shown as the second step it is vitally important that this expert advice be available during the assessment phase if at all possible. It is quite possible that the expert may be able to provide a simple and quick extrication method that is not obvious. It is also possible that the expert can prevent further injury to the victim or injury to the rescuers by preventing the rescuers from taking some action that, although appearing to be logical and effective, would cause the machine to react in an unexpected manner. A good example of this is that while turning off the power to a machine would appear to be the logical thing to do, on some punch press-type machines this would cause them to finish out their cycle.

Personnel conducting an assessment need to take the following into consideration as part of a good incident assessment:

- The apparent extent of injuries
- Current and anticipated hazards to the victim and/or rescuers
- Disentanglement requirements
- Support needs

The results of the assessment will reveal whether any additional assistance is required, such as:

- Additional medical units
- Additional or specialized extrication equipment
- Advanced medical assistance, such as an on-scene doctor
- Other specialized equipment, such as a light unit
- Hazardous materials team

Stabilization of the entrapping machinery should begin with shutting down the power (electrical, hydraulic, and pneumatic) and releasing stored energy (air or hydraulic pressure) to the machine, as soon as it is determined that this will not cause any further movement. Once the power is shut off to the machine, a firefighter should be assigned to stand by the shutoff switch to ensure that it is not turned back on inadvertently.

Generally, the next step in stabilizing the machine is to crib or block the entrapping part so that there will be no further unwanted or unplanned movement. If the machine is in midpoint of a movement, it may be necessary to crib both above and below the moving part. Although in many cases it is quite appropriate and convenient to use the same cribbing materials utilized for stabilizing vehicles, heavier and stronger materials are sometimes required since the machine may be capable of crushing regular cribbing. It may also be possible to use chains or cables to assist in stabilizing the moving part.

Accessing the patient is usually not that difficult in these types of incidents; however, accessing the entrapped appendage may be very difficult or impossible in some cases. It is very beneficial, from the medical point of view, if access can be gained to the entrapped appendage.

The three overriding factors when deciding on the method to be employed to disentangle the patient are (1) the medical condition of the patient, (2) the amount of time required to complete the extrication, and (3) the effect the extrication activities will have on the patient. While the damage to the entrapping machinery should be considered, it has a very low priority when compared to the patient's well-being.

> **Note** Amputation is a viable option in industrial entrapment incidents more than any other particular incident. This is due to the severe damage that is often done to the entrapped appendage and concern for the medical condition of a patient who may not be able to tolerate a complex and lengthy extrication operation.

When it is possible to do so without causing further injury to the patient, the simplest and fastest method of disentanglement is to operate the machinery manually through the rest of its normal cycle or to manually operate the machinery in reverse, backing the entrapped appendage out of the machine.

When these methods are not possible or practical, the entrapping machinery must be disassembled, forced, or cut. Obviously the use of on-scene expert advice and assistance is highly recommended during this phase of the operation. The tools and equipment discussed in the vehicle extrication section may also be very useful in industrial entrapment operations. However, in many instances it might be impossible to force the machinery without causing greater harm to the patient or impossible to cut the machinery components due to their size or strength. In these cases, the machinery must be disassembled piece by piece.

Elevator and Escalator Rescue

When a call is received for an elevator or escalator incident, the responding fire department unit should immediately request that the dispatcher confirm with the calling party that the service company has already been called and a technician is en route; if not, one should be requested immediately. The elevators being installed today and even those that were installed

RESCUE PROCEDURES ■ **501**

many years ago are very complex. Unless there is a compelling medical emergency requiring that the rescuers access or remove the passengers from the elevator immediately, they should await the arrival of the service technician. In many cases the service technician will be able to repair the problem immediately or at least be able to bring the car to a landing so that the passengers can be removed through normal methods. At incidents where there is no immediate need to access or remove the passengers, the fire department's responsibility is to establish communications with the passengers, ensuring them that they are perfectly safe and help is on the way.

The two basic types of elevator operating systems are hydraulic and electric/cable. Hydraulic elevators are raised and lowered by use of a hydraulic pump connected via a valve system to a multisection ram, on top of which the elevator car sits. A very basic example of this type can be found in many garages where they are utilized to lift vehicles to be worked on. The power unit for a hydraulic elevator is usually found near the lowest floor served by the elevator, **Figure 16-44**. This type of elevator is not generally used in buildings of more than five stories.

Electric/cable elevators are raised and lowered by the use of an electric motor connected to a drum or cable sheave. The cable(s) that is taken up (to raise the car) or let out (to lower the car) by the drum/sheave unit is connected to the frame of the elevator car. The motor drum/sheave unit for this type of elevator is generally located directly above the elevator shaft, **Figure 16-45**, though it can be located at the bottom or in rare cases in an adjacent room. Electric/cable elevators are in use at all high-rise buildings. Extremely tall buildings may have two or more sets of elevators, with one set serving the lower floors and another set serving the upper floors.

Elevators have a variety of doors, openings, and safety interlocks that need to be understood by rescuers. Starting with the **hoistway** door, these doors are locked in the closed position unless the car is at the correct level (the landing) in the hoistway to release the locking devices. These doors can be opened by the use of special hoistway door keys, which come in a wide variety of key shapes and styles. Unless the responding unit has an assortment of these keys, an example of which is shown in **Figure 16-46**, the

Figure 16-45 An electric/cable elevator motor and drum/cable sheave assembly in the equipment room usually located above the hoistway.

Figure 16-44 A hydraulic elevator pump and valve assembly located in the mechanical room, usually on the first floor.

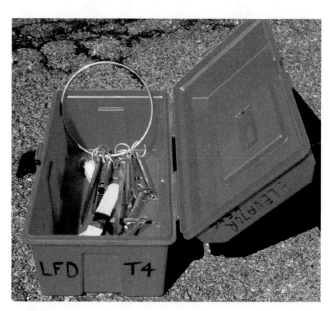

Figure 16-46 An elevator key box carried on responding apparatus shows the wide variety of keys that could be required.

rescuers will have to rely on the building management to have the proper key available when needed. When a hoistway door is opened, an electrical safety interlock will not allow the car to move.

> **Safety** Firefighters should not bet their lives on trusting an interlock switch. They must make sure the power is off.

The elevator car doors themselves do not have a locking mechanism and can be opened from either the inside or the outside of the car by pushing them apart. These doors also have an electrical safety interlock that will not allow the car to move if the slightest opening is detected.

All electric/cable elevator cars have emergency access panels either in the roof or through the side (in multiple car hoistways), **Figure 16-47**. Hydraulic elevators that have an emergency valve that can be used (by an elevator technician) to lower the car to a landing are not required to have these emergency access panels. These emergency panels are equipped with the same type of electrical safety interlock that will not allow the car to move when they are open.

Modern in-car control panels, **Figure 16-48**, generally have the floor selection buttons, an emergency stop button, and a "Fire Service" key slot. Some will have an "Independent Service" switch located behind a locked panel or a separate key slot. Many will also have a telephone or intercom-type system for emergency usage.

> **Note** Rescuers need to be aware that these telephone or intercom systems may not be answered on site and need to be very cautious relaying information or instructions through an off-site operator.

The Fire Service key slot is for use by firefighters who may want to utilize the elevator car during the investigation of a possible structure fire. When this key switch is operated the car will respond only to commands from the in-car control panel and not to any calls from other floors. The car door will not automatically open on reaching the selected floor; the operator in the car must push and hold the Door Open button.

When the Independent Service switch is operated, the car will respond only to commands from the in-car control panel and will not respond to calls from other floors. However, when the selected floor is reached the door will automatically open.

If it becomes necessary for rescuers to access or remove elevator passengers, this should only be attempted by rescue personnel with special training and experience. This can be a very dangerous process for both rescuers and passengers.

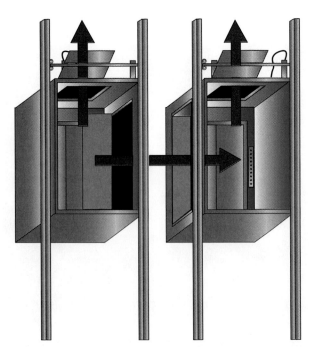

Figure 16-47 Elevator cars may have top emergency access panels, side emergency access panels, or both.

Figure 16-48 Firefighters must know and understand the operation of the various types of elevator control panels that exist within a particular response area.

The first step in accessing the passengers is to determine the location of the car in the hoistway, which can often be a time-consuming process. If rescuers are able to communicate with the passengers, they will be able to assist by telling rescuers at what floor they got on the elevator or possibly the last floor they saw indicated by the in-car floor indicator. If rescuers cannot communicate with passengers or they cannot assist, the quickest method is to go to the elevator pit below the elevator and look up, or to the equipment room above the elevator looking down, and then estimate the car's location.

While personnel are determining the car's location, a firefighter should locate the electrical control panel, which should be in the equipment room, and shut off the power to the elevator. (As with industrial entrapment procedures, this firefighter must maintain control of the electrical power.) Once an estimated location is determined, rescuers open the closest hoistway door to confirm the exact location. An open hoistway door is extremely dangerous to rescuers, passengers, and bystanders, so access to the open hoistway must be controlled. Use of a ladder or other barrier, **Figure 16-49**, is strongly recommended.

If the car is at or very close to a landing, the hoistway doors can be opened as previously discussed, the car doors can be pushed open, and the passengers assisted out of the car. Once the hoistway doors are opened, if enough of the top part of the car doors is accessible from the landing for passengers to be removed through, a ladder needs to placed through the opening down into the car, and a rescuer needs to enter the car to assist the passengers in climbing out, **Figure 16-50**. If the open hoistway doors reveal that enough of the bottom part of the car doors is accessible from the landing, a ladder needs to placed from the landing to the car, and a rescuer needs to climb into the car to assist the passengers, **Figure 16-51**. This is very dangerous; the open hoistway must be blocked.

If the car doors are not accessible from a landing, the rescuers need to go to the landing above and open the hoistway doors so that they can use a ladder to gain access to the top of the car.

> **Safety** Any rescuer entering the hoistway must have a life safety harness and lifeline.

Once on top of the car, the rescuer can open the top emergency panel and gain access to the passengers through it. If the car is close enough for a ladder to be placed from the landing above to the top of the car, the passengers can be removed up the ladder, **Figure 16-52**. If this method is used, passengers need to be placed in a life safety harness and attached to a lifeline as they begin exiting the car.

If the car is not close enough for a ladder to reach it and it is in a common hoistway with other elevator cars, it may be possible to bring one of the other cars alongside and transfer the passengers through the side emergency panel. If this method is chosen, the rescuers will need to build a makeshift bridge between the cars. Short ladders, backboards, wooden planks, and similar items work well for this. Once firefighters are ready to transfer the

Figure 16-49 An open and unprotected hoistway door is extremely dangerous. A ladder can be used to protect the opening.

Figure 16-50 Passengers can be rescued from an elevator with only the top half of the car door opening visible from the landing by extending a ladder down into the car. Firefighters then assist the passengers out.

Figure 16-51 Passengers can be rescued from an elevator with only the bottom half of the car door opening visible from the landing by extending a ladder up into the car. Firefighters then assist the passengers out. Note that the hoistway opening needs to be protected.

Figure 16-53 Passengers can be rescued by using a working elevator stopped alongside of the stuck elevator. Passengers are transferred out through the side emergency access panel. Note that the bridging in place and that a firefighter assists the passengers from both cars.

Figure 16-52 Passengers can be rescued from an elevator with the top of the car 6 to 10 feet below the landing by extending a ladder down to the top of the car and a firefighter assisting a passenger out the top emergency access panel. A Class 2 harness and safety lines need to be connected to both firefighter and passenger.

passengers, one rescuer needs to join the passengers in the stuck car in order to assist in the transfer, **Figure 16-53**.

Though not common, incidents involving escalators do occur. They usually involve a passenger getting feet caught in the area where the steps disappear into the **landing plate** or getting fingers caught under the moving hand rail. As with an elevator incident, the responding unit should request that a service technician be dispatched immediately. The other type of incident that can be encountered involving escalators is when a service technician becomes caught in the gear mechanism while working on the unit.

As with all of the other incidents that involve machinery, one of the first things to be done is to shut off and control the power supply. Most modern escalators have emergency shutoff switches located at the bottom landing on one of the end plates adjacent to the hand rail. The motor and drive gear assembly for escalators is usually located under the top landing plate, **Figure 16-54**, or near the top under the escalator itself, behind a removable panel if this area is accessible.

In cases of foot entrapment at the landing plate, the plate itself can be taken up by removing the screws that hold it down. **Figure 16-54** shows one of these plates removed. It is very difficult to

RESCUE PROCEDURES 505

Figure 16-54 The top landing of an escalator with the landing plate removed. The motor/gear assembly can be seen.

reverse the direction of travel of an escalator due to a number of safety devices designed to prevent the escalator from reversing if power is lost. In the case of finger/hand entrapment the firefighters can loosen the wheel that drives the hand rail. It is located in the same area as the motor and drive gear. If more extensive disassembly is required to extricate an entrapped passenger or repair person, it will be necessary to wait for a service technician.

Farm Equipment Rescue

Rescue of victims who are trapped, pinned, impaled, or otherwise injured by farm equipment can be very challenging. While on the surface it may appear that these types of extrications are little different from vehicle extrications, this is not the case. Everything about the farm equipment itself inherently makes extrication a much more difficult process. Farm equipment is built to withstand the rigors of very hard use on a daily basis and not bend, break, or otherwise fail. Therefore, it is routine in extrications involving this type of equipment to have to disassemble the equipment rather than use the other methods discussed in the vehicle extrication section (distortion, displacement, and severing). Couple this with the fact that these events are often located far from paved roads that allow apparatus with all their rescue equipment to be parked close, and rescue becomes all the more difficult.

If a fire department's district has the potential to be required to respond to this type of incident, specialized training should be provided to firefighters. An excellent resource for information and assistance should an incident occur is the local equipment dealer; the department should work on developing a working relationship with dealers prior to an actual incident.

Lessons Learned

This chapter covered a variety of the more common rescue situations that firefighters can expect to be confronted with from the day that they first step onto a unit. However, this does not even scratch the surface of the great variety of real-life rescue situations that firefighters confront on a daily basis. The goal of this chapter was to present a broad spectrum of situations along with the correct methods and procedures to deal with them in a safe, effective, and timely manner.

> **Note** Firefighters need to keep the skills, knowledge, and abilities learned in this chapter in their mental toolbox, building on them throughout their careers and constantly updating them as new technology and equipment is developed and put into use.

KEY TERMS

Asphyxiation Loss of consciousness or death caused by too little oxygen and too much carbon dioxide.

Confined Space A space that is large enough to be entered but is not designed for continuous occupancy.

Cribbing The use of various dimensions of lumber arranged in systematic stacks (pyramid, box, step, etc.) to support an unstable load.

Disassembly The actual taking apart of vehicle components.

Displacement The relocating of major parts (i.e., doors, roof, dash, steering column) of a vehicle.

Distortion The bending of sheet metal or components.

Engulfed To swallow up or overwhelm.

Extricate To set free, release, or disentangle a patient from an entrapment situation.

Ground Pads Sheets of plywood, planks, aluminum sheets, and so on, used to distribute weight over a larger area.

Guideline/Lifeline Rope used as a crew is searching a structure to assist them in finding their way back out.

Hardware Equipment used in conjunction with life safety ropes and harnesses (carabiners, figure eights, rappel racks, etc.).

Harnesses Webbing sewn together to form a belt, seat harness, or seat and chest harness combination.

Hoistway The shaft in which an elevator or a number of elevators travel.

Landing Plate The plate at the top or bottom of an escalator where the steps disappear into the floor.

Packaging The bandaging and preparing of a patient to be moved from the place of injury to a stretcher.

Reel Coil Memory that wire develops from having been placed on a wooden spool as it is being manufactured.

Rescue Those actions that firefighters perform at emergency scenes to remove victims from imminent danger or to extricate them if they are already entrapped.

Severance The cutting off of components (i.e., brake pedal, steering wheel) in a vehicle.

Shoring The use of timbers to support and/or strengthen weakened structural members (roofs, floors, walls, etc.) in order to avoid a secondary collapse during the rescue operation.

Spinal Immobilization The process of protecting a patient against further injury by securing them to a backboard or other rigid device designed to minimize movement.

Suspension System The springs, shock absorbers, tires, and so on, of a vehicle.

Tether Line A rope that is held by a team on shore during a water rescue to be used to haul the rescuer and victim back to shore.

Tunnel Vision The focus of attention on a particular problem without proper regard for possible consequences or alternative approaches.

Tunneling The digging and debris removal accompanied by appropriate shoring to safely move through or under a pile of debris at a structural collapse incident.

Two In/Two Out The procedure of having a crew standing by completely prepared to immediately enter a structure to rescue the interior crew should a problem develop.

Voids Spaces within a collapsed area that are open and may be an area where someone could survive a building collapse.

Webbing Sling Approximately 12 to 15 feet of rescue webbing tied end to end, forming a continuous loop.

REVIEW QUESTIONS

1. What is the biggest danger that firefighters must be aware of when involved in rescue operations?
2. What safety equipment must firefighters always be wearing and have with them prior to entering a burning structure or one in danger of becoming involved?
3. What is meant by the term *two in/two out?*
4. Name at least three of the things firefighters should be looking for when approaching a building they will be searching.
5. Why should firefighters be required to carry a forcible entry tool with them while searching a structure?
6. Describe the difference between the primary search and the secondary search.
7. What carries or drags are recommended in a single rescuer operation?
8. List the sequence of procedures, in order, that should be carried out at a vehicle crash incident.
9. List and describe each of the four disentanglement methods described.
10. List in order and describe the four methods presented for carrying out a water rescue.
11. What are the dangers presented by using a tether line in moving water?

12. Rescuers should not approach too close to a victim who has fallen through ice. Why?
13. List and describe the three types of structural collapses covered.
14. Describe the difference between *shoring* and *cribbing*.
15. Describe the actions and procedures that must be taken to make a trench collapse safe prior to rescuers entering.
16. What is the most common hazard encountered at confined space incidents?
17. Describe the differences between hydraulic and electric/cable elevators.
18. Describe the difference in how an elevator operates when in the "Fire Service" and "Independent Service" modes.

Additional Resources

Bechdel, Less and Slim Ray, *River Rescue.* Appalachian Mountain Club, 1985.

Brown, Michael G., *Engineering Practical Rope Rescue Systems.* Delmar Learning, a part of the Thomson Corporation, Clifton Park, NY, 2000.

Browne, George J. and Gus S. Crist. *Confined Space Rescue.* Delmar Learning, a part of the Thomson Corporation, Clifton Park, NY, 1999.

Downey, Ray, *The Rescue Company.* Fire Engineering Books and Videos, 1992.

Erven, Lawrence, *Emergency Rescue.* Macmillan, New York, 1980.

Frank, James and Jerrold Smith, *Rope Rescue Manual.* California Mountain Company Limited, Santa Barbara, CA, 1987.

Linton, Steven and Damon Rust, *Ice Rescue.* International Association of Dive Rescue Specialists, 1982.

McRae, Max, *Fire Department Operations with Modern Elevators.* Robert J. Brady Company, Bowie, MD, 1977.

Ohio Trade and Industrial Education Service, *Victim Rescue.* Trade and Industrial Education Instructional Materials Laboratory, Columbus, OH, 1976.

Padgett, Alan and Bruce Smith, *On Rope.* National Speleological Society, Huntsville, AL, 1987.

Setnicka, Tim, *Wilderness Search and Rescue.* Appalachian Mountain Club, Boston, MA, 1980.

Technical Rescue Program Development Manual. U.S. Fire Administration, Washington, DC.

CHAPTER 17

FORCIBLE ENTRY

Robert R. Morris, New York City Fire Department

 OUTLINE

- Objectives
- Introduction
- Forcible Entry Tools
- Safety with Forcible Entry Tools
- Maintenance of Forcible Entry Tools
- Construction and Forcible Entry
- Methods of Forcible Entry
- Windows
- Breaching Walls and Floors
- Tool Assignments
- Lessons Learned
- Key Terms
- Review Questions

STREET STORY

As a firefighter, many hours are spent training and trying to prepare yourself for any situation you could possibly encounter. After spending twenty-three years in this business, I have found there are some situations you cannot train for, especially when it comes to forcible entry. I would like to share one such incident that happened to me and how I reacted.

In my rookie year as a career firefighter we were dispatched to a structure assignment at one of our mobile home communities. En route to the scene, the dispatcher advised us that she had received a call from one of the neighbors stating that they could see flames coming from the windows, and they were unsure if anyone was home. It was around 9:00 p.m. on Saturday night so we were all hopeful that the home was unoccupied. Upon our arrival we found a mobile home approximately 50 percent involved, and the captain called for a 1¾-inch attack line and instructed us to make entry through the front door. After advancing the line, we discovered the door was locked, and we would have to make forcible entry. My partner retrieved a Halligan tool and after a quick pry the door snapped open. As the door flew open, something immediately hit me in the chest, knocking me backwards and landing on top of me. I opened my eyes and found that it was a severely burned victim who had obviously been trying to escape. After catching my breath and putting my heart back in my chest, we made entry and extinguished the fire.

When the mop-up crew arrived and started doing their thing, I found myself sitting on the tailboard of the engine with every emotion from fear to sorrow going through my head. My captain, a hardened veteran of the fire service, approached me to talk about all the events that had taken place. He told me the body was that of a seventy-year-old gentleman who had apparently fallen asleep with a cigarette.

This one incident bothered me for several years, and I will honestly admit I even considered leaving the fire service. I still find myself thinking about this fire once in a while, but now I am the hardened veteran because I have seen so much more.

The moral of this story is that forcible entry, like all of firefighting, is not only a physically and mentally challenging task, but can also be emotionally challenging, with no way to prepare. You never know what you will find behind that door.

—Street Story by Rick Townsend, Firefighter/Paramedic, Sierra Vista Fire Department, Sierra Vista, Arizona

CHAPTER 17

OBJECTIVES

After completing this chapter, the reader should be able to:

- Identify forcible entry tools by common name and use.
- Describe the inspection and maintenance procedures for each type of forcible entry tool.
- Describe building features and methods of forcible entry for doors, windows, gates, walls, and floors.
- Identify five types of locks and describe their operation.
- List the steps for the three types of conventional forcible entry.
- Demonstrate conventional forcible entry on a variety of doors.
- Describe or demonstrate the through-the-lock forcible entry method.
- List or describe four construction features of windows and methods of gaining entry.
- List or describe three considerations of breaching walls.
- Explain the three considerations of tool assignments.

INTRODUCTION

According to statistics from the Federal Bureau of Investigation over 2½ million homes are broken into every year, approximately 8,600 each day, or one break-in every 13 seconds. Because of these incidences, residents and owners install a variety of locking devices to protect both their homes and businesses. These devices present a number of challenges to firefighters who must gain access to a structure to conduct firefighting operations.

The ability to gain entry to secured areas and buildings is a primary operation at fires and other emergencies, and **forcible entry** is often one of the first operations conducted at a scene. At structural fires, a delay in gaining entry will reduce the ability to mount an aggressive fire attack, allowing the fire to extend out of control. Also, rescue operations require forcible entry to gain access to areas where victims are located. In addition, the renewed emphasis on rapid intervention teams, for the rescue of trapped firefighters, requires all firefighters to be knowledgeable and proficient with the forcible entry tools, **Figure 17-1**.

Forcible entry is a combination of knowledge and skills used to gain entry to a structure when the windows and doors are locked, blocked, or in some cases do not exist. To perform this essential task, firefighters must be knowledgeable about building construction and the operation of **locking devices** to break, remove, or bypass these elements. In addition, the firefighter must have the skills, gained by training and experience, to apply this knowledge using a variety of tools.

Knowledge

A working knowledge of the many types of locks, hardware, doors, and other assemblies is essential to successful forcible entry operations. Firefighters must be able to "size up" the quickest and easiest way to gain access to buildings, such as the doors

Figure 17-1 A typical assortment of forcible entry tools used by fire departments.

shown in **Figure 17-2**. In addition, the firefighter must know which type of tool to use and the best method to gain access through the door to the building.

> **Streetsmart Tip** As with all firefighting skills, forcible entry must be studied and practiced on a continuing basis. As security concerns increase, the types and number of locks used will also increase. Firefighters should study the locks used in their area during the development of pre-incident plans. They can visit a locksmith to learn of new types of locks or check out the Web sites of lock manufacturers. Forcible entry is more brain power than it is muscle power.

Skill

The element of skill involves a firefighter's ability to apply knowledge of building construction, lock assemblies, tools, and techniques to accomplish the necessary tasks of forcible entry. This means choosing the proper tools and applying the best techniques when using the tools. Again, skills are developed by repeated practice and experimenting with new tools, locks, and techniques.

Experience

Experience is acquired by two means. One is through drills and practice at training sessions and the other is at the scene of actual fires and emergencies. Both are the means by which skill is developed and knowledge is gained as well as reinforced. The most important experience is gained from field operations where firefighters' skills and knowledge are put to the true test.

FORCIBLE ENTRY TOOLS

Firefighters must have an understanding and knowledge of the tools available to conduct forcible entry. The selection and use of the "right" tool are essential if the task of forcible entry is to be completed as quickly as possible. Although any number of tools can be used to accomplish a task, the right tool for the right job is the quickest and easiest way to complete the operation. Because of geography or local tradition, many of the tools shown or listed may have another name or nomenclature. Firefighters must know the tools and their names as used by their particular department. Tools used for forcible entry operations are divided into several families or groups based on their intended uses as shown in **Table 17-1**.

This chapter does not attempt to show or describe *all* of the tools used for forcible entry because some tools are obsolete or used infrequently. The tools described in this chapter are the most common forcible entry tools used today.

> **Streetsmart Tip** "Try Before You Pry." The first rule of forcible entry is to attempt to open the door or window and determine if it is even locked. Firefighters should not waste time but should always try to open the door first, as this is part of the size-up of forcible entry. Appropriate eye protection is a necessity when using any and all forcible entry tools. The striking, prying, pulling, or cutting action will create sparks, metal chips, and other debris that can cause eye injury.

Figure 17-2 Firefighters should size-up the way to gain entry, "try before you pry," and think about how they would force these doors.

Striking Tools

The group or family of **striking tools**, **Figure 17-3**, is used to deliver impact to other tools, such as a Halligan tool, in order to drive it into place. Striking tools are used for impact delivery to the lock or the door itself. They may force the door or even break it down.

Flathead Ax

The most common of the striking tools is the flathead ax. Although this tool may be used for cutting, it is generally used to drive the Halligan tool. Together, the ax and Halligan tool form the "**irons**," **Figure 17-4**, and are one of the most important and useful of all forcible entry tools. The flathead ax is available in 6- and 8-pound sizes, the weight being at the head of the ax. The 8-pound ax is best for striking purposes and is commonly known as the forcible entry ax. The handle of the ax can be comprised of wood or fiberglass. Fiberglass is best for all purposes, due to its high strength and ease of maintenance.

Figure 17-3 The group or family of striking tools includes the maul, small hammer, flathead ax, and Denver tools.

> **Streetsmart Tip** A firefighter should never free-swing any of the striking tools. The firefighter using the striking tool, usually to apply force to a prying tool, should hold the tool firmly and use a strong and controlled stroke for power and accuracy.

Maul/Sledge

The maul or sledge is generally found in several sizes with the 10-pound maul being the most common and versatile. Other sizes include the 8-, 12-, and 16-pound versions. The 10-pound maul is sometimes carried with the Halligan tool to again form a set of "irons." The 16-pound maul is considered too heavy to drive other tools such as the Halligan and is sometimes used to break in doors or locks. Once again, fiberglass handles are recommended for strength and safety. Some new tool

Forcible Entry Tools

FAMILY OR GROUP	TOOLS	OPERATION
Striking	Flathead ax, maul, sledgehammer, battering ram, hammer, punch & chisel, lock breaker	Deliver impact force to break lock or drive another tool.
Prying	Crowbar, Halligan tool, hux, claw tool, pry bar, hydraulic	Provide mechanical advantage, leverage.
Cutting	Axes, saws, torches, bolt cutters	Cut material away; cut around locking devices.
Pulling	Hooks, pike poles	Limited for forcible entry; breaks glass, gypsum board, Sheetrock.
Through-the-lock	K tool, A tool, picks & key tool, bam bam, vise grip or channel lock pliers, REX tool	Remove lock cylinder.

TABLE 17-1

FORCIBLE ENTRY

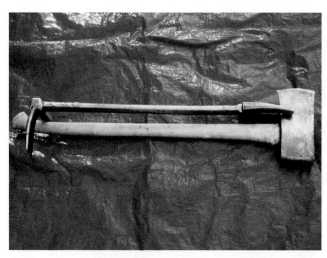

Figure 17-4 The flathead ax and Halligan tool form the tool set known as the *irons*.

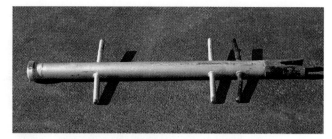

Figure 17-6 The battering ram requires brute strength to break a door, lock, or wall.

styles combine mauls with other tools such as the Denver tool, **Figure 17-5**.

The power a striking tool applies is explained by this equation: *Force = Weight × Speed*. A tool that is too heavy cannot be moved fast enough to develop proper force. Also, problems with accuracy and control can develop. On the other hand, a tool that is too light will not deliver enough force, even if it is swung hard enough.

Battering Ram

Battering rams, **Figure 17-6**, are used by two or more firefighters to break through a door or wall. The desired result is to break or push the object in. There are several types, but the type most commonly used is the old-style fire department ram. This tool is made of steel and has four handles and a round end for battering and a forked end for breaking and penetrating. Another type of battering ram is smaller and is intended for use by one or two firefighters. This tool weighs approximately 20 to 25 pounds and is used to break doors and breach walls.

Figure 17-5 The Denver tool is a maul with a cutting edge, prying end, and a pulling hook.

Prying and Spreading Tools

The group or family of **prying tools**, **Figure 17-7**, is used to spread apart a door from its jamb, move objects, or expose a locking device.

Halligan Tool

The original **Halligan tool**, **Figure 17-8**, designed by Hugh Halligan of the Fire Department of the City of New York, has proven to be the most important single forcible entry tool used in the fire service. Although the original style tool is not widely used today, the same basic concept with some improvements is capable of many forcible entry tasks.

Halligan-type tools are made in several weights and of different materials. The best type for forcible entry and structural firefighting is a 30-inch forged steel tool. The basic Halligan design is a tool that has three primary parts as shown in **Figure 17-8**. These are the adz end, the pike end, and the fork end. The size and shape of the parts are important because they determine the mechanical advantage the tool will apply to the spreading force. As noted earlier, the Halligan tool is often nested with the flathead ax, forming the "irons."

Claw Tool

The claw tool, a forerunner to the Halligan tool, was the mainstay of forcible entry operations for the previous generation of firefighters. The claw tool had two major parts: the fork end, which had a hollow, curved profile, and the hook end, which was curved and ended with a sharp point. Generally, the tool was made in two sizes. The heavy claw is 42 inches long and the standard claw is 32 inches long. The standard claw was usually carried with a striking tool such as the ax or the kelly tool. The claw tool is obsolete now, having been replaced by the Halligan tool.

Kelly Tool

The Kelly tool is a steel bar, 28 inches long, and has two main features. One end has a large adz, approximately 3 inches wide, and the other end has a large

Figure 17-7 The group or family of prying tools includes the Halligan, claw tool, hux bar, Detroit door opener, pry bar, and hydraulic spreaders.

chisel or fork. This is an older tool and has since been replaced by the Halligan tool.

Hydraulic Spreaders

Handheld hydraulic spreaders, **Figure 17-9**, are available in two different styles. One is a hydraulic pump with a spreading device attached by a length of hose and the second is a similar pump with the spreading devices attached directly to it. The spreading end is forced between the door and jamb at the location of the locking device and the pump is operated to spread the door and jamb apart. A tool is used to force the spreaders into the space between the door and jamb.

Power hydraulic spreaders such as a Hurst or Amkus system can also be used to spread objects. Their application is generally for rescue work; however, they may be used to force entry into a building.

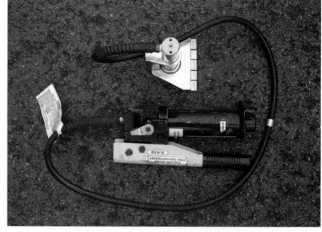

(A)

(B)

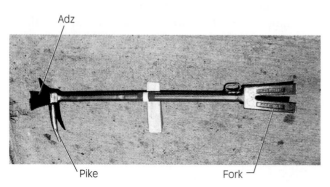

Figure 17-8 The Halligan tool is one of the most useful forcible entry tools.

Figure 17-9 Handheld hydraulic spreaders are available in two different styles but operate on the same principle. (A) Handheld pump with remote spreader. (B) Handheld unit with integral built-in spreader.

Miscellaneous Prying Tools

There are many other types of prying tools such as the crowbar, pry bar, and others, but most are of limited use. As with any tool its use depends on the type of task to be accomplished and the method of forcible entry.

Cutting Tools

The group or family of **cutting tools**, **Figure 17-10**, is used to cut away materials and expose the locking device or cut through a door or wall to accomplish forcible entry.

Ax

The fire department ax is seen in two basic configurations in **Figure 17-11**: the flathead ax and pick head ax. The flathead ax is most commonly used as a striking tool for forcible entry. In addition to cutting, this tool may be used as a prying tool, especially when equipped with a fiberglass handle. The pike head ax has a point that can be used for puncturing, pulling, and, to a limited extent, prying. It is recommended that all axes be equipped with fiberglass handles for strength and safety.

Handsaws

The most commonly used handsaw is the metal cutting hacksaw. A strong, high-quality frame with a supply of good-quality blades is used for many firefighting operations, especially vehicle rescue. Carpenter saws, both rip and crosscut, may also be carried on fire apparatus for use in areas where power saws are too large, are too noisy, or present a safety hazard.

Bolt Cutters

Bolt cutters are used to cut metal bars, cables, wires, and other hardware. These tools come in a variety of sizes; the most common for forcible entry is 36 inches in length. This size will cut steel up to ⅜ inches thick. Firefighters should avoid using bolt cutters to cut heavy-duty padlocks with case-hardened shackles. The hardness and the thickness of the metal make them difficult to cut and will damage the cutting edge of the tool.

Wire Cutters

Wire cutters are not usually used for forcible entry but are an extremely handy tool to carry. They are used to cut electric wires. They have insulated handles which reduce the risk of electric shock.

Power Cutting Tools—Saws

Power saws, **Figure 17-12**, are the most common power cutting tools used for forcible entry work. They are available in two basic types: the rotary saw with a circular blade and the chain saw. The rotary saw in **Figure 17-12A** has a carbide-tipped blade. **Figure 17-12B** shows a rotary saw with a metal-cutting disc and **Figure 17-12C** shows a chain saw. Rotary and chain saws are usually powered by a two-cycle gasoline engine and require two firefighters for safe operation.

Figure 17-10 The group or family of cutting tools includes axes, saws (both power and manual), and bolt and wire cutters.

Figure 17-11 The fire department ax is seen in two basic configurations: (A) flathead and (B) pick head.

Figure 17-12 Gasoline powered saws are available in two types. The rotary saws shown in (A) with the wood cutting blade and (B) with the abrasive disc, for either masonry or metal cutting, or, the chain saw shown in (C). The chain saw is usually used for roof ventilation.

Carbide-Tipped Blades

The rotary power saw is widely used and can be adapted to cut a variety of materials. Wood and composition materials such as roofs and flooring are cut with a carbide-tipped blade, which has 12 to 24 teeth. Light-gauge metal may also be cut with this blade. When cutting with a carbide-tipped blade, it is important to maintain full rpms to avoid having the saw bind in the material being cut.

Metal Cutting Blades

The most common metal cutting blade is an abrasive disc made of aluminum oxide. This blade is used to cut locks, hardware, steel doors, roll-down gates, and other forcible entry applications.

> **Note** With all blades, care must be taken not to bind or bend the blade while cutting. This is particularly important with abrasive discs because they can shatter or disintegrate.

Masonry Cutting Blades

Masonry materials such as concrete, brick, block, and stone can be cut with an abrasive disc made of silicon carbide or a steel with a diamond matrix blade. When cutting with these blades, a spray of water on the blade will cut down on the production of dust and keep the blade cool, thus prolonging the life of the blade.

> **Streetsmart Tip** Gasoline or other hydrocarbon fuels will break down the bonding material used in the manufacture of abrasive disc blades. To avoid this, blades must be stored separately from the fuel can for the saw. One should always refer to the manufacturer's recommendations on use and storage. Checking the fuel supply and starting the engine of gasoline-powered equipment before taking it to the location of the cutting operation will save time and frustration. Some departments start and run the saw each tour to avoid problems.

Chain Saws

Chain saws are used primarily for ventilation purposes; however, they have application as a forcible entry tool depending on the type of building construction and opening required. Several types of saws and chains are available, and many departments use a carbide-tipped chain. The chain saw can also be used to cut through wood siding, wood frame walls, certain doors, and light-gauge metal.

Reciprocating Saws

The reciprocating saw, **Figure 17-13**, is an extremely versatile tool. These saws are powered by electricity, either stationary (house current) or portable (generator). Cordless, battery-powered saws are now available and can be operated remotely from a power source, thus increasing versatility. A variety of blades are available for this type of saw depending on the material to be cut.

Cutting Torch

The cutting torch, **Figure 17-14**, uses a fuel such as acetylene mixed with oxygen to produce a flame to heat metal. A jet of pure oxygen is then applied to intensify the heat, creating a flame temperature in excess of 5,000°F, melting the metal. Many departments use these tools for heavy forcible entry in addition to other cutting tools. Use of cutting torches requires specialized training in addition to following manufacturer's recommendations and department procedures.

> **Caution** Acetylene is an unstable and extremely flammable gas. It is essential to always follow manufacturer's recommendations for storage and use of this material. In addition, cutting torches should never be used in a flammable environment.

Figure 17-13 The reciprocating saw is a versatile tool for forcible entry and rescue work.

Figure 17-14 Cutting torches use a mixture of acetylene and oxygen to generate a high-temperature flame. They cut metal by melting it.

Pulling Tools

The most common type of **pulling tool** is the **hook** or **pike pole**, **Figure 17-15**. These tools are grouped by the type of head and handle length and are used to open up walls and ceilings, to vent windows, and to pull up roof boards or other building materials.

Special Tools

A number of specialized tools are available to assist with or conduct forcible operations. Many of these have been developed by firefighters after years of experimentation and trial and error.

Bam Bam or Dent Puller

Originally designed as an automotive body work tool, the bam bam is used to pull lock cylinders. It consists of a shaft, case-hardened screw, and slide hammer. The screw is turned into the lock cylinder and the slide hammer is operated to pull the cylinder out. As lock technology has advanced, the application of this tool has become limited.

Duck Bill Lock Breaker

The duck bill lock breaker is designed to break open heavy-duty padlocks, **Figure 17-16**. The long, tapered head is placed into the shackle of the lock and driven down with a flathead ax (8-pound minimum) or a maul. This action will pull the shackle out of the body of the lock.

K Tool and Lock Pullers

The "K tool," **Figure 17-17**, is used to perform through-the-lock forcible entry. The K tool is designed to pull out lock cylinders and expose the mechanism in order to open the lock with the various key tools. The back of the tool is shaped like the letter "K" and slides over the lock cylinder. The front of the tool has a loop for the adz of the Halligan tool.

The K tool is placed on the lock cylinder and tapped into place to gain a good purchase or bite on the lock. The Halligan tool is then used to pry/pull

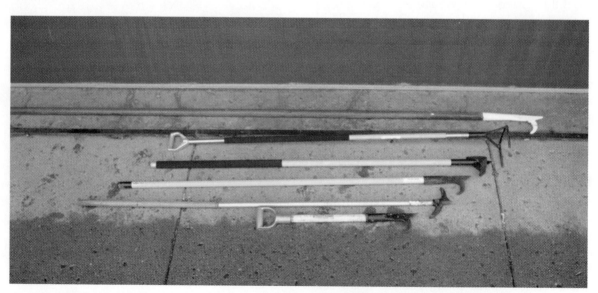

Figure 17-15 The most common type of pulling tool is the hook or pike pole, available in various styles and lengths.

FORCIBLE ENTRY 519

Figure 17-16 Duck bill lock breaker forcing a heavy-duty padlock.

Figure 17-18 The REX tool being used to pull a tubular dead bolt lock.

the cylinder out of the lock mechanism. This exposes the lock mechanism, which is operated with the proper key tool. Key tools include the bent, square, and screwdriver types, as well as a pick to slide open the shutter on some rim locks. A locking-type pliers is a good addition to the K tool kit.

A number of other tools similar to the K tool are available that can be used to perform the same operation of through-the-lock forcible entry. These are the A tool, officer's tool, or REX tool, **Figure 17-18**, to name a few.

Combinations of Tools

To perform the task of forcible entry most of the tools or groups of tools discussed are used in combination with other tools. Striking tools are used with prying tools, such as the flathead ax and Halli-gan tool, and these tools should be carried or stored together on apparatus. In addition, the firefighter must be able to size up the forcible entry task and choose the right combination of tools to provide adequate leverage or force. Experience will provide the firefighter with the knowledge to choose the right combination and to apply this knowledge to complete the task.

SAFETY WITH FORCIBLE ENTRY TOOLS

As with all tools and equipment, if misused or used for the wrong task, forcible entry tools will create safety hazards. Firefighters will become familiar with the tools, their operation, and their maintenance during training, which should result in safe

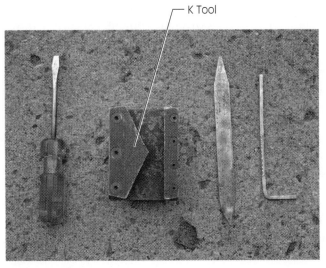

Figure 17-17 The K tool (shown with key tools) is designed to perform "through-the-lock" forcible entry.

operation. A number of general rules apply to all operations as follows:

- Always wear proper personal protective equipment including hand and eye protection.
- Follow manufacturer's guidelines for proper operations.
- Do not attempt to cut material other than that for which a blade or tool was designed for.
- Operate with regard to the safety of others in the immediate work area.
- Make sure tools are in proper operating condition before use.
- Most forcible entry operations require teamwork. Never attempt to use tools alone that require two firefighters.
- When the task is complete and if the tool is no longer needed, secure it to prevent tripping or other hazards.
- Tools should be stored and easily accessible, **Figure 17-19**.

Note The safety guidelines provided are general in nature and firefighters are reminded to read manufacturer's operating instructions for the specific tools used in their department. In addition, they should not use any power tools or other spark-producing equipment in an explosive or flammable atmosphere.

Safety Whenever forcing entry into an unfamiliar area, firefighters must be very cautious as to what they may face when creating the opening. Immediate danger to life and health (IDLH) conditions, extreme fire exposure, drug lab chemicals, booby traps, or other hazards may be present.

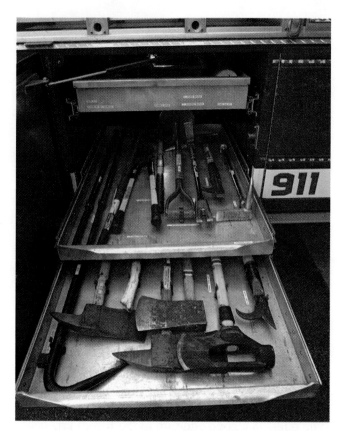

Figure 17-19 These slide-out trays provide easy access to forcible entry tools. Note the mounting brackets that hold the tools in place.

- Equip the saw with a carry strap (standard equipment with some manufacturers).
- Use the right blade for the material being cut, **Figure 17-20**.
- Never carry a running saw up a ladder or through a crowd of firefighters.
- Power saws require two firefighters: the saw operator and a guide firefighter.
- Eye protection is required when running any power equipment, especially power saws.

Rotary and Chain Saws

As the use of security gates and overhead doors increases, the power saw has become the tool of choice to remove the door or gate. The rotary saw with a metal cutting blade is an effective tool for these operations. These saws present a number of hazards, and firefighters should follow these guidelines to complete the operation safely:

- Always follow the manufacturer's instructions.
- Conduct daily checks for operation and blade condition.
- Check the saw for fuel and proper operation before proceeding to the entry location.

Carrying Tools

Many forcible entry tools have sharp or pointed ends and must be carried safely from fire apparatus to the fire scene. Firefighters should always be aware of their safety in addition to the safety of other firefighters.

- *Axes.* Carry the ax with the blade away from the body and the pick head covered. Never carry an ax over the shoulder.
- *Prying tools.* Similar to the ax, pointed and sharp edges should be carried away from the body and covered if possible.
- *Hooks or pike poles.* The tool head, the hook end, should be carried down and close to the

FORCIBLE ENTRY

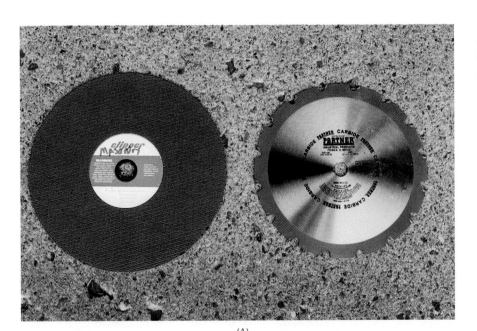 

Figure 17-20 (A) Rotary saws have three types of blades: wood, masonry, and metal. The masonry and wood blades are shown here. (B) This is a new style of wood and roofing blade for the rotary saw.

ground. Depending on handle length, beware of overhead electrical wires and other obstructions. Inside a building, carry the handle close to the body with the hooked end toward the ceiling.

- *Striking tools.* These tools tend to be heavy and the head should be carried close to the ground. When using these tools, do not use a free-swing motion. Firmly grasp the tool with two hands and use a controlled and accurate stroke to move the tool.

Hand Tools

Hand tools used for forcible entry are constructed of metal, wood, fiberglass, or some combination of these materials. All tools must be inspected regularly for condition, cracks in the handles, burrs in the metal, and loose heads.

MAINTENANCE OF FORCIBLE ENTRY TOOLS

Proper tool maintenance is the first step to tool safety. Tools must be inspected and cleaned on a regular basis and checked for wear and damaged parts. Tools should be removed from service or repaired when defects are found.

Metal Heads and Parts

- Remove any dirt or rust with steel wool or emery cloth.
- Use a metal file to maintain the proper profile and cutting edge.
- Sharpen edges and remove burrs with a file.
- Do not keep the blade edge too sharp; this may cause it to chip when in use.
- Do not grind the blade because it can overheat and cause it to lose the temper and become soft.
- Do not paint the metal parts. Keep them lightly oiled if desired.

Caution Oil should not be applied to the striking surface or face of striking tools. When used, an oil coating on the striking surface may cause the tool to slip or glance off, causing injury, and little or no force will be applied to the object being hit.

Fiberglass Handles

- Wash them with soap and water and dry completely.
- Check for damage or cracks.
- Make sure metal parts are secure.

Wood Handles

- Clean with soap and water, rinse, then dry completely.
- Check for damage and sand off any splinters.
- Do not paint or varnish the handles. A coat of boiled linseed oil may be applied if necessary.
- Ensure that the head is securely fastened to the handle.

CONSTRUCTION AND FORCIBLE ENTRY

The type and construction of the many different features of buildings, such as doors, windows, gates, walls, floors, and roofs, must be recognized and understood by firefighters to force entry. A thorough knowledge of the construction of these building features will lead to successful forcible entry operations.

Door Construction

Passage doors are manufactured in many styles as shown in **Figure 17-21** and are manufactured from plastic, wood, metal, and/or glass. The door assembly consists of the door itself, the frame or **jamb**, **mounting hardware**, and locking device mounted in a jamb, which is **rabbeted** or has a stop attached to it, **Figure 17-22**. A rabbeted jamb is formed or milled into the casing that the door closes against to form a seal. The stopped jamb has a piece of molding nailed or attached to the casing for the door to close against. The stopped jamb can be removed, allowing access to the lock assembly and hence easy forcible entry.

Types of Doors

Wood Doors

There are three types of wood doors. **Panel doors** have a solid stile and rails with panels made of wood or glass or other materials. **Flush** or **slab doors** are flat or have a smooth surface and may be of either **hollow-core** or **solid-core** construction. Solid-core doors are more resistant to fire and more secure, usually making them more difficult to force. **Ledge doors** are built with solid material, usually individual boards, and are common in barns and warehouses. Wood doors are used primarily in residential construction and are installed on a wood frame or jamb secured to wall framing with nails or screws.

Metal Doors

Metal doors manufactured either as hollow-core or metal clad, **Figure 17-23**, are common in new construction. These are usually installed in metal frames and can be very secure. The metal clad door, which has a steel surface with a wood core, is manufactured in a number of designs and architectural finishes and some may have a fire-resistive rating. Generally metal doors are used in commercial construction and as exterior doors in residential construction. Metal doors are more substantial and secured to the wall construction, **Figure 17-24**. Forcible entry may be

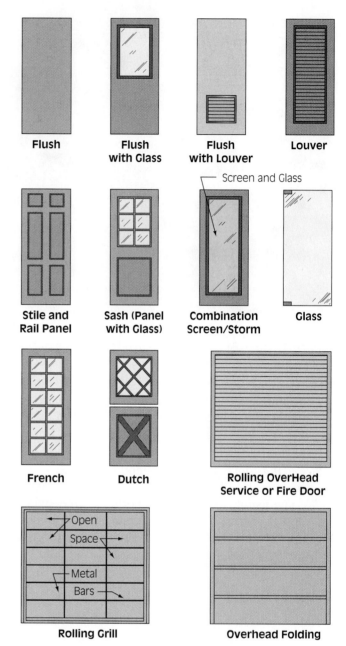

Figure 17-21 Doors are manufactured in a variety of styles and materials.

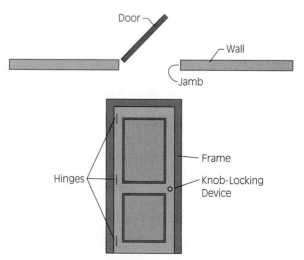

Figure 17-22 The parts of a door assembly.

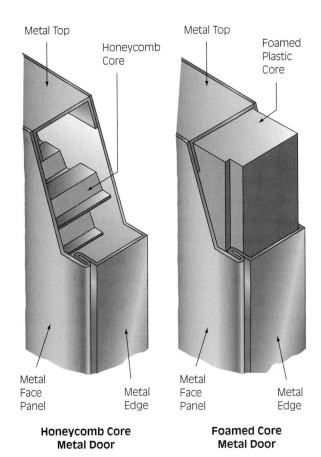

Figure 17-23 Metal doors are of both solid- and hollow-core design.

Glass Doors

Two main types of glass doors are commonly in use today: the metal or tubular frame glass door and the tempered or frameless glass door, **Figure 17-25**. The metal frame is the most common type of glass door and both styles are generally used in commercial and mercantile occupancies. Most have tempered glass, but a laminated safety glass and polycarbonate glazing (Lexan) are common and can be found in these doors.

Tempered Glass Doors

These doors are made of thick tempered glass, hung on a tubular metal frame, with hinges located at the top and bottom of the door and a locking device set in a metal rail at the bottom of the door. **Tempered glass** is plate glass that has been heat treated to increase its strength. The best method to force this type of door is the through-the-lock technique. If it is necessary to break tempered glass, firefighters should strike a sharp blow with a pointed tool such as the pike of a Halligan tool. Firefighters must be aware that the glass will shatter so they should stand to one side and strike the glass in a lower corner.

Door Swing Direction

Doors are hung in jambs with hinges, and forcible entry is accomplished by working with the direction of swing. The firefighter must determine

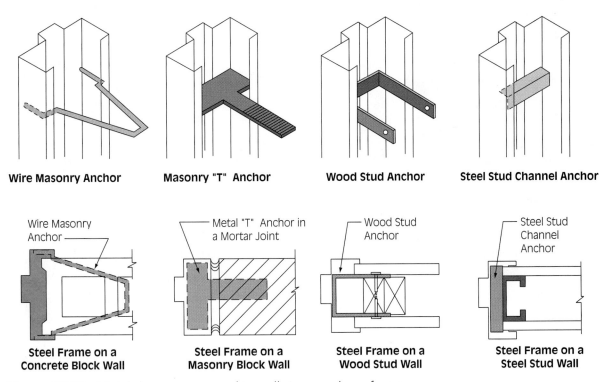

Figure 17-24 Metal doors are secured to walls in a number of ways.

which way the door swings, and the common way of describing this is in relationship to the forcible entry team. Doors with exposed or visible hinges will swing toward the forcible entry team, **Figure 17-26A**. If no hinges are visible and there is a stop on the door frame, the door will swing away from the team, **Figure 17-26B**.

Sliding Doors

Sliding doors are usually found in residential occupancies and consist of sliding and fixed sections of tempered glass in a wood or metal frame. The locking device is usually very light with a strike screwed into the jamb, **Figure 17-27**. These doors are forced by prying the door away from the jamb with a Halligan tool, breaking the lock striker out of the stationary frame. To prevent entry by intruders, many occupants will place a pipe, broomstick, or other solid object in the track to prevent the door from sliding. With these objects used as additional security devices, breaking the glass may be the only option for entry, but it should be used as a last resort.

Revolving Doors

Revolving doors are made up of four sections (doors) hung on a vertical shaft or hinge that

Figure 17-25 Typical commercial glass door.

(A)

(B)

Figure 17-26 Hinges indicate the direction of door swing and forcible entry. An outward swinging door is shown in (A) and an inward swinging door in (B).

FORCIBLE ENTRY

Figure 17-27 The strike of a residential sliding glass door is secured with two small screws and can be easily forced.

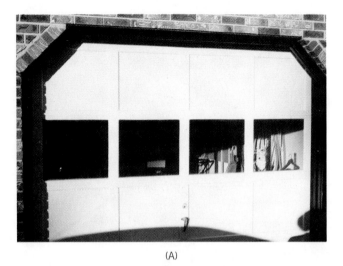
(A)

(B)

Figure 17-28 Typical overhead doors: (A) residential and (B) commercial.

allows the door to rotate. Building and life safety codes require revolving doors to collapse and allow occupants a rapid exit if needed. This is accomplished by releasing the arms that hold the door apart. The "panic-proof" type of revolving door will automatically collapse when two sections are moved in opposite directions. These doors may be locked with key cylinder locks or slide bolts. Generally it is best to avoid using revolving doors as an entry point because of lack of clearance created by the folded door and vertical shaft or hinge.

Overhead Doors

Overhead doors range from the simple garage door to the very secure roll-down steel door, **Figure 17-28**. These doors are built from wood, steel, aluminum, and fiberglass and may have a solid or insulated core. Depending on the occupancy and security requirements, they may have windows of glass or some type of synthetic material.

Residential Garage Doors

Overhead garage doors used in residential construction are typically three- to five-section folding doors of wood or metal construction, **Figure 17-28A**. Older style doors may be one-piece slabs that tilt up into the garage. Overhead doors are installed on tracks with rollers and balance springs to assist in opening. A folding overhead garage door may be forced by any of several methods:

- Break a panel or window, reach in, and unlock the securing device.
- Pull the lock cylinder and utilize through-the-lock forcible entry.
- Automatic openers hold the door in the closed position. To disconnect the opener, break out a panel near the attachment mechanism, reach in with a tool to grab the release cord, and pull as shown in **Figure 17-29**.

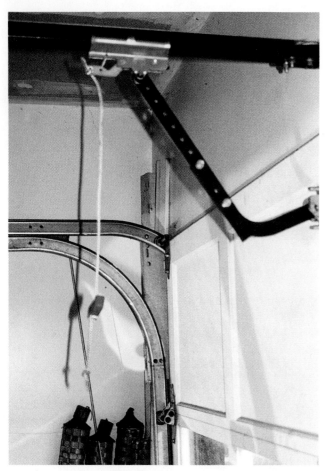

Figure 17-29 To disable a motorized residential overhead door opener, the firefighter can break out a pane of glass or door panel and pull the release rope.

> **Caution** At a recent fire, the initial attack lines advanced through the garage area of a house. The overhead door, which was equipped with an automatic operator, closed either because the balance springs were weakened by fire or the door operator shorted out. This trapped the attack team inside the garage with no means of escape. Once an overhead door is opened, a tool should be placed (a six-foot hook works well) in the track or a pair of locking pliers should be used on the track to prevent the door from closing.

Commercial Garage Doors

Commercial overhead doors are similar in operation and construction to residential doors, **Figure 17-28B**. The exception to this is the type of locking and security devices. These doors may be forced using the same methods mentioned for residential doors or by cutting the door with a rotary saw, **Figure 17-30**. After the door is opened, the same precautions used for residential overhead doors must be taken to ensure that it does not close, thus trapping firefighters.

Roll-Down Steel Doors and Gates

Roll-down steel doors, **Figure 17-31**, are of heavy steel construction to provide a higher level of security or as a rated door in a firewall or separation. Roll-down doors generally use three different methods of opening and closing:

- *Manually operated:* The steel gate is lifted by hand with the assistance of springs.
- *Chain operated:* A chain hoist mechanism is used to lift the door.
- *Electrically powered:* An electric motor connected to a switch is used to raise and lower the door.

The method used to force these doors will depend on the type and number of locks and security devices. Generally the best way to accomplish forcible entry is to:

- Cut or force the locks.
- Attack the hardware.
- Cut through the gate in a manner similar to that shown in **Figure 17-30**.

Security gates used in commercial occupancies provide security while at the same time allow for the display of merchandise. The most common means of securing these gates is to use padlocks. These padlocks can be cut or forced by use of several types of tools; the most reliable method is the power saw with a metal cutting disc. Cutting or otherwise forcing the hardware that secures the gate is a method that can be used with some success. An example of this is cutting the hasp that secures the padlock to the gate.

Locks

Locks are designed and intended to keep unwanted or unwelcome visitors out of a building or occupancy. Firefighters—although not unwelcome visitors—will face the problem of locked door and entry access on a regular basis. As security requirements increase or break-ins occur, additional locks will be placed on a door as shown in **Figure 17-32**.

The most important part of forcible entry is to know the type of lock, how the lock operates, and how to disable it. Knowing this will save valuable time and energy on the fireground and provide quicker access, rescue, and fire extinguishment.

> **Streetsmart Tip** From time to time home improvement and consumer magazines will have articles on locks and security devices. This information is an excellent resource for additional training on new types of locks and methods to force entry.

FORCIBLE ENTRY 527

Figure 17-30 On older commercial overhead doors, a large inverted V-shaped cut is made in the gate with a power saw. A slat on each side is pulled toward the center and removed. This allows all slats to be removed and a large opening created.

Figure 17-32 To gain access through this door, the firefighter must know how to force these three different styles of locks.

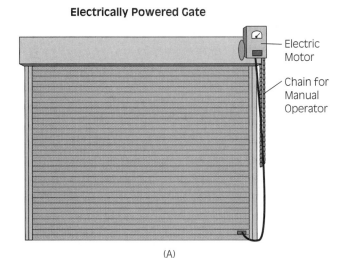

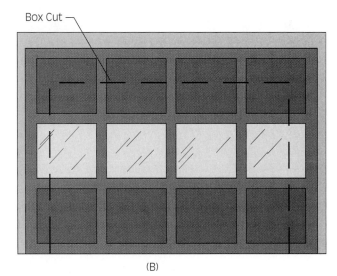

Figure 17-31 (A) Roll-down steel doors are of heavy steel construction to provide a higher level of security or as a rated door in a firewall. (B) Box cut used for this door.

The basic nomenclature of locks as shown in **Figure 17-33** will assist the firefighter in understanding the operation of most locks.

Lock mechanisms used to provide security for doors and other openings fall into the following general categories.

Key in the Knob Lock

This lock, **Figure 17-34**, is most common in residential occupancies and on interior doors in commercial occupancies. The outside of the lock will have a keyway for operation and the inside will usually have a keyway or button. The bolt on this type of lock is either a latch or dead latch bolt type. The latch bolt has a throw of approximately ½ inch, making this an easy door to force.

Figure 17-34 A key in the knob-type lock usually found in residential occupancies.

Mortise Lock

The mortise lock, **Figure 17-35A**, is designed to fit into a cavity in the edge of a door and is usually found in commercial occupancies. The mortise lock is designed with three types of operating latches: the dead bolt, the dead bolt and latch, and the pivoting dead bolt. The pivoting dead bolt may also be known as an Adams Rite lock, which is used in narrow frame/stile aluminum doors. Most mortise locks operate with a key that, when inserted in the keyway and turned, turns a cam, which moves the dead bolt or latch and opens the lock.

- *Dead bolt.* This lock has one sliding bolt and is locked or unlocked by one complete turn of the

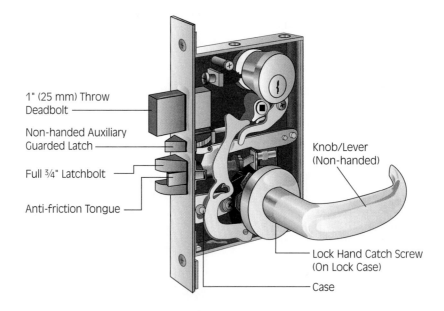

Figure 17-33 The parts of a lock device.

FORCIBLE ENTRY   529

(A)

(B)

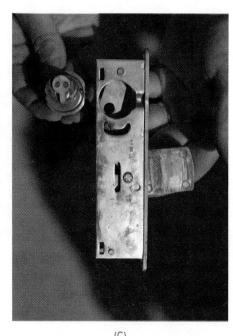
(C)

Figure 17-35 A typical mortise style lock: (A) the exterior, or normal view, (B) the locking mechanism, and (C) the pivoting dead bolt (Adams Rite lock).

key (**Figure 17-35B**) in the cylinder. This will only operate the dead bolt; additional latches or locks are operated separately.

- *Dead bolt and latch.* (**Figure 17-35B**) Similar to the dead bolt, but with an additional latch operated by a doorknob.
- *Pivoting dead bolt.* This lock, **Figure 17-35C**, is used on metal and glass doors and has a bolt that is housed vertically when retracted and pivots up to the horizontal when placed to lock. The **bolt throw** projects approximately 1½ inches into the **strike plate** when locked.

Rim Locks

Rim locks, also known as surface locks, **Figure 17-36**, attach on the inside of a door with the cylinder extending through the door and a keyway visible on the outside. There are many variations of this lock but they are all mounted on the surface of the door.

Tubular Locks

Tubular locks are mounted in a hole that has been bored into the door and are best described as a combination of the key in the knob lock and the mortise lock. However, instead of a knob, the tubular lock uses a cylinder to operate the bolt and is recognized by the cylinder's cover which protrudes about ½ inch on each side of the door. A typical tubular lock is shown in **Figure 17-37**.

Padlocks

Padlocks are portable or detachable locking devices that are manufactured for regular and heavy-duty service. This type of locking device has a movable shackle that locks into the body of the lock and is used to secure a door or gate using a hasp or chain. Padlocks come in a wide variety of shapes and sizes, **Figure 17-38**.

Regular padlocks have a shackle of less than ¼ inch in diameter and are not usually made of hardened steel. These may be cut with a bolt cutter or broken with a lock breaker. Heavy-duty padlocks have shackles larger than ¼ inch in diameter and are made

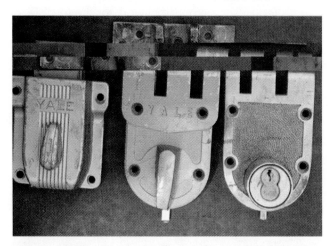

Figure 17-36 Rim locks (from left to right): a dead bolt, a vertical bolt and striker plate, and a vertical bolt with key cylinder.

Figure 17-37 Tubular dead bolts.

Figure 17-39 Cutting a padlock with a power saw. Note that both shackles are being cut at once.

of hardened steel, **Figure 17-39**. In addition the shackles on both the heel and toe are locked when the shackle is depressed into the lock case, and both sides of the shackle must be cut to open the lock.

Special Locks

Included here are all devices that do not fit into other categories. Examples of these devices are overhead door locks, magnetic locks, or card key entry systems, **Figures 17-40A** and **B**. Overhead door locks are similar to rim locks and may be forced by gaining access or through the lock. Card key entry systems are regular locking devices operated by an electric actuator or solenoid. Again, conventional or through-the-lock forcible entry is the key to access.

> **Streetsmart Tip** Many office occupancies and hotels/motels are using electronic locks for security. The fire department should contact the facility management people to arrange a procedure for obtaining a master card key if available. This reduces the amount of work for the forcible entry crew and could reduce property damage.

Figure 17-38 Padlocks are available in many shapes and sizes.

Magnetic locks use a magnetic force to hold a door secure and the system is usually disabled from the interior of the building. In most instances it is quicker and easier to find another means of entry.

Additional Security Devices

As security requirements increase, home and business owners have begun to install many varieties of locks and security devices. These may be as simple as a broom handle in the track of a sliding door or as complicated as a number of additional locks on a door. These additional devices may not provide visible signs or indication of their use to the forcible entry team, **Figure 17-41**. If these types of devices are in use the forcible entry team may need to find an alternate means of entry or use a rotary saw to gain access.

When buildings are being remodeled or demolished, the fire department should check with the owner or contractor to gain permission to salvage unwanted door locks or security devices. These make excellent training aids, especially if similar locks are in use in other buildings in the jurisdiction. In addition, new construction should be surveyed to determine the types of locking devices being installed.

METHODS OF FORCIBLE ENTRY

For fire department operation there are three standard methods of forcible entry:

1. Conventional
2. Through-the-lock
3. Power tools

Figure 17-40 (A) Residential overhead door locks are a type of rim lock and are forced easily using the through-the-lock method. (B) Card key entry systems are standard locks with electric actuators.

> **Caution** All forcible entry operations must be coordinated with fire attack and ventilation. Lack of coordination may result in rapid fire spread or a backdraft.

These methods involve the use of certain tools and the application of many techniques.

Conventional

This is an old and reliable method involving the use of leverage, force, and impact. The primary tools used in this method are the "irons," consisting of the Halligan tool and the flathead ax. This technique requires procedures that will accomplish one or more of the following:

- Force the door away from the jamb, pulling the bolt away and free from the strike plate.
- Break the lock or striker.
- Break the door and/or the frame.
- Force or remove the hinges.
- Breach the wall or door.

532 ■ CHAPTER 17

Figure 17-41 These auxiliary locking devices are usually not detected by the forcible entry team: (A) floor-mounted stop plate, (B) door blocker, (C) steel bar and brackets, and (D) sliding bolt. Note that the bar and sliding bolt are often homemade devices.

Streetsmart Tip Always size up the forcible entry task. Regardless of the type of lock, at a quick glance, the fastest way to force entry to the door shown in **Figure 17-42** is to break the glass and unlock the door from the inside. But be aware there may be additional locking devices out of view that may not be easily unlocked. At a fire situation this could magnify the problems associated with heat, smoke, and fire possibly rolling out through the broken glass. On the contrary, once the door in **Figure 17-43** is forced, the firefighter may encounter the household goods stored in front of the door.

In general, conventional forcible entry is quick and reliable and is the primary, preferred method at structural fires. Conventional forcible entry may also involve the use of the hydraulic forcible entry tools.

Conventional Forcible Entry with Door Swinging Away from the Forcible Entry Team (JPR 17-1)

Remember to always try the door to determine if it is locked or secured and forcible entry is actually necessary.

1. Size up the door to determine swing; the number, type, and location of locking devices; and the type of door and frame,

FORCIBLE ENTRY

Figure 17-42 Firefighters cannot assume that the quickest way to force entry to this type of door is to break the glass and unlock the door from the inside. Additional locking devices may be out of view.

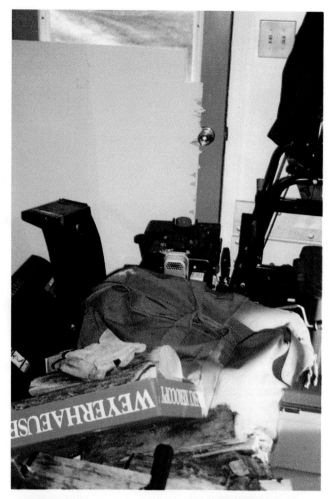

Figure 17-43 Once a door is forced, the firefighter may encounter many obstacles.

JPR 17-1A. With this type of door swing, the hinge pins are not visible. Also check to make sure the door is locked/secured.
2. Gap the door by slipping the adz between the door and the door stop 6 inches above or below center of the lock and push up or down to spread the door from the frame, **JPR 17-1B**. This will allow for placement of the fork end of the Halligan tool. Note that on some doors this may cause the door to open. On door frames with a nailed doorstop, this method will break the stop and allow easy entry of the fork.
3. Set the fork of the Halligan tool into the gapped area with the bevel of the fork toward the door no more than 6 inches above or below the center lock, **JPR 17-1C**. As the Halligan tool is struck with the flathead ax or maul, gradually bring the tool away from and at an approximately 90-degree angle to the door.
4. Set the tool using the flathead ax or maul, driving the Halligan tool in so that the tips are around the door and locked in, **JPR 17-1D**. *If difficulty is experienced, the fork may be inserted with the bevel toward the frame.*
5. Force the door; with the tool set the door is spread from the frame. The Halligan tool is pushed sharply in the direction of the door swing and opens as shown in **JPR 17-1E**. This procedure is repeated as necessary for each lock, **JPR 17-1F**.

On doors that swing away from you, try to control the door as it is forced. Generally, the door is controlled by the firefighter with the Halligan tool. As the door swings in, the firefighter must reach in with the tool and bring the door back under control.

Caution Doors leading into the fire area should be blocked or chocked open when firefighters are conducting a search or attacking the fire in that area in addition to other activities. This will prevent the door from closing behind crews and will allow rapid egress if needed.

JOB PERFORMANCE REQUIREMENT 17-1
Conventional Forcible Entry—Door Swings Away from the Forcible Entry Team

A Size up the door to determine swing; the number, type, and location of locking devices; and the type of door and frame.

B Place the adz between the door and the door stop 6 inches above or below center of the lock and push up or down to spread the door from the frame.

C Place the fork of the Halligan tool into the gapped area with the bevel of the fork toward the door no more than 6 inches above or below the center lock.

D Set the tool using the flathead ax or maul. Drive the Halligan tool in so that the tips are around the door and locked in.

E If difficulty is experienced in setting the tool, drive fork in with bevel toward frame to allow easier entry.

F This procedure is repeated for each lock.

FORCIBLE ENTRY ■ **535**

Conventional Forcible Entry with Door Swinging toward the Forcible Entry Team (JPR 17-2)

1. Size up the door to determine swing; the number, type, and location of locking devices; and type of door frame and hinges, **JPR 17-2A**. Also check to make sure the door is locked/secured.
2. Gap the door by driving the adz or fork end of the Halligan tool between the door and frame 6 inches above or below the lock, **JPR 17-2B**. Work the tool in until it contacts the doorstop. The adz end is preferred, but on some doors, especially steel, it will have a tight fit between the door and jamb and the gap may be started by using the fork end, **JPR 17-2C**. If door is tight, work tool up and down to open up gap, **JPR 17-2D**.
3. Set the tool by driving the adz or fork end past the door around the doorstop, **JPR 17-2E**. Drive in the tool until it is "locked in" around the inside of the door. Due to the configuration of the door and wall, using the adz end will allow the tool

JOB PERFORMANCE REQUIREMENT 17-2
Conventional Forcible Entry—Door Swings toward the Forcible Entry Team

A Check the door to determine swing; the number, type, and location of locking devices; and the type of door frame and hinges.

B Create a gap between the door and jamb by driving the adz or fork end of the Halligan tool between the door and frame 6 inches above or below the lock.

C On tight doors, the gap may be started with the fork end.

D If door is tight, work tool up and down to open up gap.

(continued)

JOB PERFORMANCE REQUIREMENT 17-2
Conventional Forcible Entry—Door Swings Toward the Forcible Entry Team (Continued)

E Set the tool by driving the adz or fork end past the door around the doorstop.

F This inside view shows how the adz is set.

G Force the door by pulling the tool away from the door.

---

to clear the wall. **JPR 17-2F** shows an inside view of how the adz is set.

4. Force the door by pulling the tool away from the door, **JPR 17-2G**. This will cause the bolt to pull out of the strike, the lock to break, or the door frame assembly to fail and the door will open. When dealing with several locks, this procedure may have to be repeated for each lock.

Forcing Doors with Hydraulic Forcible Entry Tools

The hydraulic forcible entry tool (HFT) is primarily used on doors that swing away and have strong metal frames.

1. Place the spreader jaw of the tool between the door and jamb as close as possible to the lock.
2. Set the jaw into the gap to the doorstop using a tool if necessary. On tight-fitting doors, a driving tool may be needed to set the jaw into the doorstop.
3. Pump the tool until the bolt is pulled out of the strike and push the door open.

> **Streetsmart Tip** If there is more than one lock present and they are relatively close together, place the jaw of the tool between the locks.

Through-the-Lock Forcible Entry

The **through-the-lock method** of forcible entry involves attacking the locking mechanism by removing the key cylinder and then operating the lock with alternative means. This method of forcible entry is best applied to mortise, rim, or tubular-type lock cylinders.

In general, the choice of through-the-lock forcible entry is made when entry needs to be gained with damage kept to a minimum. Sometimes it is also the quickest means of entry and is the best method when forcing entry of metal framed glass or all glass doors.

Through-the-Lock Entry by Unscrewing or Wrenching the Locking Cylinder (JPR 17-3)

This method is not as quick as pulling the cylinder with a lock puller and should be used only if time

JOB PERFORMANCE REQUIREMENT 17-3
Through-the-Lock Entry by Unscrewing or Wrenching the Locking Cylinder

A Using locking-type pliers, lock the pliers onto the cylinder. Turn the lock cylinder counterclockwise to unscrew the cylinder.

B Use proper key tool to open lock.

---

allows. In addition, trim work or walls may not allow the free rotation of the pliers.

1. Size up the lock to determine the type of lock and feasibility of utilizing this method. Lock cylinders with protective collars may not be able to be unscrewed.
2. Using locking-type pliers, lock the pliers onto the cylinder, **JPR 17-3A**.
3. Turn the lock cylinder counterclockwise to unscrew the cylinder.
4. Remove the cylinder and insert the proper end of the key tool as shown in **JPR 17-3B** to operate the locking mechanism as shown later in this chapter.

Streetsmart Tip Through-the-Lock Technique Modified with Rim Locks: If the cylinder crumbles and will not pull out or the lock will not unlock, the firefighter can drive the lock off the back of the door with the pike of the Halligan tool or the handle of the lock puller.

Through-the-Lock Entry Using the K Tool (JPR 17-4)

1. Size up the lock to determine the type of lock and feasibility of utilizing this method.
2. The blades of the K tool are forced over the cylinder and decorative ring. Tap the K tool into place until firmly set, **JPR 17-4A**.
3. Place the adz of the Halligan tool into the loop of the K tool, **JPR 17-4B**. If necessary, the Halligan can be tapped to firmly set the K tool.
4. Pull up on the Halligan tool and pull the cylinder, **JPR 17-4C**.

Streetsmart Tip The K tool is designed to be able to pull cylinders that are very close to the bottom or edge of the door. This is done by placing the narrow, straight blade of the K tool toward the bottom of the door to allow room for the tool.

A number of different lock-pulling tools may be used in place of the K tool. The A, officer's, and REX tool, **Figure 17-44**, are designed to be driven behind the lock cylinder to get a "bite" or "purchase" on a substantial part of the cylinder. These tools are most effective on cylinders that are recessed into a door and on odd-shaped cylinders such as tubular locks.

JOB PERFORMANCE REQUIREMENT 17-4
Use of K Tool for Through-the-Lock Forcible Entry

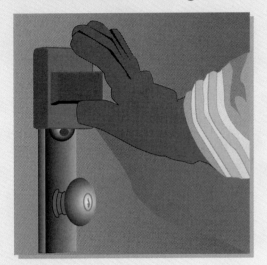

A Force the blades of the K tool over the cylinder.

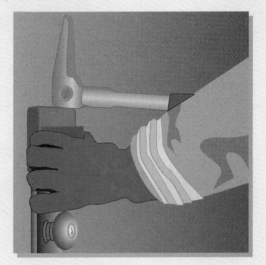

B Place the adz of the Halligan tool into the loop of the K tool.

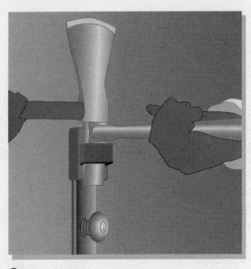

C Pull up on the Halligan tool, and pull the cylinder out.

Operating Lock Mechanisms

The final step using through-the-lock forcible entry is to manipulate the lock by using the proper key tool. Many types of key tools are used, but the most common is the two-sided, flat steel key or square key tool, **Figure 17-45**. This tool has a bent end used for mortise locks and a straight end used for rim and tubular locks.

The correct key tool to operate the locking mechanism is determined by examining the cylinder after it is removed. Lock cylinders fall into two different categories, the mortise having a cam device on the back of the cylinder and the rim lock with a flat or square blade on the back of the cylinder, **Figure 17-46**. The tubular lock has a tailpiece similar to the rim lock.

To open a mortise lock, visualize that the key hole where the cylinder was prior to pulling it out is that of the face of a clock. The keyway of the cylinder prior to removal was at the six o'clock position and the unlocking action will occur within the five to

FORCIBLE ENTRY

Figure 17-44 The REX tool is used to pull lock cylinders in the through-the-lock method of forcible entry.

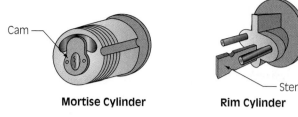

Figure 17-46 To operate the lock, look at the back of the cylinder to determine the right key tool to use.

seven or seven to five o'clock area, **Figure 17-47**. Place the bent end of the key tool into the keyway and move it to the five or seven o'clock position and engage the locking mechanism. Move it to the opposite side to retract the dead bolt. If the lock mechanism has a spring lock, push down with the key tool to depress it and then move it. To unlock the spring latch, move the key tool up to the three o'clock or nine o'clock position and engage the latch mechanism, push, and hold it in place. This will pull back the spring latch and allow the door to open.

To open a rim lock, after removing the cylinder, if a tailpiece is visible, then the lock is a rim lock. Insert the straight blade of the key tool into the slot on the rim lock and rotate the tool to unlock it.

Lock Variations

Shutter Guard

Many locks have a spring-loaded shutter guard that will close when the cylinder is pulled. This will not allow a key tool to open the lock. A shaped tool such as a pick is used to slide the shutter open. Slide the shutter horizontally toward the edge of the door and hold it open until the key tool is in place. Then turn the key tool to open.

Figure 17-45 The two types of key tools used for through-the-lock forcible entry.

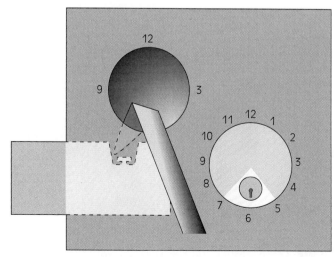

Figure 17-47 The unlocking action will occur within the five to seven or seven to five o'clock area.

Night Latch

A rim lock with the night latch engaged can only be unlocked from the inside.

Through-the-Lock Technique Modified

When the mechanism of a rim lock cannot be operated because of the presence of a night latch or shutter guard, or in any situation where the key tool cannot open the lock, firefighters should use the following procedure. Place the pike of the Halligan tool into the hole and against the back of the *rim lock*. Strike the Halligan tool with an ax and drive the rim lock off the door, **Figure 17-48**.

Tubular Locks

Locks such as the key in the knob and tubular dead bolts are opened in a manner similar to the rim lock. The knob is removed with a lock puller or knocked off with a heavy tool such as the ax or Halligan tool. This will expose the latch mechanism, which can then be operated by the key tool or screwdriver.

Removing the cylinder with a lock puller exposes the tubular dead bolt lock. The adz of the Halligan tool can be used to shear the screws, and the cylinder can sometimes be removed in this manner. The bolt mechanism is then operated with a key tool or screwdriver, depending on the shape of the hub, **Figure 17-49**.

Forcing a Padlock with a Lock Breaker

1. Place the wedge end of the lock breaker through the shackle. Depending on the size of the lock a Halligan tool may also be used for this task.
2. Strike the back or head of the lock breaker with a maul or heavy striking tool.
3. Continue step 2 until both shackles break.

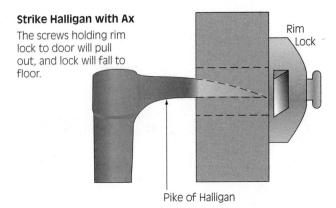

Figure 17-48 Strike the Halligan tool with the ax and drive the lock off the door. This method is used for rim locks only.

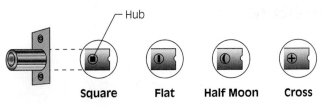

Figure 17-49 The bolt of a tubular lock is operated with a key tool matching one of the shapes shown.

Forcing a Padlock by Cutting with a Rotary Saw

1. Attach locking-type pliers to the lock case and lock the jaws. The locking pliers must have a chain or rope attached so the firefighter can hold the pliers clear of the saw.
2. A rotary saw with a metal cutting blade is used to cut the lock shackles, **Figure 17-50**.

 WINDOWS

Windows are found in many types, architectural styles, sizes, and construction as shown in **Figure 17-51**. These are installed in buildings to provide light and natural ventilation. Usually firefighters will force entry into a structure using door openings; however, windows may be used as an alternate means of entry especially for rapid intervention team operations. To successfully conduct forcible entry operations through windows, firefighters must know about the four construction features of windows: glazing, sash, frame, and security devices.

Forcible Entry of Windows

There are many types of windows and each individual one requires a specialized technique to force entry through it. There are two general reasons to force a window: to gain entry and for horizontal ventilation.

> **Streetsmart Tip** When forcing windows, an opening must be created to allow for safe entry and exit with full protective equipment in place. It is important to remove all obstructions such as air-conditioning units, security bars, or child protective gates. The opening created may be needed to make a rapid exit if conditions deteriorate.

To make a large enough opening for entry, it is often necessary to break the glazing and the sash to create the largest possible opening. A common adage heard from experienced firefighters is

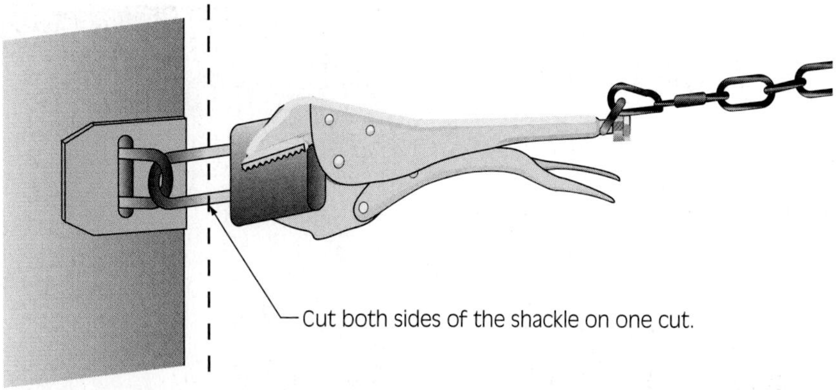

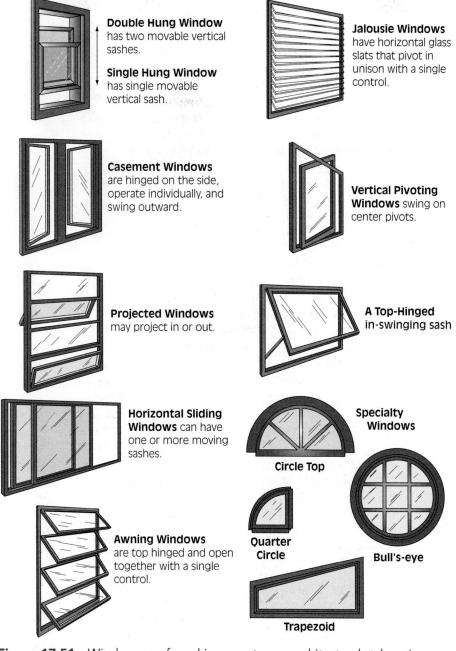

Cut both sides of the shackle on one cut.

Figure 17-50 The locking pliers must have a chain or rope attached so that the firefighter holding the pliers is clear of the saw.

Figure 17-51 Windows are found in many types, architectural styles, sizes, and construction.

"make the window a door." At structure fires, the speed of operations is more important than damage done to windows. With this goal in mind, the quickest method is to break the glass and the sash to provide a quick opening for entry, rescue, and ventilation. At the same time, firefighters must remember not to break glass unnecessarily and create a safety hazard.

Glazing

The glass or other clear material portion of the window that allows light to enter is the **glazing**. The most common glazing material is glass. There are different types of glass, of which these are just a few:

- Regular or plate glass
- Tempered glass
- Laminated (safety) glass
- Wire glass

Regular/Plate Glass

Regular glass is relatively weak and easy to break, and when struck with a tool, it breaks into very sharp, knife-like shards. Plate glass is used in larger windows and is generally thicker than regular glass, with thicknesses of ¼ to ¾ inch being the most common. This type of glass will break into large, heavy, and sharp pieces, which can be very dangerous. A long-handled tool such as a hook or pike pole works best, **Figure 17-52**. With both types of this glass, the firefighter should stand to the side and strike the window. All shards of glass that remain must be cleaned out of the window opening.

Figure 17-52 Use a long-handled tool such as a hook or pike pole to break glass out of a window.

Caution Firefighters should always use a tool such as a hook (pike pole) or other long-handled tool to clean all glass out of a window. Care must be taken that glass shards do not slide down the tool's handle.

Streetsmart Tip A unique property of tempered glass is that when it is broken, the whole pane will fracture and fragment into small and relatively dull pieces. Certain panes, even those that have been fractured, may remain in place. These will require the use of a tool to clear out the remaining glass.

Tempered Glass

This material is glass that has been heat treated to give it additional strength. The technique to use when breaking tempered glass is to strike the glass with a pointed tool, such as the pike of the Halligan tool or a pick ax. Best results are obtained when the glass is struck near a corner where the glass is more rigid than at the center of the pane.

Laminated Glass

Laminated glass is also known as *safety glass*. It is commonly composed of two or more sheets of glass with a plastic sheet between them. The purpose of the plastic sheet is to hold the glass together if broken, thus reducing the hazard of flying glass. Laminated glass is most commonly found in automobile windshields; however, it is also used for glazing in windows and doors.

Wire Glass

Wire glass has a wire mesh embedded between two or more layers of glass, **Figure 17-53**. The primary purpose is to give the glass increased fire resistance. When exposed to high heat, the glass will break. The wire will then hold the glass together and allow the glazing to remain intact to stop the horizontal spread of the products of combustion.

To break the wire glass, the firefighter strikes with a tool, an ax or Halligan, and then cleans the glass out of the frame. This is best accomplished by carefully chopping through the wire with the blade of an ax. The firefighter must use caution because the entire piece of wire glass will fall as one complete unit when cut free.

Polycarbonate Glazing

These are plastic products produced to provide additional strength as compared to regular glass. One common type is Lexan. Lexan is approximately 250 times stronger than regular glass and is available in thicknesses up to 4 inches. Polycarbonate glazing can be removed by cutting it with a carbide-tipped circular saw. It can also be removed by striking it near a corner and driving it out. A heavy striking tool is best, and progressively driving it in may work it free from the sash.

Types of Windows

Double Hung/Check Rail Windows

These windows have upper and lower sashes that both slide vertically. The sash may be composed of wood or metal and will have a locking device in place to lock the sashes to one another or to the frame. To force open this type of window, the firefighter should use a tool to force the bottom sash up and break the locking mechanism. Wood sash windows with a single lock are easier to force than those composed of metal sashes/frames or other heavily secured window frames. When this type of window is encountered, breaking the glass and sash may be the most prudent option.

Energy Efficient Windows

Most new windows and also replacement windows are equipped with double and triple layers of glass. These windows are very well sealed and have a tendency to keep products of combustion trapped inside the structure. From the exterior point of view, they will give little indication of the severe conditions inside. Extra care must be taken when using these windows for entry. Also, energy efficient windows may be more difficult to break.

Casement Windows

Casement windows are hinged on the side of the sash and generally open outward. They commonly have a metal or wood sash and can be secured by one or more latches and a crank for opening and closing. The way to open this type of window is to break the glass with a tool, reach inside, and operate the crank. The screen may have to be cut out of the way to reach inside.

> **Streetsmart Tip** The firefighter needs to size up the window and decide if it is wide enough to enter safely and exit. Often the window sash and center mullion must be removed to provide sufficient room to enter.

Awning Windows

Awning windows are hinged so that they may swing out and may have a crank-operated mechanism to move the sash. They may have a wood or metal sash. The procedure used to open this type of window is similar to that of a casement window. The window must be broken to operate the mechanism. *Many awning windows are too small to enter and the sash will have to be removed or broken out to allow for entry.*

Jalousie Windows

Similar to the awning window, the jalousie window has small sections of tempered glass that are operated with a crank mechanism and overlap when closed.

Figure 17-53 Wire glass has a wire mesh embedded between two or more layers of glass.

Projected Windows

This type of window will pivot at the top, bottom, or center and is most commonly found in commercial buildings. Pivoting windows rotate on the top or at the sides.

Fixed Windows

Fixed windows are nonoperable windows used primarily for aesthetics and the introduction of light into a structure. They can be found in inaccessible areas, such as the top of a wall near the roof of a commercial building.

> **Caution** Firefighters must be in full protective equipment before breaking *any* windows. In addition to the hazard of sharp glass and other materials, the firefighter may have to contend with the smoke, heat, and fire that can vent from a broken window.

Bars and Gates

Windows and other openings that require security measures are often fitted with gates or bars, presenting a unique forcible entry situation for firefighters. These bars and gates must be removed or forced out of the way to allow for entry and exit into the structure. The following procedures may be necessary to accomplish this task:

- *Force the locking devices:* Gates that have exposed locks can be opened by forcing or cutting the lock.
- *Attack the fastenings:* Gates and bars can be removed by breaking or cutting through the bolts or attacking the point where the bars or gates are set.
- *Cutting the gates/bars:* The firefighter can use the rotary saw with metal cutting disc (aluminum oxide) or the sawsall to cut the gate or bars.

BREACHING WALLS AND FLOORS

Emergency situations often dictate that the walls of a structure must be opened to allow entry or to remove trapped firefighters or victims. This is especially true in the event of a collapse or blocked exit. There are two main considerations when breaching walls:

- *The type of construction of the building.* Wood-framed buildings with lath and plaster or drywall are usually easy to breach. Solid brick and reinforced concrete buildings are more formidable and will be more difficult to get through and hence require more effort.
- *Tools available.* During the initial stages of an operation, regular hand tools or possibly a power saw may be the only tools available. These tools may be adequate for numerous operations, but solid masonry or reinforced concrete construction will require specialized tools and equipment.

Techniques for Breaching Walls

Breaching Wood-Framed Walls

1. Size up the wall, trying to avoid the area around doors and corners due to narrow stud spaces. *It may be beneficial to create an inspection hole using an ax handle or Halligan bar,* **Figure 17-54**, *to verify the size-up, and check for obstructions, barriers, and fire conditions. The outcome of the inspection will determine whether to move to another location.*
2. Remove the wall covering from your side first to prevent the extension of heat and smoke into the area.
3. If plumbing pipes are encountered, try to bend or break the pipes (plastic and cast iron will shatter, copper will bend).
4. When a large enough hole is made, push in the wall cover over on the other side to complete the hole. If necessary, a stud can be removed by attacking the connection at the sill or plate. If the wall is a bearing wall, do not remove more than one stud. If a larger opening is required, shoring or some other supporting techniques will be necessary for long-term operations.

Breaching Masonry Walls—Block or Brick

To breach these walls without utilizing power tools, the only option may be the 12-pound maul, the Halligan tool, the flathead ax, or a battering ram.

1. Start by removing a single unit of block or brick. Work at the mortar joints because this is usually the weak point.
2. Once the joint is weakened, use the largest striking tool available and break the masonry unit.
3. Proceed by knocking out the surrounding units or release them at the mortar joint.

masonry blade, and the diamond-tipped chain saw. An oxyacetylene torch may be needed to cut through the reinforcing rods.

> **Caution** Cutting pre- or posttensioned concrete may cause the structural member to lose its strength and collapse. Wall or floor breaching operations in buildings with this type of structural members should be managed by specialized rescue teams with structural engineering assistance.

Breaching Metal Walls

Before any cutting operations are started, the location of heavy structural members, such as columns, should be determined. The tool of choice for this operation will be the rotary saw with a metal cutting disc. The quickest cut is a triangle cut large enough to allow for safe entry, similar to the procedure for overhead doors shown earlier in **Figure 17-30**. Also, depending on the type of metal siding, it may be cut by making two vertical cuts to loosen the wall material. The metal siding can then be pushed in or pulled out depending on the situation. If necessary, the first two cuts can be joined to remove the entire piece. When light-gauge metal is encountered, the wall can be opened up with hand tools (i.e., the blade of the ax).

Techniques for Breaching Floors

Cutting Wood Floors with a Power Saw

The rotary saw with a 12-inch-diameter carbide-tipped blade will cut a maximum depth of 4 inches, **Figure 17-55**. This should be sufficient to cut through most floors in one cut. The hole must be of sufficient size and proper shape. A rectangle,

Figure 17-54 The firefighter uses an ax handle or Halligan tool to poke through a wall to determine if there are any obstructions on the other side.

> **Streetsmart Tip** When making a large opening, a triangular or diamond-shaped hole will help maintain the structural integrity of the wall.

When opening masonry walls with power tools such as a power saw with a masonry disc, cut a triangle by making two angled cuts followed by a cut parallel to the mortar line. It may be quicker to use the maul on the mortar line to break the bricks free.

Breaching Reinforced Concrete Walls and Floors

Solid concrete reinforced walls with steel reinforcing rods are extremely difficult to open up even with the proper tools. The most common tools utilized are the jackhammer, the rotary power saw with

Figure 17-55 The rotary saw with a 12-inch-diameter carbide-tipped blade will cut a maximum depth of 4 inches.

Figure 17-56 A rectangle, square, or triangle can be cut, followed by the removal of the finished flooring and the subflooring.

square, or triangle can be cut followed by the removal of the finished flooring and the subflooring, **Figure 17-56**. Carpeting and ceramic tiling should be removed before using the power saw.

Cutting Wood Floors with an Ax

When breaching a wood floor using an ax, locate the floor joists and make cuts close to them. Cut along the (parallel) joist and cut the finish floor on a bias (angle). Pull up the finish flooring to expose the subflooring. Cut the subfloor and pull up to expose the area below. Make all cuts on the subfloor first, then pull it up to confine the heat and smoke. Push down any ceiling or other obstructions.

Note these important things to consider when breaching floors:

- Always take precautions to avoid cutting wires, pipes, and conduit when performing breaching operations.
- Maintain the structural stability of the building when opening walls and floors.
- Beware of sparks produced by metal cutting tools.
- Always operate with proper protective equipment including eye, ear, and respiratory protection.

TOOL ASSIGNMENTS

The necessary tools to accomplish the tactics of structural firefighting must be carried in with first on-scene and later arriving units. The timely arrival of primary tools is as important as the placement and operation of the first hoseline and cannot be left to chance.

Tool assignments are based on the occupancy and construction of the building (i.e., multiple dwelling, private dwelling, commercial, etc.), position or task assigned, and department standard operating procedures or policies.

Lessons Learned

Forcible entry is a key tactic in structural firefighting and emergency operations, and firefighters must understand the tools, equipment, and methods used for forcible entry. As with all other fireground tactics, teamwork is an essential element. Failure to conduct effective forcible entry quickly may result in delayed search and rescue operations and unnecessary fire spread. In addition, with the renewed emphasis on rapid intervention teams, the art and skill of forcible entry techniques is a basic element for these operations.

Firefighters must continually size up buildings for firefighting and rescue operations, including how to force entry into a building.

KEY TERMS

Bolt Throw The distance the bolt of a lock travels into the jamb or strike plate. Usually ½ to 1½ inches.

Cutting Tools The group of tools used to cut through or around materials.

Flush or Slab Doors Doors that are flat or have a smooth surface and may be of either hollow-core or solid-core construction.

Forcible Entry The fire scene task of gaining entry to a building or secured area by disabling, breaking, or going around locking and security devices.

Glazing The glass or other clear material portion of the window that allows light to enter.

Halligan Tool From the prying group, a 30-inch forged steel tool with three primary parts: the adz end, the pike end, and the fork end.

Hollow-Core Door Any door that is not solid, usually with some type of filler material between face panels.

Hook A tool with a 32-inch to 12-foot handle with a pike and hook on one end. Used for pulling ceilings or separating other materials. Also known as a *pike pole*.

Irons The combination of a Halligan tool and flat-head ax or maul.

Jamb The mounting frame for a door.

Laminated Glass Glass composed of two or more sheets of glass with a plastic sheet between them. The purpose of the plastic sheet is to hold the glass together if broken, thus reducing the hazard of flying glass.

Ledge Door Door built with solid material, usually individual boards, common in barns and warehouses.

Locking Devices A mechanical device or mechanism used to secure a door or window.

Mounting Hardware Hinges, tracks, or other means of attaching a door to the frame or jamb.

Panel Doors Doors with a solid stile and rails with panels made of wood or glass or other materials.

Pike Pole See **hook**.

Prying Tools The group of tools used to separate objects by means of a mechanical advantage.

Pulling Tools The group of tools used to pull away materials.

Rabbeted A door stop that is cut (rabbeted) into the door frame. On metal door frames the stop is an integral part of the frame.

Slab Door See **flush** or **slab door.**

Solid-Core Doors Doors made of solid material such as wood or having a core of solid material between face panels.

Strike Plate The metal piece attached to a door jamb into which the lock bolt slides. Also called a *strike* or *striker*.

Striking Tools The group of tools designed to deliver impact forces to break locks or drive another tool.

Tempered Glass Plate glass that has been heat treated to increase its strength.

Through-the-Lock Method A method of forcible entry in which the lock cylinder is removed by unscrewing or pulling and the internal lock mechanism is operated to open a door. Also, the family of tools used to perform this operation.

Wire Glass Glass with a wire mesh embedded between two or more layers to give increased fire resistance.

REVIEW QUESTIONS

1. Choose an engine or truck company in a fire department and identify five forcible entry tools and describe their use.
2. List the inspection and maintenance procedures for five forcible entry tools.
3. List three different types of doors and describe a method of forcible entry for each.
4. List five types of locks and describe their operation.
5. List the steps for the three types of conventional forcible entry.
6. Demonstrate conventional forcible entry on a variety of doors.
7. Describe or demonstrate the through-the-lock forcible entry method.
8. List four construction features of windows and methods of gaining entry.
9. Describe three considerations when breaching walls.

CHAPTER 18

VENTILATION

Frank J. Miale, Retired Battalion Chief, New York City Fire Department and Lake Carmel Volunteer Fire Department

 OUTLINE

- Objectives
- Introduction
- Principles, Advantages, and Effects of Ventilation
- Heat, Smoke, and Toxic Gases
- Considerations for Proper Ventilation
- Fire and Its By-Products
- Flashover
- Backdraft (Smoke Explosion)
- Rollover
- What Needs to Be Vented?
- Air Movement
- Types of Ventilation
- Mechanics of Ventilation
- Ventilation Techniques
- Roof Ventilation
- Safety Considerations
- Obstacles to Ventilation
- Factors Affecting Ventilation
- Lessons Learned
- Key Terms
- Review Questions
- Additional Resources

STREET STORY

On July 4, 1998, we responded as the second truck to a fire on East 4th Street in New York City. As we arrived fire was blowing out five windows on the third floor of a four-story apartment building. The chief ordered me and my members to get above the fire and search the top-floor apartments for trapped occupants. As we arrived on the third floor, the first-due truck was forcing the door to the fire apartment. Thick black smoke was pouring out of the apartment. I told the first-due engine and truck that we were going to the floor above.

As we got to the half landing between the third and fourth floors, I ordered a member of my forcible entry team to totally take out the window on this half landing, knowing that smoke from the fire apartment would escape from this horizontal opening. This would buy us time to search the top floor before the roof bulkhead was opened.

As we were searching the apartments above I heard the familiar "roof bulkheads opened" transmitted from the roof firefighter over my handi talki. I knew that mushrooming would be prevented by this act and that we had a couple of minutes to search. Then I heard a saw operating on the roof and knew additional vertical ventilation was being performed. At this time, however, fire was autoexposing from the third-floor apartments into the fourth-floor apartments we were searching, forcing us to retreat to the hallway, but not before we were able to complete a thorough primary search. Although the fire extended into six apartments and the cockloft, due to the proper immediate horizontal and vertical ventilation, we were given enough time to search the top floor and see that all occupants were removed to safety without injury.

A few months later we responded to a fire in a two-story private dwelling with reports of persons trapped. As we arrived I noticed no fire, but there were black, discolored windows on the second floor. As we entered the fire apartment to search we found there was a tremendous heat condition, but no visible fire. We recognized that this fire was in the smoldering stage. My outside ventilation firefighter asked if he should take out the windows from the exterior. I immediately radioed to all members not to take out any glass until vertical ventilation was performed and the area was cooled down below the gases' explosive range.

Two different fires, two different ventilation strategies. In the first case, immediate horizontal ventilation provided us time to search apartments above the fire for occupants, while in the second case premature horizontal ventilation could possibly have caused a backdraft, seriously injuring us or worse. Proper ventilation is of the utmost importance. Learning the right ventilation operations can truly make a difference. Learn it well.

—Street Story by Mike Gala, Lieutenant, Ladder 148, FDNY

OBJECTIVES

After completing this chapter, the reader should be able to:

- Understand ventilation as a fire service tool.
- Know the principles, advantages, and effects of ventilation.
- Know the origins and effects of heat, smoke, and toxic gases.
- Identify the considerations for proper ventilation.
- Know the effects of fire and its by-products.
- Differentiate between flashover, backdraft (smoke explosion), and rollover.
- Recognize the effects of air movement in ventilation.
- Identify the types of ventilation.
- Identify the mechanics of ventilation.
- Describe ventilation techniques.
- Describe the need for roof ventilation.
- Identify safety considerations when venting operations are in progress.
- Describe the obstacles to ventilation.
- Identify the factors affecting ventilation.

INTRODUCTION

Ventilation can be defined as the planned, methodical, and systematic removal of pressure, heat, smoke, gases, and, in some cases, even flame from an enclosed area through predetermined paths. Ventilation is an essential part of the tactical and strategic objective of modern fire extinguishment. Incorrectly, in some cases ventilation is not employed until all other firefighting attacks have failed. Ventilation is a very complex subject area with many facets. Additional information can be found in books that have been written solely on the subject of ventilation.

PRINCIPLES, ADVANTAGES, AND EFFECTS OF VENTILATION

Ventilation is the relief of the products of combustion from an enclosed area. Combustion of organic material produces heat, smoke, pressure, and fire gases that quickly fill up an enclosed structure. That structure could be a one-room shed, a ten-room private dwelling, or a twenty-story building. In every case, the need to relieve the structure of these products of combustion is a very essential part of the fire suppression effort. The practice provides benefits that go beyond merely suppression to extinguish the fire.

First, by relieving the structure of heat through channeling it into the atmosphere, the fire is deprived of the ability to heat up other parts of the structure. Fire burns when the gases of a combustible substance are liberated. This liberation occurs when heat from an existing fire is applied to the unburned material. The unburned material heats up, liberates gases, and then ignites, permitting the fire to spread. Using ventilation, the heat is exhausted and dissipated into the atmosphere where its ability to spread fire through the structure is reduced.

Secondly, ventilation channels smoke out of the structure. Smoke is a combination of material, mostly unburned hydrocarbons that have a tarry consistency. Because most of these substances are microscopic in size, they are very light and can stay suspended in air. The heat emitted by the fire carries the smoke to all parts of the structure. Because smoke is made up of only partially burned solid bits of microscopic material, it obscures vision, and the more dense the smoke is, the more it will obscure. Light smoke can be like a fog in which shapes can be seen from several feet away. A heavy smoke condition can obscure light so completely that even a powerful light is rendered totally ineffective. Not only does lack of vision seriously hamper firefighting operations, but it also prevents victims from escaping.

Additionally, the unburned tarry hydrocarbons contain substances that irritate the eyes. In the natural function of the eye's protective action, anytime a foreign material is introduced, the tear ducts attempt to flush the eye with body fluids (tears) to rid it of the irritation. These tears blur vision so badly that it is nearly impossible to see.

Smoke contains many other products of combustion that are deadly substances. Many harmful compounds are mixed in with smoke, and a person that is exposed to this material will suffer ill and potentially lethal effects, **Table 18-1**. When people fall victim to smoke or its components, they usually fall into unconsciousness as life slowly ebbs away. The removal of smoke, heat, and toxic gases will add survival time to a potential victim who is unconscious, increasing the chance of successful rescue.

Gases Produced by Fire

Carbon monoxide	Takes the place of oxygen in the blood.
Carbon dioxide	Overstimulates the rate of breathing.
Hydrogen sulfide	Causes respiratory paralysis.
Sulfur dioxide	Extremely irritating to eyes and respiratory tract.
Ammonia	Extremely irritating to eyes, nose, throat, and lungs.
Hydrogen cyanide	Highly toxic and used commercially as a vermin fumigant.
Hydrogen chloride	Can become hydrochloric acid in mucous membranes.
Nitrogen dioxide	Causes respiratory distress in a delayed reaction.
Acrolein, phosgene	Gases found in certain kinds of fires; are lethal in small doses.

TABLE 18-1

HEAT, SMOKE, AND TOXIC GASES

When fire burns, air heats, expands, becomes lighter, and rises. It also begins to exert pressure on anything that surrounds it. The rising heated air becomes a means by which fire is spread or communicated to surrounding materials by convection, and, to some extent, radiation from the upper levels (see Chapter 4, Fire Behavior). Fire can also spread into other places within a structure by the pressure that is created. Pressure will take the path of least resistance within a confined area as it tries to become equalized. If that path leads to more combustible material, the heated air will spread its heat to that substance and permit it to ignite, thereby spreading the fire. Fire gases consist of many products of combustion that can contribute to the spread of fire, render a living being unconscious leading to death, or contribute the necessary ingredients for an explosion. Many substances found as the by-products of combustion are deadly.

Because of the desire to conserve energy, structures being built today and those being renovated are outfitted with heavy insulation and tight, weatherproof seams. These features make ventilation even more important because the high heat generated by the inability of the heat to escape is turned back into the compartment. These tight construction practices lead to hotter fires, early failure of structural components, and greater incidences of flashover and backdraft. These phenomena are examined in greater detail later in this chapter.

Last, in today's climate-controlled buildings equipped with windows that are unopenable or even windowless, reliance on ventilation is greater in order to move the products of combustion out of the structure. An overview of mechanical ventilation is given later.

CONSIDERATIONS FOR PROPER VENTILATION

In order for ventilation to assist in the fire extinguishment operation, the firefighter must first understand the behavior of fire gases, which include smoke, in a building. Because heat is lighter than air, it tends to rise. Smoke, when mixed with heat, also rises. As the smoke rises, it collects under any vertical obstruction and mushrooms in all directions when it meets that obstruction. Gradually, the smoke and heat fills the structure starting from the highest point in the structure and "banks down" until the entire structure is filled, **Figure 18-1**. To fully understand the concept, imagine a building. Then, pull it out of the ground and turn it upside down. While it is upside down, begin filling the cellar with water. Any channel that the water finds as it drains out of the inverted cellar into the inverted attic will be the same channels that smoke and heat will use to "drain up" out of that cellar, **Figure 18-2**, and, just as water will leak out of small holes in the windows or siding in the inverted structure, so too will smoke in the structure that is right side up. So even though smoke is coming out of the lower floors of a structure, it is only a portion of the smoke and heat that is collecting in the upper part of the structure.

Figure 18-1 Heat, smoke, and fire will follow the path of least resistance and find their way through any available opening.

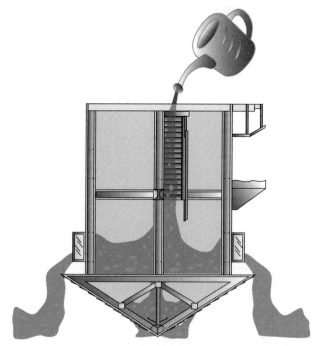

Figure 18-2 Just as water poured into the cellar of an inverted building will fill up any basins and spill over into the next void until it fills, smoke, heat, and fire will behave in a similar fashion in a structure that is right side up.

> **Note** If a large volume of smoke is coming from the windows, it is a good indicator that the smoke and heat are meeting resistance in the attempt to move vertically.

Vertical ventilation is defined as the removal of gases and smoke through vertical channels. This will prevent fire extension by convection from occurring at a remote part of the building. The opening of the structure at the top will permit the fire gases and smoke to exit the building just as punching a hole in the sample inverted building will permit water to drain out of the attic and prevent the building from filling with water.

Another type of ventilation that is often associated with fire suppression is **horizontal ventilation**, which is the channeling of smoke and heat out of the structure through horizontal openings such as windows and doors. There is a distinct difference between the needs and results of vertical and horizontal ventilation. Horizontal ventilation permits the fire's by-products to be pushed out of the structure by the advancement of the fire suppression crews. In areas remote from the fire, it will permit a reduction of smoke, and, to a lesser degree, built-up heat that will aid in a search for victims and any fire extension. Without horizontal ventilation in front of an advancing hose team, the heat, smoke, and now a new element, steam, have nowhere to go. With no outlet for the water to push the smoke, steam, and heat out of, it will be pushed over the hose team at the ceiling level. When water is applied to a fire, the fire is extinguished by the removal of heat from the fire pyramid. When raised to 212°F (100°C) water turns to steam. It takes a tremendous amount of heat to turn water into steam, and, when it does, it expands 1,700 times. That means that when 1 gallon of water is heated to the proper level, it creates 1,700 gallons of steam. If the hose team is about to apply water to a fire in an enclosed room without a ventilation hole opposite their position, the pressure created will push back out through the same opening from which the hose team entered. The fire, heat, smoke, and fire gases will push right over the hose team and superheat the area behind the nozzle team, surrounding them with the heat of the fire.

By opening a channel for the products of combustion to exit as the nozzle team moves in, the heated smoke and steam will take the path of least resistance and vent from the fire room. It is very important for the hose team to move in with a wall of water in front of them and push the fire out the window or another opening, **Figure 18-3**.

Horizontal ventilation requires some forethought before performance. If performed at the wrong time

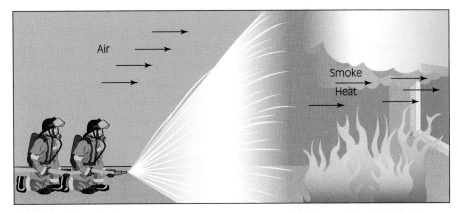

Figure 18-3 Heat, smoke, and fire will always take the path of least resistance. When pressurized by the air created by an advancing hose team, the products of combustion will seek a path that will equalize that pressure. An opening on the opposite side of the advancing nozzle team is essential.

or at the wrong place, it can accelerate fire spread. For example, if the venting is about to be performed in a part of the structure that is two rooms away from the fire room, the air currents that are created could pull the fire into a room that might otherwise not have been involved, **Figure 18-4**, but this will not occur all the time. There are many reasons why this might or might not happen. Some of the factors that will influence the air currents in a structure fire are vertical vent openings, horizontal openings, outside wind direction, the direction the hose attack team is using in relation to the fire, and the room being vented.

When properly performed, the ventilation operation can be as critical as the nozzle team applying water to the fire. It is important to note that in the confusion of a fireground operation, it may not be possible to follow a manual's every line and sentence. The firefighter must learn the conditions that cause events and use every experience to expand that knowledge.

Streetsmart Tip When access to a horizontal ventilation point is impossible, a possible recourse is to use the hose stream to break the windows from inside before mounting the attack.

In rare situations, no ventilation at all is performed at the fire extinguishment operation. Some occupancies are protected by systems that discharge a fire-inhibiting gas such as halon into the room. In this type of occupancy, an identifying placard usually warns that this gas is present. A system such as this might be used where delicate materials are found such as an

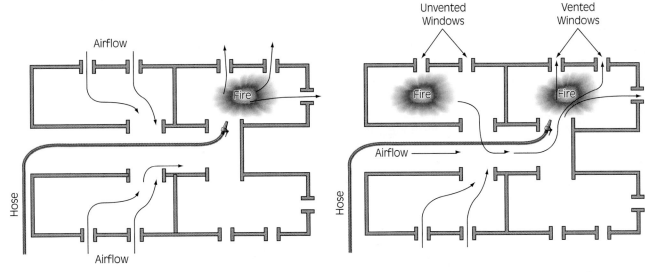

Figure 18-4 Air movement is created by water application. Openings in back of the nozzle team will create airflow from behind in the direction of the hose team. It can be a source of fresh cool air, or it can pull fire to the nozzle from behind. Indiscriminate ventilation can be a liability. Careful assessment and proper timing are important.

archiver, a museum, computer rooms, biological labs, and so on. This type of occupancy might suffer more damage from water than from the fire. With this type of extinguishing agent, the fire is extinguished by a chemical reaction with the flame production (see Chapter 4). In a structure protected by this type of system, ventilation must be avoided until instructed by the incident commander's orders. Venting this room would permit the fire-inhibiting gases to escape and permit reignition.

Many factors must be considered when venting. Some of them are access to the vent site, wind direction, weather conditions, exposures, the material burning, the height of the building, the potential for fire spread, and the escape route. Not every factor will have the same impact at every fire.

Access to the vent site can be tricky. Access to a window to be vented might be impeded by the presence of electrical wires, sloping ground, or sheer height. A window off a porch roof that needs to be vented might require the firefighter to pass windows on that porch roof that are already venting fire. This condition might well prevent an escape route and a second approach might be necessary.

Ventilation duties in the rear of the building might be difficult or impossible because of the presence of locked gates or guard animals. **Guard dogs** are trained for different levels of protection. **Watch dogs** will bark and make a great commotion. Their purpose is to scare an intruder. Guard dogs will bark and assault an intruder. Their function is to warn and injure. Attack dogs are trained not to bark and will attack an intruder with ferocious intent.

A hose stream from a hoseline or even a water extinguisher will send a guard dog running. In all cases, the best approach is to guide the dog into some kind of confined area and leave it there. Occasionally, police might have to be called to deal with the dogs by using tranquilizers or by destroying them.

The presence of electric supply lines to the structure can impede access to a vent site. Unless specifically designed, all fire service ladders, including those made of wood, are electrically conductive, and that presents a great risk when raising or positioning them in the vicinity of electricity. When the window needing ventilation is close to electric power lines, extra care must be employed. Live wires might have had their insulation burned off and energized surrounding metallic substances such as aluminum siding. A ladder resting against that energized siding can result in electrocution. When sizing up the approach, consideration must include the presence of overhead wires and, when necessary, the use of an alternate approach.

Sloping ground will significantly affect the placement of ground ladders. This is especially true if the street or road **frontage** is at a different level than the back of the building. The sides of the building will generally have some kind of slope so that the front and the back meet. Unless the landscaping is terraced with level steps, the ladders will have to be raised from uneven ground. Depending on the severity of the slope, ladders might be unavailable for use in this situation and ventilation of certain windows might be delayed.

The height of the building will also have an impact on access to a window. If the window is out of the reach of ground ladders and not reachable by aerial or tower ladders, venting might have to be performed by using adjoining windows or from windows above or below.

Wind is a factor that can dramatically alter the ventilation. The removal of a window that faces the wind could cause a complete reversal of the air current flow. If the wind is strong enough, the attack hose team can be faced with a blowtorch or wind tunnel effect that will push the fire at them. There are just too many variables to clearly establish a "go/no go" set of rules about whether to leave the glass intact or to remove it under these circumstances. Common sense and experience are factors that will help to solve this judgment call.

Humid, rainy, or foggy weather tends to cool smoke-laden, heated air and prevents the smoke from **lifting** out of the building. If the weather is humid or rainy, the smoke from the chimney does not rise in a column but tends to spread out horizontally. On a clear day, the smoke rises in a crisp column into the upper atmosphere, where it dissipates. Other atmospheric conditions such as pressure or temperature inversions alter natural air movement, affecting ventilation and hindering operations.

Chapter 4 on fire behavior explained how combustion is a process that creates its own life. If the heat produced is not large enough or dissipated quickly, the fire self-extinguishes. Through the use of ventilation, firefighters can remove some of the fire's ability to extend. If the products of combustion, especially heat and smoke, are confined in a structure, two of the three essential ingredients (heat and fuel) are present and ready to be consumed. By venting, those ingredients are removed harmlessly into the atmosphere and permitted to burn where they will do no harm to advancing firefighters.

FIRE AND ITS BY-PRODUCTS

During the pure combustion process, energy is released from an exothermic reaction as heat and light (flame), **Figure 18-5A**. Heat without smoke

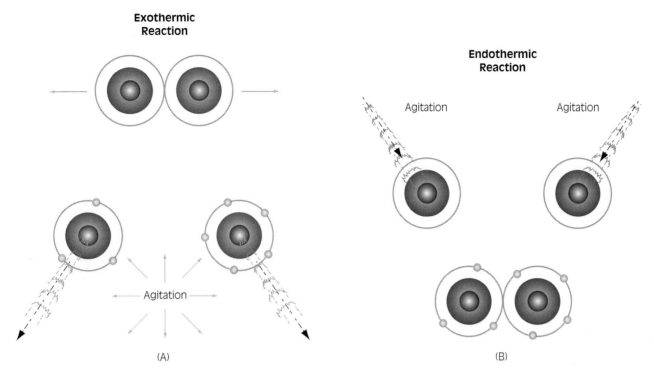

Figure 18-5 (A) In an exothermic reaction, the bond broken between two atoms or molecules creates vibration or agitation, and heat is created by that action. There are times when bonds are broken, heat is created, and the atoms or molecules find stability in another combination minus that bond. In that case, a chemical reaction releases heat, but it stabilizes as another substance. (B) Heat is present when atoms or molecules are agitated or vibrating. Very simply stated, if the loose ends of the atoms or molecules attach to another and become balanced, the agitation ceases and the heat that was being generated is absorbed by the newly created bond.

can be observed as wavy lines emanating from the fire. Other gases that are produced by the combustion process are not visible. The reuniting of molecules to form new substances is part of the endothermic reaction, **Figure 18-5B**. This process actually takes heat from its surroundings and uses it to bond the loose ends of other molecules.

Substances created by fire can be very caustic to a human. The body reacts to the invasion of foreign substances by attempting to rid itself of them. Unburned particulate matter that settles into the mucous membranes of the eye and nasal passages results in a flow of liquid from the tear ducts and nasal passages to naturally flush out the foreign material. When inhaled into the lungs, the body attempts to isolate the foreign material by coating it with phlegm. Then, the body attempts to take the collected phlegm and expel the foreign material by coughing.

When properly performed, ventilation will remove some of the harmful agents associated with the by-products of combustion. Additionally, the presence of certain chemicals and chemical compounds serves to accelerate the human respiration rate. The human body uses two mechanisms to regulate the rate of breathing. One is the level of oxygen. If the oxygen level diminishes, the rate of breathing increases. The oxygen-reading capability of the brain is actually the alternate monitor that regulates the breathing rate. The primary monitor that regulates breathing is the level of carbon dioxide in the blood. Under normal conditions, the body takes in oxygen, uses it, and produces carbon dioxide as a by-product of energy usage by the various muscles of the body. When the brain sensors read an increasing level of carbon dioxide, the body speeds up the breathing rate to increase the oxygen/carbon dioxide exchange in the lungs.

Because the by-product of combustion is carbon dioxide, the body inhales it along with the other poisonous gases produced. As the brain senses the increasing levels of carbon dioxide, it increases the breathing rate to expel the building level of CO_2. In the case of carbon monoxide, another product of combustion, which the blood cells absorb 210 times faster than oxygen, the result is a faster and faster rate of poison absorption.

> **Firefighter Fact** The faster the structure is opened and ventilated, the quicker harmful gases are replaced with fresh air, and, consequently, the better the working envrinonment for the firefighters.

Since all firefighters should be wearing self-contained breathing apparatus, one might ask if ventilation is important since a fresh air supply is available to the wearer. The answer is an emphatic *yes!* Ventilation increases the survival time of a trapped victim who might be overcome or a firefighter who has become disoriented and run out of air.

Ventilation has many other benefits beyond limiting the inhalation of contaminated air. Removal of smoke improves visibility. Victims can be discovered more quickly, and danger from a hostile environment is reduced. The chance of injury will be reduced. Dangers such as holes in floors are easier to spot, and avenues of fire extension become more obvious. With better visibility, the use of tools becomes safer.

If heat is not permitted to linger on a material long enough, it will be unable to liberate the material's combustible gases. By venting the enclosure, the heat level is kept from becoming capable of producing these phenomena:

Flashover: Everything in a confined area ignites at almost the same time.
Backdraft (smoke explosion): Unburned smoke is heated in the absence of oxygen, and, when oxygen is introduced, produces an explosive force.
Rollover: Fire begins to ignite smoke overhead in "fingers of fire" that reach out and begin to consume fuel in the gaseous state.

The mechanics associated with each of these phenomena and prevention techniques are discussed as each relates to ventilation.

FLASHOVER

Light, smoke, and heat are liberated as part of the combustion process. As the trapped heat collects at the ceiling level, its cumulative temperature increases. Across the upper area of the room, heat begins to radiate downward, heating the contents of the room. When the overall temperature reaches the **ignition point** of another substance in the room, a new chain reaction combustion site occurs and additional heat is added beyond the initial source of fire. As each item in the room follows suit, more and more heat is created and more and more items ignite. In a very short time, the entire room and all of its contents are on fire. This happens very rapidly. Slow-motion photography shows that it is not an instantaneous occurrence but a very rapid fire spread. It is important to know the mechanics of a flashover in order to recognize its development. It is even more important to know what course of action to take before, during, and after a flashover.

> **Firefighter Fact** At best, the survival time of a firefighter in bunker gear and breathing apparatus, fully encapsulated with gloves, hood, and helmet flaps down, is estimated to be between ten and fifteen seconds.

The best survival skill here is recognition and avoidance.

BACKDRAFT (SMOKE EXPLOSION)

Backdraft, also sometimes called a smoke explosion, is the rapid ignition of smoke. Sometimes confused with a flashover, it is very different.

When fire burns organic materials in the presence of oxygen, carbonaceous materials are transformed into carbon, carbon dioxide, and water. When there is a shortage of oxygen, incomplete combustion occurs, and, instead of forming carbon dioxide (CO_2), a less stable compound called carbon monoxide (CO) is formed, **Figure 18-6**. During this process, two of the three classic substances for a complete burn are present: fuel and heat.

Incomplete combustion occurs as the oxygen level decreases. Visually, this can be observed by the amount of yellow flame present. With adequate oxygen, the flame is bright yellow. As the oxygen level decreases, the yellow becomes bright orange, then dull orange, then red. As the flame approaches red, the level of oxygen is getting so low that if continued the fire would self-extinguish for lack of oxygen. As discussed in Chapter 4 on fire behavior, carbon monoxide has a very wide range of flammability. When oxygen is introduced into the mix, combustion will again occur to complete the process by converting the CO into CO_2. Under the proper conditions, this can occur with a violent reaction.

Some form of backdraft occurs at almost every fire that remains unvented prior to the arrival of firefighters. Its violence and damage will be directly commensurate with the distance from where the oxygen is introduced and the location of the flames, **Figure 18-7**.

As the fire consumes greater amounts of the oxygen in the room, the production of carbon monoxide increases. With the heat, pressure builds in the confined space. Smoke can be seen puffing from openings, and black smoke condenses on the surface of any

VENTILATION 557

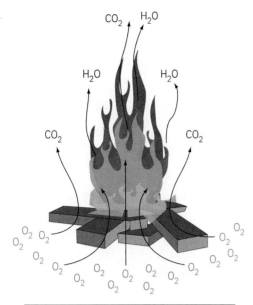

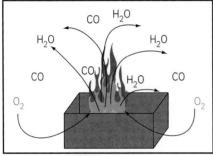

Figure 18-6 When combustion occurs, oxygen is combined with an organic material, and, through a series of bond breaking and recombination processes, carbon dioxide and water are driven off. When oxygen is lacking, instead of two oxygen atoms combining with a carbon atom to form carbon dioxide (*dioxide* meaning "two oxygens"), the carbon atom combines with only one oxygen atom, forming carbon monoxide (*monoxide* meaning "one oxygen"). In the absence of adequate oxygen, carbon monoxide is generated instead of carbon dioxide.

cooler surface, such as glass or a wall. Initially the entire room is filled with heat, carbon monoxide, and other fire gases. When an opening occurs, a billow of smoke escapes as the pressure from inside attempts to equalize. Then, what is described as "sucking" occurs where, once a large majority of the pressure in the space is released, cool outside air is introduced into the room. The cooler air causes the volume of air to contract and pull air in to attempt to equalize again. Being cool, it stays low to the floor and at some **thermal level** a mix of highly concentrated CO is mixed with the fresh, 21 percent oxygen-laden air. As the cool air snakes its way to the source of the fire to replace air that is being heated and driven upward, what could be described as a tunnel of fresh air finds its way to the fire, **Figure 18-8**. An observer might even see the smoke lift off the floor and just before ignition, actually see the fire burning across the expanse of the floor below the smoke level. In the meantime, in terms of the oxygen level on the floor, with each few inches the smoke rises, the CO and O_2 mix more. This mixture of gases at some point will permit the concentration of CO to be in the range of 14.5 to 74 percent. Once that concentration reaches the flame, the three components for combustion are present and primed for ignition. The burn begins at the seat of the fire and, like the fuse on a firecracker, burns back to the opening in the space along the perimeter of the tunnel of oxygen-laden air. This occurs so rapidly that it seems like an explosion. The burn causes the heat, which in turn causes rapid expansion of the surrounding air with such rapidity and force that it can blow out windows, knock down walls, and hurl firefighters out openings and across the street. The degree of force is dependent on several factors.

Signs of a Potential Backdraft

- Smoke-stained windows
- Puffing of smoke at seams and cracks of windows and doors
- Smoke pushing out under pressure
- No visible flame or very dull red flame in the depth of the smoke
- Heavy black smoke
- Tightly sealed building
- Large, open area structure (supermarket, bowling alley, department store)
- Can also be a large, open void (cockloft, between-space of hanging ceiling)
- Extreme heat

Consider these two backdraft conditions and the difference in the results:

- *Short distance to opening.* The atmosphere is laden with carbon monoxide. As the fresh air works in toward the thermal plume, it brings fresh oxygen. The outer perimeter mixes with the CO and achieves the proper flammability range. When the seat of the fire is reached, the tunnel perimeter ignites and follows the path to the opening. Since it is limited and relatively short, the results are unspectacular and appear merely as the fire flaring up.
- *Long distance to opening.* The same initial condition exists. However, in this case the amount of the proper concentration from the opening to the seat of the fire is much greater. When the ignition occurs, it travels a greater distance and amasses much greater force. Although technically it is moving in a chain reaction fashion, the almost

Figure 18-7 (A) When the distance between the opening and the source of ignition is short, there is little opportunity for the gases to circulate and form a mixture that is within carbon monoxide's flammable limits. When the mixture does ignite, it appears more like a flame flare-up than an explosion-like ignition. (B) A large distance between the opening and the location of the fire will permit a greater mixing of carbon monoxide and cool oxygen-laden air. The greater distance will permit the carbon monoxide concentration to be in its flammable limits over a greater area before it reaches an ignition source. A larger instantaneous ignition will result with great force. The greater the distance, the greater will be the resultant ignition force.

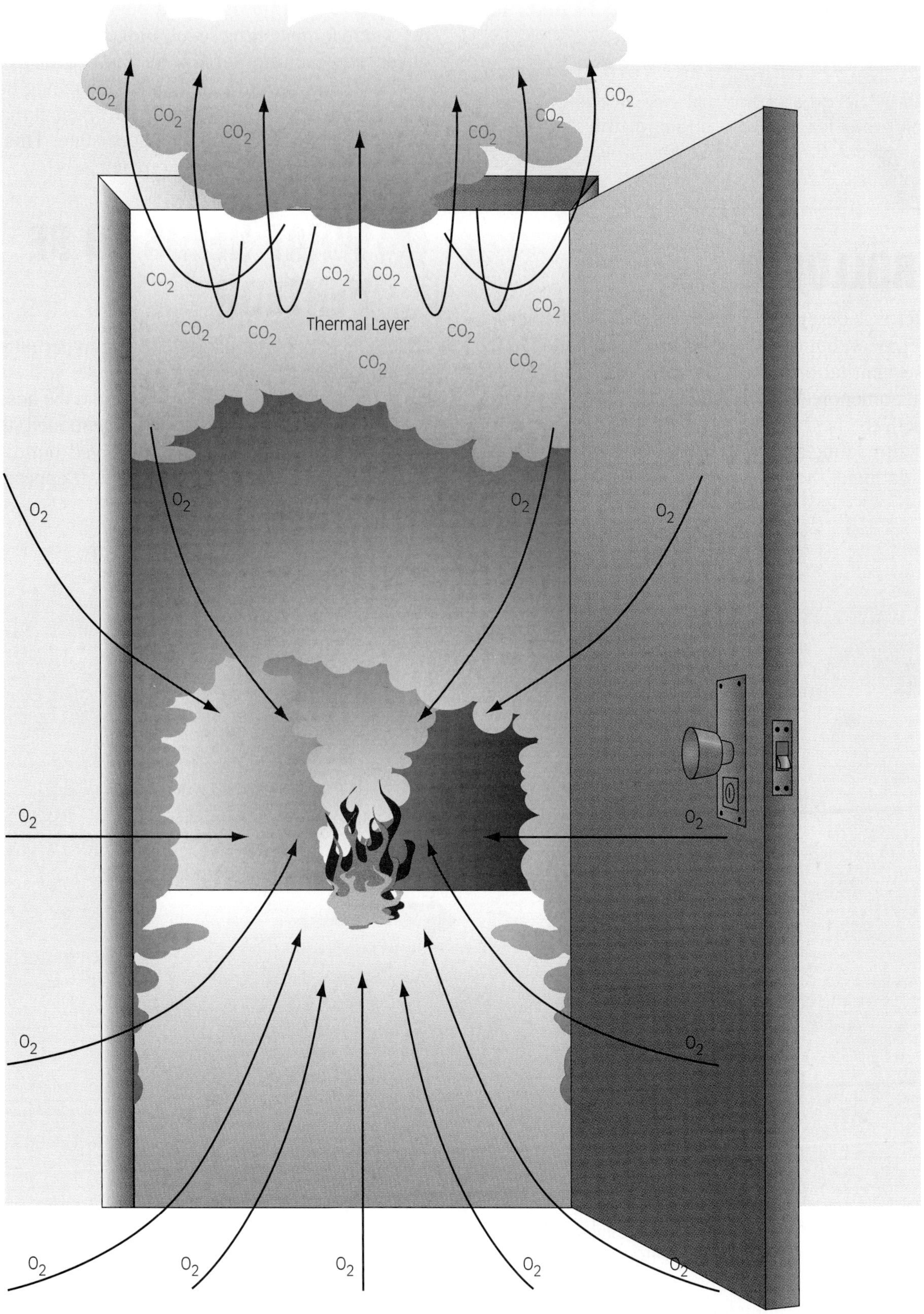

Figure 18-8 Through the open door, you can see the pressurized smoke containing carbon monoxide venting outside the upper level of the door as the oxygen-laden cooler air enters at the lower level of the thermal layer. Looking down a "tunnel" you can see the cool oxygen-laden air mixing with the oxygen-starved carbon monoxide around the perimeter of the tunnel. When it reaches the fire, the mixture will ignite and, in chain reaction fashion, follow the ignitable mixture in the tunnel right out to the opening.

instantaneous chemical reaction is like an explosion.

> **Note** The greater the distance between the smoldering fire and the opening, the greater the explosive mixture involved, and, consequently, the greater the force created.

ROLLOVER

Rollover was described in greater detail in Chapter 4, Fire Behavior, but the subject is briefly revisited here because ventilation is so very important to prevent this phenomenon. When products of combustion are produced, the heat brings them to higher levels, usually accumulating near the ceiling. The heated gases accumulating at the upper levels of the compartments reach their ignition point and begin to spread across the room at the ceiling level. Fingers of flame can be observed reaching across the room followed by a wall of flame behind it. When an advancing hoseline disrupts the upper thermal layer, the heat at that layer is forced down into an unbalanced state. This action can disrupt the rollover phenomenon.

WHAT NEEDS TO BE VENTED?

Unless there is a ventilation opening to permit escape, the expanding heated steam and smoke will roll over the wall of water and drop down behind the hose team, **Figure 18-9**. When ventilation is mentioned, the first image that appears is a fully involved building fire emitting smoke from every opening. It appears to be

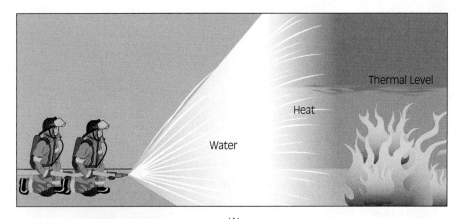

(A)

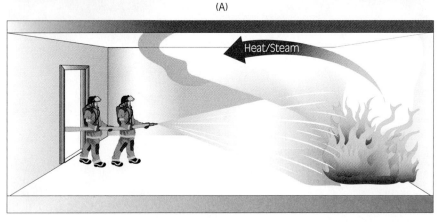

(B)

Figure 18-9 (A) Applying water to the upper levels of a thermal layer will cool and disrupt the rollover effect that is apt to occur with the proper conditions. Ventilation is critical when this is done. (B) As a hose team advances into the fire and sprays water in droplet form, it creates a wall of water and disrupts the high-heat thermal layer and cools the upper levels of the compartment. Water absorbs heat as it turns to steam, expanding 1,700 times as it does. If there is no path for the expanding water/steam conversion, it will take the path of least resistance, in this case over the wall of water and the nozzle team. The water movement will then pull any heat from the back of the nozzle team and roll over on top of it.

VENTILATION

well vented. However, long before the entire building requires venting, the smaller voids and compartments need to exhaust the increasing pressure and intensifying heat condition. If done in a timely manner, the involvement of the entire building might be avoided.

Voids and Compartments

When examining the subject of ventilation, all compartments must be treated with the same understanding. For example, a residential building is merely a very large compartment with many subcompartments (i.e., apartments). Each subcompartment can be subdivided further (i.e., rooms), then again subdivided (i.e., closets), **Figure 18-10A**. Rounding out the picture, the building might also have eaves, peaks, gables, cocklofts, and other voids, **Figure 18-10B**.

Cocklofts

Between the ceiling of the top floor and the bottom of the roof is a space called the **cockloft**, **Figure 18-10C**,

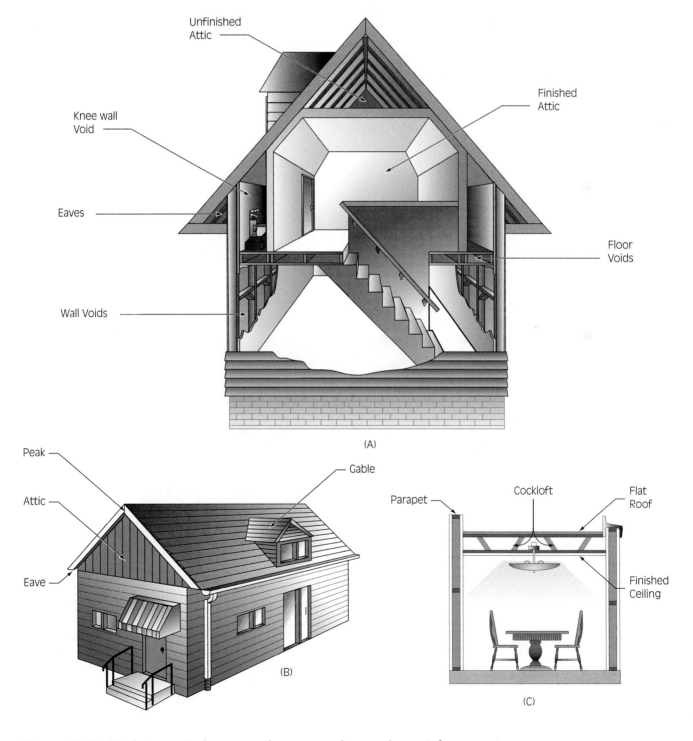

Figure 18-10 Voids in a typical structure that can trap heat and permit fire extension.

which is often the route of major fire extension. Chapter 13 on building construction dealt with this in greater detail. It cannot be overemphasized that this void is a major attack point for a ventilation crew, especially in a top-floor fire or a fire that has extended into that space.

Horizontal and Vertical Voids

Because all heat will follow the path of least resistance, an unobstructed channel could be in the form of a horizontal or vertical void. By following **pipe chases** or electrical wire pathways, heat and fire may extend without being seen. Ventilating these voids by opening them at the highest points can expose any extending fire for extinguishment. In addition, by opening the voids, the trapped heat is diffused into unheated spaces, thus minimizing the chance of heating up a new, uninvolved portion of the building, **Figure 18-11**.

AIR MOVEMENT

Defined and discussed in greater depth in Chapter 4, the attributes of the modes of heat transmission that relate directly to ventilation are revisited here:

Convection. The normal travel path of heat as it follows the convection cycle can be interrupted and challenged by opening a hole in the upper portions of the compartment. Then, instead of traveling up, across, and back down, the heated air will travel up, across, and out of the structure. By providing a path of least resistance in the normal path of heated air, the heat is carried into the atmosphere where it harmlessly dilutes with cooler air.

Conduction. Heat tends to travel along dense material. It can cause remote combustibles hidden from view to ignite. By exposing the heat conductor and ventilating the void where it exists, the trapped heat is provided with another less resistant path, thereby preventing the heat from traveling to uninvolved areas of the structure.

Radiation. In most cases, radiation is an outside function. Under normal conditions, there would be no real need to ventilate. However, in smaller applications, such as an overheated wire, the heat from the wire can radiate to nearby combustible building

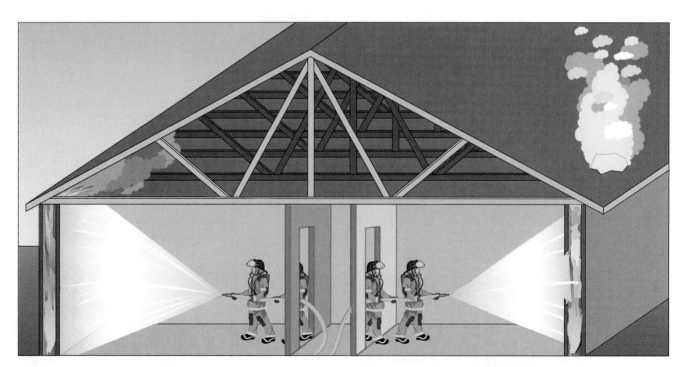

Figure 18-11 Ventilation openings are necessary to permit fire gases and steam to vent into the open air. On the left, with no openings provided, the fire is pushed into uninvolved areas of the building, and water is having no effect on the fire in the wall. On the right, an opening provides a path out through the roof to vent heat and fire gases, thus limiting extension.

components. While a combination of the three modes of heat transfer would be at work here, radiation is contributing. Ventilation would have its desired effect as heat is carried away from the recipient of the radiant heat, thereby carrying away and preventing, or at least delaying, the exposed material from reaching its ignition temperature.

TYPES OF VENTILATION

Ventilation can be performed using one of several methods either singularly or in combination with one another. Natural ventilation merely requires opening doors and windows and letting physics take care of the rest, **Figure 18-12**. Among the mechanical means are the use of smoke fans consisting of ejectors and blowers, and the use of water to create air movement.

Natural

Firefighters still use natural methods for evacuating smoke from a building. Merely opening windows and doors will provide natural ventilation. This choice is the most appropriate for incidents where time is not essential and there is no urgency to the emergency, permitting a slower venting operation. Diffusion of smoke and heat is a slow process, but it will eventually be complete when equilibrium is achieved. The natural movement of air currents is the action in this type of ventilation. This method would be appropriate for a light smoke condition where the incident is under control. It is the method of choice for removal of residual smoke from an incident such as food on the stove, overheated motor, leftover smoke from a fire some distance away in the building, or similar situations. A combination of air movement and diffusion is employed in the unassisted natural ventilation process.

Cutting a hole in the roof or breaking out windows is another type of natural ventilation when greater speed or increased volumes of smoke and heat need to be removed from a structure, **Figure 18-13**. The rising heat will take smoke and fire gases to upper levels following the path of least resistance. A large opening in the roof with an unobstructed path from the lower floor affords an express channel for ventilation using the natural method.

Mechanical

The use of mechanical devices can speed the ventilation process, especially in compartments deep within a structure or under conditions where natural air movement is not satisfactory. Mechanical aids can accelerate the air movement and even reverse the airflow against natural air current movement. Use of fans or blowers in a positive or negative mode and water from a nozzle can provide a large air movement volume.

Heating, Ventilation, and Air Conditioning

Heating, ventilation, and air conditioning (HVAC) systems can be used effectively for ventilation in sealed, climate-controlled buildings. Closing intake dampers that would draw in smoke from the outside robs the fire of fresh air. With the opening and activation of exhaust fans, smoke and heat are removed from the compartment. Uninvolved floors can be vented mechanically, and air movement in the building in general can be controlled.

Smoke Fans

Two types of fans are used for smoke and heat removal. *Ejectors* were among the first fans used by the fire service and were employed primarily to suck out the smoke and heat. Placed in an opening, the fan developed a negative pressure on the structure side of the unit and drew the heated air and smoke into the fan to be ejected, **Figure 18-14**.

The fan is placed in the window with the arrow for airflow leading to the outside and safely secured. The openings that could cause the ejected

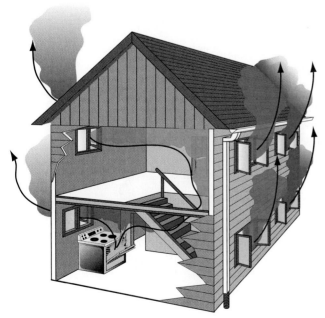

Figure 18-12 Smoke will be carried throughout the building to upper floors by normal air currents mixed in with the heat.

Figure 18-13 The hotter the air, the faster it will circulate throughout the building. Without the roof opening, the smoke will "bank down" and fill the structure. The openings provide a path to the outside for the heat and smoke.

smoke to be pulled back in and recirculated are sealed with tarps or plastic. During this time, establishment of an unobstructed path of airflow from the outside opposite the fan to facilitate fresh air introduction would be beneficial.

The disadvantage of this type of smoke removal is that the firefighting crew positioning the fan has to work exposed to the smoke and heat, **Figure 18-15**. In addition, the smoke must be pulled through the unit, subjecting it to caustic substances and deposits that could eventually cause mechanical difficulties to the motor. Deposits on fan blades can also create inefficiencies. Positioning a fan on upper level floors from

Figure 18-14 A smoke ejector exhaust fan placed in an opening will pull air through the fan as it ejects air out of the structure.

VENTILATION 565

Figure 18-15 Placement of a smoke ejector will subject the firefighters to products of combustion because in order to operate it, the firefighter must stand in the path of the airflow. In addition, it forces the firefighters to work in a hostile environment in which it is difficult to see and breathe.

the exterior is even more hazardous. In these situations, the fan is usually placed on the inside.

Positive Pressure

Positive pressure ventilation (PPV) has been around for over thirty years but only recently has it become an active tool in the arsenal. Air is introduced into the smoke-filled area through the use of PPV fans or blowers. The compartment becomes positively charged with air, and the heat and smoke are displaced through chosen avenues to the exterior, **Figure 18-16**. In addition to being used in the later stages of a fire to clear the smoke to ease overhauling, PPV has been employed with some success during the attack stage of fire suppression. Because of the narrow applicability of the practice during the attack phase, some fire officials are skeptical about introducing the practice to their operations.

> **Streetsmart Tip** Misuse of positive pressure ventilation can severely accelerate fire growth.

However, when used properly, the smoke and heat are totally removed from the access paths of the nozzle team and search team.

A distinct advantage of the PPV blower is ease of setup. The fan merely needs to be positioned at the entrance to the structure with no exposure of the setup crew to dangers from within the structure. Furthermore, some of the blowers are designed for a single-person setup. This is a particularly distinct advantage where personnel are limited. If properly applied, the blower can do the job of several firefighters attempting to open multiple holes with ladders at different locations. While not a cure-all for ventilation problems, PPV can be a material asset to an operation. The placement of the blower will depend on the fan's size and capacity and the size of the opening.

Hydraulic

The last of the mechanical modes of ventilation is to use water. Known as hydraulic ventilation, water is employed to create air movement, **Figure 18-17**. It is a quick method for expulsion of smoke and heat without the need for additional tools. Use of water can cause water damage, but it has some assets, too. If the need to evacuate a smoke-filled room is essential to search for a victim, the time needed to set up an alternative mechanical ventilation tool might be unacceptable. Water can also be used to quickly cool down a heated atmosphere in a structure by sucking and pushing the heated smoke out of the structure.

The nozzle team will quickly feel the benefits as the cooler air from the uninvolved portion of the structure is pulled to them from behind. At the same time, the air movement evacuates the heated upper portion of the compartment, **Figure 18-18**.

This method of ventilation is the most rudimentary. Tests have shown that when properly positioned, the mechanical blowers move a greater amount of cubic feet per minute (cfm) than smoke ejectors, and the use of a nozzle to move air is the most inefficient.

Using a Water Stream to Ventilate. Ventilation by hydraulic means is as simple as directing the hose stream out of the window to remove smoke and heat. Using a fog or spray stream is most effective because the water movement actually pushes and pulls the air surrounding the droplets created by the nozzle. A solid stream nozzle can also be effective by opening the nozzle partially to create a coarse, broken stream.

Whatever type of stream is used, it is placed so that the water is being directed out of the opening. A spray stream setting should be between a 30- to 60-degree angle. Fog and broken solid streams are placed so that the water discharge is entirely outside and not quite touching the sides of the opening. Ideally, the stream should fill 90 percent of the opening. The nozzle stream distance from the opening will depend on the width of the stream. The nozzle team can then move closer or back off

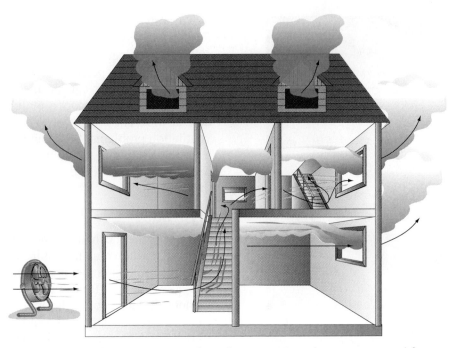

Figure 18-16 Positive pressure literally pressurizes the structure and forces smoke out any path of least resistance. Almost the same effect would occur if a light breeze were blowing directly into the structure from one side and venting out the other side.

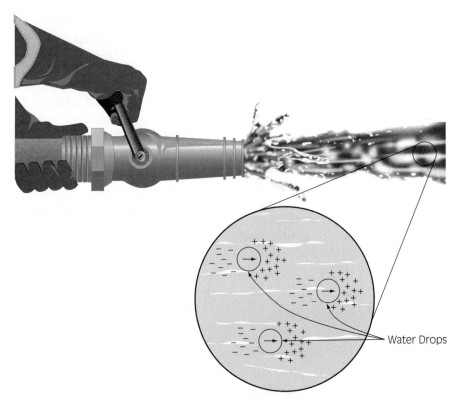

Figure 18-17 Water droplets under pressure from a hose (or even gravity) compress air in front, thus pushing air in the direction of the water flow. The vacuum created behind the droplet pulls the air along with it as the air behind the droplet attempts to fill the low-pressure void.

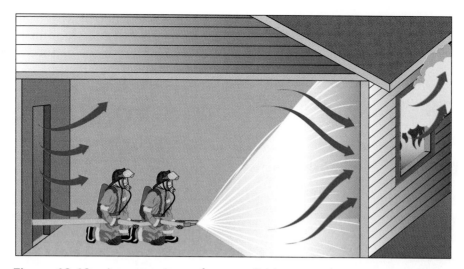

Figure 18-18 A coarse stream from a solid-bore nozzle or a spray or fog stream will create the airflow necessary to move smoke out an opening.

until the maximum beneficial effect is achieved. Because of the many variables that are always present, a certain amount of trial and error is involved with finding the best position for maximum effectiveness.

MECHANICS OF VENTILATION

The entire ventilation process, regardless of how it is achieved, is simply the movement of air from a high-pressure location to a place where lower pressure exists. Whether created artificially or naturally, air will move from a high-pressure area to a low-pressure one. Knowing this natural tendency of air to move will assist the firefighter assigned to the task of ventilation.

Vertical Ventilation

Based on the "heat rises" rule of physics, the collection of heat at the upper levels of a structure will spread fire to those upper levels if the heat level rises to the ignition temperature of the structure. By opening vertical arteries to release any pent-up gases, the odds of fire spreading to involve other parts of the structure or contents are reduced. In addition, the heated air being released will be replaced with cooler air at the lower levels. This improvement of the conditions will assist fire suppression crews attempting to locate the fire, search for possible victims, and position for suppression activities, **Figure 18-19** and **Figure 18-20**.

Horizontal Ventilation

Horizontal ventilation conforms to the same rules of vertical ventilation, which are governed by physics. Both types are actually a form of what is called **diffusion**, a naturally occurring event in which molecules keep spilling excess levels of high concentrations to areas of low concentrations. In addition to the vertical gravitation, the molecules also move laterally. Given a wall, the molecules bounce back into the mix. Given an opening, the molecules will pass into the less concentrated environment, **Figure 18-21**. That is a

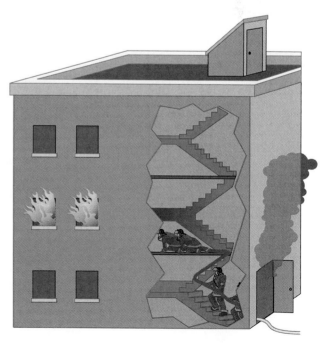

Figure 18-19 Smoke in an unvented building impedes progress. Searches are slowed. Discovery of the fire location and deployment of hoselines are delayed.

Figure 18-20 While a vented building will still have smoke in it, the difference is profound. The processes of searching, hose deployment, and fire discovery are markedly improved. In addition, the survival time of overcome victims is increased. Last but not least, potential victims might be able to self-evacuate, freeing up personnel for firefighting efforts.

scientific description of what happens, but to the firefighter on the fireground, all that is important is that the openings are made and the smoke and heat are channeled out of the structure.

VENTILATION TECHNIQUES

Many techniques are used to effect ventilation. Some are simple and require no special tools; others are more complex, are dangerous to implement, and require sophisticated tools.

Break Glass

The quickest way to open a building for ventilation is to break glass. While not always the technique of choice, it is the best investment of time for results gained if done properly. Many windows can be broken in the time it takes to force open one locked door.

> **Safety** When breaking glass, a fully equipped firefighter wearing proper protective equipment is mandatory. Slivers of broken glass can become lodged in eyes and shards of glass can act as airborne daggers.

Glass can and has penetrated skin deep enough to sever arteries and veins. The wearing of

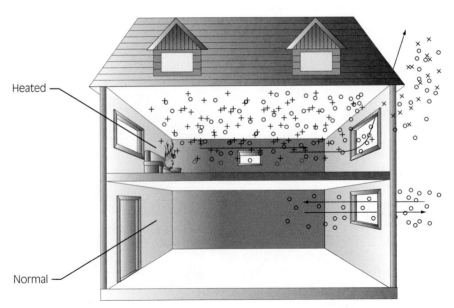

Figure 18-21 Heated air has more agitation in its molecules, causing internal pressure in a compartment. This will, in turn, create greater velocity when air exits an opening. Normal diffusion takes much longer to occur when only natural air movement and currents are employed.

protective equipment will reduce the severity or perhaps even completely eliminate an injury.

Open Doors

Merely opening a door will usually exhaust huge volumes of smoke and heat built up on the other side by pressure from the combustion. Other doors will also be helpful in opening the compartment. Keeping the door on its hinges is a good practice because the need might arise to close that door to limit fire extension or to employ a PPV blower with prechosen ventilation channels.

Effects of Glass Panes

Many windows have several panes of glass separated by wood or aluminum dividers, **Figure 18-22**. When the need to remove glass is evident, the entire window should be removed. An adage exists in the fire service vernacular: "Turn every window into a door." This refers to the possibility that the window may be needed for escape. Removal of the sashes, cross members, and any other obstacle that would impede access or egress is necessary.

When removing glass for ventilation, it is important to remove *all* glass. Shards that remain in the sash will impede airflow and reduce ventilation. A sweep around the perimeter of the sash with a tool will often produce quick satisfactory results.

Using a Tool to Take Out a Window. Regardless of the tool used, basic principles should be followed. Except for special types of glass (i.e., tempered, bulletproof), glass will break regardless of the method employed. What is important is the amount of open area created once the glass is removed.

Once the glass is broken, whatever was on the other side of that glass will be coming out. Therefore, when removing glass, it is imperative that the firefighter not be in the path created by its removal. Firefighters should position themselves off to one side and crouch underneath the window level; this is far more preferable and safer than standing directly in front of the window while removing the glass.

Using the tool to its maximum length to protect the firefighter, the tool is brought down from the upper sash through the window's midsupport and into the lower sash. Usually, this will remove a large portion of the glass. Then, the tool is used to sweep around the perimeter of the window to remove any shards that remain in the sash. Finally, the tool is used to reach into the window and unhitch any shade, blinds, or curtains that will impede airflow. If the firefighter is not able to capture the fastening, the tool can be used to hook the curtains or blinds and pull the entire unit out of the window and down until it releases. These units are held in place by small nails or screws and forcible removal is not difficult. It is essential to use care when dropping window treatments inside the structure. A victim might be lying under the window and could be inadvertently covered up. This area should be thoroughly searched as soon as possible.

The glass removal should be a two-step movement: (1) breaking through the glass from upper to lower sashes and then (2) sweeping the perimeter to remove remaining glass. Of course, not every single piece of glass will disengage. The remaining small shards will not materially affect the airflow. Good judgment must be employed when deciding whether to move on to the next window or continue to work on that opening.

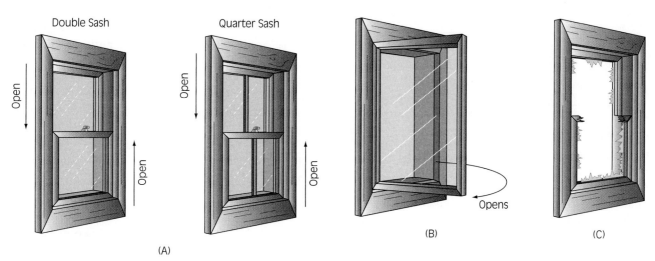

Figure 18-22 (A) Sash windows open by pulling the lower section up or the upper section down. The number of panes of glass will not change the number of sashes. (B) A casement window opens like a door, from right to left or from left to right, usually through the use of a crank or handle. (C) Regardless of the type of window, when it is vented, all the glass, including cross members, should be removed to maximize airflow with a minimum of obstructions.

JOB PERFORMANCE REQUIREMENT 18-1
Using a Tool to Take Out a Window

A Make sure the rope is securely attached to the tool using an approved slip-free knot. The tool is then lowered to the window.

B Wrap a turn of rope around your hand and toss the tool out as far as possible in a horizontal direction. The tool is then swung into the glass and will break out the window.

Note: Before performing this task, make sure there are no firefighters or victims in the building or below the building.

Rope and a Tool

A technique that has produced satisfactory results when operating off a flat roof beyond the reach of a ladder is to use a tool and a rope, **JPR 18-1**. Many windows can be taken out in a short period with this operation. The rope must be secured to the tool so there is no chance it will detach.

> **Streetsmart Tip** Some firefighters have had small rings welded to the tool and use a positive latch rope clip for quick attachment.

Using a Rope and a Tool to Ventilate Upper Floor Windows from Above. The rope is securely attached to the tool using an approved slip-free knot. The tool is then lowered to the window the firefighter desires to ventilate, **JPR 18-1A.** A turn of rope is taken around the firefighter's hand. The tool is then tossed out as far as possible from the building in a horizontal direction. As gravity takes the tool down and the end of the fully extended rope is reached, the tool will swing into the glass at the predetermined length of rope and break out the window, **JPR 18-1B.**

This technique is not the best method because it leaves shards of glass in place and does not remove screens or window shades or curtains, but it is effective when no other approach can be easily made.

Hook or Pike Pole

The use of a hook, also called a pike pole, is also an effective way to take out glass. The length of the pole keeps the firefighter a safe distance from the falling glass and it affords the firefighter the opportunity to stand off to the side of the opening. Pike poles or hooks come in assorted lengths. The use of a longer hook enables access to windows that would otherwise be out of reach without a ladder. A hook can also be used to extend the reach of a firefighter attempting to open or close a door, as a pry tool to remove roof material while standing out of the way of the opening, and for other useful purposes that will extend a firefighter's reach or remove a firefighter from dangerous positions, **Figure 18-23**.

Iron or Halligan

The iron or Halligan tool is brought down diagonally through the glass starting at the upper corner. The tool should be brought through the cross members to create one large opening. Then the tool is used to sweep around the perimeter to clean out any large shards. A disadvantage of this tool is its short length. It places the firefighter in proximity to flying glass. If prying

Figure 18-23 Making a ventilation hole requires some preplanning. (A) Firefighters should make the hole so that heat, smoke, and possibly flame do not envelop them. They must not cut off escape routes, and they should work so that the smoke is blowing away from firefighters' positions. (B) When working off a ladder, the same general precautions are necessary. Firefighters must be secured to the ladder before performing any action. This could be through use of a safety belt or a leg lock. (C) When venting from above, firefighters use the wind to their advantage and stand off to one side so that they are not standing in the path of any initial billow of heat. (D) When pulling off roof boards, firefighters should work in the clear air with the wind blowing smoke away, and be careful with roof debris. It will most likely be hidden in the smoke.

open a door or hatch, it places the firefighter almost in the path that the sudden gush of heat and smoke will take. Careful planning on the part of the firefighter must be employed to minimize exposure.

Ax

Similar to the iron or Halligan tool, the ax affords limited reach and places the firefighter in a potentially hazardous location. When using an ax for venting glass, the side of the ax is used to break the glass. Use of the blade portion or the striking head might cause the firefighter to lose a grip on the handle and lose the tool due to the manner in which a hand grasps the handle. This technique will not break tempered glass, for which a sharp pointed tool is required.

Portable Ladder

A very effective method of taking out glass in upper level windows, especially in private dwellings, is by the use of a portable ground ladder. The first thing to

Figure 18-23 (Continued) Making a ventilation hole requires some preplanning. (E) When removing a skylight, firefighters work with the wind at their backs. It is sometimes less work to lift off the entire housing than to break out each individual pane of glass. In addition, the glass for these units is usually reinforced. Depending on the amount of tar used to waterproof the unit, it might be easier to just remove the panes of glass by bending back the frame and letting each individual pane slide out. (F) When using an ax to remove window glass, the flat side of the ax head should be used, not the point or the striking surface. The position of the hands on the handle will offer better control of the ax as it breaks through the glass.

do is check for overhead obstructions. The ladder is placed to the side of the window against the siding of the house. Then the base (or heel) is arranged and measured so that the tip will fall into the glass at about two-thirds the height of the window. The ladder is then repositioned in front of the window perpendicular to the ground and shoved into the building. The tip of the ladder crashes through the upper pane and often comes down through the sash into the lower pane. The disadvantage of this technique is that sometimes only the top sash window breaks and window hangings remain in place that impede airflow. It is important to know how the interior attack is progressing and the location of the interior team before performing this operation.

Although far from totally effective, it does afford a quick method of getting some ventilation into operation in several upper level windows in a short period of time. When completed with the initial openings, ladders can be repositioned and firefighters can climb them to clean out the rest of the glass, window dressings, blinds, and so on, and to enter the building for a search if necessary.

> **Safety** When using a tool or a ladder to break glass, large glass shards can ride down the tool or the ladder into the firefighter. This shard can shatter in the face of the firefighter or slice through protective equipment into flesh.

> **Safety** When operating on a roof, there should always be a second escape route provided opposite the entry point.

Negative Pressure Ventilation

Areas of high pressure will flow to areas of low pressure. A fan moves air through the use of this physics principle. In ventilation, positive pressure is created by a fan that compresses air through the use of fan blades. That will tend to flow to an area of low pressure in the direction in which the fan is pointed. On the back side of the fan blade, negative pressure is being created. When placed in a window facing the outward flow, the positive pressure is forced out into the area outside the compartment. Along with that flow comes the heat and smoke trapped in the compartment. That is the principle under which a smoke ejector works.

Knowing this principle it is better understood that the closing of all avenues of pressure equalization from the discharge side of the fan to the input side is essential. If not properly sealed, the air will churn or recirculate and merely circle around to the intake side of the fan, **Figure 18-24**. For effective air movement, a barrier must be created that separates the positively charged air from the negatively charged air sectors. In a structure where many windows are broken and there are

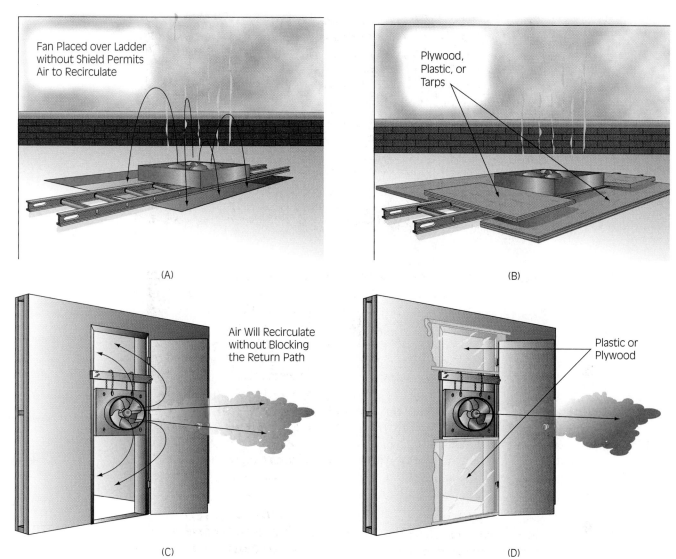

Figure 18-24 (A) When using an exhaust fan, it is important to cover the openings around the unit. Without a means to block the air from returning, it will merely circulate, reducing efficiency. (B) When covering any opening around the exhaust fan, the vacuum necessary to operate efficiently will be created and the exhausted air will not be sucked back around the fan. (C) When using an exhaust fan in a door, the air will circulate from the exhaust side into the intake side if no provision is made to block that flow. (D) Through the use of plastic, tarps, or even a piece of plywood, the air is prevented from being pulled back into the intake side of the fan.

holes in roofs and walls, this is not easily done. Therein lies the major weakness with this technique. Drapes of plastic and tarps can be employed to limit churning.

This type of setup can be very effective in limited access compartments such as cellars or basements where few openings exist. Fans can be placed over the opening in the cellar stairs or in a hole made in the first floor near a window and used to draw out the smoke through suction. This tool is primarily an overhauling aid and should not be used during the attack mode. The fan could bring fire to the location, placing the operators in a dangerous position.

Positive Pressure Ventilation

PPV uses an entirely different approach than negative pressure ventilation, but with the same principle. The positive pressure technique actually injects air into the compartment and pressurizes it. In an attempt to equalize, the smoke and heat are carried out into the areas of lower pressure outside the structure. There are many variations of this practice. The basic principle is to create a cone of air and force it from the outside to the inside, **Figure 18-25**.

Depending on the circumstances, fans can be set up to augment one another. Two fans can be placed

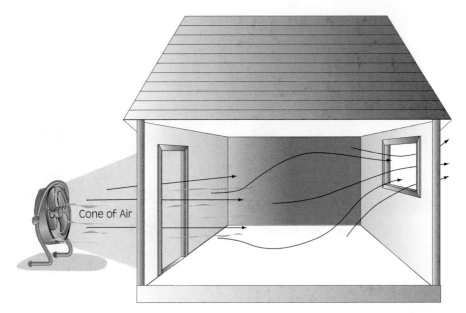

Figure 18-25 With positive pressure ventilation, the theory is to actually pressurize the compartment and then the smoke and heat will actually be pushed out another opening. To be effective, certain actions must be taken. (1) The blower or fan must be placed a short distance from the opening so that a "cone of air" is created that just barely exceeds the opening being used. (2) The exhaust opening should be smaller than the introduction opening for maximum efficiency. That opening size depends on the number of blowers and their capacity. There are many variations where this practice can be effective. Stacking fans, using them in tandem, or placing one behind the other can have varying degrees of success depending on the structure and its configuration.

side by side to cover a wide opening. They may be stacked to create an elongated cone of air. If positioned one behind the other, the rear fan can supercharge the front fan for greater efficiency. The fans can be placed in remote locations to augment the airflow from one part of the structure to another, and three or more fans can be used in conjunction with one another: one or two at the opening and others spaced throughout the structure in upper floors and along long corridors or split in halls that fork. Understanding the principle is all that is needed to create a custom application to fit an unusual situation.

Tests have shown that effective pressures can be generated up to twenty-five stories from the street opening location of a PPV blower. Augmentation beyond that level can still provide effective pressure generation. Intermediate fans will also assist in maintaining the flow.

Generally speaking, for every cubic foot of air injected into a compartment, a cubic foot of air must be ejected. The pressure loss experienced is due to the many openings in any run that will permit some air pressure to escape.

Streetsmart Tip Practical experimentation of positive pressure ventilation can be employed in any structure (even the fire station) through the use of tissue paper. With a fan or blower set up, tissue paper strips are hung in openings throughout the structure. When the fan is activated, the airflow can be demonstrated, redirected to different openings, and observed. The velocity of the air movement can be observed by the degree to which the tissues in the openings are affected.

Positive Pressure Fan/Blower Setup and Operating Principles

To effectuate positive pressure, the blower should be set up outside the structure. The cone of air that is produced must completely cover the opening. This can be achieved by moving the fan closer to or farther away from the building opening. Generally, the size of the exhaust opening must be larger than the air introduction opening. Positive pressure is most efficient when the exhaust opening is ¼ to

1½ times the size of the entrance opening. Communication is essential between the intake side personnel and the exhaust side personnel. All members engaged in a positive pressure operation must be properly trained and understand the principles of the evolution.

Many different combinations of blowers and fans can be utilized. They can be placed in tandem, stacked, side by side, in-line, or in any configuration that accomplishes the goal of pressurizing the compartment. Depending on the size, shape, and number of cubic feet in the structure, many different layouts will be successful.

- Positive pressure evacuation of smoke is successful for distances of up to 1,000 feet.
- For each cubic foot in a building, a cubic foot needs to be exhausted. Multiple floors and multiple rooms can be systematically vented by opening up a single room and window until it is vented, then closing the room off and moving to the next room. This technique is particularly effective in a school classroom-type building, motel, hotel, or similar room layout.

Depending on the size of the blower and the department's practice in using it in overhauling and/or attack modes, the novice firefighter should become familiar with the many different aspects of positive pressure techniques. The basic rules are as follows:

1. The introduction opening should be larger than the exhaust opening.
2. The air can be channeled into different parts of a building by opening and closing windows and doors.
3. A cone larger than the introduction opening is necessary to prevent recirculation of air back out of the structure.
4. Communication at the point of entry and the point of exhaust is important.
5. Practice will provide the understanding of the basic principles that can then be applied to actual fire scene operations.

ROOF VENTILATION

Vertical ventilation can be attained using several methods. One such method is to use the existing building features that allow for removal of heat, smoke, and the by-products of combustion. Those features include penthouse doors, skylights, and ventilation shafts. Care should be taken to ensure that use of these features would not spread the fire, **Figure 18-26**.

Two primary types, offensive and defensive, characterize vertical ventilation openings. The primary offensive vertical ventilation openings are generally referred to as offensive heat holes. These openings are typically placed over the fire or as close to the main body of the fire as reasonable or necessary. The primary defensive vertical ventilation openings are generally considered strip or trench ventilation operations and are broken down into subtypes from there. The goal is to place offensive openings into the structure and evaluate the need for a defensive ventilation opening. Communication with interior forces will assist in the evaluation process for vertical ventilation operations.

With snow, the hot spot might be melted away or slushy compared to the rest of the snow on the roof.

The primary hole should be cut directly over the fire if possible, and several primary holes can be cut. The more direct the heat travel to the outer air, the less potential for extension of the fire. Secondary hole cuts are any cuts that follow the primary hole. Some holes are always secondary cuts, while some types of cuts can be either primary or secondary. An example is a trench cut, also called a strip cut. It is a secondary cut and should never be employed as a primary ventilation hole. Its purpose is to create a line of defense and, as such, should not be ordered until primary holes are completed. Secondary holes are cut away from the immediate fire area and are used for inspection or to cut off fire extension.

Expandable Cut

An expandable cut is the most efficient type of cut for the time expended, **Figure 18-27**. It can produce a hole as large as is needed. Because ventilation is such a necessary operation, the roof must be opened quickly. After planning the cut so that the wind will tend to blow the smoke away from the work area, an initial cut line of about 4 feet long is made. The second cut is a small triangular purchase point. Then a third, fourth, and fifth cut are made. Using a hook (pike pole), the purchase hole is used to hook the roof material and, like a giant lid, the top is lifted off. This type of cut can produce a hole of about 16 square feet.

Care must be taken to prevent damage to support rafters or cross members when cutting a roof. A support rafter can be felt when the saw suddenly begins to slow down as it tries to cut a thickness greater than the roof sheathing. The speed of the saw will increase and the operator will actually be able to feel the difference when the rafter or support beam is passed.

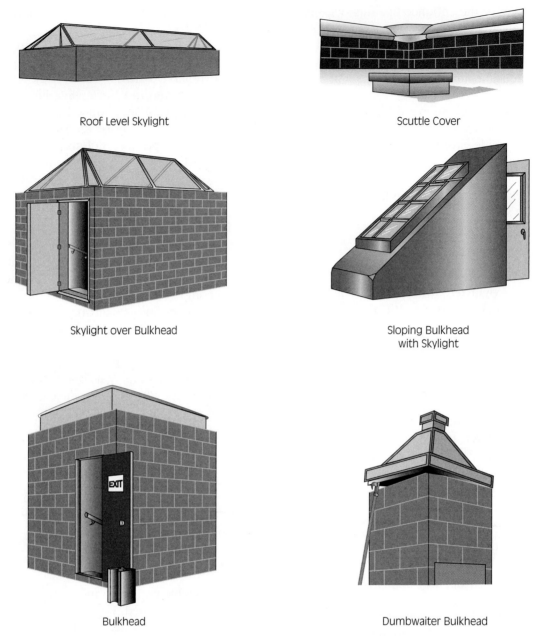

Figure 18-26 Many types of roof structures afford quick and effective initial ventilation. As is always the case, the time spent in creating a hole should be balanced by the amount of efficiency that will be gained. For example, opening a bulkhead door will be quick and effective. It should be performed before a roof cut is initiated.

If a larger opening is necessary, making three additional cuts using one of the sides of the original hole as the first cut expands the hole. With only three more cuts, the size of the hole is doubled to 32 square feet. Again, if an even larger opening is required, another three cuts will increase the hole size by another 16 square feet to a total of 48 square feet. If a larger opening is still required, the large hole can be finished off with only two cuts, **Figure 18-27**.

Occasionally, a close approach to a roof opening is impossible because of the heat condition. In this case, the expanded cut might have to be performed by making as large a hole as possible using whatever part of the hole is approachable, **Figure 18-28** and **Figure 18-29**.

By comparing the numbers, the use of this type of cut will produce a hole that is 64 square feet. In terms of manual labor, if each hole were cut individually, it would take five cuts to produce each hole. A total of four holes would require twenty-five cuts to produce 64 square feet of opening. Using the expandable cut, the same opening is produced with only thirteen cuts. In addition, one large

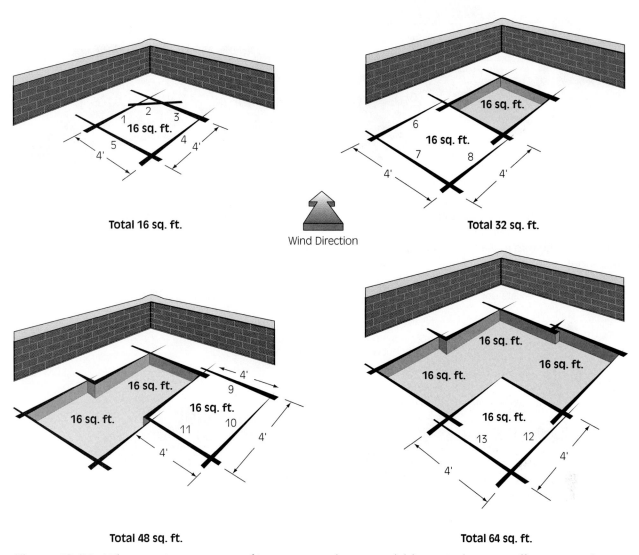

Figure 18-27 When cutting open a roof is necessary, the expandable cut is the most effective, quickest, and most efficient way to get a large opening for vertical ventilation. Each new cut is merely an extension of the previous opening. The overall number of cuts required is dramatically fewer than if each hole were individually cut.

opening produces more airflow than several smaller holes of comparable area.

Among the disadvantages of this type of cut is the amount of debris produced. With each section that is opened, roofing material, roof boards, and wood with protruding nails are left lying about. Additionally, the hole produced is large and a firefighter can fall into the opening, even with the joists that are spanning the opening after the roof material is removed.

Center Rafter Cut (Louver)

The center rafter louver cut is utilized primarily as an offensive heat hole. It is often used on building construction that utilizes plywood sheeting under paper as a roof cover. It can also be used on roofs with 1 × 6 sheeting.

The center rafter louver is based on a series of cuts that begin with an outside cut, **Figure 18-30.** The saw is placed on the roof decking and a cut is made until the roof rafter is felt with the nose of the chain saw bar. The rafter is rolled (raising the saw so as not to cut through the rafter) and the cut is continued. This "head cut" is run as far as the opening is desired, continuing over rafters as they are met.

Then the saw is taken back to the first rafter met, and a lateral cut is made from the head cut. Another cut is made parallel to the head cut until the second rafter is met. The next cut is made to "box" this cut, **Figure 18-30.** After that, the roofperson continues

Figure 18-28 Notice here that the opening had to be made in a linear fashion. In this fire, the heat was so intense that a rectangular cut was impossible. However, using the existing hole to make the next series of cuts could expand the opening. Also notice the amount of debris that accumulates. Visible in clear conditions, a severe tripping factor exists under smoke conditions.

Figure 18-29 The initial hole was cut and the "lid" lifted off in one piece. In the foreground, the cuts of the second hole can be seen along the right side of the hole. This type of roof, called an inverted roof, consists of a flimsy under roof support system while the actual weight-carrying members exist at the occupancy ceiling level. This photo illustrates how little support is available.

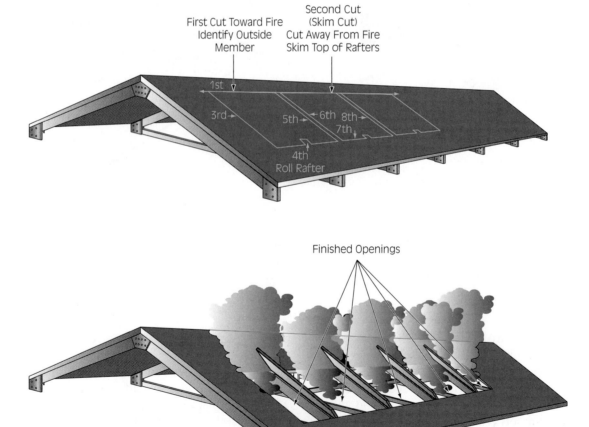

Figure 18-30 A louver opening is made by initially making several cuts parallel to and between the supporting roof joists. Then a connecting cut is made along the end of each cut perpendicular to the first set of cuts. Finally, each section is pushed down on one side and pulled up on the other. The result is a seesaw effect where panels tilt open and provide a hole for airflow. Because it is partially obstructed by each of the panels, it is not as effective as the total removal of the roof material. However, it is safer because there is less chance that a firefighter will fall into the hole. This type of cut is most effective on a plywood-sheathed roof.

duplicating this process until the cut meets the length of the head cut.

Lastly, the roofing pieces are pushed in, creating louvers along the cut and allowing the by-products of combustion to escape.

Triangular Cut

On some types of roofs, the use of the triangular cut technique is favored. A roof supported by an open web bar joist with Q-decking is the best candidate for this type of cut. Because the span of the web bar joist often exceeds 24 inches, the opening can become a funnel if opened using conventional square cuts, **Figure 18-31**.

The use of a triangular cut will help support the underlying Q-decking because it is interlocked. By cutting across the panel on a diagonal, the tendency to sag will be reduced.

> **Streetsmart Tip** Extreme care must be used in open bar truss type roofs because they are prone to failure at the early stages when this type of support system is directly involved.

Of necessity, these holes will be relatively small and might require several cuts to adequately perform the ventilation mission.

Trench Cut or Strip Cut

Purely defensive in design and execution, the trench cut is a roof opening that ventilates the cockloft area of a building where the fire is spreading under the roof, **Figure 18-32**. After the primary roof opening is made, the strategy might call for making a stop at a particular location. It is usually predicated on structural elements inherent in the building itself such as the connecting point of several wings or a place where the roof area narrows due to the presence of structural elements.

With the trench cut, the gases that are expanding horizontally under the roof are vented to the atmosphere and not permitted to pass a chosen defensive line, **Figure 18-33**. When deciding to use a trench cut, it is understood that the building on the fire side of the cut has been determined to be unsavable.

When a trench cut is being properly utilized, a hoseline should be positioned below the cut with ceilings pulled down to expose the opening. Any fire that actually reaches the opening and threatens to jump the gap should be extinguished with the hose from below. Under no circumstance should a hoseline be operated into a hole from the roof side. It will only serve to drive the heat and smoke down into the top floor environment. The purpose of this hole is to permit the laterally extending fire and heat to escape into the atmosphere where it poses little

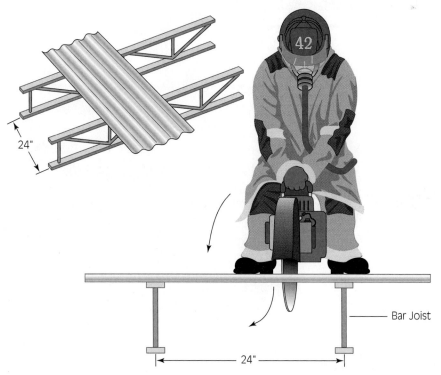

Figure 18-31 The open web bar joist-type roof structure requires extra vigilance. It is possible to cut away the supporting panel on which the firefighter is standing. Because of the large span between the supporting joists, the member can topple into the newly created hole.

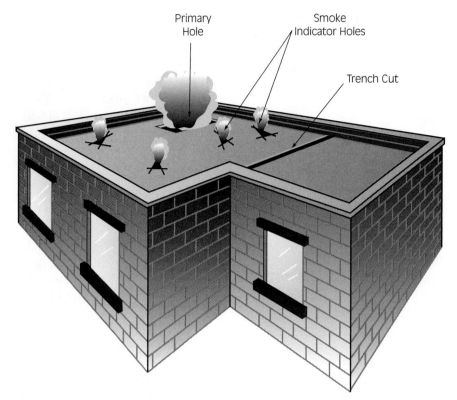

Figure 18-32 A trench cut is used to cut off fire extension. A totally defensive action, its use means that the fire side of the cut is being given up. Once opened, no firefighters should be operating on the fire side of the cut. A trench cut is (1) 2 to 3 feet wide, (2) not opened until the entire cut is complete, (3) not opened until all personnel are on the "safe side" of the cut, and (4) made using any available building features.

Figure 18-33 A trench cut is a defensive move. Ceilings should be pulled below the cut to promote vertical airflow through the trench. Additionally, a handline should be in place below the opening to cut off any horizontal extension.

threat. A hose may be operated on the roof to extinguish a roof fire, but it should never be operated into the ventilation hole.

Inspection Cut

The first operation that needs to be accomplished on a flat roof is the inspection cut. The inspection cut is placed into the roof to determine the following:

1. Roof covering and depth of covering
2. Roof sheeting material
3. Rafter direction
4. Conditions directly below firefighters

The first cut in establishing the inspection cut is made at 45 degrees to a bearing wall, followed by another cut opposite the first. Then the triangular inspection cut is completed with another cut, **Figure 18-34**. The goal is to identify the rafter and roll it when coming in contact with it. The inspection cut is larger than the smoke indicator hole and could pose a danger if placed in a path of access and egress to the ladder.

Smoke Indicator Hole

The smoke indicator hole is the only opening that will adequately determine the conditions directly below firefighters operating on the roof. The smoke indicator hole is a small triangular opening approximately large enough to push the D-handle of a roof hook through but small enough that the firefighters operating on the roof will not put a foot through the roof itself. A smoke indicator hole should be placed into the path of access and egress approximately every 15 to 20 feet of travel.

SAFETY CONSIDERATIONS

When considering the placement of a hole for ventilation, whether it is horizontal or vertical, it is of paramount importance to consider the benefit gained against the possible liability created. In some cases the best choice will be *not* to vent at that particular location, for example, if venting would expose a victim and rescuer on a ladder to danger.

Will Ventilation Permit the Fire to Extend?

There is no justification for permitting a fire to extend in order to complete a task that is assigned. On occasion, an order may be given without full understanding of the consequences of that order because early reconnaissance is incomplete and fact gathering is still taking place. Some action may still be required but the firefighter should consider the possible consequences of that action and report to the incident commander for reevaluation if conditions are changing out of sight of that incident commander. Most incident commanders in the field would gladly countermand an order given when new information shows the original order to be unsound or unsafe.

Will the Escape Route Be Cut Off?

Of paramount importance to firefighters is their individual safety. When performing a ventilation activity, an eye toward escape must always be in the forefront. When venting a series of windows, the firefighter must work toward the escape point, **Figure 18-35**. There should always be two easily recognizable ways off a roof.

Escape routes should be lighted at night if possible and the presence of such escape routes made known to everyone on the roof.

Will Ventilation Endanger Others?

An overview of the operation in progress must always be in place when operating. Careful consideration must be given to what is called **mission vision**, a "blinder" view of the work being

Figure 18-34 A triangular inspection hole cut is made by dropping the blade of the circular saw through the roof material in three intersecting cuts. If there is no support underneath, the triangle will drop into the cockloft. If there is a supporting member underneath, the roof piece can be tilted up by using a tool to depress one of the triangle points to obtain a purchase point to pull it up for removal.

Figure 18-35 Ventilate in the direction of the escape route so escape is not cut off.

performed because the mission of the individual is overriding all other activities. The activity of one firefighter performing a venting task must never endanger the position of another person.

When opening a roof, care must be taken to advise everyone else on the roof where the holes are. In dark, smoky conditions, it is easy to fall into the fire.

In addition, the torn-up roofing material should be cleared away as soon as possible to remove the tripping hazards. As the roof team pulls off the roof, they should pile it up in one place if possible. Thinking ahead of possible safety problems and working to remove them before they present themselves is safety in action.

Work in Teams

A firefighter should never work alone. The presence of a team member will afford a second set of eyes and another mind at work. It is very difficult for any one person to be aware of everything that is taking place at the same time. At times the presence of another firefighter will afford the opportunity to quickly discuss the manner in which to attack the problem at hand. A team member might make the difference between being located when a floor collapses or making a misjudgment resulting in a fall down a shaft. The presence of a team member could make the difference between being able to pry open a bulkhead door or lift a heavy skylight and not being able to do it because it is too difficult for one person.

Proper Supervision

Unfortunately, in some places, the presence of supervision is all too often considered to be a waste of personnel. There should be a person on the scene to make decisions when conflicting options are available. The presence of recognized supervision will ensure that the effort is unified. Supervision also typically brings a certain amount of experience to the activity, because supervisors are usually the more experienced personnel.

A supervisor's presence also helps to prevent team members from becoming too focused on the mission itself, because the nature of a supervisor's job is to oversee the whole operation of the sector and coordinate subordinates' activities to meet the objectives of the teams under the supervisor's command.

OBSTACLES TO VENTILATION

The importance of ventilation cannot be overstated. Because of the unpredictability of fire operations, firefighters will be confronted with many unforeseen circumstances that will delay ventilation activities. Listed next are some of the obstacles that might be encountered.

Access

Access should be one of the initial size-up considerations when arriving at the scene. The firefighter should first assess the needs of the ventilation objective, then determine the route to be employed to reach the location of the job performance task. It might be via the adjoining building, if present, by use of a ladder, or perhaps from the ground. Roof access might be delayed because of surrounding impediments such as overhead wires, elevated trains or roadways, light pole placement, or any other obstruction that might delay placement of a ladder or apparatus, **Figure 18-36**.

Access to a rear yard might be impeded by the presence of high, locked fences or by the presence of watch dogs. Alternative measures must be employed and thought out in advance before the obstacles are encountered. In routine nonemergency responses or activities, it is a good practice to map out an access strategy for a particular building and then assume that such a route would not be available. The constant practice of formulating a second and third plan on a routine basis will permit the ability to do so at a real emergency with greater ease. It is simply a matter of practicing the thinking process routine until it becomes a reflex action.

Security Devices

The presence of security devices can impede ventilation both in access and in timing. Building owners may have barricaded themselves and their possessions behind gates, screens, steel doors, and closed-up windows. In some cases, window openings have been blocked off and sealed, **Figure 18-37**.

In high-crime areas, some shop owners have even removed skylights and replaced the hole in the roof with cheap plywood attached to inferior structural support, all held in place by a few nails. Once covered with roofing, the former hole is no longer discernible and has been the cause of firefighter deaths as members walked on the covered-over holes and fell through.

> **Streetsmart Tip** These unknown alterations mandate a very cautious approach to roof operations and often result in delay of roof ventilation procedures.

Height

From multistory skyscrapers to one-story garages, the firefighter must be alert to the structure's ventilation needs. Sometimes the need to cut a hole in an area that is out of reach of the tool at hand or from a

Figure 18-36 Roof access by ladder can be slowed down when obstacles are encountered. Aerial ladder placement must take into account overhead wires, structures, and presence of utility poles before being set into place.

Figure 18-37 In the interest of security, occupants will place steel security screens or completely close off an opening that might serve as an entry point for a would-be burglar. Notice the structural material behind the steel screen on the window to the left. Any obstructions to a full opening will reduce ventilation.

particular position will be encountered. The need for a longer tool for reach or a rope tied to a tool or a ladder for access should be part of the initial size-up routine for the member assigned to ventilation duties. It must be assumed that reach will be a problem. What appears to be a single-story building from the street side can possess a topographical drop-off in the rear, placing the same floor that is at street level in the front several stories off the ground in the rear. The firefighter must practice thinking in a proactive fashion rather than reactive.

Poor Planning

One great obstacle to ventilation that is addressable is planning. Time is not a luxury. Proper on-site planning must occur without delay. A quick size-up and implementation of the plan are essential for timely ventilation. If ventilation is delayed, the interior team will suffer. Backdrafts or flashovers, decreased survival time of trapped victims, and arduous working conditions will result from delayed ventilation.

Personnel Assignment

A task that is assigned to a shorthanded or inexperienced crew will delay ventilation. If a ladder is needed to reach a setback in order to vent windows, and only one person is assigned to the task, not only will OSHA rules be violated, but objective attainment will be delayed.

Of course, there are times when adequate personnel are not yet on the scene. It is at this time that the ingenuity of the firefighter comes into play. With safety as an overriding concern the heads-up firefighter might be able to effect some ventilation while waiting for reinforcements. In a roof operation, a two-person team can open many openings relatively quickly while awaiting assistance. Structural components such as skylights, doors, roof access hatches, and ventilators can be easily opened or removed by individuals working together. A roof-cutting operation might need additional assistance but at least some openings are being produced to vent. Or two firefighters might be able to tie their tools together to get to a window that is out of the reach of just one tool. However, doing this while five other reachable windows are still not vented would be a poor use of resources.

Unfamiliar Building Layout

Especially in large buildings, the floor layout can be very confusing to the firefighter outside the building attempting to figure out which set of windows serves the fire area. This is particularly true of occupancy floor layouts that form L-shaped or U-shaped configurations in a building that appears to be square on the outside.

The building layout can also be an obstacle to a firefighter attempting to reach the rear of the building. Building wings, fences, lower floor extensions, or unusual configurations of the building's dimensions can confuse the firefighter and delay access. Some occupancies such as contemporary cluster-type construction townhouses can have door entrances on the same floor but serve three different levels. For example, there could be three entrances on the same floor. One door opens to a stairway that leads to the apartment on the floor below, one opens on the same floor as the entrance doors, and the last opens to a stairway that serves the apartment on the floor above. To the firefighter attempting to get to adjoining occupancy on the floor above the fire to vent, the need to force open three different doors might be necessary and will delay ventilation.

Observing structures while under construction, requiring building owners to identify the occupancy served, conducting familiarization drills, entering pertinent information on response ticket critical information (if such a feature is available), or simply entering an uninvolved lower floor occupancy to look at the floor layouts are just some of the ways to address this problem. Often the same floor plan exists throughout a multistory building to reduce construction costs. Utility supply voids will often dictate the "rubber stamp" floor layout of the multistory occupancies.

Ventilation Timing

Venting too early can lead to fire extension; venting too late causes unnecessary punishment to the interior forces and can even prevent forward progress. The possibilities of backdraft and flashover have already been discussed. The timing is dependent on the type of ventilation being performed.

Vertical ventilation of stairways, hallways, and any paths of egress and ingress are paramount and must be effected without delay. If fire is known to be in the area under the roof boards or on a top floor extending into the cockloft, opening of the roof must commence without delay.

> **Note** The importance of early removal of the heat and smoke through vertical channels cannot be overstated.

As with all facets of firefighting, however, nothing can be written in stone. The opening of an entrance door to a fire occupancy on a lower floor that will vent up the stairs and out the roof opening must be carefully planned. There may be situations where escaping building occupants are coming down the stairs or firefighters are moving up the stairs. In such situations, vertical ventilation through this channel must be timed properly and may even have to be delayed. Delay might be necessary when opening a roof cut because the opening might expose an adjoining building to extension. A hoseline might have to be put into place to protect the exposure before the cut can be opened.

Cut a Roof—Open a Roof

Cutting a roof and opening a roof are different processes and this distinction should be clear to firefighters. Cutting a roof is the process of making the necessary cuts to perform the ventilation. Once the cuts are made, the opening of the roof can begin. The opening refers to the actual removal of the roof material. Therefore, a roof can be cut and not yet opened. There might be situations where a delay is necessary in opening the roof because of exposures or line placement or some other reason. This does not necessarily stop the cut. The cuts can be made but the opening delayed until the time is right.

FACTORS AFFECTING VENTILATION

Several factors can affect the attainment of proper ventilation. Partial openings, screens, type of roof material, wind direction, weather, building size, and construction features are some of the more prominent elements.

Partial Openings

Research has demonstrated that one single opening of a given size has greater ventilation capability than several holes that equal the same opening area. Much of this has to do with airflow characteristics. Like water, airflow is not like a chain of molecules that stay in line and move along in single file. Air is made up of many individual molecules that will flow in generally the same direction, but will tend to drift randomly in all directions. Air will move in the direction of less resistance. The eddy currents that can be observed when smoke is released into an unconfined and unchannelled area demonstrate this. If under enough pressure, smoke appears to billow.

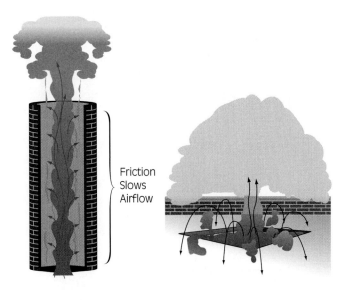

Smoke Stack **Roof Opening**

Figure 18-38 Airflow is reduced by friction.

Looking at a cross section of a chimney, it is clear that smoke rubs along the edges but tends to flow in a more uniform manner in the center. The greater the circumference of the vessel (i.e., smokestack) the less friction on the sides and the greater the flow. Once released into the atmosphere, the column of smoke continues to rise until the central core of the column is finally exposed to the reduced outside temperatures and begins to stratify. However, along the outside perimeter there is a constant peeling off of the smoke in eddy currents. The eddy currents will tend to reduce the flow, **Figure 18-38**.

Transforming the large picture of the smokestack into a roof opening, the same principles can be seen at work. With each opening, the outside perimeter of the opening will afford the airflow opportunities to eddy and slow down the overall flow. This is the underlying factor in the principle that many small holes will not be as efficient as one large hole. The large hole will have one perimeter of a given size. The many small holes will have a greater overall length of perimeters. The greater the length of perimeter, the more opportunity to reduce the speed of the airflow.

Partially Broken Windows

The same principle just described comes into play again. If a windowpane that is square is broken and many crescent-shaped shards of glass are left in place, not only is the area of the opening reduced but the presence of the shards creates more perimeter opening distance, **Figure 18-39**.

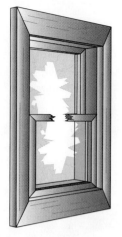 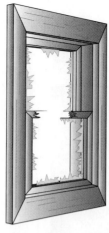

Figure 18-39 Airflow is greatest through a window where glass is fully removed. Screens, shades, curtains, and window cross members should also be removed.

Screens

The presence of insect screens in a window that has been broken out is magnified when one considers that a screen is like a solid panel with hundreds of holes.

> **Note** Conservative estimates say that the presence of a screen in a window reduces the airflow by approximately 50 percent.

Put into real-time numbers, failure to remove a screen after taking out the window is like opening 100 windows, then permitting a shutter to close on 44 of them. The effort is wasted and the ventilation is greatly reduced. Removal of the window treatments, curtains, shades, or blinds is also important. Any obstruction in the opening will reduce the airflow.

Roof Material

In some buildings, particularly those of older vintage, the roof material could be several layers thick and several layers might need to be removed before the cut is opened. When planning the cut, the firefighter must be aware that the kerf cut of the saw will usually permit a thin line of smoke to vent. If several layers of roofing must be removed, the original cut should not penetrate the under roof area. The cut should only go as deep as the roofing material that will be removed on the first pass, **Figure 18-40**. The smoke from the kerf cut can be enough to obscure vision and impede the initial roof material removal.

In some types of construction, the roof material might actually prevent the opening of a roof. The purpose of a roof is to shed water. Some materials

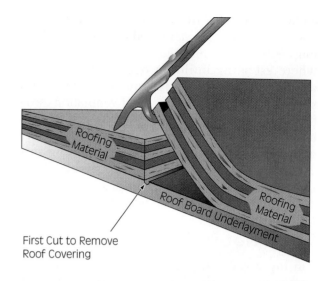

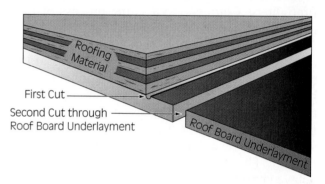

Figure 18-40 If a roof is thick with roof covering, two cuts might have to be made. The first cut is made shallow to remove the covering. It should not pierce the sheathing to prevent smoke from interfering with the completion of the first cut. With the second cut, the roof sheathing is cut and then removed.

are well designed for that purpose but are deadly in a fire. In the case of a corrugated metal roof over an open bar joist support structure, the heat from the fire melts the underside of the waterproofing material, usually some form of petrochemical product. Flammable gases are created and trapped below the cooler upper roof and tend to collect in the troughs of the correlated roof. When the heat generates enough pressure, the gases find a way into the underside of the roof through the joints of the metal. The presence of this flammable gas in the cockloft area can be explosive when reached by fire. Additionally, the unsuspecting firefighter might be providing the missing oxygen as the roof is cut. A kerf cut is all that is necessary to open an oxygen supply route. Usually without warning the roof will erupt into fire at the place of least resistance—most often the place where the roof has just been weakened by the saw cuts.

Special care must be employed and constant study of the latest innovations and techniques is essential for the firefighter to maintain and improve a margin of safety in a known hostile environment.

Other types of roofs also have their particular strong and weak points. Any type of truss roof is always suspect because of the nature of its collapse potential. Placement of any personnel on this type of roof must be very carefully considered. The full extent of the fire must be known before commitment of personnel to this type of roof for ventilation is ordered.

In many cases, one roof has been built over another, sometimes with several feet between them. Obviously, the opening of the upper roof will have no effect on the ventilation of any spaces below the under roof. The effects produced by the opening made must be evaluated by the firefighter. If there is only a little smoke and heat being vented from an opening that is over an obviously large body of fire, the incident commander should be notified and an investigation conducted to determine why. The notification lets everyone at the scene know that something is not right and the investigation determines why it is so in order to address the problem and complete the objective.

Dropped or Hanging Ceilings

Like a second roof built over the original, the installation of dropped ceilings creates a similar impediment to the ventilation process. Trapped air pockets will conceal fire and smoke, and the raging fire on the floor will not be able to vent through the opening in the roof. In older buildings, especially of the commercial type that have a regular occupant turnover, several hanging ceilings may be in place.

The space between ceilings is a haven for fire gases to collect. Venting them can be a real challenge because the dropped ceilings are (1) not always obvious, (2) often out of reach of the roof firefighter, and (3) difficult to open from either above or below. The space is usually large and is ripe for a backdraft that will cause the entire ceiling to drop into the firefighting forces below.

Advancing Under a Dropped Ceiling

The airflow in front of a nozzle can work its way into the space above a hanging ceiling, especially if of the grid and insert type. The fire is pushed into the space when the heat lifts the tiles like trap doors. Heat and fire gases fill the space and, when conditions are right, can ignite and blow down the whole grid onto the advancing firefighters, trapping them in a web of metal grid.

> **Streetsmart Tip** When entering an occupancy with a suspended ceiling, the first thing that should be done is to have a tile pushed up from the safety of the doorway into the area to determine the conditions above the ceiling.

If fire is already in the space, the tiles can usually be unseated with the hoseline merely by directing the stream upward and extinguishing fire in the ceiling while advancing.

Building Size

The building size will affect attempts to ventilate. A building's height and width both have profound effects on the ability to move air.

In tall buildings a phenomenon called a *neutral plane* can occur. It is the location where smoke will collect instead of continuing to rise. Fireground experience has found it to inhabit just a couple of floors in some cases, and other experiences have had it collect on up to ten floors many stories above the fire floor. Although certain texts give a particular number of stories above a fire floor where it will occur, the fact is that it will occur where it occurs. There is no good rule of thumb to apply because in most cases, the numbers are laboratory generated or results of carefully structured tests in existing buildings. In a fire, because so many variables are at play, there is just no way to predict accurately where this plane will occur if it occurs at all.

Factors that affect the development of a neutral plane include the presence or absence of HVAC systems and their associated ducts; wind direction around the building; presence of other buildings in the general area and height in relation to the structure on fire; outside temperature and its relative difference to the structure temperature inside; and presence or absence of smoke shafts in the building. The only way to determine what the smoke is doing in the upper floors of a high-rise building is to have reconnaissance teams report on their findings.

In large buildings, air movement can defy normal expectations. Wind direction on the outside of the building can create a negative pressure on one side of the building and cause the smoke to gravitate toward that section. This can become very confusing when smoke is being reported on a side of the building that is remote from the fire. Ventilating for such a fire can be a monumental task, and careful analysis of air movement as reported by reconnaissance teams must be conducted in order to maximize a coordinated ventilation effort.

The smaller the structure, the greater the similarity the air movement will be between buildings of the same type. Light breezes will not usually be much of a factor in a residential 2½-story wood frame building surrounded by trees and other dwellings. However, the same breeze in the same structure that sits high on an unprotected hill might exhibit very different characteristics. The same goes for a building that is facing a wide river or ravine as opposed to a structure that is surrounded by buildings of similar height, even if they are 70, 80, or more feet high. There are no absolutes in this area. So many variables factor into the formula, only experience, observation, and common sense can prepare the firefighter to observe and predict what will most likely be the result of a given situation.

Weather

Some general observation can be made about how weather will affect air movement. On cool dry days, smoke-filled structures will vent quickly. On rainy humid days, the smoke will lift slowly. Snow affects the air movement and a light breeze will tend to "pull" the smoke out of the upper vertical openings. There is no magic to this knowledge. It goes back to fire behavior and what makes smoke rise.

The heated, smoke-filled air rises because the heated air weighs less than the surrounding air sample. In humid environments, the air is actually cooled and that slows the vertical velocity of the plume. While the fire is still pumping smoke out, the column flattens quickly as it cools and might not even leave the structure at the upper levels. That will tend to make it stay low and not lift.

Rain will have the same effect with an added element. The droplets of water will, in addition to cooling the superheated gases, actually impede the vertical flow physically. There are two forces at work: (1) The heat is lifting the smoke particles, and (2) the water droplet is colliding with the airflow as gravity pulls it down resulting in a diminished vertical airflow. The ultimate result is a slowing of the venting process.

Snow affects the rising column of smoke even more profoundly. If the rising air is hot enough, the snow melts, robbing the column of heat and cooling off the air volume. The result is water that further cools down the thermal plume. Snow adds an additional factor of air resistance. Since it has weight, the snowflake will actually force the thermal plume to work harder to rise. The combination of the additional weight to lift, the cooling of the snowmelt, and the water that does further cooling works against the venting process. One can expect difficulty when trying to ventilate vertically in snow.

Horizontal venting, however, is not necessarily affected the same way because gravity or thermal plume generation does not generally affect it. Heated, pressurized air will look for a path of least resistance and will find an opening in a wall (i.e., window or door) regardless of the weather conditions. However, once outside the structure, the horizontal venting turns to a vertical direction and up the side of the structure. At that point the smoke will be affected as described earlier. Once the fire is extinguished and the source of the heat is removed, the ambient atmospheric conditions will tend to move indoors. When the humidity reaches the inside of the structure, the smoke will then tend to stay low and not lift.

Great success can be achieved with positive pressure ventilation or with the use of a hoseline stream to create negative pressure and pull the smoke out. A fog or spray stream can be used to move air by opening the nozzle and directing the stream out the window or door. The droplets of water actually push the air in front of it (positive pressure) and pull the air behind it (negative pressure). The overall effect is the creation of an artificial airflow that can be very successful.

Opening Windows

Opening windows is the simplest way to open a compartment. Not every fire requires the removal of glass. Sometimes areas unaffected by fire can become filled with smoke and merely need to be vented to remove smoke. For many years firefighters were taught to open windows two-thirds from the top and one-third from the bottom to effect ventilation. That concept has changed. Based on what was previously discussed it is now known that one large hole is more efficient than several small holes, so one full sash opening is better than two equal in area. The same holds for windows with some exceptions.

Generally, it is better to open the top sash fully and let the airflow out. The replacement air will come from the open door and the lower levels of the room. There are exceptions to this rule of thumb, **Figure 18-41**.

If the smoke condition from the door opening will make the room conditions worse, then the choice would be to close the door and open the window. If there is one window, the two-thirds/one-third rule is appropriate. The heated upper air will vent from the upper level and the cooler fresh air will enter through the lower opening.

If there are two windows, one should be fully opened at the upper sash and the other opened fully from the bottom sash. Obviously, the greater the opening at the upper level, the larger the heated air volume that can be allowed to escape.

If there are three windows, two can be opened at the top and one at the bottom. If there is a breeze or wind blowing into the outside wall on which there is a window, the window facing the breeze should be

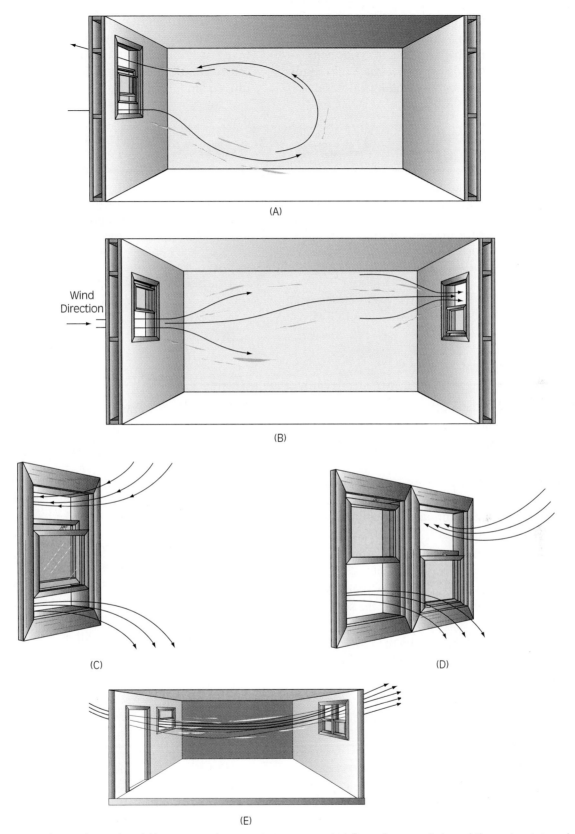

Figure 18-41 Sash windows should be opened to maximize natural airflow characteristics while maintaining the principle that large openings are more efficient than many small openings. Heat rises and cool air sinks. Depending on the number of windows and their placement, various different combinations will be effective. (A) Single window in a room. Open it two-thirds from the top, one-third from the bottom. (B) Two windows in a room. Open bottom sash of window facing wind fully. Open opposite window fully from the top sash. (C) Single window; another view. (D) Double window. Open one sash fully from the top and the other sash fully from the bottom. (E) Cross ventilation. Open window sash facing wind fully from the bottom and the others across the room fully open from the top.

opened from the bottom. The forced air will then actually pressurize the room and the vent/exit opening will be at the upper levels where the heat has collected.

The preceding scenarios represent ideal conditions. The firefighter must evaluate the effect that is being created and be prepared to adjust the plan if the desired effect is not forthcoming. Again, the many variables that tend to be present can change the rules. Nothing can be assumed and everything must be monitored until the incident is over.

Lessons Learned

Ventilation is a tool that is used in firefighting just as any other tool is used. It must be understood, manipulated to the greatest advantage, and used carefully. Its proper use can make the difference between extinguishing a fire and creating a conflagration. Its judicious use can enable a firefighter to enter a structure to make a rescue. Used inappropriately it can permit a fire to extend into uninvolved portions of a structure.

There are no hard and fast rules beyond some very simple truisms. Heat rises and cold air drops. Heated air expands and cooled air contracts. Natural air movement will follow the path of least resistance. Airflow can be artificially generated by mechanical means. Beyond these, the rest depends on the conditions.

KEY TERMS

Cockloft The area between the roof and the ceiling.

Diffusion A naturally occurring event in which molecules travel from levels of high concentration to areas of low concentration.

Frontage The portion of a property that faces and actually touches the street.

Guard Dogs Trained animals that will bark and attack an intruder.

Horizontal Ventilation Channeled pathway for fire ventilation via horizontal openings.

Ignition Point The temperature at which a substance will continue to burn after the source is removed.

Lifting A term used to describe the removal of upper level smoke and heat when cool air replaces the upper level hot air that is escaping.

Mission Vision A term used to describe a condition in which a person becomes so focused on an objective that peripheral conditions are not noticed, as if the person is wearing blinders.

Pipe Chases A construction term used to describe voids designed to house building water supply and waste pipes. The term *electrical chase* is used for wiring.

Thermal Level A layer of air that is of the same approximate temperature.

Vertical Ventilation Channeled pathway for fire ventilation via vertical openings.

Watch Dogs Trained dogs that will bark and create a commotion, but will not attack.

REVIEW QUESTIONS

1. Define ventilation.
2. What are the advantages of venting a structure?
3. Name the two types of ventilation.
4. Define and explain the terms *backdraft*, *flashover*, and *rollover*.
5. Describe an expandable roof cut.
6. Describe the concept of the two types of forced ventilation, negative and positive.
7. What is the difference between opening a roof and cutting a roof?

Additional Resources

Carter, Harry, *Firefighting Strategy and Tactics*. Fire Protection Publications, Oklahoma State University, Stillwater, OK, 1998.

Casey, James, *Fire Chief's Handbook*. Technical Publishing Company, New York, 1978.

Mittendorf, John, *Ventilation Methods and Techniques*. Fire Technology Services Publishing, 1988.

CHAPTER 19

FIRE SUPPRESSION

Dennis R. Childress, Orange County Fire Authority

OUTLINE

- Objectives
- Introduction
- Elements of Fire Control
- Tactical Considerations
- Lessons Learned
- Key Terms
- Review Questions
- Additional Resources

Photo courtesy of Central Net Fire.

STREET STORY

*The Monroeville Fire Department's Rescue 4 arrived at the scene of an apartment building in the early morning hours to find heavy smoke and fire venting out of a basement apartment. The building was a **garden apartment** structure with two stories in the front and three in the rear because of the topography, since it was built into a hillside. Each floor had four apartments between stairwells. Each stairwell was protected by firewalls or separation walls. The basement apartment was fully involved in the rear. The crew stretched a preconnected 1½-inch hoseline through the front door and down the steps to attack the fire and protect the interior stairwell.*

The other first-alarm units arrived and began normal fireground operations, search and rescue, water supply, ventilation, and secondary fire attack lines. The arriving chief quickly ordered the second alarm.

*Upon my arrival, on a second-alarm engine, the fire had **autoextended** up the rear of the building, entering the cockloft, and was venting through the roof in several areas. The third alarm was sounded. We were assigned the task of stretching a 2½-inch handline up the southern stairwell, pulling the ceiling, and supporting the firewall. By supporting the firewall, I mean we were to ensure that the fire did not get past us into the next set of apartments. We were assisted by a second engine company with a 1½-inch handline. Although an uncommon procedure today, we operated for the duration of our SCBA bottles, exited for fresh bottles twice, and returned.*

After our third SCBA bottle we exited the building and were told to take a break. By this time the fire was under control, forward extension stopped, and the fire would just require overhaul. It was daybreak. The scene in the daylight was impressive. At the height of the fire, three elevated master streams had operated over the main body of fire while six handlines supported interior firewalls, preventing further horizontal fire spread.

What I saw and experienced reinforced everything I had learned from fire academy instructors and through training. For a safe and effective outcome, the suppression of fires requires a coordinated attack, with crews working under a controlled command system and all personnel carrying out the assignments given them. Firefighters must understand the importance of fire suppression techniques and understand the role they play on the fire scene. At this fire, spread was prevented by the incident commander knowing the proper application of fireground tactics and the firefighters understanding suppression methods, application of water, and working as a team.

—Street Story by James Angle, Chief, Palm Harbor Fire Department, Palm Harbor, Florida

OBJECTIVES

After completing this chapter, the reader should be able to:

- Identify structural fire considerations to be made prior to extinguishment.
- Explain the process of fire moving from contents to structure.
- Discuss the resources a fire department considers important in fighting fire.
- Explain the fire tetrahedron in relation to wildland or ground cover firefighting.
- Discuss some of the features of topography to be considered in wildland firefighting.
- Identify some of the automotive structural dangers to firefighters in vehicle fires.
- Explain some of the components of flammable liquids and gases that affect firefighting.
- Define a plan of action for fighting fire regarding attack modes and styles.
- Explain fire stream selection considerations that must be considered in firefighting.
- Explain tactical goals to be considered when fighting fire.
- Discuss the incident management system and how it affects the way fires are controlled.
- Explain the difference between offensive and defensive modes of fire attack.
- Discuss teamwork and its part in firefighting.
- Explain the concept of the two in/two out rule and how it affects structural firefighting procedures.
- Explain why caution must be observed in the overhaul phase of fighting fire.

INTRODUCTION

This chapter gives the reader information that pulls together all of the other information presented to this point. Like a recipe, which lists ingredients and then explains how to blend and cook them, this chapter explains how to put into action the techniques and methods that have already been discussed.

This chapter looks at some of the common types of fire as well as some common techniques of firefighting in different disciplines. It opens with the more common elements of fire control, detailing most of the areas with which average firefighters will be involved during their careers. In this section, seven subjects provide information needed when working in various situations common to the average fire department. A second section details tactical considerations. At the conclusion of this chapter, the reader will have a good foundation of knowledge in fighting fire and tactical situations involving a number of elements common to today's fire service.

ELEMENTS OF FIRE CONTROL

Before firefighters can respond appropriately to an emergency involving fire suppression, they must know the basic principles involved in the processes that create and sustain fire. This section deals with those areas common to the average fire department. It discusses structural firefighting elements, ground cover or wildland firefighting elements, vehicle fires, flammable liquid and gas fires, the process of fire extinguishment, and proper stream selection.

In a previous chapter the fire tetrahedron and its importance in the suppression of fire was discussed. The elements of that tetrahedron are mentioned again here in an effort to instruct firefighters in safe, efficient, expedient fire control. A simple rule is that the earlier the fire department arrives on scene, and the more knowledge they bring with them in the fire suppression process, the lower the losses and the less risk taken.

Structural Fire Components and Considerations

When discussing structural fire components and considerations, it makes sense to look again at the fire tetrahedron, **Figure 19-1**. Most structural firefighting involves the suppression of Class A materials within the structure or as part of the structure itself. These elements are commonly suppressed by removing one side of the fire tetrahedron. The typical fire

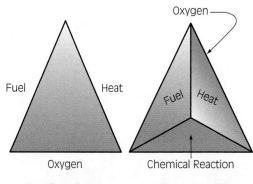

Figure 19-1 The old and new ways of visualizing the combustion process, the fire triangle and the fire tetrahedron.

department will respond with water as an extinguishing agent and suppress the fire by removing the heat, oxygen, or both as quickly as possible. This may not seem difficult in its purest form but can be extremely complicated or hazardous if not done properly and safely. A number of structural firefighting considerations must be made prior to the extinguishment, or a great number of things can go wrong.

Listed in random order are a number of factors that must be taken into consideration.

1. Length of time the fire has been burning
2. Building construction materials
3. Occupancy type and contents
4. Resources available (amount of water, staffing, equipment, etc.)

Taking these factors in order of listing and not in order of importance, recommendations can be made as to how a structure fire should be fought.

The length of time the fire has been burning is an important factor because it can determine the stability of the structure and the amount of fire the responding units will face. It will also determine which stage of involvement the fire is presenting.

As discussed in previous chapters, flashover and backdraft are extremely dangerous fire phenomena that are predicated on the stages of fire in structures. In **Figure 19-2** a flashover is shown as it occurs over the heads of a couple of firefighters. Based on length of burn, the integrity of the structure itself can be in jeopardy.

> **Caution** Most structure fires begin in contents and then spread to the structure itself. If not contained, this structural involvement can create concerns about collapse, both inside as well as outside.

This brings up the second factor: the building materials themselves. Is the structure made of highly combustible or noncombustible materials? How long will it take for the fire to become involved in the structure itself? **Figure 19-3** shows an apartment fire that began in its contents and then burned into the structure. Chapter 13 details the types of construction materials and demonstrates the dangers involved in structural firefighting.

The third factor listed is the building occupancy and contents. Building occupancy means the proposed use of the structure. Is it designed to be an automobile repair shop or a preschool? Is it a retail store or a board and care facility? Each of these occupancies presents different fire attack protocols. The life hazards, the entrance accesses, and the ventilation components are among many details associated with the occupancy types that have to be taken into consideration by the responding units. Along with the occupancy, the building contents must also be considered. In most cases, the contents will be predicated by the occupancy type. The preschool will have dramatically different contents than the repair garage. These contents will have a bearing on the suppression efforts and control problems encountered by responding units.

Figure 19-3 A contents fire has burned into the structure itself. *(Photo courtesy of Rocco Di Francesco)*

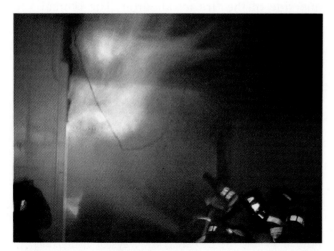

Figure 19-2 A flashover may occur in the early stages of a structure fire. *(Photo courtesy of Phill Queen)*

The fourth factor listed is resources available. Fire suppression involves all of the factors listed, as well as the resources available to the firefighting team. Some areas of the country have limited water supplies while others have practically unlimited water. Some departments have large numbers of responding apparatus while others have only one engine with the second responding unit a very long distance away. Some areas of the country have staffing problems while others do not. Some areas have unlimited training opportunities while others do not. These and many other situations are considered when dealing with response resources. These resources—or the lack thereof—have a great bearing on the tactics and strategy used in fighting structure fires.

Structural fire components and considerations are a large part of the responding agency's responsibility. Fire service personnel must be well schooled in all of these factors and situations so as to ensure efficiency, safety, and responsibility.

Ground Cover Fire Components and Considerations

In some areas of the country ground cover fires are a very large part of the fire service's work. Not all departments have jurisdictions with open areas that burn, but those that do find it to be a responsibility that is not taken lightly. In fact, large campaign fires are usually wildland (ground cover) fires. These fires, being so large and involving so many resources, usually cause the first-in departments to utilize other departments from miles around, and in some cases out-of-state departments, to assist in the attack. **Figure 19-4** shows a large wildland fire and its impact on the surrounding area. So, even if a department does not have any open land to burn in its response area, any firefighter in any department could be summoned to help in another agency's wildland area. This being the case, it is a good idea for all firefighters to know the components and considerations involved in ground cover firefighting.

In Chapter 4 the fire tetrahedron was discussed. In the previous section of this chapter the same tetrahedron was mentioned, but in the field of ground cover firefighting a fire triangle is used, **Figure 19-5**. In wildland firefighting this triangle reflects a different set of principles related to a different set of conditions when compared to structural firefighting. Each side of the triangle can be broken down into smaller components in order to understand the components and considerations of ground cover or wildland firefighting.

Weather

This component of the wildland fire triangle is the most dynamic of the group and hence is discussed first, **Figure 19-6**. This is because of its dramatic effect on wildland fire conditions. It changes all the time, including any number of times in a given day. Given similar fuels and topography, wildland fires will burn in a radically different way in different weather conditions.

> **Caution** Weather is one of the main causes of firefighter fatalities on wildland fires. In fact, whole teams of firefighters have been caught and killed on wildland fires a number of times over the years because of weather changes that were not recognized in time to change tactics or locations.

> **Caution** Frontal changes (cold and hot), high and low pressures, winds, and storms all have direct and dangerous effects on wildland fires.

Each of these elements can change in very little time on any given day. It is also a fact that some ground cover fires grow so large that they can create their own weather.

A number of safety rules or laws apply to wildland firefighting, based on years of study and practice. A great number of them pertain to the weather and its dangers to the fire environment because this area of wildland firefighting has caused so many injuries and deaths among the firefighter ranks over the years.

Figure 19-4 The Baker fire in Southern California lays out a large footprint on the area. *(Photo courtesy of Craig Covey)*

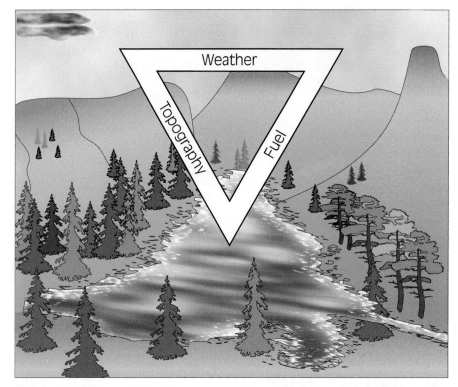

Figure 19-5 The wildland fire triangle differs from the structural fire triangle.

Figure 19-6 The components that make up the weather side of the wildland fire triangle.

Fuels

In the wildland environment, there are a great number of fuel types, **Figure 19-7**. From fine grasses to large timber, the fuels burn differently. There are two main variables in fuels and their effects on fire. These are **rate of spread** and **fire intensity**.

The rate of spread is the speed with which the fire travels across the ground or through the bushes and trees. This rate of spread is affected by a number of variables, all of which are in the wildland fire triangle. In the case of fuels, it can be said that the lighter or finer the fuel, the faster the rate of spread. Given similar weather and topography, fire will run much faster in grasses and light vegetation than in heavy brush or timber. **Figure 19-8** shows a grassy fuel fire moving quickly across the terrain. In fact, this deceptiveness has caused a great number of firefighter deaths and injuries over the years. Numbers of firefighters have been overrun in grassy fuels thinking they could "outrun" the fire when things went bad. It is a sad fact that many firefighters have been caught and burned when running from "light fuels" wildland fires.

Figure 19-8 Grasses and light fuels burn fast and must be respected. *(Photo courtesy of Rocco Di Francesco)*

The second term used here is the *fire intensity*. This is the amount of heat and flame production given off by a given fuel type. Rate of spread has an inverse relationship to fire intensity. Whereas lighter fuels have a higher rate of spread, they have a lower fire intensity. Heavier fuels have a lower rate of spread

Figure 19-7 The components that make up the fuels side of the wildland fire triangle.

and a greater fire intensity. Light grassy fuels burn fast, but not with great flame lengths, while heavy timber burns slower, but with much more intensity, releasing higher flame lengths and Btus. **Figure 19-9** shows a timber fire with long flame heights and high intensity. Flames can run up to 100 feet or more into the air over a timber or heavy brush fire.

The last fuel concept discussed here is fuel spacing or continuity. The fuel can be spaced tightly or loosely. Depending on its compactness, the fuel may or may not burn well. If air can circulate around the fuels and they are close enough to each other that they practically touch, then the fire will burn better than if the fuels are so densely packed that no air can circulate in and around them. At the other extreme is fuel that is so sparse that the fire cannot travel from plant to plant. In this case the fire may not run as well. It could even put itself out.

Topography

The last side of the wildland fire triangle is the side called topography, **Figure 19-10**. This term deals with the layout of the land. The **steepness of the slope**, the direction the slope faces, the **drainages** and **river bottoms**, and the **ridges** and **saddles** are all critical to the behavior of fire in the wildland.

It is a fact of nature that fire runs uphill faster than downhill. This is because the flame will run vertically and because the slope places the uphill vegetation close to the flame, **Figure 19-11**. The plants are preheated and burn quicker as the heat travels

Figure 19-9 Heavy timber fires burn with high intensity and flame length. *(Photo courtesy of Phill Queen)*

Figure 19-10 The components that make up the topography side of the wildland fire triangle.

Figure 19-11 The angle of the terrain will have an effect on a running fire. Note how the flame "leans" into the hill, preheating the fuel above it. If this fire was burning or "backing" down the hill, it would move much slower because the fuel is not preheated.

Figure 19-12 A fire crew enters an area with complex topographic features and stops to study the terrain. *(Photo courtesy of Rocco Di Francesco)*

upward. Conversely, it is said that when a fire is burning down a hill, it is "backing down" because the flame is leaning away from its direction of travel. When the fire backs down a hill, it naturally travels slower because the fuel ahead of it is not being preheated and has to catch fire by conduction rather than convection. It then naturally follows that the steeper the hill, the faster the fire will travel upward and the slower it will travel downward.

> **Streetsmart Tip** A concept the wildland firefighters follow here is that from flat to a 30 percent slope the fire will double its rate of spread and at a 55 percent slope it will double again.

The direction the slope faces is also very important because vegetation grows differently on different **aspects**. In the western United States, for example, the sun warms southern-facing slopes more than it does others. Because of this, the vegetation on the southern-facing slopes is dryer and generally lighter in consistency. This causes the southern-facing slopes to burn with higher rates of spread than the other aspects. Also, with the sun preheating the fuel, the chance of fire activity on these southern-facing slopes is much higher. In other parts of the country a different aspect will be dryer, but the principle is the same. Firefighters should know what aspect gets more sun in an area before planning their strategy in fighting a ground cover fire. **Figure 19-12** shows a team of firefighters entering an area that has multiple topographic features. All must be understood and monitored.

Some more common wildland firefighting terms related to topography are saddle, ridge, drainage or **chimney**, **box canyon**, and **midslope**. These are described in the Key Terms section at the end of this chapter. They all have great relevance in wildland firefighting. Each one has associated dangers when involved in fire. All but the ridge tend to draw fire to themselves. If a ground cover fire is anywhere near these topographic features, the danger level is very high because each will draw the fire to itself in a rate-of-spread increase that will astound the firefighter. These features are responsible for many firefighter deaths in wildland fires.

When discussing ground cover fire components and considerations, it becomes clear that it is a very complex science. Firefighters must understand these components to better ensure the safety and survival of those who fight fires in this environment. The numbers of firefighter deaths and injuries in wildland firefighting are very high, second only to residential firefighting. It is for this reason that ground cover firefighting must be understood and respected by firefighters the world over.

Vehicular Fire Components and Considerations

Studying vehicular fire components and considerations is not a matter of just walking up and applying water. A number of variables can be involved in this practice. This section deals with the more common variables associated with today's automotive industry.

The most logical way to discuss the automobile and its dangers when burning will be to start at the front and work toward the rear. In past years automobile models had a metal bumper held on to the car with brackets. When hit, the bumper nearly always damaged the car. The industry then developed hydraulic supports for the bumpers in order to lessen the chance of car damage when lightly struck. Today's cars have hydraulic supports in both the front and rear to protect the car.

> **Safety** Firefighters should understand that when heated, as in an automobile fire, these hydraulic fluid-filled bumper systems undergo great stress as the fluid expands. This stress can cause the bumper to be propelled off the car, traveling up to 40 feet or more. Persons standing in front of the bumper (front or rear) when this happens can be severely injured.

The point here is to avoid standing in front of any bumper after it is been heated.

The engine compartment has a mass of very hot metal filled with flammable liquids (fuel, oil, and hydraulic fluids). Any fuel leak can ignite quickly because of the heat from the engine. In many cases the fuel fire is extinguished and then keeps reigniting as the heat from the engine contacts new fuel leaking from burned lines or melted parts. These are Class B fires as opposed to Class A fires. In some cases, the engine itself is made of flammable metals such as magnesium. This is a Class D fire. Another danger in this area is the battery and its associated wiring. A battery is a plastic box filled with a hazardous material (sulfuric acid) that creates a flammable gas (hydrogen) as well as electricity. There is the potential for a Class C fire.

The recent addition of electric vehicles to the nation's highways adds another component to the firefighter's information needs. The firefighter must be able to recognize an electric vehicle and be knowledgeable about the differences it presents on the fireground. The first thing to look for will be an insignia of some type, designating electric power. The insignia will be located on the sides or rear of the body. The next step in identification comes when the hood is opened. The firefighter will not find items normally found on a gasoline engine automobile, such as an engine, a transmission, cylinder heads, an alternator, fuel injectors or spark plugs, an air filter, or accessory drive belts. Instead, the firefighter will find high-voltage warning labels on components. Although the chances of finding an electrical (Class C) fire are remote due to numerous manufacturer safeguards, firefighters may have to alter their tactics when fighting fire in this type of vehicle.

The primary dangers in the passenger compartment are the materials used in construction. Most materials found in the passenger compartment are plastic, a form of polyvinyl chloride (PVC). PVC is extremely toxic when burned. A great number of studies have been performed on this material during the past few years and all have come to the same conclusion.

> **Safety** Plastic PVC is many times more toxic than wood and cloth when burned. Firefighters must wear SCBA when working around this material in any stage of combustion. A second danger in the passenger compartment that few consider is that of the hydraulic pistons used to open the hatchback and hoods in some models. These pistons will react to heat in the same way as bumpers do.

Another item that must be mentioned here is the air bag mounted in the dash on either or both sides, and also on the side of the front seat in some models. Great care must be taken when working around these items because of the chance of an accidental discharge. Each model has its own idiosyncrasies that must be studied individually to truly understand the mechanics involved in its design. In most cases, they can be disarmed by disconnecting the battery to the car and waiting some period of time before working around the bag. The side air bag can be disconnected by cutting the power cord running up the side of the seat from the floor to the bag. It is suggested that even after disconnecting the power to air bags, caution should be used when working around them.

The next area to consider is the trunk. At first, a car's trunk might not seem to pose much of a problem but that is not the case. The trunk is used for storage. Firefighters have no idea what is stored in the trunk (without being told by the owner) until it is opened. It could be any type of flammable liquid or gas container, ammunition, explosive commodity, or other dangerous cargo. It has been known to even contain humans.

> **Safety** When opening the trunk, great care must be taken to avoid exposure to potential harm. The firefighter should stand to the side, or crouch very low, and wear full protective clothing with the eyes protected by a helmet shield pulled down or by goggles.

It may be prudent to also have a hoseline charged and ready when accessing this area of the automobile just as is recommended for all other compartments.

The last area of the automobile to consider as dangerous is the underside or fuel storage area. Action movies almost never stage a car crash without the fuel tank rupturing and exploding. In real life this is extremely rare; however, firefighters must approach with caution. What can and usually does happen in car accidents is that the fuel tank will be damaged and be leaking raw fuel on the ground around the vehicle. In this case all precautions must be taken to secure possible ignition sources until the fuel leak is stopped. Dirt or any nonflammable absorbent material will do the job of recovery until something more efficient can be obtained.

When discussing the fuel tank it is important to remember that gasoline may not be the only fuel present. Cars are run on a number of fuels today. Gasoline, diesel, propane, methanol, electricity, and steam are but a few of the fuel systems in use today. Each one carries its own risk and will dictate the tactics used in securing the fuel source.

The last consideration to be mentioned here is that of another type of vehicle, a freight truck. For the purposes of fire components and considerations, trucks (all but the trailer) are simplified as short vehicles. The compartments are the same, only compressed. The materials used are relatively the same and in the same locations. The engine compartment is larger but contains just about the same materials and hazards associated with automobiles. The same precautions are necessary.

The trailer on the truck is another story and can also be dealt with by simplifying it. For the purposes of this chapter, the trailer body is treated as a structure off the ground. It can contain just about anything a structure on the ground contains and can be just as dangerous. All of the principles of caution with structure fires will hold true here.

Studying vehicular fire components and considerations is not an easy task. A vehicle contains nearly all hazards associated with other areas of firefighting, but they are compressed into an area approximately 8 by 20 feet with possible occupants. The risks are there and the firefighter must be aware of every possibility.

Flammable Liquids Fire Components and Considerations

Unlike most Class A fires that either go out when water is applied or continue to burn at a given pace if water is not applied, flammable liquids fires can actually be complicated by improper actions of firefighters. Whereas Class A fires are somewhat predictable and follow given laws of nature when burning or when water is applied to them, flammable liquids fires are not. Each flammable liquid has its own specifications that will affect its burning characteristics and extinguishment.

Chapter 4 introduced a number of terms related to flammable liquids. Knowledge of these terms is critical. Flash point was mentioned, and that characteristic of every flammable liquid will make a very big difference in a material's storage, handling, and extinguishment.

> **Caution** The vapors given off by a flammable liquid can be that material's most hazardous profile, sometimes traveling great distances before finding an ignition source.

For instance, an automobile accident with leaking fuel and no fire can turn into a very dangerous situation if the vapors traveling outward find a distant ignition source bringing fire back to the leaking area.

Another term used in the field of flammable liquids is **solubility**. This is the degree with which a liquid will mix with water. In chemistry, the materials that are soluble in water are called **polar solvents**. Materials that are polar solvent, such as alcohol, will dilute or mix with water quite easily. Hydrocarbon-based liquids (gasoline, oil, etc.) are nonpolar solvents and do not mix with water. They either float on water or sink to the bottom of it. Most flammable liquids are not water soluble, but knowing which ones are can go a long way toward the prevention or extinguishment of a fire involving these materials. Liquids that are not water soluble do not mix with water and either float on water or sink below it. This is termed the liquid's **specific gravity**. Knowing the specific gravity of the liquid in question will tell the firefighter which method of extinguishment should be utilized. If the liquid is heavier than water and sinks, then we know water can be used to extinguish it, if applied properly. If the liquid is lighter than water, and floats on the water, then an alternate means of extinguishment must be used. **Figure 19-13** shows firefighters applying Class B foam to a flammable liquid fire by gently covering its burning surface.

Knowing the material that has spilled, or is burning, and some of its characteristics will greatly benefit the fire company, providing them with a margin of safety and the ability to effectively accomplish their goal of mitigating the hazard.

Figure 19-13 Firefighters gently apply foam to a flammable liquid fire. *(Photo courtesy of Phill Queen)*

Flammable Gas Fire Components and Considerations

Flammable gases, in many cases, can be grouped with the flammable liquids category. In fact, a great number of flammable gases are products of flammable liquids. Often when shipping a flammable gas, the manufacturer will compress the gas in order to ship a larger quantity, turning the gas into liquid under pressure.

As with flammable liquids, it is very important for the firefighter to understand the hazards associated with gases. Gases are typically stored under pressure. Therefore, when a container is ruptured or leaking, the gas escapes immediately, creating a hazard as large as the gas can spread. An example is propane. **Figure 19-14** shows a propane tank fire with firefighters cooling and containing the flames. Propane is stored as a liquid and the gas coming off the liquid is used for a multitude of purposes. Should a propane container rupture, the liquid turns into a gas, expanding over 250 times. That can be a really large problem for the responding firefighter.

The firefighter will want to know the properties of the flammable liquids so a plan of action can be determined and the hazard can be abated. One of those properties will be the gas's **flammable range**. This range will determine whether the fuel is too rich or too lean to burn in the given space it occupies. If the spill or leak is inside a structure with no windows or doors open, then there is a chance that the gas is too thick or rich to burn, as the ratio between oxygen and the gas will not support combustion. Conversely, if the spill is in a structure with open windows and a stiff breeze is blowing through the building, the gas-to-oxygen ratio may be too thin or too low to support the burning of the gas.

Figure 19-14 Firefighters apply water fog to cool a venting propane tank. *(Photo courtesy of Phill Queen)*

Another term common to flammable gases is **vapor density**. This term is very similar to the specific gravity term used with liquids. Basically, it is the weight of the gas as compared to air. Is the gas lighter or heavier than air? Knowing this fact about the material will predicate the plan of action taken because the gas may be hanging low and heavy, looking for an ignition source, or it may be light and dissipate quickly, causing far less concern.

The last term discussed here is possibly the most important: **toxicity**. Simply put, will the product kill a person who encounters it? This contact can be by a means other than breathing. It can be through skin absorption or ingestion also. Knowing this fact will determine what the responders should wear. Will turnout clothing be enough or will a hazardous materials suit have to be worn? And, in some cases, will that hazardous materials suit be enough?

The flammable gas fire components and considerations are many. Just as there are literally thousands of places where these gases can be found, there are a number of ways to deal with them when they are found. The firefighter interested in longevity will pay close attention to the details involved with this firefighting hazard.

Process of Fire Extinguishment

This section begins to develop thought processes based on the previous sections within this and other chapters. The concepts and considerations of the differing situations and classes of fire previously discussed begin to be put into the context of fighting fire.

> **Note** Before firefighters take on the beast of flame, they must first understand that fire is a multisided three-dimensional presentation of heat and chemistry. It must be considered from six sides. It has a top, a bottom, and four sides. It will follow laws of physics and applied science that only the most seasoned firefighter will understand, and even then things will go wrong on occasion. The goal in defeating this beast can only be met by extending an aggressive, fast, well-placed, adequate fire attack.

To accomplish this goal, a number of steps must be taken. The first step is to create a plan of attack. This is done by first locating the fire and determining its extent. Generally speaking, the longer it takes to find the fire, the longer it will take to put it out. Good communication becomes very important here. As Chapter 2 described, this report is built on careful and complete observation of the fire and its development to this point. Each department will usually have SOPs that determine who is first on scene and where each unit should be located. These policies are not set in concrete because each fire is different. However, they have been set up in order to maintain order and efficiency. The attack plan will be built on the observations, training, and opinions of the first-in officer after considering the facts as perceived and predicting a course of action.

The second step is then to apply the plan of action as quickly, efficiently, and safely as possible. Firefighters will take their skills, knowledge, and ability into the face of danger in an effort to mitigate the situation as quickly as possible. Three methods of fire attack are possible: the direct attack, the indirect attack, and the combination attack of fighting fire. The direct attack is quite simply the act of putting water directly onto the seat of the fire, **Figure 19-15**. This method is the most efficient use of water on free-burning fires. A solid or straight stream of water is applied in short bursts directly onto the burning fuels (Class A) until the fire darkens or is extinguished. **Figure**

Figure 19-15 A direct attack with firefighters applying a straight stream onto the seat of a fire.

19-16 shows a firefighting team about to enter a burning structure and apply a direct attack on the seat of this fire. If used indoors and done properly, the thermal balance of the room may still hold, giving the firefighting team a clear view of the fire and contents of the structure.

Another method is that of the indirect attack, **Figure 19-17**. This attack is usually utilized when the firefighting team does not see the seat of the fire. A fog stream is applied to the upper areas of the room or above the fire seat in an effort to "steam" the fire out. **Figure 19-18** shows a firefighting team applying

Figure 19-16 A firefighting team prepares for a direct attack on a contents fire. (Photo courtesy of Phill Queen)

Figure 19-17 The indirect attack has the firefighters applying a 30-degree fog into the upper heat layer of the fire in order to create a steam that will extinguish the fire.

a fog stream into the upper area of a structure in an indirect attack method. Directing the stream into the superheated upper atmosphere above the fire causes the water to "boil" or "vaporize," steaming the area and causing the fire to die. This method, if used indoors, will usually greatly disturb the thermal balance, causing a loss of visibility in the structure as heat and products of combustion are circulated downward. When using this method, the firefighting team should attempt to shield themselves from the steam vapor by backing out of the area after applying the water. They can reenter as soon as the vapor begins to dissipate.

> **Caution** Note also that the indirect method is not appropriate for occupancies with victims in need of rescue, because the steam created by the attack will burn the victims quickly.

The last method is that of a combination attack. This attack is a blend of both the direct and indirect methods. The straight stream is directed to the seat of the fire and then the nozzle is quickly changed to a fog and the upper areas of the room are covered with a quick shot into the thermal cover. This method will darken the fire and stop the growth of flashover.

In each of these methods of applying water, firefighters must remember that the goal of the fire department is to save property. If water is not applied in proportion to the amount of fire, two things can happen. If not enough water is used, extinguishment will not take place. If too much is used, then the damage by water may exceed that done by fire. The lesson here is to use only what is needed, saving the rest for overhaul if needed.

Proper Stream Selection

Fire streams were discussed at length in Chapter 11 with information regarding a number of things that are mentioned again here. The purpose of this section is to transition from knowledge of the burning characteristics of various occupancies and hazards to actual fireground use of the information.

In the case of fighting fire with water, it is important to understand that in order to be successful, sufficient water must be applied directly to the fire in order to control it. That statement may sound simple until one begins to take into account all of the mechanics of performing that task—not only the physical movement but the mental as well. Success will be based on a number of factors in the selection of the stream alone. Some of those factors, in no particular order, are proper stream type, stream size, stream placement, timing, water supply or quantity of water, stream reach, mobility needs, tactics required, speed of deployment, and personnel available.

Consideration of some of these factors provides a better understanding of the firefighter's role in proper stream selection. The first one was the proper stream type. Is a fog stream the correct choice? If so, which width? **Figure 19-19** shows

Figure 19-18 A team of firefighters applies a fog stream in an indirect attack on a fire. *(Photo courtesy of Central Net Fire)*

Figure 19-19 Firefighters use a fog pattern to shield themselves from radiant heat. *(Photo courtesy of Phill Queen)*

firefighters using a fog stream to protect themselves from radiant heat. Or is the straight stream a better choice? Each situation calls for differing stream requirements. The most basic answer to stream type may be in the type of attack required: direct, indirect, or combination. Each has its place in given circumstances.

Next to be considered is stream size. The typical response to this question may very well be "Big water, big fire and little water, little fire." It was once said that firefighting is like going to battle. The weapon must match the target. **Figure 19-20** and **Figure 19-21** show examples of the need to have the stream size match the fire size to be effective.

> **Caution** If a large line is needed, then it should be pulled right away so as not to play "catch up" with the fire. Some firefighters think the stream to start with should be the stream that will eventually be used, going smaller later, rather than ever having to go larger later.

The placement of the fire stream can make or break an attack plan. The fireground must be coordinated at all times without exception. This is a battle with an enemy that can injure or kill. When an order is given to place a fire stream in a particular location, it is part of a plan. Without that stream, the plan may fail, so this factor can be very important. An example may be a fire in a dwelling where the first line goes between the trapped occupants and the fire, with the second stream going to the seat of the fire, **Figure 19-22**. This operation will not work without both streams being coordinated and controlled. Care must also be taken to avoid opposing streams, because these streams

Figure 19-20 A firefighter applies a single straight stream onto a structure in an attempt to reduce some of the fire's heat. *(Photo courtesy of Phill Queen)*

Figure 19-21 If the line selected does not match the size needed, the fire will burn longer and hotter and can jeopardize the operation.

may drive heat and products of combustion toward the opposing firefighting teams. Still another related factor is that of timing, which will go with speed of deployment. This is self-explanatory. Timing is everything on the fireground. Stopping the progress of the fire as quickly as possible will allow the firefighter to work toward the goals of rescue and property conservation.

Many differing types of nozzles are used in the fire service. One of the commonly used types has the ability to control the amount of water it puts out. In cases of water conservation because of poor sources, a nozzle that conserves water would be advantageous. In such a case, water supply or quantity of water available plays a significant role. A different set of rules applies when the water is limited.

Another factor to consider is the mobility of the stream that will be needed. In some cases, lines are advanced quickly into the fire, whereas in other cases the line is placed in a fixed position holding the fire until another more mobile line can be set up to take over. **Figure 19-23** shows a deck gun attack on a large fire in an effort to bring the fire into handline size by reducing some of its heat. In this same area are the factors of stream reach and tactics. A line may be placed on an exposed structure that is large and

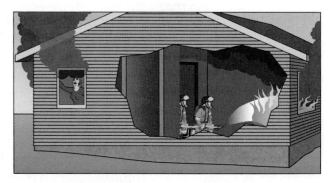

Figure 19-22 The first line pulled should be positioned between the potential victims and the fire.

Figure 19-23 Firefighters use a deck gun to hold the fire until more lines can be put into place. *(Photo courtesy of Rocco Di Francesco)*

Figure 19-24 Large stream appliances at work on a large fire. *(Photo courtesy of Brentt Sporn)*

unmoving, while a smaller, more mobile line is extended into the burning occupancy to fight the fire.

Everything taken to a fire is a resource operated by the fire department. The personnel, water, equipment, and knowledge are all resources. The selection of the proper fire stream is a significant decision. **Figure 19-24** shows the proper use of large streams on a fully involved mercantile fire. **Table 19-1** shows the differing sizes of nozzles and the effort required in order to work properly.

Firefighters carry out their part of the attack plan using water and a good basic knowledge of the requirements being predicted by the fire they are fighting. Knowledge of the factors listed, as well as many others, will arm the firefighter with an arsenal when going into battle against a fire.

TACTICAL CONSIDERATIONS

Earlier in this chapter the elements of fire control were discussed in some detail regarding the most common fire expectations of the average firefighter. Those sections detailed the components and considerations a firefighter must be knowledgeable in to be safe and successful on the fireground. This section deals with those same types of fires in a tactical firefighting capacity. It builds on the information presented in the previous section in an effort to actually use the information in fighting fire.

The actual fighting of fire on the fireground is broken down into tactical objectives. Each of these

Hose Stream Characteristics

TYPE OR SIZE OF LINE	REACH (FT)	MOBILITY	GPM	COMMON USE
1-inch or greater booster or reel line	25–50	Excellent	10–40	Very small nonstructural fires or overhaul
1½ to 1¾ inches	25–50	Good	40–175	Quick attack, one to three rooms
2½ inches	50–100	Fair to poor	125–350	One floor or more, personnel permitting
Master stream	100–200	Poor to none	350–2,000	Large, fully involved structures or exposure protection

TABLE 19-1

objectives must be accomplished in order that the safety of the firefighters, occupants, and others is a primary consideration. If there are no occupants, the overall plan still remains in order, keeping the firefighters safe and efficient. These goals are universal in the fire service, although different departments may use different terms. The priority was first set in place a great many years ago by Lloyd Layman, with the following acronym: **RECEO**—Rescue, Exposures, Confinement, Extinguishment, Overhaul.

Another way of setting this same priority is Rescue, Fire Control, and Property Conservation. Still another acronym is termed REVAS: Rescue, Exposures, Ventilation, Attack, and Salvage. There are a number more, but the point is well made that the fire service has set itself a series of goals that must be accomplished in order that the fireground remains as safe and orderly as possible.

> **Note** A second point that must be made is that the command structure must be solid and well defined in order to accomplish the set objectives in each of the goal areas.

Just as an army employs a rank structure to fight a battle, the fire service uses a rank structure to fight fire. One person will be in charge and appoint people who will command different parts of the operation under that person. **Figure 19-25** shows a battalion chief conferring with his captains in planning his objectives. This system is called the **incident management system (IMS)**. The IMS is basically designed to maintain order in any emergency operation. It can be fire, flood, rescue, first aid, mass casualty, or anything demanding a number of emergency responders. A single person will be in charge of the incident and this person is called the incident commander (IC). This person works from the goals listed earlier utilizing strategy and tactics just as a general would in the military. The IC will delegate assignments to arriving units in an effort to accomplish the goals in a safe and efficient order.

To explain the system and complexities of fighting a fire, this discussion sets the scene with an average, generic fire, because the methods and operating principles will apply to any fireground operation.

As soon as the fire is reported, and before units ever pull from the station, mental preparation takes place. These are the recall factors of past similar incidents, the recall factors of knowledge about the area where the alarm is reported to be, the recall factors regarding the fire prevention inspections of that area, the recall factors of time of day and route to be taken in response, and the factors regarding the condition of the responding units as to their capabilities. Are they fully staffed today? Are the water tanks full and the pumps working properly? Are the crews fresh and ready to go to work in possibly very dangerous conditions? All of these factors, and many more, will race through the minds of the responders. Each of these factors will eventually dictate the strategies to be considered.

Upon arrival another set of factors must be considered. All of them have to do with the environment of the response. **Table 19-2** demonstrates that the number of factors having to do with the operation can be almost overwhelming. The seasoned firefighter will go through all of these and possibly more in calculating the environment presented at the emergency. And each of them can greatly affect the outcome of the incident depending on how they are handled.

Based on the situation presenting itself to the arriving firefighters, three methods or modes of attack are possible: the **offensive attack**, the **defensive attack**, and the **combination attack**. These modes are predicated on the resources available at the emergency. If the arriving units have adequate resources to handle the situation, then they will fight the fire aggressively and offensively. They will attack the problem head-on and, following department standards, will accomplish their objectives

Figure 19-25 A battalion chief meets with his captains to set operational plans into action. *(Photo courtesy of Rocco Di Francesco)*

Fireground Factors

BUILDING

Size	Construction type	Condition
Age	Openings	Utilities
Concealed spaces	Access	Effect of fire
Extent of fire	Interior fuel load	Exterior fuel load

FIRE

Size	Location	Direction of travel
Time since ignition	Extent	Materials involved
Material left to burn	Fire load	Stage of involvement

OCCUPANCY

Type	Value	Fire load
Status (used/vacant)	Hazards of occupancy	Life hazard
Arrangement	Obstructions	

TABLE 19-2

efficiently, effectively, and safely. If they do not have adequate resources to aggressively handle the situation, then they will have to fight the fire in a defensive mode of attack. This mode will be continued until enough resources can be massed to then change to an aggressive, offensive attack. The combination mode is most often utilized when a rescue must be accomplished but not enough resources are on scene to handle the operation entirely. An example would be when a line is pulled to defend potential victims until they are removed, knowing that the line will do nothing to fight fire because it is not large enough or practical for the size of the fire.

As the operation begins to unfold, the IC calls for an attack on the fire. The IC's call will be based on the conditions presented earlier. Hoselines will be pulled, utilizing methods prescribed by the responding department's SOPs. Some common hose evolutions were covered in Chapter 10. The concept of **teamwork** will dominate the scene as each person will be responsible for his or her own assigned details of the operation. This teamwork will continue through the entire operation just as it would in a military operation. Each team member will be given an assignment based on the priority of the incident. As mentioned earlier, rescue will be the first priority, and all actions will be directed toward accomplishing that task. A line may be pulled and placed between the fire and the people needing rescue, or the situation may call for rescue first and no line pulled until it is completed, **Figure 19-26**. Each situation will dictate its own need.

A recent rule by the Occupational Safety and Health Administration sets safety standards on the fireground for rescuing occupants from burning buildings.

> **Note** The two in/two out rule tells the fire department that the teamwork concept must be followed on the fireground at all times.

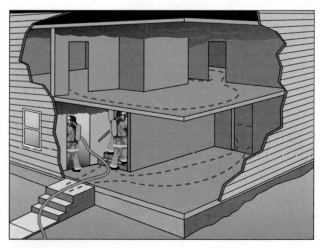

Figure 19-26 If it is not known whether victims exist, then the first line into the structure should assist the search and rescue functions.

This law is designed to protect the firefighter from being injured in rescue attempts where fire conditions are critical.

> **Firefighter Fact** In 1998 the Federal Occupational Safety and Health Administration issued a Final Rule titled *29 CFR Parts 1910 and 1926 Respiratory Protection*. This regulation requires both private industry and the fire service to practice the safe use of self-contained breathing apparatus (SCBA) by mandating a number of rules regarding face piece fit testing, physical requirements for wearers, apparatus requirements, and more. A part of this Final Rule also mandates the number of firefighters required on the fire scene when interior operations are to take place in hazardous situations. The two in/two out rule dictates that during interior structural firefighting in atmospheres that are immediately dangerous to life and health (IDLH), and that do not have lives requiring immediate rescue, at least two firefighters must enter together and remain in visual and voice contact with each other at all times. Two more firefighters must be on standby for the potential rescue of the interior team.
>
> The 1997 NFPA 1500 Standard for Occupational Safety and Health also requires this same two in/two out rule as part of its requirements. This was created in an effort to further follow OSHA laws and protect the safety of firefighters the world over.

Figure 19-27 The most common exposure problem is the closest building to the building on fire.

If the situation does not call for rescue because there are no dangers to people or animals, then the second priority is placed into action. That priority is exposure. Almost anything in the path of the fire can be called an exposure. The most common exposure is a structure next to the burning building where the fire is heating it to a point where it is close to catching on fire, **Figure 19-27**. Another example would be a car fire that is about to ignite nearby brush or tall grass. If an unburned object (i.e., exposure) is about to catch on fire, it is the responsibility of the responding firefighter to stop the spread of the fire to that object. The logic here is that the item burning has already sustained damage, but the exposed item has not. So firefighters save the exposed item before concentrating on the burning item. Exposure protection is accomplished by either removing the exposed item from the area or wetting it to cool its surface below the point of ignition, **Figure 19-28**. Radiant heat is the main cause of exposure fire, so wetting the exposed item removes that heat at the point of contact.

As soon as all exposures are deemed safe, it is time to consider the next step in the firefighting process, which is fire confinement. The fire's parameters are not always clearly defined. Visibility may be difficult at best. The members of the firefighting team must use all of their senses in order to locate the fire. Once located, water or another extinguishing agent is applied in a method that will confine the fire to its area of burn. Stopping the fire's forward progress must be done first, before the fire can be extinguished. It was mentioned earlier that after rescue came fire control. The confinement of the fire is controlling the fire. Some departments will even broadcast a "fire is under control" call to the IC when this step is accomplished. In the wildland, this point becomes even more dominant. An "under control" term may be issued days before actual fire extinguishment is accomplished. When the fire is surrounded and can burn no further outside of this confinement, then it is considered to be under control.

In many cases, to bring the fire under control in structure firefighting another task may have to be added. Ventilation (described in Chapter 18) coordinated with water application may be the only way to achieve the confinement results safely. This will be described more later in the discussion of particular structure fires.

It is common practice to attack most fires from the unburned side, **Figure 19-29**. That is to say that the line will be brought into action between the fire and its intended direction of travel. The opposite of this tactic would be to attack the fire from the rear, pushing the heat and other products of combustion forward into unburned areas, causing far more damage. This is true in almost all applications of

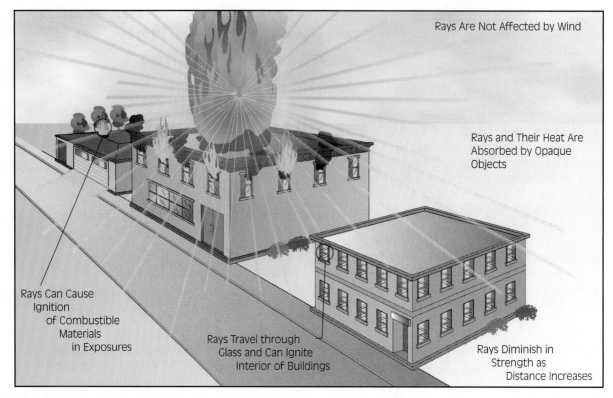

Figure 19-28 Radiant heat is the main cause of exposure fires within short distances. Radiant heat will travel in straight lines from the heat source to nearby objects.

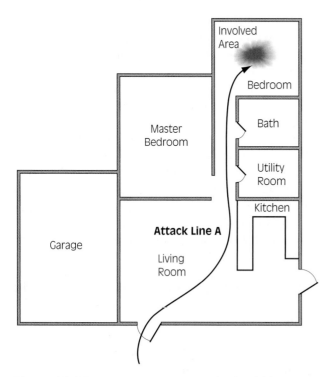

Figure 19-29 Fires most commonly should be attacked from their unburned side, pushing the products of combustion away from unburned materials and areas.

firefighting from structure to wildland or ground cover to flammable liquids and gases.

When the fire is controlled or confined, then the actual extinguishment begins. This is accomplished with direct or indirect attack strategies and resource use. This can also be thought of as property conservation. Saving the property not damaged is a very high priority here. The fire has been confined and now is extinguished utilizing the latest methods and practices in order to save the undamaged property. The use of water is restricted to only as much as is necessary, because water can cause as much and, in some cases, more damage than the fire.

As soon as the fire is extinguished, the overhaul begins. This is a methodical system of ensuring that all embers and chances of reignition are removed. The area is carefully worked by moving every item and wetting all combustibles that could carry new life to the extinguished fire. During this phase of the operation, the cause of the fire is usually determined. By carefully sifting through the fire debris the investigator looks for clues that will lead to the cause of ignition. A closely coordinated effort between these people will result in the call being closed out so the firefighters can return to their respective stations to prepare for the next call.

> **Caution** The firefighter must use extreme caution when in the overhaul phase so that the subtle clues the investigator is working with are not disturbed.

One more point must be made about firefighter safety. That point is about acts of terrorism. The fire service is called upon in nearly every conceivable emergency, representing local authority in the form of a fire department, fire district, or emergency services agency. Unfortunately the firefighter may represent the authority being targeted by an act of terrorism. For this reason it is important that the firefighter be made aware of the possibility of danger from cowardly acts of aggressiveness. Some recognized threats to firefighters are:

- Undetonated explosives: secondary devices, low-order explosions, malfunctions
- Structural instability: collapsed buildings, falling/thrown debris, live electrical lines, leaking gas mains
- Fire
- Hazardous materials (poisons, etc.)
- Biohazards

Further training in this area is highly recommended for all firefighters. Local jurisdictions, police, FBI, the Department of Justice, the State Fire Marshal, and the National Fire Academy offer this type of training. More information on terrorism is given in Chapter 24.

All of the information given in this section is next applied to tactical considerations in fighting fire in a number of given situations, bringing the information in this chapter into a fine focus for practical use.

Residential Occupancies

Fighting fire in residential occupancies has the greatest impact on national fire fatality figures, for both the public and the fire service. A direct result of this statistic was the passage of the two in/two out ruling by OSHA mentioned earlier and in Chapter 5.

> **Firefighter Fact** More people die in residential fires than in any other type of fire in the United States. This is true for both the public and the firefighters.

For this reason, the fire service must place a much greater emphasis on safety here than almost anywhere else. This subsection deals with firefighting in residential occupancies in both single-family and multifamily dwellings.

Before discussing particular residential occupancy types, it is important to discuss a number of generic considerations that hold for all residential fire emergencies. The best way to do this is to go on scene and discuss each step of consideration for the firefighter. Then, subsequent sections will detail single-family and multifamily dwelling operations.

En route to the alarm, the firefighting teams begin the size-up process discussed earlier. The team considers all factors, both silently and in discussions with each other. This practice prepares the team members for the multiple details that will be thrust on them as soon as they arrive on scene.

On arrival, the members of the firefighting team immediately scan the scene for all details remotely related to the fire. As they complete this scan they have already begun to form an attack plan based on the information gathered. The officer in charge of the team (captain, lieutenant, senior firefighter, or other) will then make a decision based on the input received, and the action will begin.

As discussed earlier, the order of attack must follow guidelines set forth in the department SOPs or training mandates. The first priority will be the rescue of occupants. Information is sought from bystanders and other sources as to the number, location, and condition of potential victims, **Table 19-3**. Determining the stage the fire is in at this time is important because it will have a direct influence on the direction the rescue or

Rescue Factors		
Number of occupants	Location of occupants	Mobility of occupants
Verification by bystanders	Condition of occupants	Firefighting required
EMS/ALS required	Manpower for search	Fire's burn time
Hazards to firefighters	Access to victims	Escape after rescue

TABLE 19-3

attack will take. In the earlier stages, the fire can be attacked aggressively and quickly in an effort to cut it off from the rescue attempt. In later stages that will be much more difficult. The firefighting team must make a decision as to the mode of attack based on the resource needs in making the rescue. Are there adequate resources, that is, enough personnel, equipment, and water to accomplish the order of attack selected? Or are the resources on scene limited, causing the team to hold as it waits for adequate help to arrive?

The two in/two out rule is important here. If a rescue is being considered, but there are no facts to substantiate that one is needed, then the rule applies that adequate manpower must be on scene before entering the structure. If the facts presented show that a definite rescue must be executed, then the rule is relaxed and that rescue can take place. As described in Chapter 16, a primary search for victims is conducted very early, as soon as the amount of personnel on scene will allow.

The sequence of priority of multiple rescues is another factor to consider. The sequence shown in **Figure 19-30**, or order of priority, is most generally as follows:

1. Those closest to the fire
2. Largest grouping of threatened people
3. Anyone else in the fire area
4. Those in the areas that will eventually be exposed

After any necessary rescues have been performed, the next priority will be exposure control. This was discussed earlier when tactical considerations were covered. Following exposure control is the step of confinement of the fire, then the actual firefighting itself. As these events begin, a number of simultaneous events must take place to ensure the safety of the firefighting team members. The utilities to the occupancy must be secured and ventilation must be considered. The utilities will consist of the electrical power, the water, and gas supplying the occupancy. A firefighter is typically assigned the task of utility shutoff. This person must find the electrical panel and shut off the breaker switches or unscrew the fuses in older occupancies. Then the gas must be secured by going to the meter or tank and turning off the flow of gas supplying the structure. After that, the company commander will decide if the domestic water to the occupancy should be shut off also. This is usually done if there has been any chance of the water lines breaking during the operation.

> **Caution** The firefighting force does not want to be electrocuted, blown up, or have water running that they do not control. Water can cause as much—or more—damage than the fire.

During this part of the operation, ventilation must be considered. Ventilation is discussed in Chapter 18. A well-coordinated ventilation effort will greatly assist the fire attack team in almost all circumstances. The extinguishment and overhaul parts of the fire attack then will take place bringing the situation to a close.

At this point, many fire departments will offer services to the occupants of the burned structure. These services consist of housing for the night, clothing, food, and, in some cases, counseling. These services are offered by the fire department through a number of volunteer and charity organizations all over the country. Two primary organizations are the American Red Cross and the Salvation Army, but there are a great many others.

This section dealt generically with residential occupancy firefighting. The next topic is the finer details of fighting fire in single-family and multi-family occupancies.

Single-Family Occupancies

The most common element of single-family dwellings is the layout. Granted, floor plans will differ more than porcupines have quills, but there is still commonality. In the one-story home, the bedrooms will generally be on one side of the dwelling, and the living, cooking, and laundry facilities on the other. In the two-story home, the bedrooms will generally be on the upper floor (not always) with possibly one of them downstairs. The other functional areas of the home will be downstairs. In the two-story home it is usually true that there are fewer rooms downstairs than upstairs also, because the functional rooms tend

Figure 19-30 When numerous rescues face the first team they must prioritize their actions based on victim exposure.

to be larger than the bedrooms. Why is this important? Rescue! In the dark, smoke-filled, hot confines of the fire environment, it is nice to have a little bearing on where the search may be most productive.

Another common consideration of two-story homes is that of heat travel. The heated, toxic products of combustion will tend to travel upward because they are usually lighter than air. They will travel toward the sleeping areas of the common two-story dwelling, many times moving up the open stairways.

It is typically the responsibility of the first-in engine to either take command of the incident and control the operation or to immediately attack the fire. As the engine arrives, the driver pulls to the far side of the structure passing by the front, giving firefighters a view of both sides and the front, and making room for other responding units, **Figure 19-31**. This gives the team three of four sides and quite a lot of information from which to draw an attack plan. Then more often than not, lines are pulled to either attack the fire or protect rescue team members. **Figure 19-32** shows an engine parked with lines pulled back to the structure it passed on arrival. In rural areas the engine may have laid a line from a water source to the fire, or they may have brought water with them in the form of a very large tank on the apparatus or a water tank truck. In metropolitan areas it is common for the first-in unit to attack the fire with the tank water in the unit and the second-in unit to bring water from the hydrant or water source, **Figure 19-33**. Departments' SOPs will differ in this area based on experience and resource availability.

As the second-in team arrives at the fire, they are commonly required to offer backup to the first-in unit. **Figure 19-34** shows two engines at a fire scene where the second-in unit has laid hose and is assisting the first-in unit. Some of the common tasks they are responsible for include these:

- Back up the initial lines laid by the first-in unit
- Protect the access and egress of the first-in company

Figure 19-32 An engine has pulled past the involved structure to pull working lines. *(Photo courtesy of Phill Queen)*

- Assist with attack while the first-in works on exposures
- Assist with exposure protection if the situation warrants more lines
- Back up the initial team in any requests they make

The truck company in most areas will arrive and secure the occupancy's utilities and begin ventilation.

Figure 19-31 As the first-in unit arrives, it should pull past the building, giving firefighters a good look at up to three different sides or views of the fire.

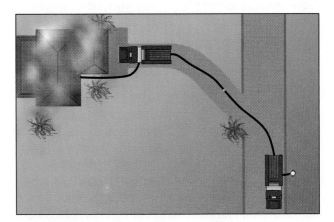

Figure 19-33 The second-due engine may bring water to the first-due unit that went into a "quick attack" using tank water.

Figure 19-34 Here the second-in engine has laid hose from the hydrant and is assisting the first-in unit. *(Photo courtesy of Phill Queen)*

Figure 19-35 A ventilation operation on a single-family dwelling. *(Photo courtesy of Rocco Di Francesco)*

In occupancies that require entry tools not common to engine companies, the truck assists with building entry, and the ladders will be used for a great many purposes as mentioned in Chapter 14. **Figure 19-35** shows a ventilation operation on a single-family residence.

The role of the firefighting teams in single-family dwellings is basic and, as mentioned before, potentially hazardous because of the need to rescue occupants in dangerous situations.

Multifamily Occupancies

Multifamily occupancy firefighting is also potentially dangerous and many times frustrating.

> **Caution** Apartment complexes, condominium developments, upstairs and downstairs tenants, common walls, common attics, poor access, confusing addressing, people or bystanders in the way, parked cars, and many more issues and items make this a hazardous environment for the arriving firefighters. The multifamily occupancy fire is more difficult, dangerous, and demanding on the firefighters than the single-family residence or single-use occupancy.

In large developments, very often the smoke may be the best indicator of the fire's location, giving the firefighters a better mark on which to focus, **Figure 19-36**.

The role of the first-in unit is basically the same as for a single-family occupancy. Size-up and the action must be quick and decisive. It will be based on the factors as perceived by the arriving team and whatever they have gathered from other sources. If the structure has a common means of egress for the occupants, they must be protected by the team, allowing escape. This can be considered a form of rescue, which is the first priority of firefighting (RECEO).

The resources needed at a multifamily occupancy fire are going to have to be upgraded because of the logistics of moving personnel and equipment into areas of need. The distance from the street alone can be great and complex. The greater number of occupants will pose a rescue situation demanding more effort. The structural design lends itself to exposure problems, and confinement will be more difficult due to access problems and larger running areas for the fire to travel. Extinguishment may be difficult because of hidden areas in which the fire will run between floors and units. **Figure 19-37** shows a large multifamily complex on fire during its construction and the size of the pending firefight.

Figure 19-36 In many cases, in larger occupancies or complexes, the smoke may be the best indicator of the fire's location.

Figure 19-37 A large multistory family complex fire will entail a lot more resources than a single-family dwelling fire. *(Photo courtesy of Central Net Fire)*

Figure 19-38 The first-in unit at a mercantile fire should consider parking in front. *(Photo courtesy of Brentt Sporn)*

Business and Mercantile Occupancies

The business or mercantile fire is fought somewhat differently than the residential fire. Here, depending on the time of day, the life hazard potential for fire victims is not as great, and the fire and life hazards to the firefighter can be greater.

In most cases the business having the fire is either closed with no occupants, or open with the occupants either evacuated or fighting the fire. In either case, the chance of having to save lives is far less than for residential firefighting. On the other hand, however, the hazards associated with the occupancy itself can be greater to the firefighter.

> **Caution** Many businesses utilize hazardous materials or dangerous processes, and sometimes storage can be piled high. These can pose a serious risk to the firefighters entering these occupancies under dark, smoky, hot conditions.

Typically, the first-in engine will go to the front of the building in an effort to locate occupants who can assist with information about the fire. They will ask about its location, spread potential (what is burning), lives endangered, time of ignition and source, hazards to watch out for, most important items or processes to protect from fire or water, and where access can be made safely. The officer will then decide on an attack plan based on these and other factors known about the area or business. **Figure 19-38** shows the first-in engine in front of the occupancy and the truck behind it utilizing a master stream. If the officer has decided to attack the fire, then the firefighter is told to pull lines from the apparatus following department SOPs in preparation for attack. If the fire can be attacked without endangering crew members, then an attack is made on the fire. If entering the building will put the crew in danger (per the two in/two out rule), the crew will prepare the attack by laying lines and placing equipment, but wait for reinforcements to assist in the interior attack.

The second-in engine has a number of options based on the conditions of the fire, the department's SOPs, or the direction of the IC. Some of these options are as follows:

1. Respond to the rear of the occupancy.
2. Supply the first-in engine with water.
3. Join the first-in unit with manpower for the initial attack.
4. Lay into the sprinkler connection and supply added water to the system.
5. Stage at the street or hydrant waiting for instructions.

If the first-in engine has not made entry due to either the business being closed or not being able to determine the extent or possibility of fire, then the second-in engine will respond to the rear of the occupancy. The second-in engine will often find the fire first, **Figure 19-39**, because many business occupancies have storage in the rear and this is where fires most often begin.

The first-due truck also has a number of options based on SOPs or need. Some of the uses for the first-in truck can be any of the following:

1. Forcible entry
2. Ventilation of structure
3. Search and rescue (primarily multistory ladder work)
4. Ground ladder operations
5. Securing of utilities
6. Salvage
7. Aerial streams

Figure 19-39 Many times the second-in unit will find the greater hazards when they respond to the rear of the structure.

This is only a partial list and if one truck cannot accomplish all of these tasks in a very short time by itself, then a second-in truck company may be called for assistance.

In these types of fires, firefighter injuries and fatalities can occur from falls and building collapse. The falls can occur from roofs either into the building or off the roof on the side of the building. Chapters 4, 13, and 18 describe safety on roofs and collapse potential.

The business or mercantile fire can be a potentially dangerous operation based on the hazards associated with the products manufactured and stored, and the structural integrity during a fire. This type of firefighting must be taken very seriously and studied well in order for the firefighter to be successful and safe.

Multistory Occupancies

The preceding discussions about fighting fires in structures apply in this section as well. Every tactic and factor considered can be utilized in attacking fires in multistory or high-rise structures. In the case of a multistory fire, it must also be expected that the resources needed for this task will be much greater than for any single-story structure. **Figure 19-40** shows a staging area for a multistory fire beginning to fill with responding units. The support structure of the firefighting system will require multiple layers because the fatigue factors alone can overwhelm smaller systems. In downtown Los Angeles, high-rise responses have been known to contain over seventy-five engines, not counting truck companies and other units.

The number of units responding, as well as the workload, will be predicated on the height of the structure, the location of the fire, and local protocol. If the structure is four stories and the fire is on the third floor, not as many units will be required as for a fire on the fifty-sixth floor of a seventy-five-floor building.

This section is intended to give a basic overview of the high-rise fire operation, because an in-depth study is beyond the scope of this book. Common courses in multistory or high-rise firefighting often run up to forty hours in length. Multiple courses are available and a number of books have been written on the subject.

In the typical high-rise response, a greater number of units will respond due to the nature of the work involved in these fires. Common practice is for the first-in unit to go to the **control room**. This room contains items that will assist in the determination of the location and severity of the fire. Some of the items found in a typical control room are:

- Alarm panels giving alarm details as to location, and so on
- Building utility controls
- Keys to all areas of the building including the elevators
- Communications to all parts of the building
- Sprinkler system activation details and controls
- Controls to auxiliary water pumping systems to supplement the building systems
- Maps and diagrams of the building layout and systems

Communications will be extremely important in this operation because the next unit to arrive typically will go to the fire's reported location. As this unit climbs the stairs or accesses the elevators they must be in communication with the first-in unit getting feedback as to that unit's findings regarding location, and so on.

> **Streetsmart Tip** In many jurisdictions, a general rule in high-rise firefighting is not to use the elevator. There is too great a chance that the crew would be taken to the fire floor, and as the door opened, they would be exposed to fire and conditions that could be dangerous to their survival, **Figure 19-41**. Because of the extreme height of many of today's high-rise structures, some jurisdictions may take elevators under certain circumstances. Two of those are if the elevator has firefighter lock-out controls built for fire department access and control or if they are part of a split bank of elevators that does not access the reported fire floor.

If an elevator is utilized, the firefighters will attempt to stop and get off the elevator two or more floors below the reported fire floor and walk the stairs the rest of the way. This way they can remain in the

FIRE SUPPRESSION 617

Figure 19-40 A staging area for a high-rise fire begins to fill with resources. *(Photo courtesy of Dennis Childress)*

Figure 19-41 Firefighters must only use elevators that are secured so that accidental fire exposure will never occur.

stairwell safely before deploying onto the fire floor. This crew will typically take a number of items with them for setting up an attack on the fire. The typical list for an engine or truck company to take with them is shown in **Table 19-4**.

Upon reaching the fire floor the attack team will set up an operation using the standpipe for water supply to

Suggested High-Rise Equipment to Be Carried Aloft

ENGINE COMPANY

One spare SCBA cylinder per member
One large hand lantern or light per company
One standpipe hose bundle to include gated wye and nozzle (see Chapter 12)
Two spanner wrenches and a small pipe wrench or adjustable pliers
One ax
One Halligan tool
One short pike pole
Portable radio

TRUCK COMPANY (FOUR-PERSON UNIT)

One spare SCBA cylinder per member
Two large hand lanterns or lights
Two axes (one flathead)
One Halligan tool
One midlength pike pole
Circular saw with fuel can
One 100-foot drop line (⅜-inch rope if possible)
Portable radio

TABLE 19-4

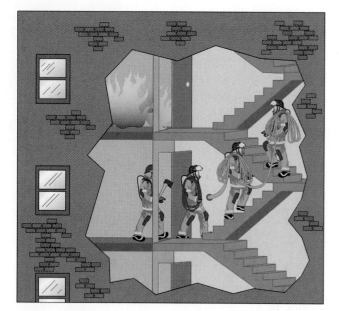

Figure 19-42 The high-rise fire is usually attacked from the stairwells. Laying hose up from the floor below the fire floor will give the firefighting team a safer and more efficient operation.

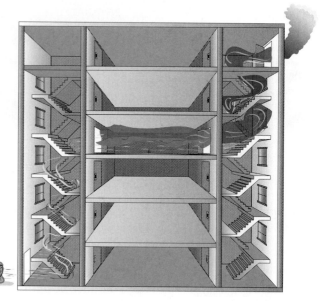

Figure 19-43 Truck company begins ventilation process.

the hose they brought with them. They will go to the floor below the fire and connect the hose to the standpipe, **Figure 19-42**, bringing it up to the fire floor in a layout that will play out as they enter the fire floor to attack the fire. Depending on the size of the fire, they may either attack or wait for a second unit for support. As in residential fire attack, if occupants are on the fire floor, then the first line must go between them and the fire in an effort to create an escape path or at least to contain the fire until sufficient personnel are on scene to focus on the rescue of the people.

The first-in truck will most usually go into a ventilation mode because ventilation will be extremely important in this situation. The products of combustion will be concentrated on the fire floor and possibly others, making rescue and firefighting very difficult at best. Their first task will be to capture the stairwells before the fire does, giving access to the fire to the fire teams as well as exits for the evacuees, **Figure 19-43**.

As other units arrive, they will begin filling in a great number of important operational positions. Some of these assignments could include search, attack team assistance, elevator control, equipment staging two floors below the fire floor (SCBAs, hose, lighting, etc.), ventilation team assistance, water supply to sprinkler or standpipe connections, salvage below the fire floor, and many more.

The multistory or high-rise fire can be one of the most taxing to personnel in the fire service. It will require more personnel, planning, practice, and other resources than almost any other type of operation. The only fire that will exceed it in resource needs will be the wildland conflagration, which is discussed elsewhere in this chapter. The firefighter assigned to a unit with the possibility of responding to a high-rise fire must be prepared for this assignment, because it is dangerous and extremely taxing physically. **Figure 19-44** demonstrates that a multistory fire requires more resources than a smaller single-story structure.

Below-Ground Structures or Basements

Fires below a structure, such as in basements, can be some of the hardest fires to fight due to the harsh conditions the firefighter may have to face. Fire and its combustion products travel upward with heat by convection and radiation. To fight one of these fires, the firefighters must travel down through this superheated air and smoke in order to reach the fire itself, **Figure 19-45**. In just about any other firefighting attack the firefighter can access the fire from below or the side, keeping low and out of the products of combustion. That is not true here.

Note The key to fighting a basement fire is to ventilate as soon as possible, releasing the heat and smoke, making access and attack easier for the firefighting team.

The basement can also be very difficult to ventilate. Chapter 18 tells of a number of ventilation techniques and that information will prove invaluable in a below-ground fire such as a basement.

Figure 19-44 More resources, and many times more water, must be available for the multistory fire. *(Photo courtesy of Rocco Di Francesco)*

The role of the first-in engine company in these fires is to get water onto the fire as quickly as possible. This is primarily because the fire itself will be traveling upward and outward, creating a very poor situation for anything or anybody above the fire in the structure. The second-in unit must quickly assist with the ventilation as well as assist with the line already pulled. A backup line must be pulled as quickly as is practical for the safety of the team that has entered the basement. The attack team must be backed up as soon as possible and this is a situation where a single firefighter will *never* enter the area alone. The team concept must be strictly followed.

In cases where access to the basement cannot be made, it may be possible to punch through the floor above the room on fire and flood the room with water or high expansion foam, **Figure 19-46**. Whatever the attack method, it must be done quickly because everything above the fire is exposed.

Structures Equipped with Sprinklers or Standpipes

Fighting fire in occupancies equipped with sprinkler systems or standpipes has been a boon to the fire service over the last decade or two. Before the mandates of these systems, firefighters were not nearly as successful as they are today fighting fires in these types of occupancies.

The sprinklered building creates a unique situation for the firefighter.

> **Firefighter Fact** In over 94 percent of the fires recorded in the recent past, in sprinklered buildings, the sprinklers have either extinguished or confined the fire prior to the arrival of the fire department.

Figure 19-45 Firefighters must travel down through superheated gases and toxic products of combustion in order to fight a basement fire.

Figure 19-46 If access to the below-ground fire is too dangerous or not practical, then other means must be used to extinguish the fire through the ceiling of the burning room.

The team entering the structure has but to find the seat of the fire and complete extinguishment. This will still not be easy because the smoke and heat will be held down over the combustibles by the cooling action of the water spray from the system. The firefighting team will have to enter the structure very carefully without shutting down the protection afforded by the operating sprinklers, and they will have to find the seat of the fire through very limited visibility to complete the extinguishment of the fire. After this is done, the system can be shut down for a short time during overhaul. Chapter 12 details the operation of sprinkler systems.

The standpipe system is somewhat different. This pipe system will carry water to the firefighting team, easing their dependence on hoselines over great distances. This system is invaluable in fighting fires in high-rise structures. It saves a great amount of time by having the water ready for the team as it arrives at the floor on fire. Team members simply hook the hose carried aloft to the system and then charge it by turning on a valve at the standpipe. Chapter 10 describes this operation in some detail.

Exposure Fires

An **exposure fire** can be described as any combustible item being threatened by something burning in another area. The fire may be carried to the exposure by three ways. Chapter 4 described these as conduction, convection, and radiation.

As stated earlier in this chapter, the second priority after rescue, in most fire scenarios, is exposure protection. This form of property protection is one of the primary focus points in most fires. A number of factors are involved in most exposure fires. In no order of preference they are:

- *Wind:* Convection carrying embers can be a problem.
- *Distance:* The nearer the exposure, the greater chance of fire.
- *Material:* The makeup of the exposed surface determines its combustibility.
- *Intensity of fire:* The intensity of the primary fire factors into its spread.

A couple of basic methods are used to defend an exposure from fire and they are based on the methods of heat transfer. In conditions of conduction, a cooling medium must be applied to the material being heated by the conduction. An example would be a water stream applied to a tank exposed to a fire burning under or on it. **Figure 19-47** shows foam being applied to a tanker that is exposed to a fire next to it. In cases of convection a downwind

Figure 19-47 Firefighters apply foam to a tank truck exposed to a nearby fire. *(Photo courtesy of Central Net Fire)*

patrol must be provided so as to ensure that embers do not ignite other materials. An example would be a downwind house with a wood shake roof, near a fire with flying brands being put into the air. Next is radiation as a medium for ignition of an exposure. The best method would be to apply water on to the exposed area to cool it, **Figure 19-48**. An example would be, as a structure is free burning, the first hoseline is placed between the burning structure and the next door unit. The water is applied to the exposed structure until another line can be trained onto the burning structure itself.

Exposure fires are common and dangerous, especially when the firefighting teams do not see the second fire and are put in danger by it.

> **Safety** It is important to put lookouts in service in order to maintain safety whenever there is any possibility of an exposure fire.

Nonstructural Fires

This subsection discusses fighting fire in nonstructural situations. It opens with wildland or ground cover firefighting, then transitions to vehicle fires and then flammable liquids and gas fires. The section finishes with a short discussion on trash and dumpster fires.

Ground Cover or Wildland Fire

The ground cover or wildland fire can be either a short pickup-type situation or a campaign fire situation lasting days to weeks or more. It can be a short, easy, nonthreatening call or a hard-fought, killer-of-firefighters call. In each of these cases the objectives remain basically the same:

- Confine the spread of the fire by surrounding it with hoselines, removing the fuel, or by another means of stopping its growth.

FIRE SUPPRESSION 621

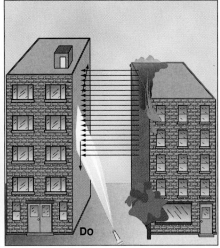

Figure 19-48 Radiation will travel through water or any opaque material. In order for water to be effective, it must be applied to the exposed surface in order to cool it.

- Guide the fire by control measures in order to keep it from burning into areas of higher intensity or faster growth.
- Operate from a position of strength and safety, never giving the fire the advantage of jeopardizing the safety of the fire crews or the citizens being protected.

The initial size-up is very important in that this type of firefighting can be very time consuming and exhausting. A number of basic facts must be considered on arrival that will dictate the attack plan for this type of fire. These facts are based on the wildland fire triangle discussed earlier and are used to determine which attack method will be most successful in fighting this fire. Based on the facts, the officer will ask two questions: What is the fire doing now and what will it be doing in the future? From this, the attack plan will be formed and the fight will begin.

The direct attack is a form of attacking the fire itself, in some fashion. It will usually involve hoselines, but not always. Hand tools or aircraft can be used in many cases as successfully as handlines. A point of attack is made on the fire called an **anchor point**. From there, the team will work up one of the **flanks of the fire** working toward the **head** or progressive end of the fire, **Figure 19-49**. Based on the factors listed in the subsection on ground cover fire components and considerations, the firefighting team will progress up the most active side of the fire until eventually the fire is surrounded and extinguished. **Figure 19-50** shows a firefighting team in direct attack of a grass fire by working up the flank while staying in the burn.

Safety By staying in or near the burn (walking with one foot in the black, or burn, and one foot in the green, or unburned area) a firefighter's safety is greatly enhanced.

If a team is progressing up the opposite flank, that is called a parallel method of attack. Both teams will meet at the head or near there and join lines, controlling the fire. Note that there is a difference between controlling and extinguishing the wildland fire. Controlling is just that. The fire is controlled and no longer burns into threatened areas. Extinguishment is the full extinguishment of all embers and fires contained within the perimeter of the control line.

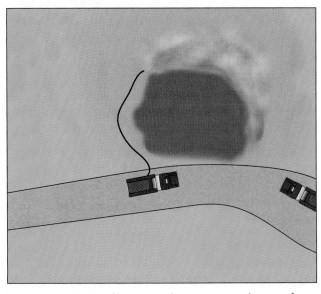

Figure 19-49 A direct attack on a ground cover fire.

Figure 19-50 A fire crew in the direct attack mode works on a grass fire. *(Photo courtesy of Phill Queen)*

Figure 19-52 An indirect attack is accomplished by burning off fuel just before the main fire arrives. *(Photo courtesy of Phill Queen)*

Extinguishment may take days to accomplish after the fire has been controlled.

Once a fire has reached such proportions that a direct attack is impractical, an indirect attack is utilized. This type of attack is common on large campaign fires that sometimes burn for days or even weeks. The indirect attack involves getting ahead of the fire in order to remove fuel from its path, **Figure 19-51**. This can be done by any of several methods. A few of the more common methods are removing fuel by hand or with bulldozers, removing fuel by utilizing natural barriers such as roads and rock outcroppings, or burning off the fuel in controlled conditions. **Figure 19-52** shows the last method as a firefighter burns off fuel before the main body of the fire can reach it. This method of firefighting is very scientific and practiced by only the most experienced fire teams. It requires the ability to read the weather, the fuels, and topography very carefully in order to remain in control of the attack. Consideration of safety and all of the safety lessons and rules must be followed very carefully on this type of firefighting.

Another type of wildland firefighting is now being utilized quite commonly in the fire service today and that is **interface firefighting**. This type of firefighting involves crews that are often oriented more toward structural firefighting. They are asked to go into the rural areas in an effort to protect structures from wildland fires, **Figure 19-53**. The skills involved with this type of firefighting are very different from urban firefighting. A number of classes and curricula are being promoted in these areas for fire service personnel that are designed to create a much safer and smarter firefighter as society continues to move out of the cities.

Ground cover or wildland firefighting is much different than structural firefighting. It requires different tools, clothing, methods of attack, and knowledge. **Figure 19-54** shows a common tool used in wildland firefighting: a helicopter with a water dropping tank. To be safe, efficient, and effective, the firefighter of today must be educated in this area as well as that of the structural fire world.

Vehicular Firefighting

Vehicle fires are very common in today's fire service. Some of the more common reasons for vehicle fires are:

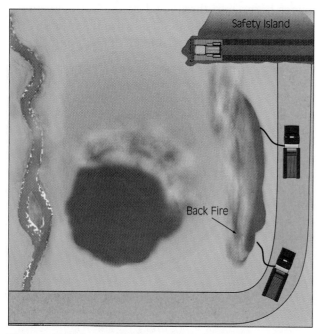

Figure 19-51 An indirect attack on a ground cover fire with two engines and a dozer.

FIRE SUPPRESSION

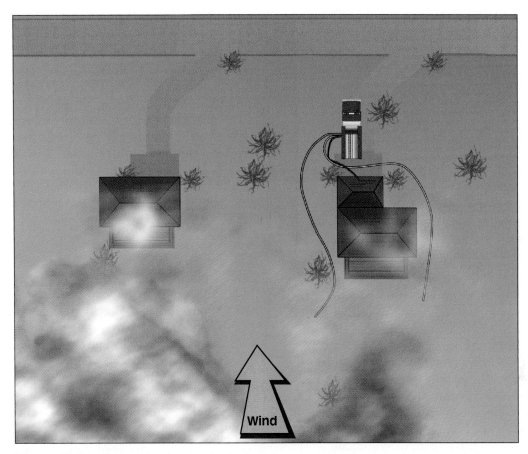

Figure 19-53 Protecting homes in the interface is a growing practice across the country for engine companies used to fighting fire in the cities.

- Collision resulting in leaking fuel and sparks
- Electrical problems causing short circuits
- Overheated components such as tires, engines, or brakes
- Muffler and catalytic converter problems
- Discarded smoking materials
- Arson

The NFPA recommends at least a 1½-inch hoseline for vehicle firefighting and preferably a combination nozzle with a partial fog of approximately a 30-degree pattern initially. The first water should be used to protect the occupants if they are still in the vehicle. **Figure 19-55** shows a firefighting team directing its initial attack on a car's interior into the passenger compartment first. After the occupants

Figure 19-54 A helicopter is readied for wildland firefighting. *(Photo courtesy of Brentt Sporn)*

Figure 19-55 The firefighters' first attack point is the passenger compartment in order to protect life. *(Photo courtesy of Brentt Sporn)*

have been rescued, then the attack on the fire itself may begin. The suggested method of attack is direct. The only time an indirect attack would take place is when the vehicle fire is exposing something else, causing concern on the part of the firefighting team.

> **Caution** As mentioned earlier in the book, backdraft and flashover are two structure fire phenomena that firefighters must be aware of. This is also true with automobile fires. A closed passenger compartment or trunk may be the perfect place for a backdraft to be set up. Or, that same area can see a flashover if the conditions are right. Care must be taken when working in these environments.

The automobile fire can also pose situations in areas not encountered in other fires—closed hood and trunk areas. It will be important to gain entry for a number of reasons. Some of those reasons are the complete extinguishment of the fire, the investigation of the fire's origin, the securing of the vehicle and all of its contents, and possibly the determination of whether the vehicle was carrying illegal materials that factored into why it burned. Gaining entry into the hood or trunk area can be extremely difficult.

In most of today's automobiles a steel cable runs from the interior dash area to the latch at the front of the engine compartment. Unfortunately, this cable burns through quickly and access to the engine area must be by force, as the cable mechanism will be out of service. Entry tools must be used as well as ingenuity in the entry effort.

The trunk is no different. While there is a cable from the interior driver area to the trunk in most new vehicles, it will be of no use due to the fire. Entry tools will also be needed here.

> **Safety** The attack should be from the upwind and uphill side of the vehicle in order to reduce the firefighter's exposure to heat and the products of combustion. This is also important in situations where leaking fuel is under the vehicle and can run toward the firefighting team members. Consideration must also be applied to chocking the tires on any vehicle burning on unlevel ground. It could be disastrous if the vehicle's brake lines burned through and the vehicle began to roll away or toward the fire engine. Full protective clothing with SCBA must be worn when fighting vehicle fires because the combustibles are toxic and sometimes explosive.

Care must be taken around the bumpers, hatchback struts, and air bag assemblies as discussed in the vehicle fire components and considerations subsection. Care must also be exercised around the tires if they are burning as well as the fuel areas of the vehicle. It is also recommended that the battery to the vehicle be disconnected by removing the cables before any overhaul begins to ensure firefighter safety.

The procedures around electric vehicles may differ due to the nature of the power source and its effect on the automobile itself. The automobile industry has gone to great lengths to educate firefighters in safe methods of fighting electric vehicle fires. The most recommended method to date consists of seven steps:

1. Approach vehicle at an angle to front or rear.
2. Identify vehicle type.
3. Secure and stabilize vehicle.
4. Power off electric vehicles.
5. Make the initial attack on vehicle fires with water and/or foam.
6. Use standard tools and cut-in areas for victim extrication.
7. Do not cut into high-voltage components. These are usually color coded by law to stand out from the rest of the components.

The action taken by an attack team on a vehicle fire in step 5 is to extinguish electric vehicle fires with water and/or foam. If it appears that the fire has extended into the battery pack, the firefighter should attempt to extinguish it following the instructions for the specific battery type. This information changes as technology changes, so it must be obtained from in-service training on an ongoing basis. If the vehicle is in a safe area, under some circumstances, firefighters may allow the fire to self-extinguish. Vehicle manufacturers are now beginning to use a number of alternative fuels in an effort to become more fuel efficient. A firefighter may find compressed natural gas (CNG), liquefied propane gas (LPG), and others. Fighting fire involving these fuels must take on the methods discussed in the next section on flammable liquids and gases fires.

As the firefighters work on the vehicle fire, caution must be considered regarding other traffic in the area, **Figure 19-56**. Unfortunately, a great number of firefighter injuries occur each year at vehicle fires when passing motorists strike firefighters working on the accident scene. Great care must be taken knowing that passing motorists will be looking at the accident more than the people working in the area. It is also important to realize that flashing lights and excitement will attract some drivers to the scene. **Figure 19-57** shows an engine being used to block traffic from the vehicle fire scene. Traffic cones, road flares, or fire engines are to be used to block traffic from the work area in order to better protect firefighters working an accident.

FIRE SUPPRESSION 625

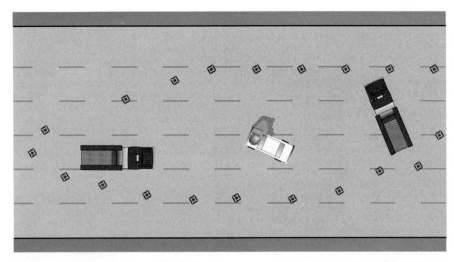

Figure 19-56 To protect firefighters on scene in the street, the engine is typically used to block traffic. Cones are used quite often to assist in this endeavor.

While fighting vehicle fires may at first seem simple, this is far from the truth. Vehicle fires are complex, dangerous situations, requiring skill in fighting each of the four classes of fires. Safety must always be considered in these situations.

Flammable Liquids and Gases Fires

Flammable liquids and gases have a number of things in common that the firefighter can use to control a fire in these commodities. The mechanisms of extinguishment are well known to the fire service. These are some of them:

- Smother the fuel, cutting off the supply of oxygen.
- Starve the fuel by removing unburned material from the area.
- Interrupt the chain reaction process with chemical agents.
- Cool the fuel, reducing vapor pressure.

A great range of possible situations can arise involving combustible and flammable liquids and gases and their burning. As the first-in unit arrives, a number of facts must be obtained in order to handle the situation effectively. Finding answers to the questions asked may be difficult or sometimes impossible. In many situations, hazardous materials teams may be needed in order to identify the materials and the recommended method of handling them.

As opposed to most of the other fire scenarios discussed thus far, flammable liquids and gases do not follow the laws of extinguishment so basic to the fire service. Water may not be the choice best suited for firefighting here. **Figure 19-58** shows a crash crew applying foam to a ground fire of flammable liquids.

Figure 19-57 The fire apparatus is used to block traffic from the firefighters' working area. *(Photo courtesy of Phill Queen)*

Figure 19-58 A team of firefighters applies foam to a pool of flammable liquid. *(Photo courtesy of Phill Queen)*

Figure 19-59 Water is applied to the heated metal surface in order to keep it cool, slowing the pressure buildup inside the exposed tank car. This will reduce the possibility of a BLEVE.

Caution In fact, in many cases, water could create more problems than the firefighting team cares to imagine. No one agent applies to all fires in this category, and, in some cases, it may not be the best solution to extinguish the fire at all. It may be better to let it burn itself out as in the case of some pesticides and gases.

Knowing this it is possible to set up some very generic principles for fighting most flammable liquid and gas fires as follows:

1. Identify the material involved and its hazards.
2. Utilize written assistance with the DOT *Emergency Response Guidebook* and other sources.
3. Evaluate the threat to nearby similar commodities such as gas cylinders and tanks. For instance, get water on exposed tanks to reduce internal pressure buildup and prevent the possibility of a BLEVE (see Chapter 4), **Figure 19-59**.
4. Determine which extinguishment agent and principle is best.
    a. For small fires consider CO_2 or dry chemicals.
    b. For larger fires consider foam suited for that particular commodity.
    c. For some fires consider letting the fire burn itself out. This is particularly true in cases where the leak, when burning, is contained, and if extinguished, may spread and reignite, causing greater hazards and situations.
    d. Look for shutoff valves or switches to shut down the flow of material to the fire. **Figure 19-60** shows a group of firefighters approaching a shutoff valve at a flammable liquid fire under the protection of a water fog.

Incidents involving combustible and flammable liquids or gases are very dangerous and have caused the fire service to bring in the concept of hazardous materials teams to deal with these special situations. The number of materials considered hazardous in today's society pose a very great threat to the fire service and its ability to fight fire. Knowledge and safety are the keys in working with these materials.

Trash and Dumpster Fires

A trash fire, whether contained in a bin (dumpster), truck, or spread on the ground, can surprise the most seasoned firefighter with a burning rate or intensity not expected. Trash can be anything discarded by anyone. Trash can produce toxic gases and, in many cases, items discarded can be hazardous and sometimes even explosive. The unprepared firefighter can be injured or killed fighting these seemingly innocent fires.

As the fire engine arrives on scene it is important for the team members to be fully dressed in their

FIRE SUPPRESSION

Figure 19-60 A team of firefighters approaches a flammable liquid fire control valve behind a fog pattern. *(Photo courtesy of Phill Queen)*

Figure 19-61 A firefighter approaches a dumpster fire, keeping below the top edge of the container. *(Photo courtesy of Brentt Sporn)*

full protective clothing including SCBA. Water is most usually the agent used in the extinguishment of these typically Class A combustibles.

> **Safety** When first applying water, the firefighter must be protected from the possible explosion of heated containers such as aerosol cans, bottles, and glass items. Using the protection of the side of the bin or the side of the engine should be considered when first attacking a trash fire.

Figure 19-61 shows a firefighter applying water while being protected from exposure.

Many times these fires will deeply seat themselves, not allowing the water to penetrate the materials. In these situations a number of things can be done, including the following:

- Use a tool such as a pike pole or rubbish hook to stir the material as water is applied.
- Use a tractor or dozer to move larger masses of trash in order to penetrate piles.
- If the fire is in a truck, consider having it dumped while water is applied.
- The use of water agents (light water) can be an effective method of achieving deeper penetration into the materials.
- In some cases the bin or dumpster may have to be filled with water to achieve better results.

The trash or rubbish fire can pose extinguishment problems as well as hazards not considered in such nonemergency-type calls. The firefighter must be knowledgeable about the possibilities and proceed with caution when dealing with the unknown contents of these fires.

Outside Stacked or Piled Materials Fires

Many times the firefighter is asked to extinguish a fire in outside storage. These may be lumber piles, used tire piles, stored pallets, bundled paper or cardboard, or just about anything else that is flammable and stored outside. **Figure 19-62** shows a typical yard storage of high piled pallets.

The hose size, stream used, amount of water, appliances for water delivery, and tools for overhaul will either make the job easy or, if not selected properly, make the job difficult.

> **Safety** Working around any high piled storage is dangerous. This danger is magnified if the materials are on fire or even wet after extinguished. Piled storage is held up by the integrity of the items stored, and as that changes with either fire or water damage, those piles may topple. Collapse zones must be maintained whenever working around high piled storage.

Logically, large piles with large fires will require large amounts of water. The delivery of that water will be the key to success. Large-diameter lines with the proper nozzles, usually smooth bore or straight stream capable, will most often be utilized for quick knockdown. As this knockdown is taking place, it will be most important to keep an eye on any exposures that may be threatened.

Figure 19-62 Typical high piled storage that can jeopardize firefighter safety.

Those may consist of nearby piles of the same material, small buildings, or even distant exposures threatened by flying fire brands. Then, after initial knockdown, the fog and/or foam nozzles will come into effect. Many times the streams will require additives, such as those described in Chapter 11, in order to add to the penetrating power of the water.

Then the overhaul begins. In most cases, firefighters can expect to have to pull all the materials out of every pile while water is being applied. In the case of used tire storage, it is not uncommon to have to bring in large tractors, cranes with buckets, or other large hauling or moving equipment to break down the piles. Hand tools will also be used as the piles are brought down to manageable sizes. In many cases the overhaul will last far longer than fighting the fire itself.

> **Caution** The products of combustion will continue to develop during the overhaul phase of the work, and it is important to continue to wear SCBA or some type of respiratory protection during this time.

Lessons Learned

The basic principles of firefighting are based on sound scientific laws as well as years of firefighting experience. Every combustible item and occupancy must be carefully studied and understood in order for firefighters to be effective, efficient, and safe in their work.

Just as both synthetic and natural items differ, so will the tactics and strategies necessary to extinguish them differ when they burn. Knowing the basic elements of fire control, and applying them safely, will result in a better environment for all.

KEY TERMS

Anchor Point A safe location from which to begin line construction on a wildland fire.

Aspect The direction a slope faces given in compass directions.

Autoextended When a fire goes out the window on one floor, up the side of the building, which is often noncombustible, and extends through the window or cockloft directly above.

Box Canyon A canyon open on one end and closed on the other. They become very dangerous when wildfire enters them.

Chimney Another term for drainage. Given because of the draw of fire as in heat going up the chimney.

Combination Attack A combined attack based on partial use of both offensive and defensive attack modes.

Control Room A room on the ground floor of a high-rise building where all building systems controls are located.

Defensive Attack A calculated attack on part of a problem or situation in an effort to hold ground until sufficient resources are available to convert to an offensive form of attack.

Drainage A topographic feature on the side of a hill or mountain that naturally collects water runoff, channeling it to the bottom of the rise. Fire is attracted to this feature.

Exposure Fire Any combustible item threatened by something burning nearby that has caught on fire.

Fire Intensity A measurement of Btus produced by a fire. Sometimes measured in flame length in the wildland environment.

Flammable Range Ratio of gas to air that will sustain fire if exposed to flame or spark.

Flanks of the Fire The sides of a wildland fire running from the start point up each side to the end of the fire running into unburned areas.

Garden Apartment A two- or three-story apartment building with common entryways and layouts on each floor, surrounded by greenery and landscaping, sometimes having porches and patios.

Head of the Fire The running top or aggressive end of the fire away from the start point.

Incident Management System (IMS) A management system utilized on the emergency scene that is designed to keep order and follow a sequence of set guidelines.

Interface Firefighting Fighting wildland fire and protecting exposed structures in rural settings.

Midslope An area partway up a slope. Any location not on the bottom or top of a slope, as in a midslope road crossing the slope horizontally.

Offensive Attack An aggressive attack on a situation where resources are adequate and capable of handling the situation.

Polar Solvent A material that will mix with water, diluting itself.

Rate of Spread A ground cover fire's forward movement or spread speed. Usually expressed in chains or acres per hour.

RECEO Acronym coined by Lloyd Layman standing for Rescue, Exposures, Confinement, Extinguishment, and Overhaul.

Ridge The land running between mountain peaks or along a wide peak. A high area separating two drainages running parallel with them.

River Bottom Topographic feature where water runs from higher elevations to lower. Can be dry or wet depending on season or recent rains.

Saddle A pass between two peaks that has a lower elevation than the peaks. Wind will pass through this area faster than over the peaks, so fire is drawn into this feature.

Solubility A liquid's ability to mix with another liquid.

Specific Gravity Weight of a liquid in relation to water. Water is rated 1.

Stairwell An enclosed stairway attached to the side of a high-rise building or in the center core of same.

Steepness of Slope The degree of incline or vertical rise to a given piece of land.

Teamwork A number of persons working together in an effort to reach a common goal.

Toxicity Poisonous level of a substance.

Vapor Density Weight of a gas in relation to air. Air is rated 1.

REVIEW QUESTIONS

1. Explain how water suppresses fire in relation to the fire tetrahedron.
2. Give at least two factors that must be taken into consideration prior to an attack plan being formed and extinguishment of a structure fire.
3. Ground cover firefighting involves a number of principles very different from those used for structural firefighting. Explain how the wildland fire triangle differs from the fire tetrahedron.
4. Vehicle firefighting involves a number of built-in dangers. Name an exterior feature and an interior feature that can injure a firefighter who is not very careful.
5. Explain why firefighting tactics will differ based on a flammable liquid's specific gravity.
6. Why does a basement fire have the potential to be so much more hazardous to a firefighter than a ground-level fire in the same occupancy?
7. OSHA has enacted the two in/two out ruling. Explain this rule and its impact on the fireground during structural operations.
8. Explain the procedures recommended if a firefighter is considering taking an elevator to the fire floor in a high-rise or above-ground fire. What does your jurisdiction recommend?

Additional Resources

Brunacini, Alan, *Fire Command.* National Fire Protection Association, Quincy, MA, 1985.

California State Fire Training, *Fire Command 1A and 1B.* California State Fire Marshal Office.

Coleman, Ronny J., *Management of Fire Service.* Daybury Press, North Scituate, MA, 1978.

Layman, Lloyd, *Fire Tactics and Strategy.* NFPA, Quincy, MA, 1953.

Lowe, Joseph, *Wildland Firefighting Practices.* Delmar Learning, a part of the Thomson Corporation, Clifton Park, NY, 2001.

NFPA 1001: Standard for Fire Fighter Professional Qualifications, 1997 Edition. National Fire Protection Association, Quincy, MA, 1997.

Richman, Harold, *Engine Company Operations.* National Fire Protection Association, Quincy, MA, 1986.

Richman, Harold, *Truck Company Operations.* National Fire Protection Association, Quincy, MA, 1986.

CHAPTER 20

SALVAGE, OVERHAUL, AND FIRE CAUSE DETERMINATION

Geoff Miller, Sacramento Metropolitan Fire District

OUTLINE

- Objectives
- Introduction
- Salvage Tools and Equipment
- Maintenance of Tools and Equipment Used in Salvage
- Salvage Operations
- Salvage Operations in Sprinklered Buildings
- Overhaul Tools and Equipment
- Overhaul Operations
- Fire Cause Determination Concerns
- Securing the Building
- Lessons Learned
- Key Terms
- Review Questions
- Additional Resources

STREET STORY

Many firefighters take more of an interest in going in and attacking the fire. But salvage is important too. Most of the damage to a person's property is caused by smoke and water, not by the actual fire. And those contents can have so much meaning for people—all they want is their private possessions. Beyond that, if there's a possibility of arson, it's your job to preserve the evidence.

I remember receiving a call for a fire at The Museum of Science and Industry—it came in through enhanced 9-1-1. By the time we responded, the museum had white smoke curling from the roof. The initial companies were locating the seat of the fire (not easy since the museum occupies four city blocks). They figured out it was in a void in the roof, and we started doing salvage. I acted as coordinator. In a house fire, we'd be likely to rope the belongings in the center of the room and put salvage covers over everything. But in this case, there were fixed displays, so we couldn't move them, but we still covered everything. We built chutes for the stairwells so that the water could be channeled without damaging the carpets, and we diked the elevators.

If we can prevent water damage and channel out some of the smoke, we'll have a better goodwill image with citizens. In this case, you may not gain the admiration of fellow firefighters, but what we did was save a lot of artifacts that couldn't have been replaced.

—Street Story by Julius Stanley, Lieutenant, Chicago Fire Department, Chicago, Illinois

OBJECTIVES

After completing this chapter, the reader should be able to:

- Explain the purpose of salvage and overhaul operations.
- Explain the importance of salvage from a "customer service" standpoint.
- Identify tools and equipment used in salvage and overhaul operations.
- Identify needed maintenance of salvage and overhaul tools and equipment.
- Perform different identified salvage cover throws, folds, and rolls.
- Arrange a room's contents into a salvageable position.
- Identify sprinkler system shutoff valves.
- Shut off a flowing sprinkler head.
- Explain how fire attack and ventilation assist in the salvage effort.
- Explain where to search buildings for hidden fires.
- Explain how to look for structural stability.
- List debris removal techniques.
- Explain the importance of evidence preservation.
- Identify how to determine the area of origin of a fire.
- Identify how to secure a building after emergency operations are complete.

INTRODUCTION

Salvage and overhaul operations are not often viewed as critical tasks within the fire service. They are not associated with the excitement of fire attack, rescue, or ventilation. This perception leads to a downplaying of importance and an "afterthought" mentality. Fire cause determination is something new firefighters will not be actively involved in, but they need to understand what actions to take so as not to make the cause determination unattainable or more difficult for the fire investigator. This chapter explains the reasons for and importance of salvage, overhaul, and fire cause determination and describes the necessary firefighter skills and abilities needed to effectively perform these operations.

SALVAGE TOOLS AND EQUIPMENT

All fire incidents involve a potential loss of material goods. Whether they are easily replaceable articles of clothing or furniture or irreplaceable photographs, heirlooms, antiques, or memories, they all have meaning to their owner. It is for this reason that firefighters can make a tremendous impact on an incident by aggressively salvaging the occupancy and making sure not to cause more damage than the emergency itself has, **Figures 20-1A** and **B**.

People usually think of the interior of a structure when salvage is considered, but items on the exterior of a house can be just as valuable. While first-in company officers are making their size-up to determine the extent of the involvement, they may be able to *quickly* move some patio furniture or arrange to have vehicle(s) covered in the front of the structure with a salvage cover. It is important that the company officer "triage" the entire scene. If quick action can save property without endangering the operation, then the officer should make an effort to do so. It is possible for any firefighter to get tunnel vision and waste time trying to save one portion of the scene while endangering the integrity of the scene as a whole. Quick thinking and experience are vital traits of a good company officer.

> **Streetsmart Tip** Company officers who are in command of a scene can often perform a small necessary or safety-oriented task if there are no other firefighters in staging and if the task will not interfere with the safety of the on-scene crews or the command of the scene. Commanding the scene is the primary role of that officer.

It is rewarding to report to owners who fear the worst possible loss that salvage crews were able to save a particular special item or memory for them. Special items or memories make a home unique, and firefighters should always consider this while working any fire scene. A quick scan followed by consideration of what the homeowner might want to save can be a great service to a person who suffers the tragedy of a fire.

The basic premise of salvage operations is to remove the harmful atmosphere from the material

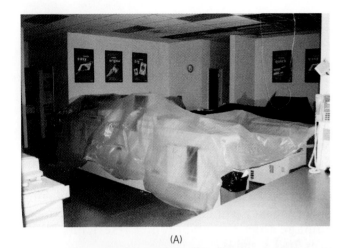

Figure 20-1 (A) Good and (B) bad salvage operations.

Salvage Covers

The mainstay of salvage is the salvage cover or a variation of the salvage cover. Salvage covers are made out of several different materials, most commonly plastic, canvas, and treated canvas.

The size of the salvage cover varies. It can be anywhere from 10 × 12 feet to 12 × 16 feet. The perimeter of the cover is ringed with grommets spaced at intervals, **Figure 20-2**. The ideal interval is 16 inches to match up with the studs in a building wall. A plastic cover has advantages in weight and water resistance. A canvas cover is durable and, if treated, can compare to the plastic cover for water resistance capability.

Salvage covers are the main materials used for constructing water chutes and catch-alls, which are discussed later.

A new tool that is making the salvage cover a less desirable option is **Visqueen** or black plastic, which can be carried easily by a firefighter, **Figure 20-3**. The Visqueen comes in rolls 120 feet long, with a width, once unfolded, of 20 feet. A roll can also be carried by putting a dowel in the center tube (with a cord attached) and hanging the cord over the neck so the roll is allowed to hang in front of the firefighter. The only additional tool needed is a razor knife or scissors to cut off the desired lengths. In some cases the plastic is precut into various sizes for quick deployment. By precutting some lengths, the weight of the roll is reduced, thus minimizing the stress on the firefighter carrying it.

Floor Runner

Another item used for salvage operations is the floor runner, which is usually around 3 feet wide and about 20 feet long. It is normally made of a lightweight canvas-type material for easy deployment. It is used just like its name implies—to cover the floor

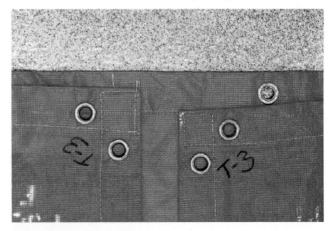

Figure 20-2 Salvage cover grommets.

or to protect the material from the harmful atmosphere. Several techniques are used to accomplish this goal and will be discussed later. Any technique is rendered useless unless the proper tools are used, and there are many tools and different pieces of equipment for salvage work. Some operations might require complex equipment while others demand a hammer and a few nails.

SALVAGE, OVERHAUL, AND FIRE CAUSE DETERMINATION

Figure 20-3 Firefighter carrying Visqueen roll.

down a hallway or along a traffic area, **Figure 20-4**. If there is a fire in one area of the building, firefighters should try not to damage the carpet in another. Remember, customer service really makes the difference and is a sign of professionalism.

Water Vacuum

A very useful tool in salvage operations is the water vacuum, **Figure 20-5**, which is available in two basic types. One is worn like a backpack and is used to remove water from areas where access prohibits squeegee use; the other is usually larger and is moved around on wheels. If SCBAs are being worn, the "backpack" water vacuum cannot also be worn. Other methods will have to be considered, or firefighters will have to wait until the air quality is safe and the area is safe to work in without SCBA.

> **Streetsmart Tip** Squeegees are very helpful where the need exists to move large amounts of water on a flat surface. A sprinkler head discharge in a large warehouse can result in many gallons of water on the floor. Once the flow is stopped, this water can be quickly moved from the warehouse by several firefighters working together with squeegees.

Figure 20-4 Floor runner in place.

The water vacuum can also be used to quickly drain catch-alls so that they do not overflow. The water vacuum usually has a limited capacity of around 5 gallons and thus needs to be emptied. If a large amount of water must be removed by a water vacuum, it is advisable to have the owner contract with a professional water removal company.

Another option carried by some fire agencies is either a submersible pump or "float-a-pump." The submersible pump can be dropped directly into water and used to pump water out through an attached line. This is a very good tool for flooded basements. The float-a-pump simply floats on top of the water and essentially drafts water and pumps it out. The submersible needs electrical power, while the float-a-pump is gasoline-driven.

Figure 20-5 Firefighter using water vacuum.

Figure 20-6 Various fastening tools.

Miscellaneous Salvage Tools

Basic hammers and nails or staple guns also have a place in salvage work. Having these tools available allows firefighters the ability to secure doors, windows, ventilation holes, or other openings to further protect the occupant's property from weather. This is an area where the use of Visqueen comes in handy. It can be left behind on a roof to cover a ventilation hole and does not have to be retrieved.

Doors and windows need to be secured with plywood or some other sturdy material to offer any resistance to vandals. This task is handled by board-up crews and is discussed later.

Sometimes salvage covers need to be used to prevent a leak from damaging an item to be salvaged or an item that is too large to move to a safer location. A hammer and nails can be used to nail the salvage cover to the wall and cover the exposed item. Firefighters also use a small fastening tool called an S-hook that can be hammered into the wall to establish a sturdy place to hang the cover, **Figure 20-6**.

A thorough knowledge of available tools to be used in salvage work will make the task much easier to accomplish. With the right tools, the job will be done correctly and the damage minimized.

MAINTENANCE OF TOOLS AND EQUIPMENT USED IN SALVAGE

After salvage and overhaul tools and equipment are used, they must be cleaned and inspected like any other piece of firefighting equipment to make sure they are ready for the next emergency. A written log should be kept on the maintenance and use of the tools so proper maintenance is done.

Salvage tools and equipment are often exposed to hazardous materials and should be placed out of service if damaged. If through the course of a salvage operation, the equipment suffers damage, the fire department should seek reimbursement from the property owners through their insurance company. Just as the insurance company is liable for the owners' property, it may also be liable for fire department equipment if it is damaged. This is particularly true for salvage equipment where loss is suffered while protecting the insured's items from damage. The department should document the damage thoroughly and then contact the insurance company through the **chain of command** to determine the means to secure reimbursement.

SALVAGE, OVERHAUL, AND FIRE CAUSE DETERMINATION

Streetsmart Tip This is one of the many reasons why it is a good idea to carry a camera on the apparatus. It provides an easy way to document items, and it can also capture some great training pictures.

This may not apply in all states, so it is important to contact a local insurance commissioner to determine if this is an option.

When salvage covers are used at an emergency scene they will end up with material or debris on them. This material should not be spilled on clean areas after it has been collected. The best way to keep this material in the cover is to do a "loose fold and roll," **Figure 20-7**. Essentially, with this technique the cover is folded into itself so the debris caught is trapped inside. Once the cover is folded into itself by bringing both sides to the middle once and then once again, the cover is then rolled up. This will keep all of the material inside until it can be disposed of. Once the cover is dumped out and brushed off, it must be washed using a mild soap solution, hung to dry, inspected for tears or holes, repaired, folded, and placed back on the apparatus. (Note that this reveals another plus of Visqueen—no maintenance.)

Drying the cover can prove difficult due to its size. Sometimes hooks are placed along high rooflines to hang the cover to dry. Another method is to rig a system to haul the cover up alongside of a building to dry, **Figure 20-8**. If neither of these is available, it can be laid out on the apparatus floor, but it must be turned over so both sides dry. In any case the cover should be dry before it is folded and returned to service.

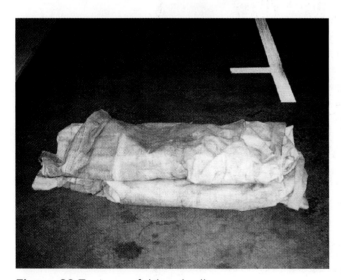

Figure 20-7 Loose fold and roll.

Figure 20-8 Salvage cover drying rack.

Once dry, the cover should be inspected for holes or tears. A simple way to inspect is for firefighters to drape the cover over themsleves and look for light shining through any holes or rips. Holes or tears should be marked with chalk and repaired.

Salvage Cover Folds and Rolls

Salvage covers are either rolled or folded for storage. The available compartment space often dictates which method to use. The roll is accomplished by placing the cover on a large flat surface; if it has a treated side, it is placed up. Sprinkling some baby powder on the cover and spreading it around with a broom will keep the cover from sticking to itself if it is not used for long periods. Two firefighters are needed to roll a cover. Place one firefighter at either end of the cover, **JPR 20-1A**. Both firefighters place one hand on the edge of the cover one-quarter of the way in and the other hand on the outside edge. Each firefighter should be on the same outside edge facing each other. While grasping the cover, with a quick flipping motion, bring the outside edge to the middle, **JPR 20-1B**.

The firefighter's inside hand should be at the fold. Repeat this fold again and again until the fold is at the middle of the cover, **JPR 20-1C**. Fold the opposite end of the cover in the same manner. The cover should now be about 3 feet wide and ready to be rolled, **JPR 20-1D**. The treated side is rolled inside and the cover is ready for deployment.

Many salvage covers are folded for deployment by one firefighter, while others are folded for deployment by two or more firefighters. Folding a

JOB PERFORMANCE REQUIREMENT 20-1
Salvage Cover Roll

A Place one firefighter at either end of the salvage cover. Both firefighters place one hand on the edge of the cover one-quarter of the way in and the other hand on the outer edge.

B While holding on to the cover, with a quick flipping motion, bring the outside edge to the middle.

C Repeat step B with the outside edge, the first fold, and again bring the outside edge to the middle.

D The folded salvage cover should be approximately 3 feet wide and is now ready to be tightly rolled.

SALVAGE, OVERHAUL, AND FIRE CAUSE DETERMINATION 639

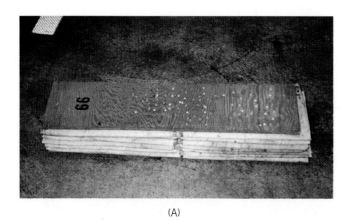
(A)

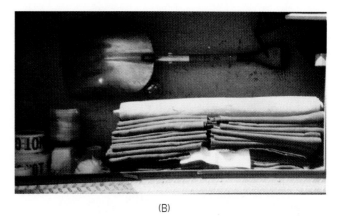
(B)

Figure 20-9 Salvage cover fold. (A) Salvage cover fold template. (B) Paper towel between salvage covers in compartment.

cover is similar to folding a bed sheet. It should be neat and orderly and take up minimal space on the apparatus. Some departments, to keep the folds neat and tidy, use a plywood template to adjust the size, **Figures 20-9A** and **B**. **JPR 20-2** shows the proper skills to prepare a salvage cover for a one-person deployment. **JPR 20-3** shows the proper skills to prepare a salvage cover for a two-person deployment. When folding a salvage cover, it is vital that the deployment method be considered. These covers often must be deployed quickly.

Other tools must be thoroughly cleaned, dried, and checked for damage. The water vacuum must be rinsed several times to remove all contaminants from the tank. Chain saws will require more than the normal amount of maintenance. They must be taken apart and cleaned thoroughly. They are run in atmospheres they are not designed to run in and used to cut things they are not really designed to cut. They must be inspected and cleaned from top to bottom. When the saw is put back into service, it should start within a few pulls and should run perfectly. Saws are not used all that often, but when they are needed they are *really* needed and are crucial to fire attack, rescue, ventilation, salvage, and overhaul.

Tool maintenance is an important aspect of firefighting. Tools need to be in good working order every time they are used.

SALVAGE OPERATIONS

Safety Considerations

The basic goal behind salvage operations is to eliminate the possibility of damage from a harmful substance to as many belongings as possible in the shortest amount of time. This may entail moving the item out of harm's way, covering the item, or removing the harmful substance from the area.

As with all emergency scene operations, the salvage group must be aware of its surroundings and work as safely as possible. If salvage operations are going on while active suppression operations are taking place—as is often the case—the crew must be appropriately dressed for the conditions; this means dressed the same as the attack crew—full protective clothing with self-contained breathing apparatus. (The individual components of full protective clothing are discussed in Chapter 6.)

The same is true for forcible entry tools. The salvage group is often sent to areas where the fire attack is not taking place, such as the apartment below the involved apartment. In this case it is necessary for group members to carry forcible entry tools with them so they make entry into the apartment. This also applies to interior doors within a business or home that are locked for security reasons.

> **Caution** The crew members must be aware of their surroundings at all times to ensure they do not find themselves in the wrong place at the wrong time. Although the content of a structure is important to its owner, it is not worth injury or a firefighter's life!

One of the most common hazards of salvage work is ceiling collapse. The salvage group may be covering material to protect it from water that is accumulating in the ceiling area due to several possible factors. The roof could have leaks or be under repair, sprinkler heads in the attic could have fused, or the attack crew might not have done a

JOB PERFORMANCE REQUIREMENT 20-2
Preparing a Folded Salvage Cover for a One-Firefighter Spread

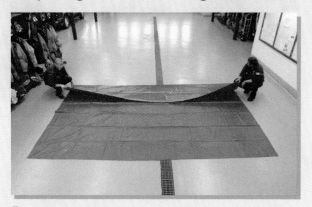

A With the cover flat on the ground, fold the outside edge to the center point of the cover. Smooth out wrinkles and align edges.

B Fold the same edge to the center point again. One fold is now beneath the other fold. Smooth out wrinkles and align edges.

C Fold the opposite side in a like manner. Smooth out wrinkles and align edges.

D Fold one end to a point just short of the center point. Smooth out wrinkles and align edges.

E Fold the same end again. One fold is now beneath the other fold. Smooth out wrinkles and align edges.

F Repeat steps to fold the opposite end. Smooth out wrinkles and align edges. The cover is ready for a one-person deployment.

JOB PERFORMANCE REQUIREMENT 20-3
Preparing a Folded Salvage Cover for a Two-Firefighter Spread

A Fold the cover in half and place it on the ground. Smooth out wrinkles and align edges.

B In similar fashion, fold the cover in half (lengthwise) again. Smooth out wrinkles and align edges.

C Fold the cover in half (widthwise). Smooth out wrinkles and align edges.

D Fold the cover end over again (widthwise). Smooth out wrinkles and align edges.

E Fold the cover in half two more times (widthwise). The cover is ready for a two-person deployment.

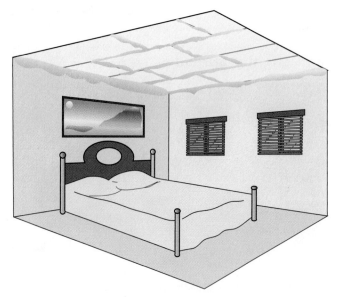

Figure 20-10 Plasterboard (Sheetrock) with water seeping through the seams.

good job of monitoring water flow. For these reasons, the plasterboard (**Sheetrock**) will become weak and sometimes fail. In most cases there will be warning signs. The most common is plasterboard seams begin to show as water seeps into them, **Figure 20-10**. With this sign comes the knowledge that water is building up in the insulation. At a weight of 8⅓ pounds per gallon, the water will make the ceiling fail. Another sign will be plasterboard that sags as if it were not nailed properly.

If these signs appear, the water must be drained before the ceiling collapses. The easiest mitigation is to stop the flow and remove the water. This can be accomplished by quickly salvaging the area below the ceiling and creating a drain hole in the plasterboard to relieve the water buildup. If it is not possible to salvage the area, a quick decision must be made: Is water damage more serious than ceiling collapse? In most cases, water damage will be the lesser of the two evils.

A safe scene involves knowing what other operations are ongoing and where they are taking place. Ventilation crews working above the salvage operation can create a dangerous situation if the cut roof or ceiling drops. Firefighters should be listening for noise and radio traffic and should be aware of crews working above and below them. The incident commander or operations section chief should be aware of all crews and their location in the building and should be contacted if the members of the salvage crew have any questions.

Stopping Water Flowing from Sprinkler Heads

Stopping the flow of water from sprinkler heads is often the main step in a salvage operation. Several types of sprinkler stops are available. The most common is the sprinkler wedge. This is usually a wooden wedge, like a doorstop, that can be jammed into place in the opening of a flowing sprinkler to stop the flow of water. The firefighter must be careful not to break the head off the sprinkler while hammering the wedge in place. This job is not easily accomplished as it is usually done from a ladder while water is flowing under a considerable amount of pressure.

The firefighter assigned this task must wear eye protection and full protective clothing. The firefighter will need to lock into the ladder in order to get some force behind the wedge and eliminate the possibility of falling. Then it is merely a matter of patience. As long as the main drain has been opened and the system is shut down, the firefighter is only working against residual pressure within the system. Getting this assignment is something most firefighters rarely forget!

Other types of sprinkler stops work on more of a mechanical theory: The stop is placed in the opening and either screwed or locked into place by the operator.

Methods of Protecting Material Goods

The "removing items from harm's way" approach can be determined very quickly by asking one simple question: Can it be moved quicker than it can be covered? If the answer is yes, firefighters should go for it! This will usually apply to quickly accessed items like patio furniture and garage items. These are easily moved from harm's way because they usually do not have to be moved very far to get them out of the smoke and heat. Items that are in the way of the fire attack crews should be moved to assist the operation and also accomplish salvage.

Arranging of Furnishings and Salvage Cover Deployment

Covering the item is usually the best choice when the item is either too large to be moved or the movement would be too time consuming. When the decision is made to cover items with salvage covers or Visqueen, the room should be surveyed for the largest item or central piece. All smaller items should be moved to

SALVAGE, OVERHAUL, AND FIRE CAUSE DETERMINATION 643

Figure 20-11 Furniture arranged for salvaging.

this area, **Figure 20-11**. Consider a bedroom for example. Items should be removed from the closets and placed on the bed, dressers should be pushed up close to the bed, and pictures should be removed from the walls and placed on the bed. If dressers are too large to be moved then the drawers can be placed on the bed.

Once everything is ready to be covered, the salvage cover or Visqueen should be deployed. This is done by one or two firefighters. This deployment method is called a *counter payoff*. With one firefighter holding the cover or Visqueen firmly by the bottom fold, the other pulls the top fold or leading edge, if it is rolled, away from the firefighter holding the cover, **Figure 20-12**. It is easier and will save time if the cover is rolled or unfolded directly onto the items needing protection. When the cover is extended, it is then placed on the furnishing and the sides are unfolded. At this point, the reason for having the treated side out is obvious—so that it is facing the ceiling to deflect any water dripping down.

If there are breakables under the cover, extra care must be taken. Depending on the room being salvaged, a glass piece should be placed on top of the cover after the room is salvaged to designate that valuables are under the cover. This signals other crews that the cover should be removed gently. The salvage cover edges should be "rolled" under so that they won't collect any water or debris. If this is a one-firefighter operation, the cover is just laid on the items and the firefighter plays it out individually.

> **Ethics** Salvage operations can provide temptations to firefighters who would like a souvenir of an incident. Sometimes items are slightly damaged and crews will determine the value to be zero and take the item back to the firehouse. This is stealing and should never be allowed to take place.

Covering items does not just apply to furnishings; it can also apply to carpeting. When attacking an attic fire, the attack crew can bring with them a salvage cover and quickly spread it out prior to pulling ceilings or making attic access cuts. This is also where the floor runner comes into play. This again is just one of those little things that create excellent customer service.

Shoulder Toss

Another way to deploy a salvage cover is to use the *shoulder toss*. This method is done by a single firefighter and is reserved for covering large unbreakable items. Rack storage in a warehouse is one example where this could be used. The firefighter takes a cover that has been folded (not rolled) to the location of deployment. The firefighter then places the cover on one arm so it can be tossed with the firefighter's throwing arm, **JPR 20-4A**. The firefighter grabs the top couple of folds that are nearest and flips them back so that the arm is in a throwing position behind the shoulder. By using a straight arm movement, the cover is tossed over the item being salvaged, **JPR 20-4B**. The cover is then spread out over the item that needs to be covered. This takes practice and some upper body strength, but is quite effective in areas where the item is too high to just spread the cover over it.

Figure 20-12 Two firefighters deploying a salvage cover.

JOB PERFORMANCE REQUIREMENT 20-4
Shoulder Toss

A Place a folded salvage cover over one arm.

B Using a straight arm movement, toss the cover over the items being salvaged.

Balloon Toss

If two firefighters are available, the balloon toss is another method of deploying a salvage cover, **JPR 20-5**. This method is used if the objective to be covered is slightly taller than the firefighters. The two firefighters place the unfolded cover on the ground alongside the full length of the base of the objective, **JPR 20-5A**. The firefighters grab the top inside edge of the cover, pick it up to about waist height, and pull with tension on the cover. Both firefighters simultaneously move the cover's edge in an upward motion so that it catches air, and while still holding the end of the cover they move to the other side of the objective while the cover is still in the air, **JPR 20-5B**. The objective is to use the air to help raise the cover over the objective. Both firefighters guide the side of the cover over the objective to the ground on the opposite side. The cover is then adjusted to help water flow off the objective, **JPR 20-5C**.

Working with Visqueen

Working with Visqueen or a rolled salvage cover is essentially the same as working with a folded cover. The limiting factor is the size of the rolled cover. In both cases, the roll is held by one firefighter and the other pays out the remaining portion. Once the material is in place, the sides are folded out and the edges are rolled under. A good example of where Visqueen is much better than a salvage cover is in a supermarket where the long rows of food would need to be covered. By starting at one end of the row, the proper length of Visqueen can be determined and cut. The Visqueen is then placed over the row and everything is covered.

Another use is covering the ventilation holes in a roof. The Visqueen can be nailed to the roof using wood strips. This is normally a task for the board-up crew, and it is discussed later.

Limiting Exposure to Smoke

As discussed in Chapter 18, firefighters should always consider the by-products of combustion—heat, smoke, and toxic gases—as a danger. These by-products should be contained from spreading into undamaged portions of a structure and should be removed from the entire structure using forced air ventilation, **Figure 20-13**. Containment of the by-products can be an action as simple as closing doors after a room has been searched or closing any nearby doors during the fire attack advance. This action can save the contents of rooms from additional smoke, fire, and water damage. Adding positive pressure ventilation not only allows firefighters better visibility and safety from the products of combustion but also affords an additional measure of protection against damaging the contents of the unburned portion of the structure.

Water Removal

Water from fire attack operations can leak through the floor and damage rooms or entire floors below the fire. It is critical to capture, divert, or remove the water to successfully salvage these areas. If there is a

JOB PERFORMANCE REQUIREMENT 20-5
Balloon Toss

A Two firefighters place the cover on the ground next to the item needing protection. The salvage cover is unfolded out of the accordion fold so that the cover is across the base of the item at its full length.

B The firefighters grab the top inside edge of the cover and pick it up to about waist height. Both firefighters simultaneously move the cover's edge in an upward motion so that it catches air, and they then move to the other side of the item while the cover is still in the air.

C The firefighters guide the cover to the floor on the opposite side. Once the items are covered, the firefighters dress the cover by folding the edges under to keep water from pooling and to lessen the chance for the edges of the cover to be disturbed by accident.

floor drain, the operation becomes a little easier. If no drain exists, firefighters should examine their options. If a toilet can be removed, a drain hole can be created. Firefighters can squeegee water down this created drain as they would a floor drain. Firefighters can be creative and use salvage covers to create dikes to channel water into the drain or out of the building. Lining a stairwell with a salvage cover can create a chute to channel the water out of the area.

Catch-All

An example of the need to use a catch-all would be a multiple-story office building. A fire may occur in an office and be extinguished by the sprinkler system. This could cause a major salvage problem on a lower floor.

Catch-alls are used to contain the water dripping through the ceiling until a system is set up to remove the water from the building (chutes, water

Figure 20-13 Positive pressure ventilation in place.

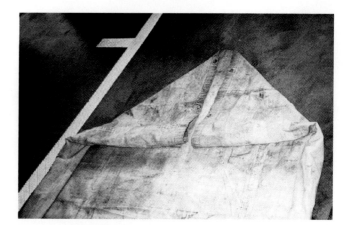

Figure 20-14 Catch-all end folds.

Water Chute

If the water volume is too much for the catch-all, then a water chute is used so the water can be diverted out of the building through a window or doorway or down a stairwell. The water chutes are a combination of ladders or pike poles and salvage covers. The salvage covers are draped on the ladders so they work like a trough and the water flows down them. Office furnishings or other equipment is used to support the ladders. The salvage cover can also be stretched out and a pike pole placed on either side. The cover is wrapped around the poles and the ends rolled in until the desired width of the chute is attained.

If the length is longer than a single cover, a leakproof seal must join the two, **Figure 20-16**. Once the direction of the flow is determined, the upper cover is put in place with the last 6 inches folded back on itself. The next cover is laid on the next portion of the ladder with its top 6 inches laid on top of the bottom 6 inches of the cover above.

Figure 20-15 Catch-all.

vacuums, squeegees, etc.). Catch-alls have sometimes been constructed using ladders and salvage covers to form a containment area. However, firefighters should consider the weight contained inside this catch-all. At 8.33 pounds per gallon, this weight can soon cause a structural integrity problem in an already damaged building, as the weight is all focused on a small area of the floor. It is much safer to consider a catch-all as a mitigation effort and construct it using a single salvage cover. A system to remove the water from the area should be constructed or devised as quickly as possible.

Catch-alls are easy to make. The firefighter places the cover with the treated side down and rolls the sides in to the desired size of the catch-all. The next step is to fold the ends as shown in **Figure 20-14,** roll the ends in, and tuck them tightly so they do not unroll. The cover is then flipped over, **Figure 20-15**. As the catch-all fills up it will not unroll the sides or ends. As the water depth will only be 2 to 3 inches, a water vacuum may be necessary to adjust the water level until the water can be directed away with a chute or other method.

SALVAGE, OVERHAUL, AND FIRE CAUSE DETERMINATION 647

Figure 20-16 Leakproof fold of water chute.

The covers are then rolled together toward the bottom cover and flattened out when there is no more to roll.

SALVAGE OPERATIONS IN SPRINKLERED BUILDINGS

Post Indicator Valve and Outside Screw and Yoke Valve

Many buildings are now equipped with sprinkler systems. These systems are beneficial at stopping or containing a fire but can cause damage to the unburned portions of the buildings if allowed to flow freely. It is important that firefighters know how to shut these systems down to prevent unnecessary damage. The incident commander will determine when the sprinkler is no longer needed for fire suppression.

Essentially two types of valves are used to shut down a sprinkler system: the post indicator valve (PIV), **Figure 20-17**, and the outside screw and yoke (OS&Y) valve, **Figure 20-18**. The PIV is generally located near the sprinkler connection for the building, but it may be just about anywhere around the building. The PIV is about 3 feet tall with a window near the top that states either "Open" or "Shut." This indicates the status of the valve. The PIV should be locked in the open position. Most PIVs are monitored so that if someone attempts to close them, an alarm sounds. If a PIV is found in the closed position it should be noted to

Figure 20-17 Post indicator valve. Note that the PIV is shut and unlocked. If found in this position at an incident, it should be opened and the condition reported to the incident commander and the fire investigator.

the incident commander and turned to the open position if active firefighting is still taking place. To close the valve, the firefighter should break the lock, which is usually a knockoff-type lock, and remove the handle, which is usually locked on the valve.

> **Streetsmart Tip** A firefighter should always take bolt cutters when sent to close a sprinkler valve in case it is locked with a regular lock rather than a knockoff lock.

The handle is used to close the valve. It helps to remember, "Righty, tighty—lefty, loosey," or clockwise to tighten and counterclockwise to loosen. Once the PIV is closed, the firefighter should also locate the system drain and open it. This will assist in bleeding the system and reducing the amount of water that will drain through the sprinkler heads.

Figure 20-18 Outside screw and yoke valve.

The OS&Y valve will be located either on an exterior wall of the building or just inside the building. It looks like a steering wheel with a large screw sticking out of the middle of it. If the screw is out then the valve is open; if it is in then it is closed. The wheel may be locked so it cannot be turned to close the valve, or it may be a supervised system that sends an alarm if attempts are made to close the wheel. Bolt cutters are used to cut the lock to shut this valve. Again, if the valve is found in the closed position, the firefighter should notify the incident commander and turn it on if active firefighting is still taking place. If there is any possibility to have the valve reopened due to fire spread then it should not be turned off. The system should only be shut down when *all* active firefighting is completed.

Sprinkler Stops

If no valves can be located, sprinkler stops will have to be used. In some cases, sprinkler stops will need to be put in place to minimize the amount of water that will flow out of the head. An A-frame ladder will reach most ceilings in sprinklered buildings. When placing a sprinkler stop from an A-frame ladder, the ladder is steadied by another firefighter as the firefighter climbs it to provide a good platform. Sometimes the sprinkler head is in a position where it cannot be reached using normal laddering techniques. In this case an auditorium raise can be used. This is a risky operation and those risks must be weighed against the amount of property damage that can be saved. The auditorium raise is discussed in Chapter 14.

Salvage Operations Lessons Learned

Salvage operations can be as important as any actions a fire service agency can perform on an emergency scene. In most cases, a majority of the damage has already been done to the structure and the main job is to make sure the emergency does not get any worse after arrival. This is where salvage operations can be very important to the success of the incident. Firefighters will learn throughout their career that the ability to make something good out of a very bad situation becomes paramount in order to find some value in what they do. With fires, that outcome is in doubt until the end of the incident. Firefighters apply all of their knowledge, skills, and abilities to aggressively attack the problem, but if all that is left is a smoldering parking lot, what have they accomplished? Salvage can immediately give firefighters that sense of purpose. It is no longer acceptable to believe that merely saving the house next door means a successful operation. Firefighters need to preserve as much of the salvageable property as possible. The operation must ensure life safety, exposure protection, and property conservation. With this thought process directing firefighters' efforts, the "customer" will agree that they accomplished what they set out to do.

OVERHAUL TOOLS AND EQUIPMENT

The process of overhaul is as important as the initial extinguishment of the fire. The process of making sure the fire is completely extinguished can be tedious but very worthwhile. The ability to check walls, floors, ceilings, and attics—especially with blown-in insulation and dead air spaces—is a skill that needs to be learned and practiced whenever firefighters attend a working incident. Due to many factors, including successful fire prevention programs, the fire service is responding to fewer working fire incidents. The need to learn good and thorough firefighting practices and make them habits is paramount to successful firefighting.

Tearing into buildings, cutting through floors, pulling ceilings, and searching for hidden fires are tasks that most firefighters enjoy. The tools used for overhaul are built for these tasks, so these tools are among firefighters' favorites. In most cases, firefighting requires precision and attention to details. In overhaul, the mission is simple—leave no area unsearched if there is potential exposure. A firefighter's knowledge of building construction and fire behavior is an asset in property conservation efforts.

Common Tools

Common overhaul tools are pike poles, pitchforks, rubbish hooks, shovels, axes, chain saws, carry-alls, and wheelbarrows, **Figure 20-19**. All of these tools meet the goals of getting into areas to determine if they were exposed to fire and removing necessary items so they will not cause any more fire. Most of these tools are common and their purpose does not need to be explained. The few exceptions are the rubbish hook and the carry-all. The rubbish hook is a heavy tool with a D-handle that is between 4 and 6 feet in length. It has two tongs spaced about 6 inches apart that run perpendicular to the handle: It looks like a rake with only two teeth. The tool is heavy enough to easily break through ceilings or strip shingles from a roof. It is a mainstay for ventilation crews on roof operations. Because of its weight and the D-handle, it is easily used to breach and pull plasterboard.

Carry-All

The other tool needing a little explanation is the carry-all. Its name makes it pretty clear as to its purpose. It is an approximately 6-foot-square piece of heavy canvas with a rope strung through the grommets for handles that is used to carry debris out of the building once it is loaded, **Figure 20-20**. It is used in areas where wheelbarrows cannot be, due to access or other restrictions.

OVERHAUL OPERATIONS

Once the determination has been made that fire investigation concerns have been met, overhaul can commence. Fire investigation concerns are discussed later in this chapter. Overhaul operations consist of breaching walls, floors, ceilings, and dead air spaces in order to confirm that fire is not present in these areas. All areas directly exposed to the fire should be stripped of coverings such as plasterboard or paneling.

Firefighters can look for obvious signs of hidden fire and utilize their senses by feeling the wall for heat, smelling around outlets and other openings for a scent of anything burning, and listening for sounds of items burning. The bottom line is if there is a chance something could be burning behind a wall covering, the wall covering must be removed to make sure it is all clear. It is much better to be safe than sorry in this instance.

Firefighters must visually observe the wooden structural members to make sure there is no smoldering fire behind the coverings. The building structural members should also be visualized in

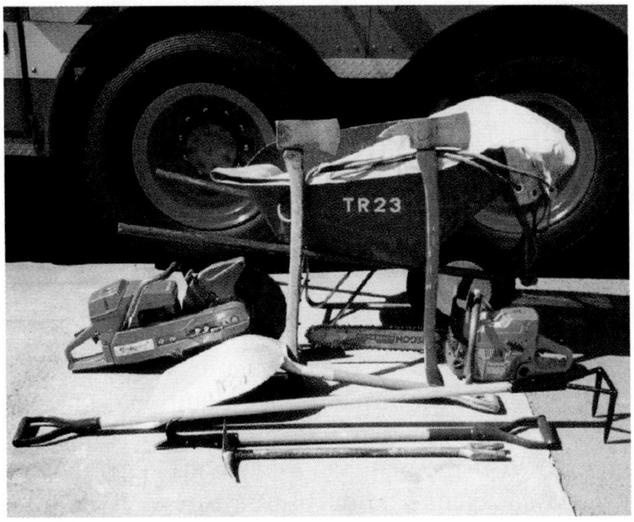

Figure 20-19 Various overhaul tools.

Figure 20-20 Firefighters carrying debris in a carry-all.

order to confirm that the building's structural integrity has not been compromised by the overhaul activities. Fire can get into these areas through electrical outlets; via conducted heat from metallic pipes or wiring; through breaches in the walls or floors where conduit runs; via heating, ventilation, and air conditioning (HVAC) ductwork; or because of a lack of fire stops in improperly constructed buildings.

Once these areas are breached, all of the insulation, if any is in place, needs to be removed so the entire space can be visualized. This is extremely important in attic spaces. Several different types of blown-in insulation can smolder for many hours prior to igniting, **Figure 20-21**. Some of the more common are cellulose or redwood insulation, which is often found in older homes. If this type of blown-in insulation is found, then the task of removing all of the exposed material becomes paramount. The smallest of embers can travel to remote areas within the attic and smolder for many hours prior to growing to a free burning state. The entire attic must be checked and revisits are recommended, as discussed later in the chapter. Newer homes have rolled-in fiberglass bats that are easily removed and for the most part noncombustible, although the paper backing the fiberglass is attached to will burn.

Overhauling Roofs

Overhauling roofs can be a very long and tedious process. Sometimes there can be multiple roofs under the most visible one and each one needs to be overhauled for possible fire extension. Sometimes flat roofs are left in place while pitched roofs are built over them, thus creating two attic spaces and a real overhaul nightmare. As with all overhaul operations, the material exposed to fire that could smolder for some time needs to be removed. This is why roofs are particularly difficult. During a hot summer day there is nothing worse than working on a roof to overhaul it because of the heat, access problems, the sheer magnitude of the operation, and the inherent safety concerns of working above ground.

> **Safety** Overhauling roofs is something that needs to be done on all working attic fires and this can turn into a very time-consuming task. This situation provides a good opportunity to bring crews in from outside the area to work a few hours on the roof while the initial attack crews rest or perform other needed tasks prior to resuming their overhaul duties. In overhaul situations, it is important to set up a "rehab" station where crew members can rehydrate and cool off.

Electronic Heat Sensors

A relatively new way to check buildings for hot spots is to use electronic heat sensors or thermal imaging cameras. These sensors determine where heat is higher as compared to the surrounding areas. The sensors' alerting mechanisms differ depending on the sophistication of the equipment. Some simply sound different when they are pointing at an area that is hotter than the surrounding areas, whereas others give the operator the ability to see and pinpoint the exact area that is showing a heat signature. Some of the more sophisticated devices are also used during searches to find people through the smoke, under water, or lost in the wilderness, and for other applications, **Figure 20-22**.

Revisits of the Involved Structure

Even after all areas have been checked, it is important to revisit the scene of the incident. Crews should never leave the scene if there is any active evidence that any part of the building is not completely extinguished. As with all good operations and focusing on customer service, it is always wise to return to make additional checks. The crews are checking for any smoldering areas and to see if the occupants have any questions or need assistance. A good rule is to revisit 2 hours

Figure 20-21 Blown-in insulation in an attic.

SALVAGE, OVERHAUL, AND FIRE CAUSE DETERMINATION

Figure 20-22 Helmet with thermal imaging camera attached.

Figure 20-23 Firefighter using wheelbarrow to remove debris.

after the last company or personnel leaves the scene and then again within 10 hours. This way the building is visited twice within 12 hours of incident termination.

In most cases crews from the next shift can go by the incident for the revisit and also learn from it by looking at the areas of involvement and problems encountered. Volunteer companies will have to work around their normal work schedules. Another option is to have the business owner hire a security company to maintain a fire watch, or local law enforcement may be able to assist if the fire service agency is unable.

Debris Removal

The need to remove the debris while searching for hidden fires is important, but as discussed later in this chapter this needs to be done in cooperation with the fire investigator. All of the loose material damaged in the fire must be removed and further extinguished outside the structure. As walls or ceilings are pulled, the material needs to be removed from the building. This is where carry-alls or wheelbarrows come in handy, **Figure 20-23**. The material needs to be placed in an area outside the structure where cleanup crews can easily access it after the firefighters leave.

The material is placed in a pile. It is *lightly* hosed down to keep from creating a bigger problem down the street with a substantial amount of runoff. Firefighters must be certain the material is extinguished. Debris placed on driveways is the easiest to clean up after the event. Although placing debris in front yards outside a main window may seem convenient for firefighters, customer service should always be considered.

> **Streetsmart Tip** Sometimes the debris needs to be placed outside a window or other opening, instead of taking it to the driveway or other hard surface. Visqueen should be placed outside the opening to protect any plants or lawn under the plastic. Remember: customer service! This effort will be appreciated by the building owner.

After the overhaul crews have completed the removal of material and are sure it is completely extinguished, they can cover the debris with Visqueen to reduce the eyesore.

Overhaul Operations Lessons Learned

As with salvage operations, the customer service aspects of overhaul are many. The detailed search for hidden fires, the ability to limit the amount of uncovered debris left behind, and the overall ability to leave the building in a state where it is safe for the owners to get in and start looking for lost articles are very important processes. Overhaul work is often the only thing the owners get to see the result of. Making sure overhaul works in concert with the fire investigator is another way to ensure the operation is successful.

FIRE CAUSE DETERMINATION CONCERNS

As the incident progresses through all the different aspects of initial dispatch, response, arrival, rescue, containment, extinguishment, ventilation, salvage, and overhaul, firefighters need to be aware of possible clues as to how the incident started. During dispatch, they should note all of the information available at that time. Information such as weather, time of day, address, type of dispatch (i.e., garage fire versus bedroom fire), and the number of reports all give firefighters a clue as to what the fire was doing prior to arrival, **Figure 20-24**.

During response, additional information may be gathered by answering these types of questions: Is the structure evacuated or still occupied? Are callers reporting explosions or different colored smoke? Were citizens seen leaving the incident at a high rate of speed? Observant firefighters on scene can provide fire investigators with reliable information as to where the fire was, what spectators said about the fire, and how the fire reacted during fire attack. Once inside the structure, firefighters should observe how the fire reacted, where the fire was, how the contents of the building were arranged, if there were signs of a break in, how the electrical breakers looked, if anybody was found in the building and how were they found, how appliances were found, how interior doors were found, and so on. All of these things need to be committed to memory so they can be relayed to the fire investigator, **Figure 20-25**, and also listed in the company officer's report. The smallest of items that seems out of place needs to be reported. Although it may not seem like much, it may be the missing piece of the puzzle for the investigator.

During overhaul, firefighters must make sure the investigator approves of the removal of any part of the structure or any piece of furnishing prior to removing it. Firefighters cannot just throw things out of the building for expediency. Each step has to be taken with knowledge and care so as not to make things worse. As investigators work through a structure, they will be able to release the different rooms for overhaul. Markings on plasterboard from the smoke may mean nothing to a firefighter, but may be that missing puzzle piece for the investigator. Experienced investigators are only as good as the clues they observe. If they show up and all they

Figure 20-24 Dispatch sheet showing various information. *(Courtesy of Town of Colonie Fire & EMS)*

Mutual Aid Dept/Unit	Disp'd	Ack'd	Enrt	Scene	I/S

Amb:	Hosp:	Enrt:	Arrv:	Remarks:	
Amb:	Hosp:	Enrt:	Arrv:		

Sig	Cover Co.	Comm. Post	Fire Inv	Op. Off	Comm. Dir.	Mob. Air	Batt. Co	Car 1
30								
40								

*Signal 20 gas leaks or hazmat incident: page fire investigation (399)

Figure 20-24 Continued

Figure 20-25 Fire investigator working a scene.

have is a bare wood floor and wooden studs for walls, they are at a disadvantage.

Preservation of Evidence

In the process of their work, firefighters may come upon something out of the ordinary and will need to make every effort to preserve it until the fire investigator can observe it, even if this means stopping their entire operation. Firefighters need to leave items where they found them and cover them up with Visqueen or mark off the area with fire line tape to keep everybody out of the affected area. Firefighters do not want to become involved in a big legal matter because they did not do their job correctly. If a room is overhauled and cleaned of all contents and later is found that an incendiary device was located in that room, it could result in a legal matter for the firefighter and the department.

Basics of Point of Origin Determination

The determination of the point or area of origin can be a very scientific pursuit in which exotic testing is done and the exact point is determined—or it can be as simple as looking at a room and knowing where the fire started based on the indicators present. The basic clues a firefighter should watch for are where the fire was first noticed or reported; where the heaviest damage was; or any markings on the walls, ceilings, or floor indicating where the fire started. The basic premise behind fire behavior is that it travels the path of least resistance, **Figure 20-26**. As fires grow, smoke and heat are forced up until they reach an obstruction, usually the ceiling. The smoke and heat will branch out, taking the path of least resistance. In a structure fire this path will be along the ceiling. As

Figure 20-26 "V" pattern in a structure fire.

the smoke and heat go in opposite directions the wall will be marked with a "V." *V pattern* is a common term in the fire investigation field. If the fire moves down a hallway from the front room, the V will start at the bottom nearest the area of origin and proceed up the wall.

Another simple manner to determine the fire's starting point and where it spread is the **depth of char**. Wherever the fire burned the longest is where the fire's depth of char will be the greatest, **Figure 20-27**. This can be illustrated by looking at a log in the fireplace. Usually the center of the log has the most "charring" and it diminishes toward each end. Wherever the fire started would have had the longest period of time to burn the wood members in that area, thus the charring would be greater in that area. The only misleading part of this indicator is when an area of the fire took longer to extinguish than others. This attack crew information can assist the investigator in determining the point of origin.

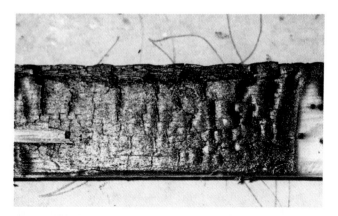

Figure 20-27 Depth of char.

Depth of char is just one of many indicators and can be influenced by many other factors. It is a simple way to determine the point of origin for a simple incident. The key to assisting the fire investigators is not to rush to complete the job and completely mess up theirs. It is imperative that firefighters pay attention to what is going on around them during all aspects of an incident and report items that fall outside the norm.

Fire Cause Determination Lessons Learned

Assisting the fire investigator is a skill that should be second nature to all firefighters. The gathering of information from the time of dispatch throughout the incident should become automatic. Every little detail should be noted and passed along to the investigator. Fire investigators are just another part of the team who are attempting to make this tragedy as tolerable as possible for the owner. Determining the cause and origin of the incident is every firefighter's duty and, like all of the other parts of the job, a very important factor. Firefighters always want to assist an investigation or criminal prosecution.

SECURING THE BUILDING

Firefighters have an obligation to make sure the building is secured after they have completed all of their operations. This would include the revisits discussed earlier in the chapter, boarding up all openings—including roof openings—so the building is secure, and securing the utilities. Some jurisdictions rely on board-up crews to handle this function, **Figure 20-28**, and the insurance carrier will pay those crews. If a building is not insured then the local public works or a local lumberyard might donate some plywood to help the owner defray costs. Again, customer service will leave a lasting positive experience for the homeowner.

These are simple items to consider prior to leaving an incident scene. But what about the displaced owners? Where are they going to stay or do business? What are they going to wear for clothes if everything was destroyed? Do they have the means to contact their insurance company? Do they have insurance? Does the incident commander need to contact the Red Cross or other charitable organization in order to get them a place

SALVAGE, OVERHAUL, AND FIRE CAUSE DETERMINATION

Figure 20-28 Building that has been boarded up after a fire.

to stay or some clothing? All of these things should be considered and the fire service should play an active role in solving these issues for the owners.

In many of these cases the owners are insured, have family or friends they can stay with and borrow clothes from, and only need on-scene crews to make sure the building is safe for them to enter so they can get on with their lives and start rebuilding. It is that other small percentage that should be assisted in any way possible to help them get back on their feet. Some fire service agencies have hotel rooms reserved if the need arises to house families that have no place to go; contacts with the local phone company for rerouting the business or homeowner phone lines; and contacts with local churches, the Red Cross, or the Salvation Army for food or clothing. This list could grow as large as the fire service wishes it to. Fire service agencies are taking a much more active role in going above and beyond when it comes to customer service and this will pay off in the long run.

Lessons Learned

This chapter dealt with areas not commonly reinforced through training. Most fire training consists of the vital roles of fire attack, building construction, search and rescue, and so on. Salvage and overhaul duties, while not glamorous, are vital to the functions of fire investigation and property conservation. These duties can often go unnoticed as firefighters are focused on the more exciting aspects of an incident. Most firefighters would rather talk about the great "stop" they made instead of the minimal damage the lower floor of the structure received. However, salvage and overhaul duties do have a direct impact on the success of an incident. They are also a great customer service for the homeowner, even if they are never recognized.

KEY TERMS

Chain of Command Common fire service term that means to always work through one's direct supervisor. The fire service is viewed as a paramilitary organization and because of this all requests for information outside the assigned workplace should go through the supervisor.

Depth of Char A term commonly used by fire investigators to describe the amount of time wooden material had burned. The deeper the char, the longer the material was burning or exposed to direct flame.

Sheetrock A trademark and another name for plasterboard.

Visqueen A trade name for black plastic. It can be used very effectively in salvage and overhaul operations. Many examples are discussed in this chapter.

REVIEW QUESTIONS

1. Why is the exterior of a building important to salvage operations?
2. What are the three types of material that salvage covers are made of?
3. What advantages does Visqueen have over customary salvage covers?
4. How can a fire department be reimbursed for equipment damaged in the course of working the incident?
5. How do firefighters check for holes in the salvage covers?
6. What are the basics of furniture arranging for salvage operations?
7. What are some indicators that a Sheetrock ceiling has a potential to fall?
8. What does PIV stand for?
9. How are overhaul and fire investigation linked?
10. What does a revisit of a structure fire accomplish?
11. What is a "V" pattern?
12. Why is knowing the "depth of char" important?

Additional Resources

NFPA 1921: Guide for Fire and Explosion Investigations. National Fire Protection Association, Battery March Park, MA, 1998.

CHAPTER 21

PREVENTION, PUBLIC EDUCATION, AND PRE-INCIDENT PLANNING

Donald C. Tully, Orange County Fire Authority

 OUTLINE

- Objectives
- Introduction
- Administration of the Fire Prevention Division
- Fire Company Inspection Program
- Home Inspections
- Fire and Life Safety Education
- Pre-Incident Management Process
- Lessons Learned
- Key Terms
- Review Questions
- Additional Resources

STREET STORY

As a fire service educator for many years, the one thing I have noticed that sets apart people in prevention from people in suppression is the time it takes to see results. In suppression, the effects of the interventions are seen in minutes, hours, or at the most several days. Prevention is different. The results may not be seen for months, or years, for that matter.

I have been fortunate in my career to see fire prevention pay big dividends. Fire deaths and severe fires have been drastically reduced in Montgomery County, Maryland, while the population and rate of building construction continue to grow. This is true not only here but throughout the country. Safety campaigns, smoke detectors, sprinklers, child-resistant lighters, school programs, and the facts that people are eating out more and smoking less are all contributing factors.

A particular case stands out in which a school fire safety program made the difference between life and death. Alena Tune of Rockville, Maryland, paid attention to the advice she got in school on fire safety and used it to save herself and her four-year-old sister when a fire broke out in their home. It was 3 a.m. on November 17, 1995. Alena woke up to the screams of her sister. She put to practice the survival techniques she learned in school. She got down on her hands and knees, got to her sister, and put her on her back. She crawled 30 feet to the front door and safety. This happened thirteen years after firefighters began teaching fire safety to fourth graders.

So, remember that your prevention message may not pay off today or tomorrow, but armed with lifesaving information, people can use it during their lifetime to save lives and property.

—Street Story by Mary K. Marchone, Program Manager, Montgomery County Fire and Rescue Services, Montgomery County, Maryland

OBJECTIVES

After completing this chapter, the reader should be able to:

- Describe the function of the Fire Prevention Division.
- Understand the purpose and value of a quality fire prevention inspection program.
- Prepare, conduct, and follow up a quality fire prevention inspection of a business occupancy.
- Identify typical violations found in business occupancies and initiate the appropriate corrective action.
- Understand the value and goals of a home inspection program.
- Conduct a fire prevention inspection of a residential occupancy.
- Identify various types and levels of fire service public education programs.
- Prepare and present a fire safety educational program.
- Understand the necessity of pre-incident management for emergencies at target hazards.
- Prepare a pre-incident management plan for a target hazard.

INTRODUCTION

What is the cardinal mission of the fire service in America? Unquestionably, it is the preservation of life and property. The scope of this lofty task is seemingly ever expanding.

> **Firefighter Fact** As recently as thirty years ago firefighters were generally expected to respond only to fires—structure, vehicle, wildland. Today's firefighters are called on to be experts in all aspects of emergency activities: medical emergencies, hazardous materials incidents, urban search and rescue, swift water rescue, high-angle and confined space rescue—and the list goes on.

With all the demands placed on firefighters by the increasing expectations of the public to perform these relatively new areas of technical expertise, it is easy to lose sight of the simplest and most effective method of achieving the goal of the preservation of life and property: prevention. Any physician will attest to the benefits of physical conditioning. Keeping in shape now helps to prevent devastating injury and illness in the future. Why do fire departments do preventive maintenance on their vehicles and equipment? Changing the oil regularly prevents costly engine rebuilds in the future. So it is with fire prevention and education. In an enlightened world, the firefighter's job must be to save lives and property from the effects of fire and other emergencies through the effective application of time-honored and innovative fire prevention and education techniques.

> **Note** Fire prevention officers have long been guided by the adage, "The Three Es of Fire Prevention—Education, Engineering, and Enforcement."

This chapter shows how, through the application of the three Es, the level of fire safety in the community can be enhanced. This information is presented to assist the firefighter/inspector in conducting a quality fire prevention inspection of both residential and non-residential occupancies. Furthermore, the chapter highlights programs available for instructing communities in proper fire prevention activities and appropriate responses to emergency situations. Finally, a discussion is included on the need and methodology for pre-incident management, or preplanning, by firefighters for emergencies at target hazards.

ADMINISTRATION OF THE FIRE PREVENTION DIVISION

The duties and responsibilities of fire prevention officers are often not well understood by suppression firefighters. These men and women are seen around the station or at headquarters, but little is known about their daily routines. Fire prevention officers play a key role in ensuring that the fire department meets its goal of preserving life and property in the community.

The Fire Prevention Division is responsible for all aspects of the fire and life safety of buildings and occupants prior to an emergency incident. This includes the "engineering" aspect of fire prevention such as interaction with architects and builders, new construction plans review and approval, **Figure 21-1**, installation of fire detection and suppression systems, inspection of newly constructed or remodeled occupancies, **Figure 21-2**, administration of hazardous materials disclosure programs, inspection of high-risk occupancies requiring specialized expertise, and public education and information programs. After a fire or hazardous materials incident has occurred, the Fire Prevention Division is often called on to

Figure 21-1 Fire prevention officer reviewing plans.

conduct the origin and cause investigation, **Figure 21-3**, and coordinate the follow-up activities of other agencies, when appropriate.

These duties may be the responsibility of one fire prevention officer, or specialized assignments may be given to a number of officers, depending on local conditions. Whichever approach is taken, it is certain that the effectiveness of the Fire Prevention Division's efforts is greatly minimized without the cooperation and assistance of suppression firefighters.

FIRE COMPANY INSPECTION PROGRAM

An effective fire company inspection program will reap immediate and future benefits for the community and fire departments. The "enforcement" pro-

Figure 21-2 Fire prevention inspector walking through new construction.

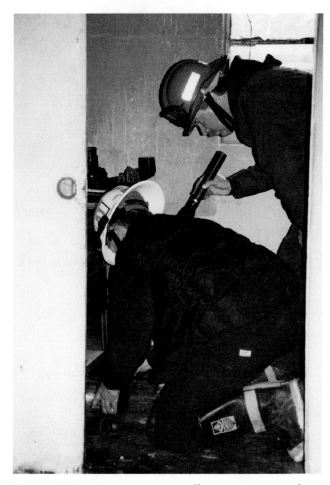

Figure 21-3 Fire prevention officer inspecting a fire scene.

gram should be approached with the following objectives in mind: reduction of fire hazards, opportunity to increase public awareness of the dangers of fire, positive public contact, and building familiarization and **pre-incident management**.

Inspection responsibilities are generally assigned to fire companies by the Fire Prevention Division. For obvious reasons, every effort must be made to ensure that such assignments do not require companies to leave their first-due response areas. The number of inspections assigned to particular fire companies will vary depending on the size and complexity of the occupancies and the level of emergency activity of the fire company.

Equipment

Any job, done right, requires the correct tools and equipment. A minimum complement of tools and equipment for the firefighter/inspector is suggested here:

- Complete standard-issue uniform, clean and pressed
- Occupancy inspection files

PREVENTION, PUBLIC EDUCATION, AND PRE-INCIDENT PLANNING 661

- Clipboard with notepad and pencil
- Flashlight
- Inspection forms
- Standardized information bulletins
- Violation notices
- Fire code reference manual

Many company officers find it convenient to keep their forms and equipment organized in a catalog case (large briefcase), **Figure 21-4**.

Preparation for Inspections

With ever-increasing demands for firefighters' time, it is important for company officers to routinely designate specific days and times for conducting fire prevention inspection. Planning and organizing an afternoon's inspections is essential. Preparation should include these steps:

1. Plan the area to be inspected. When practical, all occupancies within a given shopping center, office building, or block should be visited on the same day.
2. Review occupancy files prior to leaving the station. Note past violations, corrections, and recommendations to check for compliance or reoccurrence, **Figure 21-5**.

Figure 21-5 An officer reviews occupancy files before leaving the station.

3. Give consideration to the type of activities conducted at the business relative to the time of day chosen for the inspection. For example, it is usually not good practice to visit a restaurant during lunch hour or a small manufacturing plant after 3:30 p.m. In the case of the restaurant, the staff may be too busy to assist inspectors, and it is not unusual for a small manufacturing plant to shut down after 3:30 p.m.
4. Should the inspector make an appointment with the business owner or arrive unannounced? There is no specific rule in this regard. Generally, fire inspectors do not make "surprise inspections." Many business owners appreciate a telephone call prior to the inspection; others feel it is not necessary. Writing a note in the inspection file indicating the owner's preference is good practice.

   Note, too, that given the unpredictable nature of the firefighting business, keeping an appointment is not always possible. When making an appointment, remember to explain this to the business owner.

Conducting the Inspection

How many firefighter/inspectors are necessary to conduct a fire prevention inspection? Again, there is no magic number. This is a judgment call for the company officer. Remembering that building familiarization is an important aspect of the fire company inspection program, it would seem appropriate to include the entire company. However, consideration

Figure 21-4 Company officer preparing for inspections.

must be given to the business owner's perspective. Too many firefighters may be seen as a "show of force" or an intimidation. Too few inspectors may result in a less than thorough or inefficient inspection. Care must also be taken to minimize the disruption to business activities. It certainly does not take three or four inspectors to inspect a small barbershop, for example, but this may be the right number for a large warehouse or manufacturing occupancy.

After arriving at the occupancy to be inspected, the crew should proceed directly to the front office, **Figure 21-6**. They should not linger in front or immediately begin inspecting the exterior of the building. Although the fire department has a statutory right to inspect all businesses, permission must always be obtained from the owners or their representatives prior to beginning any inspection.

Firefighters should introduce themselves to the receptionist, state their business, and request to see the person responsible for general building and employee safety. This may be the business owner, plant safety engineer, operations manager, or a maintenance engineer, **Figure 21-7**. Whoever is designated should have the authority to speak for management and be able to correct any violations noted.

> **Note** The company officer should insist that the crew be escorted during the inspection. The escort will be able to grant access to all areas of the building and answer any questions that may arise. Additionally, the possibility of unfounded questions or impropriety on the part of the inspectors will be eliminated.

Figure 21-6 Upon arrival at the site to be inspected, inspectors should proceed directly to the office.

Figure 21-7 Introduction of a company officer to a business owner.

Occasionally, but fortunately infrequently, a business owner will deny permission to inspect. The company officer should ask for an explanation for the denial. Perhaps it is as simple as the time is not convenient for the business owner, the owner is too busy, or the safety manager is not present. Rarely is it a matter of the owner just refusing to be inspected with no rational explanation. This does occur, though. Under no circumstances should company officers engage in an argument with owners or attempt to persuade them to change their minds through the use of threats of authority or intimidation. A calm and reassuring approach is always the best tactic. "Would another day be more convenient?" or "We are not here to disrupt your business, only to help you maintain a safe working environment for you, your employees, and your customers." If all else fails, the matter should be referred to the Fire Prevention Division. Ultimately, an **administrative warrant** may have to be obtained from the local magistrate.

> **Note** There are several schools of thought regarding the proper methodology for routing through a building when conducting the inspection. Outside to inside, top to bottom, left to right are all acceptable. It is generally of little consequence in what order the occupancy is inspected as long as the method is efficient, systematic, and thorough.

Typical Violations

A primary objective of the fire company inspection program is, of course, the elimination of fire and safety hazards. A **fire hazard** can be defined as any condition, situation, or operation that could lead to the unwanted ignition of combustibles or result in proper combustion becoming uncontrolled. Fire code violations in all occupancies can be categorized in general terms: exiting, fire protection equipment, use and storage of hazardous materials, and electrical and general fire safety.

Exiting

Most fire prevention officials would agree that a building's **means of egress**, **Figure 21-8**, provides the most basic level of life safety to its occupants.

> **Note** Exits are the most often abused and neglected system in a building.

The inspector must verify that no exits are compromised by ensuring the following:

- Clear and unobstructed access is provided to all exits.
- Exits are identified and well lit.
- The proper type of door is used and it opens in the direction of travel.
- Exits are equipped with the proper opening and locking hardware.
- Exit discharges to a public way are clear.

Exit accesses must be kept clear and unobstructed by storage or other materials that may hinder their use. Service corridors in malls and hotels are easy targets for storage of recently delivered merchandise or unused tables and chairs, **Figure 21-9**.

All exit accesses and exits are required to be identified with an exit sign, **Figure 21-10**. An exception is made for the main entrance to an occupancy because it is assumed that it is obvious even to visitors. Generally, exit signs are required to be illuminated only in assembly occupancies or high-rise buildings. When the exit access is not straightforward or is confusing, as in a warehouse or office with many corridors, intermittent directional signs should be used. If the fire inspector becomes lost or misdirected while walking through the building, then it can be assumed that visitors may become lost as well. This situation should be addressed by the use of intermittent directional exit signs.

> **Streetsmart Tip** The use of a "This Is Not An Exit" sign is not considered good practice because people in a hurry to leave the building under emergency circumstances are likely to focus on the word "Exit" and may become disoriented as a result.

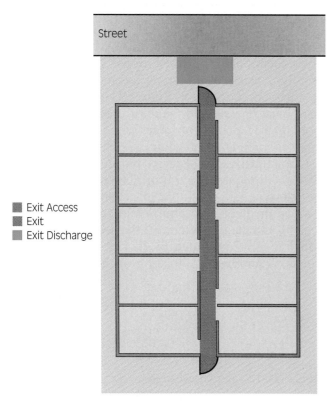

Figure 21-8 Means of egress.

Figure 21-9 Exit corridors must be kept clear of all storage.

Figure 21-10 All exits are required to be clearly identified.

The inspector must ensure that the proper type of door is used as an exit. Of the four classifications of doors—revolving, overhead, sliding, and swinging—only the swinging door is permitted to be used for a required exit. **Figure 21-11** shows the four classifications of doors.

Exits are required to be equipped with specific types of opening and latching devices depending on the occupancy classification. Approved exit hardware can be grouped in three categories: no-knowledge, panic, and special egress devices.

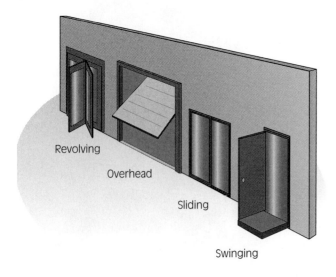

Figure 21-11 Of the four types of doors, only the swinging door can be used for a required exit.

No-knowledge hardware, Figure 21-12, is a broad category of locking devices that require no key or special knowledge to operate. Lock sets that combine the latching mechanism with a dead bolt are an example, **Figure 21-13**. The dead bolt is retracted simultaneously with the normal latching device when the knob or lever is turned. This device provides the security of a dead bolt while maintaining simplicity of operation. No-knowledge hardware is required on all exits, except the main entrance, in all occupancies when panic hardware is not required.

> **Note** A double-cylinder lock set is key operated on both sides. This type of hardware is permitted only on the main entrance to most occupancies provided a sign reading "This Door to Remain Unlocked During Business Hours" is displayed above the door. Double-cylinder dead bolts are also permitted in private residential occupancies.

> **Streetsmart Tip** Although double-cylinder dead bolts are permitted in residential occupancies by most building and fire codes, firefighters should strongly discourage their use. During an emergency evacuation of a home, there is no time to fumble for a key.

Panic hardware is required on exits in all assembly, educational, and institutional occupancies. It usually consists of a locking mechanism activated by a bar across the door, **Figure 21-14**. The mechanism must operate and the door must open with no more than 15 pounds of force. This is to allow even small children and frail adults to open the door.

Occasionally, for security purposes, a business owner will request the use of a **special egress control device**. The operators of occupancies such as jewelry or electronic equipment stores, weapons dealers, and the like often feel the need for the additional security provided by locked exits. This, of course, is an obvious violation of the fire code. To

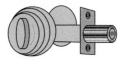

Figure 21-12 No-knowledge hardware is required on all exits not requiring panic hardware, except on the main entrance.

PREVENTION, PUBLIC EDUCATION, AND PRE-INCIDENT PLANNING

Figure 21-13 Interconnected lock set and dead bolt.

Figure 21-15 Special egress control devices often look like normal panic hardware, but must be identified with a sign.

Figure 21-14 Exit doors should be opened fully to ensure their function.

address the concerns of these owners while ensuring the safety of their employees and customers, fire and building officials have approved the limited use of special egress control devices. This type of device is allowed, provided that an approved automatic fire sprinkler system and an approved automatic smoke detection system protect the building. The door-release will unlock the door in a maximum of fifteen seconds. Pushing on the bar, **Figure 21-15,** activates the device. With the specific approval of the fire marshal, these devices are permitted in the following occupancies: general businesses, factories, institutions, mercantile businesses, storage facilities, and group care facilities and residential care facilities housing clients with various forms of dementia.

Special egress control devices must automatically deactivate whenever the sprinkler or smoke detection systems activate or whenever there is a loss of electrical power to the building or the device. There must also be posted on the door a sign that reads "Keep Pushing. Door Will Open in 15 Seconds. Alarm Will Sound" or similar wording.

Exit discharges must be kept clear to allow for occupants to continue their travel to a safe position away from the structure, **Figure 21-16**. Fire codes

Figure 21-16 Firefighters should inspect the exit discharge to ensure occupants have clear access to the public street or parking area.

usually define a "safe distance" as the nearest public way, that is, a street, alley, sidewalk, or major parking lot.

Fire Protection Equipment

> **Note** Fire protection equipment, whether portable or fixed, is the building occupants' first line of defense against unwanted fire. Firefighters must ensure that this equipment is not only present and maintained in proper working condition, but that the building occupants know when and how to use it.

During an inspection, it is most appropriate for the inspector to question the business representative: "Do your employees know where your fire extinguishers are located and how to use them?" "Do you provide any training in their proper use?" "Do you know how to manually activate your kitchen cooking equipment fire suppression system?" "Do you know where the fire sprinkler control valves are located?" "Do you know how a fire sprinkler system operates?" Many people are still under the misunderstanding that all fire sprinkler heads operate simultaneously or that they are activated by smoke!

Portable Fire Extinguishers. Firefighters should inspect for the presence of fire extinguishers, **Figure 21-17**, and ensure the following:

- Fire extinguishers are located near all exits and in exit accesses.
- The extinguisher has the proper classification and rating for its location.

Figure 21-17 Firefighter inspecting a fire extinguisher.

- The fire extinguisher is mounted on the wall and is easily visible and accessible.
- The pressure gauge indicates that the extinguisher is fully charged.
- The tag on the extinguisher indicates that it has been serviced in a time frame conforming to local requirements. (Some jurisdictions require annual servicing, whereas others permit a time span as long as five years between services.)

Automatic Fire Sprinkler Systems. Firefighters should check fire sprinkler systems for the following:

- All water supply valves are open and secured.
- The fire department connection should be free of obstructions and capped, **Figure 21-18**.
- Access should be provided to system risers, valves, and gauges, **Figure 21-19**.
- Pressure gauges on the system riser should show appropriate pressure.
- All areas of the building must be properly protected.
- Sprinkler heads should not be damaged, painted, or corroded.
- The system should be flow tested regularly, preferably quarterly, and records kept, **Figure 21-20**.
- The system should be serviced in a timely manner, usually at five-year intervals.

PREVENTION, PUBLIC EDUCATION, AND PRE-INCIDENT PLANNING 667

Figure 21-18 Outside control valves must be open and secured. The fire department connection must be clear of obstructions and capped.

Sprinkler systems in spray paint booths deserve special mention. By the very nature of the operation, sprinkler heads in spray booths quickly become covered with overspray. This condition will not only delay activation of the head but will most likely prevent its operation altogether. Therefore, keeping the heads clean is essential. To protect sprinkler heads that are likely to be ruined by overspray residue, NFPA 25, *Inspection, Testing and Maintenance of Water-Based Fire Protection Systems*, requires that they be covered with thin plastic

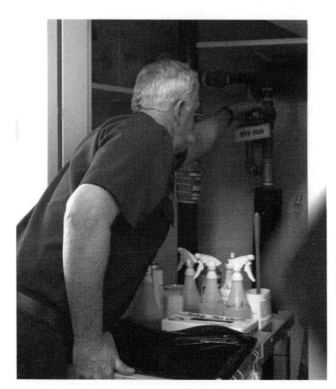

Figure 21-19 Access should be provided to fire sprinkler control valves and gauges.

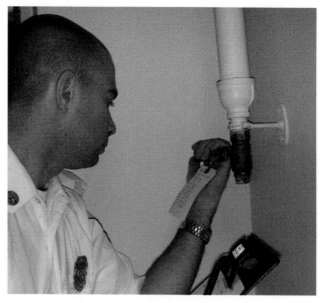

Figure 21-20 The sprinkler system should be flow tested regularly. Some inspectors flow the system themselves, while others ask the business owner to do so.

or small paper bags, as shown in **Figure 21-21**. These bags should be changed daily.

Restaurant Cooking Surface and Exhaust Hood Suppression Systems. Fixed extinguishing systems protecting the cooking surfaces and exhaust hood and ducts in restaurants are required whenever the cooking process produces grease-laden vapors. Pizza and bread ovens and stoves that are used for heating water or soup would not require a system; however, nearly all other commercial cooking would. Such

Figure 21-21 Many departments allow plastic sandwich-type bags to be placed over sprinkler heads in spray paint booths to prevent the accumulation of overspray on heads.

systems are essentially large fire extinguishers mounted on the wall, **Figure 21-22**. They contain either a wet or dry extinguishing agent.

When inspecting cooking surfaces and hood and duct systems, firefighters should ensure the following:

- The extinguishing agent cylinders are charged and armed.
- Nozzles are free of grease and are capped.
- Fusible links are free of grease.
- Hood filters are clean and in place, **Figure 21-23**. (Filters not only trap grease, but also restrict a surface fire from spreading to the duct.)
- The manual activation control is accessible, **Figure 21-24**.
- The tag on the system indicates that it has been serviced annually.

Similar systems are found in automotive, furniture finishing, and spray paint booths. They should be inspected in the same manner as restaurant systems.

Heat and Smoke Detection Systems. Beginning at the system control panel, **Figure 21-25**, the firefighter should ensure the following:

- All indicators on the panel are in a "normal" condition and not in "trouble" mode.
- The panel is receiving AC power.

Figure 21-23 Extinguishing system nozzles and missing filters.

- All detection devices are present and properly mounted—not hanging from the ceiling.
- Records are provided indicating that the system has been tested periodically.

It is often a good idea for firefighters to request that the system be tested in their presence. While not always practical given the nature of the occupancy, witnessing the activation of the system will not only ensure that it functions but will stress to the owner the importance of the system.

Figure 21-22 Fixed fire extinguishing system for cooking surfaces.

Figure 21-24 Restaurant workers should know the location of the manual release and how to activate the extinguishing system.

Figure 21-25 Firefighter checking fire alarm system control panel.

Fixed Gaseous Extinguishing Systems. These types of extinguishing systems protect building areas where water from a sprinkler system would cause massive amounts of damage to sensitive equipment, such as computers or electrical systems. The extinguishing agent is generally CO_2, halon, and halon replacement. In essence, these protection systems serve the purposes of detection *and* suppression.

Firefighter Fact Because of its danger to the environment, halon has virtually been replaced by other agents that are more environmentally friendly. Existing halon systems are "grandfathered" and can continue operation. Halon is no longer produced, and only remaining stockpiles of halon are available. Recharging systems with halon is generally cost-prohibitive. The result is that most systems are modified to use more environmentally safe extinguishing agents.

Caution The detection system should be inspected as would be any other heat and smoke detection system with the exception of the function test. Function tests should never be done by firefighters because an accidental discharge may result.

The inspection of the suppression portion of the system involves checking the cylinder gauges for proper pressure, ensuring that all discharge nozzles are unobstructed either by foreign objects or storage, and making sure access to the manual discharge control is provided.

Use and Storage of Hazardous Materials

Ensuring the proper use and storage of hazardous materials is a subject that often requires a great deal of knowledge, expertise, training, and experience. Many fire prevention officers dedicate large portions of their careers to such endeavors. However, from the firefighters' perspective, some general principles apply.

Inside Storage and Use

- Storage and use areas must be well ventilated.
- Sources of ignition must be eliminated in any area using or storing flammable or combustible liquids or gases. These include open flames and welding or grinding of metal.
- The maximum quantity of flammable and combustible liquids permitted varies according to the occupancy.
- Quantities of less than 5 gallons must be stored in approved **safety containers**, **Figure 21-26**.

Figure 21-26 Approved flammable liquid safety can.

- Small quantities should be stored in approved metal or wood storage cabinets, **Figure 21-27**.
- Larger quantities must be stored in rooms specially designed for such purposes. These rooms are constructed of fire-resistant material, have raised door sills to prevent leakage, have liquid-tight floors and walls, have explosion-proof electrical fixtures, and are well ventilated.

- Containers with a capacity greater than 30 gallons may not be stacked.
- Dispensing must be done through an approved pump or self-closing faucet, **Figure 21-28**.
- Dispensing containers must be properly bonded and grounded to protect against the discharge of static electricity, **Figure 21-29**.
- Compressed gas cylinders, whether full or empty, should be stored with protective caps in place and secured with chains or straps to prevent falling, **Figure 21-30**.
- Sources of ignition must be eliminated in any area using, dispensing, or storing flammable or combustible liquids or gases. These include open flames and welding or grinding of metal.

Outside Storage

- The selected storage area should be properly distanced from buildings and property lines to minimize the potential for fire spread or exposure.
- Storage areas must have **secondary containment**. This typically means a 6-inch curb around the storage area to contain spilled liquid, **Figure 21-31A**.

Figure 21-27 Small quantities of hazardous material should be stored in approved storage cabinets.

Figure 21-28 Approved self-closing faucet.

PREVENTION, PUBLIC EDUCATION, AND PRE-INCIDENT PLANNING   671

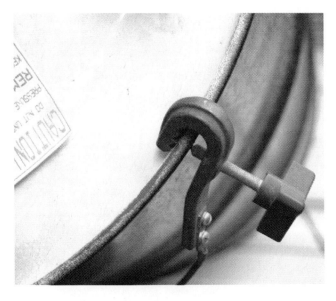

Figure 21-29   Properly bonded and grounded barrels.

(A)

Figure 21-30   Compressed gas cylinders secured with a chain and with caps in place.

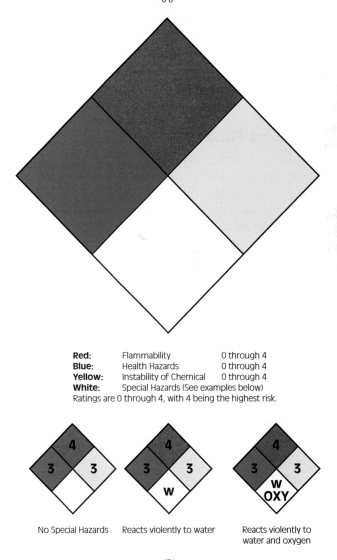

(B)

Figure 21-31   (A) Storage of flammable liquids with curb around storage area. (B) The NFPA 704 placard system identifies hazards in fixed storage facilities. The colors and numbers quickly give first responders vital information.

- Sources of ignition must be eliminated, keeping in mind the potential for the migration of vapors from spilled liquids.

Hazardous materials that are located in fixed storage facilities are not governed by the same regulations as the transportation industry. Local jurisdictions may develop a system of labeling for hazardous materials. Most local jurisdictions choose to adopt the NFPA 704 (*Standard System for the Identification of the Fire Hazards of Materials*) placard system, **Figure 21-31B.** This system provides valuable information to first responders and notifies employees and other persons using the facilities of the contents. The placard system is simple to read and provides information about health hazards, flammability, and reactivity. Firefighters inspecting facilities with these types of placards should speak to the appropriate persons at the facility to review the types of materials contained on the property and the associated hazards. These hazardous materials should be documented in the inspection report and first-due responders should be advised.

Electrical Hazards

Building electrical systems require specialized knowledge to ensure proper design and installation. The firefighter should call on the expertise of a local electrical inspector whenever questionable installation practices are noted. However, a firefighter can be effective in eliminating electrical hazards by observing the following:

- Check all fuse and breaker panels to verify that overcurrent protection devices have not been defeated and that all spaces on the panel are equipped with a breaker switch or blank to cover the opening, **Figure 21-32.**
- Access to the panels should be maintained free of storage and debris.
- All outlets and junction boxes should have an approved cover in place.
- Ensure that outlets, switches, lights, and appliances are approved for the location installed. Remember that ordinary electrical devices are not permitted in rooms designed for the use and storage of flammable or combustible liquids or explosives.
- Extension cords are not permitted to be used in place of permanent wiring. This problem is widespread and becomes more acute when light-duty "zip cords" are utilized. If the installation of additional permanent outlets is not possible, multiplug adapters with built-in circuit breakers are generally an acceptable alternative.
- No exposed wiring, approved or otherwise, should be permitted to run through doorways or under floor coverings or be stapled to a wall.

Figure 21-32 Circuit breakers must never be prevented from tripping by the use of tape or other objects. All spaces on the panel must be covered with either a breaker switch or a blank.

General Fire Safety

> **Note** Poor housekeeping practices, while seemingly a simple issue, are often the root cause of many fire hazards and eventually fires and injuries. Accumulated trash and debris are often found blocking exits or access to fire protection equipment. Haphazard storage of combustibles near open flame or hot surface devices such as water heaters or space heaters is another cause of accidental fires.

Dust accumulation is a common problem in woodworking and textile manufacturing occupancies. Care must be taken to prevent the buildup of dust on all horizontal surfaces, especially those

PREVENTION, PUBLIC EDUCATION, AND PRE-INCIDENT PLANNING 673

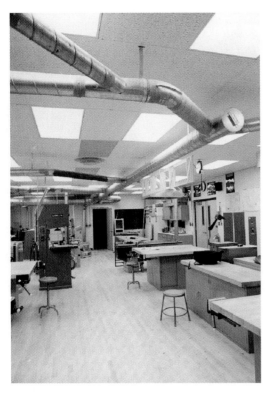

Figure 21-33 Woodworking occupancy with a dust collection system.

overhead. Dust explosions are caused when combustible dust particles suspended in the air contact an ignition source. Fortunately these devastating explosions are an infrequent occurrence. **Figure 21-33** shows the ductwork for a dust collection system.

Proper disposal of rags used with flammable or combustible fuels, solvents, and oils is easily accomplished but so often neglected. Autoignition of such rags due to their casual treatment has resulted in more than a few fires. Oily rags must be stored or discarded in an approved safety can, **Figure 21-34**.

Fire codes restrict smoking in certain occupancies. General smoking in buildings should be discouraged by firefighters. Smoking and non-smoking areas should be clearly identified and proper containers provided for the disposal of smoking materials.

Building Exterior

No fire prevention inspection is complete until the inspector has walked around the outside of the building. The following observations should be noted:

- Address numerals of sufficient size should be posted where responding emergency personnel can easily see them, **Figure 21-35**.

- All access roads should be considered fire lanes and kept clear of vehicles and storage, **Figures 21-36A** and **B**.
- Secured key boxes (also called Knox Boxes), when provided, should be checked for current keys, **Figure 21-37**. Firefighters should

Figure 21-34 Oily rags must be stored in approved safety cans with self-closing lids.

Figure 21-35 All occupancies should have plainly visible address numerals.

Figure 21-37 Key boxes on the outside of a building allow firefighters access to a locked building. They should be checked for current keys.

encourage, if not require, the use of such boxes for after-hours entry into buildings by emergency personnel.
- Trash disposal areas should be away from buildings and free of debris, **Figure 21-38**.
- Outside storage of flammable and combustible liquids should be as described earlier.
- Accumulated dried vegetation near buildings should be eliminated, **Figure 21-39**.

Concluding the Inspection

At the conclusion of the inspection, the firefighter should convey the findings to the building representative. This "wrap-up meeting," as shown in **Figure 21-40**, need not be a formal event. In fact, the efficient inspector will have pointed out any deficiencies or suggestions noted while walking through the business at the time observed. In this way, at the wrap-up, the firefighter can reiterate the

(A)

(B)

Figure 21-36 (A) Unblocked fire lane. Concrete blocks under the turf are often used to allow fire apparatus to drive in landscaped areas. (B) Blocked fire lane. Fire lanes must be kept open at all times.

PREVENTION, PUBLIC EDUCATION, AND PRE-INCIDENT PLANNING 675

Figure 21-38 Trash storage areas should be free of debris and away from buildings.

Figure 21-39 Dried grass at exterior of building should be eliminated.

findings in summary form prior to leaving the premises. The inspector should remember to thank occupants for their time and praise them for their fire safety efforts. If criticism is necessary, the inspecting officer should exercise tact.

All deficiencies, violations, and suggestions must be documented, even those corrected in the presence of the firefighter.

Note Violations that pose an imminent life hazard, in the opinion of the firefighter, must be corrected before the firefighter leaves the premises. Examples include locked or blocked exits, welding operations in the vicinity of open flammable liquids, and electrical wiring that presents an obvious shock hazard.

Many different styles of notices are available. An example of a Fire and Life Safety Inspection Notice is shown in **Figure 21-41**. At a minimum, the notice should include the name and address of the business, the date of the inspection, the building representative,

Figure 21-40 Inspector discussing findings with business owner.

FIRE SAFETY CHECKLIST

BUSINESS NAME		PHONE #
STREET ADDRESS	SUITE	ZIP CODE
BUSINESS OWNER		HOME PHONE
OWNER ADDRESS		
EMERGENCY CONTACT		
ADDRESS		EMERGENCY CONTACT PHONE #
INSPECTION CONTACT	INSPECTION DATE	INSPECTION COMPLETED
	#1 #2	( ) YES ( ) NO

#1 #2

ARTICLE 1: ADMINISTRATION
- 0136: No space heaters allowed in occupancy
- 0137: Notices & orders shall be complied with

ARTICLE 9: ACCESS
- 0901: Install approved address identification
- 0902: Maintain & post fire department access lanes
- 0903: Lock Box required for access along with current keys

ARTICLE 10: FIRE PROTECTION SYSTEMS
- 1001: Maintain fire protection systems in an operable condition
- 1002: Inspection & test reports to be made available
- 1004: FDC shall remain clear of obstructions
- 1005: 3ft. clear space to be maintained around hydrants
- 1009: All control valves & flow switches shall be monitored
- 1011: All commercial cooking equipment shall have automatic fire protection
- 1012: Provide a 40 BC fire extinguisher within 30 feet of cooking equipment
- 1013: Hood & duct system shall be free of grease
- 1014: Hood extinguishing system to be serviced and links replaced every 6 months
- 1015: Fire alarm equipment shall not be obstructed
- 1017: "Local Alarm Only" signage shall be provided
- 1018: Maximum travel distance for fire extinguishers exceeded
- 1019: Properly mount fire extinguisher
- 1020: Service fire extinguisher annually

ARTICLE 11: GENERAL
- 1103: Dumpsters shall be 5 feet from building
- 1104: Maintain good housekeeping
- 1105: Maintain storage 18 in. below sprinkler heads
- 1106: Maintain storage 24 in. below ceiling
- 1107: No storage in boiler, mechanical, or electrical room
- 1108: Drapes and other decorative material shall be fire resistive
- 1109: Aboveground gas meters & piping shall be protected
- 1110: Maintain clearance from combustibles and heat producing devices
- 1114: Maintain fire resistive construction
- 1115: Maintain fire doors and dampers
- 1117: Remove accumulation of exterior waste

#1 #2

ARTICLE 12: EGRESS
- 1201: Remove all obstructions from path of exit
- 1202: Exit door shall swing out when occupancy exceeds 50
- 1203: Remove all unapproved locking devices from exit doors
- 1205: Emergency power equipment for egress illumination shall be maintained
- 1206: Exit paths shall be identified by exit signs
- 1207: Exit signs shall be illuminated
- 1208: No storage under unprotected stairways
- 1213: Dead end corridors over 20 feet in length

ARTICLE 25: PLACES OF ASSEMBLY
- 2501: Maximum occupant load signs to be posted in occupancies over 50

ARTICLE 74: COMPRESSED GAS
- 7401: Compressed gas cylinders shall be secured

ARTICLE 80: HAZARDOUS MATERIALS
- 8002: Tanks, piping & valves shall be protected from vehicular damage

ARTICLE 85: ELECTRICAL EQUIPMENT AND WIRING
- 8501: Install cover plate on all electrical outlets, switches and junction boxes
- 8503: Discontinue the use of extension cords
- 8504: Cords shall not be subject to damage
- 8505: Cords shall be maintained in good condition
- 8506: Discontinue the use of multi-plug adapters
- 8507: Maintain 30 inch clearance around electrical panels
- 8508: Electrical panels shall be labeled as to area served
- 8509: Doors into electrical rooms shall be labeled

MISCELLANEOUS:
SEE ATTACHED SUPPLEMENTAL

COMMENTS:

REQUIRED PERMIT	ASSOCIATED FEE	NOT CURRENT	CURRENT

A routine fire inspection has been made of these premises to determine if any life or fire hazard exists. The inspection also was made to determine if any violations of the Uniform Fire Code exist. If violations have been found, this shall serve as your official notice. All violations must be corrected within the time allowed. A re-inspection will be made at a later date to verify the required corrections. Should you need assistance regarding this inspection, contact the Fire Prevention Bureau at 962-2537.

I, the undersigned, am in receipt of a copy of this inspection report and am aware of the hazards noted. I am also aware that this is a routine inspection for correction of violations and **may not encompass every possible violation.**

SIGNATURE OF BUSINESS OWNER, MANAGER, OR RESPONSIBLE PARTY	TITLE	DATE

SIGNATURE OF INSPECTOR

FIRE DEPARTMENT: WHITE   1ST INSPECTION:CANARY/BUSINESS OWNER   2ND INSPECTION:PINK/BUSINESS OWNER

Figure 21-41 Inspection notice should be given to business representatives at the conclusion of the inspection. *(Courtesy of Loveland Fire and Rescue, Loveland, Colorado.)*

PREVENTION, PUBLIC EDUCATION, AND PRE-INCIDENT PLANNING ■ 677

Figure 21-42 A business owner may affirm that minor code violations have been corrected by signing and returning a self-clearing card.

violations noted with their corresponding code references, the fire department inspector's name, and a compliance date. Most often, notices are completed prior to leaving the building. If this is not possible because further research is required, it is perfectly acceptable to return later with the completed notice. The original should be given to the building occupant and a copy should be filed in the occupancy file. Remember, this is a legal document and must be treated as such.

Reinspections

Few things can negate the validity of a fire inspection faster than the lack of follow-up. All documented violations (and they should *all* be documented) must be reinspected in some manner to emphasize their significance and ensure compliance. *If it is important enough to document, then it is important enough to follow up.* However, not all violations require the time and effort of another trip to the business site. Many departments use a self-clearing card, an example of which is shown in **Figure 21-42**. It may have different names, but the concept is the same. For a limited number of very specific infractions, the business owner affirms that the violation has been corrected by signing a postcard and mailing it to the fire department. Examples of qualifying infractions are address numerals that have been enlarged so they are clearly visible, removal of an extension cord, or recharging a fire extinguisher.

More serious violations require a reinspection. The time allotted to make the corrections will vary with the complexity and seriousness of the violation. However, once a reinspection date has been established and agreed to by the business owner, every effort should be made by the inspector to return on the specified day. Additional compliance time may be granted at the discretion of the inspector. Failure on the part of the business owner to make a reasonable effort to comply with violation notices should be referred to the Fire Prevention Division.

HOME INSPECTIONS

Firefighter Fact National fire statistics prove that year after year roughly 80 percent of all injuries and deaths in America occur in the home. To the credit of the fire service and the public, actual numbers of injuries and deaths have been decreasing in recent years. But the fact remains that eight of ten people killed or injured by fire are in the "safety" of their homes.

The overwhelming majority of fire prevention efforts are concentrated on the business community. The reasons are deep seated and fundamental but are also centered around privacy issues. Americans simply will not allow authorities legislated access to their homes, no matter how noble the reason.

This does not preclude the fire service from making an effort to conduct voluntary home inspection programs. Such programs, whether available to entire jurisdictions or only targeted areas, can be beneficial to the community as well as the fire department. Not only will the threat of fire in family homes be reduced, but much goodwill for the fire department will be fostered.

Voluntary residential inspections can be used to point out fire and life safety hazards, check and install smoke detectors and carbon monoxide detectors, and instruct families in proper emergency preparedness techniques. When granted permission to enter a home, firefighters are cautioned to limit their movement in the home to the "less private" areas. In other words, stay out of the bedrooms.

Typical hazards found in the home are not unlike those discussed previously. Exiting problems are usually centered around double-cylinder dead bolts on doors and security bars on doors and windows, **Figure 21-43**. Electrical problems involve overloaded circuits and extension cords. Storage around furnaces and water heaters is often noted. Combustibles left on top of floor furnaces continue to cause many fires at the start of each heating season. Improperly stored flammables and combustibles in basements and garages are regularly observed.

Attention should be given to issues of safety to small children. Accessible electrical outlets without protective caps, unsecured household chemicals, swimming pools without security fences, and electrical appliances in bathrooms are areas that deserve the serious attention of parents and child care providers.

Smoke and carbon monoxide detectors should be installed, **Figure 21-44**. The importance of these devices cannot be overstated and must be stressed to homeowners. Firefighters should remind residents that their chance for survival in

Figure 21-44 Smoke and carbon monoxide detectors are available separately and as one unit. *(Photos courtesy of BRK Brands, Inc. makers of First Alert® branded safety products)*

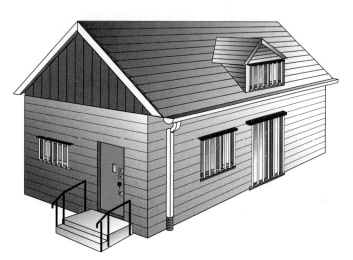

Figure 21-43 House with security bars.

case of a fire increases by 50 percent if a smoke detector is installed and functioning. However, correct installation is only a good first step. Detectors must be maintained by changing their batteries on a regular basis, **Figure 21-45**. The International Association of Fire Chiefs sponsors the annual "Change Your Clock, Change Your Battery" program to remind residents to change their smoke detectors' batteries when they reset their clocks for Daylight Savings Time.

Finally, firefighters should not fail to seize the opportunity to discuss emergency preparedness with the home's occupants. This should include first-aid firefighting techniques such as fire extinguisher placement and use, stove top fire extinguishment, **Figure 21-46**, and barbecue fires. Instruction should be given in proper home evacuation in case of fire. But this area could be expanded to include preparedness for natural disasters such as earthquakes, tornadoes, and flooding.

FIRE AND LIFE SAFETY EDUCATION

For most people, the tragedy of losing a home or a loved one to fire is something that happens to someone else in some other neighborhood or town. How many post-incident reports have appeared on television where the neighbor remarks, "I can't believe this happened on our block. This is such a quiet street"? Teaching citizens to recognize life safety hazards and to react

Figure 21-45 Batteries in smoke detectors should be checked monthly and replaced every six months. Some detectors have lithium batteries that do not require replacement.

Figure 21-46 Firefighter showing others how to place a lid on a burning pan.

appropriately is clearly a fire department function and responsibility. The problem for fire safety educators is how to make fire and safety personal issues in the community.

> **Note** National and state statistics do little to impress people about the immediacy of a problem or its potential impact on their lives. Citizens must be made aware that fire safety is an issue in their community and be taught what they can personally do to prevent becoming a statistic.

Some departments have personalized the tragedy of fire by conducting tours of burned-out homes where occupants have been injured or killed. Specially built fire safety houses, as shown in **Figure 21-47,** allow more intensive training for the general public. These fire safety houses are designed to simulate fire conditions in a safe and controlled environment. They allow citizens to experience firsthand some of the conditions present during smoke or fire conditions. The citizens learn that a basic evaluation can become a confusing scenario under adverse conditions. Educational efforts must be tailored to meet the specific needs of the community and not be limited strictly to fire-related issues. For example, water safety classes are frequently taught by fire service personnel in communities with a high concentration of backyard swimming pools.

Fire and Life Safety Program Presentations

Firefighters are frequently asked to speak to groups of all ages regarding fire and life safety issues. For a presentation to be worthwhile and effective, it should contain three main components: preparation, presentation, and practice.

The preparation phase involves the firefighter and the audience. The firefighter must be well versed with current knowledge on the subject matter to be presented and come prepared with appropriate audiovisual equipment, training aids, and literature. Once in front of the audience, the firefighter should prepare the group for the information they are about to receive. An effective method for gaining the audience's attention is to make the subject matter personal to them. For example, when speaking on the topic of drownproofing to a group of parents, it might be beneficial to recount a recent incident in the community, minus the graphic details, of course. The firefighter should let them know that they are about to receive information they can use to ensure the incident is not repeated at their homes. Getting the audience interested and personally involved is conducive to learning and the goal of the preparation phase.

Now that the audience is primed to learn, it is time to present the information. How the information is transferred from the firefighter to the individuals in the group is left to the creativity of the speaker. Showing a video and answering questions may be easy for the firefighter, but generally not much learning takes place. Professional educators know that the level of learning increases proportionally with the number of senses affected. In other words, the more the audience can see, hear, touch, or smell the subject matter, the more the information will be retained. When speaking to a group of hotel workers about evacuation of their guests during an emergency, it is important to let them hear the fire alarm sounding and see the strobe lights flashing. When speaking to children, use of fire prevention and education specialized equipment helps to keep the children's attention, **Figure 21-48**.

In the final phase of the program, the participants are asked to apply, practice, or demonstrate what they have learned. A group of small children being taught how to report an emergency could practice by using fire station telephones with a firefighter on an extension playing the role of a dispatcher receiving the call. An industrial fire brigade, having been instructed in the proper selection and use of fire extinguishers, could be allowed to extinguish small pan fires. (Appropriate safety equipment is mandatory. Firefighters must always verify local department policy regarding live fire training.) Firefighters should observe this phase closely, ensuring that the information pre-

Figure 21-47 This school bus was redesigned and specially equipped to become a fire safety house.

PREVENTION, PUBLIC EDUCATION, AND PRE-INCIDENT PLANNING   681

Figure 21-48 These fire safety characters keep the children focused and help to get the message across.

sented is being applied properly and making correction where necessary.

Forms of Fire and Life Safety Programs

Public education programs usually take three general forms: public service announcements, school programs, and adult/homeowner programs. Fire station tours also provide an excellent opportunity to heighten the public's awareness of fire safety issues.

Public Service Announcements (PSAs)

A simple, inexpensive, and timely method of spreading fire safety messages is to use the media. Radio, television, and newspapers are always willing to assist public agencies with their efforts. Short five- to ten-second television or radio spots or small news articles can be used for delivering safety messages on such seasonal topics as drowning prevention, fireworks safety, smoke detector maintenance, Halloween costumes and candy, and Christmas tree safety.

School Programs

Fire safety and education programs for elementary school children can be valuable and productive, **Figure 21-49**. They are designed to teach students the dangers of fire, how to prevent fires from happening, and what to do if they are caught in a fire. These programs can take many forms: two of the more popular are NFPA's Learn Not to Burn and the Junior Firefighter Program.

Figure 21-49 Popular public education programs are available through the NFPA.

The Learn Not to Burn program is designed for children in preschool through the third grade. It teaches children safety lessons using educational material and technical support provided by NFPA. Teachers attend workshops where they are instructed in course material by fire educators. The teachers then teach the program over a three-month span at their schools. Incorporated into the Learn Not to Burn curriculum are lessons on home evacuation using material from Operation EDITH (Exit Drills In The Home) and the ever-popular Stop, Drop, and Roll.

The Junior Firefighter Program can take many forms. It is generally presented to fifth-grade students in three stages. Phase 1 is usually an oral presentation and tabletop demonstration by a fire education specialist. Using an instructor's manual and student workbooks provided by the fire department, teachers complete phase 2 in about four weeks. Phase 3 consists of a visit to the school by an engine company. Firefighters describe their jobs and demonstrate their equipment. Additionally, they present Junior Firefighter patches or badges to students who have completed the program.

School Evacuation Drills

An effective fire prevention program in schools should be a matter of high priority not only to the fire department and school staff but to the community as a whole. A proper program will include regular and thorough fire safety inspections of the school premises by school staff and the fire department as described throughout this chapter. Additionally, the program must include emergency evacuation drills. Although such drills are usually referred to as "fire" drills, it must be emphasized that emergency evacuations can be necessary for other occurrences as well—bomb threats and natural disasters, to name two. Many states and local communities have adopted statutes requiring evacuation drills to be conducted at designated intervals, generally twice per school year. Where no such mandatory requirement exists, the fire department should work with school administrators to ensure that drills are conducted voluntarily.

Emergency exit drills are necessary to ensure that students and staff can safely, efficiently, and effectively evacuate their buildings in times of stress. Order and control are the key elements that will help to prevent injuries during the evacuation. A good emergency evacuation plan and drill will include the following points:

1. Evacuation drills should be conducted as often as necessary to ensure that all occupants of the building are thoroughly familiar with the process. The fire department should be present whenever possible but at least once per school year.
2. The focus of the drill should be placed on disciplined control and order. Speed, while important, should not be overemphasized. Speed is a by-product of properly planned and supervised evacuation.
3. Specific exits should be assigned to groups of classrooms; for example, Classrooms 1, 2, and 3 should use Exit A and Classrooms 4, 5, and 6 should use Exit B. All building

exits should be utilized to provide for the even distribution of exiting occupants. Alternative exits should also be designated.

4. Students and teachers should proceed to predesignated assembly points outside and away from the building. Playgrounds are appropriate; parking lots are not. Teachers should be required to account for all students in their classes. Taking roll call once assembled outside is the preferred method. They should report the status of their classes to the school administrator in charge. Missing and unaccounted for children must be reported to responding firefighters.
5. Emergency evacuation plans should be drawn in graphic form and posted in each classroom and at various locations throughout the school. **Figure 21-50** shows such a plan.
6. To familiarize students and staff with the sound, the fire alarm system should be used whenever conducting an evacuation drill.

Adult Programs

The types of programs and lessons taught to adults are limited only by the imagination of fire service educators. Common themes include:

- Home fire safety
- Smoke detector placement and maintenance
- Drownproofing
- Fire extinguisher demonstrations and instruction for use
- Operation EDITH (Exit Drills In The Home)
- CPR instruction
- General first aid
- Hazardous materials awareness
- Earthquake, tornado, and flooding preparedness

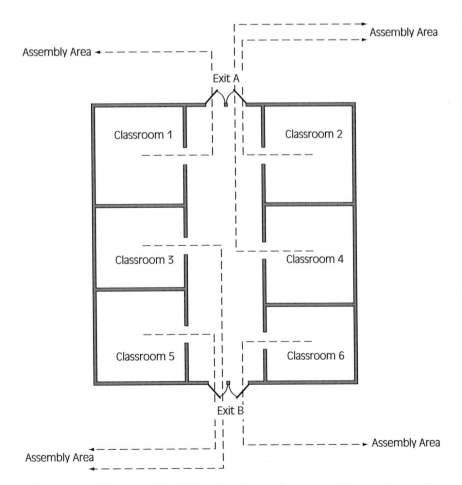

Figure 21-50 A graphic view of the school building showing exits, evacuation routes, and assembly areas should be posted in all classrooms and throughout the building.

Fire Station Tours

Visiting a local fire station has always been, and continues to be, one of the most popular field trips for all types of clubs, organizations, and schools. Station tours not only give the visitors a chance to see firefighting equipment and apparatus up close, but also give them a glimpse into life in a fire station. From the firefighters' perspective, tours provide an excellent opportunity to "spread the word" about fire and life safety.

Fire station tours should be tailored to fit the needs and interests of the visitors. Young children may be interested in trying on a firefighter's safety equipment.

> **Caution** Helmets should not be placed on very small children; neck injuries could result.

Children might also enjoy sitting in the engine's cab or watching a firefighter "slide the pole." Older children and adults may be interested in the operation of a pumper or dispatch and response procedures.

Fire safety messages should be interwoven into the station tour and directed at the appropriate level. Small children should be asked or taught how to call for emergency assistance, to "stop, drop, and roll," or how to properly react to an operating smoke detector and exit their homes safely. Older visitors should be given information regarding such issues as smoke detector and carbon monoxide placement and maintenance, home evacuation planning and drills, and general fire safety. This information should be presented orally during the tour and reinforced with written literature distributed at the end of the visit, **Figure 21-51**. It is generally not good practice to give visitors brochures and flyers during the tour because this will serve as a distraction. Firefighters should also take the opportunity to inform visitors about the specialized educational programs offered by their department. Giving visitors a printed schedule of upcoming events and a telephone number for registration or further information is an excellent method for encouraging participation.

PRE-INCIDENT MANAGEMENT PROCESS

Fire prevention activities and pre-incident management, often referred to as preplanning, are related and yet different. From a fire prevention perspec-

Figure 21-51 Fire safety brochures available through the NFPA.

tive, one assumes an incident *can* occur and seeks to take measures to ensure that it does not. Pre-incident management assumes an incident *has* occurred and utilizes tactics, strategies, and coordination of resources to minimize the impact on lives and property. The link between the two is education. Fire service educators strive to teach the public how to prevent fires and other emergencies, while at the same time teaching the proper action to take should an emergency occur.

Pre-incident management can be as simple as company officers deciding from which hydrant they will lay hose for a particular building, or as complex as a coordinated effort between many agencies and many jurisdictions. Regardless of its scope, pre-incident management must be a collaborative effort between all divisions of the fire department and involve other agencies when appropriate.

Deciding to Preplan

In a perfect world, all structures in a given jurisdiction would be preplanned with the information stored in a massive database. That information would be available to emergency responders via mobile data terminals located on all apparatus. While such systems currently exist, they are often out of the financial reach of many fire agencies—at least for the immediate future. Therefore, most data gathered during the pre-incident management process is stored in "hardcopy" form. For practical reasons—storage, staffing, and time constraints—fire agencies must prioritize occupancies to be preplanned. In so doing, consideration should be given to the following:

- Type of occupancy
- Type of incident expected
- Life hazards—civilian and firefighter
- Nature of activities conducted at the occupancy
- Exposure to surrounding areas
- Complexity of firefighting operations
- Resources required

Structures such as high-rise buildings, hotels, malls, large industrial buildings, and multistory and multi-building apartment complexes should be given high priority. Such occupancies are often referred to as **target hazards**, indicating that a greater than average life hazard or complexity of firefighting operations can be expected.

Site Visit

Generally, after deciding to preplan a structure, the first step is the on-site visit or inspection. The site visit for pre-incident management purposes differs from a fire prevention inspection in that the firefighters' purpose is fact gathering from an operational strategic perspective as opposed to ensuring compliance with fire code regulations. Certainly if code violations are noted, they should not be ignored but rather dealt with in the usual manner.

During the pre-incident management site visit, firefighters should be gathering information that will enable emergency responders to deal effectively with all levels of situations at the site. At a minimum, the following information should be obtained and documented:

- Occupancy classification
- Construction type and method
- Structure size, height, and number of stories
- Exiting systems
- Built-in fire protection
- Access points to the site and interior of the structure
- Exposure problems
- Hazardous materials usage and storage areas
- Personnel—civilian and firefighter—safety issues and features
- General firefighting concerns

Diagrams

An essential element of the pre-incident management process is the diagram of the site, structure, or occupancy. Here is where the information gathered during the site visit is displayed in graphic form. Diagrams of the site and floor plans should be included and should be presented in both **plan** and **sectional views** using a standardized set of symbols, **Figure 21-52**, and drawn to scale.

Diagrams can be hand drawn but the preferable method is to produce them using a personal computer. Sophisticated computer-aided drawing software need not be used. A simple paint or draw application will do the job. Updating the diagram will be a simple process if the original is produced in this manner.

The site plan should include the perimeter of the building, surrounding roadways, access points to the site and structures, fire hydrants, water main sectional valves, fire sprinkler control valves and connections, cross fencing, gates or other obstacles that may impede the movement of emergency vehicles, staging areas, and likely apparatus placement locations.

Floor plans should include a general layout of the interior of the building, floor by floor. Areas of high life hazard, exiting systems, fire protection features, hazardous materials use and storage areas, type of construction, roof openings, stairs and elevators, and other pertinent information can be shown on this drawing.

Seek Input from Others

Structures in need of preplanning have been identified and prioritized, site visits have been conducted, and diagrams have been completed. Now what? As suggested earlier, the pre-incident management process is a collaborative effort involving all levels and divisions of the fire department and often other agencies. All relevant information obtained should be assembled in narrative and graphic form and routed as appropriate to obtain

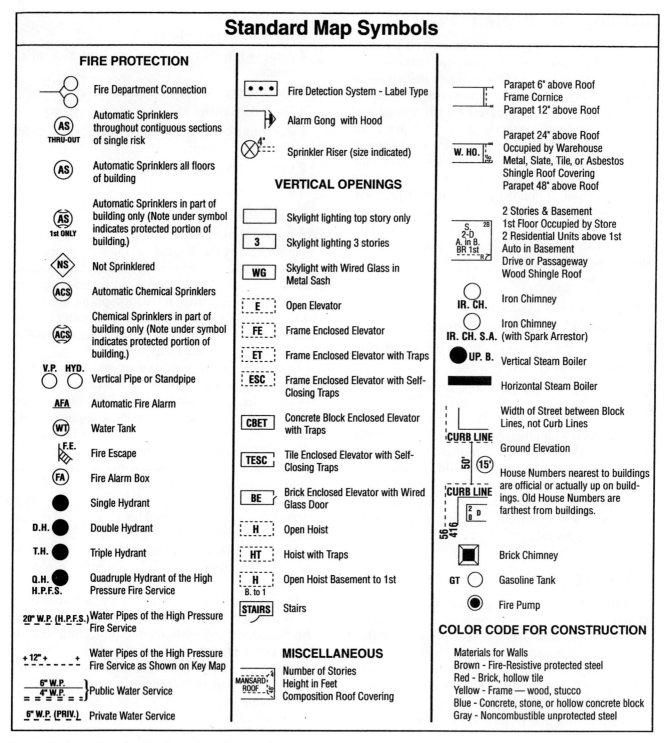

Figure 21-52 Standardized symbols for preplanning.

the perspective, expertise, and input of others. For example, in urban areas, the water department may suggest that, due to anticipated volume of fire streams during a full-scale firefighting operation, the water pressure in city mains be boosted. Law enforcement may have input regarding anticipated traffic problems. The Red Cross or other disaster relief agency may have suggestions regarding evacuation points or temporary shelter locations. In short, the pre-incident management process is not complete until everyone with potential involvement in an incident has had the opportunity for input.

The Finished Document

What should happen to the finished document? Certainly, preplans serve no purpose if they are stored on a hard drive on someone's PC or in a binder on a shelf. They should be carried in command vehicles and on all apparatus. Beyond being carried on vehicles, they should be practiced. Periodic training at specific locations using preplans enables firefighters not only to practice their skills, but to become familiar with facilities in their jurisdictions and to update pre-incident management plans when necessary.

Lessons Learned

It is hard for a firefighter to imagine anything more exhilarating and satisfying than aggressively entering a burning structure, rescuing its occupants, and defeating the sworn enemy—uncontrolled fire. Or perhaps pulling a drowning child from a swimming pool and watching him respond as a firefighter applies lifesaving techniques. Unless it would be responding to a "fire out on arrival" call because a smoke detector that the homeowner had installed at the fire department's recommendation had alerted his sleeping family to the fire downstairs in the kitchen. Or reading in the morning newspaper about the child who escaped serious injury when a fireworks sparkler ignited her clothes and she "stopped, dropped, and rolled" because the firefighters who came to her school had taught her how. There will always be emergencies and disasters. Preventing or lessening the impact of these devastating events must be the firefighter's constant goal.

KEY TERMS

Administrative Warrant An order issued by a magistrate that grants authority for fire personnel to enter private property for the purpose of conducting a fire prevention inspection.

Fire Hazard Any condition, situation, or operation that could lead to the ignition of unwanted combustion or result in proper combustion becoming uncontrolled.

Means of Egress A safe and continuous path of travel from any point in a structure leading to a public way. Composed of three parts: the exit access, the exit, and the exit discharge.

No-Knowledge Hardware Locking devices that require no key or special knowledge to operate.

Panic Hardware Hardware mounted on doors that enable them to be opened by pushing from the inside.

Plan View A drawing or diagram of a building or area as seen from directly overhead. May include a site plan or a floor plan.

Pre-Incident Management Advance planning of firefighting tactics and strategies or other emergency activities that can be anticipated to occur at a particular location. Often referred to as pre-planning.

Safety Container A storage can that eliminates vapor release by using a self-closing lid. Also contains a flame arrestor in the dispenser opening.

Secondary Containment Any approved method that will prevent the runoff of spilled hazardous materials and confine it to the storage area.

Sectional View A vertical view of a structure as if it were cut in two pieces. Each piece is a cross section of the structure showing roof, wall, horizontal floor construction, and the location of stairs, balconies, and mezzanines.

Special Egress Control Device Door hardware that will release and unlock the door a maximum of 15 seconds after it has been activated by pushing on the bar.

Target Hazard An occupancy that has been determined to have a greater than average life hazard or complexity of firefighting operations. Such occupancies receive a high priority in the pre-incident management process and often a higher level of first-alarm response assignment.

REVIEW QUESTIONS

1. List the three Es of fire prevention.
2. What are the objectives of a fire company fire prevention inspection program?
3. Identify the first step in conducting a fire prevention inspection after arrival at the occupancy to be inspected.
4. Identify the steps to be taken in the event that permission to inspect is denied by the business owner or representative.
5. Define a fire hazard.
6. Identify the three components of a means of egress.
7. Where is panic hardware required?
8. Discuss the key areas of concern to the fire inspector when inspecting a fire sprinkler system.
9. When are commercial cooking surfaces and equipment required to be protected by a fixed fire extinguishing system?
10. Is it necessary to document a deficiency noted during a fire prevention inspection if it was corrected in the presence of the inspector?
11. When conducting a home fire prevention and safety inspection, the firefighter should be aware of several typical fire hazards found in the home. Identify three.
12. Should fire department public education efforts be tailored to meet the needs of the community and not be limited to fire-related issues?

Additional Resources

Halon Recycling Corporation
http://www.halon.org

NFPA Fire Prevention Handbook
http://www.NFPAcodesonline.org

Klinoff, Robert, *Introduction to Fire Protection.* Delmar Learning, part of the Thomson Corporation, Clifton Park, NY, 1997.

United States Fire Administration, http://www.usfa.fema.gov

Occupational Safety and Health Administration, http://www.osha-safety-training.net

CHAPTER 22

EMERGENCY MEDICAL SERVICES

Andrea A. Walter, Sterling Park Rescue Squad

OUTLINE

- Objectives
- Introduction
- Roles and Responsibilities of an Emergency Care Provider
- Safety Considerations
- Assessing a Patient
- Cardiopulmonary Resuscitation/AED
- Bleeding Control and Shock Management
- Emergency Care for Common Emergencies
- Lessons Learned
- Key Terms
- Review Questions
- Additional Resources

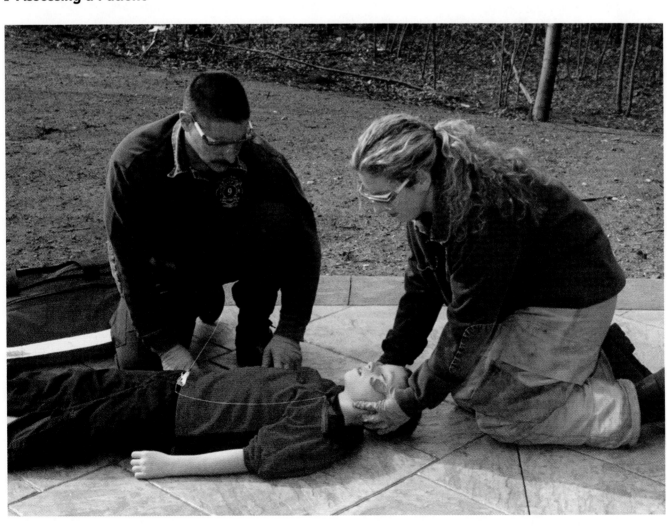

STREET STORY

During the past twenty-two years, I have seen many occasions where EMS training of firefighters has been essential to the success of incidents—especially to the patients involved. There have been many times where an engine I was on arrived at the scene of a collision before an ambulance. There were times when we pulled victims from buildings and no EMS units were available. There have been other rescue situations—in a storm drain, in a steam tunnel, in overturned vehicles—where EMS was on the scene but did not have the proper personal protective equipment to enter the hazardous area to provide care. Still, one incident stands out in my mind as a clear example of why firefighters should have EMS training.

It was a clear weekday afternoon in Fairfax County, Virginia. It had been a typical day with typical calls. It was rush hour—traffic was at a standstill; no one seemed to be moving. A vehicle fire was dispatched, a short distance from Fire Station 34. I knew that Engine 34 and Ambulance 34 were out of service for HAZMAT training and that the next due engine (Fairfax City Engine 33) was about 4 miles away and would have to battle traffic. I was about 3 miles away, just leaving headquarters, but could see the smoke—a billowing, black loom-up. I decided to roll in, just to watch.

Squad 18, a heavy squad with a crew of three, responded from Fire Station 34 and arrived within a couple of minutes. As I recall, their size-up was "a stake body truck well involved." No big deal, I thought. Then, after a minute, they requested an ambulance for a burned subject. Then they requested another, along with two medic units. Then they requested a helicopter. What seemed at the start to be "no big deal" ended up to be a serious situation with five civilians critically burned.

The crew from Squad 18 began treating the patients. I assisted when I arrived, as did the crew from Engine 33. As ambulances and medic units began to arrive, they took over patient care, but the lifesaving measures had already been taken. Yes, the fire was eventually extinguished, but our job as firefighters is life safety first, then property conservation. Thanks to our efforts, all five patients survived.

If we had not been trained to provide EMS care, there could easily have been five fatalities at that scene. Further, with rush hour traffic, there were hundreds (if not thousands) of people who could have perceived that firefighters are only trained to put "wet stuff on red stuff" and that injured victims are of no concern to us. That is not the image the fire service should portray.

The truck was a total loss even before the fire department was called. The five victims, employed by a landscaping company, were riding in the back of the truck when one tried to light a cigarette. Unable to do so because of the wind, he knelt down or bent down and lit it—right over a can of gasoline. The victims were burned from the flash fire that occurred.

Ever since this incident, I have made sure that I have kept my EMS training current, and as a chief officer I have made sure that all personnel under my command have had the opportunity and encouragement to become EMS-trained. Patient care is a key part of what we do as firefighters, and we cannot do it effectively if we are not trained.

—Street Story by Gordon M. Sachs, Chief, Fairfield Fire and EMS, Fairfield, Pennsylvania

OBJECTIVES

After completing this chapter, the reader should be able to:

- Explain the basic elements of an emergency medical system.
- Describe and practice the principles of infection control and body substance isolation for all patients.
- Perform an initial assessment on all patients, obtain vital signs, and conduct a focused history and physical exam for signs of illness and/or injury.
- List the different types of bleeding in patients, demonstrate methods for controlling the bleeding, and be able to treat patients in shock.
- Describe types of burns and ways to treat them.
- Identify the effects of ingested poisonous or controlled substances, how to contact a poison control center, and how to treat patients who have been exposed to a caustic substance.

INTRODUCTION

"When I became a firefighter, I expected to be fighting fires on every shift. I never anticipated that my engine company would respond to more emergency medical calls than fires last year. I am certainly glad that I received training in emergency care in firefighter school, because I have really helped save lives. There is more to being a firefighter than putting water on fire!"

It is true. There is more to being a firefighter than "putting the wet stuff on the red stuff." Firefighters are community defenders and trusted public saviors. When people do not know who else to turn to, they call the fire department. The community relies on firefighters to be skilled at fighting fires and to be creative problem solvers. That means firefighters need to know more than just basic firefighting techniques. That is where emergency medical care comes into the picture, **Figure 22-1**.

Emergency medical responses constitute more than 50 percent of total emergency responses for many fire departments all across the country. In some jurisdictions, emergency medical calls make up 75 to 80 percent of the fire department's total emergency responses per year.

This chapter covers very basic material that firefighters will need to know based on the NFPA 1001

Figure 22-1 Emergency medical services star of life.

> **Firefighter Fact** Emergency medical care has evolved into an essential part of the fire service.

standard and also covers standard first-aid practices. A wide variety of training courses in emergency medical services (EMS) are available that allow firefighters to increase their EMS knowledge and skills, and all firefighters are encouraged to continue their EMS training.

With the technology available today, firefighters can deliver lifesaving techniques to stabilize patients until emergency medical technicians and paramedics arrive, **Figure 22-2**.

> **Firefighter Fact** Firefighters are a critical, lifesaving link in the emergency response community.

ROLES AND RESPONSIBILITIES OF AN EMERGENCY CARE PROVIDER

As a part of the emergency response system, firefighters often act as providers of emergency care and first aid. Every emergency service delivery system is different, yet many call on firefighters to assist as the first arriving emergency unit on the scene or as a helping hand to the emergency medical technicians and paramedics. In any case, a working knowledge of the basics of emergency medical care is an important part of a firefighter's training.

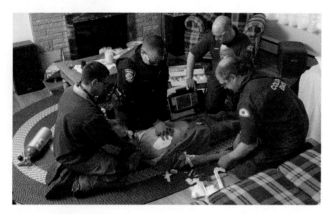

Figure 22-2 Firefighters are often called on to assist EMS crews with patient care.

Key Responsibilities

Firefighters have several key responsibilities when responding to emergency medical incidents:

- To ensure their own safety, the safety of their team, and the safety of the patient.

> **Safety** The most important factor to consider when firefighters respond to emergency medical calls is safety, for themselves, their team members, and their patient.

- To act safely from the minute they step on the fire engine until every piece of equipment is clean or sterilized and placed back on the fire engine, **Figure 22-3A–C**.
- To act in a professional manner at all times. People who have called for assistance are counting on firefighters to help them or their friends or family members. Firefighters must always be respectful and considerate, **Figure 22-4**.
- To never cause a situation to become worse or act beyond the bounds of their training.

> **Ethics** Emergency medical training is not only about what firefighters can do to assist a patient in crisis, it is also about recognizing what is beyond an individual firefighter's training and ability. If firefighters recognize that a situation is beyond their capabilities, and they call for the appropriate assistance and act within the limits of their training, they have done the best job they can for their patient.

- To practice and update emergency care and first-aid skills with training. The emergency medical field is constantly changing. That means firefighters may have to learn new skills,

(A)

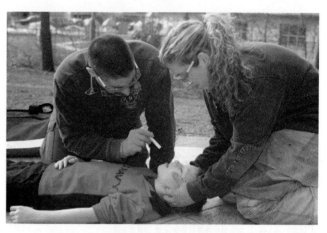

(B)

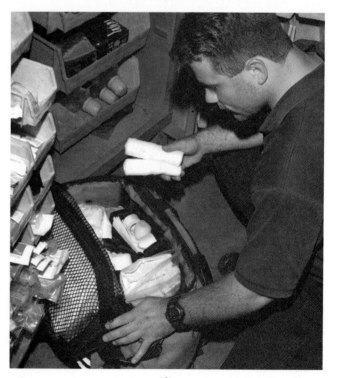

(C)

Figure 22-3 (A) EMS calls begin when firefighters leave the station, (B) continue throughout the time on the scene, and (C) do not end until firefighters have cleaned the equipment and restocked the supplies.

EMERGENCY MEDICAL SERVICES 693

Figure 22-4 Firefighters should always treat patients with respect. They should treat patients as they would want to be treated in the same situation.

Figure 22-5 Firefighters should train on new EMS skills and continue to practice skills they have already learned. Practice makes skilled responders.

as well as practice the old skills, **Figure 22-5**. Training for emergency medical responses and maintaining current knowledge and skills will allow firefighters to be ready for any emergency medical incident that occurs.

- To know the EMS equipment and maintain it properly. While emergency medical care may require the use of a firefighter's brain and hands, it also requires specialized equipment and tools. Firefighters should be familiar with each piece of EMS equipment and all supplies they carry. Also, equipment needs to be maintained properly. It should be cleaned after each use (or discarded properly if it is disposable) and checked at the beginning of each shift, **Figure 22-6**.

- To gather important information. Information about the patient on an emergency medical call

Figure 22-6 Firefighters should check their EMS equipment and supplies at the beginning of each shift.

can be critical for treatment. The firefighter must collect all important information about a patient before deciding on a course of action. Then, when EMS personnel arrive, the firefighter should transfer that information so that the EMS personnel can continue with appropriate care in the ambulance and in the hospital. Collecting patient information is discussed later in this chapter.

Legal Considerations for Emergency Care Providers

Emergency care providers should understand several important legal issues. First is the principle called **standard of care**, which is a legal term that means for every emergency medical incident, an emergency responder should treat the patient in the same manner as would another emergency responder with the same training. In short, this legal principle is what is used in court to demonstrate that the actions of an emergency care provider were adequate or inadequate. This legal issue emphasizes the importance of firefighters acting within the bounds of their emergency care and first-aid training and doing what they have been trained to do, and *only* what they have been trained to do.

Another important term to understand is **consent**. Emergency responders should always attempt to gain consent from the patient before beginning treatment. In the case of children or minors, consent should be obtained from a parent or legal guardian. Emergency responders should not withhold treatment from minors when a parent or guardian is not present. Emergency responders can begin treatment while making every effort to contact the parents or guardians to obtain permission.

Patients who are unconscious when emergency responders arrive are considered to be giving **implied consent** for emergency medical treatment. Patients who are not unconscious, and are mentally capable, have the right to refuse treatment.

Abandonment occurs when an emergency care provider begins treatment of a patient, and then leaves the patient or discontinues treatment prior to the arrival of an equally or higher trained responder.

> **Caution** If firefighters begin patient care, they must remain with the patient until another emergency responder arrives whose training is equal to or higher than theirs.

Interacting with Emergency Medical Services Personnel

As emergency care providers, firefighters interact with emergency medical service personnel while on the scene of an emergency. Emergency medical transport is done in a variety of ways. An ambulance staffed with EMTs and paramedics may come from the firefighters' own fire department, it may come from a separate rescue squad or EMS organization, or it may even come from a private transport company or a hospital.

In some cases, firefighters may be interacting with emergency medical personnel who have arrived by helicopter, commonly referred to as a medi-vac. **Medi-vac** units are staffed by nurses, paramedics, and EMTs who are usually skilled in treating patients with traumatic or serious injuries. A medi-vac unit can be seen in **Figure 22-7**.

It is also important to understand the different levels of care that are provided by EMS crews who may arrive on the scene. An ambulance may provide basic life support (BLS) care, which is the primary level of EMS care, or it may provide advanced life support (ALS) care, which involves much more aggressive treatments for patients and specialized monitoring.

In any case, the arriving EMS personnel will rely on the firefighters on scene to give them critical patient information. What is important patient information? It varies from situation to situation and requires firefighters to use their best judgment, but a few basic pieces of information are important for every call: the patient's age, the patient's medical history, known drug allergies of the patient, medications the patient is taking, the circumstances sur-

EMERGENCY MEDICAL SERVICES

Figure 22-7 Medi-vac helicopters are used to transport patients to specialty care hospitals.

rounding the illness or injury the patient is reporting, and the patient's vital signs, **Figure 22-8**.

Many other pieces of information can be helpful, and firefighters should consider gathering such information during an emergency medical call, if feasible: the patient's name and address, what condition the patient was found in, the patient's surroundings (example: condition of an automobile in a car accident), and the treatment that was given to the patient prior to the EMS personnel arriving on the scene.

All firefighters need to check with their department about their responsibility for documenting an emergency medical response. Many fire departments require that a patient care report or an emergency call record be completed for every emergency medical response. Thorough and accurate documentation of emergency medical responses will help protect the department in the event of legal action.

SAFETY CONSIDERATIONS

One of the most important considerations when performing emergency care and applying first-aid skills is the safety of the caregiver or individual firefighter performing patient care and the safety of the emergency response team, such as an engine company crew responding on an emergency medical call.

> **Safety** Safety, including **infection control**, should be the primary consideration of every emergency responder.

Safety and infection control should also remain a primary consideration for the duration of the emergency run and even afterwards in the cleanup phase of the incident. *Remember:* Emergency responders who get sick or injured are no longer a part of the solution; they are a part of the problem. Safety is the primary consideration!

Analyzing the Safety of the Emergency Scene

Firefighting training teaches firefighters to observe each and every scene for situations that may be unsafe to the public or to the firefighters present. For example, firefighters must check for overhead obstructions whenever raising a ladder to a structure and check the floor of a structure for stability before stepping on to it. Engine company officers must survey the condition of a burning building to make sure it is not on the verge of collapse before sending firefighters inside for an interior fire attack.

Figure 22-8 Firefighters should record important information for arriving EMS personnel and for their agency's patient information reports. Effective documentation is critical. *(Photo courtesy of Fred Schall)*

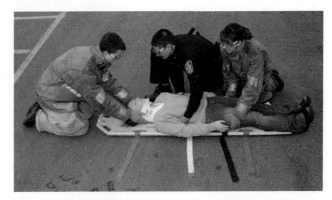

Figure 22-9 The firefighter should pay careful attention to the officer or the lead EMS provider to ensure the team is working together in patient care. Teamwork is important in all aspects of firefighting, including EMS.

While functioning on emergency medical calls, firefighters must also check for scene safety prior to entering. There are a few important considerations for EMS scene safety. Firefighters should listen carefully to all information given prior to arrival on the scene. It is essential to plan actions ahead of time! Also, firefighters should listen to the direction of the officer or the EMS provider in charge of the incident so the team is functioning as one unit, **Figure 22-9**. It may be necessary to stage away from the scene until it can be stabilized by law-enforcement personnel.

Firefighters should carefully observe the emergency scene and people present before entering. Indications of violent behavior, the use or presence of weapons, or signs of controlled substance or alcohol use may all be initial indicators of a safety concern. If after entering the emergency scene a safety concern is discovered, it is acceptable to take the team out of the situation until law-enforcement officers have secured the scene or the dangerous individual or condition has left the scene.

The firefighter should be sure to ask for additional assistance when safety concerns are present, whether in the form of additional fire and emergency medical units or a law-enforcement officer. It is also important to keep the communications center informed of the situation and the team's status to ensure continued safety throughout the incident.

Firefighter Physical and Mental Health

Firefighting is a strenuous occupation, and firefighters are encouraged to maintain a healthy lifestyle to meet the demands of firefighting work.

Emergency medical calls for assistance can also be strenuous work, both physically and mentally.

> **Note** A healthy lifestyle will help firefighters in their ability to respond to emergency medical calls for assistance as well as in typical firefighting activities.

A "healthy lifestyle" is different for each firefighter but generally includes regular exercise, a proper diet, and getting the right amount of sleep. Firefighters should contact their personal physician or department health care provider for assistance with developing a healthy lifestyle that meets their needs.

One of the most common problems for firefighters on emergency medical responses is back injuries. Firefighters should learn and practice proper lifting techniques (lifting with the legs, keeping the weight close to the body, etc.) when moving patients or emergency medical equipment.

Emergency medical calls can also take their toll on the mental health of a firefighter or emergency medical responder. Firefighters should consult their department's policies on critical incident stress debriefing or other intervention techniques available for firefighters who feel adversely affected by emergency medical responses.

Infection Control

Safety for the firefighter responding to emergency medical incidents involves not only scene safety and the prevention of injury but also protection from contracting **communicable diseases** or **infectious diseases**.

Prevention of Exposure to Infectious Diseases

The Centers for Disease Control and Prevention (CDC) is a federal organization that monitors outbreaks of infections and advises agencies on how to handle the situation and control the disease. The CDC, the Occupational Safety and Health Administration (OSHA), and the National Fire Protection Association (NFPA) have all written standards of practice or operational guidelines that firefighters should use when providing medical care.

Many laws and regulations have been enacted on the local, state, and national levels that assist in protecting emergency responders and health care providers from infectious diseases. In addition, first responders should follow any local or organizational policies for infection control.

> **Safety** Firefighters should understand and follow the infection control guidance issued by their departments, their states, and appropriate federal agencies. These guidelines and precautions are developed to protect firefighters and first responders from infectious diseases.

Any time a firefighter treats a sick or injured patient, the risk of **exposure** to disease is present, both for the firefighter and for the patient. Firefighters should protect themselves and the patient from disease. The most effective means of reducing the risk of spreading disease is hand washing, **Figure 22-10.**

Proper hand washing is a necessity in the medical profession, and many types of antibacterial soaps are available for this purpose. Although soap is a great tool, the technique of washing is more important than the type of soap.

> **Safety** The industry standard for washing hands is vigorous lathering with soap and water for fifteen seconds (or longer) followed by thorough rinsing, **Figure 22-11.** A disposable towel should be used to thoroughly dry the hands, and the towel should be used to shut off nonautomatic faucets to ensure there will be no exposure to the organisms on a dirty faucet handle.

Frequent hand washing, although critical, may have a drying effect on the skin, and a moisturizing lotion may be necessary to combat the discomfort.

Firefighters must also be aware of the possibility of transmission of disease and sickness to their family and friends. Firefighters who do not practice proper infection control procedures can carry disease-causing microorganisms home with them to their families. As a general practice, firefighters should be extra careful never to wear soiled duty uniforms home. Washing soiled items in a separate station machine designed for cleaning contaminated articles can help lower cross-contamination to other washables. It is advisable to shower and change at the station before leaving for home. Likewise, firefighters who respond from a station should keep soiled uniforms and personal protective equipment out of living areas of the station.

Body Substance Isolation

Simple precautions, called **body substance isolation (BSI) precautions**, can reduce the odds of disease transmission significantly. BSI involves wearing proper protective equipment on every call.

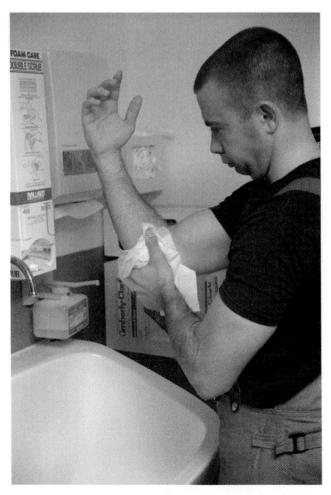

Figure 22-10 The most effective means of reducing the risk of spreading disease is hand washing.

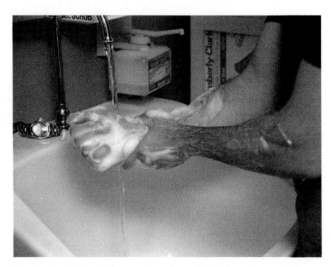

Figure 22-11 Proper hand washing involves vigorous lathering with soap and water for fifteen seconds (or longer) followed by thorough rinsing.

A person who is ill with a communicable disease will not necessarily appear ill and may not want to make the disease known to others. Therefore, firefighters should have an agency-approved face mask available and should wear the minimum of goggles or safety glasses and protective gloves any time a medical scene is entered. There is no exception to this rule!

> **Safety** Appropriate BSI precautions must be taken for every patient.

The mainstay rule of emergency medical care is to treat all body fluids as if they were infectious, **Table 22-1.** If a substance is wet, it has the potential to be an infectious substance. Again, a firefighter should wear a minimum of gloves and eye protection on every call. A face mask can be added, if necessary, for splash potential. A nonabsorbent or plastic gown may be added for medical scenes with gross contamination potential, **Figure 22-12.**

Protective gloves are an important barrier device for obvious reasons but are effective only if they fit and are used properly. Firefighters must don new gloves for each patient contact. They must never use the same pair of gloves on more than one patient. Double gloving can be used for incidents with gross contamination potential. If

Potentially Infectious Body Fluids

Blood
Amniotic fluid
Vaginal discharge
Semen
Cerebrospinal fluid
Pleural fluid
Synovial fluid
Peritoneal fluid
Pericardial fluid
Fluids with little potential to transmit bloodborne diseases:
▪ Tears
▪ Nasal discharge
▪ Vomitus
▪ Sputum
▪ Saliva
▪ Feces
▪ Urine

TABLE 22-1

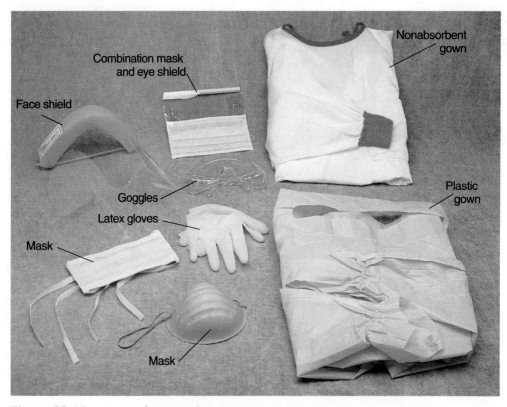

Figure 22-12 Types of personal protective equipment.

EMERGENCY MEDICAL SERVICES

gloves fail during use, firefighters should stop immediately, wash their hands, and don another pair of gloves before returning to patient care. Gloves should only be removed after all patient contact possibilities are finished and should be removed without touching the skin with the outer layer of the glove, **JPR 22-1A–C.** Firefighters should dispose of gloves in a container clearly marked with the biohazard label and wash their wash hands after every use of gloves, **JPR 22-1D.**

JOB PERFORMANCE REQUIREMENT 22-1
Removing Gloves

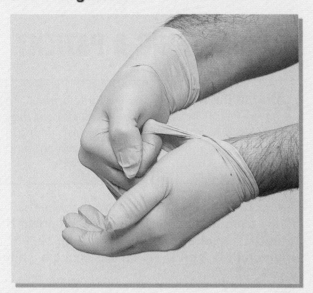

A Grasp the palm or outside cuff of the left glove with the gloved right hand.

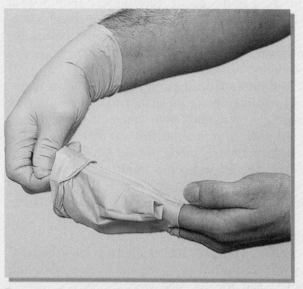

B Pull the left glove toward the fingertips. The glove should turn inside out as it is removed.

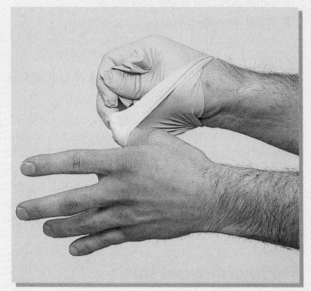

C Hold the removed glove in the still-gloved right hand. Insert the thumb or two fingers of the ungloved left hand under the cuff of the right glove, carefully avoiding any contaminated areas. Pull the right glove toward the fingertips, turning the glove inside out as it is removed. The soiled left glove should remain in the palm of the right glove as it is removed.

D Dispose of the gloves in a container clearly marked with the biohazard level and wash hands thoroughly.

Splashes of body fluid or chemicals can occur on emergency medical scenes, which is the purpose for eye protection in the BSI precautions. The eyes can provide a route for a disease microorganism to enter the firefighter's body. As stated earlier, goggles or safety glasses should also be used on every emergency call.

Any supplies or equipment used for treating a patient should either be disposed of in an appropriate biohazard container or cleansed and disinfected properly, **Figure 22-13**. Firefighters should wear gloves when disinfecting emergency medical equipment and should consult departmental policy for appropriate cleaning and disinfecting methods and procedures.

Firefighters who regularly participate in emergency medical incidents should consider receiving immunization against both common diseases and hepatitis B, an infectious disease that is the most dangerous risk to medical care providers. Firefighters who experience a significant exposure to blood or any body fluid should contact their supervisor immediately. Each agency will have policies concerning infection control and exposures.

In short, BSI precautions are important because they protect the safety of the firefighters and their families and friends. BSI precautions also protect patients with compromised disease-fighting systems from disease-causing microorganisms. A review of precautions for infection control is covered in **Table 22-2**.

For more information on infection control, BSI, and exposure to infectious diseases, firefighters can contact the CDC or OSHA. Both agencies are a part of the federal government and will provide information as requested. More information is also available on their Web sites at http://www.cdc.gov and http://www.osha.gov.

ASSESSING A PATIENT

> **Streetsmart Tip** When firefighters arrive on the scene of an emergency medical call, they can get a lot of information just by taking a quick look at the patient, the environment, and the situation.

Once a firefighter has made sure the scene is safe for the team to enter and donned appropriate BSI, an initial visual size-up of the patient should be conducted. A quick visual assessment should reveal:

- Is the patient awake?
- Is the patient in a harmful situation or environment?
- What position is the patient in?
- What people or objects are in the emergency scene that may have contributed to the patient's illness or injury?
- What is the skin color of a patient (for example, pale, red, bluish, etc.)?

Firefighters with many years of experience can often use these quick visual surveys to determine how serious a patient's condition may be. These quick visual surveys can provide important patient information in the first few seconds of an emergency responder's arrival, as well as recognize any safety concerns for the patient and the care providers.

Performing an Initial Assessment

After a quick visual survey of the patient and the scene is done, the next step is an **initial assessment**. An initial assessment is the initial investigative action taken by care providers to determine if the patient has the basic signs of life as

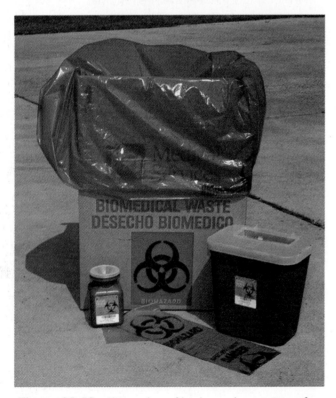

Figure 22-13 Examples of biohazard containers for disposal of infectious waste.

Standard Precautions for Infection Control

WASH HANDS (PLAIN SOAP)

Wash after touching **blood, body fluids, secretions, excretions,** and **contaminated items.**

Wash immediately **after gloves are removed** and **between patient contacts.**

Avoid transfer of microorganisms to other patients or environments.

WEAR GLOVES

Wear when touching **blood, body fluids, secretions, excretions,** and **contaminated items.**

Put on **clean** gloves just **before touching mucous membranes** and **nonintact skin.**

Change gloves between tasks and procedures on the same patient after contact with material that may contain high concentrations of microorganisms. Remove gloves promptly after use, before touching noncontaminated items and environmental surfaces, and before going to another patient, and wash hands immediately to avoid transfer of microorganisms to other patients or environments.

WEAR MASK AND EYE PROTECTION OR FACE SHIELD

Protect mucous membranes of the eyes, nose, and mouth during procedures and patient care activities that are likely to generate **splashes** or **sprays** of **blood, body fluids, secretions,** or **excretions.**

WEAR GOWN

Protect skin and prevent soiling of clothing during procedures that are likely to generate **splashes** or **sprays** of **blood, body fluids, secretions,** or **excretions.** Remove a soiled gown as promptly as possible and wash hands to avoid transfer of microorganisms to other patients or environments.

PATIENT CARE EQUIPMENT

Handle used patient care equipment soiled with **blood, body fluids, secretions,** or **excretions** in a manner that prevents skin and mucous membrane exposures, contamination of clothing, and transfer of microorganisms to other patients or environments. Ensure that reusable equipment is not used for the care of another patient until it has been appropriately cleaned and reprocessed and that single-use items are properly discarded.

LINEN

Handle, transport, and process used linen soiled with **blood, body fluids, secretions,** or **excretions** in a manner that prevents exposures and contamination of clothing and avoids transfer of microorganisms to other patients or environments.

Use **resuscitation devices** as an alternative to mouth-to-mouth resuscitation.

TABLE 22-2

 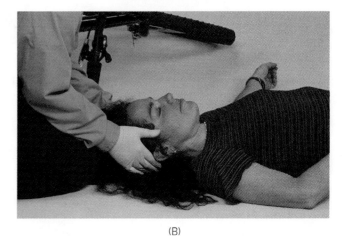

Figure 22-14 Firefighters stabilizing the head and neck of a (A) seated victim and (B) a victim lying on the ground.

well as any serious, life-threatening injuries. The initial assessment covers the following:

1. Level of consciousness of the patient (Is the patient awake?)
2. Airway (Does the patient have an open airway?)
3. Breathing (Is the patient breathing adequately?)
4. Circulation (Is the patient's heart pumping blood to the body adequately?)
5. Major bleeding (Is there any major bleeding?)

The initial assessment can be done very quickly on a conscious patient who is alert and talking. Patients who are unconscious will require closer examination to conduct the initial assessment.

When performing an initial assessment on a patient, firefighters should consider the following:

Level of Consciousness

Is the patient awake? If not, is the patient responsive to loud verbal commands or painful stimulation? After the firefighter determines the level of consciousness of the patient, another member of the team should maintain stabilization of the head and neck if the patient is either unconscious or has experienced a traumatic injury that may endanger the spine (automobile accident, fall, industrial accident, etc.).

To stabilize the head and neck of the patient, the firefighter places one hand on either side of the patient's head and holds firmly so that the head and neck are in a straight line with the body, **Figures 22-14A** and **B**. The purpose of stabilizing the head and neck is to keep them from moving so that if there is a neck injury, no further damage will be done to the neck or spinal cord.

Airway

The firefighter must ensure that the patient has an open airway. To do this for a patient without traumatic injury, the patient's airway is opened by placing the palm of one hand on the patient's forehead and the fingers of the other hand underneath the chin. The firefighter lifts the chin with the fingers and presses lightly on the forehead to roll the head into the open airway position, **Figures 22-15A** and **B.**

If the patient has a traumatic injury and there could be damage to the patient's neck, the firefighter will need to use a different method for opening the airway. To do this, one hand is placed on either side of the patient's head, placing the fingers along the curve in the jaw bone near the ear. Using the fingers, the jaw bone is pressed out to open the airway, **Figures 22-16A** and **B.**

The act of breathing brings air into the lungs where oxygen is passed into the bloodstream. Oxygen is vital to sustaining life. Many fire departments carry oxygen on their apparatus. Many patients experiencing difficulty breathing may benefit from the use of supplemental oxygen from cylinders that firefighters and EMS providers carry on their equipment. Note, however, that in very rare circumstances, the use of supplemental oxygen may worsen the patient's condition. Firefighters should consult their department's policies on the use of supplemental oxygen. Oxygen administration and the devices used to deliver oxygen to a patient are part of the curriculum of most higher level emergency medical training courses.

Breathing

When it is established that the patient has an open airway, or a firefighter has taken measures to open the airway, the next step is to check to see if the patient is

EMERGENCY MEDICAL SERVICES

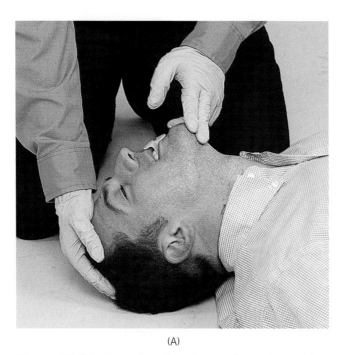 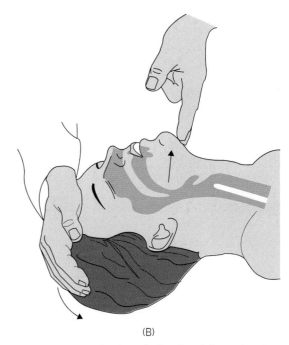

Figure 22-15 Opening the airway of a patient without traumatic injury using the head-tilt, chin-lift method.

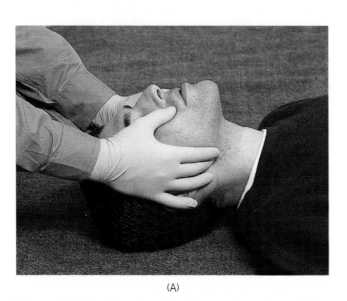 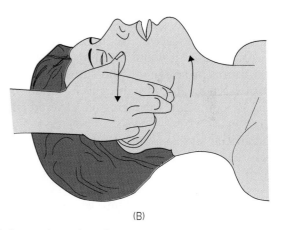

Figure 22-16 Opening the airway of a patient with traumatic injury using a jaw thrust.

breathing. If there is no breathing, the firefighter should provide rescue breathing for the patient.

Circulation

After the patient has an open airway and is breathing (or someone is breathing for the patient) the firefighter should check for a pulse in the patient. For an unconscious victim, the firefighter should locate the **carotid pulse** in the neck. On a conscious patient, the firefighter can use the **radial pulse** in the wrist to determine the pulse rate and quality. For a small child or infant, the best place to locate the pulse rate is the **brachial artery** in the inside of the upper arm. The pulse should be checked for 3 to 5 seconds to determine if it is present. The locations of these pulse points are shown in **Figure 22-17**.

> **Streetsmart Tip** Firefighters should never use a thumb to check a patient's pulse. The thumb has its own pulse, and firefighters may confuse their own pulse with the one they are trying to locate in a patient. It is best to use the index and middle fingers together to locate a patient's pulse.

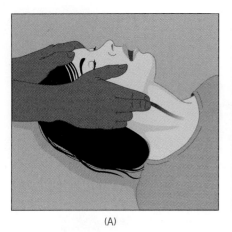 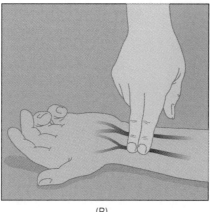 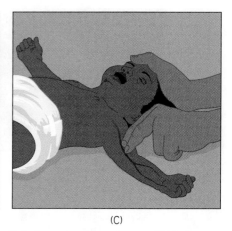

Figure 22-17 Locating pulses in (A) the carotid artery in the neck, (B) the radial artery in the wrist, and (C) the brachial artery in an infant's upper arm.

Major Bleeding

After determining that the patient has a pulse, the firefighter checks the patient from head to toe, front and back, for major bleeding. A sweeping motion is used with both hands along the entire length of the patient's body to check for major bleeding. If the patient is lying down, it is important to check underneath the patient's body. This can be done without moving the patient by simply reaching under the patient's body when doing the sweeping motion, and checking the gloves frequently for signs of blood. If a source of major bleeding is located during the initial assessment, the firefighter should attempt to stop the bleeding and bandage the wound. Bleeding and bandaging are covered later in this chapter.

The initial assessment allows firefighters to quickly identify major problems and start treatment of patients. In a mass casualty situation, where there are many patients, a triage system may be used. **Triage** is a quick and systematic method of identifying which patients are in serious condition and which patients are not, so that the more seriously injured patients can be treated first. Many triage systems are based on the same principles in the initial assessment. Firefighters can learn more about triage systems in higher levels of EMS training.

Vital Signs and the Focused History and Physical Exam

After the initial assessment of the patient is complete, the firefighter continues on to a focused history and physical exam, which is a thorough examination of the patient. This consists of three parts: (1) patient fact-finding, (2) vital signs, and (3) a head-to-toe survey of the patient.

If the patient is conscious, firefighters will need to go on a fact-finding mission for information about the patient. This includes trying to determine the age of the patient, important medical history, and allergies to medications. Firefighters should also try to establish what the patient's current situation is. Has the patient had chest pain for the past two hours? Does the patient remember exactly what happened in the automobile accident? Also, firefighters should always check patients for a bracelet, necklace, or watch that may contain vital medical information, **Figure 22-18**.

If the patient is unconscious, the firefighter can try to do fact-finding by asking questions of family members, friends, or bystanders present on the scene. If no one else is present or no one can provide medical facts about the patient, the firefighter may be able to find clues on the emergency scene to help establish some information about the patient.

After the fact-finding is done, the next step is to get a set of vital signs on the patient. For firefighters providing emergency first aid, the set of vital signs should include a pulse rate, a respiratory rate, and the color and temperature of the skin. **Figure 22-19** shows average vital sign ranges by age of the patient.

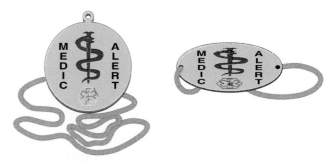

Figure 22-18 Examples of medical alert tags.

AGE	PULSE	RESPIRATIONS
Newborn	120–160	40–60
1 year	80–140	30–40
3 years	80–120	25–30
5 years	70–115	20–25
7 years	70–115	20–25
10 years	70–115	15–20
15 years	70–90	15–20
Adult	60–80	12–20

Figure 22-19 Average vital sign ranges by age.

Pulse Rate

The pulse rate is the number of times the heart is beating per minute. To establish the pulse rate, the firefighter should first locate the carotid pulse in the neck or the radial pulse in the wrist. Using the second hand on a watch or clock, the firefighter should count the number of beats for a minute to get the patient's pulse rate. With experience and practice, firefighters may choose to count the beats over a thirty-second period and multiply the number by two for the pulse rate per minute.

Respiratory Rate

The respiratory rate, like the pulse rate, is the number of times the patient is breathing in one minute. To get the respiratory rate, the firefighter observes the patient's chest and abdomen for the characteristic rise and fall that occur with each breath for one minute.

> **Streetsmart Tip** If the patient is wearing heavy clothing, or it is difficult to see the chest and abdomen rise and fall, the firefighter may lightly place one hand on the upper abdomen to feel the rise and fall of breathing.

The firefighter should also try to observe the quality of the respirations (labored, shallow, normal).

Color and Temperature of the Skin

The firefighter should carefully observe the skin color and temperature of the patient, because this can be an important indicator of the patient's condition. The skin color may be normal, ashen (gray in color), blue or cyanotic, red, yellow, or pale (white). The skin temperature may be normal, cool and clammy, hot and dry, or hot and moist. These are important signs, and should be noted and recorded on all patients.

After the patient's vital signs are recorded, it is time to do a head-to-toe examination of the patient. It is important to check the patient's vital signs periodically. A good practice to follow is obtaining vital signs every three to five minutes. To conduct a head-to-toe survey of the patient, the firefighter should follow the procedure outlined here:

Head and Neck. Gently palpate the head and neck to check for any structural damage to the bones in the head, face, and neck.

> **Caution** In the initial assessment, one of the first steps after checking the scene for safety purposes and donning BSI is to establish and maintain head and neck stabilization if the patient is unconscious or a traumatic injury is suspected. If head and neck stabilization have been initiated on a patient, the firefighter should *not* move the head or neck when performing a head-to-toe survey.

Face and Eyes. Gently palpate the bones in the face to check for swelling or deformities. Examine the mouth, nose, and ears for the presence of foreign objects, bleeding, or discharge of fluid.

Carefully examine the patient's eyes and note the size of the pupil: normal, **constricted,** or **dilated**. Also, check to make sure that the pupils are equal in size for both the right and left eye. Using a small flashlight, quickly shine the light in each eye briefly and note whether or not the pupil is "reactive" to light. If the pupil is reactive, it will shrink with the presence of light and grow larger with the absence of the light source. Refer to **Figure 22-20** for examples of different pupil sizes.

Upper Torso and Chest. Check the bones of the rib cage and breastbone (sternum) for structural integrity. Also, check to see if the chest rise and fall from breathing is equal on both the right and left sides. An unequal chest rise and fall may indicate a serious breathing problem, such as a collapsed lung.

Lower Torso, Abdomen, and Pelvis. Check the abdomen and pelvic area for any obvious signs of traumatic injury, like bruising, swelling, or pulsating masses. A pulsating mass is a sign of a rupture, or impending rupture, of the body's largest artery, the aorta, and is a very serious condition. Using one hand on top of the other, press down in each of the four quadrants of the abdomen to check for pain, tenderness, or abdominal rigidity. Next, place one hand on each hip on the side of the body, squeeze together gently to check the stability of the pelvis, then press down on the top of the hips gently to continue to check the structural integrity of the pelvis.

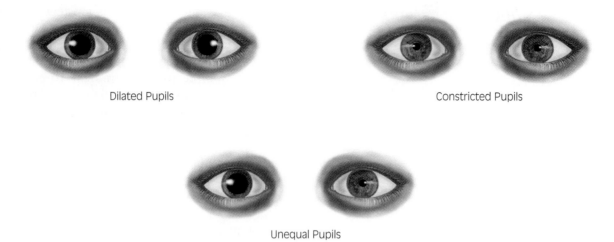

Figure 22-20   Examples of pupil size.

Arms and Legs. Gently feel the arms and legs of the patient to check for any possible damage to the extremities. Also, look carefully at the arms and legs in comparison to one another. Is one leg, or arm, longer or shorter than the other? If the patient is awake, check for movability of the fingers and toes to make sure the neurological system (the brain and spinal cord) is functioning. You can also ask a patient to grip your hand and squeeze, or place your hands at the bottom of the patient's feet and ask the patient to push against your hands. Check for the pulse in both arms and both legs to be sure that the heart is properly circulating blood to the extremities. The pulse in the foot can be felt on the top of the foot and the inside of the ankle underneath the protruding bone structure in the ankle, **Figure 22-21**.

Another very effective way of checking for circulation in the arms and legs of children under six years of age is by checking capillary refill. To do this, pinch the end of the fingers, or toes, until the skin underneath the nailbed becomes white. When you

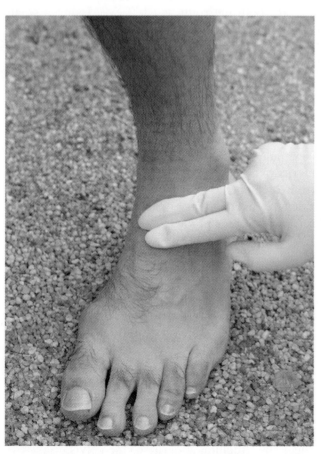
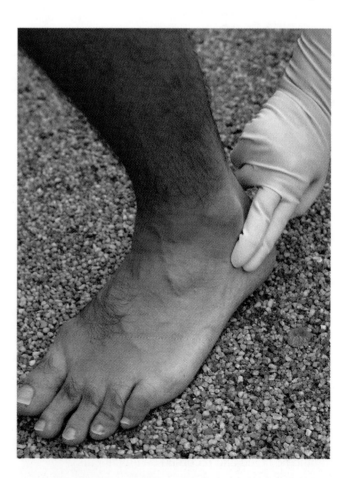

Figure 22-21   Locating pulses in the foot.

EMERGENCY MEDICAL SERVICES   707

Figure 22-22   Capillary refill.

release the pressure, watch to see how quickly the skin becomes pink or normal again. Normally, the skin will return to its original color right away. If it takes longer than a second or two for the skin under the nail to return to its original color, there may be a circulation problem in the patient. **Figure 22-22** demonstrates the capillary refill technique.

Patient Findings

If the fire department requires that firefighters complete a call record for emergency medical patients, they should be sure to document the findings of the assessment and the patient's vital signs.

> **Note** Documenting patient information is very important, and firefighters should be familiar with their department's policies concerning the recording of patient and emergency response information.

Firefighters may have found something in the survey that is important to the treatment of the patient, both by an ambulance crew and in the hospital. Everything that is found must be passed on to the emergency medical provider taking over care of the patient. That includes the initial assessment, the focused history and physical exam, the patient's vital signs, and all patient fact-finding that was done while on the scene.

CARDIOPULMONARY RESUSCITATION/AED

One of the most basic and widely learned emergency response skills is cardiopulmonary resuscitation, or CPR. This skill is learned by all types of people, from lifeguards to physicians. CPR is a critical, lifesaving skill for firefighters and emergency responders to learn and practice, **Figure 22-23**.

> **Note** CPR and techniques for helping choking victims are important knowledge to have not only while on duty with the fire department but also anywhere a firefighter may go.

Figure 22-23   Firefighters learning CPR. *(Photo courtesy of Fred Schall)*

Coronary heart disease and cardiovascular diseases are the leading killers in the United States. Heart disease, while more prevalent in the elderly, can affect anyone. Emergency care providers will most certainly be faced with situations where cardiovascular disease has caused a sudden respiratory arrest (a patient who is not breathing) or cardiac arrest (a patient who is not breathing and has no pulse). In these situations, CPR is a basic lifesaving skill.

Several different organizations support CPR training, education, and research. The two largest organizations are the American Heart Association (AHA) and the American Red Cross. CPR training is widely available in the United States. If a particular department or agency does not have CPR classes available, firefighters can contact a local chapter of the AHA or the American Red Cross and ask them about CPR training.

Safety CPR is a critical skill for firefighters to know and practice. They need this skill not only to provide care to patients on emergency medical calls but also to be prepared to help fellow firefighters on the scene of a fire or emergency incident. Heart disease is a leading killer of firefighters operating on emergency scenes every year. Firefighters should have the skills necessary to help the members of their team in need, including CPR and first-aid training.

As research on cardiovascular disease and cardiac arrest procedures continues, the skills and information regarding CPR may be changed or revised. Firefighters should follow the most current standards set forth by the AHA or the American Red Cross for education, practice, and performance of CPR. Firefighters should also be aware of recommendations concerning retraining on CPR skills, which should be done every one to two years.

Automated external defibrillators (AEDs), Figure 22-24, have become an important part of the provision of emergency medical care. Many types of first response organizations, including fire departments, use AEDs as a lifesaving tool. AEDs are becoming more common in nonemergency settings as well, including shopping malls, casinos, airplanes, and even private homes. AEDs are becoming more common because they dramatically improve the outcome of cardiac arrest.

An AED is a small machine with an internal computer that analyzes a patient's heart rhythm and delivers electrical shocks, or defibrillations, as necessary to correct a heart with a malfunctioning rhythm. It requires some basic training to place the AED on a patient and use it appropriately. Many first response agencies are training firefighters to use AEDs at the same time they are training them on CPR techniques. Firefighters should consult their agency's policies and explore training courses for AED use by firefighters.

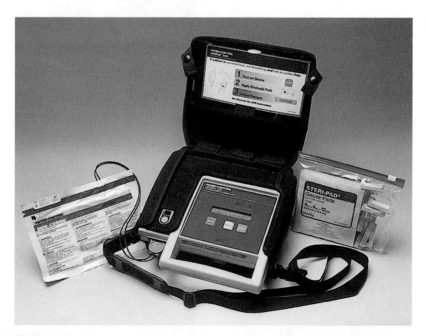

Figure 22-24 An AED is a small machine with a computer that analyzes a patient's heart rhythm and delivers electrical shocks, or defibrillations, as necessary to correct a heart with a malfunctioning rhythm.

BLEEDING CONTROL AND SHOCK MANAGEMENT

The blood in the body brings oxygen and nutrients to the cells in the body in order for the cells to survive, grow, and reproduce. The heart and the system of tubes through which blood travels (**arteries**, **veins**, and **capillaries**) are very complex and vital systems, **Figure 22-25**. The **cardiovascular system** is a closed system with a pump (the heart) and specialized tubing (the blood vessels) that provide the body with blood containing the elements humans need to live. Damage to the heart or the blood vessels may result in a lower volume of blood or an ineffective pumping system, which are serious, if not life-threatening, conditions that can cause a patient to go into shock, or hypoperfusion (a lack of oxygen and nutrients to the tissues).

> **Caution** Bleeding control and shock (hypoperfusion) management are important skills to learn in first-aid training. The firefighter should use BSI, as with all patients, but especially so in this situation because exposure to blood and body fluids is very likely.

Internal and External Bleeding

Internal bleeding occurs when there is bleeding within the body and no visible open wound is present. Internal bleeding can occur because of trauma to the body or because of illness. Internal bleeding can be a very serious condition for a patient.

Firefighters should look for the following signs, which may indicate internal bleeding:

- Bruising of the skin
- Pale skin
- Cold and clammy skin

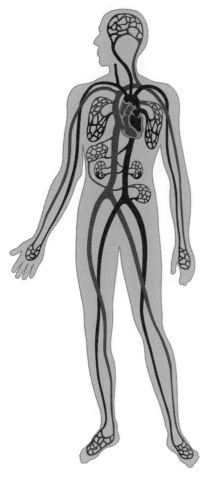
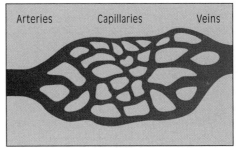

Figure 22-25 The blood vessel system in the body, which is composed of arteries, veins, and capillaries.

- Dilated pupils
- Obvious deformities to major bones (like the pelvis, the upper leg, etc.)
- A rigid and tender abdomen
- Blood in the urine or from the rectum
- Blood from the mouth or nose, or blood in vomitus

Patients who the firefighter suspects are suffering from internal bleeding should be treated for shock (hypoperfusion), as covered later in this chapter.

External bleeding is bleeding that is coming from an open wound on the body. The three types of external bleeding are discussed next.

Arterial Bleeding

Arterial bleeding is bleeding from arteries, which are the blood vessels that carry blood that has been oxygenated away from the heart to the cells in the body. Arterial blood, because it has just been supercharged with oxygen, is bright red in color. Bleeding from the arteries can be very serious. In order for the oxygenated blood to get to the cells, the heart pushes it through the arteries under pressure. When an artery is severed or damaged, this pressure can force a large quantity of blood out of the blood vessel system, depleting the system and possibly causing shock.

Firefighters providing first-aid care to patients can recognize arterial bleeding by looking for bright red blood from the wound site or blood spurting or pulsating from the wound site.

Venous Bleeding

Venous bleeding is bleeding from a vein. Veins are the blood vessels that carry blood from the cells in the body, after the oxygen and nutrients are used, back to the heart to be reoxygenated in the lungs. Venous blood, because it has been stripped of the oxygen it was carrying, is bluish red in color. Bleeding from the veins can also be very serious, especially if the bleeding is from a large vein. Unlike the arteries, however, the pressure on the blood in the veins is not as great, but can still be life threatening.

Firefighters providing first-aid care to patients can recognize venous bleeding by looking for bluish red blood from the wound site or blood flowing steadily from the wound site.

Capillary Bleeding

Capillary bleeding is bleeding from a capillary. Capillaries are the very small blood vessels connecting the arteries and the veins. These small blood vessels filter the oxygenated blood from the arteries to the cells and then take the used blood back from the cells and into the veins to return to the heart and lungs. Bleeding from the capillaries is what a person sees when he or she cuts a finger or skins a knee. It is usually not serious; however, it should be treated by the care provider.

Firefighters providing first-aid care to patients can recognize capillary bleeding by looking for blood slowly oozing from the wound site.

Caring for Patients with Internal Bleeding

Once firefighters have established what kind of bleeding is occurring in a patient, they can take measures to get the bleeding under control. Internal bleeding cannot be brought under control, however, by first-aid actions. For patients exhibiting signs of internal bleeding or shock, there are other actions a firefighter can take to provide assistance.

For minor internal bleeding or, in other words, bruising, the best treatment firefighters can give the patient is to apply a cold pack or ice pack to the affected site. The purpose of applying cold to the bruised area is to attempt to slow down the blood flow seeping into the skin by narrowing the blood vessels. The cold will also help with alleviating some of the pain of bruising as well. After applying cold to the injury site, the firefighter should attempt to elevate the part of the body that is injured, unless there is a chance there is a fracture. For example, if bruising and pain are found on the lower leg, and there appears to be no deformity of the leg structure or signs of a bone fracture, cold should be applied to the site and the leg raised.

> **Streetsmart Tip** When applying cold packs or ice packs to a patient, the firefighter should place a towel, cloth, or gauze pad between the cold pack and the skin to prevent freezing of the tissues.

Major internal bleeding is a serious problem and requires rapid medical attention. The loss of a large volume of blood can cause a patient to go into shock. The treatment for patients in shock is covered in a later section.

Caring for Patients with External Bleeding

Three methods are used when dealing with a patient experiencing external bleeding, as explained next. Note that, before acting, the firefighter should remember to wear protective gloves, since exposure to blood and body fluids is likely.

Pressure on the Site of the Bleeding, or Direct Pressure

Capillary and some venous bleeding can usually be controlled by direct pressure. Direct pressure can also help in the case of arterial bleeding. Firefighters controlling bleeding with direct pressure should first locate the wound and source of the bleeding. (They must look carefully, because a large volume of blood may be coming from a small cut or puncture.) The firefighter should place a sterile dressing or bandage over the site of the wound and apply pressure by pressing a hand against the wound site. Using a roll of gauze (preferably sterile) or a roll of bandage material, the firefighter wraps the site with the material, pulling it tight to create moderate pressure to the wound site. It is important to use a piece of medical or first-aid tape to secure the bandage and keep pressure on the wound site. If the bleeding soaks through the bandaging, it should not be removed to redo the bandage. Once a bandage is placed on a patient, it should not be removed. Instead, the firefighter should place another sterile dressing on the bandaged wound and wrap it again with a roll or gauze, this time applying a little more pressure with the bandaging, **Figure 22-26**.

Elevate the Site of the Bleeding

After applying direct pressure and a bandage to the wound that is bleeding, the firefighter should elevate the extremity or area that is bleeding. If direct pressure stops the bleeding, elevation may not be necessary. Elevation will make the blood flow to the site slower because it must travel "uphill" and, therefore, may slow down the bleeding. If the bleeding is from an arm or leg, it should be simply propped or raised above the level of the heart. If the bleeding is from the head or face, the firefighter should make sure the patient sits upright.

Use Pressure Points

The blood vessel system in the body has some main vessels, or tubes, that supply certain areas of the body. If direct pressure and elevation do not

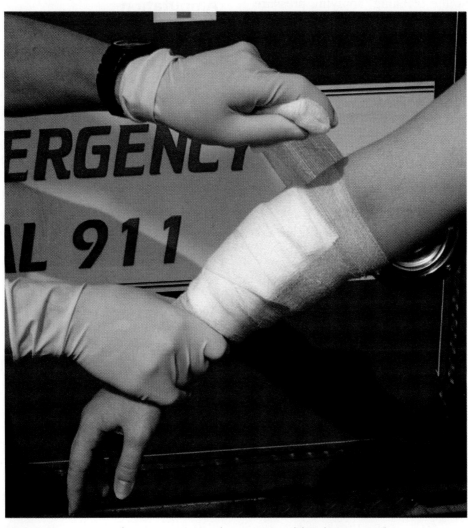

Figure 22-26 Applying a pressure dressing to a bleeding wound.

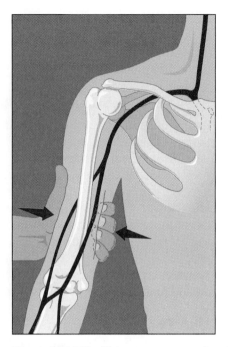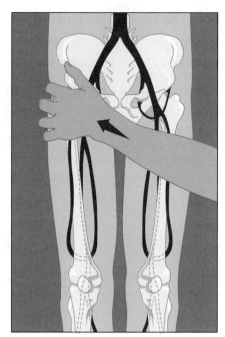

Figure 22-27 Using pressure points in the upper arm and the leg to decrease blood flow to a wound.

help stop or slow the bleeding, applying pressure to these "pressure points" may slow the flow of blood to the body area, thus slowing the flow of blood from the wound, **Figure 22-27**. The best pressure points are in the upper arm (brachial artery) and in the hip/pelvic area (**femoral artery**). In both of these areas, the major artery is near a bone where it can be compressed to slow blood flow.

Types of Wounds Requiring First Aid

There are several types of wounds that firefighters providing first aid should be able to recognize and treat.

Abrasion

An **abrasion** is a scrape or brush of the skin usually making it reddish in color and resulting in minor capillary bleeding. An abrasion is what results when someone "skins a knee," a common event in childhood. Abrasions should be treated by applying sterile bandages and, if the bleeding has not stopped, direct pressure.

Avulsion

An **avulsion** is an injury where a part of the skin is torn away, but still attached, leaving a flap or loose area hanging. Avulsions should be treated by applying sterile bandages and, if the bleeding has not stopped, direct pressure.

Amputation

An **amputation** occurs when a part of the body is severed completely as a result of an injury. This can happen to fingers and toes and even arms and legs. Firefighters should first apply sterile bandages to the site where the amputation occurred with direct pressure, using elevation or pressure points as needed. Firefighters should then locate the severed body part and wrap it completely with a sterile dressing. Once it is wrapped in the dressing, the body part should be placed into a plastic bag which is then sealed tightly. The severed body part must be kept cold and transported to the hospital with the patient. Firefighters can place the severed body part that is sealed in a plastic bag into a second plastic bag filled with ice to keep the body part cold for transport.

Laceration or Incision

A **laceration** is a cut to the skin and underlying tissues that leaves an irregular, uneven pattern. An **incision** is a cut to the skin that leaves a straight, even pattern. In both cases, the firefighter should apply sterile bandages to the wound and, if the bleeding has not stopped, direct pressure. Elevation and pressure points may be necessary for a large or deep laceration or incision.

Puncture

A **puncture** is caused by an object that has stabbed the body. If the object is no longer in the wound, firefighters can treat the puncture wound by applying a sterile bandage and, if necessary, direct pressure, elevation, or pressure points. If the object causing the puncture remains in the body, firefighters should pack the sterile dressings around the wound site, leaving the object in the body. They should not attempt to remove an object imbedded in a wound, because that could cause more harm to the patient. The object must be secured so that it will not move around and cause more damage to the skin and tissues.

What Is Shock? (Hypoperfusion)

Shock, or **hypoperfusion**, is a condition caused by a problem with or failure of the circulatory system that results in a decrease of oxygen and vital nutrients to the body's tissues. In other words, it is a lack of blood to the organs and tissues of the body. The body requires a regular and constant supply of oxygen and nutrients to function properly.

> **Caution** Shock (hypoperfusion) is a very serious condition and can cause death.

Shock (hypoperfusion) can result from a problem with the heart, which pumps the blood to the tissues. This can happen when a patient is experiencing a heart attack, and the heart has lost its ability to pump the blood adequately. Shock can also result from a failure in the blood vessels that carry the blood to and from the heart. This can be caused by a severe allergic reaction, which causes the blood vessels to expand rapidly, or by a trauma to the head or spinal cord, also causing a rapid expansion, or dilation, of the blood vessels.

Lastly, shock can result from a problem within the blood vessels, namely, a drop in the quantity of blood contained in the vessels. This can occur when the patient has an injury that has caused major internal or external bleeding. As the blood is lost from the body, the vessels and the heart have less oxygen carrying material to work with, resulting in the shock state. This type of shock can also occur in a patient who has experienced an illness and is experiencing **dehydration**. A severely dehydrated state can cause shock.

Recognizing the Signs and Symptoms of Shock (Hypoperfusion)

There are many signs and symptoms of shock, or hypoperfusion, that firefighters should look for in a patient, **Figure 22-28**:

- Pale color to the skin, or a bluish tint to the skin, especially around the lips and in the nailbeds
- A cool temperature of the skin

Signs of Shock
- Pale or Bluish Tint to Skin Color
- Cool Skin Temperature
- Moist or Sweating Skin
- Dilated Pupils
- Rapid, Shallow Breathing
- Rapid, Weak Pulse
- Nausea, Thirst, or Vomiting
- Unconsciousness or an Altered Level of Consciousness

Figure 22-28 Signs of shock.

- Sweating, or moist skin
- Pupils larger than normal, or dilated
- Rapid, shallow breathing
- A rapid, weak pulse
- Complaints of nausea, thirst, or vomiting
- Unconsciousness, or an altered level of consciousness

Firefighters should look for signs of shock (hypoperfusion) in all patients. Many different traumatic injuries or severe illnesses can result in shock, so firefighters should always be prepared for a shock state in a patient.

Caring for Patients in Shock

Firefighters who find a patient with the signs and symptoms of shock (hypoperfusion) should follow the basic treatment outlined here. Firefighters should also treat patients in this manner if they suspect that they could go into shock.

Treating a Patient for Shock (Hypoperfusion)

1. Ensure scene safety and BSI precautions.
2. Assess the patient's level of consciousness and, if needed, maintain head and neck stabilization.
3. Make sure the airway of the victim is open. This is especially important in unconscious patients.
4. Ensure that the patient is breathing and has a pulse. If either is absent, follow the procedures for cardiopulmonary resuscitation and rescue breathing.
5. Treat the injuries present and control any major bleeding.
6. Keep the patient warm. Place some blankets or a covering over the patient to prevent

any body heat loss or exposure to cold in the environment.
7. Position the patient properly. The best possible position for a patient in shock is lying on the back. If head and neck stabilization is initiated during the initial assessment it should be continued throughout patient treatment. Placing a patient on the back may not be possible if the patient is having difficulty breathing, so the patient can be placed in a reclined position. A patient who is conscious and vomiting is best placed on the side so that the vomitus does not obstruct the airway.
8. Raise the legs of the patient to allow more blood to reach the heart. The legs should be raised together and no more than a foot above the level of the heart. Do not raise the legs of a patient who has a traumatic injury to the legs or pelvis.
9. Provide reassurance to the patient, and monitor vital signs every few minutes.

EMERGENCY CARE FOR COMMON EMERGENCIES

No two emergency medical calls are the same. Firefighters must learn the basics of first-aid treatment and then adapt them to each emergency. This section lists some common emergencies, things to look out for, and first-aid treatment.

Trouble Breathing

Trouble breathing is a very common emergency that firefighters may be called on for first-aid care. There are many reasons why a patient may be experiencing trouble breathing, anything from anxiety to a major respiratory disease. Common signs and symptoms of trouble breathing include wheezing, gasping for air, shallow breathing, pale or bluish skin color, anxiety, or even unconsciousness if the patient is having serious difficulty breathing.

Care Guidelines

1. After assessing scene safety and donning BSI, perform an initial assessment and attend to any major problems found with the airway, breathing, or circulation.
2. Reassure the patient, keep the patient calm, and monitor vital signs frequently.
3. Recognize that this is a serious emergency and emergency medical care is needed right away.

Chest Pain

Chest pain is another common emergency firefighters may encounter, as shown in **Figure 22-29**. Chest pain related to a heart problem or heart attack may be described as a tightness or squeezing feeling in the chest, and the pain may be felt not only in the chest but also in the jaw, abdomen, or arm (often the left arm). Chest pain can also result from some respiratory disorders, traumatic injuries, or even stomach problems such as indigestion.

Care Guidelines

1. After assessing scene safety and donning BSI, perform an initial assessment and attend to any major problems found with the airway, breathing, or circulation.
2. Reassure the patient, keep the patient calm, and monitor vital signs frequently.
3. Recognize that this is a serious emergency and emergency medical care is needed right away.
4. Because chest pain may be a sign of a problem with the heart, there is a possibility the patient could go into shock. Look for signs and symptoms of shock and treat the patient accordingly.

Medical Illnesses

A firefigher may be called on to assist a patient with a medical illness, which can range from stomach problems like nausea and vomiting to general illnesses like fevers, the flu, or the common cold.

Care Guidelines

1. After assessing scene safety and donning BSI, perform an initial assessment and attend to any major problems found with the airway, breathing, or circulation.
2. Reassure the patient, keep the patient calm, and monitor vital signs frequently.
3. Recognize that this could be a serious emergency and emergency medical care is needed right away.
4. Because a medical illness may cause dehydration (lack of fluids) in a patient, there is a possibility the patient could go into shock. Look for signs and symptoms of shock and treat the patient accordingly.

Allergic Reactions

Allergic reactions result from the body's reaction to a substance to which there is an allergy. Many people experience allergy problems associated with hay fever or bee stings. In some cases, allergic reactions

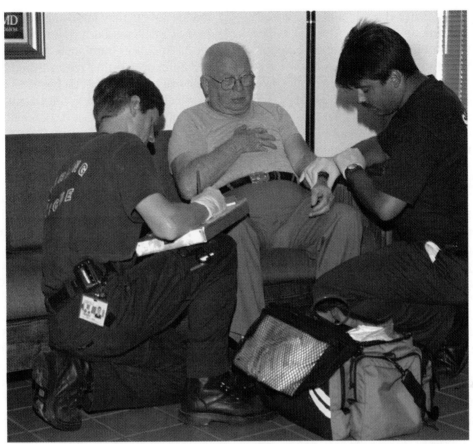

Figure 22-29 Firefighters treating a patient complaining of chest pain. *(Photo courtesy of Fred Schall)*

can be very severe, even life threatening. Allergic reactions can result from a variety of substances, such as types of food, things in the environment, bites and stings, and medications. Allergic reactions may be localized at the part of the body where the substance was introduced, and present as localized swelling, redness, and itching. Allergic reactions involving the entire body system may include trouble breathing, rapid pulse, shallow breathing, hives (red splotches or bumps on the body, typically concentrated on the torso), and anxiety, **Figure 22-30**.

Care Guidelines

1. After assessing scene safety and donning BSI, perform an initial assessment and attend to any major problems found with the airway, breathing, or circulation. Severe allergic reactions can compromise the airway and breathing of a patient because the reaction tightens the breathing passages in the lungs and airway. Be prepared to breathe for the patient if necessary.
2. Reassure the patient, keep the patient calm, and monitor vital signs frequently. Also, if the patient is still in contact with or near the substance causing the allergic reaction, remove the patient from the area or isolate and dispose of the substance.
3. Recognize that in an allergic reaction involving the body system, this is a serious emergency and emergency medical care is needed right away.
4. Because severe allergic reactions often cause the blood vessels to widen, there is a possibility the patient could go into shock. Look for signs and symptoms of shock and treat the patient accordingly.

Thermal Burns

Thermal burns, or burns caused by heat, can be as simple as a sunburn or as life threatening as severe burns across much of the body. Thermal burns are

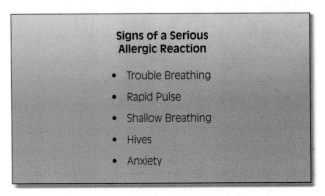

Figure 22-30 Signs of a serious allergic reaction.

very painful and can cause the patient to lose body heat and fluids, causing dehydration.

Burns are divided into three categories based on their severity. **Superficial burns**, sometimes referred to as first-degree burns, occur when the outer layer or layers of skin are burned, **Figure 22-31A.** The skin will turn red and feel hot to the touch. Depending on the severity of the superficial burn, swelling may result. These burns are often painful to the patient but will generally heal in about a week, even if left untreated. Sunburn is an example of a superficial burn.

In **partial thickness burns**, sometimes called second-degree burns, additional layers of skin are burned, but the deeper layers of the skin remain undamaged, **Figure 22-31B.** The skin will be very red, and blisters will develop. These burns can cause the patient intense pain because the nerves under the skin are sometimes affected. The major sign of a partial thickness burn is the blistering of the skin; however, this blistering does not always occur immediately. It may take several hours before blisters develop. These types of burns require medical attention, and with treatment they generally heal within a few weeks.

The last category of burn is a **full thickness burn**, often referred to as a third-degree burn, **Figure 22-31C.** In full thickness burns, all layers of the skin are burned. These burns will have the characteristics of partial thickness burns, but will also have areas of charred or blackened skin, which is the definitive sign of a full thickness burn. These burns are the most serious of the three types because the entire area of skin has been damaged, and the body is no longer able to hold fluids inside or keep bacteria and dirt outside. Many times nerves are also damaged or destroyed, and the patient may not complain of pain in the area of the full thickness burn. (However, the accompanying partial thickness burn areas will still be very painful.) Full thickness burns require medical attention, and these patients are extremely susceptible to infections. The healing process of a full thickness burn is usually long and painful and can be debilitating.

Care Guidelines

1. After assessing scene safety and donning BSI, perform an initial assessment and attend to any major problems found with the airway, breathing, or circulation. Be sure to remove the patient from the environment to stop the burning process.
2. Reassure the patient, keep the patient calm, and monitor vital signs frequently.
3. Cover the burned area with dry, sterile dressings to prevent infection. (Firefighters should consult their agency for specific treatment for burns. Some fire departments may have a different treatment for burns.)
4. Monitor the patient's airway and breathing constantly. Burn victims may also have inhaled smoke or hot gases, which can cause damage in the lungs and create airway or breathing problems.
5. Also, monitor the patient for signs and symptoms of shock and treat accordingly.

Chemical Burns

Chemical burns are caused by chemical substances that come into contact with the skin or tissues of the body, creating a caustic reaction. A variety of chemical substances can cause a chemical burn. They can

(A)

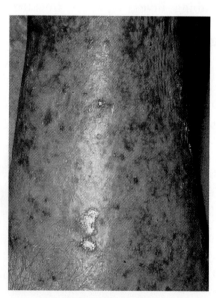
(B)

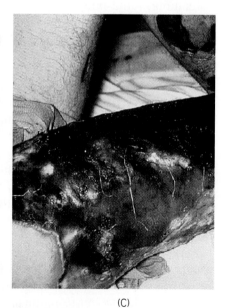
(C)

Figure 22-31 (A) Superficial burns, (B) partial thickness burns, and (C) full thickness burns.

be found in many different locations, from large, industrial plants to family residences.

Care Guidelines

1. After assessing scene safety and donning BSI, perform an initial assessment and attend to any major problems found with the airway, breathing, or circulation.
2. Reassure the patient, keep the patient calm, and monitor vital signs frequently.
3. If the chemical substance is dry or in powder form, brush off as much of the substance as you can from the patient, and remove the patient's clothing, watches, and jewelry. (Firefighters should make sure they protect themselves when removing the chemical from the patient. Do not use bare hands or brush the substance into the wind where it may become airborne and harm others. Consider additional protective gear, such as masks and gowns.)
4. If the chemical substance is not in dry or powder form (liquid chemicals, acids, etc.), flush the substance off the patient with large volumes of gently flowing water. Use a garden hose without a pressure nozzle, bottles of water, or water from a faucet to flush the substance off the patient. Remove all clothing, watches, and jewelry from the patient.
5. Monitor the patient's airway and breathing constantly. Chemical burn victims may also have inhaled a substance that can cause damage to the lungs and create airway or breathing problems.
6. If the chemical has gotten into the patient's eyes, flush the eyes with large amounts of gently flowing water for at least fifteen to twenty minutes.

Poisoning

The ingestion or inhalation of a caustic substance is considered poisoning. Poisoning can also result from the ingestion of a large quantity of a normally harmless substance, like over-the-counter medications or household substances as shown in **Figure 22-32**.

Care Guidelines

1. After assessing scene safety and donning BSI, perform an initial assessment and attend to any major problems found with the airway, breathing, or circulation.

Figure 22-32 Many types of substances and materials can be harmful to people and cause poisoning. When caring for a patient who has possible poisoning, the firefighter should make every attempt to find out exactly what substance was ingested or inhaled and how much.

2. Reassure the patient, keep the patient calm, and monitor vital signs frequently.
3. Do some investigative work to find out *exactly* what substance was ingested or inhaled. Keep that information handy.
4. *Do not* give the patient any liquid or food to ingest.
5. Contact the local poison control center. Check with your department or agency and find out what your local poison control center is and the telephone number for the center. Tell the poison control center all essential information about the patient, such as age, sex, medical history, and symptoms. Also, tell the poison control center *exactly* what substance was ingested or inhaled, and how much of it was ingested or inhaled. The poison control center will be able to suggest additional treatment for the patient before the EMS unit arrives.

Streetsmart Tip Local poison control centers are very valuable and helpful resources. Even if the exact name or dosage of the substance is unknown, they may be able to help the emergency responder figure out what it is.

Poison control centers have special references to assist them in figuring out "mystery" substances. For example, if a pill is found, a firefighter can describe it to the poison control center, and they may be able to find out what the name of the drug is and even the dosage.

Fractures and Sprains

A **fracture** is a medical term for a broken bone. A **sprain** is an injury to the ligaments that hold joints in the body together and allow them to move. Firefighters may notice swelling and tenderness to the area of a fracture or sprain. In the case of a fracture, there may be deformity to the bone and body structure.

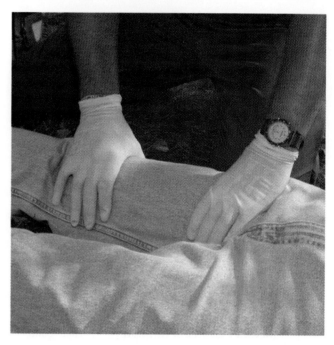

Figure 22-33 Firefighter stabilizing a leg injury until the EMS unit arrives.

Care Guidelines

1. After assessing scene safety and donning BSI, perform an initial assessment and attend to any major problems found with the airway, breathing, or circulation.
2. Reassure the patient, keep the patient calm, and monitor vital signs frequently.
3. Do not move the patient, and protect the cervical spine area in cases where the patient has fallen or been involved in a moving vehicle accident.
4. Do not move the injured part of the body. Do not attempt to straighten an arm or leg that is deformed. Carefully protect the injured area and be sure it does not move or shift, **Figure 22-33**.
5. As needed, apply cold packs or ice packs to the injured area to help relieve some of the pain and swelling.

Lessons Learned

Emergency medical care is a major part of most firefighters' responsibilities, and the treatment a firefighter gives a patient may make the difference between life and death. As the field of emergency medicine has progressed, the fire service has played an increasing role in providing first response emergency care. This chapter has provided some very basic information about first aid and emergency medical treatment, but there is a lot more to be learned.

The most important tools used to provide emergency medical first aid do not come from a trauma bag or a first-aid kit. The most important tool is the firefighter's brain. The job of a first-aid provider is a lot like that of a detective. Firefighters should use their senses to discover all that they can about a patient and the environment, and then treat the patient accordingly. A life could be saved in the process!

> **Streetsmart Tip** All firefighters should consider furthering their education to become emergency medical technicians and even paramedics.

KEY TERMS

Abandonment Abandonment occurs when an emergency responder begins treatment of a patient, and then leaves the patient or discontinues treatment prior to the arrival of an equally or higher trained responder.

Abrasion A scrape or brush of the skin usually making it reddish in color and resulting in minor capillary bleeding.

Allergic Reaction The body's reaction to a substance to which there is an allergy.

Amputation Occurs when part of the body is severed completely as a result of an injury.

Arterial Bleeding Bleeding from an artery.

Arteries The blood vessels, or tubes, within the body that carry blood rich with oxygen and nutrients away from the heart.

Automated External Defibrillator (AED) A portable computer-driven device that analyzes a patient's heart rhythm and delivers defibrillation shocks when necessary.

Avulsion An injury where a part of the skin is torn away, but still attached, leaving a flap or loose area hanging.

Body Substance Isolation (BSI) Precautions A set of precautions for emergency responders designed to prevent exposure to any body fluid or substance.

Brachial Artery A major artery in the inside of the upper arm that supplies blood to the arm. Can be used as a pressure point for controlling bleeding and for locating a pulse on an infant.

Capillaries The very small blood vessels in the body that connect arteries and veins and filter the oxygen and nutrients from the blood into the tissues of the body.

Capillary Bleeding Bleeding from a capillary.

Cardiovascular System The heart, blood vessels, and blood within the body.

Carotid Pulse The pulse located on either side of the neck.

Chemical Burns Burns caused by chemical substances that come into contact with the skin or tissues of the body, creating a caustic reaction.

Communicable Disease A disease that can be transmitted from one person to another.

Consent The acceptance of emergency medical treatment by a patient or victim.

Constricted A condition of the pupils where they are much smaller than normal and may appear almost like a "pinpoint."

Dehydration A loss of water and vital fluids in the body.

Dilated A condition of the pupils where they are much larger than normal and can take up almost the whole colored portion of the eye.

Exposure A contact with a potentially disease-producing organism; the contact does not necessarily produce the disease in the exposed individual.

External Bleeding Bleeding that is coming from an open wound on the body.

Femoral Artery A major artery in the lower body near the groin that supplies the leg with blood. Can be used as a pressure point for controlling bleeding in the lower extremities.

Fracture A medical term for a broken or cracked bone in the body.

Full Thickness Burns Burns affecting not only the skin structure but the tissues and muscles underneath. Full thickness burns may be red, white, or charred in color and will appear dry because the blood vessels in the skin are damaged extensively and are not supplying fluids to the area.

Hypoperfusion A serious condition caused by a problem or failure of the circulatory system that results in a decrease of oxygen and vital nutrients to the body's tissues. Also known as shock.

Implied Consent The assumption of acceptance of emergency medical treatment by an unconscious patient or a child with no parents or legal guardians present.

Incision A cut to the skin that leaves a straight, even pattern.

Infection Control Procedures and practices for firefighters and emergency medical care providers to follow to prevent the transmission of diseases and germs from a patient to themselves or other patients.

Infectious Disease See *communicable disease.*

Initial Assessment The initial investigative action taken by care providers to determine if the patient has the basic signs of life as well as any serious, life-threatening injuries.

Internal Bleeding Bleeding within the body when no visible open wound is present.

Laceration A cut to the skin and underlying tissues that leaves an irregular, even pattern.

Medi-vac An ambulance that transports patients by air. Typically, medi-vac units are helicopters with highly trained EMS personnel and nurses.

Partial Thickness Burns Burns affecting the entire skin structure that lies over the top of the fatty tissues and muscles causing skin to turn red and blistering of the skin.

Puncture An injury caused by an object that has stabbed the body.

Radial Pulse The pulse located in either wrist.

Shock A serious condition caused by a problem or failure of the circulatory system that results in a decrease of oxygen and vital nutrients to the body's tissues. Also known as hypoperfusion.

Sprain Injury to the ligaments that hold the body's joints together and allow them to move.

Standard of Care A legal term that means for every emergency medical incident, an emergency responder should treat the patient in the same manner as another emergency responder with the same training.

Superficial Burns Burns affecting the outermost layer of skin, which typically cause redness of the skin, swelling, and pain.

Thermal Burns Burns caused by heat or hot objects.

Triage A quick and systematic method of identifying which patients are in serious condition and which patients are not, so that the more seriously injured patients can be treated first.

Veins The blood vessels, or tubes, within the body that carry blood lacking oxygen and nutrients back to the heart.

Venous Bleeding Bleeding from a vein.

REVIEW QUESTIONS

1. What are the basic elements of an emergency medical system? What is the firefighter's role in the emergency medical system?
2. Describe the principles of BSI. Discuss how you would use BSI as a firefighter providing emergency medical care to a patient. When are BSI precautions needed?
3. What are the five major parts of an initial assessment?
4. What are the three types of external bleeding and the characteristics of each?
5. What are the three types of thermal burns, and how would a firefighter treat a patient with thermal burns? What is the difference in treatment between a patient with thermal burns and a patient with chemical burns?
6. How do you contact your local poison control center? What information do you give to the poison control center when contacting them for assistance?

Additional Resources

American Association of Poison Control Centers, http://www.aapcc.org

American Heart Association, http://www.americanheart.org

American Red Cross, http://www.redcross.org

Beebe, Richard, Med, RN, NREMT-P; and Deborah Funk, MD, NREMT-P, *Fundamentals of Emergency Care*, Delmar Learning, a part of the Thomson Corporation, Clifton Park, NY, 2001.

BLS for Healthcare Providers, American Heart Association, Dallas, Texas, 2002.

Burn Resource Center, http://www.burnsurvivor.com

Centers for Disease Control, http://www.cdc.gov

Fundamentals of BLS for Healthcare Providers, American Heart Association, Dallas, Texas, 2002.

Fundamentals of BLS for Healthcare Providers (Video), American Heart Association, Dallas, Texas, 2002.

Occupational Safety and Health Administration, http://www.osha.gov

Walter, Andrea; Edgar, Chris; and Marty Rutledge, *First Responder Handbook*, Delmar Learning, a part of the Thomson Corporation, Clifton Park, NY, 2003.

CHAPTER 23

FIREFIGHTER SURVIVAL

David W. Dodson, Lead Instructor, Response Solutions, Colorado

OUTLINE

- Objectives
- Introduction
- Incident Readiness
- Safety at Incidents
- Firefighter Emergencies
- Lessons Learned
- Key Terms
- Review Questions
- Endnotes
- Additional Resources

STREET STORY

It is one thing to talk about the life and death nature of firefighting—yet another when you are actually experiencing a situation that is about to kill you. A million thoughts go through your head and time seems to slow down. You rely on training, instinct, awareness, and even luck to get you through. My own experience happened one cold winter night—the kind of night where you didn't want to go outside. I was the second-arriving company officer to a rural, single-story farmhouse with a well-involved attic fire. The older home had a sturdy, high-mass roof structure so we elected an aggressive interior attic—that is, pulling ceilings and hitting it from underneath. Conditions inside the home were clear, although we knew there was significant fire over our heads in the attic space. In short time, we had two lines operating inside and making good progress. I noticed that the ceiling was sagging in the large "great-room" area of the house—and advised all crews to stay near walls and doorways and away from any open-span areas in case the roof collapsed. Seeing this potential increase, I thought we could simply force the collapse by pulling down the sagging area. I told my crew to take cover behind me in a doorway while I yanked down the ceiling. All was "safe" in my mind because we were eliminating a hazard and making access to the burning attic easier. Finding a good leverage point became troublesome so I moved into the room just a few feet. The roof then let go behind me.

As I turned to escape out the doorway, the roof hit me and flipped my 230 pounds over a couch. The first thought to cross my mind was dying. The second thought was a question: Where did my crew go? Training then kicked in. I took inventory of my environment: I was on my back (with SCBA—ouch), still breathing, legs working, no flames, light smoke, and a ceiling/roof structure pinning me to the floor and couch. I called for my crew—no reply. I called out for anyone—no reply. I couldn't get to my radio or PASS device manual activator. I thought about lying motionless to help activate the device but blew off that idea and went into self-help mode. I reasoned that I should try to self-extricate and get to my crew somehow. My instinct told me to move. I knew the fire was right above me with only drywall and framing separating me from it. With an energy burst, I was able to spin myself over to my side and began to crawl through a slight void created by the furniture in the room. Earlier, I was aware of three exits to the area. The way I came in was now blocked but I recalled a side door and a rear sliding glass door. Luckily, the void led to the sliding glass door. I exited and went to the entry point of my crew. They were shielded from the collapse by the doorway and they simply exited to begin looking for another way in to find me. What seemed like ten minutes under that roof was more like seconds according to my crew. They didn't even have time to report the collapse before they were out and I met them.

Training, instinct, awareness, and luck figured into this survival. Use this chapter to help build the training, instinct, and awareness to help you survive. Don't rely on the luck.

—Street Story by Dave Dodson, Lead Instructor, Response Solutions, Colorado

OBJECTIVES

After completing this chapter, the reader should be able to:

- List the three main components that lead to incident readiness.
- Define the four key checks to ensure that PPE is ready for response.
- List three types of personal accountability systems.
- Define personal size-up.
- Describe the three components that lead to "fitness for duty."
- Name three practices that lead to team continuity.
- Define risk/benefit.
- List and briefly describe the three components of rehabilitation.
- Describe the procedures that should be taken to establish and prepare for the assignment of a rapid intervention team.
- List the five steps that can lead to an organized rapid escape.
- List the three steps that should be taken when entrapment occurs.
- Compare and contrast post-incident thought patterns and critical incident stress.

INTRODUCTION

The excited radio transmission of "Mayday, mayday, firefighters injured and trapped" will present perhaps the most stressful circumstances that a firefighter will ever face. While no firefighter or fire officer expects this to happen, it would be negligence if training efforts failed to address firefighter survival and the handling of situations in which firefighters are injured, trapped, missing, disoriented, or imperiled—all firefighter emergencies.

Obviously, it makes sense to attempt to prevent a firefighter emergency from happening. Firefighter survival is accomplished through significant proficiency training and education. Daily, firefighters can help prevent firefighter emergencies through incident readiness—that is, the efforts to ensure that all firefighters are mentally and physically ready to respond. Personal protective equipment (PPE), task accountability, and fitness for duty are all issues that help prevent firefighter emergencies. At an incident, firefighter actions, inaction, and attention to hazards will help prevent a firefighter emergency. Likewise, attention to team continuity, orders and communication, rapid intervention planning, and rehabilitation all help prevent firefighter emergencies.

If a firefighter emergency were to occur, a planned, systematic process of rescue would have to be established to avoid compounding the seriousness of the emergency. Having procedures in place for rapid escape and self-extrication, accountability recall, and post-incident emotional processing is essential. This chapter discusses all of these areas and gives firefighters survival tools to help prevent firefighter emergencies and to assist them when such an unfortunate event does occur.

INCIDENT READINESS

Unlike most professionals, firefighters must perform at peak mental and physical levels with little or no notice. Constant readiness is imperative. Preparing for incident response involves more than firefighters merely making sure they know where they are going. Incident readiness is a mental process that answers a few questions:

1. Am I in a position to respond?
2. Is my protective gear available?
3. What is my relationship to the response?
4. Physically, can I respond?
5. Mentally, can I check out of my current thoughts and focus on response?

A negative answer to any of these questions may set a firefighter up for an injury. To avoid this risk, firefighters must take steps to ensure that their "system" is assembled and ready for response. Some readiness system items are addressed before an alarm is received, while others are addressed on receipt of the alarm. This "system" is the key to survival and includes some important components.

> **Note** Namely, the readiness system includes PPE, personal accountability, and fit-for-duty status.

Personal Protective Equipment

The personal protective ensemble can be considered the first step in one's ability to focus on incident handling.

> **Safety** One fire service adage claims that PPE is the first and last means of defense from risks.

Specifically, PPE is the first thing firefighters put on and the last thing they want to be left with when the incident is over, **Figure 23-1**. If firefighters rely on PPE to protect them from some hazard, they have in effect used up all other protective measures. As an example, if firefighters find that their PPE prevented a serious burn, that means the PPE became the last means of prevention. Firefighting streams, ventilation, zoning, and access/egress efforts have likely failed if the firefighter got to the point that PPE made the difference between injury and no injury.

> **Caution** At the beginning of a duty tour or shift, firefighters must ensure that their protective equipment is dry, serviceable, and ready for quick donning.

In many departments, firefighters must check *all* their various ensembles. Firefighters working at Fire Station 4 in Loveland, Colorado, must prepare their structural ensemble, aluminized proximity suits, wildland gear, and (EMS) PPE each and every shift. This readiness check ensures that the firefighters are ready for any type of incident that is likely to occur within Station 4's response district, which includes an airport, commercial and residential structures, rural areas and fields, and EMS responsibility for 10,000 people.

Figure 23-1 PPE is part of the survival system. Is it ready?

When preparing PPE for readiness, firefighters should ensure the following:

1. All clothing materials are dry. Wet clothing reduces the protective insulation of the clothing and may lead to steam burns in heated environments.
2. All PPE is present and positioned so that it may be rapidly donned. This is important for station-assigned firefighters and those who respond to a station to staff an apparatus. Firefighters who respond from home directly to an incident must pack their dry gear in such a way that it can be donned quickly and appropriately.
3. Essential "pocket tools" are available and in working order. It is not uncommon for firefighters to carry an assortment of personal tools that they feel are important. Small flashlights, trauma shears, doorstops, nylon webbing, carabiners, grease pencils, chalk, sprinkler wedges, multiple-tool pliers, knives, foam earplugs, self-escape rope, and two-way radios are all common pocket tools found in firefighters' PPE.
4. Alternative PPE items are appropriately packed and ready for use. As stated earlier, many firefighters must use (EMS) PPE, wildland PPE, proximity PPE, or lightweight rescue PPE. These should also be checked and made ready.

When an alarm is received, the time spent donning gear *before* arrival at the incident is time well invested. People waiting until incident arrival to don PPE are setting themselves up to "shortcut" their protective ensemble. These shortcuts are caused by firefighters who are concerned and distracted by the situation and the actions that need to be started rather than by completion of their ensemble. Donning structural PPE prior to incident response is preferable; however, driving in full PPE can introduce some of its own safety issues. While large fire apparatus allows some room for the bulky structural PPE, driving a personally owned vehicle (POV) or a small support apparatus may be more cumbersome. If a driver makes the choice to don PPE after driving to an incident, strict discipline *must* be exercised to fully don PPE before engaging in the incident.

> **Caution** Firefighters should not find themselves at an incident without their PPE fully donned; they must practice strict self-discipline to complete the ensemble. They must not let the urgency of the situation override prudent judgment!

Personal Accountability

Personal accountability is an essential part of firefighter readiness. Personal accountability refers not only to the established accountability system used by a particular fire department, but also to the firefighter's relationship to the response and ability to perform as trained.

Accountability System

The accountability system used by a department can take on many forms. Typically, the system will fall into one of three general types:

1. *Passport:* This is a crew-card system that is tracked on a status board by a monitor. Members of a crew give a name chip to the company officer or team leader, who places all of the team names on a card or **passport**. When given an assignment, the passport is given to the monitor, who tracks all crews. Team leaders must report any changes in their teams' location or assignment.
2. *Tag:* In this system, individual firefighters report to staging and give an identification tag to the staging manager. The staging manager groups the tags into teams and tracks the progress of each. First arriving crews and personnel will not have a staging/accountability manager to collect their tags. In these cases, individual tags are usually hooked on a ring of the first-due apparatus, and the staging/accountability manager, once assigned, will collect these.
3. *Company officer:* This system is perhaps the oldest and most used system. Here, the company officer, team leader, or other supervisor is responsible for keeping track of the crew. This system is the least preferred system in that many firefighters have been lost and many crews broken up due to the lack of formality of the system.

Regardless of the system used, each firefighter must be aware of how the system works and follow its specific guidelines. Failure to be accounted for on the incident is akin to **freelancing**—that is, performing a task that has not been assigned or performing a task alone—both are fire service taboos, **Figure 23-2**.

> **Streetsmart Tip** Not only can firefighters endanger themselves through freelancing, but may also indirectly injure others when they attempt to find the freelancer.

Figure 23-2 Accountability systems take on many forms—firefighters must know how to check in!

Worse, if freelancers become lost or trapped, nobody will come to assist if they do not know freelancers are on scene and working.

Relationship to the Response

Another part of the accountability equation is attention to each individual's relationship to the response. Specifically, firefighters' relationship to the response includes their assignments and their own personal size-ups.

- *Assignment.* In most cases, firefighters are given readiness assignments—that is, being assigned as given crew members for given positions on the apparatus. With these assignments come responsibilities to ensure that they perform preassigned or reactive duties. In cases where firefighters have no preassigned duty, firefighters must be prepared to perform a host of duties. The readiness elements entail a mental preparation to perform the tasks of that position. If firefighters are assigned positions that they have not performed recently, they may not be ready to act—leaving an invitation to injury. In some departments, a list of crew duties for a given seated position on the apparatus is listed on a plastic card—this is a great reminder. Firefighters should take the time to reacquaint themselves with the expected tasks, tools, and procedures required. This improves readiness and reduces injury potential.
- *Personal Size-Up.* **Personal size-up** can be defined as a continuous mental evaluation of firefighters' immediate environments, facts, and probabilities. This mental evaluation should have them evaluating the weather, time of day, current

chain of command, and likely assignment. Once again, they are mentally preparing themselves—thereby reducing injury potential.

Work as Trained

> **Streetsmart Tip** Personal accountability is achieved when all firefighters and fire officers are able to perform assigned essential tasks—tasks they have been trained for—and keep the chain of command advised of their progress.

Of significant importance here is everyone's ability to perform as trained. A common thread discovered in post-incident investigations is the fact that many of the injured were performing tasks they had never experienced before. In some cases, the training had been given by the individual's department, but the individual either forgot *how* to accomplish the task or chose a method that was inappropriate for the situation. How can this be prevented? Listed next are two key points to assist firefighters.

Firefighters should perform as trained. This sounds easier than it is. Most departments invest considerable cost to train firefighters, so firefighters should do their part and constantly practice and perfect their skills. For career firefighters, this should be no challenge—evenings around the station are perfect times for doing this—it does, however, take motivation. Volunteers have a bigger challenge. They should take the initiative to go to the station and practice, and get others to join in the effort.

Firefighters need to know their strengths and weaknesses. The achievement of a firefighter certificate does not mean a firefighter has all of the necessary skills to be an effective emergency responder. It is merely a minimum. Training never ends for a firefighter—ever! Firefighters must constantly test themselves and try to improve those areas where they discover weaknesses. As a matter of fact, firefighters should strive for *mastery* of their assigned tasks, **Figure 23-3**. Mastery can be defined as the ability to achieve 90 percent of an objective 90 percent of the time. Further, firefighters should expand their abilities.

> **Caution** Firefighters should not let the incident scene become an avenue for discovery and on-the-job training.

Fitness for Duty

Fitness for duty is more than just being "in good shape."

Figure 23-3 Firefighters achieve mastery of tasks through repeated training. Mastery reduces the chance of injury.

> **Note** Fitness for duty includes mental fitness, physical fitness, wellness, energy, and rest, **Figure 23-4**.

If just one of these is less than adequate, the window to being injured opens.

Mental Fitness

Being mentally ready to respond to an incident at all times rarely happens—firefighters' lives are complex, busy, and full of mental and, in some cases, emotional challenges. As has been stated

Figure 23-4 Fitness for duty includes more than just being "in shape." It also includes wellness, rest, energy, and mental fitness.

often, a firefighter's size-up begins on receipt of the incident. If a firefighter is involved in an emotional event during this phase, key size-up information may be missed. The key is for the firefighter to "check out" of the environment and "check in" to size-up. The firefighter's non-firefighting acquaintances may not understand the rapid mental departure—it is common, however, for them to forgive the firefighter.

Once again, it is easier to say a firefighter should "check in" to the incident than to actually do it. Many firefighters have a mental ritual to help them achieve the check-in state every time. They create a ritual—a physical or mental act that reminds them to check out of their emotional or current mental state. For some, it comes when they buckle their seat belt and respond. Others take a deep breath and force the exhalation. Still others perform a rapid mental checklist: What is the call? Where is the call? Is my gear ready? What is my relationship to the response—first due, second due?

Physical Fitness and Wellness

There is no doubt that fit firefighters are less injury prone than those who are unfit. Fitness is a lifestyle commitment. On receipt of the alarm, physical fitness becomes more than an issue of how much firefighters "work out." Do they have injuries? Are they recovering from injuries? Are they recovering from a cold or the flu? Any of these conditions can affect performance and fitness. The simple act of going from a resting heart rate at alarm time to a maximal heart rate on arrival at a working fire can stress the body to the point of collapse if it is not capable of accommodating the spike. A firefighter who is in "great shape" can collapse quickly at a fire if fighting off the infection of a flu. Likewise, persons taking prescription drugs may experience rapid loss of strength or fatigue easier. The lasting effects of alcohol can also impair the firefighter well after consumption.

Physical fitness and wellness are highly individual—that is, firefighters are all different and have varied fitness needs based on genetics, gender, metabolic rate, and personal history (injuries, tolerance levels, etc.). However, firefighting contains some universal demands that require a certain level of fitness and wellness. Specifically, firefighters need to create a fitness and wellness lifestyle that includes the following:

- *Cardiovascular conditioning.* Without a doubt, this is critical. Firefighters seldom operate at incidents with a resting heartbeat. Repeated conditioning at elevated heart/breathing rates strengthens the heart/lung relationship and reduces the chance of overexertion at incidents.
- *Core strengthening.* The human core can be defined as the muscle structure that supports the hips, back, and thoracic cavity. This "power center" is used in virtually all physical activity at an incident. Sit-ups, crunches, and balance-oriented exercises (using a fitness ball) help strengthen this core.
- *Flexibility improvement.* The stress on joints, tendons, and ligaments can be brutal in the firefighting world—strains and sprains are the leading firefighter injury. Following a brief cardiovascular warm-up, firefighters should apply stretching and flexing routines to help condition joints, tendons, and ligaments.
- *Resistance training.* Firefighting is labor intensive. Resistance training (weight lifting) helps build muscle that makes the firefighting labor demand easier on the body. While many strategies exist, resistance training that systematically works *all* major muscle groups is best and will help keep postural balance.
- *Nutritional balance.* Perhaps the most neglected area of fitness and wellness of firefighters is the daily intake of food. Although this highly personal topic is fodder for much debate, most fitness experts agree that the key is to eat less but eat more often. Eating a balance of carbohydrates, protein, and fats along with a programmed intake of supplements helps the firefighter find nutritional balance. Firefighters working a shift should stay well hydrated by drinking plenty of water throughout the shift. Firefighters who are on call or in volunteer systems should keep drinking water close and hydrate en route to the fire station or incident.

Energy and Rest

Safety is directly affected by the energy potential and rest a firefighter has stored. Obviously, if the firefighter is short on sleep, the physical and mental capacities are reduced. Likewise, if it has been more than three hours since a meal or the food (energy) in the stomach has been used up, the firefighter will reach a fatigued state quicker.

> **Caution** For most firefighters, the worst time to respond to an emergency is just after midnight, when they have not received much sleep after a long day, it has been five or six hours since the last meal, and the mind and body are in repair mode.

What is the solution? The answer lies in the firefighter's awareness. The firefighter should be attentive to energy and rest levels and plan to communicate an early rehabilitation need to an officer during operations.

> **Streetsmart Tip** One simple suggestion: Firefighters should keep a balanced energy bar (protein, carbohydrate, and fat) in the apparatus or car and eat part of it during their response. A half a bar of protein, carbohydrate, and fat will serve the body well if it has been more than three hours since the last meal. Hydrating prior to incident engagement, however, is essential. Drinking some water while en route helps maintain heart stroke volume when sudden work demands are required.

SAFETY AT INCIDENTS

With roughly half of all duty-related injuries and deaths happening at the incident scene, it makes sense for each firefighter to develop a system to follow to minimize risk of injury or death. It is important to note here that *nobody* tries to get injured at an incident—the circumstances of the incident have created the potential. In most cases, the individual receiving the injury or the other firefighters and fire officers have failed to proactively "see" the events that led to the injury. Additionally, injuries and death occur when firefighters allow the urgency of the situation to override prudent judgment. What is being said here is valid—injuries can be prevented. At the individual level, a firefighter can prevent injuries and even death through a series of mental and physical actions. This section shows these mental and physical actions.

Team Continuity

> **Note** The success of all incident handling is dependent on trained, assembled teams accomplishing organized tasks, **Figure 23-5**.

Departments that allow well-meaning firefighters to perform as each sees fit will ultimately experience a duty death or significant injury. The fire service has coined the term *freelancing* for this type of action. Freelancing is dangerous. Countless investigative reports cite freelancing as a significant contributing event to death or injury. To eliminate freelancing, each and every firefighter must be assigned to a team of two or more and be given spe-

Figure 23-5 Freelancing is eliminated and incident success is gained when specific tasks are assigned to teams of two or more firefighters operating from a single incident action plan.

cific assignments. On completion of a task, the team must report that the task is done and also report any update information that could be useful to the incident commander. The incident commander will then reassign the team or have the team report to staging or rehabilitation.

Additionally, the team must exercise guarded judgment when completing a task. For example, a team performing overhaul may decide that an additional tool is needed. The simple decision is to send a lone firefighter to retrieve the tool, but this sets up a situation that breaks team continuity. Rather than sending a lone firefighter, it may be prudent to send two, leaving two to continue task completion. In the case of a three-person crew, the team leader should weigh the potential of breaking the team. In most cases, it is best to keep the team together and ask for a shuttle team to bring the needed tool or to take a momentary break and have the team exit the area together, get the tool (and perhaps a drink of water), and then return. This process may seem inefficient, but it underscores a guarded, safe approach to team continuity. **Figure 23-6** outlines practices that contribute to team continuity.

FIREFIGHTER SURVIVAL

Keys to Team Continuity

- Know who is on your team and who is the team leader.
- Keep your team in *visual* contact at all times. If visibility is obscured, be within touch or voice distance of your partner.
- Communicate your needs and observations to the team leader.
- Rotate to rehab and staging as a team.
- Watch your team members—practice a strong "buddy-care" approach.

Figure 23-6 Team continuity reduces the chance for injuries.

Figure 23-7 Late-arriving teams can benefit from a quick, formal safety briefing prior to performing tasks.

Orders/Communication

The incident commander is responsible for assembling an **incident action plan (IAP)** that is implemented by teams performing tasks. These tasks are assigned to organized teams in the form of orders. Each team is responsible for carrying out the order and providing updates on a regular basis. Additionally, the team must relay information about any pertinent hazards or conditions that may be important to the overall IAP.

> **Safety** As individuals, firefighters must keep their team leader advised of conditions and hazards they find as work is performed.

The discovery of a weak stairway, of a significant weight load in an attic or on the roof, or of fire in a void space are all important facts that should be relayed to the team leader and subsequently up the chain of command of the incident management system (IMS).

Occasionally, a team performing an order is given a different order by someone else in the IMS. In this case, the team leader needs to inform the person giving the new order that they are already under orders. The person giving the new order then must either countermand the original order or find someone else to perform the new order. In these cases, it is *imperative* that both the team and the person giving the new order communicate any changes to the person giving the original order.

Typically, the first arriving group of teams and apparatus performs a prescribed set of tasks or orders based on a preplanned or standard set of procedures. In these cases, the teams should know not only the tasks, but also the tools required and any safety considerations. An official safety briefing is usually not held and teams carry on their task. This does not mean that team members should not vocalize safety thoughts among themselves—in fact, this should be encouraged. Building construction considerations, fire behavior and smoke observations, and access/egress routes should be discussed as the team approaches its task.

Later-arriving teams are given assignments based on the IAP. These teams should ascertain essential information and hold a quick, formal safety briefing based on the conditions present and the hazards that are faced, **Figure 23-7**. This practice is not new—wildland fire crews are *required* to perform a crew briefing prior to task accomplishment. Most proactive fire departments are initiating these safety briefings as a matter of habit. **Figure 23-8** lists good habits regarding orders and communications.

Good Habits Regarding Orders and Communications

- Clearly understand any order; ask questions and seek clarity.
- Assemble tools and openly discuss procedure as you approach tasks.
- Relay all pertinent hazards found during the course of task completion.
- Update crew status on a regular basis.
- Advise when tasks are complete.
- Communicate rotation to rehab and/or staging.

Figure 23-8 Practicing good reporting habits enhances communications.

Risk/Benefit

Another key to firefighter survival is the concept of risk/benefit. Simply stated, **risk/benefit** is an evaluation of the potential benefit that a task will accomplish in relationship to the hazards that will be faced while completing the task. As an example, the task of vertically ventilating a residential structure fire will benefit the fire suppression and lifesaving effort by increasing visibility, reducing heat buildup, and reducing overall damage. That benefit, however, is at the risk of roof collapse, operating on a potentially steep incline, or operating without quick egress. The team operating on the roof is often in the best place to judge this relationship. Signs of collapse, difficult footing, and obscured visibility may present a significant risk to the team. The risk/benefit evaluation in this example may cause the team leader to withdraw the team.

Risk/benefit evaluations take place at different levels within an IMS. At the command level, an overall evaluation takes place, as reflected in the IAP. At the team level, crews evaluate their immediate environment and report this back to the person giving them their orders, **Figure 23-9**. Regardless of the level of evaluation, some basic guidelines can be used to help make risk/benefit decisions:

- Firefighters will take a significant risk to save a known life.
- Firefighters will take a calculated risk, and provide for additional safety, to save valuable property or reduce the potential for civilian and firefighter injuries.
- Firefighters will take no risk to their safety to save what is already lost.

These guidelines are common within the fire service and are more clearly spelled out in NFPA Standard 1500, *Fire Department Occupational Safety and Health Program*.

Personal Size-Up

As mentioned several times in this book, firefighters must perform a personal size-up. A size-up is a continuous mental evaluation process. Specifically, firefighters should continually evaluate the safety of their environment by staying aware of the following:

- *Established work areas*. Firefighters have to be aware of collapse zones, barriers to traffic, and hazard isolation zones.
- *Hazardous energy*. Electrical equipment, pressure vessels, chemicals, and even springs and cables can suddenly release—causing injury and death to firefighters. It is important to *scan* the environment and mentally log all forms of hazardous energy.
- *Smoke conditions*. Smoke is fuel waiting to ignite. Firefighters crawling in low-visibility, hot smoke are flirting with an explosive environment. Cooling and ventilation are key.
- *Escape routes*. The firefighter who continually plans two or three escape routes while performing tasks is investing in survivability if the established work area breaks down. The firefighter should count doors and windows, scan for traffic barriers, and plot pathways to "safe havens."

Rehabilitation

> **Note** A study of firefighter duty-related injury and death statistics shows that stress and overexertion consistently rank as leading causes.

Firefighting is hot, arduous work performed in PPE that does not allow body heat to evaporate through sweat. A key health and safety concern is controlling the heat stress that occurs when a firefighter's internal core temperature rises above its normal level during incident activities. Core temperatures that have risen above normal can lead to heat- and heart-related injuries and death. Proper rehabilitation efforts (*rehab* for short) will help keep firefighters safe by reducing heat stress. Additionally, good rehab practices will provide the rest, hydration, and nourishment needed for sustained opera-

Figure 23-9 Solid risk/benefit analysis means taking no risk for that which is already lost. *(Photo courtesy of Richard W. Davis)*

tions. NFPA addresses specific practices for rehab in its document, NFPA 1584, *Rehabilitation of Members Operating at Incident Scene Operations and Training Exercises.*

> **Note** The key elements of rehab include rest, active cooling, hydration, and nourishment.

Rest

At an incident, rest is achieved during crew rotation. When a firefighter is rotated to rehabilitation, the firefighter should maximize the opportunity to rest by sitting down, by allowing medical personnel (which should be assigned) to do vital sign check, **Figure 23-10**, and by mentally disengaging from the event (which rests the mind). The firefighter who achieves these three important steps can actually work longer at any given incident—it allows a brief recovery from physical and mental stress and allows other rehabilitation elements (hydration and nourishment) to repair and prepare the firefighter for further work. The medical evaluation by a trained medical provider is an important part of rest. Many signs and symptoms indicate the need for further rest—or removal from the incident scene.

One additional point can be made here. Some firefighters do not feel the need for rehabilitation and may be reluctant to spend time in rehabilitation while others are aggressively engaged in the incident—these firefighters are potential candidates for overexertion. Remember, the time to rehabilitate is *before* a firefighter gets tired. "Rehab early and often" should be the fire service creed, especially if the incident duration will span hours, **Figure 23-11**.

Active Cooling

As mentioned at the start of this section, increased core temperatures can cause heat- and heart-related injuries and death. During rehab, firefighters should engage in active cooling to reduce their core body temperature. Typically, firefighters have used passive cooling as a strategy to reduce core temperatures. Passive cooling includes the use of shade, air movement, and rest to bring down core temperatures. Medical personnel have typically monitored heart rate and a firefighter's perceived comfort to determine when a person has recovered sufficiently to don PPE and resume firefighting operations. New studies have shown that this strategy is *not* adequate and that heart rate and perceived comfort are not good indicators of sufficient core temperature cooling.[1] Effective core temperature cooling is achieved using an "active cooling" strategy. Active cooling is best achieved using a forearm immersion technique. Simply put, firefighters should doff their coats and submerge their hands and forearms in a basin of cool water. This technique is more effective than misting fans and has a reduced tendency to cause the chills that can happen when firefighters go immediately from hot environments to cold environments.

Figure 23-11 Rehabilitation should start well before a firefighter is thirsty or tired. Failure to rehabilitate "early and often" opens the door to injury.

Figure 23-10 Effective rehabilitation includes a medical evaluation by BLS or ALS personnel.

Hydration

Hydration cannot be overemphasized in any incident environment, whether it is one of heat stress, cold stress, heavy workload, light workload, mentally taxing events, or long-duration events. Water is vital to the peak operation of virtually every body system from transport of nutrients, to blood flow, to waste removal, to temperature regulation. When the body becomes dehydrated, these systems start to shut down in order to protect themselves. With this shutdown comes fatigue, reduced mental ability, and, in extreme cases, medical emergencies such as renal (kidney) failure, shock, and death. The working firefighter must accommodate for the wearing of heavy clothing that does not allow evaporation of sweat. Additionally, the firefighter must account for strenuous physical activity under stressful situations. Hydration of firefighters should become paramount—even to the point of excess. Firefighters must not *wait until they are thirsty to drink water!*

> **Safety** As a rule, firefighters should strive to drink a quart of water an hour during periods of work—this is best delivered in 8-ounce increments spread over the hour.[2]

Substituting carbonated and/or sugared beverages or other liquids for water can slow the absorption of water into the system. For this reason, just water should be given for the first hour. For activities lasting longer than an hour, some consideration can be given to adding essential electrolytes and nutrients along with the water. Many sports drinks are available that can achieve this. These sports drinks are best diluted 50 percent with water in order to speed their absorption into the system.

Nourishment

While dehydration and thermal stress can lead to energy depletion, most firefighters associate energy depletion with the need for food. Nutrition or "fueling" of the firefighter can have the effect of rejuvenating the firefighter or putting the firefighter to sleep. Too often, rehabilitation-feeding efforts accomplish the latter. A firefighter properly nourished will work smarter and safer. The firefighter who is improperly fed will not only want to "crash," but will likely make sluggish mental calculations leading to injury.

So what is the proper way to nourish firefighters? A brief study of essential nourishment theory can help answer this question. In basic terms, maximizing energy from the human machine takes a balance of four essential elements: oxygen, water, blood sugar (from food), and insulin. When this balance is present, other essential hormones and enzymes are created that make for a well-running human machine. The element that is often misprescribed is food—the foundation for building balance. True, *all* food is fuel for the firefighter. That is, all food is converted to glucose (blood sugar) for the muscles to use as fuel. Insulin is released into the bloodstream to help convert the blood sugar into energy for muscle use. Some foods are digested faster than others though. Complex carbohydrates, such as breads, pasta, and beans, require tremendous amounts of blood, and time, for digestion. This explains why some firefighters want to go to sleep after eating—blood is needed for digestion—and less energy is available for thinking and physical tasks.

The key to providing quick energy to the firefighter is to find a balance of protein, fat, and carbohydrate. Ideally, this balance should be a 30/30/40 mix—that is, 30 percent protein, 30 percent fat, and 40 percent carbohydrate.[3] This balance provides essential elements from three food groups. A balanced approach achieves a few benefits. One, the balance will stabilize insulin release into the bloodstream, helping to reduce the rollercoaster of blood sugar levels that often lead to sporadic activity, chemical imbalance, and fatigue. Second, the balance approach stimulates the release of hormones and enzymes that optimize human performance—both physical and mental.

Choosing the best protein, carbohydrate, and fat also promotes steady, sustained performance. Protein is best derived from low-fat meats such as turkey, chicken, and fish. Eggs and cheese also offer protein. Fats should be of the monounsaturated type like olive oil, nuts, and peanut butter. Often, carbohydrates are dangerous in that so many of the foods typically found on the fire scene are rich in unfavorable carbohydrates. Candy, bread, potatoes, and bananas are all carbohydrates that are quite rich and have a tendency to slow down the worker. Good carbohydrates include green vegetables, apples, tomatoes, and oatmeal.

A good example of a quick, balanced rehabilitation meal would include water, sliced turkey sandwiches on thin-sliced bread with mustard or ketchup, apples, and a handful of peanuts. Low-sugar energy bars can be stocked on apparatus and used as a meal replacement. It is best to use energy bars that are balanced 30/30/40.

Rapid Intervention Teams

Rapid intervention teams (RIT) are being formed by many fire departments as a proactive practice to rescue lost or trapped firefighters. NFPA 1500, *Standard for Fire Department Occupational Safety and Health Pro-*

gram, and NFPA 1561, *Standard for Emergency Services Incident Management System,* both require the formation of a RIT early in the incident if firefighters are exposed to dangerous environments or tasks.

As mentioned in Chapter 19 (two in/two out rule), the initial RIT includes the two "out" team members that support the two "in" members. This requirement is specific to interior fire attack but should be conceptually used for all incidents where firefighters are taking risks. As more resources arrive, the RIT becomes a designated team for a defined task. This RIT should have no other assignment than to prepare for the rapid deployment in support of a search and/or rescue.

> **Streetsmart Tip** Firefighters assigned to the RIT should be prepared to act immediately on orders to rescue lost or trapped firefighters. This means the crew assigned the RIT task should be equipped and prepared for the task.

When preparing for the RIT assignment, firefighters should gather appropriate tools and equipment such as SCBA (donned and in the ready position), forcible entry tools (including a power saw), hand lanterns, radios, and a life rope. Additionally, the RIT crew should start preplanning a rescue operation. Information about building construction, layout, and entry/egress routes should be shared. The team should also inquire as to the IAP and current crew location and assignments and should monitor radio traffic. As a rule, the RIT should position near the main entry point of the incident hot zone and be ready for immediate deployment.

The radio transmission of "Mayday" or "Emergency traffic" should cause the RIT to size up the situation and mask up. Typically, it is best for the RIT to await orders from the incident commander to initiate the rescue/search, unless the IC has given the RIT the responsibility to self-initiate actions. Regardless, the RIT needs to communicate its actions on a frequent basis. **Figure 23-12** can serve as a RIT checklist for the firefighter assigned to a rapid intervention team.

FIREFIGHTER EMERGENCIES

So far, this chapter has focused on preventing injuries and deaths through proactive actions. Occasionally, preventive measures do not address every circumstance or the evolution of the incident presents the working firefighters with an emergency that

Rapid Intervention Team Checklist

☐ Don PPE/SCBA

☐ Assemble team tools including:
  • Hand lanterns
  • Radio
  • Prying tools/ax
  • Lifeline rope
  • Thermal Imaging Camera
  • Lifting Equipment
  • Spare Air Supply

☐ Inventory building/environment
  • Construction type
  • Egress/entrance
  • Weak links
  • Layout/hazards

☐ Receive briefing on incident action plan and crew assignments/locations.

☐ Be ready by listening to the radio and staging near primary access point.

Figure 23-12 Firefighters assigned to a rapid intervention team should increase their readiness using a "checklist" approach.

threatens their lives. In these cases, each firefighter must rely on instincts and training to help survive the situation. To help understand the actions to be taken during an actual or potential firefighter emergency, the firefighter must study procedures for rapid escape and declaring a **Mayday** for lost and trapped situations. Further, the firefighter must understand that survival includes processes for the rescue of trapped and lost firefighters as well as long-term mental survival.

Rapid Escape

During the course of task completion, firefighting teams may discover a situation that requires immediate escape or may be ordered to do so via radio and fireground warnings. The firefighting teams must then take immediate steps to exit the building or environment.

> **Caution** The two most common situations that will lead to this rapid escape need are rapid fire spread and building collapse. An event such as flashover, backdraft, or partial collapse of a building can lead to rapid fire spread, potentially trapping or endangering firefighters, **Figure 23-13**.

Figure 23-13 Rapid fire spread and partial collapse are likely to trigger the need for rapid escape.

A building being attacked by fire is always subject to collapse. Once this attack degrades the load-bearing portions of the structure, a collapse will occur. Teams that witness these events must take steps to initiate an evacuation. Teams that hear an established evacuation signal must also immediately egress. Some typical evacuation signals include a repeated series of air horn blasts from apparatus or a special radio tone followed by the evacuation order. The next section discusses some important steps that will assist in achieving immediate escape.

Rapid Escape Steps

1. *Preplan the escape.* All firefighters operating in a hazardous environment should continually look for multiple escape routes. Even in routine or nonthreatening environments, the firefighter should be constantly evaluating escape routes. This process develops good habits that will serve the firefighter well when a threatening situation does develop.
2. *Immediately report the need for rapid evacuation.* Actual collapse or signs of impending collapse can be communicated via radio while exiting the area. A delay in reporting these conditions could endanger other crews. Likewise, the witnessing of a flashover (or signs of an impending flashover or other hostile fire event) should be cause for immediate reporting.
3. *Acknowledge rapid evacuation or escape signals.* In cases where the team has not initiated the rapid escape, the team must acknowledge that it has heard the signal. Signals vary from department to department but typically consist of an audible fireground signal (a defined air horn signal or a barrage of sirens and air horns) as well as a radio broadcast (an alert tone followed by the evacuation order).
4. *Rapidly escape.* Rapid escape means just that—firefighters should leave the area immediately. They should leave hoselines and heavy tools and escape. Trying to bring these types of items with them has resulted in injuries to many firefighters. Often, it is faster for the team to use a way out that is different from the way it entered. This is where the team's escape preplan observations pay off. At times, firefighters will be faced with a situation that requires a temporary escape—in these cases, the priority is to protect the crew and plan for further escape. Here are some examples:

- *Room escape.* It may be possible to enter an uninvolved room, shut the door, and use a window for further escape. If the window is above the first floor, the firefighter may have to clear the window and wait for a ladder. These actions should be communicated if it is not obvious to crews outside. If the room has no window, the firefighter should think through the options. Often, a simple wall breach (drywall over wood studs) can get the team to rooms with alternative escape routes.

- *Roof/balcony escape.* When fire or collapse has cut off stairwells and hallways, it may be easier to get to the roof or balcony and await ladders. In these cases, it is best to close doors or create barriers that separate the team from the fire. It is essential to communicate the team's needs if the firefighters assisting the team are not in visual contact.

- *Self-rescue.* In some extreme cases, it might be possible to escape through a system of self-rescue. Many firefighters carry a personal rescue rope or strap so they can lower themselves to another floor. Firefighters equipped and trained to do this need to establish a barrier to fire spread, secure a reliable anchor point, have a device or method to descend the rope, and have a method to break windows on lower floors.

- *Ladder escape.* When a ladder is in place and available for rapid escape, firefighters should use safe and common methods to mount and descend the ladder. In hurried situations (like collapse and flashover conditions), firefighters may opt for dangerous headfirst mounting and ladder slides. It is important to note that these techniques can lead to serious injuries or death. Locally, fire departments may choose to instruct firefighters on these techniques. These skills are highly technical and require close supervision and instruction for practice. A firefighter should *not* attempt to practice these without an experienced instructor who uses protective systems.

5. *Report successful escape.* Once the team has escaped the building or area, members should report that the team is safe and accounted for. This process is usually accomplished with a fireground **personnel accountability report (PAR)**. A PAR is an organized approach to accounting for everyone on scene, **Figure 23-14**. Typically, the accountability manager runs through all the assignments given thus far and asks (usually via radio) for each crew to report its status. From this, unaccounted teams are identified and rapid intervention is initiated. Many departments perform a PAR at given intervals, such as when dispatch announces time intervals (every 10 to 30 minutes). Most departments perform a PAR after an emergency or when the incident operation changes modes or strategies (for instance, switches from offensive to defensive operations).

Lost, Trapped, and Injured Firefighters

Firefighting operations place crews in environments in which they have potentially never been—large buildings with unusual floorplans, small confined areas with minimal escape routes, and a whole host of stock, furniture, and arrangement obstacles. With this comes the potential to be trapped or lost during fire suppression, search, or rescue assignments. While most firefighters will agree that an entrapment is a true firefighter emergency, fewer believe that being lost is an emergency. They believe that, eventually, they will find a way out. This thinking needs to be altered.

> **Caution** A firefighter or crew that has lost spatial bearings in an IDLH atmosphere is experiencing a firefighter emergency!

With entrapments and lost situations, it is important to have a mental process to follow to avoid complicating the situation.

Entrapments

The first step a firefighter should take in an entrapment is to get assistance. Activation of a PASS device is warranted and the declaration of a "Mayday" should be made over the radio. Some radios are equipped with an emergency assistance button. It should be activated if the radio is so equipped. The Mayday will be followed up with radio procedures and communications to get assistance to the trapped firefighter—often via the rapid intervention team. The firefighter should follow this up with other noise-making efforts. Banging on a pipe or throwing debris may be helpful, but the firefighter must be careful not to use up energy or excessive air in doing this. Using visual signals such as a flashlight may also prove helpful. The second step in dealing with the entrapment is to size up the situation and develop a plan. Some key questions to answer include:

- What exactly is causing the entrapment?
- What is the exposure to fire/smoke/further collapse?
- How much breathing air is left in the SCBA bottle?
- What are the extent of injuries?
- Is there anything that can be done to self-extricate?
- Is there any self-first-aid that can be performed?
- How can air be preserved?

The trapped firefighter should attempt self-extrication. This must be planned and systematic as opposed to reckless and panic driven. All of these steps accomplish two important points. First, a process is created to help with rescue. Second, the firefighter's mind is kept active to help ward off panic.

Figure 23-14 A PAR is a personnel accountability report organized to check the status of all crews working an incident. PARs should take place every half-hour or after an evacuation or any firefighter emergency.

Lost Firefighters

It cannot be overemphasized that a firefighter or team lost in an IDLH atmosphere is in fact experiencing a firefighter emergency. The steps to overcome this emergency are simple and can result in a quick resolution prior to an injury. First, the firefighter or team must report the fact they are lost. This is also a Mayday situation and should be transmitted as such over the radio. The Mayday will lead to radio traffic trying to ascertain last known position and any clues that might help a RIT crew locate the firefighters. Second, the lost firefighters need to manually activate their PASS devices. Finally, the firefighters need to stop and take inventory of their surroundings. From this, they can establish direction, door and window locations, and potential paths. As with firefighter entrapments, some questions should be answered:

- What is the exposure to fire/smoke/further collapse?
- How much breathing air is left in the SCBA bottle?
- How can air be preserved?
- Are there other options that have not been explored?

It is important to maintain radio contact with the RIT members and the incident commander. A team that is lost needs to help each other remain calm and preserve air. Any member that appears to panic can have a negative effect on the others—the team should exercise steps to keep minds open.

Injured Firefighters

Perhaps the most stressful situation that will ever face a firefighter is the realization that a fellow firefighter has become seriously injured. Regardless of what caused the injury, firefighters will typically drop whatever task they have been assigned and rush to aid their fellow firefighter, **Figure 23-15**. In some cases, this rapid focus on assisting a coworker becomes heroic due to extreme dangers faced by the now-rescuers and the firefighter victim. It should also be said that the potential to injure or kill many firefighters could be risked during firefighter rescues. One example of this unnecessary risk is the all-too-often scenario where firefighters rush to a debris pile to start rescue of firefighters caught under a collapse. A secondary collapse occurs and compounds the event. To minimize this risk, firefighters must trust RITs. The RIT is designed to attempt the rescue and remove the victim firefighter from the hazard so appropriate medical care can begin.

Once a firefighter is found to be trapped or injured, the RIT should be activated. It is natural for

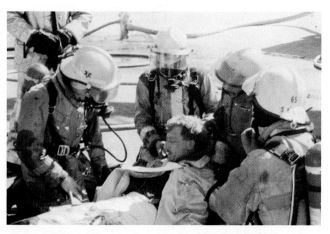

Figure 23-15 A serious firefighter injury or fatality will cause significant incident stress. Focus and use of RITs will minimize unnecessary risk during firefighter rescue and help maintain incident control.

another team that is in proximity of the trapped firefighter or crew to engage in the rescue—this must be communicated! Collapse, exposure to the fire, and smoky conditions may exist and make the rescue difficult. Furthermore, the trapped firefighter or crew will be in an area that is hard to access by rescuers. Practicing the removal of firefighters from tight enclosed areas or through small windows is an investment that can save firefighters' lives.

Firefighters who are not on the RIT or in the immediate vicinity of the rescue should resist the urge to rush in and help. History shows that most firefighter rescues require the assignment of many teams. Firefighters should let the RIT take the lead and wait for their assignment. In this way, they are part of the plan—not the problem.

Post-Incident Survival

Once an incident has reached the point where crews are starting to clean up and be released, a relaxed phenomenon encroaches on the minds of the firefighters. This phenomenon has been called **post-incident thought patterns** and is responsible for many injuries and deaths. Common post-incident injuries include strains, sprains, and being struck by objects.

> **Note** The causes of these post-incident injuries seem almost ironic in a profession where aggressive and calculated risk taking is a hallmark.

Incidents that are especially gruesome or involve significant human tragedy can easily impact firefighters in an emotional sense, leading to long-term mental and health issues. This is called *critical incident stress* (CIS). It is important that firefighters address post-incident thought patterns as well as CIS.

Post-Incident Thought Patterns

One cause of post-incident injuries has to do with the little-studied notion of post-incident thought patterns. In essence, this is inattentiveness. In cases of especially difficult, unusually spectacular, or particularly challenging incidents, firefighters will tend to reflect on their actions. The replay of the incident starts almost instantly when the order is given to "pick up." This introspection is normal. The switch from activities requiring brainpower and physical energy to an activity that is so routine so as to be dull is a hard one to adapt to. Herein lies the problem. Some signs that post-incident thought patterns are affecting crews include faraway stares, firefighters wanting to be alone, and firefighters who stop and look about as if they have forgotten their task.

> **Note** Simple safety reminders or jocularity can help firefighters regain focus and reduce injury potential.

One method to reduce the impact of these thought patterns is to take a time-out and have everyone gather for a quick incident summary and safety reminder, **Figure 23-16**. These huddles can be effective for everyone or just small groups.

Another factor to consider that leads to inattentiveness is chemical imbalance. Even the most successful rehabilitation program cannot prevent firefighters from experiencing fatigue and mental drain. With the end of an incident, especially one requiring major physical effort, comes the relaxation of the firefighter's mind, which, in turn, starts shutting down protective chemicals that stimulate performance. The adrenaline "rush" is over and the firefighter's metabolism will return to a "repair" state. This causes a mental slowdown that can lead to unclear thinking and result in injuries.

Another way to see this chemical and mind imbalance is to look at a firefighter's tools from a layperson's point of view. Most laypersons would literally have to concentrate on carrying an ax, pike pole, chain saw, roof ladder, or hose length in order to not hurt themselves or anyone around. Consider also doing these tasks in bulky, restrictive clothing and heavy boots. The firefighter does these things after incredible energy bursts under frightening conditions. Familiarity plays a certain role here, but concentration is still required. If the mind has been taxed, the body has been fatigued, and the signal to "relax" has been given—yet the concentration required to do the task has remained the same—the potential for injury rises.

Whether the issue is chemical imbalance or post-incident thought patterns, the firefighter needs to stay alert and try to pick up signs of potential injury and take steps to "survive" without injury.

Critical Incident Stress

Firefighters are expected to tolerate a certain level of incident stress given the nature of the environment that firefighting brings. Dealing with community hazards, injuries, and death can even be classified as normal for the profession. Some events, however, can trigger a significant emotional response from responders. These emotional responses may not always be external. In fact, many firefighters harbor the reaction internally. Events that typically induce an emotional response include:

- Death or serious injury to a coworker or a person known to the responder
- Mass casualty incidents
- Death or serious injury to a child, especially from an act of violence or crime
- A prolonged rescue attempt that results in a death

Firefighters will exhibit signs of CIS in many ways. **Figure 23-17** outlines a few of these signs and symptoms. It is important to note that none of these signs is abnormal—emergency responders will continue to feel the stress of critical incidents. What is important is to "survive" the CIS through a process of critical incident stress management (CISM).

CISM can take on many forms, both formal and informal. Formal CISM is usually accomplished following an incident of significant magnitude—one affecting numerous responders. Typically, formal CISM involves a critical incident stress debriefing (CISD) process that is led by peers and health pro-

Figure 23-16 A quick huddle during post-incident activities can serve to remind firefighters to continue to work safe and *smart* during pickup activities.

Selected Signs and Symptoms of Critical Incident Stress

Behavioral: Increased irritation, aggressiveness, withdrawal, flashbacks, inattentiveness, alcohol/drug abuse, and memory loss

Physical: Loss of appetite, insomnia, fatigue, headaches, muscle stiffness, and hypertension

Psychological: Guilt, sadness, depression, career introspection, claimed burnout, fear, and unsociability

Figure 23-17 Signs of critical incident stress are natural following "trigger" incidents.

fessionals. The CISD should be mandatory for all responders and is designed to help everyone with immediate and long-term mental health needs. An informal process usually includes a "defusing" that allows persons to deal with an ongoing rescue or to voice concerns and thoughts of the incident in a peer environment. Defusing should take place just prior to releasing firefighters from the scene, or back at the station just after the incident, **Figure 23-18**.

To help manage critical incident stress as well as daily incident stress, the firefighter should use CIS processes and tested stress-reduction principles such as exercise, adequate and appropriate nutrition, deep breathing, deep relaxation (massage, meditation, etc.), hobby development, and positive social interaction. With these steps, the firefighter can survive post-incident stress in the long term.

Figure 23-18 Critical incident stress management sessions can be (A) informal such as a "defusing" or (B) formal such as a process that includes peer support and mental health professionals.

Lessons Learned

Firefighter survival is dependent on many proactive and preventive actions. These actions include incident readiness, safe operations at incidents, and appropriate preparation and response to firefighter emergencies. Incident readiness includes those efforts to prepare the firefighter's personal protective equipment, personal accountability, and individual fitness for duty. Of particular note, fitness for duty relies on the firefighter's mental and physical wellness as well as the firefighter's attention to energy and rest.

Safe operations and, therefore, survival are dependent on team continuity and the elimination of freelancing. Additionally, safe incident operations are achieved when teams perform orders with attention to communications and timely updates. While operations are under way, individual teams must practice solid risk/benefit analysis for their specific job. During the incident, crews must take advantage of the rest, hydration, and nourishment offered through established rehabilitation processes.

Rapid intervention teams are formed for immediate deployment should a firefighter emergency take place. To meet this challenge, RITs must assemble tools and be briefed on essential information that will assist in a more rapid reaction should an emergency occur. Firefighter emergencies require the individual firefighter to practice a clear and concise approach to dealing with the emergency. Using the term "Mayday" can signal everyone of the need for firefighter rescue. Whether trapped or lost, the firefighter must remain calm and follow a defined series of steps to mitigate the situation.

Finally, firefighters must survive long term through an understanding of post-incident thought patterns and the effects of critical incident stress. Understanding the factors that lead to each is a useful tool in dealing with the reality of emergency responder stress and can help reduce the potential for firefighter injury or death.

KEY TERMS

Freelancing   The act of working alone or performing a task for which the firefighter has not been assigned.

Incident Action Plan (IAP)   A strategic and tactical plan developed by the incident commander.

Mayday   A universal call for help. A Mayday indicates that an individual or team is in extreme danger.

Passport   A term given to a specific accountability system where crews are tracked using a card (passport) with all members listed. An accountability manager tracks the passports on an accountability board.

Personal Size-Up   A continuous mental evaluation of an individual's immediate environment, facts, and probabilities.

Personnel Accountability Report (PAR)   This is an organized roll call of all units assigned to an incident.

Post-Incident Thought Patterns   A phenomenon that describes an individual's inattentiveness following a significant incident. Post-incident thought patterns can lead to injuries or even death.

Risk/Benefit   An evaluation of the potential benefit that a task will accomplish in relationship to the hazards that will be faced while completing the task.

REVIEW QUESTIONS

1. List the three main components that lead to incident readiness.
2. Define the four key checks to ensure that individual personal protective equipment is ready for response.
3. Describe the advantages and disadvantages of three kinds of personnel accountability systems.
4. Define personal size-up.
5. Briefly describe the three components that lead to fitness for duty.
6. Name three practices that lead to team continuity. Describe how each can increase firefighter survival.
7. Define risk/benefit.
8. List and briefly describe the three components of on-scene rehabilitation.
9. Describe the procedures that should be taken to establish and prepare for the assignment of a rapid intervention team.
10. List the five steps that can lead to an organized rapid escape.
11. Describe the reasons why a lost firefighter is considered an emergency.
12. List the three steps that should be taken when entrapment occurs.
13. Compare and contrast post-incident thought patterns and critical incident stress.

Endnotes

1. McLellan, Thomas, PhD, *Safe Work Limits While Wearing Firefighting Protective Clothing.* Defense Research and Development Council, Toronto Fire Department Grant Study Report, Toronto, Ontario, Canada, 2002.
2. U.S. Fire Administration, *Emergency Incident Rehabilitation,* FA-114. USFA Publications, Washington, DC, July 1992.
3. Sears, Barry, *Enter the Zone.* HarperCollins Publishers, New York, 1995.

Additional Resources

Angle, James S., *Safety in the Emergency Services.* Delmar Learning, a part of the Thomson Corporation, Clifton Park, NY, 1999.

Dodson, David W., *Fire Department Incident Safety Officer.* Delmar Learning, a part of the Thomson Corporation, Clifton Park, NY, 1999.

Emergency Incident Rehabilitation, FA-114. U.S. Fire Administration, Washington, DC, 1996.

EMS Safety: Techniques and Applications, FA-144. U.S. Fire Administration, Washington, DC, 1996.

Firefighter Fatality Investigations, http://www.cdc.gov/niosh

LeCuyer, John, *Designing the Fitness Program—A Guide for Public Safety Organizations,* Fire Engineering Books & Videos, a Division of PennWell Corporation, Saddlebrook, NJ, 2001.

NFPA 1500: Standard on Fire Department Occupational Safety and Health Program, 1997 Edition. National Fire Protection Association, Quincy, MA, 1997.

NFPA 1561: Standard on Emergency Services Incident Management System, 2000 Edition. National Fire Protection Association, Quincy, MA, 2000.

NFPA 1584: Rehabilitation of Members Operating at Incident Scene Operations and Training Exercises, 2003 Edition. National Fire Protection Association, Quincy, MA, 2003.

Risk Management Practices for the Fire Service, FA-166. U.S. Fire Administration, Washington, DC, 1996.

CHAPTER 24

TERRORISM AWARENESS

Christopher Hawley, Baltimore County Fire Department (Ret.)

OUTLINE

- Objectives
- Introduction
- Types of Terrorism
- Potential Targets
- Indicators of Terrorism
- HAZMAT Crimes
- Incident Actions
- General Groupings of Warfare Agents
- Detection of Terrorism Agents
- Federal Assistance
- Lessons Learned
- Key Terms
- Review Questions
- Endnote
- Additional Resources

STREET STORY

The fire service is made up of a large, tight-knit family, and the special bond we have is carried over into the other emergency services.

On September 11, 2001, our family and our nation were attacked in a manner that was unprecedented. After the attacks in New York and Arlington, Virginia, and the plane crash in Shanksville, Pennsylvania, America was at war. Terrorists had struck, and in a few precious minutes we suffered a grievous loss. America changed that day, and the emergency services took a severe hit, but we persevered. During the first few minutes of the fire in the World Trade Center my thoughts were that it was a large fire, and that it would be a tough battle, but one we could win. When the towers collapsed, my heart sank. I knew that I had just witnessed hundreds of my family members being struck down. My thoughts were immediately for my good friends who work in FDNY and for all of the other emergency responders at that scene.

The events of that day caused the deaths of many emergency responders and an enormous amount of emotional damage to the survivors, family members, and population of the United States. But on that day, when citizens across the country dialed 9-1-1, emergency responders showed up . . . we proved that we were not defeated. We are emergency responders and the keystones of our communities. We are the first line of defense against terrorism in this country. We must remain focused on our mission, which is to protect the citizens in our community. With that said, we as emergency responders must look at the real threat that terrorism presents to our communities.

There are many events that occur in this country that present substantial risks to you as a responder and to the community. The events discussed in this chapter are events that can have catastrophic consequences for your community. You must prepare for these events and become tuned in to the potential for terrorism or other criminal behavior in your community. Every community has the potential to be attacked or to be a base of terrorist operations. Read the listing of the case histories, and you will learn that all communities are at risk. As you respond in your community, remember those who served on September 11, 2001, and dedicate your career to being the most knowledgeable and prepared firefighter that you can be. . . You never know what the future may bring.

—Street Story by Christopher Hawley, Baltimore County Fire Department (Ret.)

OBJECTIVES

After completing this chapter, the reader should be able to:

- Discuss potential target locations.
- Discuss indicators of potential terrorist activity.
- Describe incident actions to be taken at a terrorist attack.
- Describe additional hazards at a terrorist attack.
- Describe other specialized resources to assist with a terrorist attack.
- Describe methods of requesting federal assistance.
- Identify common agents that may be used in a terrorist attack.

Figure 24-1 In 1993 a van was used to carry explosives to an underground parking garage in the World Trade Center. Six people were killed and more than 1,000 were injured. There were 50,000 people in the building, and the goal of the terrorist was to collapse the building into the adjacent tower.

INTRODUCTION

It is unfortunate that a chapter on terrorism needs to be included in this firefighting text, and until recent times this would not have been necessary. Although everyone has seen terrorism on the evening news, it used to occur in places such as Northern Ireland, Beirut, Israel, or somewhere other than the United States. Until recently, the United States remained for the most part immune to the reign of terrorists. This changed in February 1993 when the World Trade Center was bombed in New York City, **Figure 24-1**. Even when this bombing occurred, the fire service did not pay much attention, because the bombing was looked at as the type of incident that happened only in big cities. In addition, the persons who were found to be responsible for the bombing were controlled by an influence from outside the United States, so the thought was that it was an isolated foreign attack.

However, when the Alfred P. Murrah Federal Building in Oklahoma City was devastated by a bomb on April 19, 1995, the United States fire service took notice. The attack on the federal building, **Figure 24-2**, was brought on by a person who did not have ties to another country and was a natural born citizen of the United States. It was perceived as an attack on America from one of its own, not from some unknown citizen from a foreign nation.

And perhaps the most devastating attack on the United States occurred on September 11, 2001, when the World Trade Center and the Pentagon were hit with a total of three airplanes. Then, in Shanksville, Pennsylvania, a hijacked plane crashed into a field. As a result of the impact of the planes and the resulting fire, the 110-story twin towers of the World Trade Center collapsed. The death toll from all three sites was more than 2,600 civilians and 346 emergency responders.

Shortly following three horrific crashes, in October 2001 a flood of noncredible anthrax letters hit the country. Mixed in with the thousands of noncredible letters were several letters containing real anthrax. These letters were responsible for five deaths and up to thirteen other illnesses. The letters were sent to news media outlets and to members of Congress.

Regardless of the origin, the potential for terrorism is here in this country and has to remain in firefighters' thoughts as they respond to any incident. Other incidents are occurring on a regular basis, and, although they do not fit the exact definition of terrorism, they involve the use of large-caliber weapons or a large number of weapons. This chapter looks at terrorism, **HAZMAT crimes**, and other potentially dangerous criminal situations that are sometimes just as deadly as terrorism. Firefighters have to think outside the box in regard to terrorism. There are many events that occur in this country that place responders at risk. Terrorism, HAZMAT crimes, murders, and other criminal events all present a risk to firefighters. Someone using ricin to kill another person is committing murder, but the weapon is as deadly to the responders as it is to the intended victim. A booby-trapped drug lab is not terrorism, but it presents significant risk to responders. Many criminals are protecting themselves with

Figure 24-2 A truck bomb caused the devastation in the Oklahoma City bombing in which 167 people were killed and 759 injured. The damage extended several blocks in each direction, and 300 buildings were damaged. Fatalities occurred in 14 separate buildings. *(Photo courtesy of John O'Connell)*

body armor and fortified vehicles. Many crimes such as bank robberies involve the use of explosives and sophisticated weaponry. Street violence that can be associated with gangs is increasingly violent, and when people are killed or injured the fire service is called into action. Crimes such as assaults and murders are increasing in the school system. In recent times there have been deadly riots in Los Angeles and in St. Petersburg, Florida, usually after a sporting event. In Chicago, twenty-one people were killed and fifty were injured in a nightclub after a fight broke out. These deaths may have been caused by a lack of exits and the use of pepper spray by security personnel, which panicked the crowd. The workplace is also becoming an increasingly dangerous place to be. A firefighter in Jackson, Mississippi, entered the headquarters fire station intent on killing the fire chief. Several persons were killed, not to mention the emotional damage that occurred.

Crimes are becoming more violent, and it is perceived that this trend will continue to increase. Emergency responders are immediately placed into dangerous situations, and can get caught—literally and figuratively—in the cross fire.

HISTORY OF MODERN TERRORISM AND HAZARDOUS MATERIALS CRIMES

As a result of the terrorist attacks on September 11, 2001, the previous bombing of the World Trade Center in 1993, the Alfred P. Murrah Federal Building bombing in Oklahoma City, and other incidents, the fire service's response to some types of incidents will be forever changed. In the late 1990s there were two other bombings that are fairly well known to the fire service: the Atlanta Olympic Park bombing in 1996 and the Atlanta abortion clinic bombing in 1997. That particular abortion clinic bombing is significant because a secondary device was used. It is thought that it was strategically placed with the sole intention of harming the responders. The device was placed near the location where the incident command post was set up. Luckily, only minor injuries were received by the responders who were near the blast. In 1998 there was another bombing at a Birmingham, Alabama, abortion clinic in which an off-duty police officer employed as a security guard was killed by a device that was more accurately placed to target the responders. This secondary device was activated by the person responsible as he watched the victim approach the radio-controlled device. A brief overview of some of the recent acts of terrorism, HAZMAT crimes, use of terrorism materials, and criminal acts that presented risk to responders follows:

1980s A series of bombings targeted primarily at the Internal Revenue Service (IRS) included an attempt at a chemical bomb using ammonia and bleach. Another attempt included the use of a hot water heater as a very large pipe bomb, but the vehicle carrying the bomb caught fire while the terrorist was driving the vehicle.

1984 Dr. Michael Swango had a long history of suspicious deaths while he moved throughout the United States and abroad. He was stripped of his license to work as a doctor and got a job as a paramedic. Although his crimes do not fit the pattern of standard terrorism, he was finally arrested for attempting to poison his paramedic coworkers with chemicals.

1984 The Rajneesh Foundation was responsible for poisoning 715 people with *Salmonella* bacteria. They poisoned the salad bars in ten restaurants. The group had used other biological agents and had targeted a water supply for attack. They had previously used raw sewage and dead rodents in an attempt to poison the system. They used nursing home patients as their test targets for some of the biological attacks. The attacks were an attempt to effect a change in a local election.

1985 Members of a militant group developed a plot to poison a water supply. They were going to use 35 gallons of cyanide, which they had in their possession. Other members of the group were arrested for church arsons and attempting to blow up gas pipelines.

1993 The World Trade Center was damaged by a van bomb. Six people were killed.

1995 A man was arrested after manufacturing **ricin**, an extremely deadly toxin, with the intention of killing someone he was jealous of.

1995 The Alfred P. Murrah Federal Building in Oklahoma City was the target of a truck bomb. Several hundred people were injured and 186 people were killed in the blast.

1995 The members of a militia group known as the Patriots Council were arrested for the manufacture of ricin in the attempted assassination of a U.S. marshal.

1995 Two separate ricin incidents involved two doctors. Dr. Deborah Green killed two of her three children by burning her house, and she attempted to kill her husband three times with ricin. In Virginia, Dr. Ray Mettetal attempted to murder his boss with a syringe filled with boric acid and saline. He also possessed ricin and a number of other materials.

1995 (and throughout the 1990s) The Aum Shinrikyo are the cult group that is best known for a 1995 sarin nerve agent attack in the Tokyo subway. For a long period of time the group carried off chemical and biological weapons attacks and went to great lengths to develop their weapons program. An examination of this group yields many valuable lessons on chemical and biological attacks. The group had assets of more than $300 million. The group was also home to more than a hundred scientists whose sole function was to develop chemical and biological weapons. The group used, or explored the use of, *Clostridium botulinum, Bacillus anthracis* (anthrax), Q fever, Ebola virus, and viral hemorrhagic fever. They carried off ten chemical attacks and nine biological attacks, not to mention numerous small and large full-scale tests. They killed seven people and injured 200 in a sarin nerve agent attack in Matsumoto in 1994. The subway attack in 1995 killed thirteen but injured more than 5,000. The attack consisted of placing sarin in three subway lines, with a total of eleven sarin bags that were pierced with umbrellas. The deaths occurred to those who came in contact with liquid or were in small confined spaces. Over 4,000 people went on their own to 278 hospitals. The Tokyo Fire Department ambulances transported 688 patients. The breakdown of the injured was seventeen critical and thirty-seven severe. There were 948 patients suffering from miosis (pinpoint pupils). All of the moderately injured patients were treated and released within six hours after they arrived at the hospital. Although 85 percent of the patients did not require any treatment, they still flooded the hospitals.

1995 A train was derailed near Hyder, Arizona, and a terrorist group claimed responsibility, although no one has been charged in the attack. One person was killed and the derailment seriously injured twelve people.

1995 The Anaheim Fire Department in California was made aware of a potential sarin nerve agent attack at Disneyland. The chief was notified at midnight to be at the command post four hours later. Police agencies, federal law enforcement, and the military had known about the threat for five days. Up until notifying the fire chief and assembling the resources prior to the event, no other planning had occurred. The fire department was placed in charge of the incident and quickly developed a plan of action. Disneyland did not close and 30,000 to 40,000 people visited the park during the threat period. Luckily, the threat never materialized.

1996 Stuart Adelman, Buffalo, New York, used his employer's Nuclear Regulatory Commission (NRC) license to order several radioactive sources. The purpose of the purchase is not known, but it is suspected that foul play was being planned.

1996 Members of a paramilitary group who had access to preplanning information obtained through the fire department were arrested when they attempted to blow up a Department of Justice complex in West Virginia.

1996 and 1997 The Atlanta area was besieged with bombings, including one at the Olympic Park that killed one person. At an abortion clinic in Birmingham, Alabama, a booby-trapped device killed an off-duty police officer and wounded a nurse.

1996 Theodore Kaczynski, known as the Unabomber, was responsible for a string of bombings that lasted eighteen years. He sent sixteen bombs, which killed three people and injured twenty-three. He wrote a manifesto that appeared in the *New York Times* and was recognized by his brother, who turned him in to the Federal Bureau of Investigation (FBI).

1996 A disgruntled firefighter in Jacksonville, Mississippi, went to his department's headquarters with the intent of killing the fire chief. The fire chief was not injured, but several other fire department personnel were killed.

1996 A lab worker brought in pastries that were poisoned with *Shigella dysenterie* to coworkers. She made twelve people ill and was found guilty on five felony assault charges. Police also learned that she had attempted on more than one occasion to poison a former boyfriend with *Shigella* and other biological agents.

1997 EMS and police responded to a shooting incident, and the suspect was found to have ricin, *E. coli*, and a mixture of nicotine and dimethyl sulfoxide (DMSO).

1997 Four members of the Ku Klux Klan were arrested for plotting to blow up a hydrogen sulfide tank in order to create a diversion for an armored car robbery.

1998 In Nairobi, Kenya and Dar-es-Salaam, Tanzania, bombs exploded in two U.S. embassies, killing 224 people.

1998 Kathryn Schoonover attempted to mail 100 envelopes with cyanide disguised as diet powder all over the United States.

1998 Two men were arrested on the suspicion of **anthrax** possession. They were later released, since it was determined that they only possessed a possible anthrax vaccine. Larry Wayne Harris had previously been arrested in 1995 for the possession of plague, a biological toxin that he had ordered through the mail. He was a previous member of the Aryan Nation but was deemed too radical for the group.

1998 In Charlotte, North Carolina, a man with an explosive device held some hostages in a government building. The explosive device was thought to also contain some type of chemical agent. It was later determined that the filling agent was harmless, although the explosive was live.

1998 Abortion clinics in Florida, Louisiana, and Texas were affected by attacks using **butyric acid**, a material with a horrible, irritating odor.

1998 In Lafayette, Indiana, a pickup truck rammed into the courthouse. The bed of the pickup had flammable and combustible materials as well as several explosives.

1999 In Colorado, two students who were armed with an array of guns and explosives attacked their own high school. The suicide attack resulted in fifteen deaths and spurred a rash of bomb hoaxes throughout the country.

1999 The FBI investigated hundreds of anthrax hoaxes, none of which involved the actual use of the biological agent. Abortion clinics were the targets in most of the cases.

1999 Ahmed Ressam was arrested in Port Angeles when he crossed over the border into the United States. He was responsible for plotting the millennium "border bomb" and had the components to a large bomb in his truck. His target was the Los Angeles Airport (LAX).

2000 Police and fire officials scrambled after more than eighty vials filled with an unknown liquid were found throughout Milwaukee and its surrounding suburbs.

2000 Dr. Larry Ford's house in Irvine, California, was searched after he attempted to murder his partner. Dr. Ford committed suicide, and his house was found to have held numerous chemicals and chemical agents. The search took several weeks, and chemicals, guns, and biological materials were found.

2000 During a genetics conference in Minneapolis, Minnesota, protestors attacked three restaurants with hydrogen cyanide.

2000 Members of a militia group in Sacramento, California, were arrested for plotting to blow up a 24-million-gallon propane tank.

2001 A Chevron employee in the El Segundo, California, refinery threatened to blow up the facility. The facility produces more than 4 million gallons of jet fuel a day.

2001 Over 2,600 civilians and 365 emergency responders lost their lives at the World Trade Center in New York City; the Pentagon in Arlington, Virginia; and a field in Shanksville, Pennsylvania. Four planes were used in the attack, two hitting the World Trade Center and one hitting the Pentagon. The fourth plane crashed into a field. Members of the al Qaeda group have been held responsible for the attack.

2001 In October there was one death as a result of an anthrax-laden letter. Later, there was a total of seven deaths. The emergency services across the country were deluged with calls about white powder. A very small percentage of the calls was investigated by the FBI, but they opened more than 14,000 cases regarding the white powder events. Only five cases involved real anthrax, which was sent to members of Congress, to New York City, and to Boca Raton, Florida. In December, Clayton Lee Waagner was arrested and admitted to sending more than 500 noncredible anthrax letters to women's reproductive health centers (WRHCs). He was a member of the Army of God, a group known for pursuit of the right to life cause.

2001 The Animal Liberation Front (ALF) and the Earth Liberation Front (ELF) committed a number of terrorist acts throughout the year. The ALF admitted to 137 illegal actions at a variety of locations. Both groups target buildings and businesses that have any connection to the use of animals or perceived violation of the environment. The acts typically create inconveniences, such as glued locks, but on occasion do include arson and other violent destruction of property. For the most part these groups take great care to avoid any potential injuries to humans, but responders could be killed or injured while responding to or handling these acts.

2001 Two Jewish Defense League (JDL) members were arrested for plotting to blow up a mosque and a congressman's office. Other members of the JDL have been quoted as having a desire to continue the militant work of the two men arrested.

2001 Richard Reid was arrested on a Paris to Miami flight for attempting to detonate a PETN explosive that was located in his tennis shoes. PETN is one of the more powerful explosives. He was overpowered by other passengers and the flight crew while trying to light the fuse. He has ties to al Qaeda and is supportive of their cause.

2002 Two different bombers created fear in the United States, one in Philadelphia and the other in the Midwest. Preston Lit set pipe bombs off in U.S. Postal Service mailboxes in the Philadelphia area. Lucas Helder, a college student, set off pipe bombs in residential home mailboxes in five states. Helder's intention was to set off bombs so that the explosions when drawn on a map would form a smiley face. His bombings injured six people. He set a total of eighteen pipe bombs, six of which exploded.

2002 A fifteen-year-old stole a small plane and flew it into a high-rise building in Tampa, Florida, killing himself. His suicide note stated that he wanted to be just like Tim (McVeigh), Eric (Rudolph), and Osama (bin Laden).

2002 Joseph Konopka, a member of the Realm of Chaos group, had stashed potassium and hydrogen cyanide in a Chicago subway tunnel. The group is known for wanting to destroy public utility, water, sewage, and telecommunications systems.

2002 The leaders of two militia groups in Kentucky and Pennsylvania were arrested for possession of weapons and explosives.

2002 The FBI arrested Abdullah al Muhajir, who is associated with al Qaeda, in Chicago. He was researching the use of, and looking for materials to detonate, a radiological dispersion device. He had traveled to Pakistan and had studied methods to pull off such an attack with al Qaeda operatives. In late 2002 the FBI was still looking for 100 al Qaeda members and investigating 150 persons and groups who may have al Qaeda ties. They made two large arrests in Detroit and Buffalo, as well as other arrests throughout the United States, apprehending a number of suspects who had al Qaeda ties.

The following sections provide some information about the type of terrorism agents that currently exist. Some are unique ideas and thoughts, and some represent possible scenarios. There exists some thought within the fire service that this type of information should not be published. All of this information is readily available in the public domain and in training programs and texts throughout the United States. Most of the exact recipes and "how-to" instructions for these and any other device are easily obtained through printed texts and the Internet. This text does not provide any information that cannot be easily obtained elsewhere. If an ordinary citizen is able to obtain it in an easy, normal, and legal fashion, it is certain that the terrorist already has it. Responders need to be aware of the various chemicals and devices that someone may design to kill them and others in their community. The best defenses are education and trying to stay one step ahead of the terrorist. By being informed as to how a terrorist may operate or what some of the devices may involve, responders will be alert to a potentially fatal situation.

TYPES OF TERRORISM

The types of terrorism are divided into two distinct areas: foreign based and domestic. Until the Murrah building bombing, the fear of terrorism was aimed at a foreign source, and it was thought that any terrorist attack would be from a foreign country. To be foreign based, the motivation or supervision must come from a foreign country. Domestic terrorism originates from within the United States and is not influenced by any foreign party. From the list of terrorist acts provided earlier and from other statistics, it is clear that the largest percentage of terrorism is domestic in origin.

The FBI defines terrorism as a violent act or an act dangerous to human life in violation of the criminal laws of the United States or any segment to intimidate or coerce a government, the civilian population, or any segment thereof, in furtherance of political or social objectives. The key to this definition is the intimidation of the government or the civilian population. A militant group trying to influence the local political process sprayed *Salmonella* bacteria on a fast-food restaurant salad bar and was successful in making more than 600 people sick. The Tokyo subway sarin attack was an attempt to destroy a good portion of the police department in an effort to prevent a raid on the terrorist compound.

> **Safety** The fire service will be called to many incidents that will not fit the exact definition of terrorism, but the hazards from a pipe bomb are the same regardless of the motivation of the builder.

Many responses that would have been routinely handled in the past must now be treated much differently, and responders must always be on their guard for terrorist-style devices or potential acts of terrorism.

The terrorist's motivation is to produce fear that may be aimed at the general public or the government. Fear can be provoked by large-scale actions such as the acts on September 11, 2001, the original bombing of the World Trade Center in 1993, or the bombing of the Alfred P. Murrah Federal Building in Oklahoma City. Even acts that are not terrorism can create terror in the community. The Chicago nightclub incident, which was caused by pepper spray, and the Rhode Island nightclub fire have sparked fear and concern about safety in nightclubs. A terrorist can also incite fear just by planting the thought of potential terrorism or by devising a hoax. The latter scenario is the more likely and can be very difficult to handle from an emergency service perspective.

The thought process for determining if a threat is credible or not has five elements. If the person known or thought to be responsible is determined to have several of these capabilities, it increases the credibility factor:

1. The first of the five qualifiers involves the potential terrorist's educational ability to make a device or agent that, unlike explosives or ricin, is very difficult to attain. To truly make a biological pathogen agent, in most cases, one needs an advanced knowledge of biology. Ricin, a biological toxin, does not require any advanced knowledge compared with weaponized anthrax. Some of the threats with letters or packages have misstated the origin of the material, such as calling anthrax a virus, or have misspelled the agent's name. If the terrorists do not know the true origin of the material or

cannot spell the material correctly, they probably do not have the education necessary to make the material they state they produced. This does not take into account a person who may purchase the material.

2. The next qualifier is a person's ability to obtain the raw materials necessary to make the agents. Many of the materials necessary to make chemical warfare agents are banned for sale. Others appear on hot lists, which means they are only sold to legitimate businesses. This would not preclude someone from buying the raw materials on the black market or stealing them from a legitimate business.

3. The third qualifier is the ability to manufacture the devices or machinery required to make the agent. To manufacture chemical warfare agents requires the use of a reactor vessel, which requires about a 10-foot by 10-foot space to produce less than a gallon. There are some agents that could be produced in a bathtub using backyard chemistry, but these are not the high-end agents that attract much attention. Many people who have attempted to make agents in less than ideal conditions have died during the production process. Many criminals do not take the time to follow standard industrial safety precautions.

4. One qualifier that is often overlooked is the ability to disseminate these agents. The military conducted many tests on chemical and biological warfare agents, and although they have some good methods of dissemination, even they lack a 100 percent effective method of dissemination. The Aum Shinrikyo cult in Japan is a perfect example, as they were a group with millions of dollars in assets and full chemical and biological lab and production facilities. They employed the services of 235 scientists to develop and manufacture chemical and biological agents. The Aum Shinrikyo abandoned their biological weapons program after a full-scale release of anthrax that failed. They used sarin nerve agent twice, the first time in Matsumoto, Japan, in which seven people were killed and 200 injured. The dissemination method used in the Matsumoto attack was much more effective than the one used in the Tokyo subway attack. If they had used the same dissemination method they would have greatly altered the course of events. They would have been limited by the amount of agent that could have been produced in a short period of time.

5. The last qualifier, which is the most important, is whether the person or group has the motivation to pull off the attack. The intentional killing of one person takes significant motivation, and the intentional killing of hundreds requires a whole lot of motivation. There is always the potential for an attack, but it takes considerable education, raw materials, manufacturing, and dissemination ability to pull off a chemical or biological terrorist attack. On the other hand, explosives are easy to manufacture and do not require much education, only simple tools, and the materials required are easy to assemble. It is for this reason that explosives are used in the majority of cases and are quite successful in completing an attack. In many cases the terrorist can be successful because of the hysteria associated with a potential terrorism incident. A balance must be struck between a cautious approach and one that does not allow the terrorist to win by crippling a community and causing hysteria.

Another consideration, as we have seen in England and Ireland, is the disturbing trend of viewing emergency service personnel as targets. One theory currently under examination is that the second explosion at one of the Atlanta abortion clinic bombings was aimed at the emergency responders.

> **Safety** Responders must always be alert to the potential of a secondary device.

POTENTIAL TARGETS

Potential targets exist throughout every community in the nation and can be commercial buildings, high-rise buildings, and even residential homes. Although some incidents do not fit the definition of terrorism, the materials used are the same as those a terrorist might use. Whether the objective is murder or terrorism, the danger to the responder is the same. When looking at terrorism, potential targets can be grouped into several categories: public assembly such as the area shown in **Figure 24-3**; federal, state, and local public buildings; mass transit systems; high economic impact areas; telecommunication facilities; and historical or symbolic locations.

While obviously not an exclusive list, buildings that could be targeted include the Federal Bureau of

TERRORISM AWARENESS ■ 749

Figure 24-3 Any location is a potential target for a terrorist. Any location where large numbers of people are present, such as a mall or sports event, is a prime target.

Investigation (FBI); the Bureau of Alcohol, Tobacco, and Firearms (ATF); the Internal Revenue Service (IRS); military installations, Social Security buildings; transportation areas; city or county buildings, including fire and police stations; abortion clinics and Planned Parenthood offices; fur stores; laboratories; colleges; cosmetic production/testing facilities; banks; utility buildings; churches; and chemical storage facilities. Transportation facilities such as airports and train, bus, or subway stations are high on the potential list of targets, given the number of people who may be potential targets and the relative ease of escape. In the southeastern United States, a large number of churches have been subjected to arson fires, and in some cases explosive devices have been used. A large number of abortion clinics have been the subject of bombings, attacks, and other threats. Any incident in or near one of these facilities should be approached with caution. Responders should know the location of these facilities in their jurisdiction. Preplans for these facilities should be thought out by the company officers, but it is not recommended that these plans be committed to paper. As the battle between pro-life and pro-choice groups continues to rage, incidents at these facilities can only be expected to rise, with emergency responders caught in the cross fire.

Many of the potential targets of terrorism have not been buildings at all but events where large numbers of people are present. The Atlanta Olympic Park bombing is an example. Other scenarios involve sports stadiums, such as the one pictured in Figure 24-4, public assembly locations, transportation hubs, and fairs and festivals. First responders should have some preplans for these types of locations. One possible scenario for a stadium, devised by Captain Richard Brooks of the Baltimore County Fire Department, describes the first-in medic unit arriving at a stadium where in Section 300 there are 40 people projectile vomiting. After five minutes 200 people are projectile vomiting, and as time goes on the number increases. What happens to the responders when confronted with a situation of this nature? How many responders would be affected by this massive amount of people vomiting? How many responders would be needed to handle this incident? This act of terrorism could be accomplished by putting syrup of ipecac in the ketchup container beside one of the hot dog vendors, an easy task. Imagine the hysteria if a note was found stating that a biological agent was distributed in that section. What impact would that have on the remaining 50,000 people in the stadium if that information got out? Planners and responders involved on the national level in trying to develop response profiles to terrorism are grappling with how to plan for incidents involving 100 people, 1,000 people, 10,000 people, and 50,000 people. Terrorism incidents can very quickly overwhelm the responders and their whole emergency response system.

When dignitaries visit locations, a lengthy planning process typically takes place in which the fire department should be involved. When the Pope visited Baltimore in 1997 the planning process took

Figure 24-4 Other than special events, the most common location where large numbers of people are together is at sporting events. At this stadium, if an incident were to occur, more than 50,000 people could become part of the incident.

more than eight months. Planning for such a large event takes the cooperation of local, state, and federal agencies. Even when dignitaries visit locations such as New York City or Chicago, advance planning occurs. Other events such as political conventions or other large political gatherings all bring the potential for an incident. When one of these events comes to a community, local responders need to be prepared for not only the people arriving to the event, but also the massive federal response that may be pre-positioned. For many special events, whole task forces of federal resources may be hidden away just in case of an incident.

Certain dates have significance to several militant groups. The date of April 19 is the anniversary of the Waco, Texas, incident in which the ATF stormed a compound that housed the Branch Davidians, a group thought to be a militia group. April 19 was also the date of the Oklahoma City bombing, and the date was chosen by the bomber as a way to retaliate for the Waco incident. Other dates provoke the potential for terrorist acts. For instance, the anniversary of *Roe v. Wade*,[1] January 22, could incite a strike by antiabortion groups. Within the United States, forty states are suspected to have members of militia, patriot, or constitutionalist groups. Membership counts vary from fifty people in some states to several thousand members in other states. Groups of concern include anarchist groups and white supremacy groups such as the Ku Klux Klan, Zionist Occupation Government, skinheads, and neo-Nazi groups, including the Aryan Nation. Other groups suspected of activity or thought to have terrorism potential include patriots, New World Order militia, constitutionalists, and tax protesters. To learn which groups are active, it is relatively simple to use a Web browser and search the Internet for many of these groups, because most have Web sites.

INDICATORS OF TERRORISM

An explosion or explosive device is the most common tool of the terrorist, and police across the country have made several arrests of persons for making or storing large quantities of explosives.

> **Firefighter Fact** According to an FBI source, more than 93 percent of terrorism incidents use explosives as the weapon of choice.

The most common device is a pipe bomb such as the one shown in **Figure 24-5.** Any incident where an explosion has occurred or first responders believe that an explosion has occurred should be suspected of being a terrorist incident.

> **Safety** If one explosion has occurred, responders should always assume that there is a second device awaiting their arrival.

Figure 24-5 The most common explosive device is a pipe bomb, and it is very effective. It is a very dangerous device, not only for responders but for the builder as well.

In this day and age any suspicious package should be suspected of containing explosives and should be dealt with by a bomb technician. Firefighters should never assume that they can handle the package or remove it from the area. Just like EMS and HAZ-MAT, the handling of bombs or suspicious packages is a very specialized field and should only be done by a person trained to do that type of job.

The presence of chemicals or lab equipment in an unusual location, such as a home or apartment, is an indication of possible illegal activity. When looking at chemicals or a lab there are three possibilities: drug making, bomb making, or terrorism agent production (chemical or biological). The most likely, based on statistics, is drug making, followed by bomb making. Although very common in some parts of the country, predominantly the West Coast and the Northwest, drug production labs are not commonly located throughout the whole United States. Currently the number of drug labs is increasing in the Midwest, with responders taking down more than twenty a month in some areas. This trend is slowly moving from the West Coast to the East Coast. The most likely scenario is locating someone making explosive devices, as many individuals like to make homemade devices. It is possible, but very unlikely, that firefighters would locate a facility attempting to make a terrorism agent such as sarin. The exception to the biological agents would be for the production of ricin, as the items required to make ricin are easily obtained and the production is just as easy. Fortunately, ricin does not have the potential to easily kill large numbers of people. It is primarily an injection hazard, although it is still very toxic through inhalation or ingestion. A responder in a small community or a rural setting is the most likely to run across any of these types of production areas. Several persons a year are arrested for the possession of ricin with the intent to use it for some type of criminal act. Most of the arrests are occurring in small towns throughout the United States. No matter what agent is located, the responder should immediately isolate the area and call for assistance. The call should go simultaneously to the police, the HAZMAT team, and the bomb squad. This request would apply to all three types of labs—drug, bomb, or warfare agent—as any of these labs should be handled as a cooperative effort.

Another indicator of potential terrorism is the intentional release of chemicals into a building or the environment. Finding a chlorine cylinder in a courthouse would be unusual and should put the responders on alert to the fact that there is a high probability that a terrorist may be at work. Finding chemical containers such as bottles, bags, cylinders, or other containers in unusual locations would also be suspect. In an industrial facility in which chemicals may be common, responders may find that there is the intentional release of a chemical.

Note One of the best indicators of potential terrorism will be a pattern of unexplained illness or injury.

A response to a mall for a seizure patient is not unusual. However, a response to a mall for six people having seizures is very unlikely and could involve a chemical release from a terrorist attack. Seizures, twitching, tightness in the chest, pinpoint pupils, runny nose, nausea, and vomiting are all signs and symptoms of a warfare agent attack. EMS providers will probably be the first group to identify the use of a chemical agent. Imagine arriving at an explosion where there is a large amount of debris and twenty victims. To most the injuries would appear to be blast injuries, as they would in most cases be visible during a quick survey of a patient. Other signs and symptoms, primarily pinpoint pupils, probably would not be identified until the patient is given a more thorough exam. When confronted with victims that are unconscious or dead and now have outward signs of death, such as trauma, the responder must face the possibility that they may have been the victims of a chemical attack. This obviously has to be put in perspective. If responders in a metropolitan area of the Northeast are called to an apartment building in the winter because there are twelve unconscious people, and this happens several times each winter, it is probably carbon monoxide poisoning. However, if responders are called to a mall in the summertime because there are twelve unconscious people, it is probably not due to carbon monoxide but to a chemical release of some type. When distributing a warfare agent the explosive device will usually not have a destructive effect on the building or the surrounding area. In some cases the larger the device the more likely it is that the detonation will consume the agent. Most of the victims will not display signs or symptoms of a blast, although the one closest to the device may suffer some of those types of injuries.

Smelling unusual odors or seeing a vapor cloud may be an indicator that chemicals have been released. As may be seen with other toxic gases such as chlorine, arriving and finding dead birds or other animals should alert the responder not only to a chemical hazard but also to the potential for terrorism. It would also be unusual to respond to an office building or a house and notice security measures of the type that one would not expect in that occupancy. Items such as extra locks, bars on windows, surveillance cameras, fortified doors, guards, and other unusual protection devices may provide a clue to the responder that something out of the ordinary is at play.

HAZMAT CRIMES

Although this chapter covers terrorism, emergency responders may respond to other criminal-related events. Whether the incident is terrorism related or not, the effects on responders can be the same. Criminals are using more weapons than just guns today, and the use of clandestine labs for illicit production is on the rise. These incidents are referred to as HAZMAT crimes, since chemicals are being used in an illegal fashion. Occasionally, there will be a robbery of a convenience store where the weapon is a chemical. The robbery suspect might throw a corrosive liquid on the clerk in order to rob the cash register. There have been other robberies, attacks, or attempted murders using corrosive liquids. Whether or not the law-enforcement community or the prosecutor decides the incident is an act of terrorism, emergency responders need to recognize potentially hazardous situations and have the ability to protect themselves.

One big exposure issue for emergency responders is drug related. Many drug addicts use chemicals as part of the process to get high, and the drugs themselves are usually toxic and may present other hazards. Drug users may use a combination of chemicals to get high, and many of these items are toxic and flammable. Ether is used to assist in the heating process of several types of drugs. This material in pure form is extremely flammable, and the container may eventually become a shock-sensitive explosive. Most drugs are in solid form, which means that they present little risk unless eaten or touched with bare hands. In some cases the drugs may be stored or used with flammable and toxic liquids.

The most common situation in which emergency responders could be directly affected by drug use is when a person is huffing. While people are huffing they typically use a toxic and/or flammable material. Many common household items such as paints, glues, hair spray, and solvents provide the high that some people desire. From a HAZMAT point of view, these materials are toxic and flammable and airborne. In most cases, users will spray the material into a bag to concentrate the vapors, but when they are done the vapors remain in the room of use.

Clandestine Labs

There are several types of clandestine labs that emergency responders are likely to encounter. The most common is the drug lab, but other possible labs include explosives labs, chemical labs, and biological weapons labs.

> **Safety** All labs have inherent dangers for responders, and all are very dangerous locations to occupy. All may be booby-trapped or be set up to harm responders, and a booby trap does not know if the person coming through the door is a police officer, a firefighter, or an EMT.

Figure 24-6 Example of a possible biological lab. Shown is a microscope and an incubator; both would be used in the production of biological materials.

For the most part a biological weapons lab, shown in **Figure 24-6,** can run unattended without any major concern and may be shut down in any number of ways without major consequence. A drug, explosives, or chemical lab, on the other hand, should only be shut down by someone trained to do so, as these labs are especially dangerous.

Drug Labs

There are many types of drug labs. Just for methamphetamines, there are more than six common methods of production. In 1973 the Drug Enforcement Administration (DEA) discovered 41 labs; in 1999 they discovered 2,155 labs, and in 2001 they raided 12,715 methamphetamine labs. The map shown in **Figure 24-7** provides proof that the labs are moving eastward at a rapid rate; soon all of the United States will have to deal with this problem. Due to its popularity, meth production is becoming common but it does involve a dangerous process. The production of drugs requires the use of many chemicals. These chemicals can be purchased outright, stolen, or manufactured using other chemicals. As many drug-producing chemicals are hot listed or cannot be purchased, the producer must resort to innovative methods to produce the chemicals. The "nazi" method of meth production involves the use of anhydrous ammonia, which is usually stolen from a chemical facility. If a 150-pound cylinder of anhydrous ammonia developed a leak or catastrophically failed, people for a considerable distance downwind could be affected, and those in the immediate area would be in grave danger.

Drug labs can be very complicated setups and can be found in any number of locations such as homes, barns, hotels, storage units, and even trucks. Emergency responders routinely encounter these labs inadvertently through other responses. Responders who encounter a drug lab should notify their HAZMAT team, the bomb squad, and the local office of the DEA. The shutting down of a drug lab is a complicated and very dangerous process.

> **Safety** When chemicals are being heated or cooled there is a chance for a violent reaction.

It is this heating and cooling process that indicates the type of lab that may be present. When responders see glassware that is distilling chemicals—in other words, evaporating a certain component—and then rehydrating or condensing another portion of the original chemical, this is indicative of possible drug production. The end result of the process is a solid form, usually a powder. There will be a production line-type formation to the glassware, with some solutions being heated while others are cooled. At some point, gas cylinders may be present and the gas is allowed to mix with some part of the process. Many of the chemicals involved in the production of drugs are flammable, and although many are also toxic the predominant hazard is flammability.

> **Safety** When choosing protective clothing to enter a lab, it is important to protect responders against the predominant hazard.

The materials that are toxic cannot harm the responder as long as SCBA is worn and the materials are not touched with bare hands or eaten. It is highly recommended that personnel from the police, fire, and EMS departments receive training in drug lab awareness, with some receiving specialized training in this area.

Explosives Labs

Although not common, it is possible that emergency responders might encounter an explosives lab, which is the predominant weapon of choice for a terrorist.

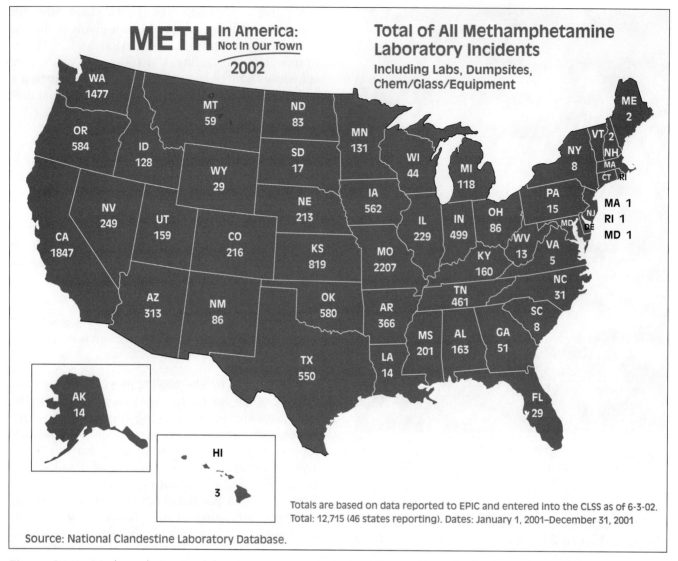

Figure 24-7 Methamphetamine lab seizures across the United States. Note the high number of labs in the Midwest. In years past, the largest numbers of lab seizures occurred in the West. The prevalence of methamphetamine labs is moving Eastward at a fast pace.

An explosives lab can be anything from a workbench pipe bomb builder to a full chemical production facility. A person building a pipe bomb does not need much equipment other than some simple hand tools, pipes, caps, and powder. A person making cyclotrimethylenetrinitramine, commonly referred to as RDX, or another more sophisticated explosive will need some more equipment. Depending on the availability of some materials, the bomb maker may have to make some chemical components as opposed to purchasing them. It is when materials are made at home that the danger increases for a responder. The processes to make the chemical components for explosives are very dangerous and present a significant risk to the builder and responders. An explosives lab differs from a drug lab in that most of the processes do not involve heating, cooling, condensing, or distilling. However, some processes used to make the chemicals do perform some of these functions, so there are no black-and-white rules for identification. The major work at an explosives lab is usually mixing of materials, in most cases solids and liquids. If gases are used, they are usually coupled with the explosive device and may be used to increase the heat of the explosion, boosting its efficiency. People making explosives will usually have a large amount of powders in their house.

Some of the other indicators of an explosives lab are the presence of ignition devices, boosting charges, or blasting caps. Many people who manufacture explosives are doing so to make fireworks and will have cardboard tubes for the explosives to be placed into. In one case, it was originally thought that an explosives maker was just making fireworks, but some other explosive devices were found to have BBs and nails taped and glued to the outside of the explosive. Neither of these have an effect on the display ability of the explosive device and are only designed to kill or maim. When an explosives lab is discovered, the local bomb squad and the local HAZMAT team should be called in to assess the materials. As the stability of many of the materials cannot be ensured, it is usually necessary to remotely destroy many of the found materials.

Terrorism Agent Labs

These are the least likely labs to be encountered in emergency response, but responders must be aware of their existence and some of the unique features of these types of labs. The two types of terrorism agent, or **weapons of mass destruction (WMD)**, labs are chemical weapons or biological weapons labs. Statistically the most likely lab is a biological toxin lab, which may be used to manufacture ricin, which is shown in **Figure 24-8**. The FBI, on average, arrests a few people a year for the possession of ricin, usually accompanied by a threat. Biological labs may be set up to attempt to make other biological materials. Other than ricin, botulinum toxin, and a few other biological materials, the manufacture of biological weapons is very difficult. The manufacture of ricin is a simple process that only requires a few items, such as castor beans and some readily available chemicals. The process to make some of the more advanced biological agents such as anthrax is more difficult and requires a higher level of education, sophisticated equipment, and access to raw materials. The development of biological agents involves culturing the material, usually in a petri dish. These petri or culturing dishes may be placed in an incubator or oven-like device to keep the material warm and at a constant temperature. Depending on the type of agent, there may be grinders, dryers, and sieves present to finish off the product. The major route of entry for many of these products is through touch or ingestion. The only inhalation concern would be during the grinding process, but a simple high-efficiency particulate air (HEPA) mask offers more than enough protection for a responder. Some of the materials used as part of the process may be flammable and have some toxicity, but the chemicals used to clean glassware and tools are usually highly corrosive and in most cases will be sodium hydroxide (lye) and/or bleach. A bioweapons lab will involve some chemicals, but may resemble more of a greenhouse than a chemical lab. Some biological materials are sensitive to light and will be kept in the dark, and the characteristic distilling and condensing glass apparatus will be missing.

Chemical weapons labs use two methods of production: the development of new chemical agents

Figure 24-8 Shown is a lab that could be used to produce a biological toxin such as ricin.

through standard production methods and the synthesis of existing materials. The development of a new chemical agent is the more difficult of the two processes and is nearly impossible except for someone with a chemistry background and access to some chemistry equipment. The recipes for chemical weapons are fairly sophisticated and require access to many raw materials that are hot listed. There have only been two arrests of persons in the United States for attempting to make a chemical weapon such as sarin nerve agent. Both were arrested for ordering the raw materials, and it is thought that the two individuals did not have the educational capacity to manufacture the agent. The production of these agents can be very risky to the producer and, without safety precautions, may result in death. Some of the off gases from the production of chemical warfare agents are extremely dangerous.

The most probable scenario for the development of a chemical warfare agent is the synthesis of an existing product. The criminal would take existing materials, which are usually in diluted form, and synthesize them or reduce them down to a concentrated product. Luckily most, if not all, of the existing products do not present much risk to humans, as they are strictly engineered to harm only insects. This would not stop a terrorist from attempting to use one of these products in an illegal manner, however. There are some pesticides on the market that are applicable to this type of scenario. The standard household pesticide usually has 0.05 to 0.5 percent of pure product mixed with an inert ingredient. A mixture that is used by a farmer may be on the order of 40 to 50 percent pure form and is then diluted in the farmer's tank. In order to be more harmful to humans the pesticide must be concentrated and not diluted. The criminal must devise a method of removing one or more of the inert ingredients. As the inert ingredient is usually flammable or combustible, this is not a difficult task. A mechanism must be devised to off-gas the inert ingredient and then capture the pesticide and collect it.

One form of pesticides, known as technical grade pesticides, is already concentrated. These technical grade pesticides are in pure form and are not diluted. These materials are not in common use but can be found in and around the United States.

INCIDENT ACTIONS

A terrorist incident combines four types of emergency response into a large incident.

> **Note** These types of incidents have these four characteristics until proven otherwise: Mass casualty EMS incident + hazardous materials release and/or explosive devices + crime scene considerations = incident management challenges.

The first-in companies at these types of incidents can easily be overwhelmed and are going to be committed to basic actions, such as life preservation. The IC will have enormous responsibilities dealing with all of the required actions. All of the components present are difficult to handle individually, and now in this type of incident they are combined and must be handled simultaneously. The handling of a 100-person **mass casualty** incident is difficult and has the potential to overload the IMS, and that may only be one-quarter of the result of a terrorist incident. Responders should examine their response systems and determine how many patients present a concern. Some systems define a mass casualty as an incident involving five victims, while in other systems it may be ten to twelve. Obviously in a system with one EMS unit, more than one patient begins to cause system problems and may involve delays in treatment and transportation.

Imagine responding to an explosion at a mall, with 100 people injured. Such a mass casualty incident (MCI) will have EMS playing a predominant role. The police department, on the other hand, will be concerned with evidence preservation and crime scene considerations. The possibility exists that the perpetrator(s) also used a chemical or biological weapon, and the explosion was the means of distribution. This situation would be an MCI, a crime scene, and a potential chemical release situation with some, if not most, of the patients contaminated. On top of this scenario, add the potential for a secondary explosive device, one that is aimed at the responders! If responders can eliminate the chemical and secondary device issues, then all they have to deal with is the simple 100-person MCI and the crime scene issues. EMS will handle the patients and the police department would handle the criminal element.

The other aspect of a suspected terrorism incident will be the massive response from the federal government, even if not requested. Later in this chapter more information is provided about federal resources that are available to respond. In most cases, the minimum response to a suspected terrorism incident would come from the FBI, initially from the local agent. Field offices are located across the country, and almost every major city has a field office. Agent(s) usually arrive one to two hours into the incident. If terrorism is suspected the FBI is the lead agency of the incident as provided by a Presidential Decision Directive known as **PDD 39**. As with hazardous materials incidents it is important to know all involved players before an incident. Knowing who the local FBI agent is can be very important, because meeting for the first time in front of an incident is not conducive to effective scene management.

Working out the "who's in charge" concerns prior to an incident is important. The fire department should liaison with its local police, EMS, and emergency management agency prior to incidents. These and many other agencies are going to be involved and have a variety of responsibilities at an incident. In general, a unified command is recommended, and although this does not mean command by consensus, input from the various agencies should be considered. The agency with the majority of the tasks to do is generally in charge. In the initial stages the fire/EMS authority would be in charge while rescuing victims, but after the victims are removed and evidence recovery becomes the next priority, the command may switch to the police department.

The magnitude of response to a terrorism incident can be the most difficult part of the incident to manage. The IC will be overtaken by a large number of federal agencies' representatives, who on a regular basis will be replaced by a later arriving supervisor. Response groups consisting of two to seventy responders may arrive uninvited, all trying to assist with the incident. Another group that can also overwhelm the system is the media. In most cases the local media will react as normal, and for the most part will cooperate. Members of the

> **Note** If an incident occurs that involves federal responders, the national media will follow close behind.

national media do not know, nor do they follow, local protocol, and they will require information. A lack of accurate information can be disastrous to an incident by causing a deterioration of the incident and creating more hysteria than is already present. Media relations are very important in these types of events.

One of the most important issues that will arise other than the life safety hazard is that of evidence preservation. As much care as possible should be taken to preserve evidence and make sure it is taken care of appropriately. The collection of evidence is primarily a law-enforcement responsibility. Unless properly trained, fire service personnel should not collect evidence but should alert the police of its presence. A whole host of issues goes along with the collection of evidence, including the chain of custody. This chain of custody, or the paper trail that follows any evidence, is crucial to the successful prosecution of the persons responsible. The failure to follow proper procedure or document the travel of evidence can result in a case being dismissed, regardless of any other evidence. There have been cases, such as the Murrah bombing, in which fire service personnel have collected evidence, but that is a very unusual occurrence. Firefighters who are assigned to collect evidence should be fire investigators or fire marshals because they are typically trained in evidence collection. Another alternative is to double up and use a firefighter and a police officer to collect evidence.

A cooperative effort is needed to combat a terrorist attack. The primary functions are rescue/life safety by fire and EMS personnel, hazard identification by the HAZMAT team, identification of possible secondary devices by a bomb technician, and incident management. It is important to communicate the hazards to all personnel, and to limit the response to essential personnel. Instead of having the whole alarm assignment report to the front of the building, it is preferable to use one or two companies to investigate while staging the other companies away from dumpsters, mailboxes, or dead-end streets. When dealing with victims it is essential to isolate them until the cause is identified. The victims can have a large amount of information and should be questioned quickly. Questions to ask include these: What did you see, hear, or smell? Was this coming from one area or was it throughout the building? Did you see anything else suspicious? What type of signs and symptoms do you have? In addition, the police will need to conduct interviews. Documentation and preservation of the evidence are essential to the successful prosecution of the terrorist.

GENERAL GROUPINGS OF WARFARE AGENTS

Terrorists could use any of a number of possible warfare agents. They are classified into three broad areas. Weapons of mass destruction are commonly used by the military. Some of the regulations that prohibit the making, storing, or using of terrorism agents are called WMD laws or regulations. Any item that has the potential to cause significant harm or damage to a community or a large group of people is considered a WMD. The other two classifications are nuclear, biological, and chemical (NBC) or biological, nuclear, incendiary, chemical, and explosive (BNICE). Both of these are descriptions of the types of materials that could be used in a terrorism attack. Although there are slight variations, they are all used to describe the various types of agents that a terrorist could use. Most of the language differences come from funding legislation or a specific federal agency.

The military has devised a naming system for many of these agents, many of which are listed in **Table 24-1**. Responders should become familiar with these names because much of the literature and help guides refer to these agents by these names. For instance, when using a military detection device, the military name is used. When dealing with terrorism, firefighters are entering another world that has its own language. The fire service has to adopt this new language to survive in this new world. The three groupings mentioned earlier are further subdivided into the categories discussed in the following subsections.

Nerve Agents

Nerve agents are related to organophosphorus pesticides and include tabun, sarin, soman, and V agent. They were designed for one purpose and that is to kill people. Although very toxic, their ability to kill large numbers of people requires that the dissemination device function correctly and a number of other critical factors be in place to be truly effective. Although several gallons of sarin agent were used in the Tokyo subway attack, the distribution method was ineffective, so out of the twelve people who died, the only people killed by the sarin

Military Designations for Agents

NAME	MILITARY DESIGNATION	UN/DOT HAZARD CLASS	NAME	MILITARY DESIGNATION	UN/DOT HAZARD CLASS
Tabun	GA	6.1	Sarin	GB	6.1
Soman	GD	6.1	Thickened Soman	TGD	6.1
V agent	VX	6.1	Mustard	H	6.1
Distilled mustard	HD	6.1	Nitrogen mustard	HN	6.1
Lewisite	L	6.1	Hydrogen cyanide	AC	6.1
Cyanogen chloride	CK	2.3	Chlorine	CL	2.3
Phosgene	CG	2.3	Tear gas	CS	6.1
Mace	CN	6.1	Pepper spray	OC	2.2 and 6.1

TABLE 24-1

itself were the two people who actually touched the liquid. The chemical and physical properties of these agents hinder their ability to be effective as a stand-alone killer. To best produce the desired effect, the agents must touch people in liquid form or be breathed in while the materials are in aerosol form. The materials will not stay in aerosol form for very long, and they have a very low vapor pressure and thus will not create vapors as standing liquid. All of the military warfare agents have a vapor pressure less than water, which means they do not evaporate quickly and unless the liquid is touched or placed on the skin it does not present a large hazard.

Signs and Symptoms of Nerve Agents

All of the nerve agents present the same types of signs and symptoms as organophosphorus pesticides, and in reality the difference is minor. Nerve agents are pesticides for humans and are a stronger, more concentrated version of commercially available pesticides. The signs and symptoms can be generally described using the acronym SLUDGEM, which stands for:

Salivation—excessive drooling
Lacrimation—tearing of the eyes
Urination—loss of bladder control
Defecation—loss of bowel control (diarrhea)
Gastrointestinal—nausea and vomiting
Emesis—vomiting
Miosis—pinpoint pupils

The term SLUDGEM describes all of the symptoms from the minor ones to the extreme signs and symptoms. A slight exposure to any of the nerve agents will cause pinpoint pupils, a runny nose, and difficulty breathing. A person who has come in contact with the liquid will be experiencing all the SLUDGEM signs in addition to convulsions. A person who is in convulsions needs immediate decontamination and medical treatment in order to survive. This treatment sequence has to occur in less than five to ten minutes. In addition, there must be sufficient medication available on scene to accomplish the treatment. Most paramedic units carry enough medication to treat one or two patients who have severe symptoms.

Incendiary Agents

For the sake of classification, **incendiary agents** are placed into the chemical classification, because chemicals are used in these devices. The most commonly used chemicals are flammable and combustible liquids. The standard Molotov cocktail is an example of an incendiary device that could be used by a terrorist. In some cases arsonists have used a mixture of chemicals, usually oxidizers, to create very fast high-temperature fires.

Blister (Vesicants)

The category of **vesicants**, or, as they are more commonly called, **blister agents,** includes chemical compounds called mustard, distilled mustard, nitrogen mustard, and lewisite. These materials were never designed to kill. They were designed to incapacitate the enemy so that if one person was affected by one of these agents several more would be needed to care for the affected person. Although at high concentrations these materials can be toxic, their biggest threat is from skin contact which causes severe irritation and blistering. Their chemical and physical properties make them less of a hazard than the nerve agents. One large concern with these agents is the fact that the effects from an exposure can be delayed from fifteen minutes to several hours. Quick identification of a blister agent is key to keeping the victims safe.

> **Signs and Symptoms of Blister Agent Exposure** One of the biggest risks with the blister agents is that they may present delayed effects. If not detected this could result in victims being released only to later have problems. In general, blister agents are not designed to kill; they were designed to incapacitate the enemy, resulting in troops being assigned to assist with the wounded. It is possible to create scenarios in which fatalities could occur, but these would be unusual cases. The effects from blister agents include irritation of the eyes, burning of the skin, and difficulty in breathing. The more severe exposure results in blisters, which may be delayed. The only real street treatment for these signs and symptoms is decontamination and supportive measures. It is important to have anyone with liquid contact blot the liquid off the skin and avoid spreading the agent. Fortunately, the chemical and physical properties of these agents make them difficult to disseminate, and coming in contact with the liquid would be the primary means of injury.

Blood and Choking Agents

The four chemicals discussed here in addition to being terrorism agents are also common industrial chemicals. The first category is **blood agents** and includes hydrogen cyanide and cyanogen chloride. Both of these materials are gases that disrupt the body's ability to use the oxygen within the bloodstream. They are also referred to as chemical asphyxiates. The **choking agents**, chlorine and phosgene, are very common in industry. Chlorine is present in almost every town, because it is used for water treatment processes and in swimming pools. Any community that has a water system or swimming pool has some form of chlorine. Chlorine comes in cylinders of 150 pounds to 90-ton railcars. It also comes in the tablet form (HTH) that is typically used in residential pools. The release of chlorine from a 90-ton railcar could result in several hundred thousand injuries and possibly an equal number of deaths, especially in an urban area. A small amount of chlorine can be very deadly or at a minimum create substantial panic in a community.

> **Signs and Symptoms of Blood and Choking Agents** Many of the blood agents are commonly found in industry and may be found at normal chemical facilities. The signs and symptoms of slight exposure to blood agents include dizziness, difficulty in breathing, nausea, and general weakness. With cyanides, the breathing initially will be rapid and deep, followed by respiratory depression and usually death. The two most common choking agents are very common in industrial use, and chlorine can be found in most communities. Signs and symptoms include difficulty in breathing and respiratory distress, eye irritation, and, in higher amounts, skin irritation. Phosgene may present delayed effects, while chlorine's effects are immediate.

Irritants (Riot Control)

The most commonly used materials that are classified as potential terrorism agents are irritants and include mace, pepper spray, and tear gas. An incident that uses an irritant often impacts a large number of people, because the usual target is a school, mall, or other large place of assembly. The use of one small container can affect large numbers of people and make them immediately symptomatic. Luckily these materials are not extremely toxic—although they are extremely irritating—and the symptoms will usually disappear after fifteen to twenty minutes of exposure to fresh air. The response to one of these incidents is difficult because patients with real medical problems need treatment and the source of the irritant is often difficult to identify.

> **Signs and Symptoms of Irritants** The signs and symptoms for a slight exposure up to a high dose are the same, with the exception of increasing severity. The signs and symptoms are eye and respiratory irritation. There is no real treatment except removal to fresh air; in fifteen to twenty minutes the symptoms will begin to disappear. Supportive care can be provided.

Biological Agents and Toxins

The most likely agents to be used in a terrorism scenario are **biological agents** and **toxins**. Some of the materials in this grouping include anthrax, mycotoxins, smallpox, plague, tularemia, and ricin. Out of all of the agents for a terrorist to make, this grouping is the easiest, especially ricin. The fatal route of entry differs with each of the agents and could occur via skin contact, inhalation, or injection. These agents are difficult to distribute effectively, and in some cases exposure to sunlight may neutralize many of these agents. Specific information about the two most popular agents is provided next.

Anthrax

Anthrax is a naturally occurring bacterial disease that is commonly found in dead sheep. It is contagious through skin contact or by inhalation of the anthrax spores. Although relatively easy to obtain, it is more difficult to culture and grow the proper grade of anthrax. To produce fatal effects, the type of anthrax required is called weapons-grade anthrax and is very difficult to produce. Even if developed it must be distributed effectively and under the right conditions.

> **Anthrax Scare 2001** The FBI is still investigating the anthrax attacks of October 2001 that started in Boca Raton, Florida, at the American Media building. It is suspected that a letter was sent and opened at this facility, which publishes the *National Enquirer*. One person died and several others fell ill at this building, but a letter was never recovered from this facility. The letters that were sent targeted the media and members of Congress. The NBC studios, the *New York Post*, and ABC all received letters. Two members of Congress, Senator Daschle and Senator Leahy, received letters. The letters killed five people and made eighteen others ill. Some of the deaths and illnesses involved people at various post offices. It is thought that the mail handling process enabled anthrax to aerosolize and get into the air. The letters contained small amounts of anthrax, which has been tested to be the Ames strain of anthrax that was produced in the United States. A lot has been learned about anthrax and its ability to be used as a weapon. The theoretical dose of anthrax in each of the recovered letters was estimated to be able to kill several hundred thousand people. The reality is that each letter typically killed one person. The other major factor in the deaths was a delay in treatment and in some cases misdiagnoses. The persons who did not seek quick treatment or who were misdiagnosed had difficult or unsuccessful recoveries. Those who sought quick medical treatment, during which the agent was recognized as anthrax, survived. The Centers for Disease Control (CDC) reported that there were ten cases of inhalational anthrax and twelve confirmed or suspected cases of cutaneous anthrax. Of the ten inhalational cases, seven occurred to postal workers in New Jersey and Washington, D.C., at mail sorting facilities. In the American Media case, one person who received the letter died, and the person sorting the mail fell ill. Six of the ten individuals with inhalational cases of anthrax survived after treatment, which is higher than originally anticipated.
>
> The signs and symptoms of inhalational anthrax are a one- to four-day period of malaise, fatigue, fever, muscle tenderness, and a nonproductive cough followed by a rapid onset of respiratory distress, cyanosis, and sweating. The recent cases also had profound, often drenching, sweating, along with nausea and vomiting.

Ricin

Although ricin is easy to make, the required distribution method leaves a lot to be desired because it must be injected to be truly effective. It is 10,000 times more toxic than the nerve agent sarin. A small amount such as one milligram, about the size of a pinhead, can be fatal. Death usually occurs several days after injection. By other routes of entry, such as inhalation or ingestion, the most likely consequence is that many people would get sick but would eventually recover. After explosives, ricin is the leading choice of domestic terrorists, and several times a year someone is arrested for possession of ricin.

> **Response to Anthrax Hoaxes** With anthrax hoaxes becoming a daily occurrence and dramatically peaking in late 2001 and early 2002, emergency responders should establish a plan for response. From the outset it would be difficult for someone to manufacture and successfully pull off an anthrax threat. The use of real anthrax is most likely an isolated case. One cannot say with absolute certainty, but a new attack is not likely and is not likely to kill or injure mass numbers of people. In the HAZ-

MAT and WMD business one should avoid the use of the words *never*, *always*, and *best*. The credibility factor and the technological difficulty make weaponized anthrax an unlikely candidate for use as a weapon. In many of the hoax incidents, letters proclaimed that the recipients had been exposed to anthrax. In some of these cases the only thing present in the envelope was a letter, which meant that there was no other material present. Anthrax is not invisible, and in order to improve the likelihood that the attack may be successful a quantity of the material must be present. In order to cause health effects the material must be inhaled, and in order to be inhaled the material needs to be distributed to put it into the air. The material, since it is a solid, requires a device to put it into the air. If this device is not present the risk to any persons in a building is very small. The only action required is to double- or triple-bag the envelope, package, or material in bags suitable for evidence collection. Procedures for the collection of evidence should be followed, and the local police as well as the FBI should be consulted. The persons who touched the material should be instructed to wash their hands with soap and water. They do not require full body decontamination nor are special solutions required such as a bleach and water mixture. The military advises that water is more than sufficient for humans. Anyone who was in the immediate vicinity (i.e., several feet) of the person does not require decontamination. The only reason for a full body wash would be if the material was splashed on a person from head to toe, but it is not required to be done immediately. The person can be taken to a shower or be provided privacy to take a shower. The person who did open the envelope should be entered into the health care system, advised of the signs and symptoms of exposure, and provided with emergency contact information. There is no need to start prophylactic antibiotics just because a person opened a letter with a powder in it. There is sufficient time to do lab analysis on the material before medical treatment is required. The threat from a solid material (no matter how toxic) is from inhaling the dust or touching the material. Gloves and SCBA are adequate protection for the collection of evidence. The FBI has a number of labs around the country set up to assist with the identification of WMD agents, and the local FBI office should be contacted for assistance.

Radioactive Agents

Nuclear agents, unfortunately, have to be put back on the list of possibilities that a terrorist could use. There are two types of radiation events, nuclear detonation and **radiological dispersion devices (RDDs)**. The use of an actual nuclear detonation device is very unlikely given the security these materials have. The amount required and the specific type also make the use unlikely. Although there is current speculation that there are some small nuclear devices missing from Russia, this has never been substantiated. There is the potential, however, for some nuclear material that could be used in an RDD coming from Eastern Europe. An RDD is a device that disperses radiological materials, usually through an explosive device. One example would be the use of a pharmaceutical grade radioactive material attached to a pipe bomb, which could cause a large amount of radioactive material to be distributed. The strength of the radiation source dictates how harmful an RDD would be. In many cases the RDD would present more of an investigative issue than a significant health issue. There are some radiation sources that would present a risk, but these are not in common use. The other factor is the explosive device and its limitations. Technologically an RDD is viable for a small area such as one-half acre of land. Attempting to distribute a significant amount of radioactive material takes a significant explosive device, not to mention the radiation source. The larger the RDD the more danger there is to the terrorist who has to assemble, transport, and detonate the device. There is an advantage to a noncredible RDD or a small RDD, and that is the public's reaction. The perception to the public and to many responders is that this would be a radioactive disaster. Radiation meters would indicate that there were radioactive materials spread around. The reality is that the amount of radiation would not be dangerous. As time passed, the danger would lessen as the radioactive material became less hazardous. Still, there would likely be tremendous concern in the community. Radiation causes fear because it is a big unknown, making it a prime weapon for a terrorist. Education on the hazards of radiation and the effective use of radiation monitors can ease this fear and allow responders to make an informed response.

Other Terrorism Agents

Although considerable emphasis is placed on warfare agents, in reality many other common industrial or household materials can be just as deadly, if not more so.

> **Note** Remember that a terrorist wants to create panic, not necessarily kill or injure massive numbers of people.

The most likely scenario is a pipe bomb, followed by the use of ricin. Responders should not be lulled into a false sense of security if they do not find a "warfare" agent or if the device is only a small pipe bomb and not a moving truck filled full of ammonium nitrate and fuel oil. Nerve agents (or any other chemical warfare agent) have never been used in the United States, nor manufactured by anyone other than the military. The FBI Bomb Data Center reports that out of the 3,000 bombings that occur each year, there have been only two large bombings, but these "small" bombs kill an average of 32 and injure 277 people each year.

DETECTION OF TERRORISM AGENTS

The detection of terrorism agents is difficult, and given the potential circumstances is done under severe conditions. The response to terrorism incidents has changed how many HAZMAT teams operate and has increased their capability to handle other situations. The confirmation that terrorist agents have been used is difficult because they are extremely hard to detect. The exceptions are the standard industrial materials such as chlorine, which is easy to detect and confirm its presence.

The detection of terrorism agents addresses three major categories of hazards: chemical agents, radiological materials, and biological agents. There are a number of devices that detect chemical warfare agents. They range from inexpensive paper strip tests (M-8 and M-9) to sophisticated electronic devices costing more than $125,000. The most common are direct reading devices, which detect nerve and blister agents, **Figure 24-9.** Some devices will also detect drugs, explosives, mace and pepper spray, and industrial chemicals. The radiological detection field has dosimeters that track the dose of radiation that one is receiving to handheld instruments. Radiation pagers and pager/dosimeters are very small. Pager-sized devices are available and can detect levels of radiation and alert the user to potential danger. Typical radiation monitors detect the dose a human is absorbing, while some newer models will determine and identify the exact type of radiation source present. The detection of biological agents is more difficult, and there are limited

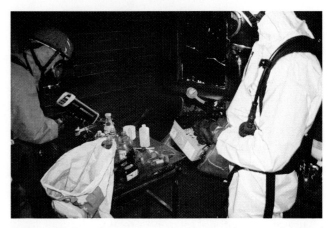

Figure 24-9 This air monitor device detects chemical warfare agents, gamma radiation, and irritants such as mace and pepper spray.

choices. One detection device that is new to the emergency response world is a polymerase chain reaction (PCR) unit, which detects the DNA of the sample and compares it to other DNA samples loaded in the library. It is the same method that is used in the laboratory to detect biological agents. The only other choice is handheld bioassays, which, depending on the brand, function with varying degrees of accuracy. Some work well and have accuracy rates in the 90th percentile, while others have accuracy rates in the 30th percentile. The major issue with any biological agent detection is that the sample collection method has to follow exacting standards without any deviation. The detection of any terrorism agent is difficult, and the HAZMAT team should take responsibility for doing the testing.

In some cases standard civilian detection devices such as photo-ionization detectors also play a role in the detection of terrorism agents. Other tests are used by HAZMAT teams that have applicability in the detection of terrorism agents. Due to the many mitigating factors the detection of these agents is going to be difficult, and to be conclusive will probably require lab tests.

FEDERAL ASSISTANCE

The federal government has established roles and responsibilities in the event an act of terrorism occurs, and these roles are provided in PDD 39. Per this document, the FBI is designated the lead agency during the emergency (crisis management) stage of an incident. The Federal Emergency Management Agency (FEMA) becomes the lead when the incident is no longer in the emergency phase (consequence management). The FBI has a Hazardous Materials Response Unit (HMRU) that provides identification, mitigation assistance, and evidence collection for potential terrorist incidents. The HMRU is a multifaceted group that not only responds to terrorism incidents, but also responds to incidents involving explosives, drug labs/incidents, and environmental crimes. A response from FEMA will vary with the incident, as will the number of FEMA personnel, but the response will be similar to any other disaster; FEMA assists in restoration and recovery issues.

The Urban Search and Rescue (USAR) teams fall under FEMA, and are activated by following the emergency management chain from the local level to the state level and then to FEMA for activation. At this time there are twenty-eight USAR teams across the country. They provide expertise in heavy rescue operations such as building collapses. This team of sixty-two people is composed of rescue specialists, dog search teams, medical specialists, communications specialists, and an engineering and rigging component. The majority of victims are rescued by the first responders and other local specialized resources. But in some cases trapped victims may be alive for many days and may require the expertise and equipment that a USAR team has available. The unfortunate thing is that the USAR teams have a delayed response, because it takes them several hours to become airborne, not to mention travel time. If there is a possibility they may be needed, their assistance should be requested early. Like many of the teams, it is not uncommon for an advance party to arrive hours prior to the arrival of the remainder of the team.

A number of other agencies may be involved in the event of an incident. The military has a couple of units that have terrorism response capabilities and responsibilities. The army has the Technical Escort Unit (TEU), headquartered at the Aberdeen Proving Grounds, Maryland. There are other units at Dugway Proving Grounds, Utah; Pine Bluff Arsenal, Arkansas; and Fort Belvoir, Virginia. The TEU is assigned to provide escort service for warfare agents and to be the troubleshooting group in the event of an incident involving warfare agents or explosives. When such an incident occurs, the TEU will respond to assist with identification and the mitigation of the incident. Team members handle chemical, biological, and explosive materials as well as other hazardous materials. TEU is a self-contained unit and has the lab resources of the Soldiers Biological and Chemical Command (SBCCOM) of the Aberdeen Proving Grounds available to assist them.

Basic Incident Priorities When dealing with an incident that involves terrorism, first responders should follow these guidelines:

- To rescue live victims, use full protective clothing including SCBA. All of the agents listed in this chapter are predominantly hazardous through inhalation. Avoid touching any unknown liquids or solids, because most are also toxic through skin contact.
- Use a quick in/quick out approach: Do not treat victims, remove them from the area. Keep in mind the terrorist may be among the injured. Watch for secondary devices.
- Request HAZMAT and the police bomb squad. The sooner responders eliminate the potential for chemical agents or a secondary device, the better off they will be. The HAZMAT team may have the ability to detect the agents listed in this chapter, and most HAZMAT teams are working with their bomb squads on a more frequent basis.
- Limit personnel operating in the hazard area.
- Establish multiple staging areas, out of the line of sight.
- Notify the local emergency management agency so that they can mobilize the state and federal resources.
- If a building has collapsed or there is potential for a building collapse, request assistance from a tactical rescue team or a USAR team.
- Isolate all victims, separating contaminated from clean victims.
- Establish a safe triage, treatment, and transport area away from the impact or hot zone.
- Notify all area hospitals of the incident.
- Remember that the incident is a crime scene and make provisions to preserve as much evidence as possible.
- If there is reason to suspect the presence of chemical agents, use the DOT ERG or other reference sources such as the *Medical Management of Chemical Casualties Handbook* to suggest safety precautions and patient treatments.

The marines have a unit known as the Chemical and Biological Incident Response Force (CBIRF), which comes from Indian Head, Maryland. This unit responds to acts of terrorism across the world and has three main components: decontamination, medical, and security. It is a self-contained unit and requires only a water source for conducting long-term operations. CBIRF is able to respond nationwide upon request to terrorist incidents and can provide detection and mitigation as well. All of the federal resources including CBIRF and TEU can be and have been pre-positioned for certain events, such as the Olympics, political conventions, and visits by dignitaries. These resources integrate within the local system, usually hidden away from the public, and are immediately available.

Other federal programs include training under the Nunn-Lugar-Domenici Legislation, which was passed in September 1996 (P.L. 104-201). This law, known as the Domestic Preparedness Training Initiative, mandated that the Department of Defense provide training to the 157 major cities and counties across the United States. It is intended to enhance the capability of the local, state, and federal response to incidents involving NBC materials. Part of this process is an assessment to determine the cities' capabilities after the training. A second component of the law provides for thirty teams to be fielded by the National Guard across the United States. These teams known as Weapons of Mass Destruction Civil Support Teams (WMD-CSTs) are being established to provide a National Guard component to the response to terrorists. Also set up around the country are Metropolitan Medical Response Teams (MMRTs), which are a group of 129 people on each team. They are trained and equipped to handle the medical component of a terrorism event. The plan is to have them in the 120 largest metropolitan areas in the country.

Lessons Learned

The response to a potential terrorism incident can be very challenging, and every responder must be alert to the possibility of such an incident. To be safe, responders should wear all PPE, not linger in the environment, and relocate to a safe area once the live victims are out. It is important to be aware of the potential for secondary devices and request HAZMAT and bomb squad assistance quickly. One to thousands of victims may be injured or killed. Scenarios involve tremendous loss of life, including a large number of responders. There exists the possibility that responders may lose and the terrorist will win, a situation that can be avoided by training, planning, and preparing for such an incident.

KEY TERMS

Anthrax A biological material that is naturally occurring and is severely toxic to humans. It is commonly used in hoax incidents.

Biological Agents Microorganisms that cause disease in humans, plants, and animals; they also cause the victims' health to deteriorate. Biological agents have been designed for warfare purposes.

Blister Agents A group of chemical agents that cause blistering and irritation of the skin. Sometimes referred to as vesicants.

Blood Agents Chemicals that affect the body's ability to use oxygen. If they prevent the body from using oxygen, fatalities result.

Butyric Acid A fairly common lab acid that has been used in many attacks on abortion clinics. Although not extremely hazardous, it has a characteristic stench that permeates the entire area where it is spilled.

Choking Agents Agents that cause a person to cough and have difficulty breathing. The terrorism agents that are considered choking agents are chlorine and phosgene, both very toxic gases.

HAZMAT Crime A criminal act that uses or threatens the use of chemicals as a weapon.

Incendiary Agents Chemicals that are used to start fires, the most common being a Molotov cocktail.

Mass Casualty An incident in which the number of patients exceeds the capability of the EMS to manage the incident effectively. In some jurisdictions this can be two patients, while in others it may take ten to make the incident a mass casualty.

Nerve Agents Chemicals that are designed to kill humans, specifically in warfare. They are chemically similar to organophosphorus pesticides and cause the same medical reaction in humans.

PDD 39 Presidential Decision Directive 39, which established the FBI as the lead agency in terrorism incidents responsible for crisis management. It also established FEMA as the lead for consequence management.

Radiological Dispersion Device (RDD) An explosive device that spreads radioactive material throughout an area.

Ricin A biological toxin that can be used by a terrorist or other person attempting to kill or injure someone. It is the easiest terrorist agent to produce and one of the most common.

Toxins Disease-causing materials that are extremely toxic and in some cases more toxic than other warfare agents such as nerve agents.

Vesicants A group of chemical agents that cause blistering and irritation of the skin. Commonly referred to as blister agents.

Weapons of Mass Destruction (WMD) A term that is used to describe explosive, chemical, biological, and radiological weapons used for terrorism and mass destruction.

REVIEW QUESTIONS

1. Describe four potential targets of terrorism.
2. Describe four indicators of potential terrorist activity.
3. What are the most readily available agents that could be used in a terrorist attack?
4. Which three main local/regional agencies or groups should be notified immediately of a suspected terrorist attack?
5. Describe the process of requesting federal assistance.
6. Explain which of the BNICE agents is the most likely to be found at an incident.
7. Describe the second most likely agent.
8. Describe which agent is designed to kill immediately.
9. What are the immediate signs and symptoms of sarin exposure?

Endnote

1. The *Roe v. Wade* decision was the Supreme Court case that allowed legalized abortions, and is the case that has created a lot of the turmoil between pro-choice and pro-life forces.

Additional Resources

Bevelacqua, Armando S., *Hazardous Materials Chemistry*. Delmar Learning, a part of the Thomson Corporation, Clifton Park, NY, 2001.

Bevelacqua, Armando S. and Stilp, Richard H., *Hazardous Materials Field Guide*. Delmar Learning, a part of the Thomson Corporation, Clifton Park, NY, 1998.

Bevelacqua, Armando S., and Stilp, Richard, *Terrorism Handbook for Operational Responders,* 2nd ed., Delmar Learning, a part of the Thomson Corporation, Clifton Park, NY, 2004.

Buck, George, *Preparing for Terrorism: An Emergency Services Guide.* Delmar Learning, a part of the Thomson Corporation, Clifton Park, NY, 1998.

Buck, George, *Preparing for Biological Terrorism: An Emergency Services Guide.* Delmar Learning, a part of the Thomson Corporation, Clifton Park, NY, 2002.

Buck, George, *Preparing for Terrorism: An Emergency Services Guide.* Delmar Learning, a part of the Thomson Corporation, Clifton Park, NY, 1998.

Buck, George, Buck, Lori, and Mogil, Barry, *Preparing for Terrorism: The Public Safety Communicator's Guide.* Delmar Learning, a part of the Thomson Corporation, Clifton Park, NY, 2003.

Hawley, Chris, *Hazardous Materials Incidents.* 2nd ed., Delmar Learning, Clifton Park, NY, 2004.

Hawley, Christopher, *Hazardous Material Air Monitoring and Detection Devices.* Delmar Learning, a part of the Thomson Corporation, Clifton Park, NY, 2002.

Henry, Timothy V., *Decontamination for Hazardous Materials Emergencies.* Delmar Learning, a part of the Thomson Corporation, Clifton Park, NY, 1999.

Medical Management of Biological Casualties. U.S. Army Medical Research Institute of Infectious Diseases, Ft. Dietrick, Fredrick, MD, 1996.

Medical Management of Chemical Casualties Handbook. U.S. Army Chemical Casualty Care Office, Medical Research Institute of Chemical Defense, Aberdeen Proving Ground, MD, Sept. 1995.

Pickett, Mike, *Explosives Identification Guide.* Delmar Learning, a part of the Thomson Corporation, Clifton Park, NY, 1999.

Scnepp, Rob and Gantt, Paul W., *Hazardous Materials, Regulations, Response & Site Operations.* Delmar Learning, a part of the Thomson Corporation, Clifton Park, NY, 1999.

Smelby Jr., L. Charles, ed., *Hazardous Materials Response Handbook,* 3rd ed. National Fire Protection Association, Quincy, MA, 1997.

Stilp, Richard and Bevelacqua, Armando. *Citizen's Guide to Terrorism Preparedness.* Delmar Learning, a part of the Thomson Corporation, Clifton Park, NY, 2003.

Stilp, Richard and Armando Bevelacqua, *Emergency Medical Response to Hazardous Materials Incidents.* Delmar Learning, a part of the Thomson Corporation, Clifton Park, NY, 1997.

GLOSSARY

Abandonment   Abandonment occurs when an emergency responder begins treatment of a patient and then leaves the patient or discontinues treatment prior to the arrival of an equally or higher trained responder.

Abrasion   A scrape or brush of the skin usually making it reddish in color and resulting in minor capillary bleeding.

Absolute Pressure   The measurement of pressure, including atmospheric pressure. Measured in pounds per square inch absolute.

Accelerator   A device to speed the operation of the dry pipe valve by detecting the decrease in air pressure. It pipes air pressure below the clapper valve, speeding its opening.

Accident   The result of a series of events and conditions that lead to an unsafe situation resulting in injury and/or property damage.

Accident Chain   A series of events and conditions that can lead to or have led to an accident. These events and conditions are typically classified into five areas: environment, human factors, equipment, events, and injury.

Acclimation   The act of becoming accustomed or used to something. Typically achieved through repeated practice within a given set of conditions.

Adapter   Device that adapts or changes one type of hose thread to another, allowing connection of two different lines. Adapters have a male end on one side and a female on the other with each side being a different thread type, for example, an iron pipe to national standard adapter.

Administrative Warrant   An order issued by a magistrate that grants authority for fire personnel to enter private property for the purpose of conducting a fire prevention inspection.

Aerial Apparatus   Fire apparatus using mounted ladders and other devices for reaching areas beyond the length of ground ladders.

Aircraft Rescue and Firefighting (ARFF)   Of or pertaining to firefighting operations involving fixed or rotary wing aircraft.

Allergic Reaction   The body's reaction to a substance to which there is an allergy.

Americans with Disabilities Act   Public law that bars discrimination on the basis of disability in state and local services. Enacted in 1990.

Amputation   Occurs when part of the body is severed completely as a result of an injury.

Anchor Point   A safe location from which to begin line construction on a wildland fire.

Anthrax   A biological material that is naturally occurring and is severely toxic to humans. It is commonly used in hoax incidents.

Application Rate   Amount of foam or foam solution needed to extinguish a fire. Usually expressed in gallons per minute per square foot or liters per minute per square meter.

Aqueous Film-Forming Foam (AFFF)   A synthetic foam that as it breaks down forms an aqueous layer or film over a flammable liquid.

Aquifer   A formation of permeable rock, gravel, or sand holding water or allowing water to flow through it.

Arson   A malicious fire or fires set intentionally by humans for vengeance or profit.

Arterial Bleeding   Bleeding from an artery.

Arteries   The blood vessels, or tubes, within the body that carry blood rich with oxygen and nutrients away from the heart.

Articulating Boom Ladder   An apparatus with a series of booms and a platform on the end. It is maneuvered into position by adjusting the various boom sections into place to position the platform at the desired location.

Aspect   The direction a slope faces given in compass directions.

Asphyxiation   Condition that causes death due to lack of oxygen and an excessive amount of carbon monoxide or other gases in the blood.

Association of Public Safety Communications Officials-Int., Inc. (APCO)   International not-for-profit organization dedicated to the advancement of public safety communications. Membership is made up of public safety professionals from around the world.

Atmospheric Pressure   The pressure exerted by the atmosphere, which for Earth is 14.7 pounds per square inch at sea level.

Atomization   The separation of atoms and molecules into an unconnected state where they are in suspension rather than in liquid form.

Attack Hose   Small- to large-diameter hose used to supply nozzles and other applicators or protective system for fire attack. Attack hose commonly means handheld hoselines from 1½ to 2½ inches (38 or 63 mm) in diameter.

Authority Having Jurisdiction (AHJ)   The responsible governing organization or body having legal jurisdiction.

Autoextended   When a fire goes out the window on one floor, up the side of the building, which is often noncombustible, and extends through the window or cockloft directly above.

Automated External Defibrillator (AED) A portable computer-driven device that analyzes a patient's heart rhythm and delivers defibrillation shocks when necessary.

Automatic or **Constant Pressure Nozzle** Nozzle with a spring mechanism built in that reacts to pressure changes and adjusts the flow and resultant reach of the nozzle.

Automatic Sprinkler System A system of devices that will activate when exposed to fire, connected to a piping system that will supply water to control the fire. Typically, an automatic sprinkler system is also supported by firefighters when they arrive on the scene.

Auxiliary Appliances Another term for protective devices, particularly sprinkler and standpipe systems.

Available Flow Amount of water that can be moved to extinguish the fire. Depends on the water supply, pump(s) and their capabilities, and the size and length of hose.

Avulsion An injury where a part of the skin is torn away, but still attached, leaving a flap or loose area hanging.

Axial Load A load passing through the center of the mass of the supporting element, perpendicular to its cross section.

Backdraft A sudden, violent reignition of the contents of a closed container fire that has consumed the oxygen within the space when a new source of oxygen is introduced.

Backflow Preventers A check valve or set of valves used to prevent a backflow of water from one system into another. Required where a building water or fire protection system connects with the public water system. Backflow preventers are being required for environmental and health reasons.

Backstretch or **Flying Stretch** An attack line lay where the engine is at the hydrant and the line is stretched back from the engine to the fire. The flying stretch is a version of the backstretch where the engine stops in front of the fire, the attack portion is removed, and the engine proceeds to the hydrant.

Balloon Frame A style of wood frame construction in which studs are continuous for the full height of a building.

Bank Down A condition in which the heat, smoke, and fire gases have reached the uppermost level in a compartment and, instead of continuing up, begin to push down from the ceiling toward the floor.

Base Radio Radio station that contains all of the antennas, receivers, and transmitters necessary to transmit and receive messages.

Basic 9-1-1 Telephone system that automatically connects a person dialing the digits "9-1-1" to a predetermined answering point through normal telephone service facilities. Number and location information is not normally provided in *basic* systems.

Beam A structural member subjected to loads perpendicular to its length.

Bed Ladder The nonextending part of an extension ladder.

Bevel The outside curve of the fork end of the Halligan tool.

Bight A doubled section of rope, usually made along the standing part, that forms a U-turn in the rope that does not cross itself.

Biological Agents Microorganisms that cause disease in humans, plants, and animals; they also cause the victim's health to deteriorate. Biological agents have been designed for warfare purposes.

Blister Agents A group of chemical agents that cause blistering and irritation of the skin. Sometimes referred to as vesicants.

Blood Agents Chemicals that affect the body's ability to use oxygen. If they prevent the body from using oxygen, fatalities result.

Body Substance Isolation Precautions A set of precautions for emergency responders designed to prevent exposure to any body fluid or substance.

Boiling Liquid Expanding Vapor Explosion (BLEVE) Describes the rupture of a container when a confined liquid boils and creates a vapor pressure that exceeds the container's ability to hold it.

Bolt Throw The distance the bolt of a lock travels into the jamb or strike plate. Usually ½ to 1½ inches.

Bond A substance or an agent that causes two or more objects or parts to bind.

Booster Hose Smaller diameter, flexible hard-rubber-coated hose of ¾- or 1-inch (19- to 25-mm) size usually mounted on a reel that can be used for small trash and grass fires or overhaul operations after the fire is out.

Bourdon Gauge The type of gauge found on most fire apparatus that operates by pressure in a curved tube moving an indicating needle.

Box Canyon A canyon open on one end and closed on the other. They become very dangerous when wildfire enters them.

Brachial Artery A major artery in the inside of the upper arm that supplies blood to the arm. Can be used as a pressure point for controlling bleeding and for locating a pulse on an infant.

Branch The command designation established to maintain span of control over a number of divisions, sectors, or groups.

Bresnan Distributors Has six or nine solid tips or broken stream openings designed to rotate in a circular spray pattern. Used to fight fire in basements or cellars when firefighters cannot make a direct attack on the fire.

British Thermal Unit (BTU) A measurement of heat that describes the amount of heat required to raise 1 pound of water 1°F.

Brush Gear Another term for a wildland personal protective ensemble.

Bunkers A slang term that is used mostly to describe the components of a structural firefighting ensemble. The original use of the term *bunkers* referred only to the pant/boot combination that firefighters wore at night and placed next to their "bunks" for rapid donning.

Butyric Acid A fairly common lab acid that has been used in many attacks on abortion clinics. Although not extremely hazardous, it has a characteristic stench that permeates the entire area where it is spilled.

Bypass Eductor Eductor with two waterways and a valve that allows plain water to pass by the venturi or through the venturi to create foam solution.

Cantilever Beam A beam that is supported at only one end.

Capillaries The very small blood vessels in the body that connect arteries and veins and filter the oxygen and nutrients from the blood into the tissues of the body.

Capillary Bleeding Bleeding from a capillary.

Carbon Dioxide (CO_2) An inert colorless and odorless gas that is stored under pressure as a liquid that is capable of being self-expelled and is effective in smothering Class B and C fires.

Carbon Monoxide Colorless, odorless, poisonous gas that when inhaled combines with the red blood cells excluding oxygen.

Cardiovascular System The heart, blood vessels, and blood within the body.

Carotid Pulse The pulse located on either side of the neck.

Cellar Nozzles Has four spray nozzles designed to rotate in a circular spray pattern for fighting fires in basements or cellars when firefighters cannot make a direct attack on the fire.

Chain of Command Common fire service term that means to always work through one's direct supervisor. The fire service is viewed as a paramilitary organization and because of this all requests for information outside the assigned workplace should go through the supervisor.

Check Valves Valves installed to control water flow in one direction, typically when different systems are interconnected.

Chemical Burns Burns caused by chemical substances that come into contact with the skin or tissues of the body, creating a caustic reaction.

Chimney Another term for drainage. Given because of the draw of fire as in heat going up the chimney.

Choking Agents Agents that cause a person to cough and have difficulty breathing. The terrorism agents that are considered choking agents are chlorine and phosgene, both very toxic gases.

Chord The top and bottom components of a beam or truss. The top chord is subjected to compressive force; the bottom chord is subjected to tensile force.

Cistern An underground water tank made from natural rock or concrete. Cisterns store large quantities of water—30,000 gallons or more—in areas without other water supplies or as a backup supply.

Class A Classification of fire involving ordinary combustibles such as wood, paper, cloth, plastics, and rubber.

Class B Classification of fire involving flammable and combustible liquids, gases, and greases. Common products are gasoline, oils, alcohol, propane, and cooking oils.

Class C Classification of fire involving energized electrical equipment, which eliminates using water-based agents.

Class D Classification of fire involving combustible metals and alloys such as magnesium, sodium, lithium, and potassium.

Class K A new classification of fire as of 1998 that involves fires in combustible cooking fuels such as vegetable or animal oils and fats.

Clipping Term associated with the use of two-way radios that is used to describe instances when either the first part of a message or the last part of a message is cut off as the result of either speaking before pressing the transmit key or releasing the transmit key prior to the end of a transmission.

Closed-Circuit SCBA A type of SCBA unit in which the exhaled air remains in the system to be filtered and mixed with oxygen for reuse.

Cockloft The area between the roof and the ceiling.

Code of Federal Regulations (CFR) The documents that include federally promulgated regulations for all federal agencies.

Collapse Zone The area around a building where debris will land when it falls. As an absolute minimum this distance must be at least 1½ times the height of the building.

Column A structural element that is subjected to compressive forces—typically a vertical member.

Combination Attack A combined attack based on partial use of both offensive and defensive attack modes.

Combination Fire Attack A blend of the direct and indirect fire attack methods, with firefighters applying water to both the fuel and the atmosphere of the room.

Combination Nozzle A spray nozzle that is capable of providing straight stream and spray patterns, which are adjustable or variable by the operator. Most fog nozzles used today are combination nozzles.

Combustion The chemical action in which heat and light are produced and the heat is used to maintain the chemical chain reaction to continue the process.

Command Vehicle Typically used by operations chief officers in the fire service.

Common Terminology The designation of a term that is the same throughout an IMS.

Communicable Disease A disease that can be transmitted from one person to another.

Communications Sending, giving, or exchanging of information.

Company A team of firefighters with apparatus assigned to perform a specific function in a designated response area.

Compound A combination of substances joined in a chemical bond that exists in a proportional amount and cannot be separated without chemical interaction.

Compressed Air Foam System (CAFS) A foam system where compressed air is injected into the foam solution prior to entering any hoselines. The fluffy foam created needs no further aspiration of air by the nozzle.

Compression A force that tends to push materials together.

Computer-Aided Dispatch Computer-based automated system that assists the telecommunicator in assessing dispatch information and recommends responses.

Concentrated Load A load applied to a small area.

Confined Space A space that is large enough to be entered but is not designed for continuous occupancy.

Conflagration A large and destructive fire.

Consent The acceptance of emergency medical treatment by a patient or victim.

Consolidated Incident Action Plan The strategic goals to eliminate the hazard or control the incident.

Constant or **Set Volume Nozzle** Nozzle with one set volume at a set pressure. For example, 60 gpm at 100 psi (227 L/min 690 kPa). The only adjustment is the pattern.

Constricted A condition of the pupils where they are much smaller than normal and may appear almost like a "pinpoint."

Continuous Beam A beam that is supported in three or more places.

Control Room A room on the ground floor of a high-rise building where all building systems controls are located.

Cribbing The use of various dimensions of lumber arranged in systematic stacks (pyramid, box, step, etc.) to support an unstable load.

Critical Incident Stress Debriefing (CISD) A formal gathering of incident responders to help defuse and address stress from a given incident.

Critical Incident Stress Management (CISM) A process for managing the short- and long-term effects of critical incident stress reactions.

Cutting Tools The group of tools used to cut through or around materials.

Database Organized collection of similar facts.

Dead Load The weight of the building materials and any part of the building permanently attached or built-in.

Defensive Attack A calculated attack on part of a problem or situation in an effort to hold ground until sufficient resources are available to convert to an offensive form of attack.

Dehydration A loss of water and vital fluids in the body.

Deluge Systems Designed to protect areas that may have a fast-spreading fire engulfing the entire area. All of its sprinkler heads are already open, and the piping contains atmospheric air. When the system operates, water flows to all heads, allowing total coverage. The system uses a deluge valve that opens when a separate fire detection system senses the fire and signals to trip the valve open.

Density The mass per unit volume of a substance under specified conditions of pressure and temperature.

Deployment Plan Predetermined response plan of apparatus and personnel for specific types of incidents and specific locations.

Depth of Char A term commonly used by fire investigators to describe the amount of time wooden material had burned. The deeper the char, the longer the material was burning or exposed to direct flame.

Design Load A load the engineer planned for or anticipated in the structural design.

Detergent-Type Foams Use synthetic surfactants to break down the surface tension of water to create a foaming blanket.

Diffusion A naturally occurring event in which molecules travel from levels of high concentration to areas of low concentration.

Dilated A condition of the pupils where they are much larger than normal and can take up almost the whole colored portion of the eye.

Direct Fire Attack An attack on the fire made by aiming the flow of water directly at the material on fire.

Disassembly The actual taking apart of vehicle components.

Discharge Flow Total amount of water flowing from the discharge side of the pump.

Displacement The relocating of major parts (i.e., doors, roof, dash, steering column) of a vehicle.

Distortion The bending of sheet metal or components.

Distributed Load A load applied equally over a broad area.

Distributor Pipe or **Extension Pipe** Devices that allow a nozzle or other device to be directed into holes to reach basements, attic, and floors that cannot be accessed by personnel. The distributor pipe has self-supporting brackets that help hold it into place when in use.

Division Command designation responsible for operations within an assigned geographic area.

Double Female Allows the two male ends of hose to be connected.

Double Male Used to connect two female thread couplings.

Drafting The pumping of water from a static source by taking advantage of atmospheric pressure to force water from the source into the pump.

Drainage A topographic feature on the side of a hill or mountain that naturally collects water runoff, channeling it to the bottom of the rise. Fire is attracted to this feature.

Dressing The practice of making sure that all parts of a knot are lying in the proper orientation to the other parts and look exactly as the pictures herein indicate.

Dry Chemicals Dry extinguishing agents divided into two categories. Regular dry chemicals work on Class B and C fires; multipurpose dry chemicals work on Class A, B, and C fires.

Dry Hydrant A piping system for drafting from a static water source with a fire department connection at one end and a strainer at the water end.

Dry Pipe Systems Air under pressure replaces the water in the system to protect against freezing temperatures. The sprinkler control valve uses a dry pipe valve to keep pressurized air maintained above with the supply water under pressure below the valve.

Dry Powders Extinguishing agents for Class D fires.

Dump Site The area where tenders are unloaded or their load dumped.

Dutchman A short fold of hose or a reverse fold that is used when loading hose and a coupling comes at a point where a fold should take place or when two sets of couplings end up on top of or next to each other. The dutchman moves the coupling to another point in the load.

Dynamic A rope having a high degree of elongation (10 to 15 percent) at normal safe working loads.

Ears Elongated folds or flaps at the ends of a layer of hose to assist in pulling that layer.

Eccentric Load A load perpendicular to the cross section of the supporting element that does not pass through the center of mass.

Eductor Device that siphons a liquid from a container into a moving stream.

Electrical Conductor Any material that will permit electricity to flow through it.

Emergency Call Box System of telephones connected by private line telephone, radio-frequency, or cellular technology usually located in remote areas and used to report emergency situations.

Emergency Communications Center Facility either wholly or partially dedicated to being able to receive emergency and, in some instances, nonemergency reports from citizens. Centers such as these are sometimes referred to as fire alarm, headquarters, dispatch, or a public safety answering point (PSAP).

Emergency Medical Dispatch System designed for use by telecommunicators to assist them in evaluating patient symptoms using predetermined criteria and responses.

Emergency Medical Services The delivery of prehospital medical treatment.

Emergency Medical Technician (EMT) An individual trained and certified to provide basic life support emergency medical care.

Employee Assistance Program (EAP) A defined program that offers professional mental health and other health services to employees.

Encoder Device that converts an "entered" code into paging codes, which in turn activate a variety of paging devices.

Endothermic Reaction A chemical reaction in which heat is absorbed, and the resulting mixture is cold.

Engine Company The unit designation of a group of firefighters assigned to a piece of apparatus designed to deliver water to the fire scene.

Engulfed To swallow up or overwhelm.

Enhanced 9-1-1 Similar in nature to basic 9-1-1 but with the capability to provide the caller's telephone number and address.

Equilibrium When referring to gas or liquids, a state where a balance has occurred in mixture or weight.

Evaporation A process in which the molecules of a liquid are liberated into the atmosphere at a rate greater than the rate at which the molecules return to the liquid. Ultimately the liquid becomes fully airborne in a gaseous state.

Exhauster A device to speed the operation of the dry pipe valve by detecting the decrease in air pressure. It helps bleed off air.

Exit Drills in the Home (EDITH) A fire survival program to encourage people to practice fire drills from their home or residence.

Exothermic Reaction A chemical reaction that releases heat, such as when two chemicals are mixed and the resulting mixture is hot.

Explosive Limits A concentration of a gas or liquid that is not too rich or too lean to ignite with force.

Exposure A contact with a potentially disease-producing organism; the contact does not necessarily produce the disease in the exposed individual.

Exposure Fire Any combustible item threatened by something burning nearby that has caught on fire.

Extension Ladder A ladder consisting of two or more sections that has the ability to be extended to a desired height through the use of a halyard.

External Bleeding Bleeding that is coming from an open wound on the body.

Extricate To set free, release, or disentangle a patient from an entrapment situation.

Federal Communications Commission Government agency charged with administering the provisions of the Communications Act of 1934 and the revised Telecommunications Act of 1996 and is responsible for nonfederal radio-frequency users.

Femoral Artery A major artery in the lower body near the groin that supplies the leg with blood. Can be used as a pressure point for controlling bleeding in the lower extremities.

Fill Site The area where tenders are filled or get their water.

Fire Alarm Notification to the fire department that a fire or other related emergency is in progess, which results in a response.

Fire Engineering The study of fire, fire behavior, fire extinguishment, and suppression.

Fire Flow Capacity The amount of water available or amount that the water distribution system is capable of flowing.

Fire Flow Requirement A measure comparing the amount of heat the fire is capable of generating versus the amount of water required for cooling the fuels below their ignition temperature.

Fire Hazard Any condition, situation, or operation that could lead to the ignition of unwanted combustion or result in proper combustion becoming uncontrolled.

Fire Hose A flexible conduit used to convey water or other agent from a water source to the fire.

Fire Hydraulics The principles associated with the storage and transfer of water in firefighting activities.

Fire Intensity A measurement of Btus produced by a fire. Sometimes measured in flame length in the wildland environment.

Fire Load The amount of heat energy released when combustibles burn in a given area or building—expressed in British thermal units (Btus).

Fire Resistive The capacity of a material to withstand the effects of fire.

Fire-Resistive Rating The time in hours that a material or assembly can withstand fire exposure. Fire-resistive ratings are usually provided for testing organizations. The ratings are expressed in a time frame, usually hours or portions thereof.

Fire Shelter A last-resort protective device for wildland firefighters caught or trapped in an environment where a firestorm or blowup is imminent.

Fire Societies Groups of people who voluntarily banded together to deal with a community's fire problems.

Fire Station Alerting System System used to transmit emergency response information to fire station personnel via voice and/or digital transmissions.

Fire Stopping Pieces of material, usually wood or masonry, placed in stud or joist channels to slow the extension of fire.

Fire Stream The water or other agent as it leaves the hose and nozzle toward its objective, usually the fire.

Fire Tetrahedron Four-sided pyramid-like figure describing the heat, fuel, oxygen, and chemical reaction necessary for combustion.

Fire Wardens Designated community individuals who walked the streets at night looking for fire and carrying large wooden rattles with which to signify a found fire.

Fire Watch An organized patrol of a protected property when the sprinkler or other protection system is down for maintenance. Personnel from the property regularly check to make sure a fire has not started and assist in evacuation and prompt notification of the fire department.

Firefighter Assist and Search Team (FAST) A company designated to search for and rescue trapped or lost firefighters. May also be called a rapid intervention team (RIT).

Firemark Signs on sheets of metal telling firefighters which company held the insurance policy on a home or building.

Flammable Limits The concentration level of a substance at which it will burn.

Flammable Range Ratio of gas to air that will sustain fire if exposed to flame or spark.

Flanks of the Fire The sides of a wildland fire running from the start point up each side to the end of the fire running into unburned areas.

Flash Point The temperature at which a liquid will liberate a flammable gas.

Flashover A sudden event that occurs when all the contents of a container reach their ignition temperature simultaneously.

Flow The rate or quantity of water delivered, usually measured in gallons per minute or liters per minute (1 gpm = 3.785 L/min).

Fluoroprotein Film–Forming Foam (FFFP) Combines protein with the film-forming fluorinated surfactants of AFFF to improve on the qualities of both types of foam.

Fluoroprotein Foam Designed as an improved protein foam with a fluorinated surfactant added.

Flush or Slab Doors Doors that are flat or have a smooth surface and may be of either hollow-core or solid-core construction.

Fly Ladder That portion of a ladder that extends out from the bed ladder. Also called *fly section*.

Foam An aggregate of gas-filled bubbles formed from aqueous solutions of specially formulated concentrated liquid foaming agents.

Fog Nozzle Delivers either a fixed spray pattern or variable combination of straight stream and spray patterns.

Forcible Entry The fire scene task of gaining entry to a building or secured area by disabling, breaking, or going around locking and security devices.

Foreman Individual designated as the leader of an early fire company; a predecessor to the modern title of fire chief.

Forestry Hose Specially designed hose for use in forestry and wildland firefighting. It comes in 1- and 1½-inch (25- and 38-mm) sizes and should meet U.S. Forestry Service specifications.

Fracture A medical term for a broken or cracked bone in the body.

Freelancing   The act of working alone or performing a task for which the firefighter has not been assigned.

Friction   Caused by the rubbing of materials against each other while in movement and converts or robs some of the movement energy into heat energy.

Friction Loss   Measurement of friction in a system such as a hoseline.

Frontage   The portion of a property that faces and actually touches the street.

Fuel Resistance   Ability to tolerate the fuel and to avoid being saturated by or picking up the fuel.

Full Thickness Burns   Burns affecting not only the skin structure but the tissues and muscles underneath. Full thickness burns may be red, white, or charred in color, and will appear dry because the blood vessels in the skin are damaged extensively and are not supplying fluids to the area.

Garden Apartment   A two- or three-story apartment building with common entryways and layouts on each floor, surrounded by greenery and landscaping, sometimes having porches and patios.

Gate Valves   Indicating and nonindicating valves that are opened and closed to control water flow.

Gauge Pressure   Measures pressure without atmospheric pressure. Normally fire department gauges do not measure atmospheric pressure. Gauge pressure is measured in psi or psig.

Girder   A large structural member used to support beams or joists—that is, a beam that supports beams.

Glazing   The glass or other clear material portion of the window that allows light to enter.

Ground Pads   Sheets of plywood, planks, aluminum sheets, and so on, used to distribute weight over a larger area.

Guard Dogs   Trained animals that will attack an intruder.

Guideline/Lifeline   Rope used as a crew is searching a structure to assist them in finding their way back out.

Gusset Plate   A connecting plate used in truss construction. In steel trusses, these plates are flat steel stock. In wood trusses, the plates are either light-gauge metal or plywood.

Halligan Tool   From the prying group, a 30-inch forged steel tool with three primary parts: the adz end, the pike end, and the fork end.

Halyard   A rope or cable that is used to raise the fly ladders of an extension ladder.

Hard Suction Hose   A special type of hose that does not collapse when used for drafting.

Hardware   Equipment used in conjunction with life safety ropes and harnesses (carabiners, figure eights, rappel racks, etc.).

Harnesses   Webbing sewn together to form a belt, seat harness, or seat and chest harness combination.

Hazardous Materials   Chemicals that are flammable, explosive, or otherwise capable of causing death or destruction when improperly handled or released.

Hazardous Materials Technician   An individual trained to meet the requirements of CFR OSHA 1910.120, *Technician Level for Hazardous Materials Response.*

HAZMAT Crime   A criminal act that uses or threatens the use of chemicals as a weapon.

Head of the Fire   The running top or aggressive end of the fire away from the start point.

Head Pressure   Measures the pressure of a column of water in feet (meters). Head pressure gain or loss results when water is being pumped above or below the level of the pump. A head of 2.31 feet (0.7 m) would equal 1 psi (6.895 kPa).

Heat Resistance   The ability of foam to stand up to the heat of the fire or to hot surfaces near the fire.

Heat Sink   The term used to denote a place where heat is drained away from a source.

Helix   The metal or plastic bands or rings used in hard suction hose to prevent its collapse under drafting conditions.

Higbee Cut   The blunt ending of the threads of fire hose couplings that allows the threads to be properly matched, avoiding cross-threading.

Hoistway   The shaft in which an elevator or a number of elevators travel.

Hollow-Core Door   Any door that is not solid, usually with some type of filler material between face panels.

Home Alerting Devices   Emergency alerting devices primarily used by volunteer department personnel to receive reports of emergency incidents.

Hook   A tool with a 32-inch to 12-foot handle with a pike and hook on one end. Used for pulling ceilings or separating other materials. Also known as a *pike pole.*

Horizontal Ventilation   Channeled pathway for fire ventilation via horizontal openings.

Hose Bed   The portion or compartment of fire apparatus that carries the hose.

Hose Bridges   Devices that allow vehicles to pass over a section of hose without damaging it.

Hose Cap   Does not allow water to flow through it. Instead, it caps the end of a hoseline or appliance to prevent water flow.

Hose Cart   A handcart or flat cart modified to be able to carry hose and other equipment around large buildings. Some departments use them for high-rise situations.

Hose Clamp   A device to control the flow of water by squeezing or clamping the hose shut. Some work by pushing a lever that closes the jaws of the device and others have a screw mechanism or hydraulic pump that closes the jaws.

Hose Jackets Metal or leather devices used for stopping leaks without shutting down the line that is fitted over the leaking area and either clamped or strapped together to control the leak.

Hose Roller or **Hoist** A metal frame, with a securing rope, shaped to fit over a windowsill or edge of a roof with two rollers to allow the hose to roll over the edge, preventing chafe.

Hose Strap A short strap with a forged handle and cinch clip attached. Used to help maneuver hose and attach hose to ladders and stair rails.

HVAC Acronym for heating, ventilation, and air-conditioning unit. HVACs are typically a rooftop unit on commercial buildings. Buildings may have one or dozens of these units.

Hydrant Valves or **Switch Valves** Valve used on a hydrant that allows an engine to connect and charge its supply line immediately but also allows an additional engine to connect to the same hydrant without shutting down the hydrant, and increases the flow of the hydrant.

Hydrant Wrenches Tools used to operate the valves on a hydrant. May also be used as a spanner wrench. Some are plain wrenches and others have a ratchet feature to speed the operation of the valve.

Hydraulic Pistons Mechanical rams that operate by pressure exerted through the use of a liquid, usually some form of oil.

Hydraulics The study of fluids at rest and in motion.

Hydrocarbon Any of numerous organic compounds, such as benzene and methane, that contain only carbon and hydrogen.

Hypoperfusion A serious condition caused by a problem or failure of the circulatory system that results in a decrease of oxygen and vital nutrients to the body's tissues. Also known as shock.

Hypoxia A deficiency of oxygen.

Ignition The point at which the need for outside heat application ceases and a material sustains combustion based on its own generation of heat.

Ignition Point The temperature at which a substance will continue to burn after the source is removed.

Immediately Dangerous to Life and Health (IDLH) The maximum level of danger one could be exposed to and still escape without experiencing any effects that may impair escape or cause irreversible health effects.

Impact Load A load that is in motion when it is applied.

Implied Consent The assumption of acceptance of emergency medical treatment by an unconscious patient or a child with no parents or legal guardians present.

Incendiary Agents Chemicals that are used to start fires, the most common being a Molotov cocktail.

Incident Action Plan (IAP) A strategic and tactical plan developed by the incident commander.

Incident Management System (IMS) A management system utilized on the emergency scene that is designed to keep order and follow a sequence of set guidelines.

Incision A cut to the skin that leaves a straight, even pattern.

Increaser Used to connect a smaller hose to a larger one.

Indirect Fire Attack An attack made on interior fires by applying a fog stream into a closed room or compartment, thus converting the water into steam to extinguish the fire.

Infection Control Procedures and practices for firefighters and emergency medical care providers to follow to prevent the transmission of diseases and germs from a patient to themselves or other patients.

Infectious Disease See **Communicable Disease.**

Initial Assessment The initial investigative action taken by care providers to determine if the patient has the basic signs of life as well as any serious, life-threatening injuries.

In-Line Eductor Eductor in which the waterway is always piped through a venturi.

Inorganic A substance that is not of any living organism.

Intake Relief Valve Required on large-diameter hose at the receiving engine that functions as a combined overpressurization relief valve, a gate valve, and an air bleed-off.

Integrated Communications The ability of all units or agencies to communicate at an incident.

Interface Firefighting Fighting wildland fire and protecting exposed structures in rural settings.

Internal Bleeding Bleeding within the body when no visible open wound is present.

Intervention The act of intervening; to come between as an influencing force. Typically a reactive action.

Irons The combination of a Halligan tool and flathead ax or maul.

Jacket The outer part of the hose, often a woven cloth or rubberized material, which protects the hose from mechanical and other damage.

Jamb The mounting frame for a door.

Jet Dump A device that speeds the process of dumping a load of water from a tanker/tender.

Jet Siphon A device that speeds the process of transferring water from one tank to another.

Joist A wood framing member that supports floor or roof decking.

Kerf Cut A quick and easily made examination hole. It is created by letting the spinning blade of a power saw cut through the material to be cut and pulling it out, leaving only a slit-like cut measuring approximately 12 inches long and only as wide as the cutting blade.

Kern A derivative of the term *kernel,* which is defined as "the central, most important part of something; core; essence."

Knockdown Speed Speed with which foam spreads across the surface of a fuel.

Laceration A cut to the skin and underlying tissues that leaves an irregular, even pattern.

Ladder Pipe An appliance that is attached to the underside of an aerial ladder for an elevated water application.

Laminated Glass Glass composed of two or more sheets of glass with a plastic sheet between them. The purpose of the plastic sheet is to hold the glass together if broken, thus reducing the hazard of flying glass.

Landing Plate The plate at the top or bottom of an escalator where the steps disappear into the floor.

Life Safety Term applied to the fire protection concept in which buildings are designed to allow for the escape of building occupants without injuries. Life safety usually makes the building more fire resistant, but this is not the main goal.

Life Safety Line According to NFPA 1983, rope dedicated solely to the purpose of constructing lines for supporting people during rescue, firefighting, or other emergency operations, or during training evolutions.

Lifting A term used to describe the removal of upper-level smoke and heat when cool air replaces the upper-level hot air that is escaping.

Liner The inner layer of fire hose, usually made of rubber or a plastic material, that keeps the water in the tubing of the hose.

Lintel A beam that spans an opening in a load-bearing masonry wall.

Live Load The weight of all materials and people associated with but not part of a structure.

Load-Bearing Wall Any wall that supports other walls, floors, or roofs.

Loaded Stream Combats the water freezing problem by adding an alkali salt as an antifreezing agent.

Loading The weight of building materials or objects in a building.

Local Application System Designed to protect only a certain or local portion of the building, usually directly where the hazard will occur or spread.

Locking Devices A mechanical device or mechanism used to secure a door or window.

Loop A turn in the standing part that crosses itself and results in the standing part continuing on in the original direction of travel.

Mantle Anything that cloaks, envelops, covers, or conceals.

Mask Confidence or "Smoke Divers" Training Training courses designed to develop a firefighter's skills and confidence for using SCBA.

Mass Casualty An incident in which the number of patients exceeds the capability of the EMS to manage the incident effectively. In some jurisdictions this can be two patients, while in others it may take ten to make the incident a mass casualty.

Master Stream or **Heavy Appliances** Non-handheld water applicator capable of flowing over 350 gallons of water per minute (1325 L/min).

Mastery The concept that an individual can achieve 90 percent of an objective 90 percent of the time.

Matter Something that occupies space and can be perceived by one or more senses; a physical body, a physical substance, or the universe as a whole. Something that has mass and exists as a solid, liquid, or gas.

Mayday A universal call for help. A Mayday indicates that an individual or team is in extreme danger.

Means of Egress A safe and continuous path of travel from any point in a structure leading to a public way. Composed of three parts: the exit access, the exit, and the exit discharge.

Medium-Diameter Hose (MDH) Either 2½- or 3-inch (63- or 75-mm) hose.

Medi-Vac An ambulance that transports patients by air. Typically, medi-vac units are helicopters with highly trained EMS personnel and nurses.

Midslope An area partway up a slope. Any location not on the bottom or top of a slope, as in a midslope road crossing the slope horizontally.

Miscible Having the ability to mix with water.

Mission Statement A written declaration by a fire agency describing the things that it intends to do to protect its citizenry or customers.

Mission Vision A term used to describe a condition in which a person becomes so focused on an objective that peripheral conditions are not noticed, as if the person is wearing blinders.

Mitigation Actions taken to eliminate a hazard or make a hazard less severe or less likely to cause harm. Typically a proactive action.

Mobile Data Computer Communications device that, unlike the mobile data terminal, does have information processing capabilities.

Mobile Data Terminal Communications device that in most cases has no information processing capabilities.

Mobile Radio Complete receiver/transmitter unit that is designed for use in a vehicle.

Mobile Support Vehicle Vehicle designed exclusively for use as an on-scene communication center and command post.

Modular Organization The ability to start small and expand if an incident becomes more complex.

Molecule The smallest particle into which an element or a compound can be divided without changing its chemical and physical properties; a group of like or different atoms held together by chemical forces.

Mortar Mixture of sand, lime, and portland cement used as a bonding material in masonry construction.

Mounting Hardware Hinges, tracks, or other means of attaching a door to the frame or jamb.

Multiple-Alarm Incident Involves the response of additional personnel.

Mutual Aid or **Assistance Agreements** Prearranged written agreements of the type and amount of assistance one jurisdiction will provide to another in the event of a large-scale fire or disaster. The key to understanding mutual aid is that it is a reciprocal agreement.

National Emergency Number Association Not-for-profit organization founded in 1982 and made up of more than 6,000 members. The association fosters technical advancement, availability, and implementation of a universal emergency telephone number system.

National Fire Protection Association (NFPA) A not-for-profit membership organization that uses a consensus process to develop model fire prevention codes and firefighting training standards.

National Institute for Occupational Safety and Health (NIOSH) A federal institute tasked with investigating firefighter fatalities and making recommendations to prevent reoccurrence.

Needed or **Required Flow** Estimate of the amount of water required to extinguish a fire in a certain type period. Based on the type and amount of fuel burning.

Nerve Agents Chemicals that are designed to kill humans, specifically in warfare. They are chemically similar to organophosphorus pesticides and cause the same medical reaction in humans.

Nested The state when all the ladders of an extension ladder are unextended.

NFPA 1001 *Standard for Fire Fighter Professional Qualifications,* a national consensus training standard establishing the job performance requirements of tasks to be performed by firefighters.

NFPA 1404 National Fire Protection Association standard created by the Fire Service Training Committee detailing the requirements for fire service SCBA programs, including training and maintenance procedures.

NFPA 1500 National Fire Protection Association standard created by the Technical Committee on Fire Service Occupational Safety and Health that addresses a number of issues concerning protective equipment.

NFPA 1981 National Fire Protection Association standard specific to open-circuit SCBA for fire service use that contains additional requirements above the NIOSH certification.

NFPA 72 National Fire Alarm Code.

NFPA Standard 1931 The standard issued by the National Fire Protection Association that governs fire service ladder testing and certification.

9-1-1 Emergency telephone number that provides access to the public safety services in the community, region, and, ultimately, nation.

NIOSH National Institute for Occupational Safety and Health, 42 CFR Part 84, sole responsibility for testing and certification of respiratory protection including fire service SCBA.

No-Knowledge Hardware Locking devices that require no key or special knowledge to operate.

Nozzle A tapered or constricted tube used to increase the speed or change the direction of water or other fluids.

Nozzle Flow The amount or volume of water that a nozzle will provide. Flow is measured in gallons per minute or liters per minute.

Nozzle Pressure The pressure required to effectively operate a nozzle. Pressure is measured in pounds per square inch or kilopascals.

Nozzle Reach The distance the water will travel after leaving the nozzle. Reach is a function of the pressure, which is converted to velocity or speed of the water leaving the nozzle.

Nozzle Reaction The force of nature that makes the nozzle move in the opposite direction of the water flow. The nozzle operator must counteract the thrust exerted by the nozzle to maintain control.

Occupancy Classifications The use for which a building or structure is designed.

Occupant Use Hose Hose that is used in standpipe systems for building occupants to fight incipient fires. It is usually 1½-inch (38-mm) single-jacket hose similar to attack hose.

Occupational Safety and Health Administration (OSHA) The federal agency, under the Department of Labor, that is responsible for employee occupational safety.

Offensive Attack An aggressive attack on a situation where resources are adequate and capable of handling the situation.

One- or **Two-Person Rope** According to NFPA 1983, a one-person rope requires a minimum tensile strength of 4,500 pounds, and a two-person rope requires a minimum tensile strength of 9,000 pounds.

Open-Circuit SCBA A type of SCBA unit in which the exhaled air is vented to the outside atmosphere.

Operational Period The time frames for operations at an incident. At large-scale or complex incidents these will usually be eight- to twelve-hour time frames.

Organic A substance derived from living organisms.

OSHA 29 CFR 1910.134 Standard establishing minimum medical, training, and equipment levels for respiratory protection programs.

Outside Stem and Yoke (OS&Y) Valve Has a wheel on a stem housed in a yoke or housing. When the stem is exposed or outside, the valve is open. Also called an outside screw and yoke valve.

Oxidizer A catalyst in the breakdown of molecules.

Oxygen-Deficient Atmosphere An atmosphere with an oxygen content below 19.5 percent by volume.

Packaging The bandaging and preparing of a patient to be moved from the place of injury to a stretcher.

Panel Doors Doors with a solid stile and rails with panels made of wood or glass or other materials.

Panic Hardware Hardware mounted on doors that enable them to be opened by pushing from the inside.

Paramedic (EMT-P) An individual trained and certified to provide advanced life support emergency medical care, including drug therapy.

Parapet The projection of a wall above the roofline of a building.

Partial Thickness Burns Burns affecting the entire skin structure that lies over the top of the fatty tissues and muscles causing skin to turn red and blistering of the skin.

Passport A term given to a specific accountability system where crews are tracked using a card (passport) with all members listed. An accountability manager tracks the passports on an accountability board.

PDD 39 Presidential Decision Directive 39, which established the FBI as the lead agency in terrorism incidents responsible for crisis management. It also established FEMA as the lead for consequence management.

Personal Alert Safety System (PASS) A device that emits a loud alert or warning that the wearer is motionless.

Personal Size-Up A continuous mental evaluation of an individual's immediate environment, facts, and probabilities.

Personnel Accountability Report (PAR) This is an organized roll call of all units assigned to an incident.

Piercing Nozzles Originally designed to penetrate the skin of aircraft and now have been modified to pierce through building walls and floors.

Pike Pole See **Hook**.

Pipe Chases A construction term used to describe voids designed to house building water supply and waste pipes. The term *electrical chase* is used for wiring.

Pitot Gauge A device with an opening in its blade-shaped section that allows water to flow to a Bourdon gauge and registers the flowing discharge pressure of an orifice.

Plan View A drawing or diagram of a building or area as seen from directly overhead. May include a site plan or a floor plan.

Platform Framing A style of wood frame construction in which each story is built on a platform, providing fire stopping at each level.

Polar Solvent A material that will mix with water, diluting itself.

Polar Solvent Type of Foam or Alcohol-Resistant Foam Foam that is compatible with alcohol and/or polar solvents by creating a polymeric barrier between the water in the foam and the polar solvent.

Polymeric Barrier A separation barrier made up of polymer or a chain of molecules linked in a series of long strands. This separates a polar solvent from an ATC foam blanket.

Portable Hydrant or Manifold Like a large water thief and may have one or more intakes and numerous outlets to allow multiple hoselines to be utilized with or without a pumper at the fire location.

Portable Water Tanks Collapsible or inflatable temporary tanks for the storage of water that is dumped from tankers or tenders. Usually carried by the tender to set up a dump site.

Positive Pressure A feature of SCBA providing a continuous supply of air, delivered by the regulator to the face piece, keeping toxic gases from entering. This pressure (1½ to 2 psi, depending on the manufacturer) is slightly above atmospheric pressure.

Post Indicator Valve (PIV) A control valve that is mounted on a post case with a small window, reading either "OPEN" or "SHUT."

Post-Incident Thought Patterns A phenomenon that describes an individual's inattentiveness following a significant incident. Post-incident thought patterns can lead to injuries or even death.

Preaction System Similar to the dry pipe and deluge systems. The system has closed piping and heads with air under no or little pressure, but the water does not flow until signaled open from a separate fire detection system. The preaction valve then opens and allows water to flow through the system to the closed heads. When an individual head is heat activated, it opens and water attacks the fire. Usually used when water can cause a large dollar loss.

Prearrival Instructions Self-help instructions intended to enhance the overall safety of the citizen until first responders arrive on the scene.

Pre-Incident Management Advance planning of firefighting tactics and strategies or other emergency activities that can be anticipated to occur at a particular location. Often referred to as preplanning.

Pressure The force, or weight, of a substance, usually water, measured over an area.

Pressure-Regulating Device Designed to control the head pressure at the outlet of a standpipe system to prevent excessive nozzle pressures in hoselines.

Primary Hole Ventilation term used to describe the first holes to be cut in a roof. They must be located as close to directly over the fire as possible to prevent laterally drawing the fire across unburned areas.

Protein Foam Made from chemically broken down natural protein materials, such as animal blood, that have metallic salts added for foaming.

Prying Tools The group of tools used to separate objects by means of a mechanical advantage.

Pulling Tools The group of tools used to pull away materials.

Pulmonary Edema Fluid filling the lungs causing death by drowning.

Pump Operator A generic term to describe the person responsible for operating a fire apparatus pump. Other commonly used titles include motor pump operator, engineer, technician, chauffeur, and driver/operator.

Puncture An injury caused by an object that has stabbed the body.

Purlins A series of wood beams placed perpendicular to steel trusses to help support roof decking.

Pyrolysis Decomposition or transformation of a compound caused by heat.

Quint A combination fire service apparatus with components of both engine company and a truck company.

Rabbeted A door stop that is cut (rabbeted) into the door frame. On metal door frames the stop is an integral part of the frame.

Radial Pulse The pulse located in either wrist.

Radiological Dispersion Device (RDD) An explosive device that spreads radioactive material throughout an area.

Rafter A wood joist that is attached to a ridge board to help form a peak.

Rapid Intervention Crew (RIC) See **Rapid Intervention Team.**

Rapid Intervention Team (RIT) A company designated to search for and rescue trapped or lost firefighters. Depending on location, may also be called a FAST.

Rate of Spread A ground cover fire's forward movement or spread speed. Usually expressed in chains or acres per hour.

RECEO Acronym coined by Lloyd Layman standing for Rescue, Exposures, Confinement, Extinguishment, and Overhaul.

Reducers Used to connect a larger hose to a smaller one.

Reel Coil Memory that wire develops from having been placed on a wooden spool as it is being manufactured.

Rehab A shortened word meaning *rehabilitation*. Rehab typically consists of rest, medical evaluation, hydration, and nourishment.

Rescue Those actions that firefighters perform at emergency scenes to remove victims from imminent danger or to extricate them if they are already entrapped.

Rescue Company The unit designation of a group of firefighters assigned to perform specialized rescue work and/or tactics and functions such as forcible entry, search and rescue, ventilation, and so on.

Rescue Specialist A firefighter with specialized training and experience in areas such as high angle rope rescue, confined space, trench, or structural collapse rescue.

Residential Sprinkler System Smaller and more affordable version of a wet or dry pipe sprinkler system designed to control the level of fire involvement such that residents can escape.

Residual Pressure The pressure in a system after water has begun flowing.

Respiratory Protection Programs Management programs designed to ensure employee respiratory protection as required by OSHA 29 CFR 1910.134 and NFPA 1500.

Respiratory System The system of the human body that exchanges oxygen and waste gases to and from the circulatory system.

Retard Chamber Acts to prevent false alarms from a sudden pressure surge in the water supply by collecting a small volume of water before allowing a continued flow to the alarm device. The water from a surge is drained from a small hole in the bottom of the collection chamber.

Ricin A biological toxin that can be used by a terrorist or other person attempting to kill or injure someone. It is the easiest terrorist agent to produce and one of the most common.

Ridge The land running between mountain peaks or along a wide peak. A high area separating two drainages running parallel with them.

Ringdown Circuits Telephone connection between two points. Going "off-hook" on one end of the circuit causes the telephone on the other end of the circuit to "ring" without having to dial a number.

Risk The chance of injury, damage, or loss; hazard.

Risk/Benefit An evaluation of the potential benefit that a task will accomplish in relationship to the hazards that will be faced while completing the task.

Risk Management The process of minimizing the chance, degree, or probability of damage, loss, or injury.

River Bottom Topographic feature where water runs from higher elevations to lower. Can be dry or wet depending on season or recent rains.

Rollover A phenomenon where the burning of superheated gases from fire extends into the top areas of the compartment in the upper thermal layers.

Rope Hose Tool About 6 feet (2 m) of ½-inch (13-mm) rope spliced into a loop with a large metal hook at one end and a 2-inch (50-mm) ring at the other. Used to tie in hose and ladders, carry hose, and perform many other tasks requiring a short piece of rope.

Round Turn Formed by continuing the loop on around until the sections of the standing part on either side of the round turn are parallel to one another.

Run Card System System of cards or other form of documentation that provides specific information on what apparatus and personnel respond to specific areas of a jurisdiction.

Running End End of the rope that is not rigged or tied off.

Saddle A pass between two peaks that has a lower elevation than the peaks. Wind will pass through this area faster than over the peaks, so fire is drawn into this feature.

Safety Container A storage can that eliminates vapor release by using a self-closing lid. Also contains a flame arrestor in the dispenser opening.

Search and Rescue Attempts by fire and emergency service personnel to coordinate and implement a search for a missing person and then effect a rescue.

Secondary Containment Any approved method that will prevent the runoff of spilled hazardous materials and confine it to the storage area.

Secondary Hole A ventilation hole that is opened only after the primary holes have been opened. It complements the primary holes.

Sectional View A vertical view of a structure as if it were cut in two pieces. Each piece is a cross section of the structure showing roof, wall, horizontal floor construction, and the location of stairs, balconies, and mezzanines.

Self-Contained Breathing Apparatus (SCBA) A type of respiratory protection in which a self-contained air supply and related equipment are worn or attached to the user. Fire service SCBA is required to be of the positive pressure type.

Setting The finishing step, making sure that the knot is snug in all directions of pull.

Severance The cutting off of components (i.e., brake pedal, steering wheel) in a vehicle.

Shear A force that tends to tear a material by causing its molecules to slide past each other.

Sheetrock A trademark and another name for plasterboard.

Shock A serious condition caused by a problem or failure of the circulatory system that results in a decrease of oxygen and vital nutrients to the body's tissues. Also known as hypoperfusion.

Shock Load A load or impact being transferred to a rope suddenly and all at one time.

Shoring The use of timbers to support and/or strengthen weakened structural members (roofs, floors, walls, etc.) in order to avoid a secondary collapse during the rescue operation.

Shoulder Load Hose load designed to be carried on the shoulders of firefighters.

Shuttle Operation The cycle in which mobile water supply apparatus is dumped, moves to a fill site for refilling, and is returned to the dump site.

Siamese A device that connects two or more hoselines into one line with either a clapper valve or gate valve to prevent loss of water if only one line is connected.

Simple Beam A beam supported at the two points near its end.

Slab Door See **Flush** or **Slab Doors.**

Slot Loads Narrow section of a hose bed where hose is flat loaded in the slot.

Small Lines or **Small-Diameter Hose** Hose less than 2½ inches (63 mm) in diameter.

Soft Suction Hose Large-diameter woven hose used to connect a pumper to a hydrant. Also known as a soft sleeve.

Solid-Core Doors Doors made of solid material such as wood or having a core of solid material between face panels.

Solid Stream Nozzles Type of nozzle that delivers an unbroken or solid stream of water to the fire. Also called solid tip, straight bore, or smooth bore.

Solubility A liquid's ability to mix with another liquid.

Spalling Deterioration of concrete by the loss of surface material due to the expansion of moisture when exposed to heat.

Span of Control The ability of one individual to supervise a number of other people or units. The normal range is three to seven units or individuals, with the ideal being five.

Spanner Wrenches Used to tighten or loosen couplings. They may also be useful as a pry bar, door chock, gas valve control, and so on.

Speaking Trumpet Trumpet used by a foreman or crew boss to shout orders above the noise of firefighting activities.

Special Egress Control Device Door hardware that will release and unlock the door a maximum of fifteen seconds after it has been activated by pushing on the bar.

Specific Gravity Weight of a liquid in relation to water. Water is rated 1.

Spinal Immobilization The process of protecting patients against further injury by securing them to a backboard or other rigid device designed to minimize movement.

Sprain Injury to the ligaments that hold the body's joints together and allow them to move.

Sprinkler Systems Designed to automatically distribute water through sprinklers placed at set intervals on a system of piping, usually in the ceiling area, to extinguish or control the spread of fires.

Staging Part of the operations section where apparatus and personnel assigned to the incident are available for deployment within three minutes.

Stairwell An enclosed stairway attached to the side of a high-rise building or in the center core of same.

Standard of Care A legal term that means for every emergency medical incident, an emergency responder should treat the patient in the same manner as another emergency responder with the same training.

Standard Operating Procedure (SOP) Specific information and instruction on how a task or assignment is to be accomplished.

Standing Part The part of a rope that is not used to tie off.

Standpipe Systems Piping systems that allow for the manual application of water in large buildings.

Static A rope having very little (less than 2 percent) elongation at normal safe working loads.

Static Pressure The pressure in the system with no hydrants or water flowing.

Staypoles The stabilizer poles attached to the sides of Bangor ladders that are used to assist in the raising of this type of ladder. Once raised, they are not used to support the extended ladder.

Steepness of Slope The degree of incline or vertical rise to a given piece of land.

Storz Couplings The most popular of the nonthreaded hose couplings.

Straight Stream A nozzle pattern that creates a hollow stream, similar in shape to the solid stream pattern, but the straight stream pattern must pass around the baffle of the nozzle. Newer fog nozzle designs, especially the automatic nozzles, only have this hollow effect from the tip on and, hence, create a solid stream with good reach and penetration abilities, some better than solid stream nozzles.

Strainers Placed over the end of a suction hose to prevent debris from being sucked into the pump. Some strainers have a float attached to keep them at or near the water's surface. A different style of strainer or screen is located on each intake of a pump.

Strategic Goals The overall plan developed and used to control an incident.

Stream Shape The arrangement or configuration of the water or other agent droplets as they leave the nozzle.

Stream Straighter A metal tube, commonly with metal vanes inside it, between a master stream appliance and its solid nozzle tip. The purpose is to reduce any turbulence in the stream, allowing it to flow straighter.

Strike Plate The metal piece attached to a door jamb into which the lock bolt slides. Also called a *strike* or *striker*.

Striking Tools The group of tools designed to deliver impact forces to break locks or drive another tool.

Superficial Burns Burns affecting the outermost layer of skin, which typically cause redness of the skin, swelling, and pain.

Supplied Air Respirator (SAR) A type of SCBA in which the self-contained air supply is remote from the user, and the air is supplied by means of air hoses.

Supply Hose or **Large-Diameter Hose (LDH)** Larger hose [3½ inches (90 mm) or bigger] used to move water from the water source to attack units. Common sizes are 4 and 5 inches (100 to 125 mm).

Surface-to-Mass Ratio Exposed exterior surface area of a material divided by its weight.

Suspension System The springs, shock absorbers, tires, and so on, of a vehicle.

Tactics The specific operations performed to satisfy the strategic goals for an incident.

Tactilely Using the sense of touch to feel for any differences or abnormality.

Tag/Guide Lines Tag lines are ropes held and controlled by firefighters on the ground or lower elevations in order to keep items being hoisted from banging against or getting caught on the structure as they are being hoisted.

Tanker The term given to aircraft capable of carrying and dropping water or fire retardant. Some departments still use the term to describe land-based water apparatus.

Target Hazard An occupancy that has been determined to have a greater than average life hazard or complexity of firefighting operations. Such occupancies receive a high priority in the pre-incident management process and often a higher level of first-alarm response assignment.

TDD Device that allows citizens to communicate with the telecommunicator through the use of a keyboard over telephone circuits instead of voice communications.

Teamwork A number of persons working together in an effort to reach a common goal.

Telecommunicator Individual whose primary responsibility is to receive emergency requests from citizens, evaluate the need for a response, and ultimately sound the alarm that sends first responders to the scene of an emergency.

Tempered Glass Plate glass that has been heat treated to increase its strength.

Tender The abbreviated term for *water tender*. A water tender is defined as a land-based mobile water supply apparatus. Some departments still use the term *tender* to describe a hose-carrying support apparatus.

Tensile Strength Breaking strength of a rope when a load is applied along the direction of the length, generally measured in pounds per square inch.

Tension A force that pulls materials apart.

Terra Cotta Tiles composed of clay and sand that are kiln fired. May be structural or decorative.

Terrorism Acts of violence that are arbitrarily committed against lives or property and intended to create fear and anxiety.

Tether Line A rope that is held by a team on shore during a water rescue to be used to haul the rescuer and victim back to shore.

Thermal Burns Burns caused by heat or hot objects.

Thermal Layering The stratification of gases produced by fire into layers based on their temperature.

Thermal Level A layer of air that is of the same approximate temperature.

Thermal Plume A column of heat rising from a heat source. A fully formed plume will resemble a mushroom as the upper level of the heat plume cools, stratifies, and begins to drop outside the rising column.

Thermal Protective Performance (TPP) A rating level, expressed in seconds, used to characterize the protective qualities of a PPE component before serious injury is experienced by the wearer.

Through-the-Lock Method   A method of forcible entry in which the lock cylinder is removed by unscrewing or pulling and the internal lock mechanism is operated to open a door. Also, the family of tools used to perform this operation.

Tidal Changes   The rising and falling of the surface water levels due to the gravitational effects between the Earth and the moon. In some areas, these changes are insignificant but in others there is more than 40 feet of difference between high and low tide.

Tip Arc   The path that a ladder's tip will take while being raised.

Torsion Load   A load parallel to the cross section of the supporting member that does not pass through the long axis. A torsion load tries to "twist" a structural element.

Total Flooding System   Used to protect an entire area, room, or building by discharging an extinguishing agent that completely fills or floods the area with the extinguishing agent to smother or cool the fire or break the chain reaction.

Tower Ladder   An apparatus with a telescopic boom that has a platform on the end of the boom or ladder. It can be extended or retracted and rotated like an aerial ladder.

Toxicity   Poisonous level of a substance.

Toxins   Disease-causing materials that are extremely toxic and in some cases more toxic than other warfare agents such as nerve agents.

Triage   A quick and systematic method of identifying which patients are in serious condition and which patients are not, so that the more seriously injured patients can be treated first.

Triple Combination Engine Company   Fire apparatus that can carry water, pump water, and carry hose and equipment.

Truck Company   The unit designation of a group of firefighters assigned to perform tactics and functions such as forcible entry, search and rescue, ventilation, and so on.

Truss   A rigid framework using the triangle as its basic shape.

Tunnel Vision   The focus of attention on a particular problem without proper regard for possible consequences or alternative approaches.

Tunneling   The digging and debris removal accompanied by appropriate shoring to safely move through or under a pile of debris at a structural collapse incident.

Turntable   The rotating platform of a ladder that affords an elevating ladder device the ability to turn to any target from a fixed position.

Two In/Two Out   The procedure of having a crew standing by completely prepared to immediately enter a structure to rescue the interior crew should a problem develop.

Type A Reporting System   System in which an alarm from a fire alarm box is received and retransmitted to fire stations either manually or automatically.

Type B Reporting System   System in which an alarm from a fire alarm box is automatically transmitted to fire stations and, if used, to outside alerting devices.

Type I, Fire-Resistive Construction   Type in which the structural members, including walls, columns, beams, girders, trusses, arches, floors, and roofs, are of approved noncombustible or limited combustible materials with sufficient fire-resistive rating to withstand the effects of fire and prevent its spread from story to story.

Type II, Noncombustible Construction   Type not qualifying as Type I construction, in which the structural members, including walls, columns, beams, girders, trusses, arches, floors, and roofs, are of approved noncombustible or limited combustible materials with sufficient fire-resistive rating to withstand the effects of fire and prevent its spread from story to story.

Type III, Ordinary Construction   Type in which the exterior walls and structural members that are portions of exterior walls are of approved noncombustible or limited combustible materials, and interior structural members, including walls, columns, beams, girders, trusses, arches, floors, and roofs, are entirely or partially of wood of smaller dimension than required for Type IV construction or of approved noncombustible or limited combustible materials.

Type IV, Heavy Timber Construction   Type in which exterior and interior walls and structural members that are portions of such walls are of approved noncombustible or limited combustible materials. Other interior structural members, including columns, beams, girders, trusses, arches, floors, and roofs, shall be of solid or laminated wood without concealed spaces.

Type V, Wood Frame Construction   Type in which the exterior walls, bearing walls, columns, beams, girders, trusses, arches, floors, and roofs are entirely or partially of wood or other approved combustible material smaller than the material required for Type IV construction.

Undesigned Load   A load not planned for or anticipated.

Unified Command   The structure used to manage an incident involving multiple response agencies or when multiple jurisdictions have responsibility for control of an incident.

Unity of Command   One designated leader or officer to command an incident.

Utility Rope   Rope used for utility purposes only. Some of the tasks utility ropes are used for in most every fire department are hoisting tools and equipment, cordoning off areas, and stabilizing objects. Also used as ladder halyards.

Vacuum (Negative) Pressure   The measurement of the pressure less than atmospheric pressure, which is usually read in inches of mercury (in. Hg or mm Hg) on a compound gauge.

Vapor Density Weight of a gas in relation to air. Air is rated 1.

Vapor Suppression Ability to contain or control the production of fuel vapors.

Vaporization The process in which liquids are converted to a gas or vapor.

Variable, Adjustable, or **Selectable Gallonage Nozzle** Nozzle that allows the nozzleperson to select the flow, with usually two or three choices, and the pattern.

Veins The blood vessels, or tubes, within the body that carry blood lacking oxygen and nutrients back to the heart.

Velocity Pressure The forward pressure of water as it leaves an opening.

Veneer A covering or facing, not a load-bearing wall, usually with brick or stone.

Venous Bleeding Bleeding from a vein.

Venturi Principle A process that creates a low-pressure area in the induction chamber of the eductor and allows the foam concentrate to be drawn into and mix with the water stream.

Vertical Ventilation Channeled pathway for fire ventilation via vertical openings.

Vesicants A group of chemical agents that cause blistering and irritation of the skin. Commonly referred to as blister agents.

Vicarious Experience A shared experience by imagined participation in another's experience.

Visqueen A trade name for black plastic. It can be used very effectively in salvage and overhaul operations.

Voice Inflection Change of tone or pitch of voice.

Voids Spaces within a collapsed area that are open and may be an area where someone could survive a building collapse.

Wall Indicator Valve (WIV) A control valve that is mounted on a wall in a metal case with a small window, reading either "OPEN" or "SHUT."

Watch Dogs Trained dogs that will bark and create a commotion, but will not attack.

Water Columning A condition in a dry pipe sprinkler system in which the weight of the water column in the riser prevents the operation of the dry pipe valve.

Water Curtain Nozzle Designed to spray water to protect exposures against heat by wetting the exposure's surface.

Water Hammer A sudden surge of pressure created by the quick opening or closing of valves in a water system. The surge is capable of damaging piping and valves.

Water Table The level of groundwater under the surface.

Water Tender The term given to land-based water supply apparatus.

Water Thief A variation of the wye that has one inlet and one outlet of the same size plus two smaller outlets with all of the outlets being gated. The standard water thief usually has a 2½-inch (65-mm) inlet with one 2½-inch (65-mm) and two 1½-inch (38-mm) outlets.

Weapon of Mass Destruction (WMD) A term that is used to describe explosive, chemical, biological, and radiological weapons used for terrorism and mass destruction.

Web The vertical portion of a truss or I beam that connects the top chord with the bottom chord.

Web Gear The term given to a whole host of personal items carried on a belt/harness arrangement worn by wildland firefighters. Items include water bottles, a fire shelter, radio, and day sack.

Webbing Nylon strapping, available in tubular and flat construction methods.

Webbing Sling Approximately 12 to 15 feet of rescue webbing tied end to end, forming a continuous loop.

Western or **Platform Framing** A style of wood frame construction in which each story is built on a platform, providing fire stopping at each level.

Wet Chemicals Extinguishing agents that are water-based solutions of potassium carbonate–based chemicals, potassium acetate–based chemicals, or potassium citrate–based chemicals, or a combination.

Wet Pipe Sprinkler System Has automatic sprinklers attached to pipes with water under pressure all the time.

Wire Glass Glass with a wire mesh embedded between two or more layers to give increased fire resistance.

Work Hardening A phrase given to the effort and physical training designed to prepare an individual to better perform the physical tasks that are expected of the individual. Work hardening is key in preventing injuries resulting from typical firefighting tasks.

Working End The end of the rope that is utilized to secure/tie off the rope.

Working Length The length of the ladder that spans the distance from the ground to the point of contact with the structure. This does not include any distance the ladder might go beyond the point of contact as would be the case when the tip extends beyond the roof.

Wye A device that divides one hoseline into two or more. The wye lines may be the same size or smaller size and the wye may or may not have gate control valves to control the water flow.

Zoning A term given to the establishment of specific hazard zones; that is, hot zone, warm zone, cold zone. Also collapse zones.

ACRONYMS

AFFF aqueous film–forming foam
AHA American Heart Association
ALS advanced life support
APCO Association of Public Safety Communications Officials—International, Inc.
ARFF aircraft rescue and firefighting
ATF Alcohol, Tobacco, and Firearms
BLEVE boiling liquid expanding vapor explosion
BLS basic life support
BNICE biological, nuclear, incendiary, chemical, and explosive
BSI body substance isolation
Btu British thermal unit
CAD computer-aided drawing; computer-aided dispatch
CAFS compressed air foam systems
CBIRF Chemical and Biological Incident Response Force (Marines)
CFR *Code of Federal Regulations*
CIS critical incident stress
CISD critical incident stress debriefing
CISM critical incident stress management
CPR cardiopulmonary resuscitation
EAP Employee Assistance Program
EDITH Exit Drills In The Home
EMS emergency medical services
EMT emergency medical technician
EMT-P emergency medical technician-paramedic
EPA Environmental Protection Agency
FAST firefighter assist and search team
FBI Federal Bureau of Investigation
FCC Federal Communications Commission
FD fire department
FEMA Federal Emergency Management Agency
FFFP fluoroprotein film–forming foam
gpm gallons per minute
HAZMAT hazardous materials
HMRU Hazardous Materials Response Unit (FBI)
HVAC heating, ventilation, and air conditioning
IAFF International Association of Fire Fighters
IAP incident action plan
IC incident commander
IDLH immediately dangerous to life and health
IFSAC International Fire Service Accreditation Congress
IMS Incident Management System
IR infrared
IRS Internal Revenue Service
kPa kilopascals

L/min liters per minute
MCI mass casualty incident
mph miles per hour
MSV mobile support vehicle
NBC nuclear, biological, and chemical
NBFU National Board of Fire Underwriters
NENA National Emergency Number Association
NFPA National Fire Protection Association
NHT national hose thread
NIOSH National Institute for Occupational Safety and Health
NP nozzle pressure
NR nozzle reaction
ORM-D other regulated material, Class D
OS&Y outside stem and yoke valve
OSHA Occupational Safety and Health Administration
PAR personnel accountability report
PASS personal alert safety system
PDD 39 Presidential Decision Directive 39
PFD personal flotation device
PPE personal protective equipment
ppm parts per million
PPV positive pressure ventilation
PSA public service announcements
PSAP public safety answering point
psi pounds per square inch
psia pounds per square inch absolute
psig pounds per square inch gauge
PVC polyvinyl chloride
RECEO rescue, exposures, confinement, extinguishment, overhaul
REVAS rescue, exposures, ventilation, attack, and salvage
RIT rapid intervention team
SAR supplied air respirator
SCBA self-contained breathing apparatus
SOG standard operating guideline
SOP standard operating procedure
SRS supplemental restraint systems
SSP standard sprinkler pendent
SSU standard sprinkler upright
TEU Technical Escort Unit
UPS uninterruptible power supply
USAR Urban Search and Rescue (team)
UV ultraviolet
WIV wall indicator valve
WMD weapons of mass destruction

OTHER FIRE SCIENCE TITLES FROM DELMAR

Fundamentals and Suppression

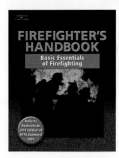

Delmar's Firefighter's Handbook: Basic Essentials of Firefighting

This briefer version of Delmar Learning's *Firefighter's Handbook* features important and up-to-date information about today's fire service without delving into the related topic of hazardous materials. It covers the critical details that apply to the job of today's firefighter, including the basic requirements of Firefighter I and II. The book is ideal for fire departments, academics, and schools in which hazardous materials are taught in a separate course with separate learning materials.

Order #: 1-4018-3582-1

Delmar's Basic Firefighting Video Series and CD-ROM Courseware

The perfect complement to *The Firefighter's Handbook*, this set of four videos leads viewers, step by step, through processes and procedures used every day by highly trained firefighters on the job. The series uses a unique blend of simulated action sequences, professional quality animations, and helpful graphics to assist viewers in acquiring vitally important technical knowledge and hands-on skills—safely and efficiently.

**Order #: Videos: 0-7668-4099-9,
CD-ROM Courseware: 0-7668-4104-9**

Introduction to Fire Protection, 2nd ed./Klinoff

This book offers a complete introduction to the field of fire protection, technology, and the wide range of services provided by both public and private fire departments of today. It covers fighting fires and the provisions of other emergency services, hazardous materials control, fire prevention, and public education.

Order #: 0-7668-4958-9

Principles of Fire Behavior/Quintiere

While explaining the science of fire with a precision found nowhere else, this text applies science to fire safety design and investigation. Using a quantitative approach, it presents an ideal introduction to the scientific principles behind fire behavior.

Order #: 0-8273-7732-0

Hydraulics for Firefighting/Crapo

Hydraulics for Firefighting leads readers throughout the principles, theory, and practical application of fire service hydraulics. This book is written in a format that will help guide the new firefighter through even the most technical hydraulic principles and complex laws of physics. The author takes care to explain theories in ways that are easily understood by anyone with knowledge of basic algebra.

Order #: 0-7668-1905-1

Introduction to Fire Pump Operations/Sturtevant

Here's the book that offers students and professional fire pump operators the updated knowledge required to efficiently, effectively, and safely operate and maintain fire pumps. With an emphasis on NFPA standards and safety, the book is logically presented in three sections: Pump Construction/Peripherals, Pump Procedures, and Water Flow Calculations.

Order #: 0-8273-7366-X

Firefighting Strategies and Tactics/Angle, Gala, Harlow, Lombardo, and Maciuba

This book is a complete source for learning firefighting strategies and tactics, from standard company responsibilities and assignments to specialized situational strategies and tactics. The reader will progress from basic concepts to the application of tactics and situational strategies for particular occupancies or types of fires. This book was written in an easy-to-follow manner. It is presented in a fashion that can be universally applied in all areas of the country, rural to urban.

Order #: 0-7668-1344-4

Explosives Identification Guide/Pickett

This is a reference guide to explosives for emergency responders such as firefighters, police officers, and EMS staff as well as security personnel. Through color photographs and short descriptions, the student can identify explosives by general type and learn the appropriate way to treat each of them. Written in a general, nontechnical style, the book is a fast and easy guide for those with little or no knowledge of, or experience with, explosives.

Order #: 0-7668-0490-9

Wildland Firefighting Practices/Lowe

The reader will learn in detail all aspects of wildland firefighting with this new, well-illustrated text. Written in a clear, how-to style by a seasoned wildland fire officer, it provides a comprehensive explanation of all the skills a firefighter needs to operate effectively against any type of wildland blaze.

Order #: 0-7668-0147-0

Terrorism

Preparing for Terrorism: An Emergency Services Guide/Buck

This text helps the reader develop the skills for dealing with terrorism on many levels: preparing and planning for a terrorist attack, mitigating its effects, proper emergency response, and recovery from terrorism disasters. It is an essential guide to the planning and implementation of antiterrorist response and operations for the overall safety of the first responder.

Order #: 0-8273-8397-5

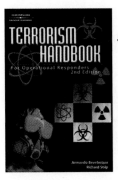

Terrorism Handbook for Operational Responders, 2nd ed./Bevelacqua and Stilp

This updated book is a guide into the most significant points that surround the emergency response processes needed to cope with terrorism incidents. It highlights new equipment and strategies that can enhance a responder's detection, monitoring, and protection capabilities against chemical and biological agents. First responders are provided with the knowledge they need to prepare for and combat acts of terrorism.

Order #: 1-4018-5065-0

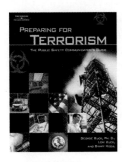

Preparing for Terrorism: The Public Safety Communicator's Guide/Buck, Buck, and Mogil

This book starts with an overview of national and international terrorism, while emphasis throughout the book is on how to prepare communications center staff and their families for a terrorist event by providing them with a well-thought-out employee emergency plans and contingencies. Solutions to communications problems, such as cellular and landline telephone overload situations, are addressed as well.

Order #: 1-4018-7131-3

Citizen's Guide to Terrorism Preparedness/Stilp and Bevelacqua

This book provides readers with facts, figures, and practical guidelines to follow as they go about their daily lives. It is designed specifically for average citizens who want to take all of the steps they can to prepare themselves for a terrorist act in their state, city, or neighborhood.

Order #: 1-4018-1474-3

Preparing for Biological Terrorism: An Emergency Services Guide/Buck

This book contains vitally important information to guide local agencies in their efforts to secure and coordinate the influx of state and federal resources before, during, and after an attack. This resource walks through the fundamental concepts of emergency planning. Subsequent chapters enable readers to immerse themselves thoroughly in specific elements of successful emergency planning.

Order #: 1-4018-0987-1

Explosives Identification Guide/Pickett

This book is a reference guide to explosives for emergency responders such as firefighters, police officers, and EMS staff as well as security personnel. Through color photographs and short descriptions, the student can identify explosives by general type and learn the appropriate way to treat each of them.

Order #: 0-7668-0490-9

HAZMAT

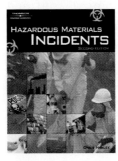

Hazardous Materials Incidents, 2nd ed./Hawley

Hazardous Materials Incidents is an invaluable procedural manual and all-inclusive information resource for emergency services professionals. Easy-to-read and perfect for use in HAZMAT awareness, operations, and technician-level training courses, this "Operations Plus" book begins by acquainting readers with current laws and regulations, including those governing emergency planning and workplace safety.

Order #: 1-4018-5758-2

Hazardous Materials Air Monitoring and Detection Devices/Hawley

This book provides HAZMAT teams with a thorough guide to effective air monitoring in emergency response situations. Each type of air monitoring devices available for emergency services is described in detail, including operating guidelines and sampling strategies. A special section discusses the latest trend in HAZMAT detection and monitoring of hazardous agents used for terrorism.

Order #: 0-7668-0727-4

Hazardous Materials Response and Operations/Hawley

While presenting an in-depth look at the response to hazardous materials releases, this book covers each class of hazardous materials and provides recommenda-

tions and guidelines for the protection of responders and victims. The text focuses on a cutting-edge response profile called Risk-Based Response, known for its progressive and aggressive approach to dealing with chemical spills. Also included is a detailed discussion of response strategies to terrorism, preparing students to be leaders in the newest area of the HAZMAT field.

Order #: 0-7668-1150-6

Hazardous Materials Chemistry/Bevelacqua

Hazardous Materials Chemistry covers the basic concepts of chemistry, emphasizing the decision-making process so that appropriate strategies and tactics will be chosen.

Order #: 0-7668-1434-3

Hazardous Materials Field Guide/Bevelacqua and Stilp

Whether the incident involves hazardous materials, a clandestine laboratory, terrorism, or a confined space operation, this user-friendly resource includes information that is consistent with the mission of all agencies. The guidebook's easy access format allows rapid identification of placards, labels, silhouettes, and common commodities that move on roadways and railways. Medical considerations are described throughout the text, identifying potential needs of an affected community.

Order #: 0-7668-0155-1

Hazardous Materials: Regulations, Response, and Site Operations/Schnepp and Gantt

This essential guide provides the student with a practical approach to the concepts of handling hazardous materials. Based on OSHA "HAZWOPER" regulations, this invaluable text addresses the specific competencies required of persons responding to a hazardous materials emergency.

Order #: 0-8273-7999-4

Emergency Medical Response to Hazardous Materials Incidents/Stilp and Bevelacqua

Medical aspects of hazardous materials response including the initial response, chemical and toxicological information, and effects on body systems—injury and treatment, physiology, and treatments of common poisonings—are explained in this book. The student will learn how to make decisions based on a scale of risk versus gain.

Order #: 0-8273-7829-7

Emergency Decontamination for Hazardous Materials Responders/Henry

This one-of-a-kind book focuses entirely on decontamination, a crucial aspect of hazardous materials emergency response. The book brings together facts about chemical contamination gathered over the last ten years and presents them in a simple, streetwise way.

Order #: 0-7668-0693-6

Hazardous Materials Air Monitoring/Detection Devices/Hawley

This text provides HAZMAT students with a thorough guide to effective air monitoring in emergency response situations. The key component to safely responding to hazardous materials is the use of a variety of detection devices for effective air monitoring. The most up-to-date information is provided along with discussion of future trends and rapidly changing technology in this field.

Order #: 0-7668-0727-4

Hazardous Materials Response and Operations/Hawley

This text presents an in-depth look at the response to hazardous materials releases, covering each class of hazardous materials, and providing recommendations and guidelines for the protection of responders and victims. The text focuses on a cutting-edge response profile called risk-based response, known for its progressive and aggressive approach to dealing with chemical spills.

Order #: 0-7668-1150-6

Inspection, Investigation, and Fire Law

Fire Prevention: Inspection and Code Enforcement, 2nd ed./Diamantes

This is a vital resource for the application of building and fire prevention codes in the inspection of buildings and facilities and for compliance through the code enforcement process. Issues such as enforcement authority, determining inspection priorities, maintenance of rated assemblies, fire protection systems, storage occupancies, detonation and deflagration hazards, and hazardous materials storage and processing are covered in depth in this comprehensive guide.

Order #: 0-7668-5285-7

Fire and Emergency Law Casebook/Schneid

Fire and emergency personnel are provided with the information about potential legal liabilities encountered every day. Actual cases are presented in detail and followed by explanations that identify the most important legal issues facing fire departments, EMS, and related organizations.

Order #: 0-8273-7342-2

Rescue

Confined Space Rescue/Browne and Crist
Confined space rescues present unique hazards to victims and emergency service workers. This new, clearly written book identifies the problems, showing users how to address them and how to rescue a victim. With this book, firefighters, police officers, EMS personnel, emergency medical staff, and industrial rescue teams can learn a simple set of skills that will provide a foundation for growth into advanced rescue operations.
Order #: 0-8273-8559-5

Engineering Practical Rope Rescue Systems/Brown
Beginners and advanced rescue technicians will learn from and enjoy this entertaining and educational book. A practical look at rope rescue systems from the point of view of an experienced professional, each chapter features exciting stories and real-life situations. The book provides a complete review of team integrity and development issues as well as team efficiency concepts that create a superior survival profile.
Order #: 0-7668-0197-7

Fire Protection Systems

Design of Water-Based Fire Protection Systems/Gagnon
A vital reference for every inspector and designer of fire protection, sprinkler, architectural, or engineering systems, this book is a must. Hydraulic calculations for the most commonly encountered water-based fire protection systems are covered in detail. Manual hydraulic calculations are thoroughly covered, and a computer disk is included to provide the reader with the opportunity to calculate a wide variety of systems.
Order #: 0-8273-7883-1

Design of Special Hazard and Fire Alarm Systems/Gagnon
As the most current guide to the design of state-of-the-art special hazard and fire protection systems, this book is essential to architects, engineers, layout technicians, plumbers, mechanical contractors, and sprinkler firms. Using the most up-to-date NFPA standards and reference data, it guides the reader through the steps needed to design a wide variety of simple and complex systems.
Order #: 0-8273-8293-6

Officer

Coming Soon! Second Edition, Fall 2004 (Order #: 1-4018-2605-9)

Company Officer/Smoke
Any firefighter that wants to gain certification as a Fire Officer will find this practical guide an excellent resource. Based on the latest information and requirements outlined in NFPA 1021, the Standard for Fire Officer Professional Qualifications, the book gives the user the information necessary to meet NFPA Standard competencies for certification as Officer I, the first-line supervisor.
Order #: 0-8273-8472-6

Going for the Gold/Coleman
Author Ronny Coleman, Chief Deputy Director of the California Department of Forestry and Fire Protection, offers a unique, must-have resource for the thousands of individuals who hope to carry the fire chief's badge. It covers the specifics of how a person actually applies leadership and decision-making concepts on a daily basis. The book provides a realistic appraisal of what it takes to aspire for, achieve, and then succeed as fire chief.
Order #: 0-7668-0868-8

Occupational Health and Safety in the Emergency Services/Angle
A comprehensive approach to program management for fire and emergency service occupational safety and health is provided in this new, practical book. Safety officers and fire department and EMS managers will make good use of this one-stop resource.
Order #: 0-8273-8359-2

Fire Department Incident Safety Officer/Dodson
This is the only book that provides a clear, focused, and detailed approach to making a difference as an incident safety officer. Company officers, battalion chiefs, safety officers, and incident commanders will benefit from the foundation material and the incident safety officer action model presented in this book.
Order #: 0-7668-0362-7

Reference

Firefighter Exam, 2nd ed./Learning Express
This new edition has been completely revised and updated to reflect changes in national firefighter tests. With this book, firefighter candidates will be able to sharpen their skills with four practice tests. Instructional chapters target the basic skills tested on firefighter exams—math, reading comprehension, grammar, vocabulary, memory, judgment, and recall.
Order #: 1-57685-440-X

A Practical Guide to Teaching in the Fire Service/Morse
Designed for the person without formal training or a degree in education who is found teaching in front of a classroom, this how-to book is the perfect resource. Its no-nonsense approach covers the day-to-day information that is needed to conduct a successful class in fire service.
Order #: 0-7668-0432-1

FIRE SCIENCE TITLES: Firefighter's Handbook

Practical Problems in Mathematics for Emergency Services/Sturtevant

This is the only math-related text specifically written for the emergency service field. Today, most certifications in the emergency service field require written exams that include related math problems. Designed for such exams, this book may be used as a preparation for certification and promotional exams, as well as a quick reference for the seasoned professional.

Order #: 0-7668-0420-8

Codes

2003 International Fire Code

The 2003 International Fire Code references national standards to comprehensively address fire safety in new and existing buildings. Topics addressed include fire department access, fire hydrants, automatic sprinkler systems, fire alarm systems, hazardous material storage, and fire-safety requirements for buildings.

Order #: 1-4018-5074-X

2003 International Building Code

The 2003 International Building Code addresses design and installation of building systems with requirements that emphasize performance. The IBC is coordinated with all the International Codes including structural, fire, and life-safety provisions covering seismic, wind, accessibility, egress, occupancy, roofs, and more.

Order #: 1-4018-5073-1

Emergency Medical Services

First Responder Handbook: Fire Service Edition/Walter, Rutledge, and Edgar

This book covers safety for first responders, patient care principles, emergency response tactics, and first responder actions for incidents involving terrorism. This street-smart book fulfills the U.S. Department of Transportation's requirements for first responder certification while offering firefighters valuable emergency medical training.

Order #: 0-7668-3919-2

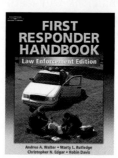

First Responder Handbook: Law Enforcement Edition/Walter, Rutledge, Edgar, and Davis

This book is for anyone in law enforcement and provides valuable emergency medical training to potential first responders to an emergency scene.

Order #: 0-7668-4191-X

INDEX

A-frame hoist, 389
A-frame ladder, 379–380
Abandonment, 694
Abrasion, 712
Absolute pressure, 290
Accelerator, 320
Access, 583
Accident, 110
Accident chain, 110
Accident prevention, 110, 111
Acclimation, 137
Accordion load, 237, 238, 239
Accountability system, 725
Acetane, 91
Acetylene, 517
Acrolein, 551
Active cooling, 731
Adams, Samuel, 9
Adams Rite lock, 529
Adapter, 228
Adjustable gallonage nozzle, 283
Administrative warrant, 662
Advanced life support (ALS) care, 694
Advancing flat load, 245, 246
Advancing hoseline at entry point of building, 260
Advancing hoselines, 251–261
    into structures, 253
    standpipe system, 255–257
    up/down stairs, 254–255
    working hose off ladders, 257–261
AED, 708
Aerial fire apparatus, 30
Aerial ladder, 373–375
AFFF, 296
Air chisels, 485
Air consumption, 169
Air cylinders, 153–155. *See also*
    Self–contained breathing apparatus (SCBA)
Air movement, 562–563
Air pressure, 87
Alarm and monitoring systems, 331
Alarms. *See* Communications and alarms
Allergic reactions, 714–715
Allied agencies and organizations, 35–36
ALS care, 694
Aluminum cylinders, 154
Ambulance stretchers, 480–481
America Burning, 18
American fire helmet, 11
Ammonia, 551
Amputation, 500, 712
Ancient history, 5–6
Angle, James, 46, 592
Anthrax, 760
Anthrax hoaxes, 760–761
Anthrax scare (2001), 760

APCO, 48
Appliance friction loss, 294
Application rate, 295
Aqueous film forming foam (AFFF), 296
Aquifer, 203
Army stretcher, 480
Arrival reports, 67–68
Articulating boom ladder, 376
Artillery, 271
Asphyxiation, 145
Assessing the patient, 700–707. *See also* Medical examination
Assistant chief, 32
Association of Public Safety Communications Officials–International (APCO), 48
Atmospheric pressure, 212, 213
Atomization, 98
Atoms, 79, 80
Attack dogs, 554
Attack hose, 222
Attack line, 244
Attic ladder, 379, 380
Authority having jurisdiction, 148
Automated external defibrillators (AED), 708
Automatic alarm systems, 56–57
Automatic hose washer, 225
Automatic nozzles, 283
Automobile, extrication from. *See* Extrication from motor vehicles
Automobile fire, 599–601, 622–625
Auxiliary appliances, 309
Auxiliary locking devices, 532
Auxiliary protective signaling system, 56
Available flow, 291
Avulsion, 712
Awning windows, 541, 543
Ax, 515, 516, 571
Axial load, 345

Backboard, 478–480
Backdraft, 100, 556–560
Backflow preventers, 209
Backpack/harness assembly, 152, 153
Backpack pump tank fire extinguisher, 189, 190
Backstretch, 244
Bagging a rope, 447–449
Baker fire (Southern California), 595
Balanced pressure demand-type foam proportioner system, 300
Balloon frame, 356–357
Balloon toss, 644, 645
Bam bam, 518
Bangor ladders, 377, 379
Bank-back technique, 303
Bank-in technique, 303
Bar joist, 361
Bars and gates (windows), 544

Base radio, 58
Basement fire, 618–619
Basic 9-1-1, 54
Basic apparatus requirements, 30
Basic figure eight knot, 433, 434
Basic life support (BLS) care, 694
Battalion/district chief, 32
Battering rams, 513
Battery-powered reciprocating saw, 486
Beam, 346, 347
Beam raise, 400, 402, 403–404
Becket bend, 429–430
Bed ladder, 372, 373
Below-grade rescue, 497–498
Below-ground fire (basement), 618–619
Bight, 426
Biohazard containers, 700
Biological, nuclear, incendiary, chemical, and explosive (BNICE), 757
Biological agents, 760
Biological weapons lab, 753, 755
Blanket drag, 469–472
Bleeding control, 709–712
BLEVE, 88–89, 626
Blister agents, 759
Blood agents, 759
Blood vessel system, 709
BLS care, 694
BNICE, 757
Body fluids, 698
Body substance isolation (BSI) precautions, 697–700
Boiling liquid expanding vapor explosion (BLEVE), 88–89, 626
Boiling point, 89
Bolt cutters, 515, 648
Bond, 80
Booster hose, 222
Boots, 129
Boston Vendome hotel fire (1972), 343
Bounce-off technique, 302
Bourdon gauge, 214
Bowline, 431–432
Brachial artery, 703, 704
Braid-on-braid rope, 423
Braided rope, 422–423
Brannigan, Francis L., 348
Breaching floors, 545–546
Breaching walls, 544–545
Break-apart nozzle, 265, 267
Breaking glass, 568–569
Breathing problems, 714
Bresnan distributor, 286
Brick, 349
British thermal unit (Btu), 77
Brush gear, 130
BSI precautions, 697–700
Btu, 77
Buddy breathing attachment, 171
Buddy system, 164

Building collapse
    collapse signs, 364
    collapse zone, 365
    factors to consider, 346
    famous collapses, 343–344
    masonry walls collapse, 350
    preparing for collapse, 364–365
    rescue, 494–497
    types, 495–496
    typical hazards, 360
Building construction, 339–368
    beams, 346
    building types, 352–357
    buildings under construction, 364
    collapse. See Building collapse
    columns, 346
    composites, 350–351
    concrete, 349–350
    connections, 347
    doors, 522–527
    fire effects, 347–352
    forces, 345
    historically significant building collapses, 343–344
    loads, 342–345
    locks, 526–530
    masonry, 350
    occupancy use, 359–360
    other construction types, 357–359
    parapet walls, 363
    roof structures, 362–363
    stairs, 363
    steel, 349
    trusses, 361
    void spaces, 362
    walls, 347
    wood, 348–349
Building loading, 342
Building size, 587
Building types, 352–357
Buildings under construction, 364
Bunkers, 127
Burning process, 91–93
Burns, 715–717
Burst hose, 268–269
Business/mercantile fire, 615–616
Butane, 91
Butterfly roof, 362
Bylaws, 35
Bypass eductor, 298
Bypass eductor bypass valve, 299

CAD systems, 51
CAFS, 299, 301
Call-taking process. See Receiving reports of emergencies
Camera, 637
Cantilever beam, 346
Capillary bleeding, 710
Capillary refill, 707
Carbide-tipped blade, 517
Carbon dioxide (CO_2), 146, 188, 333, 551, 555
Carbon dioxide extinguisher, 192
Carbon monoxide (CO), 91, 146–148, 551, 555

Carbon monoxide (CO) detectors, 313, 678
Carbon monoxide poisoning, 147
Card key entry systems, 531
Cardiopulmonary resuscitation (CPR), 707–708
Cardiovascular conditioning, 727
Cardiovascular system, 709
Carotid pulse, 703, 704
Carries, 468–469
Carry-all, 649
Carrying ladders, 395–399
Cartridge-operated dry chemical extinguisher, 193
Cartridge-operated extinguishers, 189–190
Cascade system, 156, 177–179
Casement window, 541, 543, 569
Cast iron, 349
Catch-all, 645–646
CBIRF, 764
CDC, 696
Cellar nozzle, 286
Cellulose, 79
Celsius scale, 78
Center rafter louver, 577–579
Centers for Disease Control and Prevention (CDC), 696
Centimeter, 78
Central station protective signaling system, 56
CFR, 109
Chain saws, 517
Charring, 654
Check rail windows, 543
Check valves, 209
Chemical and Biological Incident Response Force (CBIRF), 764
Chemical burns, 716–717
Chemical reaction, 83
Chemical weapons labs, 755
Chernobyl nuclear disaster, 85
Chest pain, 714, 715
Chief of department, 32
Chief officers, 32
Childress, Dennis R., 591
Chimney, 585
Chlorine, 82
Choking agents, 759
Chord, 346
Circuit breaker, 672
CISD, 737–738
CISM, 116, 737–738
Cistern, 205
Civil War, 13–14
Clandestine labs, 752–753
Class 1 harness, 491
Class 2 harness, 491
Class 3 harness, 491, 492
Class A extinguishers, 193–194
Class A fire, 100, 101, 185
Class A foam, 296–297
Class B extinguishers, 194
Class B fire, 100, 101, 185–186
Class B foam, 296, 297
Class C extinguishers, 194

Class C fire, 100, 101, 186
Class D fire, 186
Class D heavy metal fire, 100, 101
Class I standpipe system, 329, 330
Class II standpipe system, 329, 330
Class III standpipe system, 329
Class K fire, 100, 186–187
Claw tool, 513
Climbing path, 391
Clipping, 65
Closed-circuit SCBA, 158
Closet ladder, 379, 380
Clothing drag, 471, 473
Clove hitch, 427–429
CO, 91, 146–148, 551, 555
CO detectors, 313, 678
CO_2, 146, 188, 333, 551, 555
Coat method (donning SCBA), 160, 162, 163
Coats, 128
Cockloft, 561–562
Code of Federal Regulations (CFR), 109
Cohortes Vigilum, 6
Coiling a rope, 445–447
Coleman, Ronny J., 2
Collapse signs, 364
Collapse zone, 365
Column, 346, 347
Combination A-frame ladder, 379–381
Combination attack, 608
Combination fire attack, 290
Combination fog nozzle, 283, 288
Combination headsets, 134
Combination tools, 483, 484
Combination units, 31
Combined gravity-pumped water distribution system, 206
Combustible heavy metals, 101, 102
Combustion, 85, 86
Command staff positions, 41–42
Commercial cooking surfaces, 667–668
Commercial garage doors, 526
Common terminology, 37
Communications and alarms, 45–73
    arrival reports, 67–68
    communications facility, 49–51
    communications personnel, 48–49
    computers, 51
    emergency services deployment, 58–62
    MSVs, 68
    radio systems and procedures, 63–67
    receiving reports of emergencies, 51–58. See also Receiving reports of emergencies
    records, 68–70
    traffic control systems, 62–63
Communications facility, 49–51
Communications personnel, 48–49
Communications process, 47
Company, 28–29
Company officer, 28, 29
Compartments/subcompartments, 561
Composite cylinders, 154
Composite truss, 351
Composites, 350–351

Compound, 79
Compressed air foam systems (CAFS), 299, 301
Compression, 345
Compressor/purifier system, 177, 179–180
Computer-aided dispatch (CAD) systems, 51
Computers, 51
Concentrated load, 342
Concrete, 349–350
Conduction, 94–95
Conductor, 84
Confined space, 498
Confined space rescue, 498
Connections, 347
Consent, 694
Consolidated incident action plans, 38
Constant pressure nozzle, 283
Constant volume nozzle, 283
Constricted pupils, 706
Contents fire, 594
Continuous beam, 346
Convection, 95–96
Conventional forcible entry, 531–536
Core strengthening, 727
Cotton fiber rope, 420
Counter payoff, 643
Coupling/uncoupling hose, 228
Couplings, 225–227
CPR, 707–708
Cribbing, 488, 496
Critical incident stress (CIS), 737–738
Critical incident stress debriefing (CISD), 737–738
Critical incident stress management (CISM), 116, 737–738
Cross-arm carry, 469
Cutters, 483
Cutting a roof, 585
Cutting tools, 515–517
Cutting torch, 517
Cyanogen chloride, 758
Cylinder replacement procedure, 172–175
Cylinders, air, 153–155. *See also* Self-contained breathing apparatus (SCBA)

Dalmatian, 13, 14
Damaged rope, 440
De Francesco, Joseph, 419
Dead bolt, 528–529
Dead bolt and latch, 529
Dead-end water mains, 206
Dead loads, 342
Debris removal, 652
Decay stage, 92–93
Deck gun, 271
Deep wells, 204
Defense attack, 608
Deluge set, 271–273
Deluge systems, 320–321
Density, 84, 88
Dent puller, 518
Denver tool, 513

Deployment plan, 59
Deployment tables, 60
Depth of char, 654
Deputy chief, 32
Design load, 344
Designated incident facilities, 38
Detection systems, 309–314
 flame detectors, 313–314
 gas detectors, 313
 heat detectors, 310–311
 people or manual systems, 309–310
 smoke detectors, 311–312
Detector activation operations, 335
Detergent-type foams, 296
Diagram of the site, 685
Diagrammatical guides. *See* Job performance requirements
Diffusion, 87, 567
Digital logging recorders, 69
Dilated pupils, 706
Direct attack, 289, 603, 621
Direct pump water distribution system, 206
Disassembly, 489
Discharge flow, 292
Disentanglement, 489
Displacement, 489
Distilled mustard, 758
Distortion, 489
Distributed load, 342
Distributor pipe, 228
Division chief, 32
Dodson, Dave, 106, 123, 339, 343, 348, 721
Domestic Preparedness Training Initiative, 764
Door blocker, 532
Door swing direction, 523–524
Doors, 522–527
Dormers, 363
Double becket bend, 430–431
Double-cylinder lock set, 664
Double-donut roll, 231, 233
Double female, 228
Double gloving, 698
Double hung window, 541
Double male, 228
Drafting operation, 264–265, 266
Drags, 469–476
Drain and carry, 234
Dressing a knot, 426
Dropped ceilings, 587
Drug labs, 753
Dry barrel hydrants, 207, 208
Dry chemical extinguishing agents, 188
Dry chemical total flooding systems, 333
Dry chemicals, 185
Dry hydrant, 205, 207, 209
Dry pipe systems, 319–320
Dry powders, 186
Duck bill lock breaker, 518, 519
Duck's foot, 228
Dump site, 211
Dumpster fire, 626–627
Duplex radio system, 64
Dust explosions, 673

Dutchman, 237
Duty training standards, 26–28

EAP, 116
Early American history, 8–13
Ears, 245, 246
Eccentric load, 345
EDITH, 33
Educating the public, 679–684. *See also* Fire and life safety education
Eductor, 297–299
Electric/cable elevators, 501, 502
Electrical conductor, 84
Electrical hazards, 672–673
Electrical power, 83–84
Electrical situations, rescue from, 498–499
Electricity, 84
Electronic heat sensors, 650
Electrons, 80, 84
Elevated temperatures, 145
Elevator control panels, 502
Elevator rescue, 500–504
Emergency call boxes, 55
Emergency communications centers, 49–51
Emergency evacuation plan, 682
Emergency evacuation signals, 67
Emergency medical dispatch, 53
Emergency medical services (EMS), 689–720
 AED, 708
 allergic reactions, 714–715
 assessing the patient, 700–707. *See also* Medical examination
 bleeding control, 709–712
 breathing problems, 714
 BSI presentations, 697–701
 burns, 715–717
 chest pain, 714
 common emergencies, 714–718
 CPR, 707–708
 first aid, 712–713
 fractures/sprains, 718
 infection control, 696–697, 701
 interacting with EMS personnel, 694–695
 legal issues, 694
 medical illness, 714
 poisoning, 717–718
 roles/responsibilities, 691–694
 safety considerations, 695–700
 shock, 713–714
Emergency services deployment, 58–62
Employee assistance program (EAP), 116
EMS. *See* Emergency medical services (EMS)
Encoders, 61
Endothermic reaction, 80, 81, 555
Energy bar, 732
Energy efficient windows, 543
Engine company, 29
Engineered wood, 349
English measurement system, 77, 78
Enhanced 9-1-1, 54, 59

Ensembles, 127–133
Entrapment, 735
Equilibrium, 87
Equipment
   extrication from motor vehicles, 482–486
   fire extinguishers. See Portable fire extinguisher
   fire hose. See Fire hose and appliances
   fire inspectors, 660–661
   forcible entry, 511–521
   high-rise firefighting, 617
   ladder. See Ladder
   nozzle. See Nozzle
   overhaul operations, 648–649
   PPE. See Personal protective equipment (PPE)
   rope. See Rope
   safety, 112–114
   salvage operations, 633–639
   SCBA. See Self-contained breathing apparatus (SCBA)
   standpipe operations, 334
Escalator rescue, 504–505
Evacuation drills, 682
Evaporation, 89
Exhaust fan, 573
Exhauster, 320
Exit doors, 665
Exit drills in the home (EDITH), 33
Exothermic reaction, 80, 81, 555
Expandable cut, 575–577
Explosive limits, 90
Explosives lab, 753–755
Extended aerial ladder, 374
Extending hoseline, 265–268
Extension ladder, 377–378
Extension pipe, 228
External bleeding, 710–712
Extinguishing agents, 187–188
Extinguishing fires. See Fire suppression
Extremity carry, 469, 470
Extrication from motor vehicles, 482–490
   disentanglement, 489
   fluid/parts cleanup, 490
   hazards zoning, 487
   patient access, 488
   patient removal, 489–490
   scene assessment, 486
   scene stabilization, 490
   tools and equipment, 482–486
   traffic barrier, 486
   traffic calming, 487
   vehicle recovery, 490
   work areas, 486
Eye protection, 134

Face shield, 134
Fahrenheit scale, 78
Fans, 573–575
Farm equipment rescue, 505
FCC, 63
FDNY, 114
Federal Communications Commission (FCC), 63
Federal Emergency Management Agency (FEMA), 762–763
FFFP, 296
Fiberglass composite cylinders, 154
Fiberglass cylinders, 154
Figure eight, 432–436
Figure eight on a bight, 433, 436
Fill site, 211
Finish loads, 243–245
Fire
   by-products, 554–556
   gases, 551
Fire and life safety education, 679–684
   adult programs, 683
   fire drills, 682–683
   fire station tours, 684
   PSAs, 681
   school programs, 681–682
Fire attack, 289
Fire behavior, 74–105
   backdraft, 100
   BLEVE, 88–89
   boiling point, 89
   burning process, 91–93
   classes of fire, 100–102
   combustion, 85, 86
   conduction, 94–95
   convection, 95–96
   fire tetrahedron, 77
   flammable/explosive limits, 90–91
   flashover, 99
   heat transfer, 94–97
   measurements, 77–78
   oxygen, 86
   physics of fire, 78–82
   radiation, 96–97
   rollover, 99
   sources of heat, 82–85
   states of matter, 98
   structural fire, 102–103
   thermal conductivity, 97
   thermal layering, 99
   vapor pressure/density, 86–88
Fire cause determination concerns, 652–654
Fire classification, 100–102, 185–187
Fire company inspection program, 660–679
   building exterior, 673–674
   concluding the inspection, 674
   conducting the inspection, 661–662
   electrical hazards, 672–673
   equipment, 660–661
   exiting, 663–666
   fire protection equipment, 666–669
   hazardous materials, 669–673
   home inspections, 677–679
   preparation, 661
   reinspections, 677
   restaurants, 667–668
   violations, 662–669
   wrap-up meeting, 674
Fire drills, 682–683
Fire escapes, 386–387
Fire extinguisher classification symbols, 102
Fire extinguishers. See Portable fire extinguishers
Fire extinguishment. See Fire suppression
Fire flow capacity, 203
Fire flow rate, 212
Fire flow requirement, 203
Fire hose and appliances, 219–278
   advancing flat load, 245, 246
   advancing hoselines, 251–261. See also Advancing hoselines
   burst hose, 268–269
   care/maintenance, 223–224
   coupling/uncoupling hose, 228
   couplings, 225–227
   extending hoseline, 265–268
   finish loads/preconnected loads, 243–245
   fire stream, 289–295
   friction loss coefficients, 294
   hoisting, 451–454
   hose carries, 234–236
   hose lay procedures, 269–271
   hose loads, 236–242
   hose rolls, 230–233
   hose tools/appliances, 227–228
   master stream devices, 271–274
   minuteman load, 246–248
   nozzle. See Nozzle
   operating, 288–289
   service testing of hose, 274–275
   stored hose loads/packs, 248
   triple-layer load, 247–248
   types of hoses, 222–223, 288–289
   water supply connection, 261–265
   wildland firefighting hose loads, 248–250
Fire hydrants, 206–209, 213–215
Fire intensity, 597
Fire lane, 674
Fire load, 344
Fire plume, 96
Fire Prevention Division, 659–660
Fire prevention inspections, 32, 33
Fire prevention office, 32
Fire protection valves, 325
Fire resistive, 353
Fire-resistive construction, 353–354
Fire resistive ratings, 353
Fire safety brochures, 684
Fire safety characters, 681
Fire service key slot, 502
Fire service knots. See Knot
Fire shelter, 132
Fire societies, 8
Fire sprinkler systems. See Sprinkler systems
Fire station alerting system, 60
Fire station tours, 684
Fire stopping, 357
Fire stream, 281, 289–295, 604–606
Fire suppression, 591–630
   below-ground structure (basements), 618–619
   business/mercantile occupancies, 615–616

combination attack, 608
defense attack, 608
direct attack, 603
dumpster fire, 626–627
flammable liquid/gas considerations, 601–603
flammable liquid/gas fires, 625–626
ground cover considerations, 595–599
groundcover fire, 620–622
IMS, 607–609
indirect attack, 603–604
multistory occupancies, 616–618
nonstructural fires, 620
offensive attack, 607–608
outside stacked materials fire, 627–628
piled materials fire, 627–628
RECEO, 607
residential occupancies, 611–612
REVAS, 607
single family occupancies, 612–614
sprinkler system, 619–620
standpipe system, 620
stream selection, 604–606
structural fire considerations, 593–595
tactical considerations, 606
trash fire, 626–627
vehicular fire, 622–625
vehicular fire considerations, 599–601
wildland fire, 620–622
wildland fire triangle, 599–601
Fire survival programs, 33
Fire tetrahedron, 77
Fire triangle, 77
Fire wardens, 8
Fire watch, 329
Firefighter, 26
Firefighter deaths, 108, 109
Firefighter emergencies, 733–738
Firefighter fitness and health, 115–116
Firefighter injuries, 109
Firefighter readiness, 119
Firefighter safety, 106–122. *See also* Safety
  accident prevention, 110, 111
  equipment, 112–114
  firefighter responsibilities, 117–120
  injury/death, 108–109
  personnel, 114–117
  procedures, 111–112. *See also* SOPs
  safety attitude, 116–117
  safety standards and regulations, 109–110
Firefighter safety responsibilities, 117–120
Firefighter survival, 721–740
  active cooling, 731
  energy/rest, 727–728, 731
  entrapment, 735
  firefighter emergencies, 733–738
  fitness for duty, 726
  hydration, 732
  incident readiness, 723–728
  injured firefighters, 736
  lost firefighters, 736

mental fitness, 726–727
nutrition, 727, 732
on-scene rehabilitation, 730–731
orders/communication, 729
personal accountability, 725–726
personal size-up, 730
physical fitness/wellness, 727
post-incident survival, 736–738
PPE, 723–724
rapid escape, 733–735
risk/benefit analysis, 730
RITs, 732–733
safety at incidents, 728–733
team continuity, 728
Firefighter's carry, 468–469
Firefighter's drag, 476
Fireground factors, 608
Firemarks, 8
FiRP, 350
First aid, 712–713
Fishing line, 421
Fitness for duty, 726
Fixed gaseous extinguishing systems, 669
Fixed temperature heat detector, 310, 311
Flame detectors, 313–314
Flameover, 99
Flammable limits, 90–91
Flammable liquid/gas fires, 601–603, 625–626
Flammable liquid safety can, 670
Flash point, 98, 601
Flashover, 92, 93, 97, 99, 102, 556, 594
Flat carry, 396–398
Flat load, 237–238, 240, 241
Flathead ax, 513, 515
Flexibility improvement, 727
Floor-mounted stop plate, 532
Floor plans, 685
Floor runner, 634–635
Flow, 282, 291–292
Flow meters, 292
Fluoroprotein film forming foam (FFFP), 296
Fluoroprotein foam, 296
Flush doors, 522
Flush-type hydrant, 209
Fly extension raise, 402, 405
Fly ladder, 372, 373
Flying stretch, 244
Foam, 295–303
  application, 297–303
  characteristics, 295–296
  classes, 296–297
  venturi principle, 297–298
Foam eductor, 298
Foam flooding systems, 333
Foam nozzles, 300, 302
Foamed core metal door, 523
Fog nozzles, 283–285
Folding ladder, 379, 380
Follow-through figure eight knot, 433, 435
Foot-tilt method, 228, 229
Footwear, 129
Force, 345

Forcible entry, 508–547
  breaching floors, 545–546
  breaching walls, 544–545
  conventional entry, 531–536
  doors, 522–527
  glazing, 542–543
  knowledge, skill, experience, 508–509
  lock variations, 539–540
  locks, 526–530
  operating lock mechanisms, 538–539
  through-the-lock entry, 536–538
  tool assignments, 546
  tools, 511–521
  windows, 540–544
Forcing a padlock, 540
Forestry hose, 223
Formal procedures, 111
42 CFR Part 84, 149
Forward lay, 270
Four-person flat carry, 398
Fracture, 718
Franklin, Benjamin, 9
Freelancing, 119, 725, 728
Freshwater, 203
Friction, 83
Friction loss, 292–294, 294
Friction loss coefficients, 294
Fuel resistance, 296
Full escape procedure, 170
Full thickness burn, 716
Full-wrapped fiberglass cylinders, 154
Fully developed stage, 92

Gable roof, 362
Gala, Mike, 184, 549
Gambrel roof, 362
Garage doors, 525–526
Gas, 98
Gas detectors, 313
Gasoline, 91
Gasoline engine-powered hydraulic pumps, 482
Gasoline-powered saws, 516
Gate valves, 209
Gauge pressure, 290
Girder, 346
Glass doors, 523
Glazing, 542
Gloves, 129, 698–700
Go method, 493
Goldfeder, Billy, 202
Gratacap, Andrew, 11
Gravity connections, 347
Gravity-fed water distribution system, 206
Great Chicago Fire, 14
Great Fire of London, 7
Grid (looped) system, 207
Grooved and butterfly valve, 325
Grooved check valve, 325
Ground cover fire fighting, 595–599, 620–622
Ground ladder, 376–381
Ground pads, 497
Groundwater, 203–204

Growth stage, 91–92
Guard dogs, 554
Guideline/lifeline, 464
Gusset plates, 361

Half hitch, 427
Halligan tool, 513, 570
Halon, 333, 669
Halyard, 372
Hancock, John, 9
Hancock, Robert F., 417, 459
Hand in Hand, 10
Hand protection, 129
Hand washing, 697, 701
Handheld hydraulic spreaders, 514
Handline-operated medium expansion foam generator, 301
Handsaw, 515
Hanging ceilings, 587
Hard laid, 422
Hard sleeve hydrant connection, 265
Hard suction hose, 222, 223
Harness, 491–492
Hawley, Christopher, 741, 742
Hayes, Daniel, 14
Hazardous Materials Response Unit (HMRU), 762–763
Hazards zoning, 487
HAZMAT crimes, 752–757
Head pressure, 291
Head-tilt, chin-lift method, 703
Header, 346
Head's up display (HUD), 168
Hearing protection, 134
Heat detectors, 57, 310–311
Heat release rate, 94
Heat resistance, 296
Heat sink, 94
Heat transfer, 94–97
Heating, ventilation, and air conditioning (HVAC) systems, 563
Heavy-duty padlocks, 529
Heavy timber construction, 355–356
Heavy timber dimensions, 356
Heavy timber fires, 598
Heeling a ladder, 410, 411
Helix, 222
Helmet, 128–129
Hero of Alexandria, 6
HFT, 536
Hibernia Fire Company, 10
Higbee cut, 225, 226
Higbee indicator, 226
High expansion foam generator, 302, 333
High-piled storage fire, 627–628
High-pressure air bag set, 485
High-pressure air chisel kit, 485
High-pressure hydrants, 209
High-rise firefighting, 616–618
High-rise (standpipe) hose packs, 248, 251
Hip roof, 362
Historical overview, 4–17
    ancient beliefs, 5–6
    building collapses, 343–344

Civil War, 13–14
early American history, 8–13
Industrial Revolution, 14
Middle Ages, 7
Roman Empire, 6
terrorism, 914–917
World War II, 17
HMRU, 762–763
Hoist, 227
Hoisting, 449–454
    ax, 450
    hoseline, 451–454
    ladders, 412, 414, 451, 454, 455
    pike pole, 450, 451
    small equipment, 451, 454
Hoistway door, 503
Hollow-core doors, 522
Home alerting devices, 60
Home evacuation, 682
Home inspections, 677–679
Honeycomb core metal door, 523
Hood systems, 332
Hoods, 129, 130
Hook, 518, 570
Hook ladder, 378–379
Horizontal sliding windows, 541
Horizontal ventilation, 552, 567–568
Horizontal voids, 562
Horseshoe load, 239, 242, 243
Hose. *See* Fire hose and appliances
Hose bed, 223
Hose bridge, 227
Hose cap, 228
Hose carries, 234–236
Hose cart, 227
Hose clamp, 227, 268
Hose couplings, 225–227
Hose drag, 236
Hose jacket, 227
Hose lay procedures, 269–271
Hose loads, 236–242
Hose packs, 248
Hose roller, 227
Hose rolls, 230–233, 248
Hose storage rack, 226
Hose strap, 227
Hose stream characteristics, 606
Hose testing, 274–275
Hose threads, 226
Hose washer on hydrant, 224
Housekeeping practices, 672
HUD, 168
Human chain technique, 463
Humid environments, 588
HVAC systems, 563
Hydrant connections, 262–265
Hydrant valves, 228
Hydrant wrench, 227
Hydration, 137, 732
Hydraulic elevators, 501
Hydraulic forcible entry tool (HFT), 536
Hydraulic spreaders, 514
Hydraulic ventilation, 565–567
Hydraulics, 290
Hydrocarbon, 79, 80, 297
Hydrogen chloride, 146, 551

Hydrogen cyanide, 143, 146, 551
Hydrogen sulfide, 551
Hypoperfusion, 713–714
Hypoxia, 145, 147

IAP, 729
IC, 607
Ice rescue, 494
Ice rescue ensemble, 132, 133
ICF, 357
IDLH atmosphere, 125, 143
Ignition, 91
Ignition stage, 91
Immediate danger to life and health (IDLH), 125, 143
Impact load, 342
Implied consent, 694
Imposition of loads, 344–345
IMS. *See* Incident management system (IMS)
In-line eductor, 298
Incendiary agents, 758
Incident action plan (IAP), 729
Incident command designations, 41–42
Incident commander (IC), 607
Incident engagement checklist, 120
Incident management, 36–37
Incident management system (IMS), 37–42
    command staff positions, 41–42
    components, 37–38
    fire suppression, 607–609
    firefighter survival, 729
    functions, 39–41
    safety, 118
    unified command, 42
Incision, 712
Increaser, 228
Indirect attack, 603–604, 622
Indirect fire attack, 289–290
Induction heat, 84
Industrial entrapment rescue, 499–500
Industrial Revolution, 14
Informal procedures, 111
Information half-life, 17–18
Infrared flame detectors, 314
Initial assessment, 700, 702
Injured firefighters, 736
Inorganic, 80
Insect screens, 586
Inspection cut, 581
Inspection notice, 676
Inspections. *See* Fire company inspection program
Intake relief valve, 228
Integrated communications, 38
Interconnected lock set and dead bolt, 665
Interface firefighting, 622
Internal bleeding, 709–710
Intersecting roof, 362
Intervention, 110
Ionization-type smoke detector, 311–312
IR detectors, 314
Irons, 513
Irritants, 759

Jacket, 221
Jalousie windows, 541, 543
Jamb, 522
Jaw thrust, 703
Jefferson, Thomas, 9
Jet siphon valve, 211
Job performance requirements
   carries, 468–471
   coupling/uncoupling hoses, 229–230
   drags, 472–476
   forcible entry, 534–538
   hoisting, 450–455
   hose carries, 234–236
   hose loads, 238–243
   hose rolls, 231–233
   knots, 427–432
   ladders, 395–414. *See also* Ladder skills
   removing gloves, 699
   returning sprinkler system to service, 327–328
   reverse horseshoe load, 244
   rope, 446–449
   salvage operations, 638–645
   SCBA, 159–180. *See also* Job performance requirements–SCBA
   taking out a window, 570
   victim removal, 468–477
Job performance requirements–SCBA
   changing out a cylinder (other firefighter), 175–176
   cylinder replacement procedure, 172–175
   daily inspection, 172, 173
   donning–coat method, 160, 162, 163
   donning–face piece, 166, 167
   donning–over the head method, 159–160, 161
   donning–seat-mounted apparatus, 164, 165
   servicing cylinder (cascade system), 177–179
   servicing cylinder (compressor/purifier system), 179–180
Joist, 346
JPR. *See* Job performance requirements
Junior firefighter program, 682

K tool, 518–519, 537, 538
Kelleher, Mike, 75
Kelly tool, 513–514
Kernmantle rope, 423
Kerosene, 91
Kertzie, Peter F., 370
Key in the knob lock, 528
Key tools, 539
Kilometer, 78
Knee-press one-person uncoupling method, 230, 231
Knights of Malta, 7
Knockdown speed, 296
Knot, 424–439
   becket bend, 429–430
   bowline, 431–432
   clover hitch, 427–429
   double becket bend, 430–431
   elements of, 425
   figure eight, 432–436
   half-hitch, 427
   overhand, 427
   rescue, 437–438
   water, 438–439
Knox boxes, 673–674
Koonce, Ric, 183, 201, 219, 307

Laceration, 712
Lach, Bernard, 142
Ladder, 369–416
   A-frame, 379–381
   aerial, 373–375
   articulating boom, 376
   breaking glass, 571–572
   certification/testing procedures, 395
   extension, 377–378
   fire escapes, and, 386–387
   folding, 378, 379
   ground, 376–381
   hoisting, 412, 414, 451, 454, 455
   hook, 378–379
   maintenance, cleaning, inspection, 381–383
   painting, 394–395
   parking, 393–394
   parts, 371–373
   Pompier, 381
   roof level, and, 386
   safety, 383, 390–393
   selection criteria, 384–387
   skills. *See* Ladder skills
   special uses, 387–390
   storage, 393
   straight, 376–377
   terminology, 374, 375
   tower, 375–376
   truck-mounted, 373–376
   uses, 383–384, 387–390
   windows, and, 385–386
   working off, 392–393
Ladder belt, 491
Ladder commands, 400
Ladder company, 29
Ladder escape, 734–735
Ladder load, 391–392
Ladder painting, 394–395
Ladder pipe, 271
Ladder pipe operations, 274
Ladder positioning terminology, 374
Ladder raising terminology, 375
Ladder selection, 384–387
Ladder skills, 395–414
   carrying ladders, 395–399
   carrying tools, 408, 409
   engaging hook on hook ladder, 412, 413
   flat carry, 396–398
   heeling a ladder, 410, 411
   hoisting ladders by rope, 412, 414
   leg lock, 408
   lower ladder into building, 404, 406
   mounting/dismounting, 410–412
   raising ladders, 399–408
   roof ladder deployment, 412
   shoulder carry, 396, 397
   single carry, 398–399
   suitcase carry, 395, 396
   window dismount, 410, 412
Ladder truck, 272
Ladder uses, 383–384, 387–390
Laid rope, 422
Laminated glass, 542
Landing plate, 504
Large-diameter hose (LDH), 222
Layman, Lloyd, 17
LDH, 222
Lean-to collapse, 495
Learn Not to Burn program, 682
Leather lace-up boot, 129
Leather pull-up boot, 129
Ledge doors, 522
Left-hand search, 463
Leg lock, 408
Lewisite, 758
Liaison officer, 41
Life safety rope, 424, 425
Light obscuration detector, 312, 313
Light obscuration photoelectric smoke detector, 313
Light-scattering detector, 312, 313
Light-scattering photoelectric smoke detector, 313
Lightweight floor truss systems, 362
Lightweight wooden I beams, 350, 351
Line-of-duty deaths, 26
Line-type heat detector, 311
Lintel, 346
Liquid, 86, 98
Live loads, 342
Load, 342–345
Load-bearing wall, 347
Loaded stream, 187
Local application system, 332
Local protective signaling system, 56
Lock cylinders, 538, 539
Lock variations, 539–540
Locks, 526–530
Logging recorders, 69
Logistics, 40
Long spine boards, 478
Loop, 426
Loose fold and roll, 637
Lost firefighters, 736
Louver opening, 577–579
Low/medium-pressure bag, 485
Low-pressure carbon dioxide extinguishing system, 333
Lug style butterfly valve, 325

Mace, 758
Magnetic tape logging recorders, 69
Manifold, 228
Manila rope, 420, 421, 442
Mansard roof, 362
Manual incident card, 70
Manual pull box, 57
Manufactured buildings, 358
Marchone, Mary K., 658
Mask rule, 143

Masonry, 349, 350
Masonry cutting blades, 517
Masonry wall collapse, 350
Mass casualty incident (MCI), 756
Master streams, 271–274, 288
Matter, 78
Maul, 512
MCI, 756
MDH, 222
Means of egress, 663
Measurements, 77–78
Mechanical ventilation, 563
Medi-vac, 694, 695
Medical alert tags, 704
Medical examination, 700–707
 airway, 702
 bleeding, 704
 breathing, 702
 circulation, 703
 color/temperature of skin, 705–707
 fact-finding, 704
 level of consciousness, 702
 vital signs, 705
Medical illnesses, 714
Medium-diameter handlines, 289
Medium-diameter hose (MDH), 222
Meltdown, 85
Mental fitness, 726–727
Mental health, 116
Mercantile fire, 615–616
Metal cutting blade, 517
Metal doors, 522–523
Meter, 78
Methamphetamine laboratory incidents, 754
Methane, 82, 88
Metric system, 77, 78
Metropolitan medical response team (MMRT), 764
Miale, Frank J., 74, 369, 548
Microphones, 66, 67
Microwaves, 84
Middle Ages, 7
Mill construction, 356
Miller, Geoff, 631
Minuteman load, 246–248
Mission statement, 4, 24
Mission vision, 581
Mitchell, David, 722
Mitigation, 110
MMRT, 764
Mobile data computers, 63
Mobile data terminals, 62
Mobile microphone, 66
Mobile radio, 58
Mobile support vehicles (MSVs), 68, 69
Mobile water supply apparatus, 30, 31, 204–205
Model Procedures Guide for Structural Firefighting, 204
Modified Gasner bar pack, 250, 251, 252
Modular organization, 38
Molecules, 79
Monitor pipe, 271, 272
Montagna, Frand, 280
Morris, Robert R., 508

Mortar, 350
Mortise lock, 528–529, 538
Motor vehicle, extrication from. *See* Extrication from motor vehicles
Motor vehicular firefighting, 599–601, 622–625
Mounting hardware, 522
MSVs, 68, 69
Multifamily occupancy firefighting, 614
Multiple ladder removal technique, 387
Multipurpose dry chemical agent, 188
Multisite trunked radio systems, 64, 65
Multistory occupancy firefighting, 616
Multivalved wet barrel hydrant, 207
Municipal fire alarm systems, 55
Mutual aid or assistance agreements, 36

National Board of Fire Underwriters (NBFU), 15
National Emergency Number Association (NENA), 54
National Fire Protection Association (NFPA), 26
National Institute of Occupational Safety and Health (NIOSH), 110
Natural gas, 91
Natural ventilation, 563
NBC, 757
Needed flow, 291
Negatively charged, 84
Nerve agents, 757–758
Neutral plane, 587
Neutrons, 80
New York City Fire Department (FDNY), 114
NFPA, 26
NFPA 72, 50
NFPA 220, 352
NFPA 704, 671, 672
NFPA 704 placard system, 671
NFPA 1001, 26, 27
NFPA 1021, 28, 32
NFPA 1031, 33
NFPA 1041, 34
NFPA 1061, 48
NFPA 1221, 49
NFPA 1404, 144, 148, 149
NFPA 1500, 109, 126, 144, 148, 609, 730, 732
NFPA 1500 series, 110
NFPA 1561, 733
NFPA 1581, 136
NFPA 1901, 28, 29, 30
NFPA 1931, 381, 395
NFPA 1932, 382
NFPA 1975, 136
NFPA 1981, 148, 149, 1506
NFPA 1983, 128, 419, 424, 439, 491
Night latch, 540
9-1-1, 54
NIOSH, 110
Nitrogen, 333
Nitrogen dioxide, 146, 551
No-knowledge hardware, 664
Nomex, 86
Non-load-bearing wall, 347

Noncombustible construction, 354
Nourishment theory, 732
Nozzle, 281–288
 cellar, 286
 defined, 281
 fire stream, 289–295
 foam, 300–302
 fog, 283–285
 operations, 287–288
 piercing, 286
 playpipes, 286, 287
 pressure, 293
 shutoff, 286, 287
 solid tip, 282–283
 straight steam, 285
Nozzle flow, 282
Nozzle pressure, 282, 293
Nozzle reach, 282
Nozzle reaction, 282, 293
Nuclear, biological, and chemical (NBC), 757
Nuclear energy and heat, 85
Nucleus, 80
Nunn-Lugar-Domenici Legislation, 764
Nutrition, 115, 727, 732
Nylon ropes, 420–421

Occupancy classifications, 359–360
Occupant use hose, 223
Occupational Safety and Health Administration (OSHA), 109
Offensive attack, 607–608
Offensive heat holes, 575
Oily rags, 673
Old-style sprinklers, 315, 316
On-scene rehabilitation, 730–731
One-person foot-tilt method, 228, 229
One-person over-the-hip method, 229
One-person raise, 406–408
One-person rope, 424
Open-circuit SCBA, 151, 152
Open hoistway door, 503
Open web steel joists, 362
Opening a roof, 585
Opening doors, 569
Opening windows, 588–590
Operating lock mechanisms, 538–539
Order of Saint John of Hospitaliers, 7
Ordinary construction, 354–355
Organic substances, 79, 81
Organizational structure, 24–25
Oriented strand board (OSB), 351
OSHA, 109
OSHA 29 CFR 1910.134, 148, 149, 156
OSHA CFRs, 109
OS&Y valve, 318, 319, 324, 325, 648
Outside stacked materials fire, 627–628
Outside stem and yoke (OS&Y) valve, 318, 319, 324, 325, 648
Over the head method (donning SCBA), 159–160, 161
Over-the-hip method, 229
Overhand safety knot, 427
Overhaul, 610, 611
Overhaul operations, 648–652
Overhauling roofs, 650

Overhead doors, 525
Overhead garage door, 525–527
Oxidation process, 81–82
Oxidizer, 81
Oxygen, 86, 90
Oxygen-deficient atmosphere, 145

Padlock, 529–530, 540
Pagers, 61
Pancake collapse, 495
Panel doors, 522
Panic hardware, 664
Pant, 128
PAR, 735
Parallel chord truss, 361
Parallel method of attack, 621
Parapet, 363
Parmalee, Henry, 14
Partial openings, 585
Partial thickness burns, 716
Partially broken windows, 585, 586
PASS devices, 135–136, 144
Passport, 725
Patient, examining the. *See* Medical examination
PCR, 762
PDD 39, 756
Pendent sprinkler, 316, 317
Percent drop, 215
Personal accountability, 725–726
Personal alert safety system (PASS), 135–136
Personal flotation device (PFD), 492, 493
Personal protective equipment (PPE), 123–140
    care/maintenance, 136
    eye protection, 134
    firefighter survival, 723–724
    hearing protection, 134
    ice rescue ensemble, 132, 133
    PASS devices, 135–136
    proximity ensemble, 129–130, 131
    standards/regulations, 126
    street smarts, 137–138
    structural ensemble, 127–129
    swift-water ensemble, 133
    technical rescue ensemble, 133
    wildland ensemble, 130–132
    work uniform, 136
Personal size-up, 725–726, 730
Personnel accountability report (PAR), 735
Peshtigo (Wisconsin) forest fire, 14
Pesticides, 755–756
PFD, 492, 493
Philadelphia Contributionship, 9, 10
Phoenix from the ashes, 5, 6
Phosgene, 146, 551, 758
Photoelectric detectors, 312
Physical fitness/wellness, 727
Physics of fire, 78–82
Pick head ax, 515, 516
Piercing nozzle, 286
Pike pole, 518, 570
Piled materials fire, 627–628

Pinned connections, 347
PIO, 41
Pipe bomb, 761, 762
Pitot gauge, 214
PIV, 324–325, 647
Pivoting dead bolt, 529
Plastic PVC, 600
Plate glass, 542
Platform framing, 357
Playpipes, 286, 287
Plug, 206
Plywood, 349
Point of origin determination, 653–654
Poisoning, 717–718
Polar solvents, 297, 601
Pole ladders, 377, 379
Policies, 34
Polycarbonate glazing, 543
Polyester rope, 421
Polyethylene rope, 421–422
Polymerase chain reaction (PCR), 762
Polymeric barrier, 297
Polypropylene rope, 421
Polyvinyl chloride (PVC), 600
Pompier ladder, 381
Ponds, 205
Popcorn, 89
Portable deluge sets, 272–273
Portable fire extinguishers
    care/maintenance, 197
    classes, 193–194
    extinguishing agents, 187–188
    fire classification, 185–187
    inspection requirements, 197–198
    kinds, 188–193
    limitations, 194–195
    operation, 195–197
    rating system, 193–194
Portable hydrant, 228
Portable ladders, 376–381
Portable radio, 66
Portable water tanks, 210–211
Positive pressure ventilation (PPV), 565, 566, 573–575
Positively charged, 84
Post and beam, 347
Post-incident survival, 736–738
Post-incident thought patterns, 737
Post indicator valve (PIV), 324–325, 647
Power hydraulic combination tools, 484
Power hydraulic cutters, 483
Power hydraulic rams, 483
Power hydraulic spreaders, 483, 514
Power hydraulic tools, 482–484
Power saws, 515, 516
PPE. *See* Personal protective equipment (PPE)
PPE ensemble, 127–133
PPV, 565, 566, 573–575
Pre-incident management process, 684–687
Pre-incident management site visit, 685
Preaction systems, 321, 322
Prearrival instructions, 48

Precast concrete, 350
Preconnected combination loads, 245
Preconnected loads, 243–245
Pressure
    absolute, 290
    atmospheric, 212, 213
    defined, 86
    gauge, 290
    head, 291
    nozzle, 282, 293
    pump discharge, 292
    residual, 213
    static, 214
    units of measure, 282
    vacuum, 291
    vapor, 86–88
    velocity, 291
    water distribution systems, 212–213
Pressure dressing (bleeding wound), 711
Pressure points, 712
Pressure-regulating device, 329
Primary search, 465
Printers, 62
Procedures, 35, 111. *See also* Job performance requirements; SOPs
Projected windows, 541, 544
Propane, 88
Proprietary protective signaling system, 56
Protection plates, 373
Protective gloves, 698–700
Protective hoods, 129, 130
Protective systems, 307–338
    alarm and monitoring systems, 331
    detection systems, 309–314
    fire department operations, 333–336
    local application and hood systems, 332
    sprinkler systems, 314–329
    standpipe systems, 329–331
    total flooding systems, 332
Protein foam, 296
Protons, 80
Proximity PPE, 129–130, 131
Prying and spreading tools, 513–515
PSA, 681
Public education, 679–684. *See also* Fire and life safety education
Public fire/life safety education, 32
Public information officer (PIO), 41
Public service announcement (PSA), 681
Pulling tools, 518
Pulmonary edema, 145
Pulse rate, 703, 704, 705, 707
Pump discharge pressure, 292
Pump operator, 281
Pump panel gauges, 214
Pump pressure gauges, 291
Pump-type extinguishers, 188–190
Pumper, 30
Puncture, 712–713
Pupil size, 706
Purlin, 346
PVC, 600
Pyrolysis, 83

Quarter-turn thread, 225
Quick response sprinkler, 316
Quint, 30, 31

Rabbeted jamb, 522
Radial pulse, 703, 704
Radiant heat, 610
Radiation, 96–97
Radio microphone, 66
Radio systems and procedures, 63–67
Radioactive agents, 761
Radiological dispersion device (RDD), 761
Rafter, 346
Rain, 588
Raindown technique, 303
Raising ladders, 399–408
Rams, 483
Rapid escape, 733–735
Rapid intervention crew universal air connection (RIC/UAC), 168
Rapid intervention team (RIT), 732–733
Rate-of-rise heat detector, 310
Rate of spread, 597
Rattlewatch, 8, 9
RDD, 761
RDX, 754
Reach, 282
Reach method, 492
Readiness assignments, 725
Reading buildings, 360
Reading smoke, 102–103
Rearmount aerial ladder, 373
Receiving reports of emergencies, 51–58
    automatic alarm systems, 56–57
    cellular telephone, 54–55
    municipal fire alarm systems, 55
    still alarm/walk-ups, 58
    TDD, 58
    telephone, 54
RECEO, 607
Reciprocating saw, 485, 486, 517
Records, 68–70
Red Rover, 11
Reducer, 228
Reel coil, 499
Reflective trim, 128
Regular glass, 542
Regulations, 34
Rehabilitation, 730–731
Reinforced concrete, 349
Reinspections, 677
Remote station protective signaling system, 56
Required flow, 291
Rescue, 437–438
Rescue company, 29
Rescue procedures, 459–507
    backboards, 478–480
    below-grade rescue, 497–498
    carries, 468–469
    confined space rescue, 498
    drags, 469–476
    electrical situations, 498–499
    elevator rescue, 500–504
    escalator rescue, 504–505
    extrication from motor vehicles, 482–490. *See also* Extrication from motor vehicles
    farm equipment rescue, 505
    ice rescue, 494
    industrial entrapment rescue, 499–500
    rescue of firefighter wearing SCBA, 476–477
    searching burning structures, 461–467
    stretchers, 480–481
    structural collapse rescue, 494–497
    TICs, 466–467
    trench rescue, 497–498
    tunnel vision, 461
    vertical rescue, 491–492
    victim removal, 467–476
    water rescue, 492–494
Residential garage doors, 525, 526
Residential occupancy firefighting, 611–612
Residential overhead door locks, 531
Residential sprinkler systems, 321, 322
Residual pressure, 213
Resistance training, 727
Resource management, 38
Respiratory hazards, 144–148
Respiratory protection policy, 143
Respiratory rate, 705
Rest, 727–728, 731
Restaurants, 667–668
Restricted openings, 170
Retard chamber, 318, 319
REVAS, 607
Reverse horseshoe load, 244
Reverse lay, 270
Revolving doors, 524–525
REX tool, 519, 539
RIC/UAC, 168
Ricin, 760
Riding the fire apparatus, 120
Right-hand search, 463
Rigid connections, 347
Rim lock, 529, 539
Ringdown circuits, 58
Risk/benefit analysis, 730
Risk/benefit philosophy, 118
Risk management, 108
RIT, 732–733
RIT checklist, 733
Roll-down steel doors, 526, 527
Rollover, 99, 560
Roman Empire, 6
Roof/balcony escape, 734
Roof ladder, 378–379
Roof ladder deployment, 412
Roof structures, 362–363
Roof ventilation, 575–581, 586–587
Room escape, 734
Rope, 417–458. *See also* Knot
    cleaning, 441–444
    coiling, 445–447
    construction methods, 422–423
    drying, 443–444
    hoisting, 449–455
    inspection, 439–441
    lagging, 447–449
    maintenance, 441
    parts, 425
    storage, 444
    strength, 421
    tying, between two objects, 454–457
    types, 420–423
    uses, 423–424
    washing, 442–443
Rope hose tool, 227
Rope strengths, 421
Rope washer, 442, 443
Rotary power saw, 517
Rotary saws, 520, 521
Round turn, 426
Row method, 492, 493
Rowdies and Rum era, 11
Rubber boot, 129
Rubber-coated fire hose, 222
Run card system, 59
Rung raise, 400–402
Rungs, 373
Running end, 425
Rural water supply, 210–212
"Rushing to the Conflict," 12
Rusting metal, 82

Sachs, Gordon M., 690
Safety
    air bags, 484
    air chisels, 486
    ambulance stretchers, 480
    EMS, 695–700
    equipment, 112–114
    firefighters, of, 721–740. *See also* Firefighter survival
    forcible entry tools, 511–521
    generally, 106–122. *See also* Firefighter safety
    ladders, 383, 390–393
    personnel, 114–117
    power hydraulic tools, 484
    reciprocating electric saws, 486
    salvage operations, 639, 642
    SCBA, 168–169
    scene assessment, 486
    water rescue, 492
Safety attitude, 116–117
Safety glass, 542
Safety glasses, 134
Safety officer, 41
Safety standards and regulations, 109–110
Safety triad, 111
Salvage cover, 634
Salvage cover fold, 639
Salvage cover roll, 637, 638
Salvage operations, 633–648
    balloon toss, 644, 645
    catch-all, 645–646
    ethics, 643
    limiting exposure to smoke, 644
    protecting material goods, 642–643
    safety, 639, 642
    shoulder toss, 643, 644
    sprinkled buildings, 647–648
    tools and equipment, 633–639

Visqueen, 644
    water chute, 646–647
    water removal, 644–645
SAR, 158
Sash windows, 569, 589
Saws, 515–517
Scaling ladder, 381
SCBA. *See* Self-contained breathing apparatus (SCBA)
SCBA cylinders, 153–155
SCBA face piece, 157
SCBA field maintenance sheet, 174
Scheerer, Randy, 124
School evacuation drills, 682–683
Scott, Ralph, 15
Screens, 586
Seat carry, 469, 471
Second-degree burns, 716
Secondary search, 465
Secured key boxes, 673–674
Securing the building, 654–655
Security bars, 678
Security devices, 583
Selectable gallonage nozzles, 283
Self-clearing card, 677
Self-closing faucet, 670
Self-contained breathing apparatus (SCBA), 141–182
    air quality, 155–156
    backpack/harness assembly, 152, 153
    changing cylinders, 172–176
    closed-circuit SCBA, 158
    cylinders, 153–155
    daily maintenance checks, 159, 172, 173
    design/size, 150
    donning/doffing, 158–168
    emergency procedures, 171
    face piece assembly, 157
    hostile environment, 170
    JPRs. *See* Job performance requirements - SCBA
    legal requirements, 148, 149
    limitations, 148–150
    maintenance, 159, 172, 173
    new designs, 168
    open-circuit SCBA, 152
    regulator assembly, 156
    respiratory hazards, 145–146
    restricted openings, 170
    safe use, 168–169
    SARs, 158
    seat-mounted apparatus, 162, 164, 165
    servicing cylinders, 176–180
    toxic gases, 146–148
    types, 151–158
Self-rescue, 734
Sensor label, 372
Set volume nozzle, 283
Setting a knot, 426
Severance, 489
Shallow wells, 204
Shear, 345
Shears, 483
Shed roof, 362
Sheet bend, 429

Shock, 713–714
Shock load, 419
Shoring, 496–497
Shoulder carry, 396, 397
Shoulder load, 237
Shoulder loop carry, 234–235
Shoulder toss, 643, 644
Shouldis, William, 308
Shutoff, 286, 287
Shutter guard, 539
Siamese, 228
Silencing activated detector, 335
Simple beam, 346
Simplex radio system, 64
Single-donut hose roll, 231, 233
Single hung window, 541
Single occupancy firefighting, 612–614
Single-person ladder carry, 398–399
Single-person raise, 406–408
Single-section street drag, 236
SIP, 351
Sisal rope, 420
Sit and drag method, 473, 475
Site plan, 685
Skeletal frame, 347
Slab doors, 522
Sledge, 512
Sliding bolt, 532
Sliding doors, 524, 525
Slot load, 246–248
SLUDGEM, 758
Small-diameter headlines, 288
Small-diameter house, 222
Small lines, 222
Smith, James P., 340
Smith, Mike, 22
Smoke, 145
Smoke color, 103
Smoke density, 103
Smoke detectors, 57, 311–313, 678, 679
Smoke ejector, 563–565
Smoke explosion, 556–560
Smoke fans, 563–565
Smoke indicator hole, 581
Smoke velocity, 102
Smoke volume, 102
Smokestack, 585
Smooth bore nozzles, 282–283
Snorkel basket, 376
Snorkel ladder, 376
Snow, 588
Snowflake technique, 302
Sodium bicarbonate, 188
Sodium chlorate, 82
Sodium chloride, 82
Sodium chlorite, 82
Soft sleeve, 223
Soft sleeve hydrant connection, 264
Soft suction hose, 223
SOG, 35
Solid, 98
Solid-core doors, 522
Solid stream nozzle, 282–283
Solid tip nozzle, 282–283
Solubility, 601
Soman, 758

SOPs, 35, 112, 113
Sources of heat, 82–85
Spalling, 350
Span of control, 36, 38
Spanner wrenches, 227
Special appliance friction loss, 294
Special egress control device, 664, 665
Special locks, 530
Special-purpose nozzles, 286
Specialized ensembles, 133
Specialty hydrants, 209
Specialty units, 32
Specific gravity, 601
Split lay, 271
Sprain, 718
Spray sprinkler, 316
Spreaders, 483, 514
Springs, 204
Sprinkler heads, 315–317
Sprinkler piping diagram, 323
Sprinkler stops, 642, 648
Sprinkler system operations, 335
Sprinkler systems, 314–329
    connections/piping, 321–324
    control devices, 324–325
    deluge systems, 320–321
    dry pipe systems, 319–320
    fire department operations, 335
    fire protection valves, 325
    fire suppression, 619–620
    preaction systems, 321
    residential systems, 321
    returning system to service, 325
    sprinkler heads, 315–317
    sprinkler stops, 642, 648
    water pipe systems, 318–319
Sprinkler tongs/wedges, 326
Sprinkler wedge, 642
Squad, 29
Squad company, 29
Squeegees, 635
SRS devices, 488
SSP, 315
SSU, 315
Staging, 39
Stairs, 363
Standard map symbols, 686
Standard of care, 694
Standard operating guideline (SOG), 35
Standard operating procedures (SOPs), 35, 112, 113
Standard response sprinkler, 316
Standing part, 425
Standpipe hose packs, 248, 251
Standpipe operations, 334–335
Standpipe pack equipment, 334
Standpipe system, 255–257, 329–331
Standpipe valves, 331
Stanley, Julius, 632
States of matter, 98
Static electricity, 84
Static pressure, 214
Status reports, 68
Staypoles, 377
Steel, 349
Steel cylinders, 153

Steel trusses, 361
Steele, Michelle, 220
Step-by-step procedures. *See* Job performance requirements
Stiff-arm method, 230, 231
Still alarm, 58
Stone, 349
Stops, 373
Storage hose roll, 230, 232
Stored pressure AFFF/FFFP extinguisher, 191
Stored pressure dry chemical extinguisher, 191, 192
Stored pressure foam extinguisher, 191
Stored pressure water extinguisher, 190, 191
Storz coupling, 225, 226
Straight bore nozzle, 282–283
Straight finish load, 243
Straight hose lay, 270
Straight (storage) hose roll, 230, 232
Straight ladder, 376–377
Straight stream nozzle, 285
Strainer, 228
Stream application, 289–295
Stream pattern, 282
Stream shape, 282
Stream straighter, 273
Stress, 737–738
Stretchers, 480–481
Striking tools, 512–513
Strip cut, 579–581
Structural collapse. *See* Building collapse
Structural collapse rescue, 494–497
Structural fire, 102–103
Structural firefighting ensemble, 127–129
Structural fires, 102–103
Structural insulated panel (SIP), 351
Structural steel, 349
Suitcase carry, 395, 396
Suitcase ladder, 379, 380
Sulfur dioxide, 551
Superficial burns, 716
Supplemental restraint system (SRS), 488
Supplied air respiratory (SAR), 158
Supply hose, 222–223
Surface locks, 529
Surface-to-mass ratio, 346
Surface water, 204
Survival, firefighter. *See* Firefighter survival
Suspended ceiling, 587
Swift-water ensemble, 133
Swift-water rescue, 493
Switch valves, 228
Symbols
  fire extinguisher classification, 102
  labels. *See* Labels
  placards. *See* Placards
  preplanning, 686

Tabun, 758
Tactical worksheets, 40
Tag, 725
Tag/guide line, 449, 450
Tanker, 204
Target hazards, 685
TDD, 58
Team continuity, 728
Tear and run printers, 62
Technical Escort Unit (TEU), 763
Technical grade pesticides, 756
Technical rescue ensemble, 133
Technological obsolescence, 18
Telecommunicators, 48, 51, 52
Tempered glass, 542
Tempered glass doors, 523
Tender shuttle operation, 211–212
Tension, 345
Terrorism, 611, 741–766
  clandestine labs, 752–753
  credibility of threat, 747–748
  detection of terrorism agents, 762
  drug labs, 753
  explosives labs, 753–755
  federal assistance, 762–764
  groupings of terrorism agents, 757–762
  HAZMAT crimes, 752–757
  historical overview, 744–747
  incident actions, 756–757
  incident priorities, 763
  indicators of, 750–752
  potential targets, 748–750
  terrorism agent labs, 755–756
Terrorism agent labs, 755–756
TEU, 763
Theory of catastrophic reform, 15
Thermal burns, 715–716
Thermal conductivity, 97
Thermal imaging camera (TIC), 466–467, 650–651
Thermal layering, 99
Thermal plume, 95
Thermal protective performance (TPP), 128
Third-degree burn, 716
Threaded couplings, 225, 226
Three Es of fire prevention, 659
Three-person flat carry, 396, 398
Through-the-lock forcible entry, 536–538
Through-the-lock technique modified, 540
Throw method, 492, 493
TIC, 650–651
Tidal changes, 204
Tie-rods, 373
Tilt-up construction, 350
Tools. *See* Equipment
Topography, 598–599
Tormentor poles, 377
Torsion load, 345
Total flooding systems, 332
Tower ladder, 375–376
Townsend, Rick, 509
Toxic gases, 146–148
Toxicity, 602
Toxins, 760
TPP, 128
Traffic barrier, 486

Traffic calming, 487
Traffic cones, 487
Traffic control systems, 62–63
Training, 114–115
Training division, 33–34
Transfer of command, 36–37
Trash storage areas, 675
Trench cut, 579–581
Trench rescue, 497–498
Triage, 633, 704
Triangle Shirtwaist Factory fire, 314
Triangular cut, 579
Triangular inspection hole cut, 581
Triple combination engine company, 16
Triple-layer load, 247–250
Trousers, 128
Truck company, 29
Truck-mounted ladders, 373–376
Truss, 346, 361
Tubular locks, 529–530, 540
Tully, Donald C., 657
Tunnel vision, 461
Tunneling, 497
Turntable, 374
29 CFR 1910.134, 148, 149, 166
29 CFR Parts 1910 and 1926 Respiratory Protection, 609
Twin donut roll, 231, 233
Two in/two out rule, 110, 462, 609, 612
Two-person beam raise, 402, 403–404
Two-person coupling method, 229, 230
Two-person rope, 424
Two-person rung raise, 400–402
Two-person stiff-arm method, 230, 231
Two-way radios, 67
Type 1 manila rope, 420
Type A reporting system, 56
Type B reporting system, 56, 57
Type I fire-resistive construction, 353–354
Type II noncombustible construction, 354
Type III ordinary construction, 354–355
Type IV heavy timber construction, 355–356
Type V wood frame construction, 356–357

Ultraviolet flame detectors, 313–314
Ultraviolet-infrared flame detectors, 314
Uncoupling hose with spanners, 230
Underwriters' playpipe, 286, 287
Undesigned load, 344
Unequal pupils, 706
Uninterruptible power supply (UPS), 50
Units of measurement, 77–78
Unity of command, 36
Upright sprinkler, 316, 317
UPS, 50
Urban search and rescue (USAR) teams, 763
UV detectors, 313–314
UV-IR detector, 314

V agent, 758
V pattern, 654

V-type collapse, 495–496
Vacuum (negative) pressure, 291
Vapor density, 87–88, 602
Vapor pressure, 86–88
Vapor suppression, 296
Variable combination fog nozzle, 284
Variable gallonage nozzle, 283
Vehicular firefighting, 599–601, 622–625
Velocity pressure, 291
Veneer, 350
Venous bleeding, 710
Vent for life strategy, 465
Ventilation, 548–590
   air movement, 562–563
   backdraft, 556–560
   benefits, 550
   breaking glass, 568–569
   building size, 587–588
   defined, 550
   dropped/hanging ceilings, 587
   factors to consider, 551–554
   flashover, 556
   horizontal, 552, 567–568
   obstacles to, 583–585
   opening doors, 569
   partial openings, 585
   partially broken windows, 585
   PPV, 573–575
   rollover, 560
   roof, 575–581
   roof materials, 586–587
   safety, 581–582
   screens, 586
   types, 563–567
   vertical, 552, 567
   weather, 588
   when needed, 560–562
   windows, 569–572, 585, 588–590
Ventilation timing, 584
Venturi principle, 297, 298
Vertical indicator post, 325
Vertical pivoting windows, 541
Vertical rescue, 491–492
Vertical ventilation, 552, 567
Vertical voids, 562
Vesicants, 759
Vesta, 6
Vicarious experience, 116
Victim removal, 467–476
Vigiles, 6
Visqueen, 634, 635, 644
Visual guides. *See* Job performance requirements

Void spaces, 362
Voids, 561, 562

Wafer check valve, 325
Wagon pipe, 271, 272
Walk-ups, 58
Wall, 347
Wall-bearing buildings, 347
Wall fire hydrant, 209
Wall indicator post, 325
Wall indicator valve (WIV), 324–325
Wall ladder, 376–377
Walter, Andrea A., 3, 689
Washington, George, 9
Watch dogs, 554
Water, 438–439
Water-based foam extinguishers, 187–188
Water chute, 646–647
Water columning, 319
Water curtain nozzle, 286, 287
Water distribution systems, 205–206, 209–210, 212–213
Water hammer, 216
Water motor gong, 331
Water rescue, 492–494
Water-steam conversion, 99
Water supply, 201–218
   fire hydrants, 206–209, 213–215
   groundwater, 203–204
   mobile water supply apparatus, 204–205
   obstructions, 215–216
   percent drop, 215
   pressure, 212–213, 214
   rural areas, 210–212
   surface water, 204
   tanks, ponds, cisterns, 205
   water distribution systems, 205–206, 209–210, 212–213
Water supply connection, 261–265
Water table, 203
Water tanks, 205
Water tender, 204
Water thief, 228
Water vacuum, 635, 636
Waterflow alarms, 331
Weapons of mass destruction (WMD), 755, 757
Weapons of mass destruction civil support team (WMD-CST), 764
Weather, 588, 595
Web, 346
Web gear, 132

Webbing sling drag, 472–473, 474
Weight lifting, 727
Well pump with storage tanks, 204
Wells, 204
West, Mike, 466
Wet barrel hydrants, 206–207
Wet chemical agents, 188
Wet chemicals, 187
Wet pipe sprinkler system, 318–319
Wildland fire shelter, 132
Wildland fire triangle, 595–599
Wildland firefighting, 595–599, 620–622
Wildland firefighting hose loads, 248–250
Wildland hose advancing and extension, 269
Wildland PPE, 130–132
Wildland web gear, 132
William Stieger Act, 109
Wind, 554
Windisch, F. C. (Fred), 107
Window dismount, 410, 412
Window search, 465
Windows
   bars and gates, 544
   forcible entry, 540–544
   glazing, 542–543
   opening, 588–590
   partially broken, 585
   types/styles, 541, 543–544
   ventilation, 569–572, 585, 588–590
Wire cutters, 171, 515
Wire glass, 543
Wisko, Mike, 460
WIV, 324–325
WMD, 755, 757
WMD-CST, 764
WMD laws, 757
Wood, 348–349
Wood doors, 522
Wood frame construction, 356–357
Wood sash windows, 543
Wood trusses, 361
Work hardening, 115, 116
Work uniform, 136
Working end, 425
Working length, 399, 400
World War II, 17
Wound requiring first aid, 712–713
Wutz, Thomas J., 141
Wye, 228